Coastal Research Library

Volume 36

Series Editor
Charles W. Finkl
Department of Geosciences
Florida Atlantic University
Boca Raton, FL, USA

The aim of this book series is to disseminate information to the coastal research community. The Series covers all aspects of coastal research including but not limited to relevant aspects of geological sciences, biology (incl. ecology and coastal marine ecosystems), geomorphology (physical geography), climate, littoral oceanography, coastal hydraulics, environmental (resource) management, engineering, and remote sensing. Policy, coastal law, and relevant issues such as conflict resolution and risk management would also be covered by the Series. The scope of the Series is broad and with a unique cross-disciplinary nature. The Series would tend to focus on topics that are of current interest and which carry some import as opposed to traditional titles that are esoteric and non-controversial.

Monographs as well as contributed volumes are welcomed.

More information about this series at http://www.springer.com/series/8795

Laith A. Jawad
Editor

Southern Iraq's Marshes

Their Environment and Conservation

Springer

Editor
Laith A. Jawad
School of Environmental and Animal Sciences
Unitec Institute of Technology
Mt Albert, Auckland, New Zealand

Every effort has been made to contact the copyright holders of the figures and tables which have been reproduced from other sources. Anyone who has not been properly credited is requested to contact the publishers, so that due acknowledgment may be made in subsequent editions.

ISSN 2211-0577 ISSN 2211-0585 (electronic)
Coastal Research Library
ISBN 978-3-030-66237-0 ISBN 978-3-030-66238-7 (eBook)
https://doi.org/10.1007/978-3-030-66238-7

This Springer imprint is published by the registered company Springer Nature Switzerland AG
The registered company address is: Gewerbestrasse 11, 6330 Cham, Switzerland

Night at the southern marshes of Iraq

Houses and rest home in the marshes of Iraq.

Marsh Arabs in part of their daily life.

Upper photo, Iraqi bread; lower photo, water buffalo grazing.

Dedicated to every member of my delightful family who wonders if I'm devoting this work for them.
I am

Preface

I am always fascinated by the southern marshes of Iraq and the way of life that the Marsh Arabs are living in. This captivation has augmented when I read the works of the British explorer Wilfred Thesiger, a retired soldier, who spent several years in the marshes between 1951 and 1957 and published his memoirs, *The Marsh Arabs* and Gavin Young, a British journalist, visited the region and befriended Thesiger with whom he photographed the area during the 1950s. Young published his remembrances about the marsh areas in his well-known book "*Return to the Marshes*". These works have created an idea in my mind of writing a book about the marshes of Iraq and their inhabitants, but in different ways that the previous writers have done. This was sometimes in the early 1970s and since then and for more than 50 years I am embracing this idea that accompanied my travels through the countries until I settled down in New Zealand and decided to put this idea on paper.

The new technology of the internet and fast communication have assisted my goal in producing my book on the southern marshes of Iraq. Inputting the proposal of the book and the preliminary contents table, I visited hundreds of literature, videos and television programs about the marshes of Iraq. Such materials enrich my idea with several aspects of the marshes ranging from the environment to the Marsh Arabs lifestyle.

When I decided to edit a book about the southern marshes of Iraq, the idea that I had in mind is to see this book having information that not have been dealt with anybody previously. Therefore, I decided the contents of the book should include wide range spectrum of aspects about the marshes of Iraq including the historical, geological, environmental, the fauna and flora, the fisheries, the impacts and challenges that the marshes experiencing, the conservation of the marshes and finally the daily life particulars of the Marsh Arabs.

The longest chapter in the present book is that on the daily life of the Marsh Arabs. Although several authors have written on the habits and habitats of the Marsh Arabs in the form of books and articles in magazines and journal, I decided this chapter to hold information that has never been told before. Such differences are placed in attaining facts about the daily life of the dwellers of the marshes accompanied by figures. A step that never been presented in any book about the marshes of Iraq before. This chapter takes more than 4 months to finish as I used to relate each daily act and habit of the Marsh Arabs to the ancient Mesopotamian and see whether there is any similarity can be derived. In doing so, I explored an extensive number of kinds of literature on ancient Mesopotamian and retrieved from these kinds of literatures what are comparable to the life of their present-day descendants that inhabit the marsh area in Iraq. Besides, I included several facts about the daily life of the Marsh Arabs that I assume to be documented for the first time. Among these are answers for many questions such as, why the number of the main pillars of the guesthouse "Al-Madhif" of the Marsh Arab and pot of coffee in this guesthouse is always odd and even? Are the Marsh Arabs recognise several types of water buffaloes? If so, what are the names of these types? Are the Marsh Arabs specialised in making different food items such as bread, milk products and sweets? In order of not ruining the surprise that readers of this book in general and the chapter about the daily life of the Marsh Arabs in particular, I shall not give the answers for these questions here and leave it for the readers to discover them while they are reading this book.

I would like to convey my thanks to all the contributors of this book that agree to share their scientific work as chapters are written about different aspects of the southern marshes of Iraq in order make it accessible by the readers in all over the world. Also, my sincere thanks should go to Springer Publisher that agrees to publish this book and make my dream a reality.

Auckland, New Zealand Laith A. Jawad

Acknowledgements

First, the editor would like to thank each one of the authors, who contributed their time and expertise to this book. Without their support, this book would not have become a reality.

Second, the editor wish to thank the staff of Springer, The Netherlands in general about the efforts they put in publishing this book and for all the consultant staffs for their distinguished service, valuable advice and suggestions that they delivered throughout the period of checking the text of this book.

Contents

1 **Introduction** 1
Laith A. Jawad

Section I Historical Perspectives

2 **Cities in the Water: Waterscape and Evolution of Urban Civilisation in Southern Mesopotamia as Seen from Tell Zurghul, Iraq** 15
Davide Nadali

3 **Reed-Swamps in the Sumerian Material Culture: Archaeological, Archaeobotanical and Experimental Insights from the Abu Tbeirah Excavations** 33
Licia Romano, Alessandra Celant, and Maria Virginia Montorfani

4 **Human Ecology of the Marshes** 55
Ariel I. Ahram

5 **Role of Women in Ancient Mesopotamia and the Southern Marshes of Iraq: A Comparative Account** 77
Laith A. Jawad

Section II Environmental Factors

6 **Physical and Chemical Characters of Mesopotamian Marshes: A Short Review** 95
Bahram K. Maulood and Fikrat M. Hassan

7 **Hydrologic Structures in the Tigris-Euphrates Basin and Their Impact on the Vitality of the Marshes** 113
Toon Bijnens

8 Importance of Hydrological and Hydrogeological Studies in Wetlands: Examples from Turkey 127
Melis Somay-Altas

9 Use of Multispectral and Hyperspectral Satellite Imagery for Monitoring Waterbodies and Wetlands 155
Mahdi Hasanlou and Seyd Teymoor Seydi

10 Usage of Satellite Technology in Monitoring the Wetlands of Turkey, Tigris, and Euphrates Watershed 183
Gordana Kaplan, Zehra Yigit Avdan, and Ugur Avdan

Section III Geology

11 Sedimentology and Mineralogy of Quaternary Sediments of Marshes, South of Iraq 201
Badir N. Albadran

Section IV Major Biotope

12 Phytoplankton and Primary Production in Iraqi Marshes 217
Bahram K. Maulood and Fikrat M. Hassan

13 Distribution Patterns, Diversity Centers, and Priorities for Conservation of Aquatic Plants in Iran 233
Ahmadreza Mehrabian and Farzaneh Khajoi Nasab

14 Compiled Checklist of Aquatic Invertebrates of the Southern Marshes of Iraq 251
Laith A. Jawad

15 The Indian Shad *Tenualosa ilisha* (Hamilton 1822) in the Southern Marshes of Iraq: A Revision and Evaluation of a Compiled Data 271
Laith A. Jawad

16 The Presence of *Gambusia* Fish in the Southern Marshes of Iraq: Bad or Good 285
Laith A. Jawad

17 Checklist of Freshwater Fishes of Southwestern Wetlands of Iran 295
Hamid Reza Esmaeili

18 Checklist of Fishes of the Caspian Sea Basin: Land of Wetlands 319
Hamid Reza Esmaeili and Keyvan Abbasi

19 The Ornithological Importance of the Southern Marshes of Iraq 351
Mudhafar A. Salim, Salwan Ali Abed, and R. F. Porter

Section V Biodiversity Aspects

20 Northern Gulf Marine Biodiversity in Relevance to the River Discharge 379
Faiza Al-Yamani, Igor Polikarpov, and Maria Saburova

21 A Possible Threat to the Fish Biodiversity in the Southern Marshes of Iraq: A Mini-Review 439
Laith A. Jawad

22 Biotic Homogenization: A Process That Is Happening in the Southern Marshes of Iraq 449
Laith A. Jawad

Section VI Natural Resources

23 Fish, Fishing Methods and Fisheries of the Southern Marshes of Iraq 459
Laith A. Jawad

24 The Artisanal Fishers in the Southern Marshes of Iraq and the Traditional Ecological Knowledge 479
Laith A. Jawad

Section VII The Marshes and Human Health

25 Fish Consumption of Mothers and Their Children in Fishermen Society in the Marsh Areas 487
Laith A. Jawad

26 Ingestion of Fish Bones: Clinical Cases from the Marsh Areas of Iraq 497
Laith A. Jawad

Section VIII Environmental Challenges

27 The Status of Pollution in the Southern Marshes of Iraq: A Short Review 505
Nader A. Salman, Hamid T. Al-Saad, and Faris J. Al-Imarah

28 The Effects of Man-Made Noise on the Fishes in the Marshes of Iraq 517
Laith A. Jawad

29 Heavy Metals in Wetlands in Turkey 527
Onur Can Türker and Jan Vymazal

30 Locals' Awareness of Ecotourism in the Southern Marshes of Iraq 551
Laith A. Jawad

31 The Effects of Thermal Pollution on the Aquatic Life in the Southern Marshes of Iraq 559
Laith A. Jawad

Section IX Conservation

32 What Ecological Principles Required for a Proposed Establishment and Management of National Parks in the Southern Marshes of Iraq 575
Laith A. Jawad

33 The Possible Difficulties and Outcomes of the Biodiversity Conservation of the Southern Marshes of Iraq 579
Laith A. Jawad

34 Co-management Scheme to Protect the Southern Marshes of Iraq: A Proposal 591
Laith A. Jawad

35 Eradication as a Method to Manage the Non-native Fish in the Marsh Environment: A Proposal 599
Laith A. Jawad

36 Freshwater Commercial Bycatch in the Southern Marshes of Iraq: The Unexploited Aquatic Wealth and Understated Conservation Problem 611
Laith A. Jawad

Section X Socio-economic Aspects

37 Ornamental Fish Farming: A Proposal for a Successful Small-Scale Aqua Business Project in the Southern Marshes of Iraq 621
Laith A. Jawad

38 Migration of Youth to the Big Cities from the Southern Marshes of Iraq: Solutions to Keep Locals with an Upgrading for Their Skills and Livelihood 631
Laith A. Jawad

39 The Daily Life of the Marsh Arab of the Southern Marshes of Iraq Through a Camera Lens 641
Laith A. Jawad

40 Socio-economic Status Comparison of Fishermen Community in Two Marsh Areas in Southern Iraq 809
Laith A. Jawad

Contributors

Keyvan Abbasi Inland Waters Aquaculture Research Centre, Iranian Fisheries Sciences Research Institute, Agricultural Research, Education and Extension Organization, Bandar Anzali, Iran

Salwan A. Abed College of Science, University of Al-Qadisiyah, Al Diwaniyah, Iraq

Ariel I. Ahram Virginia Tech School of Public and International Affairs, Arlington, VA, USA

Badir N. Albadran Department of Geology, University of Basrah, Basrah, Iraq

Hamid T. Al-Saad College of Marine Sciences, University of Basrah, Basrah, Iraq

Faris J. Al-Imarah College of Marine Sciences, University of Basrah, Basrah, Iraq

Faiza Al-Yamani Kuwait Institute for Scientific Research, Salmiya, Kuwait

Ugur Avdan Earth and Space Sciences Institute, Eskisehir Technical University, Eskisehir, Turkey

Zehra Yigit Avdan Environmental Engineering Department, Eskisehir Technical University, Eskisehir, Turkey

Toon Bijnens Genk, Belgium

Alessandra Celant Department of Environmental Biology, Sapienza University of Rome, Rome, Italy

Hamid Reza Esmaeili Ichthyology Research Laboratory, Zoology Section, Department of Biology, College of Sciences, Shiraz University, Shiraz, Iran

Fikrat M. Hassan Department of Biology, College of Science for Women, University of Baghdad, Baghdad, Iraq

Mahdi Hasanlou School of Surveying and Geospatial Engineering, College of Engineering, University of Tehran, Tehran, Iran

Laith A. Jawad School of Environmental and Animal Sciences, Unitec Institute of Technology, Mt Albert, Auckland, New Zealand

Gordana Kaplan Earth and Space Sciences Institute, Eskisehir Technical University, Eskisehir, Turkey

Bahram K. Maulood Department of Biology, College of Science, University of Salahaddin, Erbil, Iraq

Ahmadreza Mehrabian Department of Plant Sciences and Biotechnology, Faculty of Life Sciences and Biotechnology, Shahid Beheshti University, Tehran, Iran

Maria Virginia Montorfani Department of Classics, Sapienza University of Rome, Rome, Italy

Davide Nadali Sapienza Università di Roma, Rome, Italy

Farzaneh Khajoi Nasab Department of Plant Sciences and Biotechnology, Faculty of Life Sciences and Biotechnology, Shahid Beheshti University, Tehran, Iran

Igor Polikarpov Kuwait Institute for Scientific Research, Salmiya, Kuwait

R. F. Porter Adviser to Middle East Programme, BirdLife International, Cambridge, UK

Licia Romano Department Institute of Oriental Studies, Sapienza University of Rome, Rome, Italy

Maria Saburova Kuwait Institute for Scientific Research, Salmiya, Kuwait

Mudhafar A. Salim Natural Heritage Division, Arab Regional Centre for World Heritage (ARC-WH), Manama, Kingdom of Bahrain

Nader A. Salman Al-Manara College of Medical Sciences, Misan, Iraq

Seyd Teymoor Seydi School of Surveying and Geospatial Engineering, College of Engineering, University of Tehran, Tehran, Iran

Melis Somay-Altas Geological Engineering Department, Dokuz Eylul University, Buca, Izmir, Turkey

Onur Can Türker Faculty of Science and Letters, Department of Biology, Aksaray University, Aksaray, Turkey

Jan Vymazal Faculty of Environmental Sciences, Czech University of Life Sciences Prague, Praha, Czech Republic

About the Editor

Laith A. Jawad obtained a degree (MSc) in fish taxonomy from the Zoology Department, University of Bristol, UK, in 1980. He continued as fish taxonomist at Basrah University, Iraq, where he worked for more than 20 years before he immigrated to New Zealand in 1997. During this time, he started the biochemical taxonomy of fishes of Iraq and published over 385 scientific papers and book reviews in leading scientific journals. He is the author and co-author of several textbooks in biology published in Arabic. Recently, he contributed five chapters to a book about coastal fishes, Coastal Fishes: Habitat, Behavior and Conservation, published by Nova Publishers, Canada. And authored a book about Dangerous Fishes of the Eastern and Southern Arabian Peninsula published by Springer in 2017. He served as fish biodiversity expert and consultant at the Ministry of Agriculture and Fisheries in Oman for the period 2008–2012 during which he co-authored two papers describing a new fish species from the Omani waters and reported over 80 fish species as a new record to the Omani waters. He authored a guide to the fishes of the southern coasts of Oman published by the Ministry of Agriculture and Fisheries in Oman in 2018. He also published over 90 papers on fish fauna of Oman, Iraq, Kuwait, and Saudi Arabia. In 2013, he broadened his scientific contact and started to collaborate with over 60 scientists from more than 50 countries around the world in researches dealing with different aspects of fish taxonomy and ichthyology.

Chapter 1
Introduction

Laith A. Jawad

Abstract The Near Eastern wetlands are the vast marshes of southern Iraq (al-ahwār) in the lower Mesopotamian region, an impression shaped tectonically as a consequence of the Arabian plate being subducted beneath the Iranian or Eurasian plate. These are considered the famous among the lower lands in the Middle East. These wetlands covered in 1970 an estimated area ranging from 15,000 to 20,000 square kilometres. The eastern margins of the marshlands spread over the boundary into south-western Iran, and they therefore create a transboundary ecosystem under shared responsibility. Euphrates River is the prime supplier of marshes with water, with contributions from tributaries of Tigris River. The fundamental region of the marshes is located in the area around the convergence of the Tigris and Euphrates. Therefore, the whole marsh area is divided into three major areas: (i) Al Hammar Marshes; (ii) the Central Marshes and (iii) Al Hawizeh Marshes. These three chief marsh zones have been at the centre of the great fluctuations that have been in process over the years. Marsh inhabitants who experienced a mixture of pastoralist, inactive and marsh existence grounded on the periodic growth, and collapse of the marsh waters have been recognized including the Maʿdān (Marsh Arabs) of southern Iraq. The majority of the area of the marshes in Iraq is covered with aquatic plants that are dominated by reed (Pharagmites communis) and reed mace (Typha augustata) in the transient seasonal zone. Situated on the routes of the migratory birds, the marshes are chiefly significant for birds. The marshlands set up a main wintering and staging area for waterfowl travelling between breeding grounds. The environment of the marsh area in Iraq has experienced several kinds of influences. Among such impacts was the influence of the changes in the water supply due to building dams in the upper Mesopotamian plain in Turkey and Iran and the ecocide that Saddam Hussein has implemented when he ordered to dry the marshes. The results of drying the marsh areas have impacted the life of the inhabitants of the marshes in early 1990s; accordingly, the Marsh Arabs have been compulsory to flee their areas. Besides to the engineering workings, their birthplace converted to be one

---

L. A. Jawad (✉)
School of Environmental and Animal Sciences, Unitec Institute of Technology, Mt Albert, Auckland, New Zealand

L. A. Jawad (ed.), *Southern Iraq's Marshes*, Coastal Research Library 36,
https://doi.org/10.1007/978-3-030-66238-7_1

of the chief areas of fighting that overwhelmed southern Iraq in 1991–1993. Owing to the huge loss in the marsh areas, the wildlife and biodiversity have severely influenced. Such impacts were extended outside the borders of Iraq and showed effects on both the regional and the international levels.

Keywords Marshes. Iraq · Basrah · Mesopotamia · Marsh Arabs

1.1 Geological Perspectives

In general, the area of any water body on this globe is under the rules of geology (Kornfeld 2009). Prior to 10,000 BCE, a glacial ice sheet enclosed a considerable area of the Northern Hemisphere, containing the area north of the Arabian Sea (University of California 1995). At 6500 BCE, the ice sheet had retreated from Europe and the Middle East, and the climate warmed similar to what is now (Dellapenna 1996).

Maybe the most famous of the Near Eastern wetlands are the vast marshes of southern Iraq (*al-ahwār*) in the lower Mesopotamian region, an impression shaped tectonically as a consequence of the Arabian plate being subducted beneath the Iranian or Eurasian plate (Baeteman et al. 2005).

Located for the main part in southern Iraq (29° 55′ to 32° 45′ N and 45° 25′ to 48° 30′ E), the wetlands covered in 1970 an estimated area ranging from 15,000 to 20,000 square kilometres. The eastern margins of the marshlands spread over the boundary into south-western Iran, and they therefore create a transboundary ecosystem under shared responsibility.

Geomorphologic investigations in the area have acknowledged the development and enlargement of the marshes as an advanced trait that augmented over time (Baeteman et al. 2005). The preliminary development intricate natural and physical variations in the climate and sea level, pre- and during the Holocene period, prepare the site for wetland creation in several phases (Asa 2011). Originally, throughout the Last Glacial Maximum (16,000 BCE), the marshes were not existing owing to the very low groundwater, and the Arabian Gulf was rather shallow. At that time, the sea level ascended during the Postglacial (12,000 BCE), carrying the coastline more inland (Sanlaville 2002). In the third phase, the aridity of the weather and descent in sea level, inaugurating in the third millennium BCE and enduring into the second millennium, alleviated the deltaic progradation of the coasts and instigated the waters of the Arabian Gulf and coastline to retreat considerably to their present location, sending off a clean basin that gradually filled in with saline-brackish lagoons (salt marshes) and tidal flats (*sabkha*) over the extended delta land (Sanlaville 2002). Ultimately, these established into small, everlasting freshwater marshes and lakes by the first millennium BCE

1.2 Hydrology and Precipitation Aspects

The hydrology of the Euphrates and Tigris Rivers and precipitation causing more stream flow and flooding further added to marsh creation. Euphrates River is the prime supplier of marshes with water, with contributions from tributaries of Tigris River (Hritz 2005). Subsequently, the steadiness of the coasts and sea level, flooding and sedimentation of the lower courses of the river were the chief reasons of the formation of marshes (Sanlaville 2002). The water flow from the highland at Zagros Mountains brings with it a huge amount of sediments that precipitate in the marsh area.

Soils of the marsh area made from sedimentation are another significant physical factor to marsh formation. Sedimentation can be produced by natural or anthropogenic issues causing erosion from the surrounding sandstone and limestone uplands ((Heyvaert and Baeteman 2007). Though less important, sedimentation can also be carried by dust storm (aeolian) activity or stuck sediment in seawater in the salt marshes and *sabkhas* from tidal activity (Heyvaert and Baeteman 2007). Throughout avulsion, the stiffest sediments established near the riverbanks and formed high levees. It is in these regions where silty clay was dumped, creating waterlogged beds for marshes and lakes (Sanlaville 2002). By the seventh century CE, the development of marsh creation had alleviated into an enormous wetland (370 × 90 km). Additionally, the passage of the Tigris and Euphrates Rivers into the Gulf (the Shatt al-Arab River) established in the eighth–tenth centuries (Aqrawi 2001). It is probable that this channel delivered both a drain for the marshes and access from the Gulf through them to Baghdad. The heart of the wetlands of the lower Euphrates and Gulf outlets (the Ahwār marshes), the enduring freshwater Lake Hammar (Hawr al-Hammar) and the secondary Lake al-Azim (Hawr al-Hawiza) were relatively recent creations, as was the Shatt al-Arab (Pournelle 2003).

1.3 The Major Wetland Divisions

The marshlands of lower Mesopotamia extend from Samawa on the Euphrates and Kut on the Tigris (150 km south of Baghdad) to Basrah on the Shatt al-Arab. The wetlands create a series of almost interrelated marsh and lake developments that overflow one into another (UNEP 2001). Throughout periods of high floods, large areas of desert are under water. Thus, some of the previously discrete marsh units combine together, forming larger wetland complexes. In the marsh area, there are wetlands and lakes separated by small islands. These comprise stable and periodic marshes, shallow and deepwater lakes and mudflats that are frequently flooded during periods of elevated water levels (UNEP 2001). The fundamental region of the marshes is located in the area around the convergence of the Tigris and Euphrates. Therefore, the whole marsh area is divided into three major areas: (i) Al-Hammar Marshes, (ii) Central Marshes and (iii) Al-Hawizeh Marshes. These three chief marsh zones have been at the centre of the great fluctuations that have been in process over the years.

1.3.1 Al-Hammar Marshes

The set of marshes that are known as Al-Hammar are located completely south of the Euphrates, traversing from near Al-Nasiriyah in the west to the outskirts of Basrah City on the Shatt al-Arab River in the east (UNEP 2001). This marsh broadened at its south end, with the presence of broad mud shoreline. The estimated area of this marsh ranges between 2899 km^2 and 4.500 km^2. The water of this marsh is slightly brackish owing to its location near the Arabian Gulf, eutrophic and shallow. It reaches a maximum depth of 1.8 m and about 3 m at high water mark (Maltby 1994). Euphrates River is the main supplier of water to this marsh. A substantial amount of water from the Tigris River, spilling over from the Central Marshes, also sustains the Al-Hammar Marshes. The set of marshes of Al-Hammar support one of the most imperative waterfowl areas in the Middle East, both in terms of bird numbers and species diversity (UNEP 2001). The enormous and thick reed beds offer ideal environment for breeding populations, while the mudflats sustenance for shorebirds. International important focusses of migratory waterfowl have been reported during winter, and although not correctly surveyed, the area is likely to host similarly high numbers during the spring and autumn seasons (Scott 1995).

1.3.2 The Central Marshes

This marsh is situated just north of the convergence of Euphrates and Tigris Rivers. The Central Marshes are at the focal point of the Mesopotamian wetland ecosystem (UNEP 2001), and the region is coarsely bordered by a triangle between Al-Nasiriyah, Qalat Saleh and Al-Qurnah. These marshes obtain water directly from the Tigris River (branches of Shatt al-Muminah and Majar al-Kabir in addition to the Euphrates from its southern side). These marshes extend over 3000 km^2 and during flood, they encompass to about 4000 km^2. Included with these marshes are smaller marshes such as Al Zikri Marsh and Hawr Umm al Binni, which are situated around the middle of the Central Marshes (Thesiger 1964).

1.3.3 Al-Hawizeh Marshes

These marshes are located at the eastern side of the Tigris River, overlapping the Iran-Iraq border (UNEP 2001). The Iranian fragment of the marshes is known as Hawr Al Azim. They receive their water supply from branches of Tigris River near Al-Ammarah City known as the Al-Musharah and Al-Kahla. An additional water supply enters these marshes through the Karkheh River in the east. These marshes showed to have a length of about 80 km from north to south and a width of 30 km from east to west and with an approximate area of not less than 3000 km^2. These marshes are characterised as having water in the north and middle parts around the year, and the southern part is seasonal. Water of these marshes links Shatt al-Arab River 15 km south of Al-Qurnah via the Al-Swaib River.

1.4 Settlement

The economy in Mesopotamia is based on agriculture mainly grains and became the basis of civilisation, agreeing development of humans and the growth of villages (Leick 2002). By 3000 BCE, the Sumerians had built a mature civilisation, with prosperous town centres (Leveson 1980). Leveson (1980) proposed that the upsurge of civilisation and cities that took place in the Middle East was not by chance. In the southern plains, between modern-day Baghdad and the Shatt al-Arab, as well as the Persian Gulf, the Sumerians settled and built the cities of Eridu, Kish, Lagash, Nippur, Ur, Uruk and Susa (Leveson 1980). The alluvial soils that the floodwaters of the Tigris and Euphrates Rivers discarded here were not homogeneous, comprising diverse minerals, unlike the fertile soils in the north—the land of the Akkadians and Assyrians. Though the southern plains were flat and possibly productive, they got slight rain. Therefore, these soils remain unplanted until the Sumerians learnt how to adapt to this environment, considerably over control of the waterways by means of canals and dykes (Leick 2002). With this, the first large-scale societies started to progress, and people increased beyond existence farming to produce an excess, expand their cultural actions and live in increasingly large numbers in a new form of cooperative communal, the city (Leick 2002).

It is challenging to accept an inert settlement form that evaded the marshes and adhered to the fringe dry zones (Asa 2011). Wetlands are crucial ecological means both for settlement and existence, their yearly converted resources given that reeds and other plant material for building supplies (along with animal fodder) and wildlife such as fish and waterfowl for food (Pournelle 2003). Settling in the wetland can be performed alongside the two old-style Near Eastern existence policies of wheat/barley cultivation and sheep/goat breeding. 91

Marsh settlement happens in two places: either on the many ancient rise area that upsurge above the waters or on small islands (*īshān*) built of stacked reeds and mud and surrounded by reed barriers. 95 The huts and almost the whole thing else are made of reeds merged with clay and water buffalo dung. In Iraq, these frequently take up as intricate buildings of thick reed pillars and barrel-curved reed roofs. The door and window openings are of adaptable reed mats. Throughout floods, the floor planes are effortlessly elevated or low barriers are built around the islands (Asa 2011).

Conveyance between huts and across the marshes is and always has been by small boat known locally as *masshuf* or *ṭarrāda*. These are built with wood and waterproofed with tar, a significant supply in ancient Mesopotamia (Asa 2011). Marsh inhabitants survive by raising water buffalo, fishing, hunting and growing rice. Water buffalos are used mostly for their milk (and milk products such as butter, *kaymar*, *labban* and cheese) and dung, which provides fuel and mortaring material (Asa 2011).

Marsh inhabitants who experienced a mixture of pastoralist, inactive and marsh existence grounded on the periodic growth, and collapse of the marsh waters has been recognised including the Maʿdān (Marsh Arabs) of southern Iraq. Marsh actions such as fishing, hunting and reed cutting are accomplished all year round; the reeds are used as fodder for the grazing sheep of the pastoralists, cattle/water buffalo of the farmers and mats for settlements.

It may be expected that the establishment of human in ancient times trailed analogous routes and therefore was not limited only to the dry edges. The dependence on short-lived means promotes the question of their conspicuousness within an archaeological landscape (Hritz 2007). In southern Iraq, early sign for wetland settlement and land use originates from a small site (H3) dated to the Ubaid (5500–5300 BCE) near the sea on the Kuwaiti coast northwest of Failaka Island. 108 Besides the remains of the marine fish and fowl, wetland catfish and freshwater shells were characteristic of those from the southern Mesopotamian marshes (Asa 2011). In addition, faunal mark for nurturing livestock and pastoralism in sheep/goat and cattle bones was also available. Among other artefacts that are obtained is a boat made of reeds and coated with tar thoroughly like Maʿdān's watercraft (Asa 2011). Establishment of human in the southern marshes of Iraq occurred in the Jemdet Nasr period (3200–2900 BCE) settlements, where detached chains of elevated land alienated by canals, similar to the huts style of Maʿdān settlements (Asa 2011). The signs for the settled groups of human in the marshes of Iraq can be traced through the figures of water buffalo that are depicted on clay tablets in the third millennium BCE (Adams and Nissen 1972).

1.5 The People of the Marsh Area

The people settled in the marsh area of Iraq are locally recognised as the Ma'dan or "Marsh Arabs", whose population is estimated to range from 350,000 to 500,000. Successors of the Sumerians and Babylonians, the inhabitants of the marshes act as a relic of the ancient Mesopotamia (UNEP 2001). From the ethnological perspective, the Marsh Arabs communities have been heavily subjective by arrivals and intermarriages with the Persians to the east and Arab Bedouins to the west (Thesiger 1964). The marsh people are Shi'ite Muslims, and their way of life is mostly grounded on the traditions of the Arab Bedouin (UNEP 2001).

The people of the marshes have developed an exclusive sustenance way of life that is decisively originated from their aquatic environment. For the Marsh Arabs, water buffalos are considered members of the family as they contribute to their socio-economic life, a relationship similar to that of the Bedouin and their camels (Thesiger 1957). The buffalos eat young reed shoots and give milk, butter and yoghurt, as well as energy and crop fertiliser in the form of fuel dung and manure (UNEP 2001). Fishing, waterfowl hunting and partially farming are the main professions of the Marsh Arabs.

The Marsh Arabs were confined to their marshes, and no connections with the people at the urban areas were made until the start of the First World War in 1914, where they got mixed with the remaining population of Iraq after the encouragement of the central government extended to remote parts of the country. In the early 1930s, Marsh Arabs started to move out of their environment and inhabit slum areas near the large cities such as Baghdad, Basrah, Ammarah and Nasiriyah (Batatu 1978).

1.6 Flora and Fauna

The majority of the area of the marshes in Iraq is covered with aquatic plants that are dominated by reed (*Phragmites communis*) and reed mace (*Typha angustata*) in the transient seasonal zone (UNEP 2001). Other types of aquatic plants are found in mudflats such as *Carex* and *Juncus* spp. and *Scirpus brachyceras*. Submerged plants were usually available in deeper areas of the marsh, and these are represented by hornwort (*Ceratophyllum demersum*), eel grass (*Vallisneria* sp.) and pondweed (*Potamogeton lucens* spp.), as well as bottom vegetation such as stonewort (*Chara* spp.). Floating vegetation are present in smaller lake, and among these are water lilies (*Nymphaea* and *Nuphar* spp.), water soldier (*Pistia stratiotes*) and duckweed (*Lemna gibba*) (Scott 1995; Rechinger 1964).

Situated on the routes of the migratory birds, the marshes are chiefly significant for birds. The marshlands set up a main wintering and staging area for waterfowl travelling between breeding grounds in the Ob and Irtysh river basins in western Siberia to wintering quarters in the Caspian region, Middle East and northeast Africa (UNEP 2001). Two-thirds of West Asia's wintering wildfowl, estimated at several millions, are believed to reside in the marshes of Al-Hammar and Al-Hawizeh (UNEP 2001).

Among the birds that are chiefly reliant on the marshlands are the Dalmatian pelican, pygmy cormorant, marbled teal, white-tailed eagle, imperial eagle, slender-billed curlew and an endemic subspecies of the little grebe (*Tachybaptus ruficollis iraquensis*) (UNEP 2001). The Goliath heron, sacred ibis and African darter, whose world population has been progressively dwindling, are also known to propagate in the marshes (UNEP 2001). The marsh area supports almost the entire global population of two species, the Basra reed warbler and Iraq babbler, as well as most of the world population of grey hypocolius (Maltby 1994; Scott 1995).

Mammals have been under massive impact. For example, lions were completely removed from the marsh area once the rifle was introduced during the First World War. The last lion shot in the area is stated to have been in 1945 (UNEP 2001). Among the threatened mammal species that inhabit the marsh area are the grey wolf, the long-fingered bat and a subspecies of the smooth-coated otter which is endemic to the marshes. Other large animals, notably the honey badger, striped hyena, jungle cat, goitered gazelle and Indian crested porcupine, have been recorded in the marsh area (UNEP 2001), in which they became very rare species and probably became extinct in 1980. Formerly, the utmost well-known mammal in the marshes was the wild boar, which sits as the main threat to the marsh dwellers' crops and was their chief opponent, and it declined in number drastically. Other usually seen mammals are the small Indian mongoose, the Asiatic jackal and the red fox. The usual reptiles seen in the marshes are the Caspian terrapin, a softshell turtle and a variety of snakes.

The fish accommodate a rich fauna in the marsh area; however, the invasive fish species such the species of carp have competed with the native species and later disappear from the marshes or their number has been reduced significantly. The dominant species living in the marsh area of Iraq are those belonging to the family Cyprinidae "the carp family". Members of other families such the catfish family Siluridae are represented in several species, with *Silurus triostegus* as the main species. Individuals of this species reach large sizes. Among the marine fish species that enter the marsh area for spawning are the well-known hilsa fish, *Tenualosa ilisha*, and to a lesser extent the pomfret *Pampus argenteus*. Many other marine fish species ascend in the Shatt al-Arab River reaching the marsh areas for feeding and reproduction purposes especially in recent years when the salinity of water started to increase notably. Together with the marine fish species, the penaeid shrimp *Metapenaeus affinis* was also among the attendees of the environment of the marsh area (Banister et al. 1994).

1.7 Impacts on the Marsh Area

The environment of the marsh area in Iraq has experienced several kinds of influences. Among such impacts was the influence of the changes in the water supply due to building dams in the upper Mesopotamian plain in Turkey and Iran and the ecocide that Saddam Hussein has implemented when he ordered to dry the marshes. In this section, a short notification will be given about the impacts of these two factors on the marsh areas.

The impacts on the marshes of Iraq were huge and comparable to those big events that happened somewhere else in the world such as the deforestation rates of Amazonia and the desiccation of the Aral Sea (UNEP 2001). The main reason behind such devastations is the constructions of dams in the upper reaches of Mesopotamia, i.e. Turkey, Syria and Iran. Such drastic changes in the environment of the marshes of Iraq had led to a noteworthy loss in one of the world's largest and most important wetland ecosystems. Among the marshes that has been severely affected is Al-Hammar, which is formerly considered as the largest lake south of the Euphrates River (UNEP 2001). Most of Al Hawizeh Marsh in Iraq has been diminished and changed into unfertile land. Only a small northern section remains and its shorelines are in steady retreat. The following are the major issues that happened to the environment of the marshes owing to drainage changes (UNEP 2001).

1.7.1 Habitat Loss

This section is too small to give the complete picture about the habitat loss in the marsh area; therefore, highlights of what happened will be given below:

1. At least 7600 km^2 of main wetlands (leaving out the seasonal and temporary flooded regions) disappeared between 1973 and 2000. Most of the change, however, happened between 1991 and 1995 (UNEP 2001).
2. The most extremely impacted are the Central and Al-Hammar Marshes. Of the original 3121 km^2 area of the Central Marshes in 1973, only 98 km^2 or 3% remained in 2000 (UNEP 2001).
3. Al-Hammar has diminished to 6% of its original extent. Once more, the enduring area is largely concentrated around the canals and does not constitute a real part of the original wetland system (UNEP 2001).
4. Al-Hawizeh has experienced a comparatively less severe decrease in its surface area. However, it has also diminished by 2000 km^2, leaving in place only a third of the original coverage. The shoreline of Hawr Al Hawizeh/Al Azim has been in stable withdrawal throughout the last decade (UNEP 2001).
5. The problem of shortage of water is likely to hasten as a consequence of considerable water retaining by the Karkheh Dam and plans to transfer water from its reservoir to Kuwait (UNEP 2001).
6. The impacts of the changes in the drainage influenced not only the marsh environment itself but also the north-western part of the Arabian Gulf. The reduction in the freshwater discharge from Shatt al-Arab River and the cessation of the filtering action of the marshes have led to drastic changes in the marine environment around Warbah Island at the Iraq-Kuwait border, and the water quality degraded with possibly harmful impacts on regional fish resources (UNEP 2001).

1.7.2 Refugees

The results of drying the marsh areas have impacted the life of the inhabitants of the marshes in the early 1990s; accordingly, the Marsh Arabs have been compulsory to flee their areas. Besides the engineering workings, their birthplace was converted to be one of the chief areas of fighting that overwhelmed southern Iraq in 1991–1993 (UNEP 2001). Many marsh villages were surrounded and harmed and fire was set in their huts. Different chemicals were reported to be added to the water of the marshes (Human Rights Watch 1993, 1994; United Nations 1993, 1994, 1999; Wood 1993). For those who remained inside Iraq, poor documentations about their life status were provided in which high percentage of them were dispersed in the outskirts of the big cities (AMAR ICF 2001; UNHCR 1996).

The Marsh Arabs who have forced to leave their homeland are entitled to have the designation of "environmental refugees". Such term was demarcated in a UNEP-commissioned study "as those who had to leave their habitat, temporarily or permanently, because of a potential environmental hazard or disruption in their life-supporting ecosystems" (El-Hinnawi 1985).

1.7.3 Wildlife Decline and Extinction

Owing to the huge loss in the marsh areas, the wildlife and biodiversity have severely influenced. Such impacts were extended outside the borders of Iraq and showed effects on both regional and international levels (UNEP 2001). Wildlife specialists agree that devastation of the wetlands would virtually positively lead to the global disappearance of the endemic smooth-coated otter subspecies and the bandicoot rat as they were exclusively reliant on this special habitat for their life. It can also lead to the vanishing in the Middle East of the African darter and the sacred ibis and the extermination in Iraq of the pygmy cormorant and Goliath heron (Scott and Evans 1994).

The great effects of the drying of the marshes were severe on the migratory birds. The event of drying of the marshes affected bird's populations from the Arctic region to southern Africa (UNEP 2001), where about 66% of the bird's species visiting the marsh area are an internationally significant number and they are at risk (UNEP 2001). On the other hand, the global populations of the Iraq babbler and the Basra reed warbler and the regional population of the Dalmatian pelican are likely to fall and may be diminished as a result. It is assessed that the world populations of Harrison's gerbil, an endemic subspecies of the little grebe and the marbled teal, may have declined by 50%. Moreover, chief drops in the regional populations of the eastern white pelican (*Pelecanus onocrotalus*, 30–60%), purple heron (*Ardea purpurea*, >10%), little bittern (*Ixobrychus minutus*, >10%), glossy ibis (*Plegadis falcinellus*, >10%), tufted duck (*Aythya fuligula*, >20%), marsh harrier (*Circus aeruginosus*, >10%), purple gallinule (*Porphyrio martinicus*, >50%) and coot (*Fulica atra*, 10–20%) are expected (Scott and Evans 1994).

1.7.4 Regional Weather Alteration

Speedy drying of over 9000 km^2 of wetlands and lakes is destined to have noteworthy consequences on the local microclimate (UNEP 2001). As the controlling part of the wetlands is removed, evapotranspiration and humidity levels will abruptly decrease. Rainfall designs will be altered. Temperatures will always increase, chiefly throughout the hot and long summers. Strong and dry winds attaining temperatures of over 40 °C, formerly broken by the reed beds, will blow unrestricted (Maltby 1994). The dry marshes that contain impurities will be blown to thousands of kilometres outside the border of Iraq and inflict their effects on people and environment in other neighbouring countries.

References

Adams RMC, Nissen H (1972) The Uruk countryside. University of Chicago Press, Chicago
AMAR ICF (2001) Iraqi marshlands: prospects (draft report). AMAR ICF, London
Aqrawi AA (2001) Stratigraphic signatures of climatic change during the Holocene evolution of the Tigris–Euphrates delta, lower Mesopotamia. Glob Planet Chang 28(1–4):267–283
Asa A (2011) The swamps of home: marsh formation and settlement in the early medieval Near East. J Near East Stud 70(1):55–79
Baeteman C, Dupin L, Heyvaert V (2005) Geo-environmental investigation. Akkadica (126, fasc. 1):5–12
Banister K, Backiel T, Bishop J (1994) The present state and likely future of the fishes of the Tigris-Euphrates Basin. Unpublished report, Wetland Ecosystems Research Group, University of Exeter
Batatu H (1978) The old social classes and the revolutionary movements of Iraq: a study of Iraq's old landed and commercial classes and of its Communists, Baathists, and Free Officers. Princeton University Press, Princeton
Dellapenna JW (1996) Rivers as legal structures: the examples of the Jordan and the Nile. Nat Res J 217–250
El-Hinnawi E (1985) Environmental refugees. UNEP, Nairobi
Heyvaert VMA, Baeteman C (2007) Holocene sedimentary evolution and palaeocoastlines of the Lower Khuzestan plain (Southwest Iran). Mar Geol 242(1–3):83–108
Hritz C (2005) Landscape and settlement in southern Mesopotamia: a geo-archaeological analysis. Ph.D. dissertation, University of Chicago, p 43
Hritz C (2007) Appendix III: remote sensing of sites in and around the Hawr al-Hammar and Hawr al- Hawiza. In The Persian Gulf Shorelines and the Karkheh, Karun, and Jarrahi Rivers: a geo-archeological approach. A Joint Belgo-Iranian Project. First Progress Report – Part 3, ed. Herman Gasche, Akkadica 128 (2007):63
Human Rights Watch (1993) Human Rights Watch world report 1993 (events of 1992). Human Rights Watch, New York
Human Rights Watch (1994) Human Rights Watch world report 1994 (events of 1993). Human Rights Watch, New York
Kornfeld IE (2009) Mesopotamia: a history of water and law. In: The evolution of the law and politics of water. Springer, Dordrecht, pp 21–36
Leick G (2002) Mesopotamia: the invention of the city. Penguin, London
Leveson D (1980) Geology and the urban environment. Oxford University. Press, New York
Maltby E (ed) (1994) An environmental and ecological study of the marshlands of Mesopotamia. Draft consultative bulletin. Wetland Ecosystems Research Group, University of Exeter. Published by The AMAR Appeal Trust, London
Pournelle JR (2003) Marshland of cities: deltaic landscapes and the evolution of early Mesopotamian civilization. PhD dissertation, University of California, San Diego. https://doi.org/10.6067/XCV8J67GK3
Rechinger K (1964) Flora of lowland Iraq. Hafner Publishing Ltd., New York
Sanlaville P (2002) The deltaic complex of the lower Mesopotamian plain and its evolution through millennia. The Iraqi Marshlands:133–150
Scott D (ed) (1995) A directory of wetlands in the Middle East. IUCN/IWRB, Gland/Slimbridge
Scott D, Evans M (1994) Wildlife of the Mesopotamian marshlands. Unpublished report, Wetland Ecosystems Research Group, University of Exeter
Thesiger W (1957) Marsh dwellers of Southern Iraq. Nat Geogr CXIII:204–239
Thesiger W (1964) The marsh Arabs. Penguin, Harmondsworth
UNEP (2001) Partow, H. The Mesopotamian Marshlands: Demise of an Ecosystem Division of Early Warning and Assessment United Nations Environment Programme Nairobi, Kenya, 47p
UNHCR (1996) Background paper on Iraqi refugees and asylum seekers. UNHCR Centre for Documentation and Research, Geneva

United Nations (1993) Situation of Human Rights in Iraq (E/CN.4/1993/45). United Nations Commission on Human Rights, Geneva
United Nations (1994) Situation of Human Rights in Iraq (A/49/651). United Nations Commission on Human Rights, Geneva
United Nations (1999) Situation of Human Rights in Iraq (E/CN.4/1999/37). United Nations
Wood M (1993) Saddam drains the life of the Marsh Arabs. The Independent, Saturday 28 August 1993

Section I
Historical Perspectives

Chapter 2
Cities in the Water: Waterscape and Evolution of Urban Civilisation in Southern Mesopotamia as Seen from Tell Zurghul, Iraq

Davide Nadali

Abstract Ancient canals, marshes and proximity of the sea heavily characterised the landscape and environment of the ancient State of Lagash in southern Iraq, from the mid-fifth to the second millennium BC: indeed the diachronic changes that can be analysed thanks to geological and archaeological observation and investigation show how this waterscape definitely influenced the shapes of settlement and the organisation of ancient societies from a cultural, economic and biological point of view.

Recent excavations at Tell Zurghul in southern Iraq are giving the possibility to test, in the field, the presence of water: ancient cuneiform sources, from the mid-third millennium BC, show the intense programme of the rulers of the State of Lagash in managing water through the construction of canals and the regulation of marshes characterised by marine water due to the proximity of the sea. In this respect, human actions (such as the digging of canals) and natural conditions (such as the reduction in the fifth millennium and the progressive growth in the fourth millennium BC of water level) are recognisable in the field, and they of course explain the morphology of the site in the past and the changes it suffered even in the present: water in fact is doubtless a fundamental resource for suitable conditions of formation and growth of a urban centre, but it also limits the possibility of extending occupation on the entire surface (as, e.g. the exploitation of lands for agricultural purposes).

Keywords Marshes · Canals · Lagash · Mesopotamia · Tell Zurghul · Waterscape · Urbanisation · Water

The study of the presence and management of water in southern Mesopotamia recently increased, thanks to the research that principally focused on the analysis

D. Nadali (✉)
Sapienza Università di Roma, Rome, Italy
e-mail: davide.nadali@uniroma1.it

L. A. Jawad (ed.), *Southern Iraq's Marshes*, Coastal Research Library 36,
https://doi.org/10.1007/978-3-030-66238-7_2

Fig. 2.1 Tigris and Euphrates alluvial watercourses, 8000–1720 BC. (After Pournelle 2003: fig. 26)

of the morphology of landscape and the detection of ancient canals on the surface, specifically between the main Mesopotamian rivers, Euphrates and Tigris (Fig. 2.1).[1] Indeed, the two rivers strongly characterised and shaped the Mesopotamian landscape, particularly if we think of the changes their riverbeds suffered:[2] at the same time, changes to the morphology of the landscape have also been caused by the anthropic action of digging new canals, a feature that allowed the regulation of water, the irrigation of fields as well as the transportation and movement of both people and goods.[3]

Up to now, little research has been conducted in the field: with the few exceptions of surveys made in southern Iraq (Hammar district, Dhi Qar, and Basra)[4] and

[1]Pournelle 2013; Wilkinson 2013; Wilkinson and Hritz 2013. The present resarch stems from the 3-years research "Fluid Crescent. Water and Life in the Societies of the Ancient Near East", funded by the Italian Ministry of University and Research (PRIN 2017, no. 2017NMK5FE).

[2]Garzanti et al. 2016: 112–113; Pournelle 2013: 14–16; Ur 2013: 132; Wilkinson 2003: 82; Wilkinson and Hritz 2013: 20–21.

[3]Liverani 1990; Widell 2013: 58–59; Widell et al. 2013: 66–73.

[4]Hritz et al. 2012a; Al-Hamdani 2014a.

excavation of palaeochannels aimed at their reconstruction and dating,[5] studies have been so far limited to the analysis of CORONA satellite images:[6] on the one hand, if this kind of research allows the possibility of detecting traces of canals on the surface, it does not allow on the other hand the complete understanding of the formation and use of those waterways in antiquity. In particular, the real question concerns the time: that is, when have those canals been dug? How long have they been used? Assumptions merely based on the observation of satellite images are not sufficient: a careful analysis and survey directly in the field must necessarily complete the preliminary considerations. At the same time, cuneiform sources give very important and useful information on the system of canals and the water environment of ancient Sumer: excavations of canals, constructions of dams and the existence of an organised bureaucracy for the management of this complex water network are extensively documented in the texts of the third millennium BC, showing not only the high degree of technology but also the constant and necessary care of the waterways that were used for irrigation, transport and communication.[7] In this respect, it is interesting to point out that the first real developed system of irrigation network can be precisely identified, in the Early Dynastic IIIb Period, in the ancient Sumerian State of Lagash that included the cities of Girsu, Lagash, Nigin (Tell Zurghul) and the yet unidentified Gu'abba.[8]

Extensive surveys on terrain – that offer the possibility of matching the archaeological evidence with the information provided by texts – have been possible only in recent times: the resumption of archaeological investigations in southern Iraq in the provinces of Basra, Dhi Qar and Al Diwaniyah provides the opportunity to shed light on the morphology and formation of the ancient waterscape, taking into consideration the presence of rivers, the artificial excavation of canals and the area of the marshes (that must also encompass the study of the ancient marine shore that should be considered more as a threshold rather than a proper border).[9]

The archaeological project – a joint expedition of Sapienza University of Rome and the University of Perugia established since 2014 in the region of Dhi Qar at the site of Tell Zurghul, ancient Nigin (Fig. 2.2) – also encompasses the study of the territory and the morphology of the area focusing on how landscape has been transformed and affected by water. Actually, the entire area of the site (with the sole exception of the two main mounds, labelled A and B) has been deeply characterised and transformed by the presence of water: in this respect, the landscape of Tell Zurghul can rightly be labelled as a waterscape, and one might even advance the hypothesis that the growth and formation of the settlement strongly depended on

[5]See in particular in Jotheri et al. 2016 and Jotheri et al. 2018.

[6]Hritz and Wilkinson 2006; Hritz 2010, 2014.

[7]Bagg 2017.

[8]Steinkeller 2001; for a detailed analysis of the "canal which goes to Nigin" that, with its total length of about 50 km, was the largest canal in the ancient state of Lagash, see Carroué 1986 and Rost 2011.

[9]Pournelle 2007, 2013: 19–20.

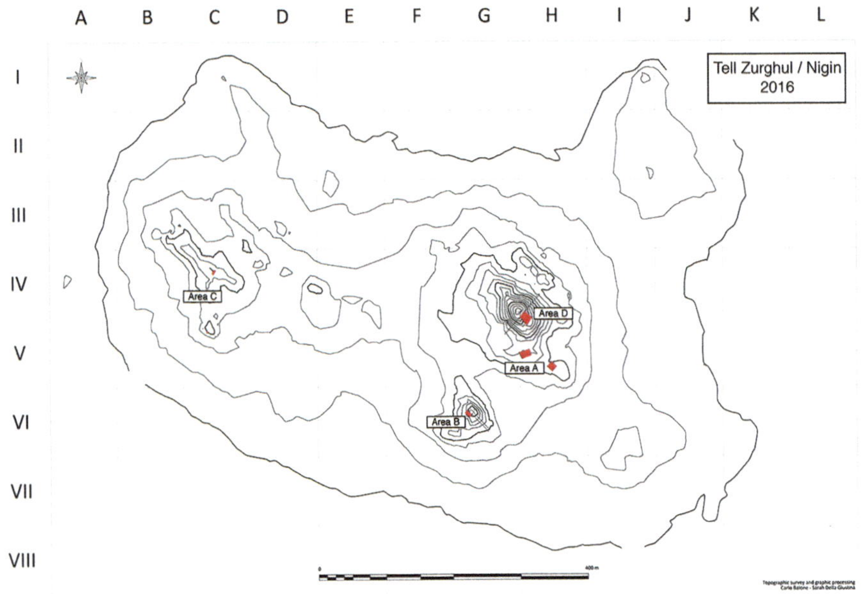

Fig. 2.2 Topographic map of Tell Zurghul (© MAIN)

the presence/absence of water.[10] In this respect, for example, the sea ingression, with the sea level gradually rising during the Holocene Period (about 6000 years BP), led to the advance of the coastline to close proximity to sites such as Ur, Eridu and Tello (ancient Girsu);[11] in the Ubaid Period (firth millennium BC) – as it has been possible to recognise in southern Mesopotamia near Tell Oueili and Larsa[12] – the sea level was slightly higher than today, and early Ubaid villages and settlements mainly occupied exposed surfaces of Pleistocene "turtlebacks":[13] the later process of sediment accumulation therefore hid the earliest occupation levels;[14] conversely, the heavy aeolian degradation revealed parts of archaeological sites dating to even earlier phases.[15] Environmental and natural conditions heavily affected the archaeological landscape of the south plain of Mesopotamia,[16] particularly taking into

[10]Wilkinson and Hritz 2013: 18.

[11]Sanlaville 1989; Pournelle 2013: 19.

[12]Sanlaville 1989; Geyer and Sanlaville 1996.

[13]Hritz et al. 2012a: tab. 1; Pournelle 2003, 2007, 2013: 22.

[14]As has been recognised in the levels Ubaid 0 and 1 at Tell Oueili (Vallet 1996; Pournelle 2007: 46–48).

[15]Pournelle 2013: 20; Wilkinson and Hritz 2013: 19.

[16]The study of the morphology and topography of Mesopotamian sites in the southern plain heavily depends on the analysis of natural conditions, environment and landscape in a diachronic perspective, showing changes across time. In this respect, it seems useful to point out the altitude of Tell Zurghul with the highest point at 12 m asl (on the main Mound A) and the lowest point of −1 m

consideration the role of water and the consequent possible seasons of inundation of the area (Fig. 2.3).[17]

Not only do cuneiform sources of the third millennium BC (Early Dynastic IIIb and Neo-Sumerian Period) testify to the intensive and programmatic actions of the rulers of the first and second dynasty of Lagash in excavating canals (the "canal which goes to Nigin" is only one of the most documented and well known).[18] Archaeological evidence on the site also shows a massive and almost constant presence of water that can be clearly recognised in the quite large white bands and pools (made of sandy soil);[19] therefore the presence of water has been long lasting, and it might be inferred that it deeply affected the morphology of the settlement as well as the stratigraphy and integrity of archaeological deposits.

Excavations, carried out at Tell Zurghul in the recent years (2015–2017), precisely showed and pointed out this aspect: on one hand, water was a fundamental resource for the growth and transformation of the urban centre; on the other, speaking from a strictly archaeological perspective, water was, indeed it has been until recent times, extremely invasive: the formation of small rivers during the heavy winter rains erodes the archaeological deposits, resulting in the presence of strata of accumulated debris and archaeological materials (namely pottery) that are dragged downstream from the main mounds and elevations (Fig. 2.4). As previously mentioned, erosion by wind and heavy rains, moreover, contributed to the disappearance of the upper layers of occupation of the archaeological sites:[20] taking the original environment of the archaeological sites of southern Mesopotamia into consideration, one might evaluate how the presence of water affected the growth and formation of ancient settlements with the consequent result that if occupation of places has been favoured by water, it has also been adapted to water; as suggested by Carrie Hritz and Jennifer Pournelle,[21] the system of southern Mesopotamia can be described as resilient due to the ability of coping and adapting to change. New archaeological

---

below sea level: this last datum shows the very special morphological nature of Mesopotamian sites in the southern alluvium characterised and shaped by the natural actions of water, wind and accumulation together with anthropic activities. In particular, the morphology of Tell Zurghul reflects the processes of sedimentation and the Gulf ingression with the continuous changes of sea level (increasing in the Holocene period, with a retreat to approximately modern sea level at the beginning of the second millennium BC). See Potts 1997: 33–4; Hritz et al. 2012b; Ur 2013: 132.

[17] Wilkinson 2003, 76–80, 87. The extensive presence of water covering the surface of Tell Zurghul has been also documented by some of the first visitors to the site: the American scholar Raymond Dougherty (1926) could in fact reach the site only by boat; again, at the end of the 1980s, Jeremy Black (1989–1990) says that to reach Zurghul from al-Hiba/Lagash, it was necessary to travel around the marshes.

[18] On the "canal which goes to Nigin," see the analysis by Rost 2011.

[19] Soundings in Operation A1 (to the south-east of Mound A) in fact revealed the presence of a thick deposit of sand corresponding to a place originally occupied by water, maybe one of the streams that crossed the city. It is interesting to point out that the northern limit of the strata of sand presented consistent layers of accumulation with materials (pottery) carried by water from the main mound.

[20] Wilkinson 2003: 81–83.

[21] Hritz and Pournelle in press.

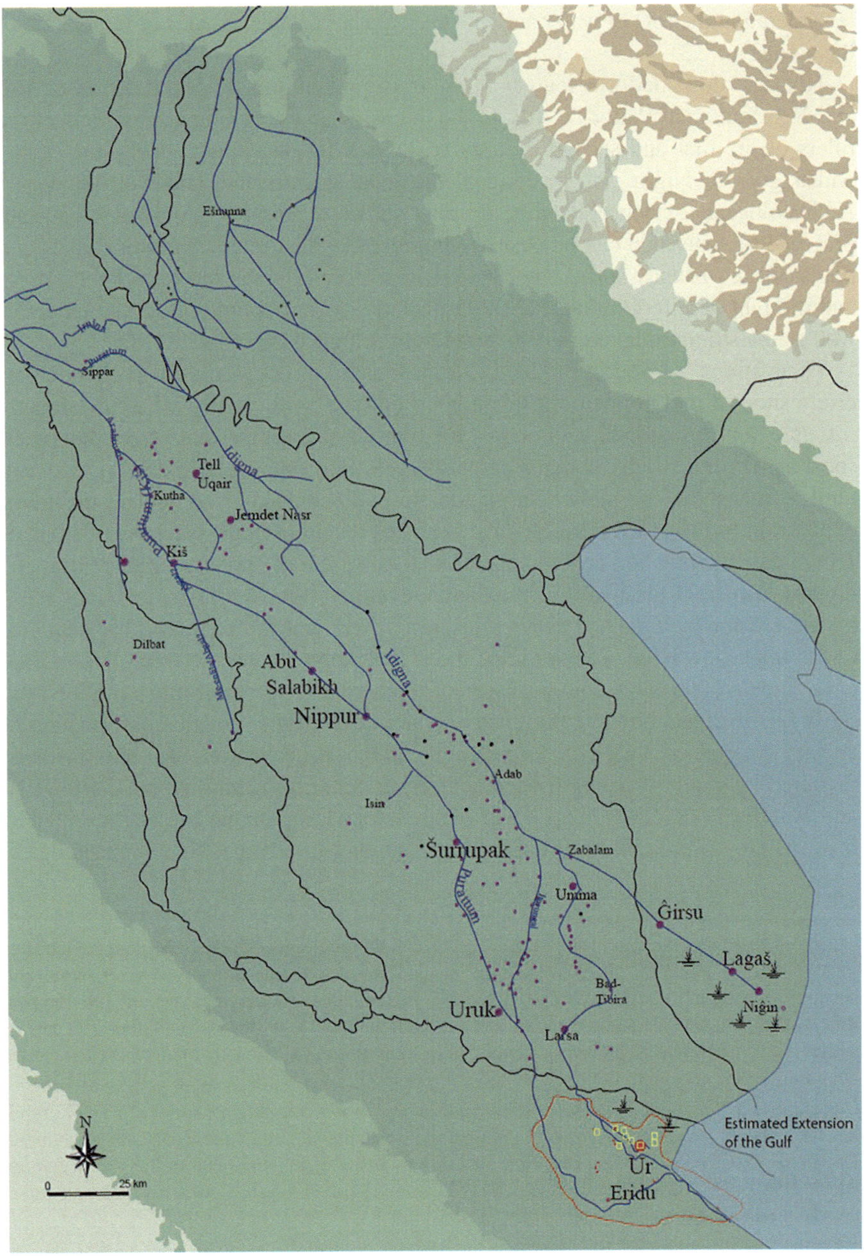

Fig. 2.3 Third millennium BC southern Mesopotamia. (After Benati 2015: fig. 2.1)

Fig. 2.4 Action of rain and traces of salt on the surface (© MAIN)

excavations in southern Iraq should also start from this premise: shapes and extension of sites change according to the features and evolution of natural ecosystems, and water is doubtless the most prominent element to which ancient Mesopotamian inhabitants and rulers paid specific attention with the construction of new and the care of old canals for the regulation of hydric resources. This is in fact a precise duty of Mesopotamian rulers: the care of ancient canals is prerogative for the good management of cities and territory they received from gods. The management and control of waterways was so important for Mesopotamian kings that hydraulic warfare occurred in the mid-third millennium BC between the cities of Lagash and Umma:[22] for a long period, the two cities confronted each other for the control of Gu'edenna, a very fertile and well-irrigated portion of land. These events explain how water was important for the economy of the cities of southern Mesopotamia, particularly in the area where cultivation of arable lands was possible only, thanks to artificial irrigation,[23] a situation completely different from the northern Mesopotamia, the region of Babylonia – as defined by Piotr Steinkeller – to make a distinction from the southern alluvium that can properly be labelled as the land of Sumer.[24]

The archaeological excavations at Tell Zurghul concentrated on three areas with operations that delivered layers of occupation dating to the Late Ubaid Period

[22] I owe the definition of "hydraulic warfare" to Ingo Schrakamp who used these terms in a communication presented at the workshop "Ancient Lagash– a workshop on current research and future trajectories" held at Vienna in 2016 within the 10th ICAANE. See also Selz 1998: 312.

[23] Widell 2013.

[24] Steinkeller 1999: 290.

(Area B),[25] the final phase of the fourth and beginning of the third millennium BC (Area A), and finally to the late third millennium BC (Neo-Sumerian phase) (Area D) (Fig. 2.2). At the same time, an extensive survey has been conducted along the western edge of the site (Area C) with a collection of pottery sherds showing that the site has long been occupied up to the very beginning of the second millennium BC (Isin-Larsa Period in common Mesopotamian chronology): the nature of the soil and the high percentage of slags so far recovered and registered seem to confirm that this area has probably been devoted to working activities with the presence of workshops and kilns for the production of pottery and bricks.[26]

As pointed out, the morphology of the site and the nature of the soil clearly show that water was a natural element that deeply characterised and, as we might infer, transformed the settlement: beyond the canal quoted in the inscriptions of Urukagina and largely described by Gudea,[27] the site has been largely and extensively covered by water (flooded) in the last century. This situation lasted for years, and only recent reclamation and climatic changes made the site drier with the presence of water pools (swamps and lagoons) localised along the margins.[28]

Since 2017, an agreement with geologists from Sapienza University of Rome and the University of Perugia has been specifically initialled for the study and analysis of the geological and hydrogeological aspects of the site and, more in particular, of the white areas made of sandy soil to detect the persistence of water as well as the period of formation of either natural or artificial aquatic surfaces (Figs. 2.5 and 2.6). However, information gathered from geological soundings can then be compared with the data of cuneiform sources of the first and second dynasty of Lagash – thus having the possibility of mapping in the ground the public works of digging canals since the mid-third millennium BC. At the same time, results of geological analyses can supply important and fundamental hints for the comprehension of the development of the settlement across time:

- How and when was the site occupied?
- Was it extensively occupied in all periods?
- Did water prevent the occupation of some areas of the site?
- Did water force and therefore guide the process and progress of occupation and growth of the urban centre?

These are only a few questions that new archaeological and geological investigations at Tell Zurghul and in the area of the ancient State of Lagash can answer. In few words, as a very preliminary assumption, the extensive presence of marshes and

[25]Nadali and Polcaro 2016.

[26]Area C is also extensively characterised by the presence of drain pipes that could in fact be attributed to domestic structures no longer visible and preserved or, in fact, to workshop and activities related to the use and management of water (drainage). See McMahon 2015; see also George 2015.

[27]See Heimpel 1998, with references to aquatic environment and canals of the ancient city of Nigin.

[28]Still today in winter time, pools of water can be seen on the western and north-western margins of the site, but not directly touching and covering the site.

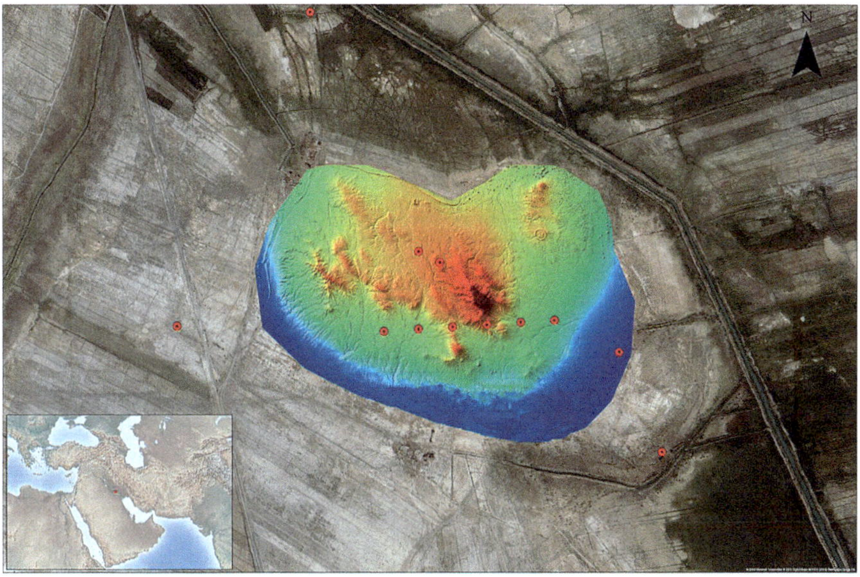

Fig. 2.5 Elevation core locations, Tell Zurghul (© MAIN)

Fig. 2.6 Sounding, Operation A1, showing the sediment of sand in the lower city (© MAIN)

Fig. 2.7 Area B, the rest of shark vertebras in situ, sector B North, 2017 (© MAIN)

regularly excavated canals might have shaped the region more generally, and the city more specifically, with emerging places and hills occupied by buildings. Geological investigations are essential to verify and clarify where and how water was present at the site. Marshes of the southernmost alluvium of Mesopotamia were characterised by marine water, with an impact on the ecosystem (flora and fauna) and the possible use of water for irrigation.[29]

In the northern sector of Area B, excavations brought to light a structure made of three rooms connected to each other. Two filling layers of soft sandy grey soil covered the floors: those two layers are characterised by the presence of many fishing net clay weights, shells and fish vertebras and some preserved fish skins. Two sickles and two grinding stones, almost complete, were also found together with a small worked crystal. This archaeological evidence from Area B confirms that Zurghul, during the Chalcolithic Period, was largely covered by marshes of brackish-marine water. This explains in fact the recovery of fish vertebras (Fig. 2.7) of "bull shark" (*Carcharhinus leucas*), one of the few species that are able to go back upstream in the rivers and that could perfectly live in a brackish-marine environment.[30] At the same time, the presence of the typical Ubaid sickle (Fig. 2.8) can be also connected with the waterscape of the region: sickles in fact, instead of being used for agricultural purposes (at least not exclusively), were employed to cut reeds in the marshes, thus making the movement on boats and the fishing easier. Interestingly enough, at

[29]Wilkinson and Hritz 2013: 23–4, 27–8.

[30]Jawad 2018.

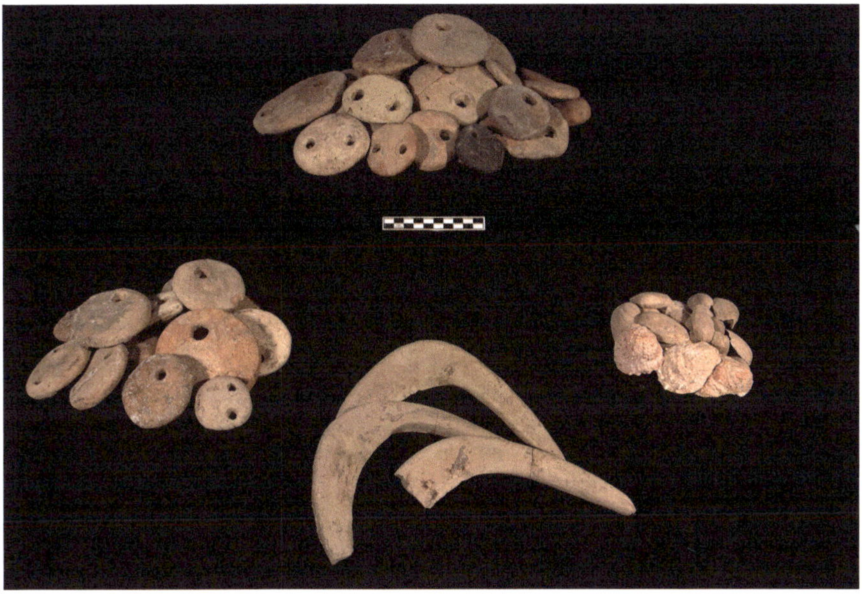

Fig. 2.8 Area B, clay sickles and group of fishing net clay weights from the sector B North, 2017 (© MAIN)

Tell Zurghul, clay sickles have been found in association with the vertebras and some fishing net clay weights.

Salinisation of soil affects the exploitation of lands for agriculture:[31] the surface of Tell Zurghul is characterised by a thick stratum of salt, and it is interesting to note that this does not completely cover the surface of the site, but is located on precise spots – most precisely, salt is present in the western sector of the site, on the inner and outer side of the elongated elevation identified and surveyed in Area C. Once salt is removed, thick strata of silt (humid clay), often without consistent evidence of archaeological materials, are found, and they can be interpreted as the result of alluvial deposits created by the movement of water across the ancient city.

An inscription of Urukagina (Early Dynastic IIIb) states that "For Nanshe [the patron deity of Nigin] he constructed the Nigin-Going-River, built the Eninnu at its beginning, Sirara House at its end" explaining that he further "extended it for her into the midst of the sea".[32] The words of the ruler of Lagash show that the ancient city of Nigin, still in the third millennium BC, laid on or was very close to the sea: the indication of Urukagina is the proof of a wide work of hydraulic engineering with the canal entering into the sea or, as we might suppose, flowing into the marine delta of the southern alluvium that was characterised by marine incursion into the swamps of

[31]Powell 1985; Sanlaville 1989: 8; Widell 2013: 58; Wilkinson 2013: 37, 40, 43.

[32]Heimpel 1998: 153.

the region. Also Gudea refers to the Canal Going to Nigin: the waterway linked the northern city of Girsu to the southernmost city of Nigin of the ancient State of Lagash – used for transportation and on occasion of religious ceremonies,[33] the canal surely shaped the landscape of the entire region, being a connection between the three cities of the State and even reaching the sea or seashore as it was during the time of Urukagina. Again, texts from the period of the Third Dynasty of Ur show the existence of a harbour and shipyard at Nigin or in the surroundings of the city.[34] The location of the ancient Gu'abba, the harbour of the State of Lagash and the trade centre where merchants and goods from the Indus Valley reached Mesopotamia, is still a matter of debate.[35] Based on the analysis of the description and measures given in the cuneiform sources, one might be suppose that the ancient canal excavated by Gudea coursed on the eastern side of the city, coming from the North: observing the satellite image of Tell Zurghul, it is possible to detect a canal on the eastern side, although it is difficult to ascertain whether it corresponds exactly to the ancient one that presumably ran into the city reaching, as Gudea says, the mountain – presumably corresponding to the main mound of the site – where the temple Sirara of the goddess Nanshe was built.[36]

For that reason, the programmed geological and hydrogeological analysis in the field aims to precisely detect the real extension of waterway, the ancient trace and the identification of the levees that surely were quite impressive in dimension if one thinks of the total length of about 50 km of the canal, indeed the largest one in the ancient State of Lagash.[37]

The dense network of canals crossing the landscape of central and southern Mesopotamia is not only clearly visible from aerial and satellite images, with the traces of waterways and pools of water; direct observation in the field is a necessary integration of the distance analysis, with the possibility of verifying the implications on the environment and morphology of ancient settlements. The excavation of canals required a huge involvement of large investment and labour:[38] the operation was a social, political and cultural activity having different levels of function and impact. The excavation of canals was a duty of Mesopotamian rulers, but indeed it was also a real necessity for the management and exploitation of arable lands: texts of the third millennium BC, from the Early Dynastic Period to the time of the Third Dynasty of Ur, clearly point to the attention Mesopotamian rulers paid to the care of old and the excavations of new canals. Texts also enlist the names of the positions of appointed employees, and they give the total amount of specialised workers that were specifically devoted to this office. Moreover, cuneiform sources also show that the general

[33]Rost 2011.

[34]On the Sumerian mar-sa, its function and management, in the time of the Third Dynasty of Ur, see Alivernini 2013a, b.

[35]Vermaak 2008.

[36]Huber Vulliet 2009–2011.

[37]See the reconstruction in Rost 2011.

[38]Steinkeller 2013.

idea of a system exclusively based and depending on the central management and administration of the regulation of water sources must be questioned, having texts and data that in fact document the existence of different levels of action and people that were involved in the planning of the irrigation and regulation of the correct flow of waters (from the main rivers to smaller canals, with the construction of ditches, dykes and dams).[39]

The analysis of the environmental landscape where the cities of the ancient State of Lagash arose and developed allows researchers to reconsider the role and importance of the presence (one might even say the predominance) of water around the sites, at least for the area of southern Mesopotamia with the very interesting and special environmental condition of the presence of rivers, artificial canals and the sea. However, the old view of the hydraulic society of Karl A. Wittfogel applied to ancient Mesopotamia should not be reformulated according to the archaeological evidence:[40] in this respect, the complex hydraulic system of southern Mesopotamia is not (at least not only) a derivation of city-states and political powers that transformed and used water to build and support their socio-economic system; in fact, it might be said that the contrary is true, that is, the waterscape of southern Mesopotamia worked as a prerequisite or even a fundamental condition to make city-states grow and enlarge their power to the level of regional and territorial states. The pre-existing hydraulic landscape (or waterscape) of ancient southern Mesopotamia contributed to the birth and the economic growth of cities, marking the development of social and political complexity.[41]

The operation carried out at Tell Zurghul since 2015 on the top and south-western slope of Mound B led to the identification of phases of occupation that can be dated to the Ubaid 4 Period (ca. 4800–4500 BC):[42] although evidence is still too scanty to present comprehensive conclusions, the data so far unearthed in the area of Mound B are intriguing if historical period (Ubaid), morphology of the mound and archaeological evidence (architecture and material culture) are taken into consideration and matched together.

Mound B has a surface of nearly 0.7 ha, with an altitude of nearly 4 m on the surrounding level of the lower city (Fig. 2.9). It has been very briefly investigated by Robert Koldewey, with a trench on the south-eastern slope and a square sounding on the top.[43] The presence of well-preserved and undisturbed Ubaid layers just below the surface of the mound is not surprising: wind and heavy rains can contribute to the remodelling of ancient mound.[44] Interestingly, the preliminary data pointing to the occupation of Mound B of Tell Zurghul in the Late Ubaid Period fits in with the

[39]On the administrative organisation and technical terms of the management of water in ancient Sumer, lastly see Schrakamp 2017.

[40]Wittfogel 1955, 1957. See also the critical reconsiderations by Liverani (2013: 162–8, 280).

[41]Algaze 2001; Wilkinson et al. 2015.

[42]Nadali and Polcaro 2016.

[43]Koldewey 1887: 416, 429.

[44]Pournelle 2013: 20.

Fig. 2.9 General view of Mound B, from North-East (© MAIN)

results of the survey of Abdulamir al-Hamdani that revealed that the area of Lagash was occupied in the Late Ubaid Period with settlements of no more than 3 hectares, mainly located around the city of Girsu and Bad-Tibira, as well as al-Hiba itself, and the southern city of Tell Zurghul, and having the shape of turtlebacks emerging from the water.[45]

The settlement dated to the Late Ubaid Period so far excavated at Tell Zurghul seems to have the characteristic of a small turtleback mound: it might be inferred that the proximity of Tell Zurghul to the sea, within a waterscape characterised by marshes, definitely forced the occupation of the pre-existing natural mounds; at the same time, earlier phases of occupation,[46] as well as levelling and rebuilding, led to the increasing and enlargement of both the mound and the settlement, with a second lower hill to the south-west. The large presence of clay cones on the surface, mostly found within gullies, indicates the existence of public buildings, the pottery being coherent with the Ubaid phase.

Surrounded by marine marshes, the mound has been extensively occupied, and further research will be conducted to verify the real extension of Ubaid layers, encompassing the area at the foot of the mound: in fact, from a general observation and first investigation of the area, Mound B, at least according to the stratigraphy of the area,[47] has not been reoccupied and covered by later structures, although Koldewey collected later materials on the surface and in the deep soundings he excavated during his brief exploration to the site in 1887.[48]

Excavation in Area A, to the south of the main mound, revealed layers of occupation from Late Uruk to the Early Dynastic I, thus showing an uninterrupted

[45]Al-Hamdani 2014b; Hritz et al. 2012a; Nadali and Polcaro 2016: 82.

[46]Sherds dating to the phase Ubaid 2 have been collected on the surface of Mound B: this datum therefore points to the previous occupation of the area.

[47]Nadali and Polcaro 2016.

[48]Koldewey 1887; Huh 2008: 245–6, 252, 752.

sequence from the end of the fourth to the third millennium BC: earlier occupation has not been detected, and one might therefore wonder and infer whether Tell Zurghul could thus have been made of several hills (turtleback mounds) or spots emerging from the surrounding waterscape of marshes.

Further study of the morphology and occupation of the site on its whole surface might in fact show that settlements moved from one spot to another according to the extension, presence and absence of water in each period: not only climate changes and natural sedimentation, but also anthropic actions (dams, canals, irrigation and reclamation) heavily affected the environment and the presence of water around and within the site. In summary, water on the one hand favoured the birth and growth of the settlement(s), but on the other hand, it might have forced and conditioned people to choose available and emerging places adapting to the changes, either natural or generated by man.

References

Algaze G (2001) Initial social complexity in Southwestern Asia. The mesopotamian advantage. Curr Anthropol 42:199–233

Al-Hamdani A (2014a) Dirasa Maidaniya lil mawaqei al-Athariya fi Hair al-Hammar [A field survey for the archaeological sites in Hawr al-Hammar]. Sumer 59:63–115. (in Arabic)

Al-Hamdani A (2014b) Kingdom of reeds: the archaeological heritage of Southern Iraqi Marshes. TAARII Newslett 9(1–2):15–20

Alivernini S (2013a) La struttura amministrativa del mar-sa nella documentazione della III dinastia di Ur (Supplemento n. 1 alla Rivista degli Studi Orientali, Nuova Serie 86). Fabrizio Serra editore, Roma

Alivernini S (2013b) Some Consideration on the Management of an Administrative Structure in Ur III Mesopotamia: The Case of mar-sa. In: Garfinkle SJ, Molina M (eds) From the 21st century B.C. to the 21st century A.D. Proceedings of the international conference on Neo-Sumerian Studies Held in Madrid, 22–24 July 2010. Eisenbrauns, Winona Lake, pp 105–113

Bagg A (2017) Watercraft at the beginning of history: the case of third-millennium Southern Mesopotamia. In: de Souza P, Arnaud P (eds) The sea in history. The ancient world. The Boydell Press, Woodbridge, pp 127–137

Benati G (2015) Re-modeling political economy in early 3rd millennium BC Mesopotamia: patterns of socio-economic organization in Archaic Ur (Tell al-Muqayyar, Iraq). Cuneiform Digital Library Journal 2015/2. http://www.cdli.ucla.edu/pubs/cdlj/2015/cdlj2015_002.html

Black JA (1989–1990) A note on Zurghul. Sumer 46:71–73

Carroué F (1986) Le "Cours-d'Eau-Allant-à-NINA^{ki}". Acta Sumerol J 8:13–57

Dougherty RP (1926) Searching for ancient remains in lower 'Irâq: report of an archaeological survey made in Southern Babylonia during the first quarter of 1926. Annu Am Sch Orient Res 7:1–93

Garzanti E, Al-Juboury AI, Zoleikhaei Y, Vermeesch P, Jotheri J, Bal Akkoca D, Obaid AK, Allen MB, Andó S, Limonta M, Padoan M, Resentini A, Rittner M, Vezzoli G (2016) The Euphrates-Tigris-Karun River system: provenance, recycling and dispersal of quartz-poor Foreland-Basin sediments in arid climate. Earth Sci Rev 162:107–128

George AR (2015) On Babylonian lavatories and sewers. Iraq 77:75–106

Geyer B, Sanlaville P (1996) Nouvelle contribution à l'étude géomorphologique de la région de Larsa-Oueili (Iraq). In: Huot J-L (ed) Oueili. Travaux de 1987 et 1989. Éditions Recherche sur les Civilisations, Paris, pp 391–408

Heimpel W (1998) Nanše. A. Philologisch. RlA 9:152–160
Hritz C (2010) Tracing settlement patterns and channel systems in Southern Mesopotamia using remote sensing. J Field Archaeol 35(2):184–203
Hritz C (2014) Contributions of GIS and satellite-based remote sensing to landscape archaeology in the Middle East. J Archaeol Res 22(3):229–276
Hritz C, Pournelle J (in press) Feeding history: deltaic resilience, inherited practice, and millennial-scale sustainability in an urbanized landscape. In: Foster HT II, Goldstein DJ, Paciulli LM (eds) The future in the past: historical ecology applied to environmental issues. University of South Carolina Press, Columbia
Hritz C, Wilkinson TJ (2006) Using shuttle radar topography to map ancient water channels in Mesopotamia. Antiquity 80:415–424
Hritz C, Pournelle J, Smith J (2012a) Revisiting the sealands: report of preliminary ground reconnaissance in the Hammar District, Dhi Qar and Basra Governorates, Iraq. Iraq 74:37–49
Hritz C, Pournelle J, Smith J (2012b) Mid-Holocene dates for organic-rich sediment, palustrine shell, and charcoal, Southern Iraq. Radiocarbon 54(1):65–79
Huber Vulliet F (2009–2011) Sirara. II. RlA 12(7–8):552–553
Huh SK (2008) Studien zur Region Lagaš. Von der Ubaid- bis zur altbabylonischen Zeit (AOAT 345). Ugarit-Verlag, Münster
Jawad LA (2018) Dangerous fishes of the Eastern and Southern Arabian Peninsula. Springer, Berlin
Jotheri J, Allen MB, Wilkinson TJ (2016) Holocene avulsions of the Euphrates River in the Najaf area of Western Mesopotamia: impacts on human settlement patterns. Geoarchaeol Int J 31:175–193
Jotheri J, Altaweel M, Tuji A, Anma R, Pennington B, Rost S, Watanabe C (2018) Holocene fluvial and anthropogenic processes in the region of Uruk in Southern Mesopotamia. Quat Int 483:57–69
Koldewey R (1887) Die altbabylonischen Gräber in Surghul un El Hibba. Zeitschrift für Assyriologie 2:403–430
Liverani M (1990) The shape of Neo-Sumerian fields. Bull Sumer Agric 5:147–186
Liverani M (2013) Immaginare Babele. Due secoli di studi sulla città orientale antica. Laterza, Roma/Bari
McMahon A (2015) Waste management in early urban Southern Mesopotamia. In: Mitchell PD (ed) Sanitation, latrines and intestinal parasites in past populations. Ashgate, Surrey, pp 19–39
Nadali D, Polcaro A (2016) The early stages of the Sumerian City at Tell Zurghul: new results from recent excavations. Origini XXXIX:75–100
Potts DT (1997) Mesopotamian civilizations. The material foundations. The Athlone Press, London
Pournelle J (2003) Marshland of cities: deltaic landscapes and the evolution of early Mesopotamian Civilisation. PhD Dissertation, Department of Anthropology, University of California, San Diego
Pournelle J (2007) KLM to Corona. A bird's-eye view of cultural ecology and early Mesopotamian civilization. In: Stone EC (ed) Settlement and society: essays dedicated to Robert McCormick Adams. UCLA Cotsen Institute of Archaeology and Oriental Institute, Los Angeles/Chicago, pp 29–62
Pournelle J (2013) Physical geography. In: Crawford H (ed) The Sumerian world. Routledge, London, pp 13–32
Powell MA (1985) Salt, seed, and yields in Sumerian agriculture: a critique of the theory of progressive salinization. Zeitschrift für Assyriologie 75:7–38
Rost S (2011) Irrigation management in the Ur III period: a reconstruction based on a case study of the maintenance of the íd-NINA-šè-DU Canal of the Province on Lagaš. In: Selz GJ (ed) The empirical dimension of the ancient near Eastern Studies (Wiener Offene Orientalistik 6). LIT Verlag, Wien, pp 211–269
Sanlaville P (1989) Considerations sur l'évolution de la Basse Mésopotamie au cours des derniers millénaires. Paléorient 15(2):5–27
Schrakamp I (2017) Das Bewässerungssystem des präsargonischen Staates von Lagaš (ca. 2475-2310 v. Chr.). Untersuchungen zu den technischen, administrativen, sozioökonomischen und rechtlichen Aspekten eines der ältesten regionalen Irrigationssysteme Südmesopotamiens.

Habilitationsschrift, dem Fachbereich Geschichts- und Kulturwissenschaften der Freien Universität Berlin vorgelegt im Juli 2017
Selz GJ (1998) Über Mesopotamische Herrschaftkonzepte. Zu den Ursprüngen mesopotamischer Herrscherideologie im 3. Jahrtausend. In: Dietrich M, Loretz O (eds) Dubsar anta-men. Studien zur Altorientalistik. Festschrift für Willem H.Ph. Römer zur Vollendung seines 70. Lebensjahres mit Beiträgen von Freunden, Schülers und Kollegen (AOAT 253). Ugarit-Verlag, Münster, pp 281–343
Steinkeller P (1999) Land-tenure conditions in third-millennium Babylonia: the problem of regional variation. In: Hudson M, Levine BA (eds) Urbanization and land ownership in the ancient Near East, vol II. Peabody Museum of Archaeology and Ethnology, Harvard University, Cambridge, MA, pp 289–329
Steinkeller P (2001) New light on the hydrology and topography of Southern Babylonia in the third millennium. Zeitschrift für Assyriologie 91:22–84
Steinkeller P (2013) Corvée Labor in Ur III times. In: Garfinkle SJ, Molina M (eds) From the 21st century B.C. to the 21st century A.D. Proceedings of the international conference on Neo-Sumerian Studies Held in Madrid, 22–24 July 2010. Eisenbrauns, Winona Lake, pp 347–424
Ur J (2013) Patterns of settlement in Sumer and Akkad. In: Crawford H (ed) The Sumerian world. Routledge, London, pp 131–155
Vallet R (1996) L'architecture des phases Obeid 0 et 1. Travaux de 1989. In: Huot J-L (ed) Oueili. Travaux de 1987 et 1989. Éditions Recherche sur les Civilisations, Paris, pp 103–139
Vermaak PS (2008) Guabba, the Meluḫḫan village in Mesopotamia. J Semitics 17(2):454–471
Widell M (2013) Sumerian agriculture and land management. In: Crawford H (ed) The Sumerian world. Routledge, London, pp 55–67
Widell M, Fritz C, Ur J, Wilkinson TJ (2013) Land use of the model communities. In: Wilkinson TJ, Gibson MG, Widell M (eds) Models of Mesopotamian landscapes. How small-scale processes contributed to the growth of early civilisations (BAR-IS 2552). Archeopress, Oxford, pp 56–80
Wilkinson TJ (2003) Archaeological landscapes of the Near East. University of Arizona, Tucson
Wilkinson TJ (2013) Hydraulic landscape and irrigation system of Sumer. In: Crawford H (ed) The Sumerian world. Routledge, London, pp 33–54
Wilkinson TJ, Hritz C (2013) Physical geography, environmental change and the role of water. In: Wilkinson TJ, Gibson MG, Widell M (eds) Models of Mesopotamian landscapes. how small-scale processes contributed to the growth of early civilisations (BAR-IS 2552). Archeopress, Oxford, pp 9–33
Wilkinson TJ, Rayne L, Jotheri J (2015) Hydraulic landscapes in Mesopotamia: the role of human niche construction. Water Hist 7(4):397–418
Wittfogel KA (1955) Developmental aspects of hydraulic societies. In: Steward JH (ed) Irrigation civilizations: a comparative study. Pan-America Union, Washington, DC, pp 43–52
Wittfogel KA (1957) Oriental despotism. A comparative study of total power. Yale University Press, New Haven

Chapter 3
Reed-Swamps in the Sumerian Material Culture: Archaeological, Archaeobotanical and Experimental Insights from the Abu Tbeirah Excavations

Licia Romano, Alessandra Celant, and Maria Virginia Montorfani

Abstract Excavations at the site of Abu Tbeirah are giving new evidence of the long cultural continuity that existed (and in a way still exists) between the life in the Iraqi Marshland and that of Sumerian people inhabiting the Land of the Two Rivers during the third millennium BC.

This continuity is related to several life's aspects, first of all in the material culture. This study focusses on the use of reed-swamps in artefact production and building technique. The evidence brought to light during Abu Tbeirah's excavations are analysed and compared with the ethnographic data from the Iraqi marshes.

Keywords Reed-swamps · Marshes · Sumer · Mesopotamia · Experimental archaeology · Abu Tbeirah

3.1 Introduction

The reprise of archaeological activities in southern Iraq is offering the opportunity of analysing the Sumerian material culture from contexts excavated with modern techniques and documented with up-to-date methodology. The present paper analyses the use of reed-swamps from the Sumerian site of Abu Tbeirah, a third millennium BC *tell* located in southern Iraq, in the S-E periphery of Nasiriyah city (Romano and D'Agostino 2019).

Abu Tbeirah 42 ha site (Fig. 3.1) was surely an important city in the constellation of city-states in which Mesopotamia was divided during the so-called Early Dynastic

L. Romano (✉)
Department Institute of Oriental Studies, Sapienza University of Rome, Rome, Italy

A. Celant
Department of Environmental Biology, Sapienza University of Rome, Rome, Italy

M. V. Montorfani
Department of Classics, Sapienza University of Rome, Rome, Italy

L. A. Jawad (ed.), *Southern Iraq's Marshes*, Coastal Research Library 36,
https://doi.org/10.1007/978-3-030-66238-7_3

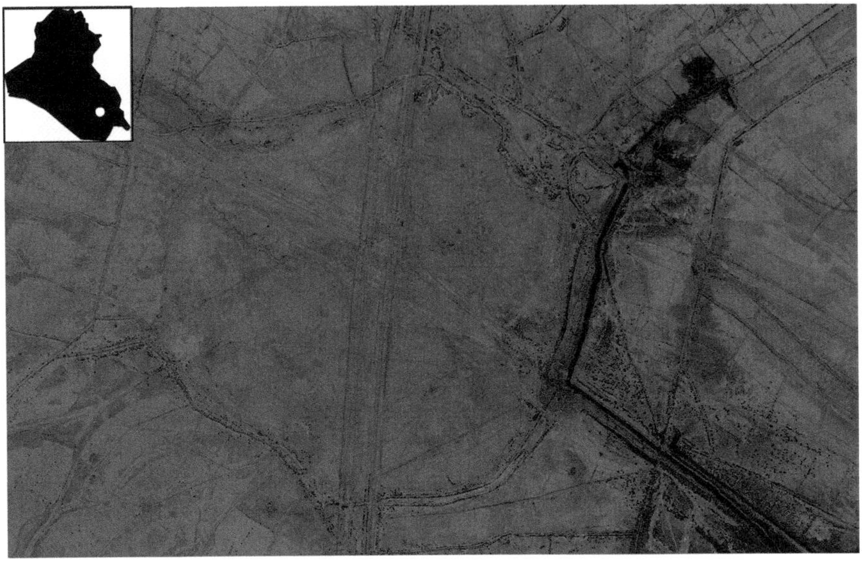

Fig. 3.1 Satellite view of the archaeological site of Abu Tbeirah (Nasiriyah). (Courtesy of E-Geos)

period (2900–2350 BC), before the unification of the territory in the First World Empire under the Akkadian King Sargon.

Since 2012, seven excavation campaigns were carried out at Abu Tbeirah. The archaeological activities led by Sapienza University of Rome and by the Iraqi State Board of Antiquities and Heritage are bringing to light a Sumerian city with housing quarters and cemeterial areas, dated from the Early Dynastic up to the Ur III period (2900–2000 BC). The multidisciplinary team involved in the research is aiming at highlighting the presence of changes, in particular in the material culture, revealing the impact of the political and presumably also climatic changes characterizing the end of the third millennium BC on the local population and in its everyday life.

Sumerian material culture was based mainly on the resources locally available: the so-called Cradle of Civilization was a rich territory for agriculture and breeding, but poor of main construction materials such as wood and stones. However, the great Mesopotamian civilization grow up using clay and reeds, together with bitumen and few other materials, as main elements not only for their constructions but also for the realizations of most of their tools and common objects. Southern Iraq is indeed an ideal environment for marshes and for an extremely diversified variety of Gramineae, Juncaceae and Cyperaceae.

3.2 Abu Tbeirah and the Marshes

Abu Tbeirah is characterized by the presence of two artificial canals that cross the city from SE to NW and that connected the site to the complex canalization system of the Sumerian third millennium BC city-states. In the western part of the city, an artificial basin, interpreted as a harbour, is visible and during the rainfall period is still characterized by the accumulation of water. As for all the city-states of the third millennium BC, Abu Tbeirah was clearly connected to the regional system of canalization that crossed the Lower Mesopotamia. The environment that surrounded the ancient city of Abu Tbeirah is comparable to those of the modern Marshland.

The Mesopotamia Fluvial Basin belongs to the so-called Zagros Fold-Thrust Belt (ZFTB), comprised between the Arabian Platform and the Zagros Mountains, filled during the Holocene by sediments of various origins (Milli and Forti 2019). The gulf shoreline, due to the delta progradation, is now located southwards, if compared with its position during the third millennium BC. At the same time, also the location of the marshes was different, being now shifted south-eastwards.

The Marshes constituted the perfect environment in which the Mesopotamian civilization and its cities developed. The natural channels were modified by Sumerians, and other artificial canals were excavated or natural channels were modified in order to ensure agriculture and transportation among the settlements. Also, nowadays, the Marshes constitute a rich and unique ecosystem: its peculiarity, its role in the development of one of the first civilizations and the ethnographic continuity of way of life of its inhabitants are at the basis of the admission in 2016 of the southern Iraqi Marshes in the UNESCO WHL, together with the archaeological sites of Ur, Uruk and Eridu.

3.3 Reed-Swamps in Abu Tbeirah's Material Culture

3.3.1 State of Preservation and Excavation Methodology

Abu Tbeirah's silty-clay soil allows a quite good macroscopic preservation of vegetal twines that were found in several excavated areas. These artefacts are often preserved as mummified, flattened and compressed remains or as impressions left on the clay soil or on bitumen. The peculiar taphonomy allows in several cases to fairly reconstruct the original shape of the objects. Notwithstanding this, their fragile nature makes the excavation and preservation quite difficult. In some cases, the twines were covered with bitumen, and this practice allowed a better preservation but hampered also any possibility of ^{14}C dating of these short-lived remains.

In order to allow a better preservation of these artefacts, excavations were preferably realized in a lab: the object and the surrounding soil were cut and lifted after being temporarily consolidated using cyclododecane spray. Then the excavations were carefully and meticulously conducted in a lab with brushes and the use of soft material pointed tools. Each finding was drawn through a CAD software and extensively documented. When possible, the artefacts were consolidated through Paraloid-B72.

3.3.2 Basketry

According to the terminology defined by J.M. Adovasio (1977), the term basketry indicates a wide range of artefacts including different kinds of recipients (e.g. baskets) and reed-mats.

Basketry excavated at the site of Abu Tbeirah comes from both funerary and domestic contexts and is reported in Table 3.1. Here a summary of the typologies of artefacts is presented with a focus on the archaeological context. The manufacturing techniques are explained in detail in § 6. The studies made by Lemorini et al. (2008) and Di Lernia et al. (2012) on other archaeological contexts represent a fundamental methodological base for the present paper and the ongoing research on Abu Tbeirah's vegetal twines.

3.3.2.1 Baskets and Other Containers

Containers in twined vegetal fibres are attested in two of the areas excavated so far, in contexts dated to the end of the third millennium BC.

As far as domestic contexts are concerned, baskets were found in the ED-Akk, Transition (2400–2300 BC) Building A in Area 1 and in apparently later Building D in Area 4 (Romano and D'Agostino 2019).

In the second and earlier phase of Room 4 in Building A, two baskets were discovered at both sides of the door that connected the space with the rest of the building. Both baskets (IDs 1–2 – Fig. 3.2) are of medium dimensions and are realized in simple plaiting (D'Agostino et al. 2015; Montorfani 2019). They were found compressed on the pavement and both resulted to be coated with bitumen. The function of the two containers is unknown: no caryopses or seeds were found inside them.

Inside Room 5 of Building D in NE Area 4, a medium size basket AbT.17.95 (ID 3 – Fig. 3.3) was found lying against the NW wall on the supposed housing surface and with some pottery shards under it. It was realized in simple plating too.

Apparently, another simple plaited big container was used for the inhumation of the grave of an infant (Grave 101) in Area 2 in the NE part of the site (ID 4 – Fig. 3.4). The infant was deposed in a semi-crouch position with the legs bent towards north and the head on the right shoulder, possibly due to a post-depositional dislocation. Near the head a plain rim jar was the only object deposed as funerary equipment (D'Agostino et al. 2015). The body and the jar were laid on a simple plaited mat that has a circular shape. The mat was raised vertically near the head, laying against a wall. It is not clear if the shape assumed by the mat is due to the cut

Table 3.1 Catalogue of the reed-swamps artefact found at Abu Tbeirah

ID	Context		Type of find	Technique	Plant taxa	Function	Diameter (cm)	Length (cm)	Width (cm)	Surface (cm^2)	Weft/element wide (cm)	Preservation	Bitumen coating
1	Residential	Area 1. Building A – Phase 1. Room 4. On the west side of the door	Basket	Simple plating	*Arundo donax*	Container	30	–	–	706.9	1.5–2.5	Good	Yes
2	Residential	Area 1. Building A – Phase 1. Room 4. On the east side of the door	Basket	Simple plating	*Arundo donax*	Container	25	–	–	490.9	1.5–2.5	Good	Yes
3	Residential	Area 4. Building D – Phase 2. Room 5. Against the north-western wall	Basket	Simple plating	*Arundo donax*	Container	–	28.9	28.1	516.2	1.2–1.3	Good	n.d.
4	Funerary	Area 2. Grave 101	Cradle?	Twill plating	*Arundo donax*	Used to host the body of an infant	–	49.5	33.54	1351.9	1.5–1.7	Good	Yes
5	Funerary	Area 1. Cemetery Grave 23	Basket	Close twining	*Typha* sp.	Container	10	–	–	78.5	0.2–0.3	Average	n.d.
6	Funerary	Area 1. Building A – Phase	Basket	Close twining	*Typha* sp.	A cosmetic shell	10	–	–	78.5	0.2–0.3	Good	Yes

(continued)

Table 3.1 (continued)

ID	Context		Type of find	Technique	Plant taxa	Function	Diameter (cm)	Length (cm)	Width (cm)	Surface (cm^2)	Weft/ element wide (cm)	Preservation	Bitumen coating
		1. Room 4. Grave 12.				was found inside it							
7	Funerary	Area 1. Building A – Phase 2. Room 3. Grave 9.	Basket	Close twining	*Typha* sp.	Container	10	–	–	78.5	0.2–0.3 ca.	Bad	n.d.
8	Residential	Area 1. Building A – Phase 2. Room 1.	Reed-mat	Twill plaiting	*Arundo donax*	Covering the pavement. Presence of a hole for a post sustaining the roofing	–	224	88	7650.0	2–2.3	Good	Yes
9	Funerary	Area 1. Cemetery Grave 1	Reed-mat	Twill plaiting	*Arundo donax*	"Winding sheet"	–	17	8	118.0	1.5 ca.	Bad	n.d.
10	Funerary	Area 1. Cemetery Grave 25. Preserved only on jar AbT.15.385.9	Reed-mat	Twill plaiting	*Arundo donax*	"Winding sheet"	–	7.8	5	22	1.4 ca.	Average	n.d.
11	Funerary	Area 1. Cemetery Grave 24	Reed-mat	Twill plaiting	*Arundo donax*	Used to lift the coffin's lid	–	n.d.	n.d.	n.d.	n.d.	Bad	n.d.
12	Funerary	Area 1. Building A – Phase	Reed-mat	Twill plaiting	*Arundo donax*	Used to keep in	–	n.d.	n.d.	n.d.	1.2–1.5	Bad	n.d.

		1. North-western out-side Grave 4 + 5 + 13				place the remains of a funerary banquet							
13	Residential	Area 1. Build-ing A – Phase 1. Room 4.	Reed-mat	Twill plaiting	*Arundo donax*	Covering a bitumen ingot	–	19	18	238.6	2.3–2.5	Average	Yes
14	Residential	Area 1. Build-ing A – Phase 1. Room 14–15	Reed panels with reed bundles		*Arundo donax*	Panel used to divide the space of the room and connected with a fir-ing struc-ture. Reed–bun-dles used as poles	–	n.d.	n.d.	n.d.	1.5	Average	No
15	Funerary	Area 1. Build-ing A – Phase 2. Room 22 Grave 28	Reed impressions		*Arundo donax*	Reed stretcher for trans-port and deposing the body	–	67	30	1121	1.2–1.4	Bad	No

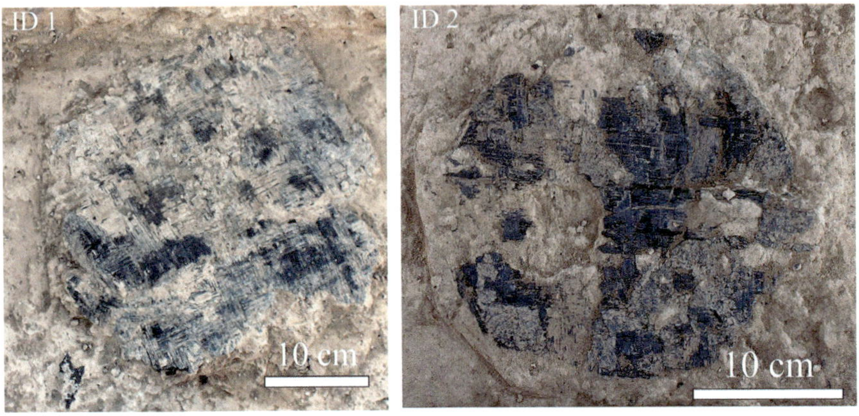

Fig. 3.2 The two baskets IDs 1–2

Fig. 3.3 ID 3 basket

of the grave adjacent to an earlier structure. However, the impression is that the mat formed originally a sort of a container, possibly a kind of cradle.

Small containers were common also in SE Area 1 inhumations, both in sub-pavement graves of Building A and in the burials of the later cemetery, cutting the earlier building.

Inside Area 1 Cemetery, a small twined basket was found in infant burial Grave 23 (ID 5 – Fig. 3.5). The basket was located in front of the skull though at some

Fig. 3.4 Infant inhumed in a cradle-like basket (ID 4)

Fig. 3.5 IDs 5–7 small close-twined baskets

distance, near the limits of the Grave's cut. The body of the infant was probably wrapped in reed-mat.

Among the sub-pavement burials of Building A – Phase 1, the only adult grave (Grave 12 – Room 4) had as equipment several pottery vessels and stone tools and a small twined basket (ID 6 – Fig. 3.5) containing a shell with cosmetic paste inside. In

Building A – Phase 2 Room 3 inside Grave 9, another small twined basket (ID 7 – Fig. 3.5) was found upside-down, located near the left hip.

3.3.2.2 Reed-Mats

Inside Building A – Phase 2 Room 1, the remains of a reed-mat covering the pavement were preserved near the NE wall (ID 8 – Fig. 3.6). Near the wall, the NW part of the reed-mat was pierced in order to host a small post. Part of the reed-mat was bent inside the hole, indicating that the post was set after the reed-mat was placed on the pavement.

As far as the funerary contexts are concerned, some graves show clear evidence of reed-mat on and surrounding the bones: notwithstanding the general bad state of preservation, it is clear that the bodies were wrapped in reed-mat before being deposed inside the burials. The best-preserved evidence of this kind of practice are those of Grave 1 (ID 9 – Fig. 3.6) and Grave 25 (ID 10 – Fig. 3.6) of Area 1 Cemetery. The 6–7-year-old body inhumed in Grave 1 showed traces of reed-mat near the upper part of the body, while inside Grave 25, the reed-mat was preserved only on the external surface to the jar AbT.15.385.8 located near the back of the 5–6-year-old subadult.

Grave 24 shows a particular situation. The inhumation in the coffin showed reed-mat traces (ID 11) covering both the inner and the outer side of the pottery sarcophagus: it might indicate that the reed-mat was used to lift the lid and laid it over the coffin.

Fig. 3.6 Reed-mats IDs 8–10 and 12

Fig. 3.7 Reed bundle ID 14, bitumen ingot with reed-mat cover ID 13 and the traces of the reed stretcher ID 15

Outside Building A – Phase 1, in an area that was plausibly considered as belonging to the structure, a huge multiple inhumation was found (Grave 4 + 5 + 13). Here the evidence of a funerary banquet is preserved: three subadults were inhumed here, each one flanked by a heap of pottery shards mixed to animal bones. These concentrations of pottery fragments were kept in place by a reed-mat cover with bricks over it. A similar finding is attested in the Royal Cemetery of Ur (Woolley 1934: 91 PG/105, Pl. 46a).

Abu Tbeirah's evidence shows also that reed-mat was used in the transportation and perhaps storage of bitumen. Several pieces of bitumen covered with reed-mat were often found in our contexts. The best-preserved one (ID 13 – Fig. 3.7) was discovered in the filling of Room 4 (Building A – Phase 1). It is an item with convex base and rounded corners, showing in the upper part a reed-mat impression and with the nucleus constituted by a mix of finely crushed straw/vegetal temper, soil and bitumen.

3.3.3 Reeds as Architectural Elements

The only evidences of the use of reeds in Abu Tbeirah third millennium BC architecture are related to a peculiar movable structure discovered inside Room-14-15 of Building A – Phase 1 (ID 14 – Fig. 3.7). The structure was realized with reed panels sustained by four posts and partially set in the ground surface inside a long and narrow foundation pit (Romano and D'Agostino 2019: 250–252). The reed panels and posts were probably partially covered (at least at their base) with clay: a huge quantity of clay lumps hardened by fire and with reed impressions were dispersed both in the filling of the room and along the foundation pits. This kind of structure was similar to those already attested in the Marshland, with posts made by reed bundles. One of the post-holes showed indeed clearly the impressions of one of these reed bundles: the clay hardened by fire presents the signs of the leaves, nodes and internodes/culms. A second one had the impression of a palm stem. The reed panels were linked to a sort of long firing structure, probably vaulted, located along the south-west wall and partially dug in the pavement. The reed structure had an opening near the northeast wall, supported by other posts. In front of the opening the remains of what was interpreted as a reed door was discovered collapsed on the floor. Indeed, a fourth post-hole was connected to this reed concentration to the door mechanism.

3.3.4 Further Evidences of the Use of Swamp Herbs

Inside Room 22 of Building A – Phase 2, the impression of at least eight reeds laying parallel was highlighted under the body (ID 15 – Fig. 3.7). This evidence might be interpreted as a sort of reed stretcher, possibly used to transport and deposit the body into the grave's cut, or as a sort of matting used to cover the bottom of the grave. The hypothesis of a wooden coffin, as attested in the Royal Cemetery of Ur, was considered too (see Woolley 1934: e.g. PG/1648 Pl. 81b). However, there was no evidence of a vertical structure as those excavated by Woolley. A similar evidence is also the reed "table" discovered inside PG/1847 (Woolley 1934: 194 Pl. 84b).

Additional indication of the use of reed-swamps is connected to artefacts in which crushed reeds, sedges and rushes were used as temper. Coarse pottery vessels show evidence of vegetal temper, plausibly the same used in the realization of the nucleus of bitumen ingots and other bitumen objects (loom weights, small pyriform objects, etc.) largely found in Abu Tbeirah's contexts. The addition of vegetal temper to bitumen is a common expedient to get a lower melting temperature limit (Schwartz and Hollander 2000: 86). In particular, the mixture of vegetal temper and bitumen is also frequently found under pestles and grindstones.

3.4 Archaeobotanical Analyses

The archaeobotanical identification of plant fibres used for weaving is not always easy, sometimes due to the lack of precise morphological-biometric elements, sometimes due to their poor state of preservation. Often, only impressions or superficial traces of parallel ribs of monocotyledonous angiosperms leaves are left, so that the only elements useful for identification are size and shape of the tissues, which may be studied through the stereomicroscope. Even observations under the SEM are not always decisive, both because impressions may be barely perceptible and because the materials bearing the impressions may be incoherent or with a more or less coarse grain size. Thus, biometric and anatomical comparison of the fossil remains with modern reference materials and interwoven fibre artefacts is fundamental.

In Table 3.1, the results of fibre analysis from the site of Abu Tbeirah are reported. These are 15 samples found in funerary contexts or used in domestic environments. Most of the finds belong to the species *Arundo donax*, whose leaves and culm can be distinguished from other reed species (e.g. *Phragmites australis*) based on their size. Experimental research (see Sect. 3.6) may provide complementary suggestions as to which reed may have been profitably used by the ancient communities. For example, differently from *Arundo*, which was used for various purposes (baskets, reed-mats, panels and coffins), small baskets were exclusively produced in *Typha* or a similar softer vegetal fibres.

The abundance of reed remains at Abu Tbeirah documents a type of environment that is not any more present in the area of the site. Marshes represent an extensive, allochthonously fed wetland complex in a very arid area (100–300 mm annual precipitation). From the ecological point of view, they host numerous endemic species of birds and mammals, besides contributing as staging posts for migrating birds. The geographical location of the Mesopotamian Plain allows a unique combination of Saharo-Arabian, Irano-Turanian and Mediterranean floristic elements (Al-Hilli et al. 2009). Wetland vegetation includes different forms, according to dominant species and average seasonal depth of water, ranging from permanently submerged to seasonally or occasionally submerged to intermittently wet. In particular, *Arundo* and *Typha* can be classified as herbaceous tall emergent weeds, typical of permanently flooded areas together with *Schoenoplectus litoralis* and *Phragmites australis*, while in seasonally flooded areas, low sedges and rushes (*Carex* spp., *Juncus* spp., *Scirpus brachyceras*) dominate, forming ephemeral and salt-tolerant vegetation in temporarily flooded areas (Hussain and Alwan 2008).

Pollen analysis from the marshlands in southern Iraq (Al-Ameri and Jassim 2011) confirms that, during the last few thousands of years, plants belonging to the families Poaceae and Typhaceae were very abundant and fully available for human activities (Al-Ameri and Jassim 2011; Celant and Magri 2019). The regional vegetation was characterized by Poaceae and Arecaceae, while marshlands formed either permanently flooded area, with abundant *Typha* and reeds, associated with the deposition of peat and/or organic clay, or partially dry marshlands, where abundant chenopods were found, indicating a salt-rich substratum. A transition from a dominance of *Typha*, palms and Poaceae, including cereals, to rising percentages of chenopods, indicating increasingly saline and arid conditions, is often found in pollen records from the southern Mesopotamia. This climatic trend may have had severe repercussions on the ancient societies, being the possible cause for the abandonment of the settlement of Tell Umm al-Aqarib around 2100 BC (Al-Ameri and Jassim 2011).

3.5 Abu Tbeirah's Evidence Between Sumerians and Ma'dan

The life of the Ma'dan, the so-called Marsh Arab, was object of several ethnographic studies (e.g. Fulanain 1927; Maxwell 1957; Salim 1962; Thesiger 1964) and still today shows interesting features that can find direct comparisons with the life in the Sumerian epoch and Abu Tbeirah archaeological evidence. In this sense, it is of invaluable help the work of E.L. Ochsenschlager (2004) that reports crafts and practices of the Ma'dan. In particular, he focussed also on the use of reeds, rushes, sedges and other grasses in different craft activities (from building techniques to the realization of musical instruments).

The persisting importance of reeds in the Ma'dan life is evident from the numerous terms used to describe the different kinds and their growing phases (Ochsenschlager 2004: 129–130). Sumerian texts were full of references to the use of reeds, as a further evidence of the continuity between Sumerians and Ma'dan. In Sumerian, "gi" translates the word reed, and is a term used also as determinative, in order to specify that the associated word belongs to a particular semantic group (e.g. Goetze 1948; Pomponio 1989; Sallaberger 1989, 1992; Waetzoldt 1992; Streck 2009). Sumerians distinguish several categories of reeds on the basis of their qualities (e.g. gi-zi/gi-ne indicates a young and fresh reed), and numerous are the words that indicate objects, building parts or actions connected with the use of reeds.

As well as the Ma'dan, Sumerians used to cut reeds and other reed-swamps during all the year (Sallaberger 1989: 314; Streck 2009: 184), to carry them by boat (e.g. Goetze 1948: 165) and join them into bundles (Pomponio 1989: 230; Sallaberger 1989: 317–318; Streck 2009: 185). The period of the year obviously influenced the kind of reed gathered (Sallaberger 1989: 314; Streck 2009: 184). Reeds were used by Sumerians and partially are still used by the Ma'dan for a wide range of purposes. In what follows, a summary of these uses will be presented, highlighting in particular those attested in the archaeological record from Abu Tbeirah.

Several kinds of baskets are reported in Sumerian texts, but, due to the limited archaeological evidence, their identification is still unknown. Marsh Arabs however realized both plaited and coiled baskets, with techniques comparable to those highlighted in Abu Tbeirah's excavations: see respectively IDs 1–4 and IDs 5–7 (Ochsenschlager 1992: 65–66). As previously said, some of Abu Tbeirah's baskets were coated with bitumen, a practice attested also in the Marshes (Ochsenschlager 1992: 66), aiming at waterproofing the containers. It is interesting that also in early texts (e.g. in Ebla archives), reed artefacts are recorded together with bitumen (Sallaberger 1989, 321 baskets; Civil 2008, 317–320 mats; Stol 2012, 51 mats and baskets). Baskets in Sumerian culture are also well-known from ancient iconography: e.g. the king is often represented in his role of temple constructor carrying on his head a basket filled with bricks.

Plaited mats, such as IDs 8–13, were largely used by Sumerians: several are the accounts concerning the time and work required for their production (Goetze 1948; Streck 2009: 186). Some of them were also coated with bitumen as testified by texts (Stol 2012: 51) and by Abu Tbeirah ID 8. The use of reed-mat wrapping bodies in funerary contexts is attested also in other Mesopotamian sites (Potts 1997: 230). In the Marshes, reed-mat is largely used for several purposes: in the construction of building, as covers for the pavement or as roofing (Ochsenschlager 2004: 130–131).

ID 13, as said, represents an example of the bitumen ingots discovered at Abu Tbeirah and probably used to bitumen transport (Schwartz and Hollander 2000: esp. Fig. 5) from the sources located in Iraq (Stol 2012: 58–59). The same kinds of findings were discovered also in the Uruk Mound at Abu Salabikh: here several pieces were recovered during the excavation of a dump pit and were impressed with reeds, weaved reed-mats and ropes (Pollock 1990: 70).

Reeds are largely used in the marshes as building material, a practice that clearly shows a continuity with the Sumerian architecture: though the archaeological literature shows mostly evidence of the use of mud-bricks and bricks, the cuneiform texts attest clearly the use of reeds as building material both for shelter, fences and partitions (Moorey 1999: 361–362). The few archaeological evidences of the use of reeds in architecture were summarized by Aurenche and Moorey (Aurenche 1981; Moorey 1999: 361–362): the most outstanding evidence are the still visible reeds used as building material in the Ziqqurat of Uruk. Further indication of the continuity of building techniques between Ma'dan and Sumerians is the Late Uruk iconography depicting structures identical to the mudhif houses of the Marshes (Potts 1997: 116–117).

The peculiar structure discovered in Building A Room 14–15 at Abu Tbeirah (ID 14) shows the use of reeds to realize a light partition inside the space made of lattice work sustained by poles of different nature (Romano and D'Agostino 2019: 250–251, Figs. 8.71–72). In particular, reed bundle is used as a sort of column/post, a practice common in the marshes but well-known also in Sumerian iconography, as demonstrated, e.g. by the already quoted mudhif architecture in the late Uruk iconography and by the reed bundle connected to the symbolism of the goddess Inanna. ID 14 was connected to an oven: this gives the opportunity to make a further comparison with the Ma'dan evidence. Ochsenschlager describes in detail the realization of the *sarifa*, a simple dwelling realized with lattice works or reed-mats as side walls, sustained by bundles or palm stems set in the soil at a distance of ca. 4 forearms (Ochsenschlager 1992: 58; 2004: 161–162). This distance is possibly comparable to that of the post-holes in the structure with the reed bundles (ID 14). Though apparently not appropriate for security reasons due to its inflammable nature, the connection of a reed structure to an oven is very common in the Marshes, where reed shelters were realized over tannur and other firing installations as a protection against the rain or the sun (Ochsenschlager 1992: 62; 2004: 100, 107–108). The use of reeds for the realizations of doors, similar to that found on the pavement of Room 14–15, is commonly attested in cuneiform texts (Streck 2009: 187).

Many other uses of reeds and reed-swamps are attested in the Marshes and in Sumerian cuneiform texts: as materials for the realization of dams, fences, arrows, rods, spears, flutes, rods, writing instrument, fodder and fuel, as ingredient for the Sumerian beer, etc. (Stretch 2009: 185–187; for the Marshes, see Ochsenschlager 2004: *passim*).

The last evidence from Abu Tbeirah to be discussed is the use of crushed reeds and swamp herbs as vegetal temper for the realization of mud-bricks and bricks, pottery and bitumen ingots. The widespread use of crushed reeds in the Marshes (Ochsenschlager 1992: 64) has its counterpart in the cuneiform texts: according to Pomponio, the word gi-gaz should be translated as "crushed reeds", while gi-saḫar as "reed with earth" (Pomponio 1989: 231). Mixing bitumen with crushed reeds was a practice aimed at reducing the melting temperature: the mixed bitumen had a higher value than the pure one in Ur III Mesopotamia (Schwartz and Hollander 2000: 86). The frequent presence of reeds and bitumen under pestles and stone tools at Abu Tbeirah might find an explanation from the Marshland ethnographic record: bitumen-covered pestles were used to reduce reeds and other swamp herbs in pieces and stripes suitable for the different final products (Ochsenschlager 1992: 64).

3.6 Experimental Reproduction of Selected Artefacts

As part of the research on Abu Tbeirah's material culture, an experimental study on plaited baskets and mats was carried out. Experimental archaeology replicates the past in a controlled environment with the aim to prove or disprove archaeological data and obtain new hypothesis and information on ancient crafts and technologies.

Selected mats and baskets from Abu Tbeirah were, thus, reproduced in order to obtain more information about the artefacts structure and the use in their context. The experiments were based on archaeological, textual and ethnographic evidence (see Sect. 3.5), considering also the raw material available to the ancient inhabitants of Abu Tbeirah (see Sect. 3.4) and the limits due to the technical aspects of basketry production.

From a technical point of view, there are three main types of plaiting techniques, defined by archaeologist James Adovasio (1977): plaiting, twining and coiling. These three techniques are nowadays used by the Ma'dan in order to realize mats and baskets (Ochsenschlager 2004). In particular, plaiting is involved in the production of reed-mats using *Arundo donax* and in basketry using different kinds of rushes; coiling is used with rushes or reeds to produce three traditional basketry shapes, *tabaq*, a wide tray; *quffa*, a deep plate; and *saba*, a jar-shaped basket used to milk water buffalos (Ochsenschlager 2004).

The best-preserved vegetal twines discovered at Abu Tbeirah were selected and their structure analysed in detail. The high-resolution photographic documentation allowed to obtain a technical drawing of the structure of every basket or mat. Hypotheses on the structure of every artefact were formulated on the basis of the botanical identification, the technical details and the ethnographical and archaeological evidence. Following these hypotheses, the experimental reproductions were realized with the help of C. Lemorini from the Laboratory of Technological and Functional Analyses of Prehistoric Artefacts (LTFAPA) of Sapienza University of Rome and of the expert basket maker and craftsman Giovanni Morra from Parco Naturale dei Monti Aurunci. All the steps of the reproduction were documented and photographically recorded.

Selected material consisted in two big baskets from Building A (IDs 1–2), two little baskets from two different graves (IDs 5–6) and two reed-mats from the floor of Building A (ID 8) and from the bottom of a grave (ID 4).

3.6.1 Simple Plaited Baskets IDs 1–2

IDs 1–2 baskets (Fig. 3.8) had a diameter of approx. 30 cm. ID 1 had a more squared shape than ID 2 basket. As it can be seen in the drawing obtained, it is clear that the two baskets were plaited using simple plaiting technique. ID 2 basket is rounder and has a thicker part that was tentatively interpreted as a handle. According to the botanical analysis, *Arundo donax* was used to realize these baskets.

For the experimental reproduction of IDs 1–2 baskets, a squared base of 26 × 30 cm was realized using plaiting; *Arundo donax* was harvested and treated

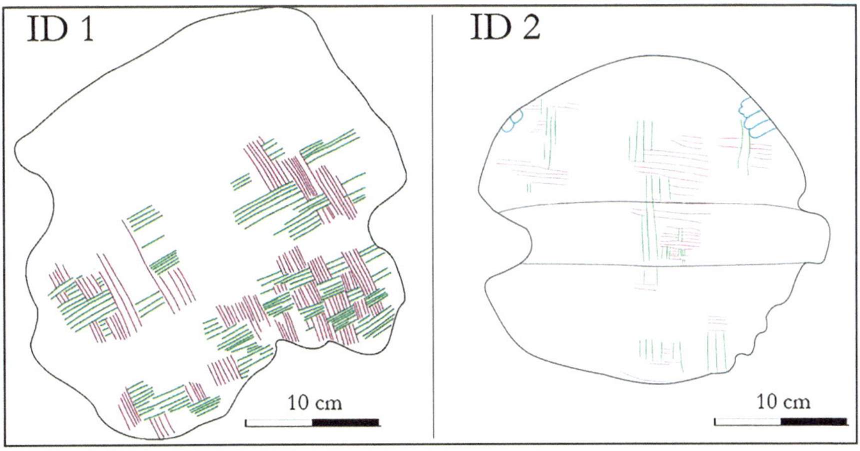

Fig. 3.8 Technical drawings of IDs 1–2 baskets

Fig. 3.9 Experimental reproduction of big basket from Building A (ID 2) as a ring basket

by beating it to make it softer, according to the Ma'dan technique. The attempt to soak it into water, and also by leaving it overnight, did not change the hardness of the fibre. Every reed was cut lengthwise and in ca. 50 cm length segments. The base was plaited with two segments crossing two segments, i.e. with a 2:2 ratio. Even though the reed was repeatedly beaten, the segments were no soft enough to be plaited, breaking repeatedly during the work.

ID 1 basket had clearly a squared shape; therefore, the exceeding material was simply cut, and the plaiting was fixed with a *Typha* rope. The experimental reproduction led to three different interpretative hypotheses about ID 1 basket: (1) it was the cover of a box, maybe made with a perishable and not preserved material; (2) it was a small mat or a tray; and (3) it was the base of a not preserved container with the walls made in another material.

ID 2 basket had a round shape and, apparently, a handle: though plaiting always leads to a squared object, the realization of a round shape was attempted starting from the same squared base made for ID 1 basket. A ring in *Arundo donax* was realized and sewed to the round shape, in order to realize a sort of "ring basket", a flat based container with a ring sieved on it (Adovasio 1977). The attempt was unsuccessful, and the thread, made with *Arundo donax* and used to sew the ring, broke immediately. Hence, the original basket was probably a "ring basket" realized using a softer tread, e.g. *Typha*, to sew the base to the ring (Fig. 3.9). The same joining technique can be hypothesized for the handle, though its actual identification in the flattened and compressed remains of the basket is uncertain.

3.6.2 Twined Baskets IDs 5–6

IDs 5–6 baskets were discovered in two different graves, and both were round shaped and had a diameter of ca. 10 cm. Thanks to the accurate observation and technical drawings, a twined structure was identified in both baskets: the twining elements are at a distance of no more than 0.2–0.4 mm, shaping two baskets of a very thin structure. Due to the uncertainty of the identification of the vegetal fibre used (see Sect. 3.4), the original material was only hypothesized. However, a material that can be reduced in filaments or have very thin stems was surely used to realize the thin twined structure of the two baskets. The possible candidates are thus the *Juncus*, whose stems are about the desired thickness, or rushes that can be shredded in thinner filaments, such as *Typha*. The use of *Arundo* was excluded because of its hardness, unsuitable for the realization of thin baskets.

Three different experimental replicas were realized for IDs 5–6 baskets. The first and the second were twined using in one case *Typha* fibre soaked in water and cut lengthwise and in the other using *Juncus* boiled in water (*Juncus* is traditionally treated by boiling it, as suggested by Mr. Morra). The third replica was realized sewing a seven-strand braid made by *Typha*. In every reproduction, the plaiting was as close as possible. In all the three cases, three soft and light small baskets were obtained.

The comparison between the experimental results and the archaeological remains IDs 5–6 confirms that the small baskets were realized through simple twining: in both the modern and the ancient artefacts, the concentric circle structure is indeed evident. As far as the material used is concerned, the vegetal fibre was plausibly *Typha* (or *similia*): the tube-like structure typical of *Juncus* is not indeed visible in the archaeological remains (Fig. 3.10).

3.6.3 Twill Plaited Mats IDs 4 and 8

IDs 4 and 8 mats' fibres were botanically determined as *Arundo donax*. These vegetal twines were realized clearly trough twill plaiting as evident from the comparison of the archaeological remains and the ethnographic examples from the Iraqi Marshes. The experimental reproduction of the mats confirmed this interpretation.

3.7 Conclusions

The rich vegetation of the Marshes was widely exploited by Sumerians, as well as by the Ma'dan, that produced and partially still produces a huge variety of artefacts and containers in vegetal fibres. The evidence for the use of reed-swamps at Abu Tbeirah

Fig. 3.10 Reproduction of twined baskets (IDs 5–6): three different hypotheses

confirms on the ground the importance of this natural resource, widely documented by cuneiform administrative texts. Among the several uses attested in Abu Tbeirah's record, basketry deserves indeed a special attention due to the important social implications that might be hidden in this peculiar part of the material culture. Two-dimensional classes of baskets are so far attested in the excavated areas, but cuneiform sources suggest that a higher variety of shapes and dimensions may be hopefully revealed by future excavations. As shown through experimental research, producing weaved containers and reed-mats implies a high practical knowledge not only of several plaiting techniques but also of the raw material and its characteristics. Were these craft skills commonly owned by Sumerians? Or were there specialists that produced these common artefacts for the community? The extramasticatory wear on some adult skeletons, both male and female discovered in Abu Tbeirah graves (Tafuri 2019: *passim*), might indicate that basketry production was a common skill.

This seems to be confirmed by the ethnographic and ethnohistorical evidence. Indeed, in the Marshes, according to Ochsenschlager, all the men were able to realize weaved artefacts, but also women were often involved in basketry production (Ochsenschlager 1992: 64). Cuneiform texts of the Ur III period seem to depict a different picture, referring to basketry mainly as a male occupation (Lafont 2016). However, as stated by Lafont, the textual evidence studied up to now should not be considered as definitive because "for a proper evaluation of the situation, references to an occupation are not enough" (Lafont 2016). Further discoveries and further research will hopefully contribute in the near future to a better understanding of this peculiar and important craft and its social implications.

Archaeological and archaeobotanical evidence from Abu Tbeirah once again testify, thanks also to the experimental research, the long continuity of life in the Iraqi Marshes with the Sumerian civilization. This continuity represents a unique treasure that must be preserved both as tangible and intangible heritage. The inscription of the marshes in the WHL UNESCO is surely an important step towards the protection of this unique heritage, but the main role in the maintenance of these millennial traditions is fundamentally in the hands of the Iraqi stakeholders.

Author Contributions L. Romano is author of Sects. 3.2, 3.3 and 3.5; A. Celant of Sect. 3.4; M. V. Montorfani of Sect. 3.6. Introduction, conclusions and the data reported in Table 3.1 are common work of the authors.

References

Adovasio JM (1977) Basketry technology, a guide to identification and analysis, Chicago

Al-Ameri TK, Jassim SY (2011) Environmental changes in the wetlands of Southern Iraq based on palynological studies. Arab J Geosci 4(3–4):443–461

Al-Hilli MR, Warner BG, Asada T, Douabul A (2009) An assessment of vegetation and environmental controls in the 1970s of the Mesopotamian wetlands of Southern Iraq. Wetl Ecol Manag 17(3):207–223

Aurenche O (1981) La Maison orientale. L'Architecture du Proche-Orient ancien des origines au milieu du IVe millénaire, Paris

Celant A, Magri D (2019) Palaeoenvironment, climate and land use in southern Mesopotamia/Nasiriyah Area. In: Romano L, D'Agostino F (eds) Abu Tbeirah excavations I. Area 1 last phase and building A – Phase 1, Rome, pp 39–47

Civil M (2008) The early dynastic practical vocabulary A (Archaic HAR-ra A)(=Archivi Reali di Ebla Studi – ARES. 4), Rome

D'Agostino F, Romano L, Ghanim AK (2015) Abu Tbeirah, Nasiriyah (Southern Iraq). Preliminary report on the 2013 excavation campaign. In: Biga MG, Córdoba JM, del Cerro C, Torres E (eds) Homenaje a Mario Liverani, fundador de una ciencia nueva (II)/Omaggio a Mario Liverani, fondatore di una nuova scienza (II) (=ISIMU 13). 2011(2015), Madrid, pp 209–221

di Lernia S, Massamba N'siala I, Mercuri AM (2012) Saharan prehistoric basketry. Archaeological and archaeobotanical analysis of the early-middle Holocene assemblage from Takarkori (Acacus Mts., SW Libya). J Archaeol Sci 39:1837–1853

Fulanain SEH (1927) The Marsh Arab – Haji Rikkan, Philadelphia

Goetze A (1948) Umma texts concerning reed mats. J Cuneif Stud 2(3):165–202

Hussain DA, Alwan AA (2008) Evaluation of aquatic macrophytes vegetation after restoration in East Hammar Marsh, Iraq. Marsh Bull 3(1):32–44

Lafont B (2016) Male and female professions in Sumer at the end of the IIIrd millennium BC. Orient 51:63–67

Lemorini C, Cristiani E, Massussi M, Piccol I (2008) L'attività di cesteria: tradizione artigianale, sperimentazione e archeologia. In: Lugli F, Stoppiello AA (eds) Atti del 3° Convegno Nazionale di Etnoarcheologia, Mondaino, 17–19 marzo 2004/Proceedings of the3rd Italian Congress of Ethnoarchaeology, Mondaino (Italy), 17–19 March, 2004 (=BAR-IS 1841), Oxford, pp 58–64

Maxwell G (1957) People of the reeds, New York

Milli S, Forti L (2019) 3. Geology and Palaeoenvironment of Nasiriyah Area/Southern Mesopotamia. In: Romano LD, Agostino F (eds) Abu Tbeirah excavations 1. Area 1 last phase and building A – Phase 1, Rome, pp 19–38

Montorfani MV (2019) Vegetable plaiting materials from the site of Abu Tbeirah (Southern Iraq, Third Millennium BC): experimental approach. Exarch 2019/1. https://exarc.net/ark:/88735/10399

Moorey PRS (1999) Ancient Mesopotamian Materials and Industries: The Archaeological Evidence, Winona Lake, Indiana

Ochsenschlager EL (1992) Ethnographic evidence for wood, boats, bitumen and reeds in southern Iraq. Ethnoarchaeology at al-Hiba. In: Powell MA (ed) Trees and timber in Mesopotamia, Bulletin on Sumerian agriculture, vol 6, pp 47–78

Ochsenschlager EL (2004) Iraq's marsh Arabs in the garden of Eden, Philadelphia

Pollock S (1990) Political economy as viewed from the garbage dump: Jemdet Nasr Occupation at the Uruk Mound, Abu Salabikh. Paléorient 16(1):57–75

Pomponio F (1989) A Neo-Sumerian account about reeds. Orientalia NS 58(2):230–232

Potts DT (1997) Mesopotamian civilization: the material foundations. Cornell University Press, Ithaca

Romano L, D'Agostino F (eds) (2019) Abu Tbeirah excavations 1. Area 1 last phase and building A – Phase 1, Rome

Salim SM (1962) Marsh dwellers of the Euphrates Delta, London

Sallaberger W (1989) Zum Schilfrohr als Rohstoff in Babylonien. In: Scholz B et al (eds) Der orientalische Mensch und seine Beziehungen zur Umwelt: Beiträge zum. Graz, Grazer morgenländischen Symposion, p 2

Sallaberger W (1992) Anmerkung zu den neusumerischen Texten über Schilfrohr. In: Powell MA (ed) Trees and timber in Mesopotamia, Bulletin on Sumerian agriculture, vol 6, pp 123–124

Schwartz M, Hollander D (2000) Annealing, distilling, reheating and recycling: bitumen processing in the ancient Near East, La pyrotehnologie à ses débuts. Évolution des premières industries faisant usage du feu/Early Pyrotechnology. The evolution of the first fire-using industries. Paléorient 26(2):83–91

Stol M (2012) Bitumen in ancient Mesopotamia. The textual evidence. Bibliotheca Orientalis LXIX (1–2):48–60

Streck MP (2009) Schilf. Reallexikon der Assyriologie und vorderasiatischen Archäologie 12:182–189

Tafuri MA (2019) The human remains. In: Romano L, D'Agostino F (eds) Abu Tbeirah excavations I. Area 1 last phase and building A – Phase 1, Rome, pp 389–418

Thesiger W (1964) The Marsh Arabs, New York

Waetzoldt H (1992) 'Rohr' und dessen Verwendungsweisen anhand der neusumerischen Texte aus Umma. In: Powell MA (ed) Trees and timber in Mesopotamia, Bulletin on Sumerian agriculture, vol 6, pp 125–146

Woolley CL (1934) Ur Excavations II. The Royal Cemetery of Ur, British Museum, London. https://www.britishmuseum.org/collection/term/BIB1124

Chapter 4
Human Ecology of the Marshes

Ariel I. Ahram

Abstract This paper places the fate of the Iraqi marshes within the deeper history of man-made environmental change in the Middle East. Irrigation was crucial to every society inhabiting the Tigris-Euphrates basin. Modern engineering promised to overcome Iraq's natural hydrological limitations. Environmental manipulation became the lynchpin in wider projects of social engineering. Using a theoretical perspective inspired by James C. Scott's work on the tensions between societal resistance and state-building processes, the paper argues that the marsh ecology endured as a kind of anti-state space, immune to the surveillance and domination of central governors. Marsh dwellers, in turn, sought refuge from political domination. As the Iraqi states accumulated scientific and economic prowess, the ability to resist or evade declined. Engineers and technocrats deployed a variety of techniques to alter the human ecology of what they saw as wasteland. Saddam adopted and deployed these plans in an even more desperate gambit to augment state control. The final solution to the problem of the marshes came with eviction of the marsh inhabitants and the obliteration of the marsh habitat.

Keywords Saddam Hussein · Marshes · Iraq · Hydrology · Rebellion · Anarchy

> He created the Tigris and the Euphrates and set them in place
> Their names he appropriately proclaimed
> He created the grass, the rush of the marsh, the reed, and the woods...
> Lord Marduk piled up a dam at the edge of the sea
> [...] a swamp he made into dry land –*The Babylonian Genesis*[1]

At the confluence of the Tigris and Euphrates rivers in southern Iraq historically lay one of the largest wetlands in southwest Asia, known in Arabic as *al-Baṭiḥa* (the

[1] Heidel 1963, p. 63.

A. I. Ahram (✉)
Virginia Tech School of Public and International Affairs, Arlington, VA, USA
e-mail: ahram@vt.edu

L. A. Jawad (ed.), *Southern Iraq's Marshes*, Coastal Research Library 36,
https://doi.org/10.1007/978-3-030-66238-7_4

marshlands). In the late twentieth century, al-Baṭiḥa consisted of three distinct wetlands – Hawr al-Amarah, Hawr al-Hammar, and Hawr al-Hawizeh – and covered some 15,000 square kilometers in total. The marshes were the home of the *maʿdan* or *maʿdi*, a group whose traditional way of life consisted of shifting agricultural production, pastoralism, and fishing.

The marshes were an epicenter of the March 1991 uprising in the wake of the 1990–1991 Gulf War. After crushing the uprising, the Iraq government attacked the marshes, destroying whole villages and deporting the population to government-built towns in the desert. The following years, a series of barrages and dams cut off nearly all of the flow of water to marshes. The assault on the marshes and their inhabitants figures prominently in any bill of particulars against Saddam Hussein and the Ba'th regime (Joffe 2000; Morley 1993; Furr 2010; Moumin 2007). Iraq's 2005 constitution lists the destruction of the marshes among the "pains of sectarian oppression inflicted by the autocratic clique" upon the Iraqi people. Since the removal of Saddam in 2003, there have been efforts to revive the marsh ecology, but to marginal effect.

This paper seeks to place the fate of the marshes within the deeper history of man-made environmental change in the Middle East (Burke 2009). Irrigation, of course, has been crucial to every society inhabiting the Tigris-Euphrates basin. Karl Wittfogel famously posited that emergence of ancient despotism in Mesopotamia, Egypt, China, and India to a distinctive form of "hydrological civilization" where control over irrigation allowed rulers to amass enormous power over society (Wittfogel 1957). But tides, rains, erosion, and other natural phenomena could stymie even the most ambitious of ancient ruler. By contrast, the advent of modern scientific knowledge spurred ever greater confidence in the ability to shape human ecology (Richards 2003). As in other arid regions, modern engineering promised to overcome Iraq's natural hydrological limitations (Worster 1992). Environmental manipulation became the lynchpin in wider projects of social engineering.

From this historical perspective, the destruction of al-Batiha was not just a singular act of political repression, but the culmination of a much longer effort to render the marshes and the marsh inhabitants as legible and productive political subjects. Using a theoretical perspective inspired and informed by James C. Scott's work, the paper argues that the marsh ecology endured as a kind of anti-state space, immune from the surveillance and domination by central authorities (Scott 1998, 2009, 2017). Marsh dwellers, in turn, were typically those who could exploit the marsh ecology to avoid state control and find political refuge. As the state accumulated scientific and economic prowess through the long twentieth century, it steadily degraded the autonomy of the marshes ecologically and politically. Seeing the marshes as a wasteland and obstacle to development, engineers and technocrats deployed a variety of techniques to alter and undermine the human ecology of the marshes. The destruction of the marshes and displacement of its population, then, came incrementally and in direct connection to the state-led development plans (Dwivedi 2002; Cernea 2000).

In the late 1980s and early 1990s, battered by combined internal and external security crises, Saddam adopted and deployed these plans in an even more desperate gambit to augment state control. The final solution to the problem of the marshes came with the eviction of the marsh inhabitants and the obliteration of the marsh habitat. As with many other plans for social transformation that matched technocratic means with totalitarian ambitions, the ultimate redemption of the marshes and the ma‘di came only through their elimination.

4.1 State and Anti-state in Ancient and Medieval Iraq, c. 3000 BC to 1858

If Egypt is the gift of the Nile, Mesopotamia has had a more uneasy relationship with its twin benefactors. Salinity, siltation, and unpredictable and untimely flooding can render the waters of the twin rivers as much a curse as a blessing. Early evidence of channeling date back at least to 3000 BC. Yet the marshes still dominated the landscape. The early city-states of Ur formed an archipelago within the midst of wider wetlands. The flora and fauna of the marshes likely yielded a surplus equal or greater to that available from sedentary agriculture, but without the need for centralize coercive control (Oates 1960; Pournelle and Algaze 2014).

It was not until the first millennium BC that animal power was used to move water, creating the first large-scale networks of branching canals to provide water for irrigation. The Nahrawan canal, connecting the Diyala and the Tigris rivers, was built in the sixth century CE under the Sassanid, evidence of a particularly strong central government capable of impounding the labor of peasants or slaves to construct and maintain the canals (Adams 1965). Indeed, Iraq's history of massive irrigation projects, as Peter Christiansen observes, exemplified a "most drastic form of coercion and manipulation of nature" (Christensen 1993, p. 4).

But while hydrology sustained the state in Mesopotamia, the state too had to sustain hydrology. These networks repeatedly fell into disrepair, allowing the marshes to return to their natural state. Baladhuri reports that when the rivers overflowed in 629 CE, the Sassanid king tried to stop the breaches:

> but the water got the better of him, and turned towards al-Bata'ah and overflowed the buildings and plants, drowning many cantons that were there. [The king] rode out in person to block the breaches; he scattered money right and left, put many workmen to death and, according to a report, crucified on certain breaches forty dam builders in one day; but all that was of no avail against the force of the water. (Hitti and Murgotten 1916, p. 454)

By the time of the Muslim conquest of Mesopotamia 11 years later, the Sassanid irrigation grid was already in disrepair due to plague and political infighting within the empire, compounding problems of flooding and food shortages. It was only with the Umayyad ascension that the canals, dams, weirs, and floodgates were somewhat repaired. Periodic neglect and over-manipulation led to problems of sedimentation and siltation on the canals alternatively inundating or desiccating previously productive agricultural land (Eger 2011; S. W. Cole 1994). With the central state in abeyance, the populous itself launched smaller scale efforts to harness the rivers continued. Tax system gave special dispensation for the acquisition of "new" lands reclaimed from the swamp (Morony 1984). Tribes who often proved the bane of the central governments still invested in irrigation sufficient for their own more limited agricultural purposes (Fernea 1969, pp. 26–29).

Al-Baṭiḥa figures as a redoubt of lawlessness, rebellion, religious heterodoxy, and otherwise uncivilized behavior in the accounts of courtly chroniclers, geographers, and occasional travels. It served, as the Encyclopedia of Islam put it, as a "hiding place for all sorts of robbers and rebels, and an asylum for discontent" (Streck and Saleh 2012). The Prism of Sennacherib, named for the Assyrian king who conquered Babylonia between 703 and 689 BCE, discusses numerous rebellions emanating from the marshes (Luckenbill 2005; Jwaideh 2007). Under Umayyad rule, the Zuṭṭ, a group of northwest Indian descent who had migrated to the area under the Sassanids and been incorporated into Muslim Arab tribes, were deported after a series of uprisings. The Zuṭṭ continued to cause problems for the Abbasids into the ninth century. Al-Jahiz names the Zuṭṭ "of the thicket" as brigands allied with a "beggar chief." Caliph Ma'mun (r. 813–833) repeatedly dispatched boats and soldiers to subdue the Zuṭṭ and their allies, including a number of runaway slaves. Twenty-seven thousand were reportedly deported to the Cilicia (Morony 1984, p. 211).

The Zanj Revolt (869 to 883), often called the Islamic Spartacus, also emanated from the marshlands. The uprising was led by a Persian freeman who claimed a lineage to the Prophet Mohammed. His message drew massive support from the black slaves who worked the salt fields in southern Iraq. At their zenith, the Zanj controlled the city of Wasit and launched attacks on Basra, Ahwaz, Ramhurmouz, and Ubulla (Waines 1977; Popović 1999; Clarence-Smith 2006, p. 63). The Zanj, according to Tabari, fished in the marshes and ferried their catch in small boats along the canals to their outposts. Eventually, Abbasid forces encircled the marsh and blocked off the flow of supplies, starving the rebels into submission (Yarshater 2007, pp. 68).

The marshes remained a lawless zone almost 500 years later. Journeying between Najaf and Basra around 1325, Ibn Battuta mentions encountering "a water-logged jungle of reeds, inhabited by nomad Arabs known as *al-ma'di*. They are brigands of the Rafidi [=Rejectors, i.e., Shi'i] sect." After describing how these ruffians waylaid his fellow travelers, Battuta concludes that "they fortify themselves with this jungle and are able to defend themselves in it against all attacks" (Battuta et al. 2010, pp. 271–272). In the 1430s, the Mushaʿshaʿ, a Shi'i millenarian sect founded by the adopted son of a Shi'i jurist, established a base on the eastern edge of the marshes in Huwayza. The Mushaʿshaʿ drew on the support of Arab tribesmen from Ahwaz to the Tigris (Moojan 1985, pp. 101–102; Soucek 1984, pp. 203–204).

With the crystallization of Iranian-Ottoman frontier in southern Mesopotamia in the sixteenth century, the marshes became a seam in which peoples could flee to avoid state control. In 1566, the Ottomans sent 450 ships down river from Mosul, 2000 Janissaries, and 6000 local Kurdish and Arab soldiers into the marshes to put down a tribal revolt. The Ottoman force cut down the marsh palm groves and leveled villages to bring the uprising to a heel. Still, the Ottomans could only force the marsh tribes into some kind of negotiated settlement, not subjugation (Matthee 2006). On the Iranian side of the marshes, the Mushaʿshaʿ became Iranian lieges and hereditary princes of Huwayzah. The region remained the domain of Arabic-speaking, buffalo-rearing tribes (Moojan 1985, p. 102; Soucek 1984). The Bani Kaʿb tribe of southern Khuzestan played off Iranian and Ottoman governors, avoiding more than a nominal tribute to either government between the seventeenth and nineteenth centuries. They raided the shipping on the Shatt al-Arab and even plundered Basra. The Iranians launched an expedition in coordination with the Ottomans and the English East India Company to encircle the Bani Kaʿb. The tribe took to the marshes, hopping from island to island before finally escaping to the open sea (Floor 2006). Even into the early twentieth century, the Bani Kaʿb continued to resist incorporation under Iranian state rule under the leadership of Sheikh Khazaʿl of Mohammareh, champion of an independent Arabistan (Strakes 2011). The marshes displayed many of the same characteristics as Scott observed of the uplands of inland Southeast Asia, a non-state zone, a "location where, owing largely to geographically obstacles, the state has particular difficulty establishing and maintaining its authority" (Scott 2009, p. 13).

Chroniclers rarely consider the more mundane but still critical question of the interior of marsh society. To offer any historical claims about the human ecology of the marsh dwellers themselves requires extrapolation and triangulation among the older sources and the ethnographic studies conducted in the marsh region since the late nineteenth century. This is an intellectually risky undertaking (Tapper 1997, pp. 18–19). The problem begins with the seemingly simple question of the name *al-maʿdi* itself. The etymological roots of the word remain obscure (Dauphin 1960, p. 39). Ibn Battuta's use of the term, though, is suggestive of an exonym, a name used by outsiders to refer to the group rather than the name a group calls itself. It could denote an ethnological grouping, but it could also be functional, applying only to brigands in marsh, not to marsh dwellers as a whole. Seen in this light a likely candidate for the origin is the Arabic root *ʿayn-dal-waw*, which has connotations of enmity, aggression, and hostility as well as avoidance and transgression.[2]

Outsiders' attempts to account for the maʿdi's origins are inconsistent. The fact that Western ethnographers, travelers, and imperialists referred to those residing in the marshes as *maʿdi* does not mean that these are the same people. Some physical anthropological studies report the maʿdi as robust and towering, others as malnourished and diseased. Some portray the maʿdi as having Indian or Persian physical features, others pure Arab.[3] Since the mid-1800s, ethnographers have tried to differentiate between the "real" Marsh Arabs, who lived in the reed huts and raised buffalo and cultivating and pastoral tribes that surrounding the marshes (Thesiger 2007; Ochsenschlager 2014). Most of these efforts, though, remain unpersuasive. As P.A. Buxton and V.H.W. Dowson presciently observed in the 1920s, ma'di is a largely occupational category, not an ethnological one (C. L. Cole 2016; Buxton 1922).

---

[2]I thank Mohammad al-Masri for bringing this possibility to my attention.

[3]In this regard, recent genetic studies of the marsh dwellers report the predominance of genetic markers typical of the rest of the Middle East, alongside some Southwest Asian and African genetic markers. (See Al-Agidi and Roberts 1984; Al-Zahery et al. 2011).

A more fruitful approach to the ma'di than assuming biogenetic continuity is to conceive of a flexible social category defined subjectively in relation to the state and other foci of power. In the case of the maʿdi, the dominant element in the relationship with the state has been evasion. Flight has been a persistent feature in Mesopotamian peasant life. The marshes, like the desert and the steppe, were historically inaccessible and thus fairly protected from the rapacious state (Nieuwenhuis 1982, pp. 119–120). The marshes thus became a sort of maroon colony, an accumulation point for those fleeing various forms of subjugation (Jwaideh and Cox 1988).

With the state at a distance, tribal solidarity in the marshes provided a ready alternative system of protection. Indeed, throughout Iraqi history, tribalism has seemed to rival the state. Modern ethnographers report that the maʿdi deem themselves to be bedouin, descendants from the illustrious Arab tribal confederations. Such a genealogy is no less improbable than any other tribal kinships. While espousing a myth of unified tribal ancestry, Middle Eastern tribes often formed as composites of multiple lineages that coalesced for particular political purposes. Underlining this point, linguistic anthropologists note that the maʿdi speak the nomadic dialect rather than the dialect common to sedentary peasants (Ingham 1976).

But tribalism entails its own forms of domination and hierarchy. Tribal culture held nomadic camel-herding as the purest and highest form of life, and the bedouin formed a kind of aristocracy in the great tribal confederations of Iraq. The shepherding and cultivating clans incorporated into the tribal confederation were generally regarded as inferiors (Batatu 2012, p. 16). Particularly in southern Iraq, the tribes took on the trappings of mini-states. Lower ranking members were impounded to work in irrigation maintenance and repair, a status verging on serfdom (Nieuwenhuis 1982, pp. 155–156). Those who lived in the marshes, then, were considered especially debased. Superior tribes refused to give their daughters in marriage to the marsh dwellers. Still, the divisions within the tribes were not always clear-cut, and there was often a fusion between different groups (Batatu 1982; Nakash 1994). This meant that the boundary between the maʿdi and the more highly esteemed categories of tribalism was permeable. Individuals and households could move into and out of the marshes depending on circumstances (Drower 1947).

The marsh economy thrived precisely when irrigation-fed agriculture, premised as it was on the state's ability to impound labor, was at its nadir. This is not to say that the maʿdi were self-sufficient—indeed far from it. The maʿdi do not appear to have fashioned their own canoes or fishing spears, acquiring them and other forms of metalwork or consumer commodities from Sabaean boat-builders (Drower 1937, pp. 51–52; Buxton 1922, p. 294). But the marsh economy in many ways kept the state at arm's length. Riparian piracy and highway robbery clearly flourished whenever that state was weak. Similarly, hunting and fishing probably provided a baseline of subsistence. Marsh fishing, particularly of carp (*Barbus sharpeyi* and *B. escimus*), dates back to 5000 BC. Pre-Islamic deities were devoted to fishing. City dwellers ate salted or fermented fish (Sahrhage and Lundbeck 2012, pp. 34–41). Buffalos (*Bubalus bubalis*) are mentioned in nearly every description of the marshes. The buffalo produced milk and cheese, staples of the marsh diet, and formed the core asset of every marsh household. Another mainstay of the marsh economy was cultivating reeds. The reeds could grow 6–9 meters in height and 5 centimeters in diameter. The reeds were a useful raw material for matting, irrigation, boat building, and home construction and for styluses. Medieval Basra hosted a market devoted to trading in marsh flora (Pournelle and Algaze 2014, pp. 8–9; Naji and Ali 1981).

The marshes provided a human ecology allergic to political hierarchy, seemingly designed to foster what might be dubbed an *anti-state* (Scott 2009, p. 3). Eventually, all the minor rebellions in the marshes were co-opted, overtaken, or destroyed. What could not be stopped, though, was the use of the marshes as a sanctuary by those whose primary political ambition was to be left alone. The motley of ex-slaves, freemen, and religious zealots who joined the Zanj and other rebels never established the kind of durable logistics and internal governance system necessary to occupy towns or cities. They were more interested in plunder than rule (Waines 1977, pp. 301–303). The same could be said of the amphibious Bani Kaʿb, who melted away rather than fight. The maʿdi seem to purposefully shrink to the margins of power tribal confederations that alternatively protected and dominated them. This type of devolved social system could allow marsh residents to scatter and escape when confronted with power (Tapper 2009). Such strategies are especially effective in frontiers, deserts, mountains, swamplands, and other terrains where surplus is not readily expropriated. It rendered the marsh society the antithesis of the centralized and hierarchical society that rulers envisioned.

4.2 High Modernism, Enclosure, and Displacement, 1858 to 1980

For much of the Islamic era, the management of Iraq's twin rivers was dominated by what Alan Mikhail calls "coordinated localism" (Mikhail 2013, p. 33). In contrast to the elaborately engineered controls that Wittfogel and others imagined necessary for agricultural success in Mesopotamia, it was the local tribes that initiated, financed, and supplied most of the irrigation projects using reed dams, animal or human-powered water wheels, and small channels (Fernea 1969, pp. 36–37; Murphey 1987). As such, the marshes and maʿdi society—the unintended spillover of the state's technological and coercive overreach—were integral features in the Iraqi landscape. Yet by the nineteenth century, both the Ottomans, the nominal imperial sovereign over Mesopotamia, and Great Britain, whose imperial purview increasingly extended to the Persian Gulf, devoted ever-expanding resources to controlling al-Baṭiḥa. For different reasons, both sought to strengthen central state authority and facilitating economic development in Iraq. Both evinced what Scott calls the high modernist ideology, a muscle-bound confidence in:

> scientific and technical progress, the expansion of production, the growing satisfaction of human needs, the mastery of nature (including human nature), and, above all, the rational design of social order commensurate with the scientific understanding of natural law (Scott 1998, p. 4).

Engineers, scientists, economists, and other specialists in impersonal—and therefore more scientific—methods of fostering change play a privileged role in high modernist endeavors, as do the capitalists who employ their services. As in Egypt, irrigation was a singular focus of modernization projects, a metric by which Britain could measure itself against empires past (Mikhail 2013; T. Mitchell 2002). High modernist technologies provided the state for the first time the means to overcome the obstacle posed by the marshes and their inhabitants. Though the maʿdi continued to evade and frustrate the state's ambitions throughout the nineteenth and mid-twentieth century, the marshes were enclosed, isolated, and diminished by state power.

An essential first step in the progress of high modernism was mapping Iraq's human and physical geography. The British began surveying the Persian Gulf in the late 1700s, including the Shatt al-Arab up to Basra. In the 1830s, the Ottomans authorized Francis Chesney's survey team north of Basra, the first serious attempt to devise navigational aids for the rivers. Chesney's team created not just a physical map, but also a catalog of the peoples they encounter (Yapp 1993). This expedition cleared the way for regular steamer transit between Basra and Baghdad. As Camille Cole notes, though, cartographers found that rivers and streams shifted with rain, drought, or sluicing—often times with calamitous results for the steamers. Bedeviling ethnographers, tribes and villages alighted and withdrew from the marshes with similar fickleness (C. L. Cole 2016; Satia 2007).

The Ottoman's nineteenth-century reforms seemed specifically to address this perceived lack of social and physical fixity (Kasaba 2009). The traditional collective tribal holdings prevalent in Iraq were regarded as an antiquated relic and the persistence of nomadism a drag on the empire. Reformists sought to promote a system of private property that would foster the creation of a body of small-holding peasants (Farouk-Sluglett and Sluglett 1983; Jwaideh 1984; Batatu 2012). Midhat Pasha, the energetic governor of Baghdad (1869–1872), recognized the linkage between land reform and hydrology. A robust and region-wide irrigation network delivering the water to modern farmsteads was essential. The rivers were now conduits allowing Iraqi grain, dates, and other products to reach India and beyond. Midhat oversaw the building of a transverse canal between the Tigris and Euphrates and officiated the launching of a steamship line to Mosul. Subsequent Ottoman governor did their best to implement the program of hydraulic management (Murphey 1987, p. 25; Ceylan 2011, pp. 187–204).

In 1908, the Ottomans hired Sir William Willcocks to develop the first comprehensive hydrological plan for the Tigris and Euphrates. Willcocks had extensive experience in the water authority in India and had overseen the construction of the first Aswan dam. Willcocks, like many British imperial servants, saw his place in the Orient through the lens of the biblical and classical history. He was convinced that that the historical location of Eden lay between the twin banks of the rivers. Irrigation could return Mesopotamia to its ancient splendor (Satia 2008). Willcocks called for the establishment or rehabilitation of a number of transverse canals and a massive reclamation of floodland around Basra (Wilensky-Lanford 2011, pp. 100–110). Willcocks was personally involved in the first stage of the construction, digging the Hindiyya canal near Hilla. But Willcocks expressly ruled out the possibility of destroying the marshes entirely. He noted that although it was possible to gain land through drainage, the result would "be disastrous. We should convert 800,000 hectares of swamps, worth something, even today, into desert," and deny fresh water to the farms on the Shatt al-Arab (Willcocks 1917, p. 35). This insight was lost of subsequent planners, though. They viewed the "unregulated" marshes as wasting precious fresh water and arable land (Ionides 1937, p. 201; Howell 1922).

Though Iraq gained formal independence in 1932, British technical advising continued. Land reclamation was a primary concern, buttressing an approach to agriculture that focused on gains from field expansion. A 1944 Iraqi government report estimated that only 20% of Iraq's potentially cultivable land was being utilized and that current usage still fell below the Sassanid and early Islamic era. Still, the report noted "marshes, swamps, and lakes which are slowly but surely silting up and will in time add still further to the already great area suitable for perennial irrigation" (Sousa 1944, p. 2). In 1951, the technocratic Iraq Development Commission released a comprehensive hydrological study. Citing Willcocks' schemes as its model, the Development Commission report set the explicit goal of "reclamation of the marshes." The plan called for new regulators at Khan Beni Saad, Amarah, Gharraf, and other infrastructure sufficient to divert some 3000 meters per second of water from the marshy wasteland (Sassoon 1987, pp. 142–146; Lebon 1955). Over the next decades, the Iraqi government retained British, Russian, American, Dutch, and Pakistani firms to execute different components of the scheme (Qubain 1958, pp. 59–64).

Though land and irrigation reforms were meant to transform the mass of tribesmen into tax-paying freeholders, it was the sheikhs and urban notables who realized the opportunity of land registration. By choice, trickery, or duress, many illiterate tribesmen surrendered their individual claims to their tribal superiors or urban notables. Almost overnight, the sheikhs became landlords and the lesser cultivating tribesmen tenants. Customary usages like encampment (*naqsha*), which allowed individual tribesmen to exclusive use of land for rice cultivation if they endeavored to maintain irrigation infrastructure, were disallowed. In Kut and Amarah, much of the land was actually claimed by the state and granted on long-term leases to the tribal paramounts and urban landlords. In 1906, the largest landholders in Amarah, for instance, were tribal sheikhs. These new landlords preempted tribal grazing rights. Once peripatetic tribesmen became sharecroppers. People in and around the marshes chose to flee the land entirely, while marsh dwellers were drawn to work as wage laborers (Haj 1997, pp. 18, 26–19; Jwaideh 1984, pp. 338–350; Batatu 2012, p. 120).

British colonial rule in Iraq, beginning with the invasion of 1914, intensified many aspects of the Ottoman high modernization project. Technological improvement, particularly hydrology, was the shibboleth of British colonial rule. The British repaired many of the old canals and built new embankments on the Tigris and Euphrates to control the floods and reclaim agricultural lands. They dispatched a corps of irrigation officers to adjudicate local usage disputes (Haj 1997; Fernea 1969, 136–8).

Unlike the Ottomans, the British deliberately bolstered the sheikh-cum-landlords, creating what might be dubbed a new irrigator class. A 1926 law sought to encourage the replacement of the traditional human or animal-powered waterwheel with mechanical pumps. Only urban investors, though, had the capital to invest in such devices. A follow-up law 6 years later allowed pump owners to seize land in case of default, further accelerating peasant alienation. The British also introduced new strains of barley and wheat and set up model farms for cultivating cotton and other cash crops (Haj 1997, pp. 49–51; Batatu 2012, pp. 121–125; Fisk 1952).

Other new technologies allowed the British to penetrate Iraq's difficult geography to an unprecedented degree. Aerial cartography proved the ideal tool to supplant the sketchy Ottoman era maps. RAF surveillance sorties charted the locations of thousands of huts, herds, and reed patches (Omissi 1990, pp. 93–95; Satia 2008, pp. 159–161). Aerial surveillance and bombardment played a critical role in suppressing tribal uprisings. The emergent colonial doctrine held that "air policing" was a more surgical, humane, and economical. Just as they had done with farming, the British set up model Iraqi towns in which to practice, perfect, and demonstrate the effects of aerial bombardment. Defending the hard-won gains in irrigation was among the missions of air policing, equating tampering with hydrological infrastructure with rebellion (Cox 1985, pp. 168–169).

The maʿdi drew on their customary repertoire of resistance by evasion, at least initially, in response to these changes. The first Ottoman gendarmerie posts appeared on the marshy banks in the 1860s, but troops, census takers, or other government agents still did not venture into the marshes themselves. Tribes like the Albu Muhammed remained heavily armed and fiercely independent, with some serving as armed retainers for prominent urban magnates (Salim 1970; Visser 2005, pp. 15–18). Uprisings occurred sporadically from the 1890s through the 1900s. Tribesmen destroyed weirs and dikes in order to create floods that would impede the government forces dispatched to suppress them. In response to one episode of piracy in 1899, a British official recommended holding tribal headman responsible for the piracies committed by anyone in his village. Ottoman troops repeatedly deported unruly tribes from the water's edge. Still, this was no commonplace brigandage. The pirates seemed to have deliberately avoided attacking British ships, calculating that by threatening only Turkish vessels, they could spur enough concern from the British that they would intervene with Ottoman authorities on their behalf. Complicating matters further, tribes straddling the Persian-Ottoman border were considered Persian subjects under the tribal leadership of the Sheikh of Muhammareh and could appeal to the Persian government for support as well (Çetinsaya 2006, pp. 93–98; Lorimer 1970, pp. 1499–1529).

When the British troops conquered Qurna in December 1914, some of the marsh sheikhs declared their loyalty, while others joined the calls for jihad against infidel rule. Some nurtured hopes that the British would accede to creating an independent state centered in Basra. One Amarah-born Shi'i intellectual, for instance, wrote romantically about the marshes as the core of a southern Mesopotamian state. Sheikh Khazail of Muhammareh put himself forward as a candidate for the throne of Iraq (Visser 2005, p. 124). Though these political plans proved for naught, the marshes were still a haven for tax evasion, snipping, theft, and attacks on the telegraph. When the sheikh of Chibaysh, a town on the edge of the marshes, defied the government, RAF aircraft dropped leaflets warning that bombers would destroy the town unless the disobedience ceased. Eventually, the sheikh was deported to Mosul, a testament to the increasing capacity of state coercion. Avoidance became the only option. Villagers in nearby Madina a few years later simply withdrew with their animals and other moveable property into the marshes to avoid taxes and rent (Salim 1970, pp. 31–32; Visser 2005, pp. 55, 100).

The maʿdi also found ways to take advantage of the economic opportunities provided by irrigation and mechanization. Through the nineteenth century, much of the steamer cargo was buffalo milk *ghi*. Reeds and reed matting remained a major trading commodity throughout the markets of central and southern Iraq. The British occupation spurred a building boom in Basra, increasing the demand for reed matting. Even the marsh canoes, the emblem of maʿdi society, endured. Reeds were too bulky to be profitably shipped by steamer, so they continued to be transported in the traditional vessels. Moreover, when steamers stalled or ran aground, queues of reed canoes offered passage to the stranded (C. L. Cole 2016).

More and more, though, the maʿdi were pulled into the orbit of the state and capital. With private landholding becoming entrenched and protected by an immensely coercive state apparatus, the pillars of the marsh society became increasingly shaky. The richest of the maʿdi chieftains, those who traveled in the grandest canoes, boasted the largest herds, and most wives were positively impecunious when compared to town dwellers. They still relied on Sabaean and Jewish middlemen to acquire metalwork, coffee, tea, and guns. Amarah and Kut had severe land concentrations well into the 1950s, with landholdings dominated by the sheikhs of the Albu Muhammed and other tribal aristocracy. Landholders exploited the ample labor pool. There were deadly confrontations between tenants and the sheikh's armed foremen. Malaria was brought under control by the government's DDT campaign in the 1950s, but bilharzias, hookworm, and syphilis remained endemic (Qubain 1958, pp. 233–235; Lanoix 1958; Al-Dewachi 2017). Similarly, though the government funneled money to education, the effects didn't reach the deeper parts of the south. Some sheikhs blocked the construction of schools near their property (Warriner 1962, pp. 136–157; Qubain 1958, p. 206; Haj 1997, pp. 37–38).

Escaping to the cities became a new form of evasion. By the 1920, Baghdad and other cities were rimmed with suburbs of reed huts. Marsh residents brought dairy, fowl, and sheep to the town market. In the 1950s, 800 families reportedly had fled from one of the Albu Muhammed sheikhs (Drower 1947; Dauphin 1960, pp. 46–48). The government built new middle-class neighborhoods featuring health clinics, cinemas, and shopping centers. Rural migrants, meanwhile, swarmed to sewage-laced slums like Baghdad's Sheikh Omar (Qubain 1958, pp. 246–247). They were mercifully free of bilharzias and mosquitoes, not to mention debt peonage, but ridden with trachoma and dysentery (Warriner 1962, p. 181; Ishow 1983).

The high modernist project launched by the Ottomans in the nineteenth century and carried forward by the British and then independent Iraq in the twentieth century deeply influenced Iraq. It dramatically altered the physical and human geography.

Baghdad boasted new construction reflecting Iraq's aspiration to become a modern and international capital city.[4] The transformation in rural areas, although less aesthetically dramatic, was perhaps even more crucial. Yet these projects rarely functioned in the ways their designers intended or anticipate. In the rural areas, nearly all of schemes to expand arable land, introduce new crops, or develop new political inclinations in Iraq fell short of their goals. For planners, the fault tended to lie with the Iraqi people themselves, who seemed culturally incapable of appreciating the benefits of state-led development or to recognize that their displacement came for the benefit of the greater good (Ireland 1970, pp. 123–125; Main 1935, p. 205).

Reflecting on a century of reform projects in 1957 (a year before a military coup toppled the Iraqi monarchy), Doreen Warriner described how thinking about developing in Iraq had:

> concentrated on the control of the environment. It has been carried out by consulting engineers... The engineer's approach is two dimensional, a blue-print in land and water, which money can turn into dams and dykes and drains. Except in so far as they provide laborers, the people of Iraq do not enter into it.... (Warriner 1962, p. 159)

Oil revenues further emboldened plans for massive infrastructural and development infrastructure. But while engineers dreamed of resurrecting Babylonia, Iraq—especially the marsh regions—acquired what S.A. Nahi, an Iraqi jurist, dubbed the "feudalism of the Hammurabi" (Nāhī 1955).

There were some voices of dissent along the way. In 1955, in what was essentially a minority report to the Haigh commission, Lord Salter warned that mega-scale hydrology and other imported technologies were no panacea. The rivers responded unpredictably to human interference and "the administration of the control made possible by the great dams presents a problem at once more difficult and more lasting than the construction of the dams themselves" (Salter 1955, p. 39; Warriner 1962, pp. 130–131). Indeed, even the International Bank of Reconstruction and Development doubted the feasibility of land reclamation. The turn to state socialism following the 1958 revolution put only a small dent in the south's endemic poverty. The apparent failures of both capitalist and socialist development models further highlighted the need to premise development plans on traditional ecological and social practices and indigenous institutions (Fernea 1969).

Heedless of these warnings, the state's high modernist project induced drastic ecological changes. In the 1970s, Syria, Turkey, Iran, and Iraq all built new dams that stifled the marshes' inflow. In 1968, the river flow was estimated at 68×10^9 cubic meters, and the Amarah and Hammar marshes spanned an estimated 7970 square kilometers. By 1984, the flow had diminished by over 40% and the marshes shrank by over a third (Jones et al. 2008). Combined with the diversion of water to Al-Tharthar Lake in Iraq, poor drainage, and the runoff of nitrate-tainted fertilizers, the waterways of southern Iraq were stricken with salinization and pollution. Even

---

[4]For broader discussion of high modernism in Iraq, see Harrington (2014), Mehdi (2008), and Bahoora (2013).

with water treatment, by 1979, the total dissolved solids in the river water at Nasiriya were already above the 1000 ppm standard for potability (by 2001, it was four times the standard). Consequently, official estimates in the mid-1970s held that 20–30% of Iraq's cultivated land had to be abandoned. Another 20–50% had lost yields due to salinization, especially in the reclaimed areas around the marshes (Al-Layla 1978, pp. 200–201; Hillel 1994, pp. 99–101).

The state also encroached on the marshes politically and socially. By the 1930s, once restive marsh villages had a government-run dispensary and school (both housed in reed huts) and by 1950s a post and veterinary office as well. Ethnographer Shakir Salim observed in the 1950s, though, that the average resident of Chibaysh was "more conscious of the government than he is of tribal authorities… He can lodge complaints against his fellow tribesmen and even against his *sirkal* [foreman] at the government house" (Salim 1970, pp. 15–18, 41). Fishing, buffalo, canoes, and reed endured as scenic vestiges. Western orientalists began espousing romantic notions of the marshes as a lost Edenic realm exactly at the moment that its unique ecology became most precarious (Adriensen 2004). Iraqi nationalist and leftist intellectuals embraced the marshes as a symbol of renewal and emancipation, most famously in Jawad Salim's massive Monument to Freedom sculpture series in Baghdad's Liberation Square (*Nasb al-Hurriyah*) (Al-Musawi 2006, p. 124; Pursley 2019, pp. 200–227). Yet the marsh regions' distinctive human ecology, characterized by illegibility and impenetrability, was wearing progressively thinner. In 1967, Communist guerrillas moved into the marshes, but were quickly crushed by government forces (Ismael 2008). This would be the last significant rebellion to emerge from the marshes for the next quarter century.

4.3 War, Development, and the Destruction of the Marshes, 1968–1995

The ascendance of Saddam Hussein and the Ba'th Party in 1968 brought to power a group that was at once highly authoritarian and deeply committed to high modernist designs. Although the Ba'th did not pursue a full state control of their economy, their commitment to absolute political domination was modeled upon and reminiscent of Stalinist Russia. The Ba'th used every possible source of information to penetrate, monitor, and control the population, including censuses, detailed questionnaires, identity cards, work, educational and health records, and membership lists of associations and clubs. Such measures were deemed essential for transforming the furthest reaches of Iraqi society (Sassoon 2012; Blaydes 2018).

Hydrological manipulation was integral of this drive. New irrigation and hydrology infrastructure was built in Hindiyya, Haditha, Habbaniya, the Tharthar Lake, and around Kirkuk in order to reclaim land. Many of the larger development project suffered setbacks or failed to deliver expectations, often attributed to the shortcoming of Iraqi peasants. But the Ba'th remained committed to use technology to spur development and remake Iraq (Springborg 1981, 1986).

The pressures of war, first the Iran-Iraq War (1980–1988) and then the First Gulf War, brought increasingly coercive and destructive measures to bear. Saddam launched his invasion of Iran on fall of 1980 in the hopes of exploiting Iran's postrevolutionary turmoil, gaining a quick military victory, and forcing Iran to cease its instigation of Iraqi Shi'is. Iranian and Iraqi troops slogged through the marshy frontier. Iraq's military leaders had initially thought the marshes were an impassable defensive barrier. Iranian forces, though, proved adept at infiltrating in small boats. Iraq set up mines and electrical wiring to block Iranian advances. Toward the end of the war, as Iraq launched its own advances on Iranian territory, military engineers poured gravel in the marshes and set up pontoon to allow their infantry to advance (Razoux 2015). For Iraqis, the war transformed the marshes from a symbol of liberation to one of devastation and exhaustion (Khoury 2013, pp. 188–90).

The state's understanding and perception of the marsh residents themselves also changed in the course of the fighting. As Iran advanced, the Iraqi government set out to deport the population from the border near the Hammar and Huwaizeh marshes. Internal documents reveal how the Ba'th leadership and government technocrats saw this relocation as a step toward societal transformation, getting the ma'di to abandon fishing and animal husbandry in favor of modern agriculture. The governor of Maysan wrote that the ma'di represented a primitive but pure form of Arab tribalism. This was an essential distinction in Ba'th thinking. Although the Ba'th avoided explicit discussion of the Sunni-Shi'i cleavage in Iraq, they often used terms like "Iranian origin" as a euphemism to question the loyalty and patriotism of Iraq's Shi'i citizens in the south. Unlike other Shi'is in Iraq, in the Ba'th Party's analysis, the ma'di were innocent of Iranian influences. But the human ecology of the marshes remained problematic. It remained too easy to survive by fishing and foraging and thus disappeared in the marshes. Army deserters had found succor in the marshes and may even have assisted Iranian forces (Blaydes 2018, p. 270; Helfont 2018, pp. 82–3).

The government took steps to purge deserters, insurgents, and anyone else found near the marshes. Thousands of marshes homes were destroyed. Hunting and fishing were restricted. Government-issued food rations were revoked, making it illegal to sell food to those living in the marshes (Khoury 2013, 113–114). The coincidental outbreak of cattle plague in the mid-1980s further diminished food security (ALsaedy 2010; Juma 1997). At the same time, the state renewed its effort to siphon water toward the defensive moats parallel the eastern Tigris (C. Mitchell 2001, pp. 68–69). With the marshes shrinking to a mere 1500 square kilometers, Iraqi officials debated what to do with the remainder. The Iraqi oil ministry recommended further desiccation in order to facilitate oil exploration. Military planners concurred. The agriculture ministry objected that the loss of the marsh reservoirs would hurt agricultural productivity in the south. The idea of full destruction was tabled, at least temporarily (Ahram 2015, p. 454).

The government further tightened its grip on the marshes even further during the Gulf War (1990-91). The introduction of new ration cards in 1990 effectively reinforced the economic siege that had been begun during the Iran-Iraq War. Similar to how government-issued identification cards were used to identify targets during the Kurdish Anfal, the ration cards were another important step laying bureaucratic groundwork for annihilation (C. Mitchell 2001, pp. 89–90; Sassoon 2012, p. 339).

Iraq's military buckled and began to disintegrate under the onslaught of the Allied air and ground offensive that began in early 1991. In March 1991, Iraqi army units mutinied in Basra, sparking an insurgency in both southern and northern Iraq. The uprising divided the marsh tribes. Some marsh tribal leaders joined the rebellion, while others publically avowed fealty to the government (Khoury 2013, 138). The insurrection, though, never reached critical mass. In a speech on March 16, 1991, Saddam claimed that in the south:

> traitors. . . were carrying counterfeit Iraqi identification cards infiltrated our country to spread destruction, sabotage, and looting in the a number of southern cities and villages of Iraq. They were supported by mobsters who lost the righteous way.

Likening the uprising to the Mongol's sack of Baghdad in 1258, he claimed that the rebels were anarchists who destroyed public and private property and burned civil, marriage, and real estate registries (Ahram 2015, p. 456).

By the end of March, the government had regained control of all of the major southern cities, leaving the marshes as the last resort. Tens and perhaps hundreds of thousands of Iraqis fled to the marshes or across the border. International observers warned that Saddam was preparing to ethnically clean the Shi'i. A UN delegation visiting the town of Hammar in July found a Potemkin village, with well-fed residents adulating Saddam. As soon as the delegation left, all UN operations in the marshes were ordered closed (Goldstein 1992, pp. 22–23).

The state began plans to drain the marshes as the rebellion smoldered through 1991. In late March, Saddam claimed in a rambling soliloquy that "the majority of the people in the south played a role in preparing the requirements for the mutiny and the creating of a psychological condition, but they did not want the mutiny to happen the way it did" (Woods et al. 2011, p. 201). Iraqi intelligence surveys in 1992 zeroed in on over 50 villages in Basra, Dhi Qar, and Maysan provinces were more an estimated 7000 traitors, saboteurs, and criminals who were suspected of hiding out. Still, this almost certainly conflated the activities of the guerrillas with those of common criminals. Western reporters found a mixture of native marsh dwellers, army deserters, and insurgents. Most were intent merely on survival. They could conduct little more than small-scale attacks (Ahram 2015, p. 546). The Iraqi state seized the opportunity to eliminate the obstreperous marshes in totality. In the previous decade, Ba'th officials had regarded the ma'di as political neophytes but still nascent members of Iraq's Arab community. Revised assessments following the uprising deemed the marsh dwellers as inveterately hostile. Government-run media pilloried the ma'di as "monkey faced" descendants of "African slaves" in the midst of a broader and more blatant attack on Iraqi Shi'is (Makiya 1993, pp. 102–3). Internal discussions among party and military functionaries were equally damning, if less crude. A 1992 document discussing the draining of the Hawizeh marsh deemed the marsh dwellers as not just uncivilized, but incapable of civilized behavior. The marshes "represent a national treasure," the report concluded, but "the favorable way to eliminate the sabotage is draining the area of the marshes and transforming it into arable lands." This was the only way to introduce residents to "new pattern of life based on production and direct trade with the government" (cited in Ahram 2015). Saddam's cousin, Ali Hassan al-Majid, was dispatched to oversee military operations in the marshes. Nicknamed 'Chemical Ali' for his use of chemical weapons against Kurdish villages during the genocidal Anfal campaigns of the 1980s, he deployed similar techniques for all-out war in the marshes. Water flows to individual villages were diverted and reed huts burned and then shelled. There were repeated reports of chemical weapon attacks on civilians and of poisoning food supply. Some 2500 residents of Chibaysh were deported and eventually executed. Through 1994, four divisions conducted search and destroy operations in the triangle between Nasiriya, Qurna, and Basra (C. Mitchell 2001, pp. 79–86).

At the same time, though, the Iraqi state continued to frame this destruction as part of a larger development plan. In December 1992, Iraqi television devoted an entire day to covering the ceremonies opening the so-called "Third" or "Saddam" River. Saddam personally inaugurated the opening events, which featured a parade of engineers and military officers. This 550-kilometer-long networks of dikes and sluices diverted much of the water from the marshes toward modern agricultural settlements. The Iraqi National Assembly announced plans to relocate the marsh residents to government-built housing along the Basra-Qurna highway. The ma'di would not have a choice to move or stay. Speaker of Parliament Sadi Mahdi Salih stated that the objective was to take the ma'di "from deep inside [the marshes]... to the fringes, [where] we can reach them and provide amenities" like electricity, schools, and running water. Today, he said, the ma'di were afflicted by bilharzia and cohabitated with animals. Salih ominously likened resettlement to the bloody forced relocation of Kurds in the north (Anderson 1992). The government offered tribal leaders special status and favors, including fishing licenses, to induce them to comply with the plans and force their followers along. Relocated residents were offered a small stipend. Amnesties were also announced for deserters who returned to the national fold (Khoury 2013, pp. 142–3; Cockburn 1992).

Iraq blamed its upstream neighbors for precipitating ecological disaster in the marshes. Turkey had impounded a considerable amount of the Euphrates river flow as part of its own hydrological reengineering strategy in southern Anatolia (Hommes et al. 2016). Nevertheless, Turkey's diversion had only a secondary impact compared to Iraq's own Third River. Nearly the entirety of the Amarah marsh and 80% of the Hammar marsh were desiccated by the mid-1990s. The Hawizeh marsh, fed partially from waters on the Iranian side of the borders, was the only component of al-Batiha that remained even close to its pre-1980 stature (Munro and Touron 1997). Yet the radical approach to land reclamation project failed to yield the kinds of agricultural gains Iraqi technicians had expected. The reclaimed land had residual salinity and other chemical defects. The flora and fauna of the marshes declined dramatically (Al-Yamani et al. 2007).

The denuded marsh landscape left the insurgents fully exposed to state repression. Elites Republican Guard units were dispatched to patrol the area, replacing regular army troops that tended to cavort with civilians. Like hundreds of thousands of other Iraqis, marsh dwellers faced a choice to join the urban underclasses or seek safety abroad. Iran resettled an estimated 40,000 former Iraqi marsh dwellers in refugee camps near Dezful and elsewhere, but many simply squatted on the Iranian side of the marshes (Le Roy 2001). By the middle of the 1990s, the once restive marshes were entirely neutralized as a locus of resistance.

4.4 Conclusion

The restoration of the marshes was supposed to provide validation to the US invasion of Iraqi and the removal of Saddam Hussein in 2003. American and Iraqi leaders touted their plans to reverse Saddam's hydraulic engineering and return the marshes to their pristine and Edenic condition. International nongovernmental organizations made grand designs for turning the Iraq marshlands into a global ecotourism destination. In 2013, the Iraqi parliament designated the marshes as a national park (Guarasci 2015).

But these steps taken to reverse the desiccation, just like the original initiative, fell far short of expected goals and yielded unintended side effects. Iraq's upstream neighbors refused to release impounded water, leaving the marshes ecologically precarious. Region-wide drought late in the 2000s exacerbated the problem. Even when water was abundant, the marshes had fundamentally changed, becoming in effect man-made ponds, not unregulated natural space. Reviving ma'di society proved equally problematic. In 2003, some marsh Arab tribes rose up in advance of the American invasion force, claiming to liberate Amarah from Saddam's control singlehandedly. Abu Hatem (Abd al-Karim al-Muhamedawi), scion of the powerful Albu Muhammed, who had fought the regime since the 1980s, joined the US-backed interim governing council. Refugees from Iran and elsewhere came back to southern Iraq to reclaim their homes and territories. But the era of the marshes seemed definitively over. Few people chose to reside in the marshes full time and resume fishing and foraging. Continuing the centuries-long pattern of rural-urban migration in Iraq, former marsh dweller preferred demi-urban slums of Baghdad, Basra, and Amarah. Many of the followers of Muqtada Sadr's Mahdi Army, centered in the slums of eastern Baghdad, bore surnames indicative of ma'di origins. Urban-based political movements quickly eclipsed the influence of marsh tribal leaders in the south (J. Cole 2003, 2008; Rasan 2005; Le Roy 2001).

Rulers of Iraqi ancient and modern shared a common the goal of mastering the marshes. Marshes stood out as an indomitable landscape, a human ecology resistant to legibility and centralization. Just like the perennial ebb and flow of the marshes themselves, the ma'di society defied the state's vision and its grasp. When states were weak, the marshes expanded and so too did the political autonomy of those who inhabited them. When states were strong, though, the marshes shrunk back, and the marsh dwellers scattered deeper into the impenetrable space of the wasteland.

It was only in the twentieth century, however, that states obtained the technological means to enact the enclosure and taming of the marshes. In this respect, the destruction of the marshes was incremental but also inexorable. It involved combined efforts of multiethnic empires powers and nation-states, indigenous governments, foreign powers, and global capitalists, each working for their agenda but toward a common goal. Warnings about the political, economic, and social impact of this destruction regularly went unheeded. The cost of this develop-induced displacement seemed not only justified, but often ignored.

As Scott argues, this calamitous logics took especially violent form when rulers combined fear of overthrow and hubristic confidence in their own capacities. The experience of war from the 1980s and early 1990s set the stage for Saddam and the Ba'th to undertake ever more radical steps to address what had seemed as the marshes' indomitable human ecology. These final blows at once continued and escalated the destruction already underway.

References

Adams RM (1965) Land behind Baghdad: a history of settlement on the Diyala Plain. University of Chicago, Chicago

Adriensen HK (2004) What happened to the Iraqi Marsh Arabs and their land. DIIS Working Papers, Copenhagen. Retrieved from https://www.files.ethz.ch/isn/18548/Iraqi_Marsh_Arabs.pdf

Ahram AI (2015) Development, counterinsurgency, and the destruction of the Iraqi marshes. Int J Middle East Stud 47(3):447–466

Al-Agidi SK, Roberts DF (1984) A note on the genetics of the Marsh Arabs of Iraq. *Garcia de Orta*. Serie Antropobiologia 3(1–2):63–65

Al-Dewachi O (2017) Ungovernable life: mandatory medicine and statecraft in Iraq. Stanford University Press, Stanford

Al-Layla MA (1978) Effect of salinity on agriculture in Iraq. J Irrig Drain Div 104(2):195–207

Al-Musawi M (2006) Reading Iraq: culture and power in conflict. IB Tauris, New York

ALsaedy JK (2010) Iraqi Buffalo now. Ital J Anim Sci 6:1234–1236

Al-Yamani FY, Bishop JM, Al-Rifaie K, Ismail W (2007) The effects of the river diversion, Mesopotamian Marsh drainage and restoration, and river damming on the marine environment of the northwestern Arabian Gulf. Aquat Ecosyst Health Manage 10(3):277–289

Al-Zahery N, Pala M, Battaglia V, Grugni V, Hamod MA, Kashani BH et al (2011) In search of the genetic footprints of Sumerians: a survey of Y-chromosome and mtDNA variation in the Marsh Arabs of Iraq. BMC Evol Biol 11(1):288

Anderson J (1992) Iraq to uproot isolated Marsh Arabs. Reuters

Bahoora H (2013) Baudelaire in Baghdad: modernism, the body, and Husayn Mardan's poetics of the self. Int J Middle East Stud 45(2):313–329

Batatu H (1982) Iraq's underground Shi'i movements. MERIP Rep 102:3. https://doi.org/10.2307/3010792

Batatu H (2012) The old social classes and the revolutionary movements of Iraq: a study of Iraq's old landed and commercial classes and of its communists, Bathists and free officers. Saqi, London

Battuta I, Gibb HAR, Beckingham CF, Hakluyt S (2010) The travels of Ibn Battuta, AD 1325–1354. Volume II (1 online resource (xvii, 269 pages) Illustrations, maps). Retrieved from https://search.ebscohost.com/login.aspx?direct=true&scope=site&db=nlebk&db=nlabk&AN=508167

Blaydes L (2018) State of repression: Iraq under Saddam Hussein. Princeton University Press, Princeton

Burke E III (2009) The transformation of the Middle Eastern environment, 3000 BCE–2000 CE. In: Burke E III, Pomeranz K (eds) The environment and world history, 1500–2000. University of California Press, Berkeley
Buxton PA (1922) The Marsh Arabs of lower Mesopotamia. British India Press, Bombay
Cernea MM (2000) Risks, safeguards and reconstruction: a model for population displacement and resettlement. Econ Polit Wkly:3659–3678
Çetinsaya GK (2006) Ottoman administration of Iraq, 1890–1908. Routledge, London
Ceylan E (2011) The Ottoman origins of modern Iraq : political reform, modernization and development in the nineteenth-century Middle East. I.B. Tauris, London
Christensen P (1993) The decline of Iranshahr: irrigation and environments in the history of the Middle East, 500 BC to AD 1500. Museum Tusculanum Press, Copenhagen
Clarence-Smith WG (2006) Islam and the abolition of slavery. Oxford University Press, New York
Cockburn P (1992) Tide turns against the Marsh Arabs. The Independent, May 7
Cole SW (1994) Marsh formation in the Borsippa region and the course of the lower Euphrates. J Near East Stud 53(2):81–109
Cole J (2003) The United States and Shi'ite religious factions in post-Ba'thist Iraq. Middle East J 57:543–566
Cole J (2008) Marsh Arab rebellion: grievance, mafias and militias in Iraq. Fourth Wadie Jwaideh Memorial Lecture. Department of Near Eastern Languages and Cultures, Indiana University, Bloomington. https://deepblue.lib.umich.edu/bitstream/handle/2027.42/150664/iraqtribes4.pdf
Cole CL (2016) Precarious empires: a social and environmental history of steam navigation on the Tigris. J Soc Hist 50(1):74–101
Cox JL (1985) A splendid training group. J Imp Commonw Hist 13(2):157–184
Dauphin J (1960) Les Ma'dan de Basse-Mésopotamie. Annales de géographie 69:34–49
Drower L (1937) Mandaeans of Iraq and Iran. Brill, Leiden
Drower L (1947) Marsh people of South Iraq. J R Cent Asian Soc 34(1):83–90
Dwivedi R (2002) Models and methods in development–induced displacement. Dev Chang 33 (4):709–732
Eger AA (2011) The swamps of home: marsh formation and settlement in the early medieval Near East. J Near East Stud 70(1):55–79
Farouk-Sluglett M, Sluglett P (1983) The transformation of land tenure and rural social structure in central and Southern Iraq, c. 1870–1958. Int J Middle East Stud 15(4):491–505
Fernea RA (1969) Land reform and ecology in postrevolutionary Iraq. Econ Dev Cult Chang 17 (3):356–381
Fisk B (1952) Dujaila: Iraq's pilot project for land settlement. Econ Geogr 28(4):343–354
Floor W (2006) The rise and fall of the Banū Kaʿb. A borderer state in Southern Khuzestan. Iran 44 (1):277–315
Furr KA (2010) Environmental degradation in the Mesopotamian marshlands: a case study in legal deficiencies. Seattle Environ Law J 19:271
Goldstein E (1992) Endless torment: the 1991 uprising in Iraq and its aftermath. Human Rights Watch, New York
Guarasci BL (2015) The National Park: reviving Eden in Iraq's marshes. Arab Stud J 23 (1):128–153
Haj S (1997) The making of Iraq, 1900–1963: capital, power, and ideology. SUNY Press, Albandy
Harrington, L. (2014), Architecture and nation-building in mid-20th century urban
Heidel A (1963) The Babylonian genesis, 2nd edn. University of Chicago Press, Chicago
Helfont S (2018) Compulsion in religion: Saddam Hussein, Islam, and the roots of insurgencies in Iraq. Oxford University Press, New York
Hillel D (1994) Rivers of Eden : the struggle for water and the quest for peace in the Middle East. Oxford University Press, New York
Hitti PK, Murgotten FC (1916) The origins of the Islamic State, being a translation from the Arabic, accompanied with annotations, geographic and historic notes of the Kitâb Fitûh Al-buldân of Al-Imâm Abu-l Abbâs Ahmad Ibn-Jâbir Al-Balâdhuri. Columbia University

Hommes L, Boelens R, Maat H (2016) Contested hydrosocial territories and disputed water governance: struggles and competing claims over the Ilisu Dam development in southeastern Turkey. Geoforum 71:9–20

Howell EBS (1922) River control in Mesopotamia. Q Rev

Ingham B (1976) Regional and social factors in the dialect geography of Southern Iraq and khūzistán. Bull Sch Orient Afr Stud 39(1):62–82

Ionides MG (1937) The régime of the rivers. E.F. Spon, Ltd., London

Ireland PW (1970) Iraq; a study in political development. Russell & Russell, New York

Ishow H (1983) The rural exodus in Iraq and its economic and social consequences. Afrique et l'Asie Modernes 136:27–44

Ismael TY (2008) The rise and fall of the Communist Party of Iraq. Cambridge University Press, New York

Joffe AH (2000) The environmental legacy of Saddam Husayn: the archaeology of totalitarianism in modern Iraq. Crime Law Soc Chang 33(4):313–328

Jones C, Sultan M, Yan E, Milewski A, Hussein M, Al-Dousari A et al (2008) Hydrologic impacts of engineering projects on the Tigris–Euphrates system and its marshlands. J Hydrol 353 (1–2):59–75

Juma KH (1997) Present status in buffalo production in Iraq. Buffalo J 13(2):103–114

Jwaideh A (1984) Aspects of land tenure and social change in lower Iraq during late Ottoman times. In: Khalidi T (ed) Land tenure and social transformation in the Middle East. American University of Beirut, Beirut, pp 333–356

Jwaideh A (2007) The marsh dwellers of Southern Iraq: the habit, origins, society, and economy. Retrieved from Bloomington, Ind. http://www.indiana.edu/~nelc/lectures/jwaideh%20documents/Jwaideh06_AlbertineJwaideh_2007.pdf

Jwaideh A, Cox JW (1988) The black slaves of Turkish Arabia during the 19th century. Slav Abolit 9(3):45–59

Kasaba R (2009) A moveable empire : ottoman nomads, migrants, and refugees. University of Washington Press, Seattle

Khoury DR (2013) Iraq in wartime: soldiering, martyrdom, and remembrance, New York

Lanoix JN (1958) Relation between irrigation engineering and bilharziasis. Bull World Health Organ 18(5–6):1011

Le Roy J (2001) Statistical outline of the situation of the South Iraqi refugees in Iran, including the Marsh Dwellers. In: Magee S, Clark P (eds) The Iraqi marshlands: a human and environmental study. AMAR International Charitable Foundation, London

Lebon JHG (1955) The new irrigation era in Iraq. Econ Geogr 31(1):47–59

Lorimer JG (1970) Gazetteer of the Persian Gulf, 'Omān, and Central Arabia. Superintendent of Government Printing, Calcutta

Luckenbill DD (2005) The annals of Sennacherib. Wipf and Stock Publishers, Eugene

Main E (1935) Iraq from mandate to independence, vol 86. Routledge, New York, p 543

Makiya K (1993) Cruelty and silence: war, tyranny, uprising and the Arab world. W.W. Norton, New York

Matthee R (2006) Between Arabs, Turks and Iranians: the town of Basra, 1600–1700. Bull Sch Orient Afr Stud 69(1):53–78

Mehdi S (2008) Modernism in Baghdad. DC PAPERS, revista de crítica y teoría de la arquitectura 1:81–89. Middle East J 57(4):543–566

Mikhail A (2013) Nature and empire in Ottoman Egypt: an environmental history. Cambridge University Press, New York

Mitchell C (2001) The assault on the marshes. In: Magee S, Clark P (eds) The Iraqi marshlands: a human and environmental study. AMAR International Charitable Foundation, London

Mitchell T (2002) Rule of experts: Egypt, techno-politics, modernity. University of California Press, Berkeley

Moojan M (1985) An introduction to Shi'i Islam. Yale University Press, New Haven

Morley D (1993) Iraq: ecological and human disaster. Lancet 341(8859):1528

Morony MG (1984) Landholding and social change: lower al-'Iraq in the early Islamic period. Land Tenure Soc Transforma Middle East:209–222
Moumin MA (2007) Mesopotamian marshlands: an ecocide case. Geo Intl Envtl L Rev 20:499
Munro D, Touron H (1997) The estimation of marshland degradation in southern Iraq using multitemporal Landsat TM images. Int J Remote Sens 18(7):1597–1606
Murphey R (1987) The Ottoman centuries in Iraq: legacy or aftermath? A survey study of Mesopotamian hydrology and ottoman irrigation projects. J Turk Stud 11:17–29
Nāhī S (1955) Muqaddimah fī al-iqṭa' wa-niẓām al-arāḍī fī al-'Irāq. Dār al-Ma'ārif, Baghdād
Naji AJ, Ali YN (1981) The Suqs of Basrah: commercial organization and activity in a medieval Islamic City. J Econ Soc Hist Orient 24(3):298
Nakash Y (1994) The Shiis of Iraq. Princeton University Press, Princeton
Nieuwenhuis T (1982) Politics and society in early modern Iraq : Mamluk Pashas tribal Shayks and local rule between 1802 and 1831. M. Nijhoff Publishers, The Hague
Oates J (1960) Ur and Eridu, the prehistory. Iraq 22:32. https://doi.org/10.2307/4199667
Ochsenschlager EL (2014) Iraq's Marsh Arabs in the garden of Eden. University of Pennsylvania Press, Philadelphia
Omissi DE (1990) Air power and colonial control. Manchester University Press, Manchester
Popović A (1999) Revolt of African slaves in Iraq in the Third/ninth century. Markus Wiener Pub
Pournelle J, Algaze G (2014) Travels in Edin: deltaic resilience and early urbanism in greater Mesopotamia. In: McMahon A, Crawford HEW, Oates J (eds) Preludes to urbanism: studies in the Late Chalcolithic of Mesopotamia in Honor of Joan Oates. McDonald Institute for Archaeological Research, Cambridge
Pursley S (2019) Familiar futures: Time, selfhood, and sovereignty in Iraq. Stanford University Press, Stanford
Qubain FI (1958) The reconstruction of Iraq, 1950–1957. Stevens, London
Rasan D (2005) 'Prince' loses favor in Amara. Institute for War and Peace Reporting, 21 February 2005. http://iwpr.ne/eport-new/rince%E2%80%9D-losesfavour-amar
Razoux P (2015) The Iran-Iraq war. Harvard University Press, Cambridge, MA
Richards JF (2003) The unending frontier: an environmental history of the early modern world. University of California Press, Berkeley
Sahrhage D, Lundbeck J (2012) A history of fishing. Springer
Salim SM (1970) Ech-Chibayish: an anthropological study of a marsh village in Iraq, 2nd edn, Baghdad
Salter JAS (1955) The development of Iraq: a plan of action. (Place of publication not identified)
Sassoon J (1987) Economic policy in Iraq, 1932–1950. Frank Cass, London
Sassoon J (2012) Saddam Hussein's Ba'th Party: inside an authoritarian regime. Cambridge University Press, New York
Satia P (2007) Developing Iraq: Britain, India and the redemption of empire and technology in the First World War. Past Present 197:211–255
Satia P (2008) Spies in Arabia: the great war and the cultural foundations of Britain's covert empire in the Middle East. Oxford University Press, New York
Scott JC (1998) Seeing like a state: how certain schemes to improve the human condition have failed. Yale University Press, New Haven
Scott JC (2009) The art of not being governed: an anarchist history of upland Southeast Asia. Yale University Press, New Haven
Scott JC (2017) Against the grain: a deep history of the earliest states. Yale University Press, New Haven
Soucek S (1984) Arabistan or Khuzistan. Iran Stud 17(2–3):195–213
Sousa A (1944) Iraq irrigation handbook. Pt. 1, The Euphrates. Iraqi Ministry of Agriculture, Baghdad
Springborg R (1981) Baathism in practice: agriculture, politics, and political culture in Syria and Iraq. Middle East Stud 17:191–209

Springborg R (1986) Infitah, agrarian transformation, and elite consolidation in contemporary Iraq. Middle East J 40(1):33
Strakes J (2011) Between two masters: Khuzestan, southern Iraq, and dualities of state making in the Arab/Persian Gulf. Arab World Geogr 14(4):336–361
Streck M, Saleh A (2012) al-Batiha. In: Encyclopaedia of Islam, 2nd edn. Brill, Leiden, p I:1093b
Tapper R (1997) Frontier nomads of Iran: a political and social history of the Shahsevan. Cambridge University Press, New York
Tapper R (2009) Tribe and state in Iran and Afghanistan: an update. Études rurales 184(2):33–46
Thesiger W (2007) The marsh arabs. Penguin, New York. Turkey and Iraq, MA thesis, University of Washington
Visser R (2005) Basra, the failed gulf state: separatism and nationalism in southern Iraq. Transaction Press, New Brunswick, NJ
Waines D (1977) The third century internal crisis of the Abbasids. J Econ Soc Hist Orient 20 (3):282. https://doi.org/10.2307/3631960
Warriner D (1962) Land reform and development in the Middle East; a study of Egypt, Syria, and Iraq, 2nd edn. Oxford University Press, London
Wilensky-Lanford B (2011) Paradise lust: searching for the Garden of Eden. Grove/Atlantic, New York
Willcocks WS (1917) Irrigation of Mesopotamia, 2nd edn. E. & F.N. Spon, New York
Wittfogel KA (1957) Oriental despotism: a comparative study of total power. Yale University Press, New Haven
Woods KM, Palkki DD, Stout M (2011) The Saddam tapes: the inner workings of a tyrant's regime, 1978–2001. Cambridge University Press, New York
Worster D (1992) Rivers of empire: water, aridity, and the growth of the American West. Oxford University Press, New York
Yapp ME (1993) The Euphrates expedition. Middle East Stud 3:585
Yarshater E (2007) History of al-Tabari, SUNY Press, vol XXVIII. Cambridge University Press, York

Chapter 5
Role of Women in Ancient Mesopotamia and the Southern Marshes of Iraq: A Comparative Account

Laith A. Jawad

Abstract Women in the ancient Mesopotamia have performed several kinds of professions that cover a wide range of skills. Such professions were portrayed in tablets and recovered through the archaeological investigations. It was interesting to learn whether there are any similarities and differences in the jobs the women performing at the present time in the Marsh Arabs society and those accomplished by women in the ancient Mesopotamia. The comparison revealed that there are some jobs that women did in ancient Mesopotamia that no longer exist in the society of the Marsh Arabs and vice versa. Further excavations in the lower Mesopotamia might reveal evidences showing more professions where women have achieved.

Keywords Lifestyle · Housewives · Modern Iraq · Jobs

5.1 Introduction

Women continued to take a leading and a significant role in the life of human societies since the ancient time. Such activity is not different in Iraq whether in the ancient or present time.

Otto (2016) in her study on the role of women in ancient Mesopotamia has shown that women participated in several aspects of the daily life of ancient Mesopotamia since the third millennium, when the image of women was portrayed as statues or on monuments and seals. In the Old Babylonian era, the situation was different, and depiction of women was more common and appeared in on the terracotta tablets.

For the representation of women on monuments not later than the third millennium, i.e., in the second millennium, Otto (2016) suggests that such depiction may already exist, but the fragile materials that made on to were the reason for not enduring the long years of history. On the other hand, Otto (2016) claims that

L. A. Jawad (✉)
School of Environmental and Animal Sciences, Unitec Institute of Technology, Mt Albert, Auckland, New Zealand

L. A. Jawad (ed.), *Southern Iraq's Marshes*, Coastal Research Library 36,
https://doi.org/10.1007/978-3-030-66238-7_5

reasons that made the evidence to last as one piece and not fragmented and reaching archaeologists at present time are those artifacts that were made of stone, a material that is unavailable in the lower reaches of Mesopotamia but is available in the northern Mesopotamian locations.

Otto (2016) also suggests another factor that may affect the decrease in the number of appearance of women images on artifacts. This factor is the discrimination between different types of women according to their social rank. The depiction was made according to the type of event, and accordingly, the daily life activities were not depicted on artifacts, while the ceremonies and religious rituals were the events that the Mesopotamian artists were looking for. If women were included in such events, then their images were portrayed.

On the other hand, the depiction on Terracotta was aimed to illustrate daily life, and hence few of women images appeared on these artifacts, but not in large number.

In ancient Mesopotamia, women have taken several roles, ranging from the high ranking such as the queen down to the simple workers. In this aspect, Otto (2016) classifies the professions that women took into nine main groups that encompass different jobs.

Leaving the women of the ancient Mesopotamia and turning to that of the Marsh Arabs in southern Iraq, here, we find women, in addition to their feminine responsibilities, who are accountable for several professions that some of them belong to men. Therefore, the idea of this chapter was born, and it was to compare the profession of women in the ancient Mesopotamian time and the present southern marsh areas in Iraq. Women of the marshes were selected for this comparison in this chapter as the Marsh Arabs may represent the descendants of Sumerians and also due to their significant role in the Marsh Arab's society.

5.2 How the Comparison of the Professions of Women in Ancient Mesopotamia and Present Marsh Arab Society Is Discussed in the Present Chapter?

In this section, the professions that Otto (2016) has listed will be taken as a base for the comparison with those professions taken by women in the present time Marsh Arabs in the south of Iraq. A short information about each profession in the ancient Mesopotamia will be given first, and then description of a similar profession in the present Marsh Arabs societies will be presented for comparison.

5.2.1 Queens and High-Ranking Women

5.2.1.1 Royal or High-Ranking Women Engaged in Cultic Activities

In Mesopotamia and in the third millennium, women from the upper class of the society have their images depicted (Otto 2016). The role of women as a queen has

Fig. 5.1 Mother and daughter shown on Stela of Ur-Nanše. (Image courtesy of Otto 2016, Fig. 1)

been described by Suter (2008) as a representation in the form of figurines fixed in temples and portrayed on the public icons such as seals. A perfect example on such representation is that performed on the stela to commemorate the opening of the Ibgal Temple (Fig. 1 = Fig. 5.1 in Otto) (Borker-Klahn 1982). In this depiction, the image of the goddess is portrayed on the side, while on the other, the images of the royal family of Ur-Nanše, $ensi_2$ of the first dynasty of Lagaš, are shown including his wife and daughter. There is some discrimination in the type of cloths women were wearing, e.g., the headband and hairstyle, in the depiction, suggesting differences in social rank (Boese 1971; Otto 2016). Women who appeared in the ritual tablet (Fig. 5.2) could be seen wearing a smooth cloak, which could be made of a fine material. Such habit of head cover might be related to the event, which could be either an official feast or a funerary event (Dolce 2008).

Similarities and Differences with the Societies of Marsh Arabs

Rank of queen does not exist in the Marsh Arabs society, but high ranks are present and many women playing this role in the present southern marshes of Iraq. High-ranked women in the marsh areas can be either the wife/wives of the sheik (the head of the tribe) and his daughters or the wife/wives of the religious man (Saeed). These two ranks are present in every human inhabitant in the marsh areas as both the sheik and the religious man are two social icons that have to be present in any society. Women of these important characters are usually dress in good quality cloths and wear a noteworthy amount of gold, and they do not perform the daily works that other women do. Here, we find similarity with the high-ranked women of

Fig. 5.2 Woman wearing a cloak from Mari, Ninni-zaza temple. (Image courtesy of Otto 2016, Fig. 5)

ancient Mesopotamia. The other similarity is the head cover, which is always worn by all women in the marsh society, but those covers used by high-ranked women are different because they are soft and made of good materials. In the marsh areas, head covers are usually worn on everyday events, but in the happy and sad events, something special is usually worn.

5.2.1.2 The Priestess Women

The role of high-ranked women or priestess was present in ancient Mesopotamia according to the archaeological evidences (Parrot 1956; Evans 2003; Otto 2016), but this role disappeared with the end of the third millennium (Otto 2016). Women taking such role usually dress differently from the other women of the society as it has been described by Otto (2016) that the high-ranked women put a cloak on their head, which looks made of expensive materials.

Similarities and Differences with the Societies of Marsh Arabs
In the Marsh Arabs society, such a woman can be the wife/wives of the religious man (Saeed) as described above. But in some societies, there is a profession that women take, which is related to telling the future by reading the shells. The woman performing this profession is known as "Fatahat Al-Fal or Arafa" Ordinary women usually visit her and ask her to read their fortune, and in many cases, she gives them amulets or charms in the form of a written paper enclosed in a piece of cloth that the women should attached to their cloths or used in the form of a drink. Most of the problems that women came to visit Fatahat Al-Fal about are social matters such as wife-husband relationship or asking to give birth to a boy.

5.2.2 *Court Ladies and Female Attendants*

From the archaeological evidences from the third millennium in Mesopotamia, a women profession was depicted on seals. This profession is related to assist high-ranking characters of the society such as queens and kings. Although these women hold such low-ranking jobs, they are considered high ranking to the common individuals of the society (Otto 2016) (Fig. 5.2). The depiction on tablets and seals also showed women performing other jobs such as musicians and dancers (Figs. 5.3 and 5.4) (Barrelet 1968). The scene depicted in Fig. 5.5 showed women moving the lower parts of their bodies (as portrayed by a wavy long skirt), clapping and raising one foot off the ground.

Similarities and Differences with the Societies of Marsh Arabs
Comparable to the royal court found in ancient Mesopotamia, women as servants can be seen as workers in the women section of the living area of the head of tribe "Sheikh" in the Marsh Arab society. These women can prepare food and do other household jobs. The gypsies or what they are known for the Marsh Arabs as "Kawlia" are comparable to the musicians and dancers of the courts in ancient Mesopotamia. The Kawlia are usually hired to perform dances and songs at happy occasions such as marriages. Women of Kawlia performing a dance are usually shaking the lower parts of their bodies and clapping, with untied long hair.

5.2.3 *The Profession of Mother and Wet Nurse*

Motherhood is a priceless profession that any woman could have and is by nature not learned. Scenes of motherhood such as women sitting with their children on their laps are frequently obtained from archaeological sites in Mesopotamia (Otto 2016). Among these are those recovered from Tell Mozan for the Akkadian period (Buccellati and Kelly-Buccellati 1995–1996; Kelly-Buccellati 2015) (Fig. 5.6) and

Fig. 5.3 Queen of Mari. (Image courtesy of Otto 2016, Fig. 6)

on old terracotta tablets showing women breastfeeding a child (Fig. 5.7) (Genouillac 1936; Otto 2016).

The profession of wet nurse was also performed by women in the ancient Mesopotamia. This profession is mainly about breastfeeding the child in case the mother died or got sick and cannot feed her child. This profession was clearly depicted on several seals from the Akkadian period (Suter 2008; Otto 2016) (Fig. 5.8) (Genouillac 1936).

Similarities and Differences with the Societies of Marsh Arabs

As with any other human society in the world, motherhood profession or instinct also exists among women of the Marsh Arabs. In Islamic belief, motherhood has been mentioned in Quran as a mother breastfeeding her child for the first 2 years of her/his life. Therefore, a strong similarity is present here for this profession.

The other similarity is related to the profession of wet nurse. This is an ancient profession and could well be taken after the ancient Mesopotamia. The woman that performs this profession is known as "Murtheaa" and as in the ancient Mesopotamian societies, the service of this woman is needed when the mother of a child got sick or died. The prophet Mohammed himself was given to a wet nurse to feed him while he is a newborn because his mother died after giving birth to him. In addition, Quran set some rules about marriages of boys and girls who are fed from

Fig. 5.4 A dressed woman portrayed on an old Babylonian terracotta tablet. (Image courtesy of Otto 2016, Fig. 16)

the same woman. In Islamic society, they are known as breastfeeding brothers and they are not allowed to get married. The responsibility of the wet nurse is to tell the families of both boys and girls about this incidence so they keep it their mind and tell the kids when they grow.

5.2.4 *Female Kitchen Personnel, Female Brewers, and Women in a Garden*

Cooks and chief cook were employed in ancient Mesopotamia in their palaces to prepare meals and banquets for a large number of guests and for the members of the palace in the ordinary days (Otto 2016). Such events are depicted in numerous seals recovered from Tell Mozan, and among them is a seal showing a chief cook and his assistant working in a kitchen preparing food (Fig. 5.9) (Kelly-Buccellati 2015; Otto 2016).

Events of women who were active outdoor, e.g., gardening in the field, were portrayed in the Akkadian seals (Fig. 5.10) (Boehmer 1965). In these seals, women were shown taking dates from a short date palm. Otto (2016) is wondering whether

Fig. 5.5 Women and musicians depicted on a seal from the Great Death Pit at Ur. (Image courtesy of Otto 2016, Fig. 15)

Fig. 5.6 Queen Uqnitum of Urkeš with her family depicted on a seal. (Image courtesy of Otto 2016, Fig. 20)

this is a real agricultural practice, where women can easily reach the cluster of dates or a symbolic act to show how to pick the dates. In southern Iraq, where there are thousands of date palms of over 250 varieties, a date palm tree can have fruits early in its life and is short enough to be able to be reached by humans. Therefore, what

Fig. 5.7 Image of a mother or wet nurse with a child portrayed on Terracotta tablet from Isin. (Image courtesy of Otto 2016, Fig. 23)

has been depicted is a reality of what is happening in some cases of collecting dates from the date palm tree.

Similarities and Differences with the Societies of Marsh Arabs

In regard to the cook profession, there are differences between what was found in ancient Mesopotamia and the present society of the Marsh Arabs. Usually, the families of the Marsh Arabs are small comprising not more than 15 individuals. Thus, they do not need a private cook to prepare food for the family, but this profession becomes a reality in happy and sad occasions that happened to any family in the Marsh Arabs society. In such instances, women from all neighborhood will attend the occasion to help in cooking and preparing food for the guests.

As to the profession of farming, women of the Marsh Arabs have pronounced and leading role in collecting reeds and other plants in addition to plantation.

5.2.5 *Female Textile Workers*

The profession of weaving textile and rugs was not very well portrayed in the ancient Mesopotamia (Otto 2016). The best of those depictions are the mosaic panel from

Fig. 5.8 Image of a mother or wet nurse breastfeeding a child depicted on Terracotta tablet from Girsu. (Image courtesy of Otto 2016, Fig. 24)

Fig. 5.9 Image of male and female cooks portrayed on a seal of Tuli. (Image courtesy of Otto 2016, Fig. 25)

Fig. 5.10 Image of women working in a garden depicted on Akkadian seal. (Image courtesy of Otto 2016, Fig. 27)

Fig. 5.11 Reassembled mosaic board from Mari displaying female and male individuals holding goods for a woman newly giving birth. (Image courtesy of Otto 2016, Fig. 33)

Mari (Fig. 5.11) (Kelly-Buccellati 2015) and the seal of ED III from the Ištar temple at Mari (Fig. 5.12) (Parrot 1956).

Similarities and Differences with the Societies of Marsh Arabs

The similarity in regard to this profession is very high with the women of the Marsh Arabs. The latter are keen in weaving to make rugs and cushions made of reed or wool. They used horizontal and vertical looms for this purpose.

Fig. 5.12 Image of a woman weaving portrayed on a cylinder seal from the Ištar temple at Mari. (Image courtesy of Otto 2016, Fig. 28)

Fig. 5.13 Image of a woman giving birth illustrated on a cylinder seal. (Image courtesy of Otto 2016, Fig. 30)

5.2.6 *Midwives*

The profession of midwives was portrayed on cylinder seals since the Sumerian time (Asher-Greve and Goodnick Westenholz 2013; Otto 2016) (Fig. 5.13). This seal shows a woman giving birth with the assistance of a midwife. It also portrayed some animals including scorpion, which is considered a symbol of fertility (Otto 2016). The other seal from the Akkadian period (Fig. 5.14) (Buchanan 1981) shows a woman lying on a bed, with bull's legs and a scorpion below the bed. The seal is decorated with many plants and trees that show seven lancet-shaped, which are considered as feminine touch (Selz 1983). Also there are frequently associated depictions of plants and vessels because they were thought to promote pregnancy and birth (Stol 2000).

Fig. 5.14 Image of a woman resting showed on a cylinder seal. Also showing oil vessels and the birth plant. (Image courtesy of Otto 2016, Fig. 31)

Similarities and Differences with the Societies of Marsh Arabs

Similarities here are high in all aspects of the process of giving birth with the help of midwives. The profession of midwives is still being performed by some women that have no medical education, but they attain the skill through practice over the years. With the location of the marshes being far away from any village or city that has a medical center, Marsh Arabs depend on local midwives in helping women in delivering babies.

The rituals that are followed on the occasion of giving birth have a great similarity to what are happening in the Middle East and in the marsh areas. Among these rituals as has been described above in the ancient Mesopotamia is bringing plants and parts of trees in the room where the woman is giving birth. In most countries in the Middle East, one of the plants brought when a woman is giving birth is known in Arabic as "Palm of Mary" and its scientific name is *Vitex agnus-castus*. This plant is usually brought and made available in the room where a woman is giving birth, and once the woman has given birth, the plant starts to stretch and open its leaves and stems. However, there is no scientific connection between the opening of the leaves of the plant and the process of giving birth by the woman. What happened is when they bring the plant, they put some water on it. The plant has a distinctive character that once it absorbed water, the leaves start to open, but this will take some time and coincide with the completion of the process of giving birth.

5.3 Remarks

Except for the profession of being a queen and being on a certain high-ranking social category, women in the southern marshes of Iraq have perform similar or nearly similar professions that are also accomplished by women in ancient Mesopotamia, and they are listed by Otto (2016).

It is interesting to see how many great similarities are there between the two cultures. Here, the question must be, are the professions of women in the marshes have inherited these professions? After going through all the archaeological evidences given by Otto (2016) on the professions of women, it is believed that people of the marsh areas in Iraq have inherited the performance of these professions through passing them from one generation to another.

There is an important point here that needs to be discussed. Otto (2016) has listed and discussed a range of professions that women in the ancient Mesopotamia have performed. Are these all the professions that women have performed in their daily lives? Or are these the most important professions? Could there be more professions that were not depicted on tablets because these tablets were completely destroyed and not retrieved? It is important to inquire as there are other professions that women in the southern marshes of Iraq usually accomplished but were not included in the list of Otto (2016).

Other professions that women of the Marsh Arabs are performing are:

1. Making tattoo: making tattoo has been fully discussed in another chapter of this book; therefore, readers who are interested to learn more about this subject need to refer to chapter "Daily Life of the Marsh Arabs." Men and women alike are doing this profession. Tattoo also has been mentioned to be used in ancient Mesopotamia.
2. Removing hair from the face and body of women: this profession is common in all human societies, and its presence in the marsh society is not unusual. Usually, there are specialized women who perform this type of job and earned the experience from their mother or any women in the family. It is interesting to see whether there are any archaeological evidences to show this type of job being performed in ancient Mesopotamia.
3. Being a tailor: as the marsh areas are remote and away from any modern living area, women usually made their clothes by themselves. Not all women have the skill to make a dress, but certain women have these skill and they are called as tailors or what they are known in Arabic as "Khaita."

Probably, further excavation in Mesopotamia may reveal evidences for more women professions that have a connection with the present people living in the area.

References

Asher-Greve JM, Westenholz JG (2013) Goddesses in context. On divine powers, roles, relationships and gender in Mesopotamian textual and visual sources, OBO 259. Academic Press/ Vandenhoeck & Ruprecht, Fribourg/Göttingen

Barrelet M-T (1968) Figurines et reliefs en terre cuite de la Mésopotamie antique I, Potiers, termes de métier, procédés de fabrication et production. Geuthner, Paris

Boehmer RM (1965) Die Entwicklung der Glyptik w. hrend der Akkad-Zeit, UAVA 4. de Gruyter, Berlin

Boese J (1971) Altmesopotamische Weihplatten, Eine sumerische Denkmalsgattung des 3. Jahrtausend v. Chr, UAVA 6. de Gruyter, Berlin

Börker-Klähn J (1982) Altvorderasiatische Bildstelen und vergleichbare Felsreliefs, BaF 4. Zabern, Mainz am Rhein

Buccellati G, Kelly-Buccellati M (1995–1996) The royal storehouse of Urkesh. The glyptic evidence from the southwestern wing. Archiv für Orientforschung 42/43: 1–32

Buchanan B (1981) Early near eastern seals in the Yale Babylonian collection. Yale University Press, New Haven/London

de Genouillac H (1936) Fouilles de Telloh II, époque d'Ur IIIe Dynastie et de Larsa. Paul Geuthner, Paris

Dolce R (2008) Ebla before the achievement of palace G culture: an evaluation of the early Syrian Archaic Period. In: Kühne H, Czichon RM, Kreppner FJ (eds) Proceedings of the 4th International congress of the archaeology of the ancient near east. Harrassowitz, Wiesbaden, pp 65–80

Evans JM (2003) Female figures wearing tall headdresses, pp 153–154 in Aruz 2003

Kelly-Buccellati M (2015) Power and identity construction in Ancient Urkesh, pp 111–130 in Ciafardoni and Gianessi 2015

Otto A (2016) Professional women and women at work in Mesopotamia and Syria (3rd and early 2nd millennia BC): the (rare) information from visual images. In: The role of women in work and society in the ancient near east, studies in ancient near eastern records, vol. 13, pp 112–148

Parrot A (1956) Le temple d'Ishtar, Mission Archéologique de Mari II/3. Geuthner, Paris

Selz G (1983) Die Bankettszene, FAOS 11. Franz Steiner, Wiesbaden

Stol M (2000) Birth in Babylonia and the Bible. CM 14. Styx, Groningen

Suter CE (2008) Who are the women in Mesopotamian art from ca. 2334–1763 BCE? KASKAL 5:1–55

Section II
Environmental Factors

Chapter 6
Physical and Chemical Characters of Mesopotamian Marshes: A Short Review

Bahram K. Maulood and Fikrat M. Hassan

Abstract Physical and chemical statuses of Mesopotamian Marshes have been studied in the pre- and post-draining processes. It was found that light penetration was ranged between 25 and 155 cm and is mainly affected by the nature of the marsh bottom and boating activity in addition to sedimentation that lead to increase the turbidity. The pH value in the area is generally ranged between 7.1 and 9.18 during 1980, whereas alkalinity had ranged between 68 and 137 mg $CaCO_3$/l, which was mainly bicarbonate alkalinity. In the marshes, water was very hard and was characterized by slightly saline to oligohaline water during 1970–1980, and it is well aerated with high level of oxygen. The main form of inorganic nitrogen was nitrate, and both nitrite and ammonia were also recorded in the environment of the marsh areas. No confidence can be placed in the accuracy of phosphorus measurement as its value is much less than that of the Shatt al-Arab River, whereas silicate value has been varied slightly between 7.7 and 9.8 mg SiO_3/l. A clear seasonal variation was evident in southern marshes before draining. Water character after reflooding showed slight value of suspended matter, pH value was 6.6–8, and EC was 1120–2410 μs/cm, and TDS was between 467 and 944 mg/l. All these figures have been recorded in Abu-Zirig Marsh; therefore, the hydrological regime had a clear effect on physicochemical properties of water in the marshes. Electrical conductivity ranged from 910 to 4413 μs/cm in Central Marshes after reflooding. A temporal variation of salinity was observed, and it ranged between 0.8 and 2.25‰ in Central Marshes. Temperature was found to be the main reason of variation in the values of pH, dissolved oxygen, organic matter values, and salinity. Plant nutrient, particularly nitrate, was found to be with much more concentration up to 141 μg/l in comparison to the previously recorded values (24 μg/l). Phosphate levels have also been increased, and the reason might be rainfall and fertilizer. Other parameters such

B. K. Maulood (✉)
Present address : Scientific consultant/precedency of Kurdistan regional government, Erbil, Iraq

Department of Biology, College of Science, University of Salahaddin, Erbil, Iraq

F. M. Hassan
Department of Biology, College of Science for Women, University of Baghdad, Baghdad, Iraq

L. A. Jawad (ed.), *Southern Iraq's Marshes*, Coastal Research Library 36,
https://doi.org/10.1007/978-3-030-66238-7_6

as chloride, sulfate, total hardness, calcium, and magnesium have been also measured in some stations within marshes.

Keywords Environment · Seasonal variation · Localities · Iraq · Freshwater

6.1 Introduction

Physicochemical and biological processes are known to be wetland functions that will in one way or another interact with society, ultimately ending up with quite many direct benefits and services to the community or indirectly to the whole society; however such functions reflect as wetland values (Kooser and Lund 1996). The present chapter will be confined to physicochemical processes and factors within Iraqi Marshes for pre- and post-draining processes of the marsh areas, which represent completely two different environmental regions according to their physical and chemical characteristics.

The marshes in the south of Iraq had attracted several studies on both national and international levels (Table 6.1). These studies had focused on different aspects of the areas such as the limnological characters of marshes (Maulood et al. 1979, 1981; Pankow et al. 1979; Al-Saadi and Antoine 1981), seasonal studies of Al-Hammar Marsh (Al-Saboonchi et al. 1982), and general other environmental studies (Rzoska 1980; Al-Lami 1986; Kassim 1986; Maulood et al. 1986).

Pankow et al. (1979) studied the marshes near Qurna (southern Iraq), while Al-Hammar Marsh was dealt with by Al-Saadi and Antoine (1981) and Nurul-Islam (1982). Garma Marshes by (Al-Saboonchi et al. 1982), on the other hand, Rise-field Soil from marshes was dealt with by (Al-Kaisi 1976; Al-Mousawi and Whitton 1983), Tigris Marshes (Hinton and Maulood 1982), different areas of marshes (Maulood et al. 1979, 1981), Qurna Marshes (Al-Zubaidi 1985), Al-Hammar Marsh (Al-Lami 1986; Al-Aarjy 1988; Hassan 1989) (Table 6.2).

From the abovementioned studies, it becomes evident that air and water temperature within marshes are ranged from 15 °C to 40 °C and 12 °C to 32 °C, respectively, as reported by Al-Zubaidi (1985) and 13–43 °C and 12–31 °C as given by Al-Lami (1986). On the other hand, Maulood et al. (1981) reported that the Central Marsh has a characteristic of having a wider range of temperature variation. Generally, water temperature of the marsh area ranged between 10.2 and 34.1 °C, which is considered high in summer and spring, while low values have been recorded in autumn and winter.

The present chapter gives a short review for the physical and chemical characteristics of the southern marshes of Iraq. Such a review will be the baseline for further studies on the ecology of the marsh areas in Iraq in the future.

Table 6.1 Different studies on the physical and chemical characteristics of the environment of the southern marshes of Iraq during the period 1977–1979

Parameters	Pankow et al. (1979)	Maulood et al.	Al-Saadi and Antoine (1981)	Al-Saboonchi et al. (1982)
Temp air °C	–	–	30.73–32.94	–
Temp water °C	26	15–28.6	23.95–27.97	12.5–31.0
Light penetration	160–195	1.0–3.5	–	
pH	–	6.8–8.2	7.19–8.20	8.1–8.7
Cond	–	–	–	
S‰	0.54–0.56	–	0.34–0.46	
DO	3.50–7.00	3.5–7.0	5.80–9.80	6.8–9.8
TA	7.5–3.5	270.451–1487.48	–	110–127
PO_4 (μg/l)	–	10–26	–	0.02–0.044
NO_3 (μg/l)	–	36–300	300–1700	0.145
NO_2 (μg/l)				0.370
NH_4-N (μg/l)	–	<10–655	90–320	
Reactive Si (mg/l)	–	0.72–0.8	–	
Sediment composition (μg/g dry weight and μg/g ash)	–		–	
Na	–	827, 966	–	
K	–	6.25, 7.32	–	
Mg	–	23,300, 25,750	–	
Ca	–	136,000, 160,000	–	
Al	–	10,180, 11,940	–	
Mn	–	732, 857	–	
Fe	–	17,100, 20,070	–	
Co	–	38.7, 44.0	–	
Ni	–	112, 130	–	
Zn	–	470, 543	–	
Cd	–	2.5, 2.	–	
Pb	–	79.2, 92.0	–	

6.2 The Review of the Physical and Chemical Characteristics of the Southern Marshes of Iraq Before the Draining

6.2.1 Light

Light penetration may reach to the bottom of Al-Hammar Marsh (about 3 m depth) (Al-Saadi and Antoine 1981), while Al-Lami (1986) showed light penetration ranged between 25 and 115 cm; moreover, at one station, light reached the bottom

Table 6.2 Physical and chemical characteristics of the environment of the marsh areas of Iraq during the period 1984–1989

Parameter	Al-Zubaidi (1985)	Al-Lami (1986) and Kassim (1986)	Al-Aarjy (1988) and Hassan (1989)	Mohammed and Barak (1988)
Temp air °C	15–40	13–42.6	19.8–25.0	–
Temp water °C	12.0–32.0	11.8–30.7	18.66–22.25	12.5–31.0
Light penetration	–	Till bottom–115	115–280	–
pH	7.38–9.16	7.63–8.62	7.16–7.28	8.1–8.7
Cond	–	–	1.73–3.06	–
S‰	0.61–21.45	1.3–7.33	1.11–1.98	–
DO	1.67–10.06	3.07–11.95	8.00–8.97	6.8–9.3
TA	26–355	68.0–136.8	102–115	110–127
PO_4 (μg/l)	0.14–1.88	0.068–0.899	0.01–0.12	20–44
NO_3 (μg/l)	0.37–3.39	0.046–9.238	0.01–0.55	145–270
N/P (inorganic form)	3:1	1.23:1–23.92:1	2.6:1–267:1	–
TN/TP	–	–	9.8:1–27.7:1	–
NH_4-N (μg/l)	–	–	–	–
Reactive Si (mg/l)	10.79–325.27	9.8–190	11.6–77.0	–
TH	157.88–13765.16	468–1808	64–753	530–873
Ca	11.88–249.02	94.6–310.9	128–159	64–172
Mg	31.67–3264.64	56.3–250.6	77–95	458–715

of the marsh. In Qurna Marshes and other marshes, light penetration ranged from 155 cm to 195 cm (Pankow et al. 1979; Maulood et al. 1979). The fluctuation in light penetration may be due to the nature of the marsh bottom and boating activity in addition to sedimentation, which might lead to increase in turbidity (Maulood et al. 1979; Al- Saadi et al. 1986).

6.2.2 pH Value

The pH values recorded in most of the studies that have been done in the marsh areas of Iraq ranged from 7.1 to 9.18 during the 1980s (Al- Zubaidi 1985; Al-Lami 1986; Kassim 1986; Al-Aarjy 1988; Hassan 1989). The total alkalinity recorded in the marshes ranged from 68.0 to 136.8 mg $CaCO_3$/l. The type of alkalinity studied in the marshes is known to be mainly bicarbonate, which might be related to the high buffer capacity of whole Iraqi water systems (Al-Mousawi et al. 1994; Hassan 2004). The hydrological process of the marshes has a vital role on the temporal and spatial variation, which might be explained by the aforementioned studies.

6.2.3 *Salinity*

The previous studies on Iraqi Marshes had reported that their salinity value is ranging from slightly saline to oligohaline during the 1970s and 1980s, respectively. The water of the marshes found to be very hard because of the concentrations of calcium and magnesium and the calcareous nature of their sediment (Maulood et al. 1981). Beside that, presence of shells, such as mollusks shell such as mollusk's shells in the marsh might contribute to this increase (Saad 1974). The marsh's water is characterized as well aerated with a high value of dissolved oxygen without anoxia status which was recorded in most of the studies (Tables 6.1 and 6.2).

6.2.4 *Nutrients*

The nutrient concentration had showed quite a wide range of variation within Iraqi Marshes in both spatial and temporal manners. The main nutrients (nitrate, phosphate, and silicate) had reflected a clear seasonal variation throughout most of the published papers. Generally, the availability of nutrients depends on many environmental factors in the marsh, in which some of those are related to the sediment, watershed, hydrology, and redox potential (Cronk and Fennessy 2001; Mitsch and Gosselink 2015). In general, the highest values of nutrients were recorded in autumn and winter of the year, while lower values were observed in spring and summer.

6.2.5 *Nitrogen*

Most studies had reported that the concentrations of nitrate and nitrite were generally low in the marshes in comparison to other water systems in Iraq (Maulood et al. 1981; Talling 1980; Al-Saboonchi et al. 1982; Al- Zubaidi 1985; Al-Lami 1986; Kassim 1986; Al-Aarjy 1988; Hassan 1989). Farthermore Maulood et al. (1979) had interpreted the temporal variation of nitrogen concentration within marsh area depends on the existence of a huge amount of aquatic plants and their growth seasons. In fact, nitrate was the dominant form of inorganic nitrogen. Still, nitrite and ammonia were also recorded, with quite low values and less variations in most of the studies on marshes. However, Al-Mousawi et al. (1994) were concerned about the determination of the filtered particulate and total nitrogen, which were reported to have low concentrations in comparison to other aquatic ecosystems (Lean and Knowles 1987; Hassan et al. 2001).

6.2.6 *Phosphorus*

Reviewing the published papers on phosphorus in southern marshes up to the 1980s of last century (Talling 1980), the results showed that all values for PO_4 are given as 0.1 mg/l and no confidence can be placed in their accuracy. Such no-confidentiality may be due to the fact that there is no estimation of the total phosphorus content that appears to be found for any Mesopotamian freshwater sites. In contrast, phosphate contents vary between 0.12 and 0.875 mg/l, with the highest value in winter reported by Rzoska (1980), which does not agree with Al-Sahaf (1976) nor with Antoine and Al-saadi (1982). However, Maulood et al. (1979) reported that phosphate values in the marshes are much less than that of Shatt al-Arab or the Arabian Gulf environments as it was estimated to be less than 20 μg/l.

The phosphors were measured in the marshes, according to different methods which reflected the different values. The reactive phosphorus was the dominant form of phosphors, and Al-Mousawi et al. (1994) reported that reactive phosphorus was about 70% of the total phosphorus in the marsh area. A temporal variation in the phosphorus concentration was noticed, where certain studies have shown that the marshes suffer from deficiency of phosphorus according to the ratio of nitrogen/phosphorus (both fractions and total) (Al-Lami 1986; Al-Mousawi et al. 1994).

6.2.7 *Silicate*

Silicate also plays an important role as a plant nutrient especially in aquatic ecosystem. Silicate is one of the significant nutrients to planktonic algae and particularly to diatoms which are the most predominant group in southern Iraqi marshes qualitatively. Antoine and Al-Saadi (1982) indicated that reactive silicate concentration in water exceeded diatom requirements. Talling (1980) referred to silicate concentration in Mesopotamian Marshes and showed that SiO_2 varied rather slightly between 7.7 and 9.84 mg SiO_2/l and the water is rich with silica. However, a quite variable level of silicate concentration has been reported for the environment of the Iraqi Marshes. The concentration of silicate can range between 0.72 and 0.8 mg/l (Maulood et al. 1981) and from 9.8 mg/l to 190 mg/l (Al-Lami 1986; Kassim 1986; Al-Mousawi et al. 1994).

A clear seasonal variation for silicate concentration was evident in southern marshes; low concentration was recorded in winter and spring, while a high concentration was recorded in summer and autumn (Al-Saadi and Antonie 1981; Al-Lami 1986; Kassim 1986; Al-Mousawi et al. 1994). The algal growth and blooming, flooding, and temperature are among the factors that play a big role in the temporal variation of silicate in Mesopotamian Marshes (Maulood et al. 1979; Al-Saadi et al. 1993).

6.3 A Review of the Physical and Chemical Characteristics of the Southern Marshes of Iraq After the Draining

6.3.1 Abu-Zirig Marsh

During the rehabilitation regime, many scientific institutions were joined with an Iraqi scientific group to follow up the response of the region as an ecosystem (marshes) to the reflooding effect and outcome after 2004 (Al-Kinzawi et al. 2010). An early ecological study after 2003 was conducted by an initiative group of Canadian and Iraqi scientists through a group of postgraduate students from different universities. UNEP on the other hand also cooperated and supported studies on the environmental management of the Iraqi marshlands through the Iraqi Marshlands Observation System (IMOS). They all assisted these scientific teams financially after 2003 (Partow et al. 2005).

A study on Abu-Zirig Marsh was to monitor the restoration of this marsh for a period of 10 months starting in May 2004 up to March 2005 (Al-Obaidi 2006) (Fig. 6.1). That it is one of the small marshes that faced drying during 1991, it is about 30 km east of Al-Nassiriah City which gain its water source from Tigris River through its branch known as Al-Gharraf River (Yasir et al. 2018).

Light penetration ranged between 63 and 94.75 cm in the Abu-Zirig Marsh, in which the lowest value was recorded in summer and the highest value in other seasons. Water flow ranged from 0 to 42.06 m^3/s. Usually, the low water flow in the marsh habitat is due to poor mixing. Never the less, high water flow in Abu-Zirig Marsh is noticed in winter and summer, while low water flow in spring and autumn.

High values of suspended particulate matter were recorded in three seasons (summer, autumn, and winter), whereas the lowest values in spring. Flood was a key factor for these differences, and also the existence of a large number of macrophytes played a biofilter role (Brenner et al. 1999). Conductivity, salinity, and total dissolved solids are all measures of the concentrations of materials dissolved in water (Tobin et al. 2001; WASC 2002). However, conductivity ranging between 1208 and 1417 μs/cm for this marsh was considered to be slightly saline (Reid 1961).

The pH in this marsh ranged between 7.4 and 9.1. Dissolved oxygen concentration on the other hand ranged from 0.3 to 18.4 mg/l, with the lowest value recorded in spring and summer, whereas the highest value in autumn and winter.

Plant nutrients were also studied in Abu-Zirig Marsh. The inorganic nitrogen (nitrate and nitrite) concentration had showed a temporal variation. The lowest value was recorded in the mid-summer and spring, while the highest value in autumn and winter. Nitrate concentration ranged from 1 to 159.2 μg/l, whereas nitrite

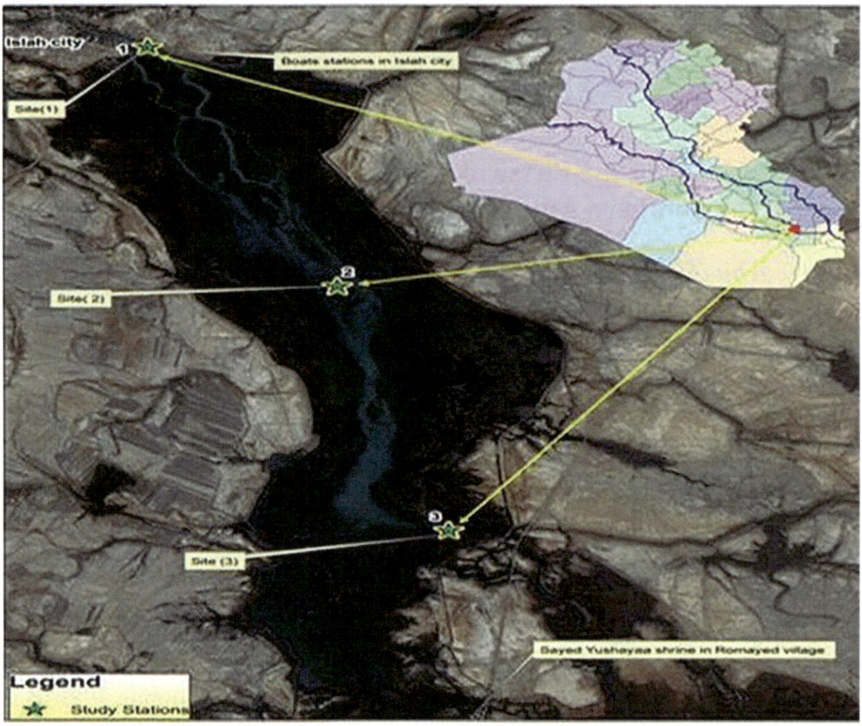

Fig. 6.1 Map showing the location of Abu-Zirig Marsh. (Courtesy of image Abu-Hadal and Al-Hassany 2020)

concentration ranged between 0.3 and 57.75 µg/l. Such variation may be due to several factors such as discharge by runoff and rainfall that acts also on the decomposition processes by bacteria (WASC 2002; Al-Saadi et al. 1986) and to biological or chemical factors that affect on such large variation in nitrogen level.

Inorganic phosphorus concentration had showed a temporal variation but in a different manner with that of nitrogen concentration, where the lowest concentration (0.2 µg/l) was recorded in wet season (autumn and winter) and the highest concentration (122.4 µg/l) in spring and summer. These values were found to be contradictory with the previous studies on marshes; however, such increase of phosphate values may possibly be due to the discharge of domestic and agricultural activities into marshes; moreover, the dryness process had showed its effects (Bass and Potts 2001; Behar and Cheo 2004a, b).

Reactive silicate concentration had recorded high values in Abu-Zirig Marsh and ranged from 127.3 to 1672 µg/l. The seasonal variation of this type of silicate was similar to the pattern of variation of phosphate. However, the dryness processes of this marsh lead to increase the releasing process of silicate from the sediment, particularly through a higher temperature season (Wetzel 2001).

Another study in Abu-Zirig Marsh lasted from December 2008 to May 2009, where a lower water temperature was recorded (1.15–27.00 °C). This is different from what was recorded by Al-Obaidi (2006). Electric conductivity ranged from 1299 to 1750 μS/cm with a slightly higher value than the previous study, and they had considered the marsh as fresh water (0.5–0.8 S‰) and the water was alkaline (100–120 mg/l). The pH values (7.6–8.5) were still within natural water ranges. The dissolved oxygen recorded a high value (6.3–9.67 mg/l) in contrast with the previous study. Nutrient values were less than what was recorded by Al-Obaidi (2006), i.e., nitrate, 0.34–0.57 μg/l; phosphate, 0.03–0.25 μg/l; and reactive silicate, 17.83–71.20 μg/l. These differences may be related to the hydrological status of the marsh that has been affected by reflooding (Qgaili and Al-Kubaisi 2014).

Al-Suhili et al. (2012) had reported that the water quality in Abu-Zirig Marsh may be within acceptable limit and recorded the following values for the physicochemical parameter of the water during the year 2009–2010: TDS, 784 mg/l; electric conductivity, 1330 μS/cm; turbidity, 30.3 NTU; pH, 8.05; dissolved oxygen, 5.95 mg/l; NO_3, 1550 μg/l; and PO_4, 85 μg/l. They concluded that the evapotranspiration in the marsh may be the main cause of water loss (22.33%). However, they suggested a few scenarios for the inundation area such as water depth and velocity distribution.

These investigations showed quite different values, particularly in respect to nutrient values for phosphate and nitrate that might be not acceptable and possibly because of the method of analysis used. Recently, two projects were carried out in Abu-Zirig Marsh, and one of these studies lasted from May 2017 to November 2017 and was concerned about the bloom of *Eichhornia crassipes* (Mart) Solms in addition to the measurement of the physiochemical parameters (Qgaili and Al-Kubaisi 2014). According to this study, the water temperature was found to range between 11 and 34 °C, depth from 40 cm to 300 cm, light penetration from 10 cm to 73 cm, pH value from 6.6 to 8.0, and EC from 1120 to 2410 μS/cm. This investigation reported minimum concentration of DO (3.3 mg/l), while the maximum concentration was 11 mg/l and TDS value ranged between 467 and 944 mg/l. On the other hand, nutrient concentrations were as follows: nitrate, 1.12–7.52 ppt, and phosphate, 0.04–64.3 ppt.

6.3.2 *The Other Marshes in South of Iraq*

Rehabilitation projects were conducted in the three main marshes in southern Iraq with Canadian support during 2006. Several postgraduate programs in Iraqi universities in collaboration with Waterloo University in Canada were affiliated to these projects.

Water temperature in all marshes under investigation (Al-Hammar, Al-Hewaizeh, and Central Marshes) ranged between 12.18 and 32.42 °C. The Iraqi climate being identical as characterized by desert climate (CSO 2012–2013; Abbas and Hassan 2018) showed a huge variation between day/night temperature, day light, rainfall, and drought periods as the seasonal variation was quite clear in these marshes. The lowest air temperature value recorded was in February and January 2006 in all marshes, ranging from 12.18 °C to 14.9 °C, whereas the highest value was recorded in summer of 2006 (27.1–32.42 °C). In Al-Hammar Marsh, the highest water temperature was in August 2006 (32.42 °C), whereas the lowest water temperature was in February 2006 (12.18 °C) (Table 6.2). In Al-Hewaizeh Marsh, the water temperature ranged between 12.23 and 31.35 °C, while in Central Marsh, the water temperature ranged between 12.35 and 29.93 °C (Al-Kinzawi 2007; Al-Haidarey 2009; Talib 2009).

Many researchers had confirmed that water temperature is well affected by air temperature and sampling time in these areas which are very important factors (Munawar 1970; Saad 1976; Antoine 1977; Al-Zubaidi 1985; Kassim 1986; Al-Saffar 2006; Al- Obaidi (2006); Hussein and Taher 2007; Al-Kinzawi 2007; Al-Haidarey 2009; Talib 2009). These studies also showed that the marsh water is known to be shallow and poses a big surface expansion in contrast to its water volume, which causes the variation of temperature values that appeared during these studies. Talib (2009) showed the effect of water temperature on primary productivity for surface and 1 m depth, electrical conductivity, chlorophyll-a, NO_3, PO_4, dissolved oxygen, oxygen saturation rate, and phytoplankton total count, whereas water temperature in the marshes statistically as well had only a negative and highly significant correlation with pH.

It is a well-known fact that hydrological regime has an effect on many physico-chemical properties of the aquatic ecosystem, so the marshes' water level and flood pulses may cause the disruption of marshes' ecosystem (Bayley 1995; Woltemade 1997). The reduction in water depths in all three marshes (Al-Hammar, Al-Hewaizeh, and Central Marshes) was observed after reflooding of Iraqi Marshes, for example, Al-Hammar Marsh's water depth reduction was from 5 m to 3.25 m, and this reduction was also noticed with other studied marshes (Al-Lami 1986; Talib 2009). The water depth values varied among the studied marshes; in the Central Marsh, it ranged from 0.9 m in October 2006 to 2.0 m in January 2006, and in Al-Hammar Marsh, from 1.15 to 3.25 m in October 2006, November 2005, December and January 2006. On the other hand, in Al-Hewaizeh was Marsh, it ranged between 1.82 and 2.4 m in September 2006 and both February and April 2006. Seasonal variations in water depth were evident, which assume that light has an important value in the marsh environment.

In the two marshes (Al-Hewaizeh and Central Marshes), minimum values obtained for light penetration were in autumn (September and October 2006), while in Al-Hammar Marsh, in summer (August 2006). In contrast, the maximum values were recorded in July and June 2006 in the two marshes (Al-Hewaizeh and Central Marshes). The minimum values were 6.7 cm, 75.0 cm, and 82.5 cm at Al-Hammar, Al-Hewaizeh, and Central Marshes, respectively. On the other hand, the maximum values were 140 cm, 150 cm, and 200 cm. These results were found to be low in contrast to that recorded before the draining process (before 2004). Therefore, the hydrological regime has a clear effect in this aspect.

The electrical conductivity ranged from 910.7 μS/cm in June 2006 at Al-Hewaizeh Marsh to 4413.3 μS/cm in July 2006 at Central Marsh. However,

Al-Hewaizeh Marsh recorded the lowest EC values in all months except for March 2006 as compared with other marshes, with the highest values recorded in Al-Hammar Marsh except in summer. In Central Marsh, the lowest value was 1897.0 μS/cm in January 2006, and the highest value was 4413.3 μS/cm in July 2006. A noticeable variation was observed, and the highest EC values had been reported in summer and spring for both marshes (Al-Hewaizeh and Central Marshes) and the lowest in winter, whereas inversely at Al-Hammar Marsh. Temperature and tide are the most important factors affecting the values of EC (Behar and Cheo 2004a, b). These results agreed with the previous studies on these marshes prior to the draining (Al-Zubaidi 1985; Al-Aarjy 1988) and also with studies after reflooding period (Al-Obaidi 2006; Al-Kinzawi 2007).

The salinities of these marshes reflected quite different values after draining in 2004. Salinities of the most studied marshes before 2004 are confined to Al-Hammar, Qurna, and Central Marshes, and these studies have revealed that the values of salinity do range from 1.5‰ in west Al-Hammar Marsh to 4.5‰ in Al-Hammar Marsh (Al-Mousawi et al. 1994), while in the Central Marsh, it ranged from 0.27 to 0.54‰ (Maulood et al. 1979) and in Qurna Marsh was only 1.02‰ (Al-Zubaidi 1985). Recent investigation after 2004 showed that water salinity values for the main marshes ranged from 0.25‰ in June and July 2006 at Al-Hewaizeh Marsh to 2.1‰ (2.1‰) in November 2005 at Al-Hammar Marsh. Talib (2009) reported that salinity values in Al-Hewaizeh Marsh had recorded minimum values in most months as compared with other marshes and showed significant differences with both Central and Al-Hammar Marshes.

Temporal variations of salinity were observed, the lowest value (0.8‰) being recorded in March 2006 (spring) at Central Marsh and the highest value (2.25‰) in October 2006 (autumn). It appears that salinity values in this marsh, in general, were exceeding 1‰. However, at Al-Hammar Marsh, salinity values were higher than other recorded values among other marshes. The lowest values always exceed 1‰, and this clearly explains how much is the impact of tide on the marsh as it reaches more deeply inside the marsh area more than before 2004. Hence, there are different environmental consequences of this rise in salinity on other properties of marshes. Also there are many scenarios that if the sea level exceeds to more than 7 m, it may lead to the disappearance of all these marshes.

The three marshes, Al-Hewaizeh, Al-Hammar, and the Central Marshes, have been known to have oligohaline-brackish water after and before 2004 as was mentioned in the studies after 2004 (Al-Kinzawi 2007; Al-Haidarey 2009; Talib 2009; Hussein 2014) and before draining (Maulood et al. 1979). All these studies suggested that the increasing and decreasing of salinity values are results of rainfall, hydrological regime, and temperature (Saad 1976; Al-Saadi et al. 1977; Maulood et al. 1979; WASC 2002; Al-Saffar 2006; Adam et al. 2007; Talib 2009).

Most studies on Iraqi aquatic ecosystems in the south (Hinton and Maulood 1982; Al-Saadi and Antoine 1981; Maulood et al. 1981) reported that Iraqi Marshes tend to be alkaline within a normal range of natural aquatic systems because of their higher buffer capacity almost generally its value are above 6 and less than 9 both the two hydrological regime before/after 2004. The pH values was ranged from 6.95 in

September 2006 and 8.02 in March 2006 at Al-Hewaizeh Marsh, whereas pH values ranged from 6.68 in September 2006 to 8.78 in February 2006 at Central Marsh and from 7.10 in September 2006 to 8.25 in both November 2005 and December 2006 at Al-Hammar Marsh. These results showed and indicated that Al-Hammar Marsh is more alkaline than other marshes. A clear seasonal variation was evident in respect to pH values in all three marshes under investigation which was possibly caused by photosynthetic activity, aquatic plant degradation, and temperature (Sreenivassan 1971; Goldman and Horne 1983; Van Dolah et al. 2002; Al-Essa 2004; Al-Obaidi 2006; Al-Kinzawi 2007; Al-Haidarey 2009).

The lowest values of DO (less than 5 mg/l) were recorded in the main marshes. In the Central Marshes, the DO concentration dropped to 1.14 mg/l in May 2006, and such values were also observed by Talib (2009) when the lowest DO concentrations were recorded during April–October 2006 (1.36, 1.14, 2.70, 2.67, 1.80, 3.23, and 1.89 in Al-Hewaizeh, Al-Hammar and the Central Marshes, respectively), while its concentrations suddenly rise and ranged from 8.36 mg/l in November 2005 to 11.89 mg/l in January 2006. It seems that the reason behind such changes might be the effect of temperature, existence of higher organic matter, and increasing salinity (AL-Saffar 2006; Al-Kinzawi 2007). Al-Kinzawi (2007) in his report on Al-Chubaish (Central Marshes) recorded the lowest concentration in April (1.4 mg/l) and the highest concentration (9.9 mg/l) in December, with noticeable seasonal variations where the maximum values were at the winter and the minimum values were at the summer season.

The DO concentration value at Al-Hewaizeh Marshes dropped to lowest values (4.35 mg/l and 4.77 mg/l) in August and September 2006, while the highest concentration (10.14 mg/l) has been recorded in February 2006. On the other hand, in Al-Hammar Marsh, the lowest recorded value (2.88 mg/l) was in October 2006, and other lowest values (<5) were noticed during the period between April and August 2006. In contrast, the highest value (10.35 mg/l) was recorded in January 2006. These results come in accordance with the studies before 2004.

Talib (2009) had noticed that the general concentrations of NO_3 in Al-Hewaizeh Marshes were higher than 100 μg/l in most of the study periods except for winter. On the other hand, Al-Hassany (2010) had also recorded low concentrations of NO_3 (<20 μg/l) during the study period in winter 2008, and these values were less than those given and estimated by Talib (2009). These results show the presence of an autochthonous and allochthonous sources. Al-Hassany (2010) results might be referred to and explained by the effect of recovery of the marsh to restoration, and the two studied sites do not undergo the desiccation regime which might also be considered here.

In Central Marshes, on the other hand, the concentration of NO_3 ranged from 2 > μg/l to <13 μg/l (Al-Kinzawi 2007). Talib (2009) had recorded undetectable values of NO_3 concentration in February and March 2006, whereas an increase of NO_3 concentrations had been recorded in 2006 (141.08 μg/l and 92.77 μg/l). A lower concentration, which did not exceed to more than 24 μg/l, had been observed. Such results were much more than those recorded before 2004 (3.9–6.0 μg/l) and reported by Maulood et al. (1979) and recently given by Hussein (2014).

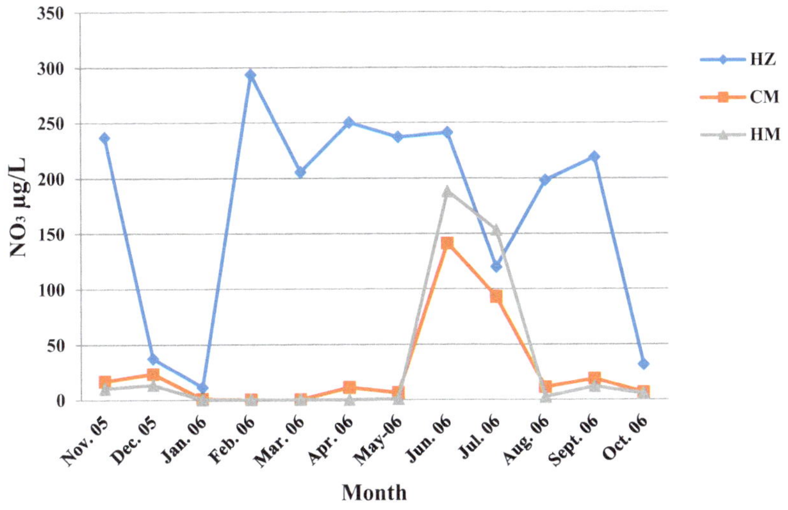

Fig. 6.2 Monthly reactive nitrate value variations for the studied marshes, Al-Hewaizeh (HZ), Central Marshes (CM), and Al-Hammar (HM) (followed Talib 2009)

However, clear seasonal variations were observed from all studies on the marsh areas after 2004. Al-Hassany (2010) reported a low concentration of NO_3 in summer and a high concentration in winter at Al-Hewaizeh Marshes, whereas Talib (2009) showed that the seasonal variation in these marshes was not clear. In contrast, the lowest values for NO_3 had been reported in winter and spring, whereas the highest concentration was recorded in summer at the Central Marshes (Al-Kinzawi 2007). In fact, only one peak of nitrate values has been recorded in June 2006 in the same marshes as indicated by Talib (2009). The same trend has been noticed in Al-Hammar Marsh, but with two peaks in Jun and July 2006.

Seasonal changes of NO_3^- concentration in the environments of the marshes come in accordance with other areas, when the maximum values were recorded in the summer season, while the minimum values were recorded in the spring season, that was found in Al-Hewaizeh Marshes two studied marshes (Central and Al-Hammar Marshes) (see Fig. 6.2); that's mean they have only one peak in spring and beginning of summer during the investigation. Although high values have been recorded in Al-Hewaizeh and Central Marshes, wide fluctuations in their values have also been observed. This may be due to fact that marsh water had been affected by water that originated from adjacent agricultural fields which is enriched with nitrogen fertilizers (El-Wakeel and Wahby 1970; Seenaya 1971).

Nitrite concentration generally was found to be low in all published studies on the marshes of Iraq, except for the recorded high value of nitrite (44.94 µg/l). The study on Al-Hewaizeh Marshes showed the highest values in February 2006, and the lowest concentration was (3.04 µg/l) in November 2005 (Talib 2009), which is suggested as an input from runoff and rainfall. In addition to reduction in the uptake of nutrients in cold season, this explanation might be not match with the results of

DO and its oxygen saturation rate (>10). The study of Al-Hassany (2010) recorded a high concentration value (5.40 μg/l) in cold season, but its value varied from that of Talib (2009) on the same marsh (summer) (Al-Kinzawi 2007). In Central Marsh, the nitrite concentrations ranged from 0.656 μg/l in March 2006 (spring) to 2.32 μg/l in August 2006 (summer) (Al-Kinzawi 2007). Another report showed a small value of 3.42 μg/l for NO_2 in August 2006 and a high value of 4.24 μg/l in January 2006 (Talib 2009). These variations may be due to the different sampling locations. The NO_2 concentration at Al-Hammar Marsh ranged from 0.94 μg/l in September 2006 to 4.08 μg/l in August 2006. However, these parameters need more detailed studies in order to find out the role of NO_2 in the marshes, particularly in respect to pollution and productivity.

Phosphate (PO_4^{-3}) is an essential nutrient for algae and other organisms and is important for metabolism (Jarvis 2000). The concentration of phosphate (PO_4^{-3}) after 2004 at Al-Hewaizeh Marsh had been shown to range from 9.57 μg/l in December 2005 to 63.31 μg/l in September 2006, which is below 10 μg/l (Talib 2009). Al-Hassany (2010) showed lower values of phosphate in comparison to that recorded by Talib (2009) and these ranged from 0.13 to 3.97 μg/l in autumn and winter 2008. An increase in the value of up to 7.10 μg/l was observed in one site (Umward Marsh) in winter 2008, it has been suggested that the rainfall that was the agricultural lands, which are rich in phosphate fertilizer was standing behind such an upsurge in the value (Sharpley 2001).

The PO_4^{-3} concentrations at Central Marshes were 0.194 μg/l in April 2006 and 2.613 μg/l in October 2006 according to Al-Kinzawi (2007). Different values were shown in another study, 1.05 μg/l in January 2006 and 15.50 μg/l in October 2006, while an increase of 32.13 μg/l was noticed in May 2006 (Talib 2009). These results have shown a remarkable increase of PO_4^{-3} concentrations in comparison with the studies before 2004.

Within Al-Hammar Marshes, the concentration of PO_4^{-3} was 19.3 μg/l in July 2006; however, no data was available in January 2006, and there were three high concentration values, 10.50, 18.06, and 19.30 in May, August, and July 2006, respectively (Talib 2009). Phosphate in the aquatic systems of the marshes originates from the decomposition of macrophytes, organic matter, nearby runoff from agricultural lands, and exchangeable reaction in marsh sediment (El-Wakeel and Wahby 1970; Horne and Goldman 1994).

A few studies are concerned with the determination of silicate in marshes. Al-Hassany (2010) recorded seasonal variations of silicate concentrations: a low value of 107.56 μg/l in spring 2008 and a high value (>300 μg/l) in winter, summer, and autumn 2008 at Al-Hewaizeh Marsh.

Other parameters such as chloride, sulfate, calcium, and magnesium concentrations and total hardness were measured only in some studies such as by Al-Kinzawi (2007), who reported that chloride values in Central Marshes ranged from 246.93 mg/l in March to 1417.9 mg/l in June 2006. Seasonal variation was observed in this study, where low values were recorded in spring and high values in summer. Also, sulfate concentrations varied temporally in the same trend as chloride concentrations ranged from 304.54 mg/l in April 2006 and 825.12 mg/l in July 2006. Total hardness, calcium concentration, and magnesium concentration ranged as 210.93 mg/l–1460.9 mg/l, 58.26 mg/l–380.46 mg/l, and 37.16 mg/l–280.4 mg/l in April (for low values) and in July and August (for high values), respectively.

6.4 Conclusions

Such review clearly indicates the presence of a big gap in the knowledge in respect to physicochemical parameter's role in Mesopotamian Marshes. Never the less, quite good and valuable efforts that have been carried out by local and international scientist, still much more endeavors are needed in respect to the role of phosphate, nitrate, and silicate in particular and trace element in general. However, the concentration of Zn has been shown to be high (Maulood et al. 1979). Almost certainly, the local scientist will attempt to fill up the gap in the near future in order to show the exact status of the Mesopotamian Marshes, which is one of the most attractive and unique ecosystems of the world.

References

Abbas AAA, Hassan FM (2018) Water quality assessment of Euphrates river in Qadisiyah Province (Diwaniyah river), Iraq. Iraqi J Agric Sci 48(6)

Abu-Hadal L, Al Hassany JS (2020) Using diatom indices to evaluate water quality in Abu-Zirig Marsh Thi-Qar Province/south of Iraq. Baghdad Sci J 17(2 Special Issue):599–603

Adam RS, Al-Shawi IJM, Al-Imarah FJM (2007) Distribution of some chemical elements in the marsh lands of southern Iraq after rehabilitation. Marsh Bull 2(1):11–17

Al Hassany JS (2010) A study of the ecological and diversity of epiphytic algae on some aquatic plants in Al-Hawizeh marshes. PhD thesis. College of Science for Women, University of Baghdad, Iraq, 178p

Al- Obaidi GSA (2006) A study of phytoplankton community in Abo Zirig Marsh, Southern Iraq. MSc thesis, College of Science, University of Baghdad

Al-Aarjy MJ (1988) An ecological study on phytoplankton and nutrients in Al-Hammar Marsh, Iraq. MSc thesis, College of Science, University of Basrah

Al-Essa SAK (2004) Ecological study of the aquatic plants and epiphytic algae in Shatt Al-Arab River. PhD thesis, University of Basrah, Iraq, 191p

Al-Haidarey MJS (2009) Assessment and Sources of some heavy metals in Mesopotamian Marshes. PhD thesis, College of Science for Women, University of Baghdad

Al-Kaisi KA (1976) Contributions to the Algal Flora of the Rice-fields of Southern Iraq. Nova Hedwigia 27:813–827

Al-Kinzawi MAH (2007) Ecological study of aquatic Macrophytes in the central part of the Marshes of southern Iraq, MSc thesis, College of Science for Women, Biology Department, University of Baghdad, Iraq

Al-Kinzawi MAH, Asada T, Hassan FM, Warner BG, Douabul A, Al-Hilli MRA, Alwan AA (2010) Vegetation response to re-flooding in the Mesopotamian wetlands. Southern Iraq J of Wetland 30:177–188

Al-Lami AA (1986) Ecological study on phytoplankton on some marsh regions southern Iraq. MSc thesis, College of Science, University of Basrah
Al-Mousawi AK, Whitton BA (1983) Influence of environmental factors on algae in rice-field soil from the Iraqi Marshes. Arab Gulf J Sci Res 1:237–253
Al-Mousawi AHA, Al-Saadi HA, Hassan FM (1994) Spatial and seasonal variations of phytoplankton population and related environmental in Al-Hammar Marsh. Iraq J Sci B 12(1):9–19
Al-Saadi HA, Antoine SE (1981) Primary productivity and phytoplankton population dynamics in polluted Ashar canal and Shatt al-Arab, Basrah, Iraq. Vern Internat Verein Limnol 21:880–885
Al-Saadi HA, Saad MAH, Hadi RA, Hussien NA (1977) Further investigations on some environmental characteristics of North-West Arab Gulf. J Indian Nat Ac Proc 43(3):183–192
Al-Saadi HA, Al-Daham NK., Al-Hassan LA (1986) Aquatic ecology. Dar Alkutub Directorate for Printing and Publishing/University of Mosul
Al-Saadi HA, Al-Mousawi AHA, Al-Aarajy MJ (1993) Physico-chemical features of Al-Hammar Marsh, Iraq. J Coll Edu For Women Univ Baghdad 4:35–39
Al-Saboonchi AA, Mohammed ARM, Barak NA (1982) A study of phytoplankton in the Garma Marshes, Iraq. Iraqi J Mas Sci 1(1):67–78
Al-Saffar MAT (2006) Interaction between the environmental variables and benthic macro invertebrates community structure in Abu Zirig Marsh, Southern Iraq. MSc thesis, University of Baghdad, Iraq, 156p
Alsahaf M (1976) Water resources in Iraq. Ministry of Irrigation, Baghdad
Al-Suhili RH, Ebrahim SE, Al-Khazaali HJ (2012) Optimum water allocation for Abo-Ziriq Marsh ecological restoration. J Eng 18(5):683–692
Al-Zubaidi AJM (1985) Ecological study on the Algae (phytoplankton) in some marshes near Qurna-Southern Iraq. MSc thesis, University of Basrah, Iraq, 131 pp
Antoine SE (1977) Seasonal variations of the environmental characteristics and phytoplankton blooms of Tigris, Iraq. MSc thesis, University of Basrah, Iraq
Antoine SE, Al-Saadi HA (1982) Limnological studies on polluted Ashar canal and Shatt al-Arab at Basrah (Iraq). Int Rev Ges Hydrobiol 67(3):405–418
Bass D, Potts C (2001) Invertebrate community composition and physicochemical conditions of Boehler Lake, Atoka County, Oklahoma. Proc Okla Acad Sci 8:21–29
Bayley PB (1995) Understanding large river-floodplain ecosystems. Bioscience 45(3):153–158
Behar S, Cheo M (2004a) Hudson Basin river watch guidance document. River Network. www.hudsonbasin.org
Behar S, Cheo M (2004b) Hudson Basin river watch guidance document: helping to coordinate monitoring of freshwater wadeable rivers throughout the watershed, river network – river watch program, revised draft. Retrieved June, 2000, from river network (www.hudsonbasin.org)
Brenner M, Keenan LW, Miller SJ, Schelske CL (1999) Spatial and temporal patterns of sediment and nutrient accumulation in shallow lakes of the Upper St. Johns River Basin, Florida. J Wetl Ecol Manage 6:221–240
Cronk JK, Fennessy MS (2001) Wetland plants biology and ecology. CRC Press LLC, Boca Raton, 462 pp
CSO (Central Statistical Organization). (2012–2013). Annual statistical abstract. Issued by the Iraqi Ministry of Planning
El-Wakeel SK, Wahby SD (1970) Hydrography and chemistry of Lake Manzala. Egypt Arch Hydrobiol 67:173–200
Goldman CP, Horne AL (1983) Limnology. McGraw-Hill, New York, 464 p
Hassan, F. M. (1989) An Ecophysiological and qualitative study on phytoplankton in the Al Hammar Marsh, Iraq. MSc thesis, College of Science, University of Basrah
Hassan FM (2004) Limnological features of Diwanyia River, Iraq. J Um-Salama For Sci 1 (1):119–124
Hassan FM, Al-Saadi HA, Mohamed AAK (2001) On the ecological features of Razazzah Lake, Iraq. Natl J Chem 2:549–565

Hinton GCF, Maulood BK (1982) Contributions to the algal Flora of Iraq: the non-diatom Flora of the Southern Marshes. Nova Hedwigia 37:49–63
Horne AJ, Goldman CR (1994) Limnology, 2nd edn. McGraw-Hill, New York
http://www.dnr.sc.gov/marine/scecap/PDF%20Files/SCECAP%2019992000%20technical%20report%2010_29_02final%20with%20append.pdf
Hussein AH (2014) Biotopes of Iraqi marshlands. Difaf Publication, Basra-Iraq, 432p
Hussein NA, Taher MA (2007) Effect of daily variations, diurnal fluctuations and tidal stage on water parameters of East Hammar marshland, Southern Iraq. J Marsh Bull 2(1):32–42
Jarvis PJ (2000) Ecological principles and environmental issues. Pearson Education, (T.J. International), Padstow, Cornwall, 361 pp
Kassim TI (1986) An ecological study on benthic algae for some marshlands southern Iraq. MSc thesis, University of Basrah
Kooser JC, Lund P (1996) Water quality guideline for wetlands; using the surface water quality standards for activities involving wetlands. Shorelands and Water Resources Program, Washington State Department of Ecology, Olympia, WA, Publication #96–06
Lean DR, Knowles R (1987) Nitrogen transformations in Lake Ontario. Can J Fish Aquat Sci 44 (12):2133–2143
Maulood BK, Hinton GCF, Kamees HS, Saleh FAK, Shaban AA, Al-Shahwani SMH (1979) An ecological survey of some aquatic ecosystems in southern Iraq. J Trop Ecol 20(1):27–40
Maulood BK, Hinton GCF, Whitton BA, Al-Saadi HA (1981) On the algal ecology of lowland Iraqi marshes. J Hydrobiol 80:269–276
Maulood BK, Sulaiman NI, Farj S (1986) A comparison ecological study on algae in Yousefia and Rashdia Baghdad. Research biology. Biol Cent Sci Council Res Baghdad 17(2):13–30
Mitsch WJ, Gosselink JG (2015) Wetlands, 5th edn. Van Nostrand Reinholdk, New York
Mohammed ARM, Barak NA (1988) Growth and condition of a cyprinid fish. Barhus sharpeyi Gunther in Al-Hammar marsh. Basrah J Agric Sci 2:18–25
Munawar M (1970) Limnological studies on fresh water ponds of Hyderabad-India. I The biotope. J Hydrobiol 35(1):127–162
Nurul-Islam AKM (1982) Marsh algae from southern Iraq. Int Rev Gesamt Hydrobiol 67 (2):245–260
Pankow H, Al-Saadi HA, Huq MF, Hadi RAM (1979) On the algal flora of the marshes near Qurna (Southern Iraq). Willdenowia 8:493–506
Partow H, Jaquet JM, Allenbach K, Schwarzer S, Nordbeck O (2005) Iraqi marshlands observation system. Nairobi, Kenya, 71 p
Qgaili JW, Al-Kubaisi ARA (2014) A study of the physical and chemical characteristics of the waters of the central marshes in southern Iraq after restoration. Baghdad Sci J 11(2 special issue):991–998
Reid GK (1961) Ecology of in landwater and estuaries. Rhien hold crop, New York, 375 pp
Rzoska J (1980) Euphrates and Tigris, Mesopotamian ecology and destiny. Junk, W., The Hague, 122 pp
Saad MA (1974) Calcareous deposits of the brackish-water lakes in Egypt. Hydrobiologia 44 (4):381–387
Saad MAH (1976) Some limnological investigations of Lake Edku, Egypt. J Arch Hydrobiol 77 (4):411–430
Seenaya G (1971) Ecological studies in the plankton of certain freshwater ponds of Hyderabad-India, I. Physic-chemical complexes. Hydrobiologia 37(10):7–31
Sharpley A (2001) Managing phosphorus agriculture and the environment college of science, The Pennsylvania State University, 8 pp
Sreenivassan A (1971) Shallow aquatic biotopes Limnological models. J Hydrobiol 37 (3–4):553–564
Talib AH (2009) Ecological study on the phytoplankton and primary productivity in southern Iraqi Marshes. PhD thesis, College, of Science for Women Baghdad University, 142 pp

Talling JF (1980) Water characteristics in Euphrates and Tigris. Mesopotamia ecology and destiny, 38

Tobin M, Yakub G, Buys M, McCartney B, Meyer S, Earth E (2001). Three rivers – second nature water quality report. Retrieved march 5th 2005, from the College of Fine Arts, Carnegie Mellon University website, Pittsburgh PA 15213. Available: 3r2n.cfa.cmu.edu/Year1/reports/aquatic/waterquality/6.pdf

UNEP (2010) Support for environmental management of the Iraqi marshlands 2004/2009, 100p

Van Dolah RF, Jutte PC, Riekerk GHM, Levisen MV, Zimmerman LE, Jones JD, Lewitus AJ, Chestnut DE, McDermott W, Bearden D, Scott GI, Fulton MH(2002) The condition of South Carolina's Estuarine and coastal habitats during 1999-2000: technical report. South Carolina marine resources division, Charleston. Technical report no. 90. 132p.

WASC, Waterwatch Australia Steering Committee (2002) Water watches Australia National technical manual. Environment Australia, Commonwealth of Australia

Wetzel RG (2001) Limnology, lake and river ecology, 3rd edn. Academic Press, 1006 pp

Woltemade CJ (1997) Water level management opportunities for ecology. Benefit, pool 5 Mississippi RWER 1. J Am Water Resour Assoc 33(2):443–454

Yasir RA, Rahi KA, Abudi ZN (2018) Water budget for Abu Zirig Marsh in southern Iraq. J Eng Sustain Dev 22(1):25–33

Chapter 7
Hydrologic Structures in the Tigris-Euphrates Basin and Their Impact on the Vitality of the Marshes

Toon Bijnens

Abstract Hydrological conditions of the Iraqi Marshes have changed over the course of the past decades. The hydrological system of the marshlands lost many of its components during the draining in the 1980s and 1990s. Some of its integrity returned after the partial restoration since 2003. In current times, the hydrology of the marshlands remains susceptible to droughts and decreased water flows from upstream. The impact of dam construction on the hydrology of the Marshes has been considerable: due to reduced water levels from the Tigris and Euphrates Rivers, different areas of the Marshes remain hydrologically disconnected and fragmented, compared to earlier levels. Salinization caused by changes in the hydrological regime has damaged the ecology of the marshlands. Restoration of the hydrological regime of Iraqi Marshes depends on the water quantity and quality that feed the marshlands. Re-flooding alone has not restored the hydrology of the marshlands; instead, further comprehensive data on water discharge and quality is still needed. A management plan would have to find suitable alternative water supplies, especially with regard to discharge which is of low quality, or establish suitable treatment so that salinity levels are reduced before it enters the marshland area. To better determine the quality, quantity, and hydroperiod of the hydrological regime of the Marshes, further research is needed. Technology could facilitate hydrological conditions in the Marshes, provided it is based on reliable water allocations and target environmental variables.

Keywords Hydrology · Dams · Hydrological regime · Hydroperiods · Discharge · Salinity

T. Bijnens (✉)
Save the Tigris, Amsterdam, Netherlands
e-mail: bijnenstoon@gmail.com

L. A. Jawad (ed.), *Southern Iraq's Marshes*, Coastal Research Library 36,
https://doi.org/10.1007/978-3-030-66238-7_7

7.1 Introduction

The Iraqi Marshes were once considered the largest system of wetlands in West Asia. It consists of Hammar, Huweizah, and Central Marshes, with its depths fluctuating between 0.5 and 3.5 meters (Al-Ansari et al. 2012). The hydrological past of the marshlands is complex. From the 1970s up to 2003, over 90% of the original Marshlands were drained, whether on purpose (political reprisal) or not (mismanagement and overexploitation).[1] The first extensive dam construction on the Euphrates River in Syria and Turkey in the 1970s reduced the discharge of the Tigris and Euphrates Rivers feeding the marshlands. What followed was desiccation of parts of the area and increased salinity levels, causing deterioration of the natural ecosystem. Salinization in the Marshes increased as salinity started to rise in the Tigris and Euphrates Rivers in the early 1980s (Schwartzstein (2015), leaving an impact on the ecology of the wetlands. From the 1980s, the Iraqi Marshes were drained by the Saddam regime. Since those decades, water shortages and rising salinity levels have remained the main obstacles to the restoration of the Marshes. After the 2003 invasion and the fall of Saddam, the Marshes were partially restored with the assistance of communities, NGOs, and authorities. With their help, over 40% has been rehabilitated.[2] Meanwhile, the work continues to restore the Marshlands.

The Marshes are fed by the Tigris and Euphrates Rivers. The main channel of the Tigris River feeds Huweizah and Central Marshes; however, as it nears the marshlands, its flow capacity decreases significantly.[3] The Euphrates River flows downstream south of Nasiriyah into Hammar Marsh, including secondary channels. Its main branch flows toward the Tigris at Qurnah. Embankments constrain the Euphrates River from Nasiriyah to Qurnah. An embankment upstream of Qurnah blocks the overflow into the wetlands. Secondary channels connect the Euphrates to the Central Marsh through a series of breaches along the Euphrates north embankment.[4] From Hammar Marsh, the water feeds into the Shatt al-Arab and Shatt al-Basra Rivers through Qarmat Ali, while the Central Marsh feeds the upper reach of the Euphrates and consequently the Shatt al-Arab at Qurnah. The Tigris and Shatt al-Arab receive water that flows back from Huweizah Marsh. The Marshes are completely dependent on upstream water supplies, and as water flows are

---

[1] *Managing Change in the Marshlands: Iraq's Critical Challenge*, United Nations White Paper: Report of the United Nations Integrated Water Task Force for Iraq, 2011, http://www.hydrology.nl/images/docs/ihp/2011.08_Marshlands_Iraq.pdf, 2.

[2] *Managing Change in the Marshlands: Iraq's Critical Challenge*, United Nations White Paper: Report of the United Nations Integrated Water Task Force for Iraq, 2011, http://www.hydrology.nl/images/docs/ihp/2011.08_Marshlands_Iraq.pdf, 2.

[3] *Physical Setting of the Marshlands*, Fanack Water, 30 May 2017, https://water.fanack.com/specials/iraqi-marshes/physical-setting

[4] *Physical Setting of the Marshlands*, Fanack Water, 30 May 2017, https://water.fanack.com/specials/iraqi-marshes/physical-setting

decreased because of dams or diversions, the lack thereof will impact the marshlands.

An estimated 105 tons of pollutants and sediment annually are filtered by tidal flows and its ecosystems (Partow 2001). Embankments – such as dams and dikes – have reduced the water flows to and from the Marshes into the Gulf. Human interference has disrupted the natural flows. A large amount of sediment such as suspended particles is trapped behind dams, reducing its load. Marsh water will be disrupted because of the change in tidal flows and cause rising salinity levels, while decreasing plankton and organic carbon levels. This in turn could reduce soil fertility and reduce the amount of freshwater.[5]

The hydrological system of the Iraqi Marshes lost many of its components during the draining in the 1980s and 1990s. Some of its integrity returned after the partial restoration since 2003. However, the hydrology of the marshlands remains susceptible to droughts and decreased water flows from upstream. This article will explore how the hydrological conditions of the Marshes changed over the course of the past decades and what is required to restore the hydrological regime of the marshlands.

7.2 Hydrologic Setting of the Marshes: Natural Hydrologic Conditions (Pre-restoration)

Until the 1970s, the Marshes consisted of natural channels, which was a highly interconnected system in which rivers merged. Upstream of the marshlands, the Tigris and Euphrates Rivers formed a delta, feeding the wetlands. Human actions have changed the original hydrological system of the Marshes since then through the construction of (agricultural) canals, drainage systems, and embankments which have altered the hydrology of the entire wetlands.[6] In Southern Iraq, north of Qurnah, the Tigris and Euphrates consisted of many branches which fed the marshlands over an area of 15,000 km^2 in the early 1970s (Fawzi et al. 2016). Hammar Marsh borders the Southern Desert and several saline lakes and is fed by the Euphrates River and its tributaries. It also received water from the Tigris River via Central Marsh and further input from groundwater. The length of this Marsh used to be 120 km^2 with a width of 25 km^2 and depth ranging between 2 and 3 m (Al-Ansari et al. 2012, p. 66). It permanently covered 2800 km^2 approximately and would increase to more than 4000 km during periods of inundation (Al-Ansari et al. 2012, p. 66). Banks and islands emerged during the hot seasons (Al-Ansari et al. 2012, p. 66). The Central Marshes are located in between Nasiriyah and Qurnah, fed

[5] *Managing Change in the Marshlands: Iraq's Critical Challenge*, United Nations White Paper: Report of the United Nations Integrated Water Task Force for Iraq, 2011, http://www.hydrology.nl/images/docs/ihp/2011.08_Marshlands_Iraq.pdf, 29.

[6] *Physical Setting of the Marshlands*, Fanack Water, 30 May 2017, https://water.fanack.com/specials/iraqi-marshes/physical-setting

by the Tigris River tributaries southward from the city of Amara. Until the 1970s, the Central Marshes covered an area of 3000 km^2 (Al-Ansari et al. 2012, p. 67). Huweizah Marsh lies on the Iraq-Iran border, east of the Tigris River. Two tributaries from the Tigris River near Amara, Al-Musharah and Al-Kahla, feed the Iraqi side of the Marsh, while on the Iranian side, the Kharkeh River flows into the Hoor al-Azim. The area covered about 2300 km^2, 80 km in length and 30 km width (Al-Ansari et al. 2012, p. 67). The south part of Huweizah Marsh is very seasonal, while the rest of the marsh is permanent, with vegetation and open stretches of water (Al-Ansari et al. 2012, p. 67).

During tidal flows, the high areas within the marshlands were flooded by water of the Tigris and Euphrates. The main channel of the Tigris River feeds water into Huweizah and Central Marsh. The carrying capacity of the channel decreases from 6000 m^3/s to 70 m^3/s as it passes from Kut to Kalat Salih-Kassarah and then increases again through inflow from Huweizah Marsh (Al-Ansari et al. 2012, p. 68). The Iraqi Marshes used to consist of a natural system of natural channels with interior deltas formed by the Tigris and Euphrates Rivers. Over the past 1000 years, this area was changed by man-made agricultural canals, some of which flow into the Central Marsh (Gharraf, Butaira, Arid, Mujar Al-Kabir) and Huweizah Marsh (Mujarrah, Qurnah). Hammar Marsh used to be occasionally fed by the Tigris River during high floods. Embankments have halted this process. Hammar Marsh is now mostly fed by the Euphrates through several channels south of Nasiriyah, some of which include Aglawin and Haffar.

In the twentieth century leading up to the 1970s, the Marshes were altered by hydraulic projects, the first major human interventions in the Tigris and Euphrates Rivers to control the flooding. Irrigation channels altered the course of the rivers. In 1913, Hindiya Barrage on the Euphrates and in 1938 Kut Barrage on the Tigris were constructed, diverting the course of the rivers for agricultural means. This meant that water flowing from the Tigris to Central and Huweizah Marshes was decreased. In the 1970s, Syria constructed several dams on the Tigris River, as did Turkey in the 1990s. As a consequence, the discharge downstream of Hindiya Barrage decreased by 250 m^3/s (Al-Ansari et al. 2012, p. 69).

7.3 Hydrologic Setting of the Marshes: Current Hydrologic Conditions (Recent Water Resources Development)

During the Iran-Iraq War of the 1980s, Iranian troops advanced inside Iraq. To block the advancement of Iran and to combat opposition, troops were moved inside the Marshes. To facilitate their movement, segments of the Marshes were drained. After the 1991 rebellion, opposition against Saddam took refuge in the Marshes. Major drainage projects took place in order to further desiccate the wetlands. Water from the Tigris and Euphrates Rivers was obstructed from flowing into the Marshes, the hydroperiod of the rivers was reduced, and polders appeared in the Marshes.

Permanent marshland area decreased by 84%, while seasonal marshlands increased by 48% (Garstecki and Amr 2011).

The flow of the Euphrates was decimated by half after the 1990s, while the Tigris lost almost two-third (Wilson 2012). After Kut Barrage was constructed, the discharge of the Tigris went down from 945 m^3/s to 368 m^3/s by 1991 (Al-Ansari et al. 2012, p. 69). Huweizah Marsh, dependent upon the Karkheh River for its water supply, receives only 2 MCM currently; before 1990, this was 5 MCM (Al-Ansari et al. 2012, p. 69). Flows were decreased due to the Kharkeh Dam which was constructed in 2002 (Bijnens 2017). If the Marshes had shrunk by one-third between 1968 and 1984, by 1992, the Marshes had diminished even further[7] due to the drainage program of Saddam coupled with Turkish dam construction upstream on the Tigris and Euphrates Rivers. In addition, oil fields were exploited within the marshland area. About 300 km^2 of the Marshes were drained to facilitate the oil exploration (Al-Ansari et al. 2012, p. 70). As a consequence of these actions, Central and Hammar Marshes were almost completely dried up. The hydrologic regime of the Marshes was completely altered. By 1993, two-third of the marshes were without water (Pearce 1993) and by 2000 less than 10% remained (Wayman 2009).

When the Saddam regime fell in 2003, hydrologic structures, embankments, and levees were torn down in order to direct the water to the Marshes again. Since then, the Iraqi authorities with the help of international NGOs and experts have partially restored the wetlands. As a consequence of the breaching of embankments and more than sufficient amounts of rainfall after 2003, the marshlands were partially recovered (Garsteck and Amr 2011). With the help of UN organizations and countries such as Italy, Japan, the USA, and Canada, the government managed to restore parts of the wetlands (Shafy 2010). The aim was to restore the Marshes up to 75% of its original size, which would imply 1800 km^2 for Huweizah Marsh and for Hammar Marsh and 2425 km^2 for Central Marsh, with a required water quantity of 5495, 3263, and 4128 CM, respectively (Al-Ansari et al. 2012, p. 75). By 2008, the marshlands covered 4950 km^2 again, an increase of 58%. A year later, by July, its size was reduced again to 2313 km^2 due to lack of water supply (Al-Ansari et al. 2012, p. 73).

According to the United Nations Environment Programme (UNEP), the Marshes from 2003 expanded at a rate of 900 km^2 per year to 3000 km^2. By December 2005, the Marshes had expanded to 41% of its former size, with Huweizah and Hammar at 50% of their original size. By 2006, the marshlands were at 58% of their original size (Shafy 2010). However, due to a major drought in 2008–2009 and reduced flow from Iran, the marshland area shrunk again to 30% (Muir 2009), demonstrating that the recovery was not permanent and restoration would take additional time and efforts. By early 2010, the wetlands had partially recovered again (Becker 2014). In 2018, Iraq experienced water shortages across the country. The Mosul Dam

[7] *Lessons from the Destruction of Iraq's Marshes*, Washington Post, 17 August 2015, https://www.washingtonpost.com/gdpr-consent/?destination=%2fnews%2fmonkey-cage%2fwp%2f2015%2f08%2f17%2flessons-from-the-destruction-of-iraqs-marshes%2f%3f&utm_term=.24c8757af414

reservoir contained only 3 billion m3 compared with 8 billion in 2017 (Aldroubi 2018). The level of the Tigris River fell to alarmingly low levels, from 21 billion m^3 to 10 (Al-Samarai 2018). Such low water levels in the rivers that feed the Marshes can be disastrous for the water supply.

The uncoordinated and fragmented approach to re-flooding the Marshes had been disorganized and not completely successful. It became clear that it would not be possible to completely restore the Marshes again, due to the high salinity levels in the soil and polluted water (Richardson 2018). In addition, landowners are not willing to return certain areas of the marshlands which are used as agricultural areas, and oil extraction is taking place in about one-third of the former marsh areas (Garstecki and Amr 2011). Therefore, the Marshes are not expected to regain its former size.

Current initiatives for restoration focus on water regulation of the Marshes, in particular storing and preserving water from the Euphrates and Tigris Rivers upstream and flows downstream, through expansion of irrigation channels or diversion of wastewater. An example is the diversion of wastewater through the Main Outfall Drain into Hammar Marsh.[8]

Experts agree that the restoration of the remaining areas of the marshes requires action on different fronts. In the Strategy for Water and Land Resources in Iraq (SWLRI) from 2014, the Iraqi government identified the minimum flows for restoration of the Marshes.[9] These minimum flows were not met in 2017 and 2018.[10] To support the supply of water, the distribution should be continuously maintained.[11] In addition, the water quality of the Marshes requires mitigation measures supported by contaminant hydrology. Water pollution in the wetlands has been driven by hydrologic processes. Studies on the Euphrates River done since 1996 have found that the river's waters reach alarmingly high levels of salinity as the river progresses southward from Turkey (Bachmann et al. 2019, p. 4). Overpumping of wells has made the problem worse, and irrigation development projects have caused greater pollution and higher salinity from returning agricultural drainage water (Bachmann et al. 2019, p. 4). This drainage water should be prevented from entering the Marshes, while irrigation techniques should be modernized. In the case of the Tigris River, its salinity continues to increase as the river flows southward. The impact of

[8]*Managing Change in the Marshlands: Iraq's Critical Challenge*, United Nations White Paper: Report of the United Nations Integrated Water Task Force for Iraq, http://www.hydrology.nl/images/docs/ihp/2011.08_Marshlands_Iraq.pdf, 2011, 8.

[9]*UNESCO Should Address the Impacts of Upstream Dam Construction on the Ahwar of Iraq and Support Transboundary Water-Sharing Agreements*, Save the Tigris, 2 July 2019, https://www.savethetigris.org/unesco-should-address-the-impacts-of-upstream-dam-construction-on-the-ahwar-of-iraq-and-support-transboundary-water-sharing-agreements.

[10]*Second State of Conservation Report Addressed by the Republic of Iraq to the World Heritage Committee on the Ahwar of Southern Iraq: Refuge of Biodiversity and Relict Landscape of the Mesopotamian Cities*, World Heritage Property no.1481, February 2019, 5.

[11]*UNESCO Should Address the Impacts of Upstream Dam Construction on the Ahwar of Iraq and Support Transboundary Water-Sharing Agreements*, Save the Tigris, 2 July 2019, https://www.savethetigris.org/unesco-should-address-the-impacts-of-upstream-dam-construction-on-the-ahwar-of-iraq-and-support-transboundary-water-sharing-agreements.

decreasing water quality on the marshlands can inflict unpredictable changes to the hydrology of the marshlands.

Since the Marshes became a World Heritage Site in 2016, progress on the preservation of the wetlands has been monitored by UNESCO and its advisory body IUCN. The World Heritage Committee set forth a series of recommendations which the Iraqi authorities should implement, including securing sufficient amount of water flows to the marshlands.[12] This is hindered by the lack of an adequate management system within Iraq, and the issue is set to worsen in the future when Turkey starts operating Ilisu Dam on the Tigris River. Iran built an embankment on the Iraq-Iran border of Huweizah Marsh in 2009 which remains to this day. The Marshes also continue to remain vulnerable to oil and gas development. The Iraqi government has not made permanent commitment not to explore or exploit oil and gas within the marshlands or on the borders of the area.[13]

7.4 Impact of Dam Construction on the Hydrology of the Marshes

The Marshes depend on upstream water supplies which are necessary to preserve the wetlands. Approximately 70% of the waters of the Tigris and Euphrates Rivers are supported by the headwaters in Turkey, Syria, and Iran (Fawzi et al. 2016). Over the past decades, water flows have been diverted or diminished to store water upstream. Historically, the Marshes depended on flood flows in the spring for soil fertility and elimination of salt in the surface of the Marshes. Upstream alteration has diminished the seasonal flooding which drove the dynamics of the Marshes. Water supplies were diverted during the big drainage of the 1980s and 1990s in which diversions and embankments desiccated the marshlands. Dam construction upstream of the Marshes started from the 1970s on the Euphrates River in neighboring Syria (Tabqa Dam) and later on Turkey. This was followed by the damming of the Tigris River in recent decades, in particular Ilisu Dam. Construction of this dam started in 2006 and was finished in 2018 (Warner 2012, p. 231). Turkey has started filling its reservoir in 2019, which is expected to have the size of 10.4 billion m^3 (Warner 2012, p. 231). The Ilisu Dam is part of the Southeastern Anatolian Project (GAP), which includes further dams to be constructed, such as Cizre. The complete GAP project consists of 22 dams.[14]

[12] *Decision: 42 COM 7B.66 The Ahwar of Southern Iraq: Refuge of Biodiversity and the Relict Landscape of the Mesopotamian Cities (Iraq) (C/N 1481),* UNESCO World Heritage Committee, 4 July 2018, https://whc.unesco.org/en/decisions/7295.

[13] *Decision: 42 COM 7B.66 The Ahwar of Southern Iraq: Refuge of Biodiversity and the Relict Landscape of the Mesopotamian Cities (Iraq) (C/N 1481),* UNESCO World Heritage Committee, 4 July 2018, https://whc.unesco.org/en/decisions/7295.

[14] *Ilisu Dam Turkey*, Banktrack, 1 November 2015, https://www.banktrack.org/project/ilisu_dam.

In Iran, from 2001, the Karkheh Dam on the Karkheh River and other water infrastructure on the Karun and Karkheh Rivers have reduced water flowing into Huweizah Marsh. Before the construction of an embankment on the Iraq-Iran border by the Iranians in 2009, Huweizah Marsh and Hoor al-Azim formed one Marsh. Ideally, the embankment should be taken down in order to restore the Marsh. Huweizah used to receive part of its water from the Karkheh River. Additional hydrological linkages between the two parts still need to be explored. The Huweizah Marsh was registered as a site "where changes in ecological character have occurred, are occurring, or are likely to occur as a result of technological developments, pollution or other human interference" in April 2010 on the Montreux Record of wetland sites of the List of Ramsar Wetlands (Hashim et al, 2019, p. 199), a year after the construction of the embankment. Yet Huweizah Marsh remains the marshland area that is best preserved and therefore one of the most important parts of the Marshes as a whole. Experts have recommended removing the Iranian-built embankment, or the construction of culverts (Garstecki and Amr 2011, p. 99). The Marshes as a whole, both parts, require joint management, or at least joint coordination. To this aim, dialogue has been facilitated by the Ramsar Convention, without any tangible results so far. IUCN has recommended supporting an agreement between Iraq and Iran on joint management of Huweizah Marsh and Hoor al-Azim, without the formal recognition of the area being transboundary (Garstecki and Amr 2011, p. 124).

Within Iraq water infrastructure on the Euphrates in Ramadi and Samarra diverted peak floods into depressions and established reservoirs. This altered the hydroperiod downstream. Most recently, the SWLRI states unambiguously that "No new large dams are required to assist Iraq in achieving its 2035 objectives. [. . .] Therefore, due to a reduction of the incoming water to Iraq, revolutionary changes in reservoir operation, and improved efficiency in all water sectors, Iraq does not need any new large dams, such as those proposed at Bekhma, Mandawa or Fatha."[15] The SWLRI recommends no further dams should be built until agreements have been reached with upstream states on downstream flows.[16] Yet the Kurdistan Regional Government (KRG) in its most recent master plan envisions 18 dams being built in the Kurdistan Region, and the KRG has recently announced the construction of three dams in the Greater Zab basin.[17] The long-term future of the Marshes is threatened by lack of coherence in Iraq's water strategy, particularly in relation to the impact of dams on current and future flows to the Marshes. Such policy incoherence must be addressed by the Iraqi authorities in order to meet minimum flow requirements to the Iraqi Marshes.

---

[15] *UNESCO Should Address the Impacts of Upstream Dam Construction on the Ahwar of Iraq and Support Transboundary Water-Sharing Agreements*, Save the Tigris, 2 July 2019, https://www.savethetigris.org/unesco-should-address-the-impacts-of-upstream-dam-construction-on-the-ahwar-of-iraq-and-support-transboundary-water-sharing-agreements.

[16] Ibidem.

[17] Ibidem.

The effects of all these hydraulic structures are large. Dams in riparian countries will allow Iraq's neighbors to control the amount of water flowing downstream the Tigris and Euphrates to the marshlands. As such, Iraq would become dependent on its upstream neighbors to release sufficient amounts of water to sustain the wetlands in Southern Iraq, as seasonal flood periods of the rivers are eliminated. Dams affect the quality, quantity, and timing of the water flows to the Marshes. It is impossible to predict the extent of the impact of dams, but one can predict the type of changes it can inflict on the water that reaches the Marshes. In case of successive dams, the effects on the hydrology are cumulated by the time the marshlands are reached. Taking into account that there are no fixed bases for accurate and specific prediction, current and future dams on the Tigris and Euphrates Rivers (will) certainly have a big impact on the Marshes. Upstream dams impact downstream use. As less surface water is available, populations would have to deplete groundwater reserves, hereby intensifying saltwater intrusion from the Shatt al-Arab and polluting freshwater. The city of Basra already suffers from alarmingly high levels of salinity.[18] Dams discontinue the upper parts of the rivers with its lower parts. The amount of water discharged by dams is often not sufficient for species, such as migratory fish, to migrate. Settling of sediments upstream will alter the biologic system downstream of the dam as a result of lack of nutrients. The process of photosynthesis and the subsequent release of oxygen are obstructed by dams, hereby enhancing the depletion of oxygen in the water discharged from dams, impacting the river flows. This oxygen deficit issue can be mitigated as water is released through the turbines, but it is unlikely to be sufficient because the time water passes through the turbines is limited. Aquatic organisms in the river downstream of the dam are affected by rising salinity levels, caused by evaporation upstream of the dam. The water downstream of dams is of different temperature, different from the natural river temperature. The thermal system of the river changes downstream of each dam, affecting aquatic organisms. The difference in water temperature upstream and downstream of a dam causes further oxygen deficits in the water that reaches the Marshes. River water that suffers from a lack of oxygen will not be able to completely purify itself and dilute pollutants that are collected as the Tigris and Euphrates pass through some major cities before they reach the Marshes. By that time, the rivers could be contaminated by waste.

Basin-wide transboundary water-sharing agreements between riparian states of Iraq, Iran, Turkey, and Syria are a necessity to safeguard the future of water supply and water quality to the Marshes. Currently, there is not enough water available to meet the restoration goals of the Marshes, let alone when dam construction upstream intensifies. Upstream dam construction is now a permanent threat to the hydrology of the marshlands. The Iraqi government maintains a dialogue with Turkey, but is weak as a downstream country. Joint management of the Tigris-Euphrates basin has

---

[18] *Fact-Finding Team Report on the Humanitarian Situation in Basra*, Save the Tigris, 8 November 2018, https://www.savethetigris.org/fact-finding-team-report-on-the-humanitarian situation-in-basra.

not been agreed upon by the riparian countries due to the complexity of the political context, political changes, and the complexity of the hydrologic regime (Fawzi et al. 2016).

7.5 Future Scenario for the Marshes Hydrology Within the Framework of Existing and Planned Hydraulic Structures in the Tigris-Euphrates Basin

Initially, the desiccation of the Marshes occurred in partial due to the use of upstream water for consumption and irrigation. Water shortages were compensated for by the use of water from reservoirs and other ways to regulate water. The hydrological structure of the marshlands was further altered by the drainage inflicted by the Saddam regime, which closed the major water inputs of the Marshes and constructed embankments. Since then, shortage of water supplies has been the main hindrance to the restoration process of the Marshes. In 2008, the Iraqi Ministry of Water Resources reported that 43% of the former Marsh area had been restored (Al-Maroofi 2015, p. 29), while the remaining area was used for agriculture and oil exploitation. The decrease of the marshlands size following the 2008–2009 drought demonstrated that water supply is challenged and the current situation for the Tigris and Euphrates Rivers, lifelines of the Marshes, is not sustainable.

To what extent then can the hydrological regime of Iraqi Marshes be restored? This depends on the water quantity and quality that feed the marshlands. In recent years, water shortages have continued, exacerbated by climate change, dam construction, and uncoordinated management of the Tigris and Euphrates Rivers. Salinization is still a real threat, obstructing the full recovery of the wetlands (Al-Maroofi 2015, p. 32). These two factors should be included in any serious recovery plan for the Iraqi Marshes.

As per SWLRI, which is adopted by the Iraqi government, Iraq should abstain from any new dam construction, including in the Kurdistan Region, until a Regional Environmental Assessment (REA) has been undertaken with the assistance of UNESCO and its advisory bodies. The REA should be basin-wide and include existing as well as proposed dams on the Tigris-Euphrates Rivers and their tributaries, with a specific mandate of assessing future impact flows on the marshlands. The REA should be conducted in compliance with the IUCN Guidance for World Heritage Environmental Assessments. Note that the Iraqi government has not published its SWLRI; it is therefore not available for public comment. At the same time, to safeguard water supplies to the Marshes, the Iraqi government should intensify its efforts to establish long-term water-sharing agreements with Turkey and Iraq. A dialogue has been ongoing between Turkey and Iraq, and in 2014, the "Memorandum of Understanding in the Field of Water between the Ministry of Forestry and Water Affairs of the Republic of Turkey and the Ministry of Water Resources of the Republic of Iraq" was agreed upon, which is still to be ratified by both parliaments.[19] This memorandum commits both Iraq and Turkey to equitable shares of water, but it does not specifically address minimum water flows to the Marshes (Dawood 2019), despite Iraq and Turkey being parties to the World Heritage Convention. Such minimum water flows need to be agreed upon in a binding legal agreement.

The re-flooded marshlands are showing different speeds and levels of recovery. The Marshes that are located close to water inputs have shown positive signs of recovery. Yet the flooding process remains poorly managed, and some of the marshlands receive poor water quality, while others do not. Initially, before the SWLRI, the Iraqi Ministry of Water Resources developed a Water Management Plan for the Marshes based on incorrect theoretical assumptions (Al-Maroofi 2015). First the plan assumed that 200 m^3/s would be the received water inflow from Turkey, incorrectly. Second, it was assumed that Huweizah Marsh would continue to receive water input from Iran, but since the construction of the embankment, inflow has ceased. Third, the plan incorrectly expected extra water input from Iran to maintain the water level of the Shatt al-Arab. The plan also did not take into account upstream dam construction or climate change and the use of water for other consumption and industries. The minimum flows of water required to sustain the Marshes were established only once the Marshes were inscribed in the UNESCO World Heritage List and have not been met yet.[20] Meanwhile, to sustain the marshlands, authorities used polluted water discharge for supplies, through the Main Outlet Drain whose water had been used for agriculture and contains high salinity levels and is thus of poor quality (Al-Maroofi 2015, p. 30). In addition to agriculture, oil exploitation and military activities are other causes of desiccation which lead to increased salinity. This is now a major problem in the marshlands, due to the accumulation of salt on the surface of the dried sediment as a consequence of evaporation (Al-Maroofi 2015, p. 30). Salinization caused by changes in the hydrological regime damages the ecology of the marshlands.

Re-flooding alone has not restored the hydrology of the marshlands. The level of input from the Euphrates and Tigris Rivers continues to reduce, causing further decrease in the seasonal periodicity of the rivers' discharge into the Marshes. The flushing effect of seasonal floods has diminished which has left an impact on the sediment of the marshlands. Less sediment reaches the Marshes, due to sedimentation upstream as a consequence of dam construction. Due to reduced water levels, different areas of the Marshes remain hydrologically disconnected and fragmented, compared to earlier levels (Garstecki and Amr 2011).

---

[19] *UNESCO Should Address the Impacts of Upstream Dam Construction on the Ahwar of Iraq and Support Transboundary Water-Sharing Agreements*, Save the Tigris, 2 July 2019, https://www.savethetigris.org/unesco-should-address-the-impacts-of-upstream-dam-construction-on-the-ahwar-of-iraq-and-support-transboundary-water-sharing-agreements.

[20] *Second State of Conservation Report Addressed by the Republic of Iraq to the World Heritage Committee on the Ahwar of Southern Iraq: Refuge of Biodiversity and Relict Landscape of the Mesopotamian Cities,* World Heritage Property no.1481, February 2019, 5.

Taking into account hydrological constraints in Iraq, experts agree that conservation efforts should focus on the Marshes that are closest to water inputs, since these have the most sustainable ecological foundations (Al-Maroofi 2015, p. 137). Huweizah Marsh contains parts of the marshlands that have recovered best, and recovery plans focusing on such areas will likely lead to best results. The amount of nutrients that enters the hydrologic system of the Marshes should be controlled through the treatment of agricultural discharge, in order to reduce the impact of nitrification (Al-Maroofi 2015, p. 137). Salinity levels and nutrients can be identified through the monitoring of phytoplankton assemblage.

Minimum water flows needed to sustain the marshlands have been identified by the Iraqi government. For the future, it should consider the impacts of reduced water flows from Turkey into the marshlands and Iran into Huweizah Marsh, reduced water levels of the Shatt al-Arab, the impacts of climate change on water resources, and the impacts of oil exploitation. Further comprehensive data on water discharge and quality is needed, especially within Iraq. A management plan would have to find suitable alternative water supplies, especially with regard to discharge which is of low quality (Main Outfall Drain in Hammar Marsh), or establish suitable treatment so that salinity levels are reduced before it enters the marshland area (Al-Maroofi 2015, p. 137). To better determine the quality, quantity, and hydroperiod of the hydrological regime of the Marshes, further research is needed. Technology could facilitate hydrological conditions in the Marshes, provided it is based on reliable water allocations and target environmental variables.

References

Al-Ansari N, Knutsson S, Ammar A (2012) Restoring the Garden of Eden. J Earth Sci Geotechnical Eng 2(1): 53–88. International Scientific Press, p 74

Aldroubi M (2018) Turkey dam sparks renewed water shortage panic in Iraq. The National, 3 June 2018. https://www.thenational.ae/world/mena/turkey-dam-sparks-renewed-water-shortage-panic-in-iraq-1.736439

Al-Maroofi SS (2015) Ecological assessment of re-flooded Mesopotamian Marshes (Iraq), PhD thesis. University of Waterloo. https://pdfs.semanticscholar.org/1f72/fed49ead6064f33de9cc78750313dafd564b.pdf, p 29

Al-Samarai M (2018) Ilisu Dam crisis proves the failure of Iraq's successive governments. The Arab Weekly, 9 June 2018. https://thearabweekly.com/ilisu-dam-crisis-proves-failure-iraqs-successive-governments

Bachmann A, Tice V, Al-Obeidi LA, Dicle TK (2019) Tigris-Euphrates river ecosystem: a status report. Mesopotamian Water Forum 2019, March 2019. https://www.savethetigris.org/wp-content/uploads/2019/03/Paper-Challenge-C-Ecosystem-FINAL-to-be-published-1.pdf, p 4

Becker R (2014) The stalled recovery of the Iraqi Marshes. Remote Sensing 6, 30 January 2014. https://esajournals.onlinelibrary.wiley.com/doi/full/10.1002/ehs2.1207, p 1260

Bijnens T (2017) The border wetlands of Iran-Iraq: an environmental crisis with regional consequences. Iraqi Civil Society Solidarity Initiative, 2 July 2017. https://www.iraqicivilsociety.org/archives/7330

Dawood I (2019) Does the Memorandum of understanding in water management between Turkey and Iraq guarantee Iraq its water rights? Save the Tigris, 18 April 2019. https://www.

savethetigris.org/does-the-memorandum-of-understanding-in-water-management-between-turkey-and-iraq-guarantee-iraq-its-water-rights
Fawzi N, Goodwin KP, Stevens ML, Medhi B (2016) Effects of Mesopotamian Marsh (Iraq) desiccation on the cultural knowledge and livelihood of Marsh Arab women. Ecosyst Health Sustain 2(3):1
Garstecki T, Amr Z (2011) Biodiversity and ecosystem management in the Iraqi Marshes – screening study on potential world heritage nomination. IUCN Regional Office for Western Asia, p 25
Hashim SJ, Borisovna VK, Schamilevich DR (2019) The International Legal Status of Marshes of Iraq as Protected Natural Areas. Academic Journal of Interdisciplinary Studies 8(4). Academic Journal of Interdisciplinary Studies, p 199
Muir J (2009) Iraq's Marshes face grave new threat. BBC, 24 January 2009. http://news.bbc.co.uk/2/hi/middle_east/7906512.stm
Partow H (2001) The Mesopotamian Marshlands: demise of an ecosystem: early warning and assessment, Technical Report UNEP, p 7
Pearce F (1993) Draining life from Iraq's marshes: Saddam Hussein is using an old idea to force the Marsh Arabs from their home. New Scientist, 17 April 1993. https://www.newscientist.com/article/mg13818691-800-draining-life-from-iraqs-marshes-saddam-hussein-is-using-an-old-idea-to-force-the-marsh-arabs-from-their-home
Richardson CJ (2018) Mesopotamian Marshes of Iraq. In: Finlayson C, Milton G, Prentice R, Davidson N (eds) The wetland book. Springer, Dordrecht, p 1685
Schwartzstein P (2015) Iraq's famed Marshes are disappearing – again, National Geographic, 9 July 2015. https://news.nationalgeographic.com/2015/07/150709-iraq-marsh-arabs-middle-east-water-environment-world
Shafy S (2010) Iraq's Garden of Eden: restoring the paradise that Saddam destroyed. Der Spiegel, 3 August 2010. https://www.spiegel.de/international/world/iraq-s-garden-of-eden-restoring-the-paradise-that-saddam-destroyed-a-709866.html
Warner J (2012) The struggle over Turkey's Ilısu Dam: domestic and international security linkages. Int Environ Agreements Politics Law Econ 12(3):231
Wayman E (2009) Lack of water threatens "Garden of Eden". Earth Magazine, 15 April 2009. https://www.earthmagazine.org/article/lack-water-threatens-garden-eden
Wilson R (2012) Water-shortage crisis escalating in the Tigris-Euphrates Basin. Future Directions International, 28 August 2012. http://www.futuredirections.org.au/publication/water-shortage-crisis-escalating-in-the-tigris-euphrates-basin

Chapter 8
Importance of Hydrological and Hydrogeological Studies in Wetlands: Examples from Turkey

Melis Somay-Altas

Abstract Providing the necessary water and nutrients for human beings to survive, and having a wide variety of plant and animal species within its body, has always made wetlands important throughout human history. Wetlands are not ordinary water bodies. It is a common life union that includes the smallest microorganism, the fish and birds that form the top link of the food chain, and the human communities that benefit from the system and live in dependence on it. Hydrogeological and hydrogeochemical studies help us to determine the type of water in wetlands, recharge–discharge limits, and wetland water storage capacities. Interpretation of wetlands by hydrogeological tools begins with determining the hydrological basin of wetland. Other important steps are determining the classification, hydrological components, water budget, detailed geological map, and hydrogeological properties of rocks and recharge–discharge relationship with water table maps, analyzing hydrogeochemical tools, and building a hydrogeological conceptual model. If we evaluate wetlands hydrogeologically in Turkey, besides drought, excessive pumping of groundwater for agricultural, industrial, and urbanization purposes has caused large drawdowns of the water table in and around wetlands. Apart from these, groundwater and surface water pollution are one of the biggest problems in the wetland's hydrological basins due to agricultural and industrial facilities.

Keywords Hydrology · Wetlands · Turkey · Nutrients · Geology · Rocks

I want to dedicate this chapter to the memory of my advisor: Prof. Dr. Sevki FILIZ.

M. Somay-Altas (✉)
Geological Engineering Department, Dokuz Eylul University, Buca, Izmir, Turkey
e-mail: melis.somay@deu.edu.tr

L. A. Jawad (ed.), *Southern Iraq's Marshes*, Coastal Research Library 36,
https://doi.org/10.1007/978-3-030-66238-7_8

8.1 Introduction

Over the years, wetlands have been an important heritage for civilizations and have played a role in establishing much larger civilizations where water is available. Only 2.5% of the water on earth is fresh water (Shiklomanov 1993). Thirty percent of this fresh water is groundwater and 1% is surface water. The remaining amount is glaciers and ice caps that cannot be used now. Freshwater areas are lakes, rivers, marshes, and groundwater associated with these surface waters. In this context, "wetlands" are of great importance in terms of the supply of fresh water necessary for the future. According to FAO (2011), approximately 3900 km^3/year of 42,000 km^3/year of global renewable water resources obtained from rivers and aquifers are for human use.

Turkey, in terms of wetlands, is one of the region's most important countries due to the reasons mentioned below:

1. Two of the four bird migration routes in the Western Palearctic Region pass through Anatolia.
2. Turkey is situated on the transition point between the continents of Europe, Asia, and Africa.
3. The effect of plant migration that occurred in the glacial period continued in the interglacial period and increased biological diversity in Turkey. This biodiversity is caused by factors related to latitude, latitude, degree of altitude, elevation, location, topography, geological structure, and soil cover differences.
4. There are altitude differences exceeding 5150 m above sea level.
5. Three sides are surrounded by seas of different ecological character.
6. There are climate variations that emerge because of these features.

Until the 1970s, enough importance was not given to wetlands in Turkey. These areas were dried to combat malaria, to open agricultural areas, and to prevent floods. Turkey has been a party to the Ramsar wetlands of international convention in 1994 and began to be held the same year national legislation. In 1999, national wetland planning was initiated by the Ministry of Forestry and Water Affairs. One of the biggest tools for the protection and rational use of wetlands is strategy. In Turkey, the first strategy for wetlands was drafted in 2002 and implemented in 2003. Today, there are a total of 135 identified wetlands in Turkey. Fourteen of these wetlands are Ramsar sites. Many of them are at risk of drying out.

The purpose of this study is to show the hydrogeological work steps and to compile the hydrogeological study be done in the wetlands of Turkey.

8.2 Hydrological Cycle and Hydrogeological Benefits of Wetlands

Wetlands are the world's most important genetic reservoirs. It is estimated that it contains approximately 40% of all species in the world and approximately 15% of all animal species.

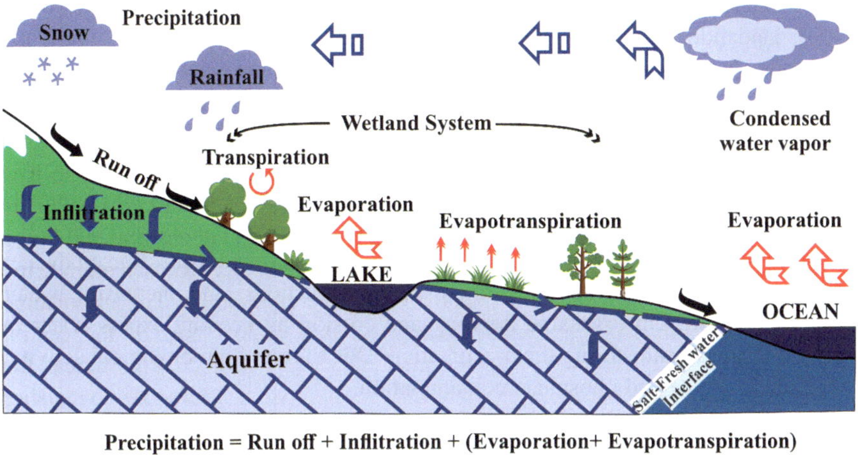

Fig. 8.1 Hydrological water cycle. (Courtesy of Melis Somay-Altas)

Water is the primarily needed factor, which provides nutrients for both animal and plant species in the wetlands (Somay et al. 2008). Based on wetland definition, it is seen that wetlands are of great importance in the hydrological water cycle.

In the hydrological cycle, the sun that activates the water cycle heats the water in the oceans, and the warmed water evaporates into the atmosphere. Rising air currents carry the water vapor up in the atmosphere, and the cooler air there causes condensation in the clouds (Fig. 8.1).

Cloud particles come together, and they grow and fall from the sky as precipitation. Some precipitation returns to the earth as snow. Most of the precipitation falls on the oceans or soil and flows as a runoff under the influence of gravity. Part of the stream mixes into the rivers in the valleys, and from there, it moves to the oceans through the rivers. Surface flows and springs are collected in lakes and rivers as fresh water. Not all runoffs reach the rivers. Most of the flow infiltrates underground. Some groundwater discharges as freshwater sources from the openings they find on the ground surface. Shallow groundwater is taken up by plant roots and returns to the atmosphere by transpiration from the leaf surface. Evaporation and transpiration events occur together, and it is not possible to separate these two events. So, this event is called as "evapotranspiration." Some of the water that leaks underground goes deeper and recharges aquifers that can store large amounts of fresh water for a very long time. Over time, this water also moves, and some of it mixes into the oceans where the water cycle begins and ends. When all these processes are taken into consideration, it is observed that the wetlands have a very important place in providing fresh water, filtering the water that can leak dirty from the surface, in the groundwater recharge–discharge process.

Wetlands are themselves water tanks. Also, these are recharging the groundwater that we use as drinking and irrigation water. The United Nations Water (UN-Water) estimates that 90% of natural disasters are related to water. If a society has the power to protect itself from natural hazards like water surplus and/or water deficiency (e.g., floods, landslides, drought etc.), this society has also "water security". Wetlands protect us from floods, droughts, and other natural disasters, as well as provide us with fresh water. Wetlands are normally found in low-energy environments – that is, in areas where water normally flows with a slow velocity. This results, in part, because the land surface is relatively flat in these areas (Orme 1990). Because wetlands lie in relatively flat landscapes, their surface area expands and contracts as the water stage changes, allowing for the storage of large volumes of water (EPA 2008). It is of great hydrogeological importance especially in terms of groundwater recharge and discharge, surface–groundwater interactions, and increasing water quality. They prevent excessive flooding and erosion by keeping excess water in excessive rainfall and storing the precipitation. Due to their high biological activity, they reduce nitrate and phosphate contamination.

Wetlands contribute positively to the local climate by making the area where they are more humid and preventing seawater from entering the inner regions, thus salting the soil. They also provide millions of people with food and livelihoods, support rich biodiversity, and store more carbon than other ecosystems. They help maintain the level of water required for agricultural activities. Deltas and river mouths have a very intense biological activity since they are constantly fed with mineral substances carried by streams.

Since the quality of water affects all wetland characteristics, wetlands should be investigated not only for surface water but also for groundwater and their relations with each other. The quality of water in wetlands depends on lithology, distance from the sea, industrialization, agricultural activities, and urbanization. Generally, lithology controls the storage capacity, porosity, and permeability of rocks and major and minor ion content. In some special cases, lithology causes pollution in and around mining areas. Thus, hydrogeological and hydrogeochemical investigations help us to identify the recharge–discharge areas, wetland water type, and water–rock interactions in the wetlands (Somay and Filiz 2005).

The value of wetlands is largely unknown among politicians and decision-makers around the world.

There are three important elements in wetland studies:

1. Presence of water: The existence, quantity, quality, and potential of water are the most important factors that directly affect wildlife and vegetation in wetlands.
2. Wetland vegetation: They are communities of water-resistant plants in flooded and water-saturated regions called hydrophytic.
3. Wetland soil: These are dark gray-colored soil groups that are exposed to water for an average of 21 days or 1 month, indicating an aerobic environment condition.

In this regard, hydrogeology, hydrogeochemistry, and water–rock relationship gain importance in the groundwater recharging of wetlands (mixing of surface waters from ground and rocks into the groundwater). These issues help determine the type of water in wetlands, recharge–discharge limits, and wetland water storage capacities.

8.3 Hydrological and Hydrogeological Work Methods in Wetlands

In Turkey, especially coastal wetland areas are the discharge points of the groundwater. This indicates that the recharging of wetlands in arid–semi-humid areas is from groundwater. It is precisely for this purpose to examine the hydrogeology of wetlands, which will enable the determination of freshwater potentials in arid–semi-arid areas, especially in terms of global climate change.

The International Union for Conservation of Nature (IUCN) reported that more than 50% of wetlands in the Mediterranean Region are dry. As mentioned in the section above, hydrology section, Turkey is also under threat of drought in the western and southeast Anatolia. This includes the threat that wetlands in these areas will dry out in the future.

Hydrogeological and hydrogeological studies were researched and evaluated on a regional scale in the study areas. Hydrogeological/hydrogeochemical studies planned to be carried out in a wetland are given in the flowchart in Fig. 8.2.

When working in a wetland, the *hydrological basin*, which determines the recharging area of the wetland superficially, is determined with the help of topographic maps (Fig. 8.3).

Changes in the hydrological structure will primarily change the physical and chemical properties, amount, quality, level, etc. of the waters in these areas and will threaten the existence of wetlands and directly and/or indirectly affect its benefits. A water resource management carried out within the boundaries of the basin also reveals the natural water resources and living relationships. One of the purposes here is to determine the recharge area and the physical and chemical properties of the waters coming into the wetland. In this way, it is foreseen how the enterprises (such as landfill, wastewater treatment plant, industrial zone) planned to be built in this area in the future can change the chemistry of the water. The fact that the hydrological cycle is interconnected means that changes occurring at a distance from the wetland can negatively affect the wetland (Meriç et al. 2010).

Determination of wetland types is the second step for a hydrogeological study in wetlands. There are several classification types all around the world. In Turkey, especially Ramsar classification is used to identify wetlands. In Fig. 8.4, wetland types were identified both by Ramsar (2004) and Cowardin (1979) classifications for Küçük Menderes coastal wetland in the study of Somay et al. (2008).

Determining hydrological components and climate types is the next step. Various climate types in Turkey was interpreted by using the Thornthwaite method. In Thornthwaite climate classification, monthly average temperature, monthly average precipitation, and monthly evapotranspiration values are used to determine the climate of a place. With this method, moisture values, accumulated water in the

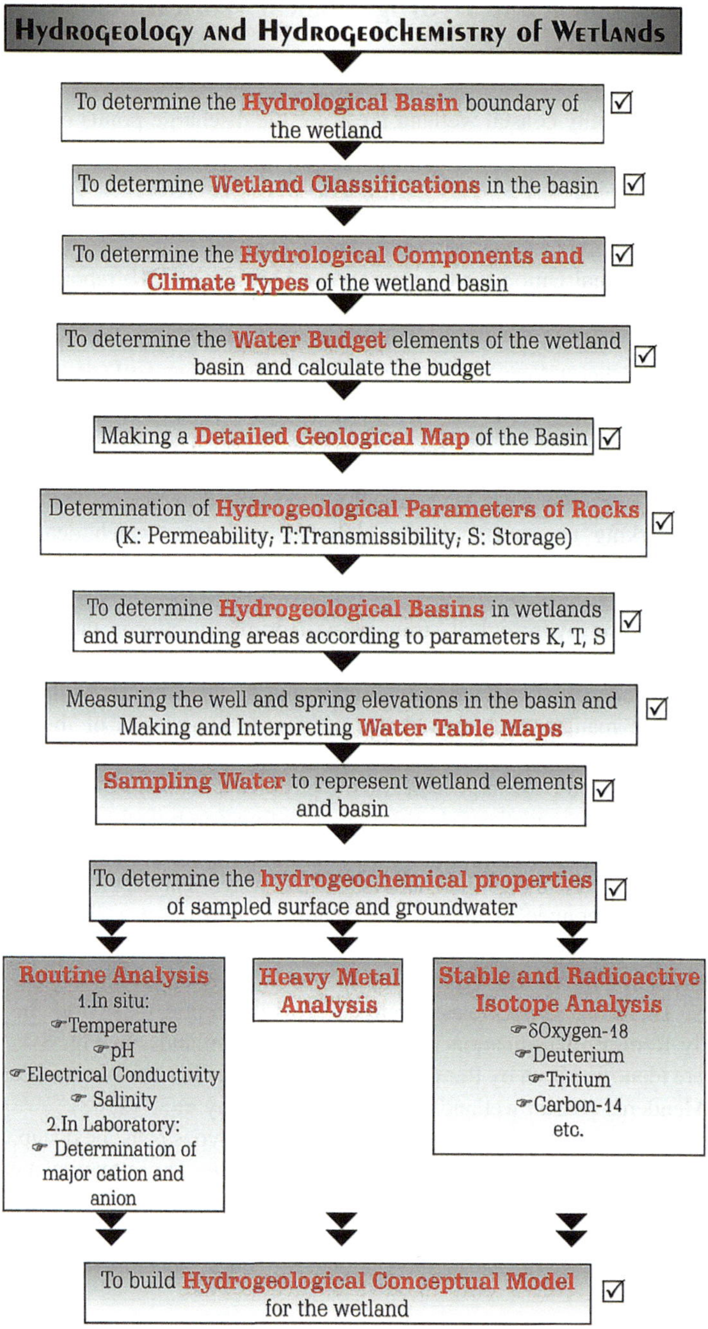

Fig. 8.2 Flowchart for the hydrogeological studies in wetlands. (Courtesy of Melis Somay-Altas)

Fig. 8.3 An example of hydrological basin from a coastal wetland–Küçük Menderes coastal wetland (Izmir). (Courtesy of Melis Somay-Altas)

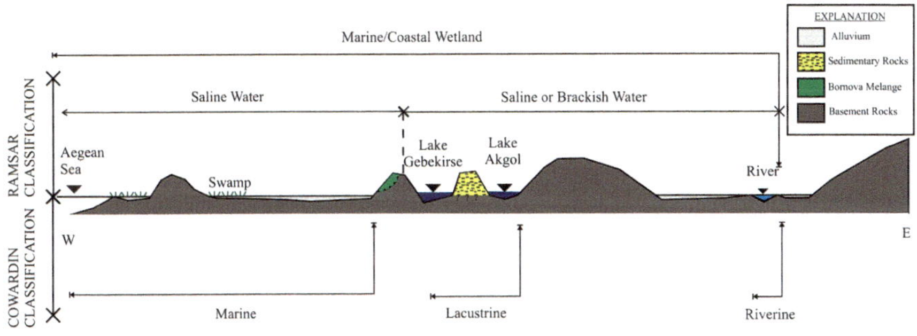

Fig. 8.4 Wetland classifications for Küçük Menderes coastal wetland in its hydrological basin. (Courtesy of Somay et al. 2008)

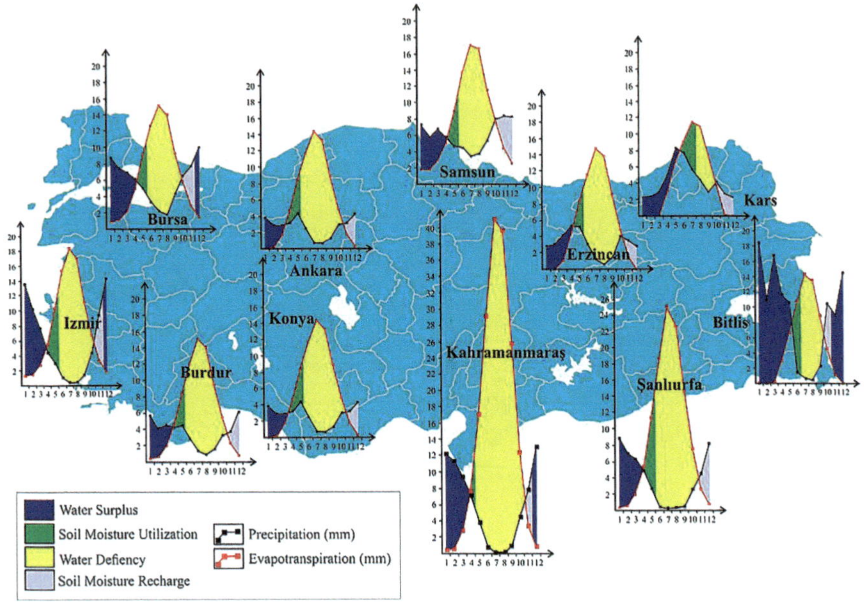

Fig. 8.5 Thornthwaite evapotranspiration–rainfall diagram for Ramsar regions in Turkey. (Courtesy of Melis Somay-Altas)

soil during the year, monthly change of accumulated water, annual real evapotranspiration value, water deficiency, and excess water and flow are calculated. According to Thornthwaite method, where precipitation is more than evaporation, the soil is saturated, and there is excess water in these places. So, the climate of this place is *humid*. On the contrary, where precipitation is less than evaporation, water does not accumulate in the soil and the soil cannot supply the water that plants need. In such places, there is a water deficiency. So, the climate of this place is *dry*. Thornthwaite diagrams of the Ramsar sites in Turkey were given in Fig. 8.5.

Turkey has 26% humid, 31% semi-humid, 33% dry–semi-humid and 8% of semidry regions according to Yilmaz and Cicek (2016) by using the Thornthwaite method. Areas that attract severe water deficiencies in humid areas in summer are seen both on the coasts and mountainous areas of the Mediterranean Region (Burdur), wide plains of the Aegean (Izmir), the Eastern Anatolia Region (Erzincan), and Marmara regions (Bursa). The areas with moderate water deficiency in humid areas are seen in the Black Sea (Samsun) Region. The areas with no surplus water defined as drought are between Ceylanpınar and Harran in the Southeastern Anatolia Region; between Konya, Ereğli, and Upper Sakarya plains in the Central Anatolia Region; and in the Black Sea Region along the Kızılırmak Valley, in Afyon in the Aegean Region, and in Iğdır and Doğubeyazıt plains in Eastern Anatolia (Yilmaz and Cicek 2016).

For example, in Izmir Bird Paradise (Gediz Delta), which is one of the Ramsar sites of Turkey, the climate is dry–subhumid, with a third mesothermal season of deficiency and with a large seasonal surplus and a temperature-efficiency regime third mesothermal (C1B13S2W2b13) according to the Thornthwaite classification system (Somay and Filiz 2003). From January to March, water surplus from rainfall recharges the surface and groundwater. Between June and September, there is water deficiency in the wetland. After November, soil moisture is recharged by water surplus.

As to the next step, "water budget" calculations are made in wetlands. In water budget calculations, conservation of mass is the continuity equation that states water neither disappears nor exists from the system. In this context, we need to determine the water budget elements of the wetland in the field in detail in order to be able to calculate the changes in the reservoir of the system + entering the system + leaving the system with the most accurate approach. Changes in the hydrological structure will primarily change the amount, quality, level, physical and chemical properties, etc. of the waters in these areas and will threaten the existence of wetlands and affect their benefits (functions, products, and qualities) directly and/or indirectly (Aydın et al. 2015).

For the further step, it is very important to make a detailed *geology/tectonism map* of the wetland. In this way, we provide the necessary data especially on the recharge–discharge relationships of groundwater, its origin, and relations with surface waters. A geological/tectonism map example was given from coastal wetlands in western Turkey (Fig. 8.6).

Determining the hydrogeological properties (permeability, conductivity, storage coefficient) of rocks in the hydrological recharge area of the wetland will make us knowledgeable about the aquifer and impermeable rocks in the basin (Fig. 8.6). In western Turkey, the basement rocks are formed by Menderes Massif Metamorphics. The permeability of the marble unit ranges from $2.16*10^{-5}$ to $4.87*10^{-2}$ m/s with an average of $1.53*10^{-2}$ m/s (Yazicigil et al. 2000). The marbles of Paleozoic could eventually constitute important aquifer (Somay et al. 2008). Lycian nappe complex which overlies the autochthonous Menderes Massif consists from the base to the top of meta-sediments, a melange unit and an ophiolitic sequence (Rimmele et al. 2003). Lycian nappes have very effective and widespread karstic features due to the distancing from long distances and being affected by epeirogenic movements. Transmissibility (T) of this unit was calculated to be 395 m^3/day/m (Önhon and Nazik 1998). Especially in the northern coastal area, Menderes Massif Metamorphics is overlaid by Bornova Flysch Zone. This melange consists of matrix of flysch-like sedimentary rocks and splits, separated by strongly sheared serpentinite slices in and around the study area (Güngör and Erdoğan 2002). Only limestone, sandstone levels, and fractured shales are aquifers in the area. Neogene sedimentary rocks that contain limestone–claystone–clayey limestone alternately overlay the Bornova Flysch Zone with an unconformity. Limestone, which has relatively high permeability due to its porous and fractured structure, is a good aquifer for water (Somay and Filiz 2003). Neogene volcanic units that consist of tuff, agglomerates, and andesites have a low primary porosity. But tectonic activity has resulted in secondary porosity, which gives rise to a significant increase in permeability (Somay and Filiz 2003). Quaternary alluvium in the coastal areas is a very important aquifer and irrigation area. The permeability and porosity of the alluvium are very high. Especially in the western part of Turkey, the urbanization on the alluvium deposits is the major problem for groundwater.

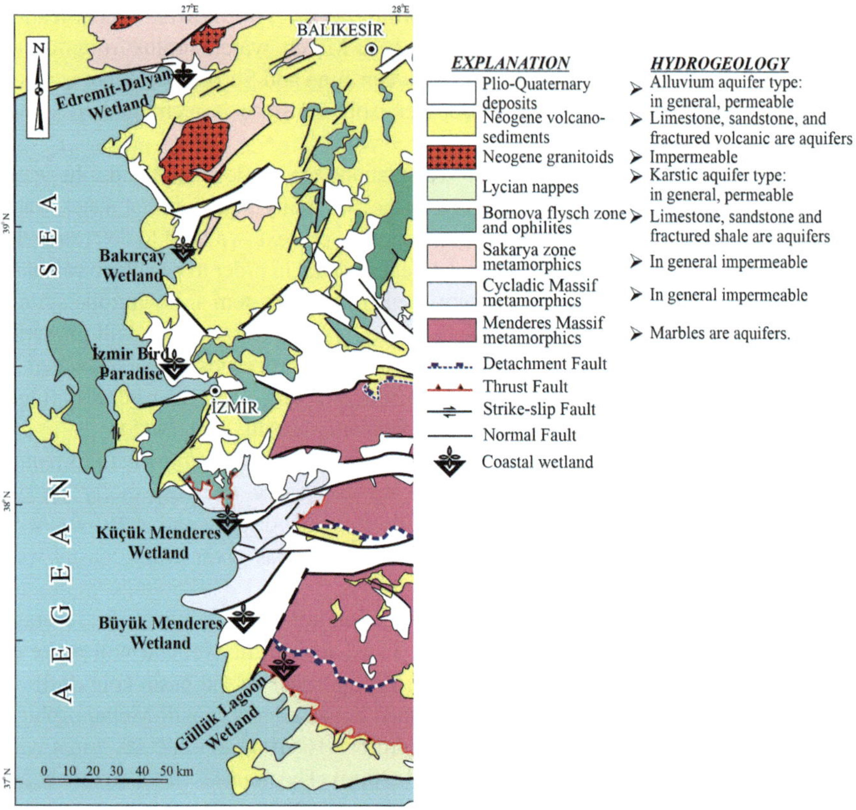

Fig. 8.6 Geological map of the western Anatolian coastal wetlands. (Modified from Ersoy et al. 2011)

Following these studies, the *hydrogeological basin boundaries* of the wetland have been determined, and it will shed light on us in the regions where groundwater recharging is also involved.

As a next step, measuring the spring and well elevations in the hydrological basin of the wetland and drawing groundwater level maps reveal the recharge–discharge relations and ground–surface water interactions (Fig. 8.7). Data of groundwater depths around Izmir Bird Paradise, which were measured by DSI in 1998, was used to draw water level maps for Menemen Plain (Izmir). It is clearly seen that the general flow direction of the groundwater is from Aegean Sea to Menemen Plain (Fig. 8.7). This phenomenon is called as "seawater intrusion." Long-term drought and excessive pumping in the study area have resulted in these large drawdowns of the water table in the unconfined aquifer of the coastal area (Somay and Filiz 2003). Many groundwater-related coastal wetlands occur in areas with a shallow water table, where groundwater can discharge to the surface, but some are linked to deep groundwater that ascend to the land surface in the coastal zone (Manzano 2019).

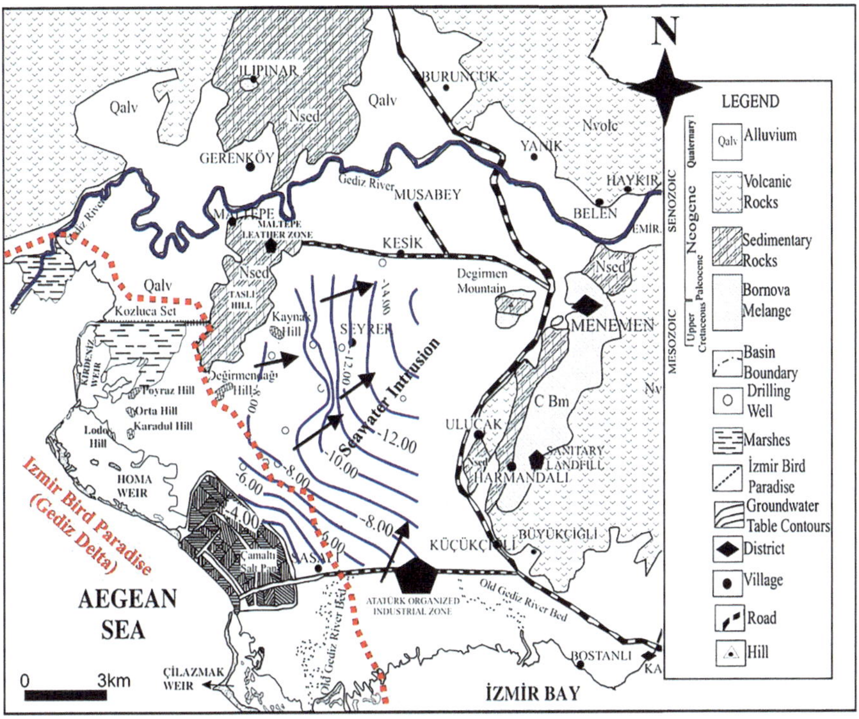

Fig. 8.7 Geology and water-level map of the Gediz Delta–Menemen Plain. (Modified from Somay and Filiz 2003)

Part of the precipitation falling on the surface infiltrates underground. It fills all porous gaps, and it can reach and goes down to the impermeable layer. The upper level of accumulated groundwater is defined as a water table. The water table is the water surface that limits the upper part of the unconfined aquifer. These maps indicate the location of the groundwater level (water table) relative to a specific reference plane (sea level). Water table maps are prepared to determine usable water, aquifer properties, and the direction of groundwater movement. Water levels in wetlands serve as an indicator of the dissolved oxygen state of the soil water system. Since the groundwater is in a dynamic process, there are changes in these maps depending on the recharge and discharge conditions over time. The water table rises and falls in natural conditions without human influence. Especially in rainy times in winter, the water table can rise and recharge the rivers and lakes (Fig. 8.8).

Finally, in order to determine the quality of the water in the hydrogeological conceptual model of the wetland (Fig. 8.9), groundwater and surface water sampling should be done from the wetland elements (such as swamp, freshwater lake, saltwater lake, lagoon) and regions representing the basin. Routine analysis, heavy metal analysis, and environmental isotope analysis of these samples taken according to the

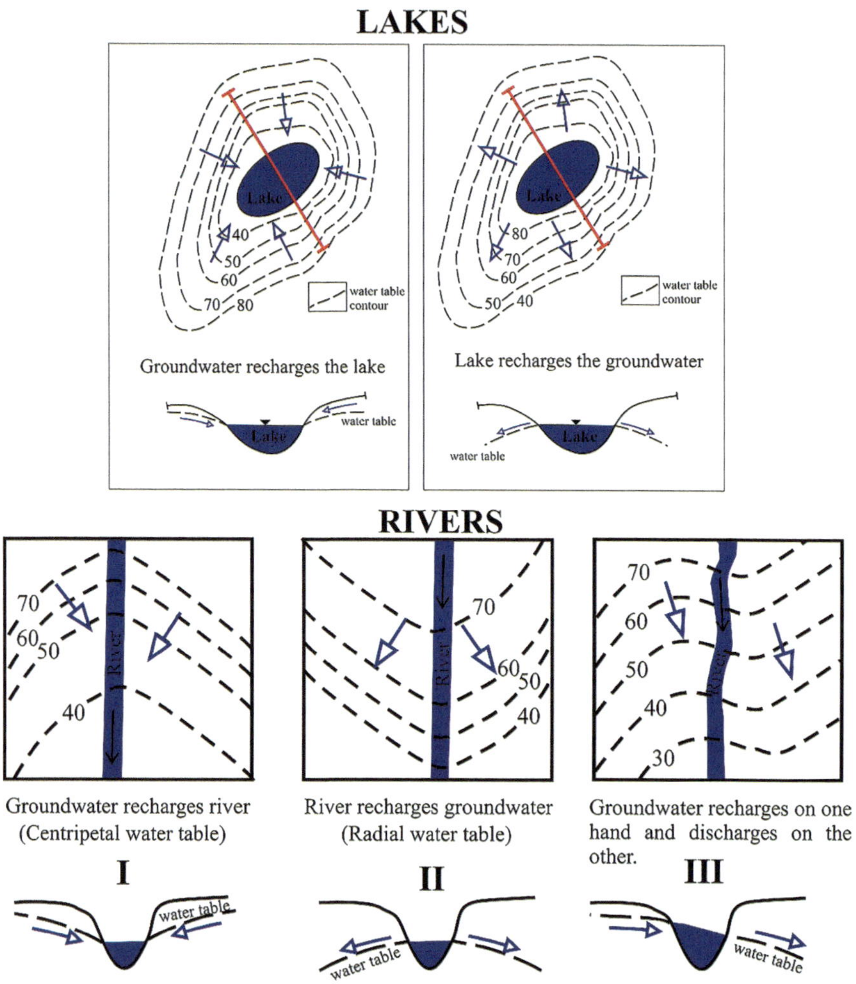

Fig. 8.8 Recharging and discharging mechanism of the wetlands. (Courtesy of Melis Somay-Altas)

purpose will give us information about the origin of the water in the region and recharge and discharge area characteristics.

Hydrogeochemical functions in coastal aquifers are also important for the sustainability of water quality in the freshwater wetland. Land use for various purposes in coastal areas affects water quality in coastal wetlands recharged by groundwater streams.

Hydrogeochemical facies were used to characterize groundwater evolution along different flow paths of groundwater in an aquifer. The facies are the function of lithologies, solution kinetics, and flow patterns of the aquifer (Back et al. 1979). According to IAH (International Association of Hydrogeologist) (1979) water-type

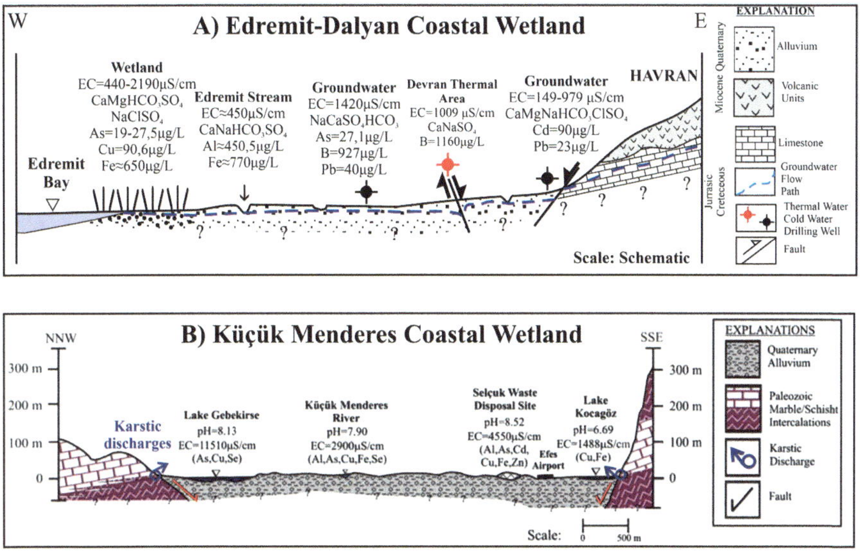

Fig. 8.9 Hydrogeological conceptual models of some coastal wetlands in western Turkey. (Courtesy of Somay et al. 2008; Somay 2016)

classification, coastal areas have Na–Cl, central parts of hydrological basin have Na–Ca–Mg–Cl–HCO_3 (Na–Cl mixed)/Ca–Na–Mg–HCO_3–Cl (Ca–Mg–HCO_3 mixed), and the hilly areas have Ca–HCO_3 water types.

Piper diagram was used to classify surface and groundwater and describe their hydrogeochemical functions (Fig. 8.10). With this diagram, the water samples taken from different regions in the hydrological basin are classified and compared with their dominant ions.

The line moving from right to left is the theoretical seawater mixture line in the abovementioned equilateral quadrilateral. Especially in the coastal areas, water is evolving from Ca–HCO_3 water type in the hilly areas to Na–Cl water type in the coastal zones. The waters on the far right of the diagram are Ca–HCO_3 waters that were taken from the hilly area of the hydrological basins. As it moves toward the sea, salinity is observed in the waters depending on seawater intrusion. In the middle of the diagram, the waters are represented as "mixed waters" due to the seawater effect. On the far left of the diagram are water samples from the coastal wetlands that have Na–Cl water types.

Schoeller semi-logarithmic diagram can be used to identify the origin of the waters (Fig. 8.11). In this diagram, waters of different origin can be immediately distinguished. Also, in this diagram, waters of similar origin form similar broken lines. Waters from same aquifer and recharge areas reflect the similar pikes in Schoeller diagram (Gemici et al. 2016). When we compare the wetlands in western Anatolia, we see that the waters near the sea show a peak like that of seawater

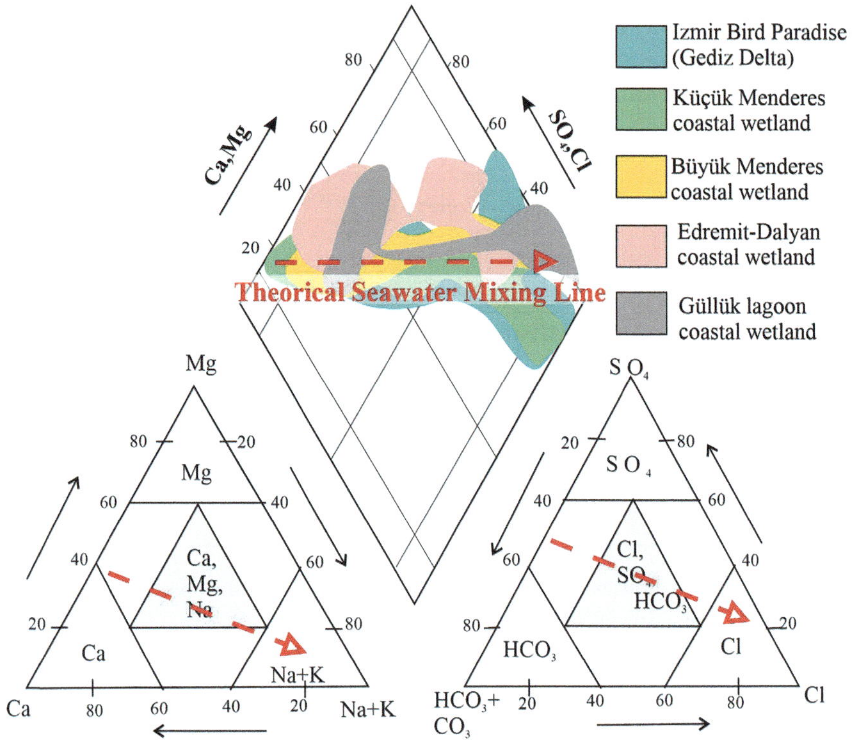

Fig. 8.10 Piper diagram for the coastal wetland basin waters. (Data from Somay and Filiz 2003; Somay et al. 2008; Somay and Gemici 2009, 2012; Somay 2016; Somay-Altas 2021)

(Fig. 8.11). On the other hand, the waters from recharging area have much less mineralization than the coastal area. According to Fig. 8.11, the wetland least affected by seawater is the Edremit–Dalyan coastal wetland.

UNESCO-IHP and Med Partnership have prepared an interactive map called "Hydrogeological and ecosystem services classification of representative Mediterranean groundwater-related wetlands" in 2015. With this map, it is aimed to make a hydrogeological classification for coastal Mediterranean wetlands connected to groundwater for the management of coastal aquifers and groundwater. This map, of which 11 were from Turkey (42%) in a total of 26 wetlands, covers the wetlands.

After all these hydrogeological studies, parameters have been proposed to create inventory in the hydrological and hydrogeological evaluations of wetlands, according to the 2019 report by UNESCO and the International Hydrological Programme (IHP). These parameters are as follows (Manzano 2019):

- *Mean rainfall* (mm/year), *mean temperature* (°C), *and mean evapotranspiration-*ET (mm/year): *Hydrology*
- *Underlying lithology*: *Geology* (e.g., sedimentary, magmatic, etc.)

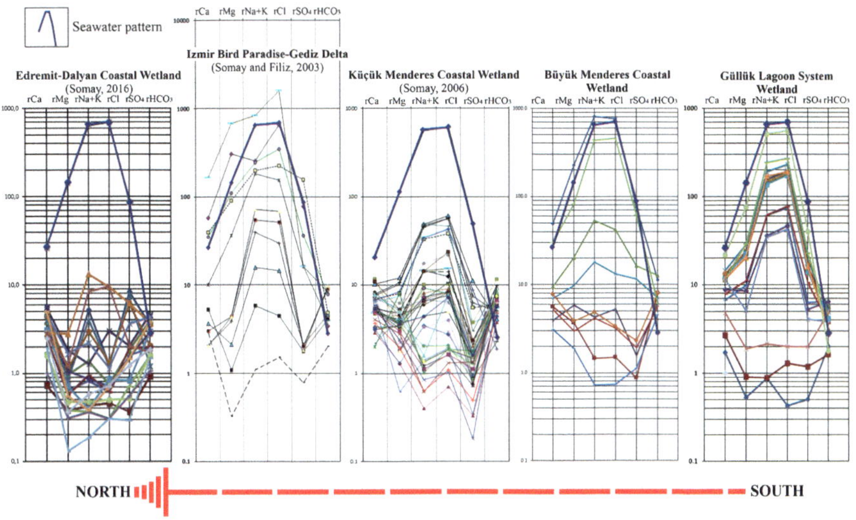

Fig. 8.11 Schoeller semilogarithmic diagram for the coastal wetland basin waters. (Courtesy of Somay and Filiz 2003; Somay 2006, 2016; Somay-Altas 2021)

- *Wetland genesis*: *Geological structure type* (e.g., volcanic, tectonic, etc.)
- *Wetland sediments*: e.g., silty, muddy, etc.
- *Water source*: *Hydrogeology* (recharging mechanism of the wetland)
- *Groundwater flow type*: *Hydrogeology* (recharge area, discharge area, etc.)
- *Groundwater dependence*: *Hydrogeology* (dominant, shared, or secondary)
- *Hydrogeochemistry*: e.g., salinity and water types
- *Resource exploitation*: *Hydrogeology* (water abstraction, biological exploitation, or mineral exploitation)
- *Modification of hydrological cycle*: *Hydrology* (e.g., inputs and outputs of the wetland, drainage, and artificial recharge)
- *Pollution*: *Hydrogeology and hydrogeochemistry,* (e.g., anthropogenic, agricultural, etc. sources)

8.4 Hydrological and Hydrogeological Studies in Turkey

Wetlands in Turkey, three sides of the different climate that our country is surrounded by sea, topography, and elevation, show different characteristics depending on the soil structure and permeability. Turkey has animal and plant species diversity because of the presence of different types of wetlands. Lagoons, flooded forests, and deltas are found by the sea (Fig. 8.12). There are shallow lakes in the Central Anatolia Region. It has many different types like glacial lakes when rising to higher elevations. Coastal wetlands and wetlands in the inner parts of Anatolia differ significantly in terms of climatic conditions. It is much more suitable

Fig. 8.12 Locations of some important wetlands in Turkey. (Courtesy of Melis Somay-Altas)

for sheltering, feeding, and protecting waterbirds in terms of coastal wetlands, year-round water availability, plant and nutrient richness, and climate conditions.

To mention briefly from Turkey by hydrogeology, there are four main groups showing aquifer characteristics (Yazicigil and Ekmekci 2020): (1) alluvial deposits (especially deltas and plains), (2) karstic aquifers, (3) fractured volcanic rocks, and (4) fractured metamorphic bedrock aquifers. All the wetlands in Turkey are recharging and/or discharging from all these aquifer types. Wetlands are divided into six different groups for their geological and hydrogeological properties, according to the classification of the Ministry of Forestry and Water Affairs, General Directorate of Nature Conservation and National Parks (Meriç and Çağırankaya 2013) in Turkey:

1. Formed by tectonism: These are the lakes formed by the accumulation of water in the bowls formed as a result of the movement of the earth's crust. Tuz, Sapanca, Iznik, Uluabat, Manyas, Beyşehir, Eğirdir, Burdur, and Seyfe lakes are the best examples of tectonic-formed lakes in Turkey.
2. Formed by volcanic activities: These types of wetlands formed in calderas of the volcanoes or by the lava that erupts to the earth as a result of volcanic eruptions, blocking or closing the mouth of the valley. Nemrut Caldera and Meke Maar Lake are the unique samples of the caldera type of wetlands in Turkey. Erçek and Nazik lakes that formed in Van city are the good samples for lava sets lakes. On the other hand, Köyceğiz–Dalyan wetland system can be classified as geothermal spring-based wetland type.
3. Formed by karstification: Karstification is a process that leads to the formation of a special landform formed by the rocks affected by chemical dissolution. The main rocks in which karst terrain forms occur are limestone, dolomite, and gypsum. Examples of karstic structure are the Kızören Lake, Eğirdir Lake, and Beyşehir Lake.
4. Formed by river sediment accumulation: They occur as a result of blocking up of valleys, bays, or gulfs with embankments formed by alluviums carried by streams. Deltas are the best structures for this classification. Deltas are flat areas

where a stream reaches a stagnant river mass, formed by collapsing the deposits it drags and grew forward in time. Examples of deltas are located in the western part (Meriç, Gediz (İzmir Bird Paradise), Küçük Menderes, and Büyük Menderes deltas), northern part (Kızılırmak and Yeşilırmak deltas), and southern part (Göksu delta) of Turkey.

5. Formed by wave sediment accumulation: Lagoons and estuaries are the best examples for this type of wetlands. Lagoons are wetlands rich in biodiversity, which are connected to the sea by a natural narrow waterway, often separated from the sea by a narrow land. Estuaries are defined as wetlands where the river mouth expands into marine ecosystems. Yumurtalık, Güllük, and Köyceğiz lagoon wetland systems are the best examples in Turkey.
6. Formed by landslide sets: Such wetlands occur when the materials sliding as a result of a landslide event obstruct a valley and water accumulates behind this embankment. This type of wetlands is observed in the Northern Anatolia Region where precipitation is very high. Sera Lake and Uzun Lake are the best examples of this classification.

Because of its complex geologic, geomorphic, and climatic settings, Turkey has many rivers that enter the surrounding seas and neighboring countries of Iraq, Iran, and Armenia (Akbulut et al. 2009). A total of 26 main drainage basins occurred in Turkey. All rivers with large discharge rates originate from huge karstic springs in Turkey, e.g., Dicle and Fırat rivers, SE Anatolia; Seyhan, Ceyhan, Manavgat, and Aksu rivers, southern Anatolia; Sakarya River, central Anatolia; and Gediz River, western Anatolia (Yazicigil and Ekmekci 2003). As a result of the studies carried out by the General Directorate of Nature Conservation and National Parks, there are 320 natural lakes in Turkey (DSİ 2020). The usable surface water of Turkey is about 94 billion m^3, and the annual withdrawable amount of groundwater is 18 billion m^3 according to DSI data. In Turkey, 77 percent of the water is used in agriculture. The other part is used for drinking and industry (DSİ 2020). Most of the agricultural irrigation is done by traditional methods, which leads to waste of water. Large projects such as dams and highways can negatively affect wetlands if not well planned.

Ahlat Marshes (Bitlis), located in the shore of Lake Van, is a freshwater area. Fresh water flowing from the marshes is mixed with the lake. The recharge of this wetland is from precipitation, surface, and groundwater (Akyüz 2018). Discharge from the wetland takes place toward Lake Van (Nergiz et al. 2018).

Bulanik Plain Wetlands are located in the province of Muş. It covers a wide area of about 340 km^2. This wetland is a natural wetland area consisting of surface waters and reeds formed by the Murat stream, which is one of the main tributaries of the Firat River, and its tributaries. The main sources of wetlands and groundwater recharge in the Bulanik plain are the Murat stream and the surface waters brought by the tributaries. It provides the main discharge with the Murat stream. According to DSİ (1978) data, there is a large alluvial cover with a thickness of 5–25 m in the lowland where wetlands are located. This alluvium unit acts as aquifer in the area. The transmissibility of the alluvial aquifer is 500 m^3/day/m, and the electrical conductivity value of the groundwater is around 1000μS/cm.

Gâvur Lake Wetland, formed by tectonic movements, is an important wetland in Kahramanmaraş. The sources that recharge the Gâvur Lake Wetland are precipitation, surface runoffs, and stream sources and karst resources (Korkmaz 2008). All the streams flowing into wetlands are all connected to the drying channels today. For this reason, the water of the streams in question is normally drained through these channels, and the recharge of the wetland is prevented (Okumuş 2011). The discharge of the wetland takes place through streams.

The river gets its main resources from the mountains of the Eastern Anatolia Region and from the Hazar Lake near Elâzığ through infiltration from the bottom. Hevsel Gardens, which covers an area of 700 hectares and is in the south of Diyarbakir, is in the delta formed when the waters of the Tigris River decrease. An application will be made for the international Ramsar Convention in 2020 for the protection of this wetland. Not only this area, but also about 500 m from the Tigris River in Diyarbakir, Batman, the areas are wetlands. Tigris River is an important resource that is used for potable and utility water, as well as agricultural irrigation, with dams constructed in the regions of the river basin (Çelik 2019). In this region, it is not acceptable to use groundwater too intensively for agriculture. Especially in the climate change process, it is possible to see drought around the river in the future.

Hazar Lake Wetland (Elâzığ) is in the Upper Euphrates Section of the Eastern Anatolia Region. As a result of the closure of Maden Stream, which is the only discharge channel of the lake and connected to the Tigris River, with the construction of HEPP, the lake took the form of a closed basin. Kavak Creek is connected to a canal to prevent the level drop in the lake. Water types are Ca–Mg–HCO3 and Mg–Ca–HCO3 in and around Hazar Lake (Öztekin Okan and Güven 2019). According to the isotopic tracers of ^{18}O, ^{2}H, and ^{3}H, the spring water discharging at higher altitudes from the fractured igneous aquifer reflects rapid circulation and recharging from recent precipitation compared to alluvial aquifer groundwater (Öztekin Okan and Güven 2019). If we examine the Hazar Lake basin, it can be seen that it is polluted with domestic, agricultural, and industrial wastes. Pollutants from sand pits and brick factories around the lake also reach the lake and cause pollution. According to Özatlar (2019), the order of heavy metal concentrations in water was measured as follows: Cr > Zn > Cu > Cd > Co > Ni.

Nemrut Caldera is a caldera that was formed as a result of the explosion of Mount Nemrut and contains the most characteristic and unique landforms, showing the structural features of a caldera collapse (Dizdaroğlu 2019). This wetland is an Internationally Important Wetland-RAMSAR Area. Nemrut Crater Lake is approximately 6 km in diameter. There are also four more lakes. Ilık Lake is the second largest lake. The other three lakes dry in the summer. This wetland is the amazing sample of the geothermal source wetland in the world.

Erzurum Swamp is a seasonal wetland. The wetland is an important point on the bird migration route. The River Karasu, one of the two branches of the Euphrates, passes through the wetland. According to Kuzulugil (2017), it has been determined that the wetland has lost its wetland quality due to climate change, has shrunk in size, and has turned into a meadow.

Kuyucuk Lake in Kars, the Ramsar area, is in the category of "continuous freshwater lake and swamp." It recharges on small springs and streams, and its deepest place is 13 m (Sevindi 2013). The water level in the lake, which was 13 m deep in 1997, has decreased continuously due to the wrong water usage. The water that accumulated in the ponds and dams built around the lake, which came to the drying point in 2014, could not be given, and a permanent solution could not be found. Since the lake could not be fed, the remaining water evaporated due to the effect of the hot summer. In 2019, the State Hydraulic Works (DSİ) drilled a water well with a depth of 140 m, a static water level of 11 m, a dynamic water level of 80 m, and a well yield of 20 l/s. The ecosystem will be revitalized by pumping 200 tons of water per day from Kuyucuk Lake from the borehole. According to the physical and chemical analysis results of the surface waters of Kuyucuk Lake, the pH value of the lake water was determined as 9.5, the salinity was 0.30%, and the temperature was 18 °C. 0.2 mg/L of nitrate, 70 mg/L of sulfate, 0.25 mg/L of phosphorus, 0.77 mg/L of phosphate, and 0.92 mg/L of nitrogen were detected. Besides the pH of the lake waters (alkaline water), the amount of phosphorus and nitrogen in the water was found to be higher than the water quality evaluation criteria (Meriç et al. 2010).

Lake Uluabat at the eastern end of the Manyas–Karacabey depression has a maximum water depth of 2.50 m and a surface area of c. 138 km^2 according to Kazancı et al. (2006). Uluabat is a large, shallow freshwater lake on the south side of the Sea of Marmara. There are four islands in the lake. The lake lies in a large east-to-west-oriented tectonic depression. It is principally recharging from Mustafakemalpaşa River. Other small sources of water are karst springs at the bottom and the surroundings and small streams and agricultural fields during rainy periods. Lake water is used to irrigate surrounding fields. The metals lead, copper, nickel, and zinc had levels higher than the Turkish Environmental Guidelines limits in water (Arslan et al. 2010). According to Hacısalihoğlu and Karaer (2016), Lake Uluabat has moderately polluted sediment, and its Cr and Ni values have been determined to be higher than upper effect threshold concentrations.

Manyas Lake Wetland is a shallow, freshwater, and tectonic lake. It is recharged by several streams, as well as by groundwater (the Kocasu Stream provides 70% of the water input). Where streams flow into the lake, small deltas have formed with extensive marshes and tree-lined riverbanks. The surface area of Manyas Lake is 164 km^2. The height of Manyas Lake from the sea level is 15 m and the deepest place of the lake is 5 m. The pH values of the lake water are between 8 and 9.5. According to Karafistan and Arık-Çolakoğlu (2005), one of the largest borax and boric acid production industries in that region (Etibor) has been estimated to discharge wastes about 6300 m^3/day to the wetland. Boron contamination is the biggest problem in the area.

Göksu Delta is a very important wetland on the migration route of many waterbird species. The Göksu Delta includes Paradeniz Lagoon and Akgöl Lake, which were formed because of bed displacements of the Göksu River and sea movements. Paradeniz Lagoon is relatively saline owing to its opening to the sea. In many parts of the delta, the water is artesian in depth, or the units closer to the stuck surface are also pressurized. Akgöl Lake contains fresh water. The region has abundant surface water sources and underground (karst) water sources, which are used for irrigation. Groundwater is found at 5–15 m depth. Groundwater pH value is between 7.5 and 8 in Göksu Delta. The recharging of the aquifers takes place in the coarse-grained units of the surface waters draining from the river in the upper codes of the delta and the accumulation of rain and melting snow waters in the limestones under these units. The discharge takes place in the form of groundwater flow into the Göksu River, flow into the lakes, and withdrawal of water by pump for irrigation and drinking water supply (Güner 2016). According to Demirel et al. (2011), it is inferred that the excess concentration of Fe, Ni, Mn, Mo, and Cu at some locations is the cause of undesirable drinking water quality. Nitrogen in delta groundwater was greatly affected by agricultural activities but was mainly sourced from domestic wastewater (Gülçiçek and Demirel 2019). Kavşut (2019) concluded that there is a continuous and dynamic interaction between underground and groundwater.

Akyatan Lagoon is one of the most important Mediterranean lagoons. It includes a large coastal lagoon with varying salinity, which is connected to the Mediterranean Sea. Hyperhaline lagoon occurred with high seawater and evaporation content of the wetland. The lagoon is surrounded by brackish marshes, sandy shores, natural wet meadows, and sand dunes. Most of the salt marshes of the region have been turned into agricultural land in recent times (Lécuyer et al. 2012). This wetland recharges from precipitation, Seyhan and Ceyhan rivers, and irrigation channels. Demir-Yetiş et al. (2014) indicated that the drainage channels had the primary impact on the spatial distribution of the water quality parameters. The drainage channels collect all water in the catchment and the agricultural areas and discharge the collected water into the lagoon as a point source (two points) instead of a nonpoint source. The pH values ranged between 7.8 and 9.8. Salinity value in the region shows a great difference. It varies between 2.5‰ and 93‰. According to Kjerfve (1994) classification, water with a salinity value greater than 40‰ is called *hyperhaline*. Hydrological dynamics of the Akyatan Lagoon are far from steady state, and one consequence concerns the response of biotic communities to rapid salinity changes that prevail in the northwestern part of the lagoon during the warmest seasons (Lécuyer et al. 2012).

Burdur Lake is a tectonic lake and the one of the deepest lakes in Turkey. This wetland is recharged by precipitation, rivers, groundwater, and seasonal streams. The water discharge of these rivers and streams is low and some of them dry up in summer. Burdur Lake is a close basin and has no outlet. Thus, water loss only occurs through evaporation. Wetland waters have hydrochemical facies such as $Ca–HCO_3$, $Ca–Mg–HCO_3$, $Mg–Ca–HCO_3$, $Na–Mg–SO_4$, and $Na–SO_4–HCO_3$ (Davraz et al. 2019; Şener et al. 2019). Groundwater chemistry was controlled by reverse ion exchange, dissolution of carbonates, and silicate weathering (Şener et al. 2019). Burdur Lake is the first wetland that protected by the Ramsar Convention in Turkey. Although it is under protection, it has lost almost half of its total volume due to drought, especially in the last 40 years. According to Davraz et al. (2019), the lake area decreased from 210 km^2 to 131 km^2 between 1975 and 2016, and the shrinking area was 37%. Their results of hydrologic, climatic, and human activities data analyses suggest that the change of lake levels might depend more on human effects than on climatic factors.

Sultansazlığı Wetland is located in Erciyes pull-apart basin in Kayseri. It covers the lower part of the Develi Plain. It consists of the saline Lake Yay, nearby salt steppe, eutrophic freshwater marshes with small islands, small lakes, and wet meadows around the marshes. The wetland recharges from groundwater, streams, and springs. Due to being close to a basin, there is no output in the wetland. A simplified hydrogeologic framework of the basin includes three aquifers: the karstic aquifer in the southern heights; the volcanic–sedimentary aquifer in the west, north, and east; and the alluvial aquifer in the central part of the basin (Bayari and Yildiz 2012). According to Yildiz (2007), the mean annual groundwater recharge was 90.5 million m^3 in and around the wetland. The wastewater disposal pollutes Sultansazlığı Wetland. EC, nitrate, orthophosphate, and ammonium concentrations are increased and dissolved (Unsal et al. 2010). On the other hand, according to Yalcin et al. (2007), high concentrations of the elements Fe, Pb, Zn, Sb, W, Mo, Co, Cu, Hg, Ni, Cr, Mn, and Cd were found in Sultan Marsh, surrounding rocks (geogenic sources), mines of Fe and Pb/Zn, industrial facilities, residential and agricultural areas, and major traffic routes (anthropogenic sources). In the region, excessive groundwater is drawn. This causes groundwater to drop to around 55–60 cm every year around Yeşilhisar.

Seyfe Lake is a tectonic and saline lake. It is recharged by a stream, runoff, and karstic springs in the west. This wetland is located in a close basin. In the basin, Paleozoic aged marbles and crystallized limestones have features of an aquifer. In addition, Neogene sediments in the basin also carry groundwater. The groundwater in the basin headed toward the lake. The amounts of precipitation and excessive groundwater pumping are quite effective on the lake's level and area. The pH of the waters taken from different parts of the Seyfe Lake varies between 8, 9, and 13 (Çiftçi 2013). According to Çelik et al. (2008), groundwater in the alluvium formation along Lake Seyfe has a salty character in the Na–Cl facies. The main reasons of the formation and change of the groundwater salinity and hydrochemical facies in the Seyfe basin are causing various (a) lithogenic pollutions and heterogeneities, (b) salinities of the upper soil zones, and (c) evaporations of the trench and channel waters open to the atmosphere.

Kizilirmak Delta was formed by sediments deposited by the Kizilirmak River. The lagoon lakes of Karabogaz Lake in the west of the coastal plain, Liman Lake in the east, and Lake Gernek and Balık Lake are important morphological elements. These lakes were formed by the progression of the coastal sand sets as a result of storing the fine clastic material (alluvium) brought by the Kızılırmak River to the east with coastal longitudinal currents. Large parts of the wetland are flooded during wet seasons. The wetland suffers pollution from agricultural runoff and untreated sewage from the town of Bafra. The study of Özturk and Sesli (2015) showed that the lagoons tended to shrink in the periods when the Kizilirmak River carried abundant amounts of sediment, prior to the construction of the Altinkaya and Derbent dams between 1962 and 1987. However, after the construction of the dams, the amount of material reaching the Black Sea has decreased by 98%, and nowadays, the

Kızılırmak Delta area is getting smaller due to currents, wind and wave erosion, and rises of sea level. The accumulated amounts of various trace metals in bottom sediments and wetland plants were found in the following order: Fe > Mn > Zn > Ni > Co > Cu > Pb and Fe > Mn > Zn > Ni > Co (Engin et al. 2017). Black Sea is comparatively more polluted than the river and that the river behaves as a diluting water resource (Samsunlu et al. 2002). According to the hydrogeophysical and hydrochemical study of Demirci et al. (2017), it shows that there is saline water intrusion up to 3.5 km inland. Fresh water can also be taken from the aquifers up to the first 10 m from the surface. After the first 10 m deep in the plain, saltwater intrusion occurred.

Yumurtalık Lagoon is a highly complex wetland system consisting of lagoons, freshwater and saltwater marshes, wide barren plains, mud flats, reeds, wet meadows, dunes, and pine forest between the points where Ceyhan River flows into the sea. Yumurtalik Bay is the only known wintering area of the endangered green turtle (*Chelonia mydas*) in the Mediterranean. Yumurtalik Lagoon is a lagoon open to the sea. The old bed of Ceyhan River and Yapı, Ömer, Avcıali, and Eşemen lakes are the main water sources that recharge the Yumurtalik Lagoon. In the lagoon, the groundwater is very close to the surface and the alluvium consisting of clay, sand, and gravel; the aquifer has a feature (Altan et al. 2004). In areas close to the sea and lakes, groundwater is salty, and in remote areas, it is salt-free or less salty. The unplanned construction of the area affects the heavy industry facilities and the Sugözü Thermal Power Plant in the east. The Baku–Ceyhan Pipeline passing near the area should also be monitored for any leakage possibility. Water samples had pH values between 7.99 and 8.48, and EC values range between 700 and 161,500μS/cm (Aydin et al. 2015). According to Aydin et al. (2015), in addition to less freshwater input, high evaporation rates and the structure of the inlet channel between the lagoon and the Mediterranean are determined as major causes of high salinities. The authors calculated the monthly required flow rates of fresh water from the Ceyhan River to the Yelkoma Lagoon were to be in the range of 0.199 m^3/s to 3.698 m^3/s.

Kızören Lake is a karstic formation lake in Konya. 145 m of the sinkhole with a diameter of 228 m and a depth of 171 m is filled with water. The lake surface, which is 20 m below, becomes even lower due to excessive groundwater use. Kızören Lake is the biggest sinkhole in Turkey. Kızören Lake, which is recharged by groundwater, shows a level difference of 5 m during the year. Excessive water withdrawal with the pump for irrigation lowered the lake level. Water withdrawal from the lake for agricultural irrigation and domestic use was stopped. Between 1996 and 2006, the level was determined to drop by 10 m. In this area, the increase in the cultivation of agricultural products caused the number of thousands of deep irrigation wells in the basin to increase. As a result, various problems arise, such as lowering of the groundwater level and the formation of new sinkholes, salting in the soil, drying of many swamps and springs, and lowering of the surrounding lakes (Yılmaz 2010). Doğdu et al. (2007), in their study, calculated that the groundwater decreased between ~0.2 and 0.9 m/year in Konya Plain and ~0.7 m/year in Karapınar between 1982 and 2007. Göçmez et al. (2008) state that the change in groundwater which is about 60% in Konya and 40% in Karapınar is related to the remaining excessive withdrawal with climatic variables.

Meke Maar is located in Konya, where young volcanic shapes are intertwined. A very young cone of basaltic slag and tuff is located in the middle of the Meke Maar, which is approximately 800 m long and 500 m wide. There is a 25 m deep crater on the cone. The lake between the cone in the center and the steep slopes of the maar is Meke Lake. In a phase after the bottom of the recess which was covered with a lake, a new cone was formed as a result of new eruptions. This wetland is recharged by groundwater and precipitation. The salt content of Meke Maar water, which is a shallow lake, is very high. The EC value is about 65,000μS/cm (Şilliler Tapan et al. 2008). But in recent years, due to the unconscious consumption of groundwater for irrigation purpose in the Konya Basin, it is completely dry in the summer months. On the other hand, although it is prohibited from the surrounding area, pyroclastic materials were taken for briquetting in the past years (Somuncu 1986).

Beyşehir Lake is a tectonic lake, its length is 50 km, and its width perpendicular to this direction is 18–20 km. Its depth is at most around 8–9 m. The elements that recharge the Beyşehir Lake Basin are surface flow, groundwater recharging, and precipitation falling on the lake surface. The elements of discharge are irrigation, evacuation, evaporation, and karstic leaks. This leakage is of the order of 5 m^3/s, reaching up to 29 m^3/s as the water level of the lake rises during wet periods (Ekmekçi 1990). The groundwater flow direction in the Beyşehir Lake Basin is toward the lake (Soyaslan and Hepdeniz 2018). Irrigation water is drawn from the lake for surrounding agricultural land. Wastewater from settlements and industrial facilities is generally discharged into the lake without treatment. Especially toward the end of spring, inorganic matter and food salt are transported to the lake. In addition, since Beyşehir provides drinking water from the lake, water is drawn in the lake for the purpose of drinking water for 19,60% of the basin population (Hoşafçıoğlu 2007).

Köyceğiz Lake is located in southwestern Turkey. The Iztuzu beach area in the region is one of the most important breeding grounds of the sea turtles (*Caretta caretta* and *Chelonia mydas*) and the Nile turtle (*Trionyx triunguis*). It is a meromictic lake, comprised of waters of both thermal and cold-karstic origin. The section between Köyceğiz Lake and the sea is covered with four small lakes and numerous canals. The lake is principally fed from surface flow, groundwater recharge from alluvium, and thermal discharges located at the bottom of the lake (Bayarı et al. 1995). There are two different water types. The deeper layer is saltier. According to Bayarı et al. (1995), cold-karstic waters have Ca > Mg > Na > K and $HCO_3 > SO_4 > Cl$ composition that indicates water–carbonate rock interactions. The chemical compositions of the thermal springs are Na > Mg > Ca > K and $Cl > SO_4 > HCO_3$ that indicate the seawater mixing. Gökgöz and Tarcan (2006) have measured temperatures of 24–41 °C and specific EC of 14,310–45,600μS/cm, dominated by Na (1550–8500 mg/kg) and Cl (2725–15,320 mg/kg). The ionic composition of the lake water samples suggests that they could be formed by the mixing of karstic and thermal discharges (Bayarı et al. 1995).

Meriç (Enez) Delta is located in Greece–Turkey border. Meriç River is a *transboundary* and *border-forming* river between the two points where it was risen and flowed into sea. This river rises in Bulgaria, creates the Turkey-Greece border and flows into the Aegean Sea at the end of the Meriç (Evros) Delta. It is a wetland system consisting of large swamps, seasonal marshes, sweet and salty lakes, and agricultural areas. The most important of the remaining lakes and lagoons within the borders of Turkey are as follows: Small and Large Gala, Sığırcı and Pamuklu lakes, and coastal lakes near Enez. The lakes in the delta are recharged by the water carried by small streams and the precipitation falling on the surface. Meriç lagoons are controlled by seawater–freshwater inputs and by salinity (Altınsaçlı et al. 2018). The recharge basin of these two lakes is 469 km^2 (Atınayar et al. 1986). The annual total amount of water coming to the Gala and Pamuklu lakes is 198.7 hm^3/year on average. According to Zal et al. (2006), only 9.7% of the waters coming to the Gala Lake Basin are fresh water; the remaining 90.3% of the waters are waste waters that come from agricultural, domestic, and industrial areas.

8.5 Conclusion

The existence of water is the most important and the priority component for wetland sustainability. The quantity and quality of the water where it interacts with recharge and discharge areas of wetlands should be investigated to understand the system. At this point, the water crisis can be solved by hydrogeological research in wetlands.

If we evaluate all the wetlands in general in Turkey, almost all are under the threat of both drought and pollution from domestic, agricultural, and industrial activities. For example, Burdur Lake lost almost half of its water volume in the last 40 years and entered the drying period. Ramsar areas such as Kuyucuk Lake, Sultansazlığı Wetland, Seyfe Lake, and Meke Maar Lake are either completely dry or are suffering from seasonal water accumulation. Uluabat Lake is also at risk of pollution and drought. Correctly planned hydrogeological studies gain importance in order to regulate all these problems.

The fundamental requirement for wetland conservation and wise use of wetlands is assigned to the desired quality and quantity of water they need at the right time. Therefore, the determination of hydrogeological relationships between groundwater and surface waters recharging and discharging the wetland has great importance. Not only ecological and hydrological studies but also hydrogeological and hydrogeochemical studies should be done regularly to ensure groundwater sustainability and to control elements that can damage and pollute the wetland. Also, hydrogeological studies in wetlands will make great contributions to the country's economy.

References

Akbulut NE, Şahin Y, Bayarı S, Akbulut A 2009 Rivers of Turkey. In: Tokner, Uehlinger, Robinson (eds) Rivers of Europe. Academic Press, pp 643–672. https://doi.org/10.1016/B978-0-12-369449-2.00017-5

Akyüz F (2018) Ahlat sazlığı (Bitlis) potansiyel kirlilik düzeyinin yerüstü su kalitesi bakımından araştırılması. Bitlis Eren Üniversitesi ve Fırat Üniversitesi Fen Bilimleri Enstitüsü. MSc thesis (In Turkish)

Altan T, Artar M, Atik M, Çetinkaya G (2004) Çukurova Deltası Biyosfer Rezervi Yönetim Planı. LIFE- Çukurova Deltası Biyosfer Rezervi Planlama Projesi. Çukurova Üniversitesi Peyzaj Mimarlığı Bölümü, 372 Sf. (In Turkish)

Altınayar G, Başkan M, Ertem B (1986) Gala Gölü Limnolojik Araştırma Raporu. DSİ Genel Müd, p 120, Ankara (In Turkish)

Altınsaçlı S, Paçal FP, Altınsaçlı S (2018) Assessments of environmental variables affecting the spatiotemporal distribution and habitat preferences of living Ostracoda (Crustacea) species in the Enez Lagoon Complex (Enez-Evros Delta, Turkey). Ecologica Montenegrina 19:130–151

Arslan N, Koç B, Çiçek A 2010 Metal contents in water, sediment, and Oligochaeta-Chironomidae of Lake Uluabat, a Ramsar site of Turkey. Sci World J, v:10, 1269–1281

Aydin H, Karakuş H, Erdem O (2015) The salinity problem at Yelkoma Lagoon (Yumurtalık-Adana) and its restoration by mixing with freshwater from Ceyhan River. Fresenius Environ Bull 24(1a):235–243

Back W, Hanshaw BB, Pyle TE, Plummer LN, Weidie AE (1979) Geochemical significance of groundwater discharge and carbonate solution to the formation of Caleta Xel Ha, Quintana Roo, México. Water Resour Res 15(6):1521–1535

Bayarı S, Yıldız E (2012) Effects of the North Atlantic oscillation and groundwater use on the contraction of Sultansazlığı Wetland, Turkey. Hydrogeol J 20:369–383. https://doi.org/10.1007/s10040-011-0809-z

Bayarı CS, Kazancı N, Koyuncu H, Caglar SS, Gökçe D (1995) Determination of the origin of the waters of Köycegiz Lake, Turkey. J Hydrol 166:171–191

Çelik R (2019) Evaluation of groundwater potential by GIS-based multicriteria decision making as a spatial prediction tool: case study in the Tigris River batman-Hasankeyf Sub-Basin, Turkey. Water 11:2630. https://doi.org/10.3390/w11122630

Çelik M, Ünsal N, Tüfenkçi OO (2008) Assessment of water quality and pollution of the Lake Seyfe basin, Kirşehir, Turkey. Environ Geol 55:559–569. https://doi.org/10.1007/s00254-007-1007-0

Çiftçi E (2013) Seyfe Gölü havzası'nda(Kırşehir) doğal ortam- yeraltısuyu ilişkisi. Fırat Üniversitesi Sosyal Bilimleri Enstitüsü. MSc thesis (In Turkish)

Cowardin LM, Golet FC, LaRoe ET (1979) Classification of wetlands and deepwater habitats of the United States. US Department of the Interior, U.S. Fish and Wildlife Service, Office of Biological Services, Washington. FWS/OBS-79/31

Davraz A, Sener E, Sener Ş (2019) Evaluation of climate and human effects on the hydrology and water quality of Burdur Lake, Turkey. J Afr Earth Sci 158:103569. https://doi.org/10.1016/j.jafrearsci.2019.103569

Demir Yetis A, Selek Z, Seckin G, Davutluoglu OI (2014) Water quality of Mediterranean coastal plains: conservation implications from the Akyatan Lagoon, Turkey. Environ Monit Assess 186(11):7631–7642

Demirci I, Candansayar ME, Soupios P, Vafidis A, Arslan H, Gündoğdu NY (2017) Interpretation of geophysical data for saltwater intrusion on Bafra plain guided by hydrogeological and Hydrochemical data. In: 9th Congress of the Balkan geophysical society 5–9 November 2017, Antalya, Turkey

Demirel Z, Özer O, Özpinar Z (2011) Investigation of groundwater pollution in a protected area in Turkey, The Göksu Delta. Gazi Univ J Sci 24(1):17–27. https://dergipark.org.tr/tr/pub/gujs/issue/7418/96738

Dizdaroğlu NC (2019) Nemrut krater gölü'nün sediment ve bazı sucul bitki örneklerinin ağır metal konsantrasyonlarının incelenmesi. Bitlis Eren Üniversitesi ve Fırat Üniversitesi Fen Bilimleri Enstitüsü. MSc thesis (In Turkish)

Doğdu MŞ, Toklu MM, Sağnak C (2007) Konya Kapalı Havzası'nda Yağış ve Yeraltı Suyu Değerlerinin Irdelenmesi, I. Türkiye Iklim Değişikliği Kongresi, Bildiri Kitabı:394–402, 11–12 Nisan 2007, Istanbul (In Turkish)

DSİ (Devlet Su İşleri – State Hydraulic Works) 2020. http://www.dsi.gov.tr/toprak-ve-su-kaynaklari

DSİ (Devlet Su İşleri- State Hydraulic Works) 1978 Malazgirt, Bulanık ve Patnos Ovaları Hidrojeolojik Etüt Raporu, DSİ Matbaası, Ankara (In Turkish)

Ekmekçi M 1990 A conceptual model for Beysehir Lake Karst system. Hydrogeological processes in Karst Terranes. In: Proceedings of the antalya symposium and field seminar, October 1990. IAHS Publication no. 207

Engin MS, Uyanik A, Cay S (2017) Investigation of trace metals distribution in water, sediments and wetland plants of Kızılırmak Delta, Turkey. Int J Sediment Res 32(1):90–97

EPA (Environmental Protection Agency) (2008) Methods for evaluating wetland condition #20 wetland hydrology. http://www.epa.gov/waterscience/criteria/wetlands/

Ersoy YE, Helvaci C, Palmer MR (2011) Stratigraphic, structural and geochemical features of the NE–SW trending Neogene volcano-sedimentary basins in western Anatolia: implications for associations of supra-detachment and transtensional strike-slip basin formation in extensional tectonic setting. J Asian Earth Sci 41:159–183. https://doi.org/10.1016/j.jseaes.2010.12.012

FAO (Food and Agriculture Organization of the United Nations) (2011) Aquaculture production 1950–2009

Gemici U, Somay MA, Akar AT, Tarcan G (2016) An assessment of the seawater effect by geochemical and isotopic data on the brackish karst groundwater from the Karaburun Peninsula (Izmir, Turkey). Environ Earth Sci 75:1008. https://doi.org/10.1007/s12665-016-5808-x

Göçmez G, Dıvrak BB, Galena I (2008) Konya Kapalı Havzası'nda Yeraltı Suyu Seviyesinin Değişiminin Tespiti Özet Raporu. WWF, Istanbul. In Turkish

Gökgöz A, Tarcan G (2006) Mineral equilibria and geothermometry of the Dalaman–Köyceğiz thermal springs, southern Turkey. Appl Geochem 21:253–268

Gülçiçek, Demirel (2019) Applying isotope techniques and Modeling to identify the nitrogen source in Göksu Delta. Pol J Environ Stud 28(5):3179–3195

Güner ED (2016) Göksu deltasındaki yeraltı su kalitesi düzeylerinin saptanması ve dağılımlarının belirlenmesi. Çukurova Üniversitesi Fen Bilimleri Enstitüsü. PhD thesis (In Turkish)

Güngör T, Erdoğan B (2002) Tectonic significance of mafic volcanic rocks in a Mesozoic sequence of the Menderes massif, West Turkey. Int J Earth Sci (Geol Rundsch) 91:386–397

Hacısalihoğlu S, Karaer F (2016) Relationships of heavy metals in water and surface sediment with different chemical fractions in Lake Uluabat, Turkey. Pol J Environ Stud 25(5):1937–1946

Hoşafçıoğlu S (2007) Beyşehir Gölü havzasında noktasal ve noktasal olmayan kirletici kaynakların değerlendirilmesi. Selçuk Üniversitesi Fen Bilimleri Enstitüsü. MSc thesis (In Turkish)

IAH (International Association of Hydrogeologists) (1979) Map of mineral and thermal water of Europe scale: 1:500,000. IAH, UK

Karafistan A, Arık-Çolakoğlu F (2005) Physical, chemical and microbiological water quality of the Manyas Lake, Turkey. Mitig Adapt Strateg Glob Chang 10:127–143

Kavşut ME (2019) Yer altı suyu-yer üstü suyu etkileşim modeli Silifke-Göksu deltası örneği. Çukurova Üniversitesi Fen Bilimleri Enstitüsü. PhD thesis (In Turkish)

Kazancı N, Toprak O, Leroy SAG, Oncel S, Ileri O, Emre O, Costa P, Erturac K, McGee E (2006) Boron content of Lake Ulubat sediment: a key to interpret the morphological history of NW Anatolia, Turkey. Appl Geochem 21:134–151

Kjerfve B (1994) Coastal lagoon processes. Elsevier, Amsterdam, The Netherlands

Korkmaz H (2008) Antakya – Kahramanmaraş Graben Alanında Kurutulan Sulak Alanların (Amik Gölü, Emen Gölü ve Gâvur Gölü Bataklığı) Modellerinin Oluşturulması. Mustafa Kemal Üniversitesi Sosyal Bilimler Enstitüsü Fen Edebiyat Fakültesi Coğrafya Bölümü Dergisi, Cilt:5–9 Antakya-Hatay (In Turkish)

Kuzulugil AC (2017) Sulak alanların sınırsal değişiminin kent iklimine etkisi Erzurum sulak alanı örneği. Atatürk Üniversitesi Fen Bilimleri Enstitüsü. MSc thesis (In Turkish)
Lécuyer C, Bodergat AM, Martineau F, Fourel F, Gürbüz K, Nazik A (2012) Water sources, mixing and evaporation in the Akyatan lagoon, Turkey. Estuar Coast Shelf Sci 115:200–209
Manzano M (2019) Environmental, social and economic relevance of groundwater-related wetlands in Mediterranean coastal areas: management and protection of Mediterranean groundwater-related coastal wetlands and their services. UNESCO
Meriç BT, Çağırankaya S (2013) Sulak Alanlar. Orman ve Su İşleri Bakanlığı, Doğa Koruma ve Milli Parklar Genel Müdürlüğü, Hassas Alanlar Dairesi Başkanlığı, Sulak Alanlar Şube Müdürlüğü. P: 160. (In Turkish)
Meriç BT, Çağırankaya S, Erden SH, Köycü Y (2010) Kuyucuk Gölü Sulak Alan Yönetim Planı 2011–2015. Çevre ve Orman Bakanlığı Doğa Koruma ve Milli Parklar Genel Md. Doğa Koruma Dairesi Başkanlığı Sulak Alanlar Şube Müdürlüğü (In Turkish)
Nergiz HD, Yetiş A, Yetiş R (2018) Su Kalitesinin Su Kuşları Üzerindeki Etkisi: Ahlat Sazlığı Örneği, International Conference on Stem and Educational Sciences 03–05 May, Muş (In Turkish)
Okumuş M (2011) Gâvur Gölü Havzası (Gâvur Gölü Nostajisi; Sağlık Ovasında Kanayan Bir Yara). GNO Ofset, Kahramanmaraş (In Turkish)
Önhon BE, Nazik M (1998) Selimiye-Milas-Ekinambarı Karst Havzası Hidrojeolojik Etüd Raporu. Devlet Su İşleri XXI. Bolge Müdürlüğü. Aydin (in Turkish)
Orme AR (1990) Wetland morphology, hydrodynamics and sedimentation. In: Williams M (ed) Wetlands: a threatened landscape. The Institute of British Geographers, Alden Press, Oxford, pp 42–94
Özatlar E (2019) Hazar gölü kıyı sularında yetişen sucul bitkilerde, göl suyunda ve sedimentinde ağır metal konsantrasyonlarının belirlenmesi. Fırat Üniversitesi Fen Bilimleri Enstitüsü. MSc thesis (In Turkish)
Öztekin Okan Ö, Güven A (2019) Hydrochemistry of groundwaters from alluvial and fractured igneous aquifers at the western region of Lake Hazar (Elâzığ, Turkey). Arab J Geosci 12:52. https://doi.org/10.1007/s12517-018-4209-8
Ozturk D, Sesli FA (2015) Shoreline change analysis of the Kizilirmak lagoon series. Ocean Coast Manag 118:290–308
RAMSAR (2004) Convention on wetlands of international importance especially as waterfowl habitat, Ramsar, Iran
Rimmele G, Jolivet L, Oberhänsli R, Goffé B (2003) Deformation history of the high-pressure Lycian nappes and implications for tectonic evolution of SW Turkey. Tectonics 22:1007–1027
Samsunlu A, Akca L, Kinaci C, Findik N, Tanik A (2002) Significance of wetlands in water quality management- Past and present situation of Kizilirmak Delta, Turkey. Water Sci Technol 46(8):145–152. Conference Paper
Şener Ş, Davraz A, Varol S (2019) Hydrogeological and hydrochemical investigation in the Burdur Saline Lake Basin, southwest Turkey. Chemie der Erde, 125592. https://doi.org/10.1016/j.chemer.2019.125592
Sevindi C (2013) Ekoturizm ve Kuş Gözlemciliği Açısından Kuyucuk Gölü Kuş Cenneti (Arpaçay-Kars). Türk Coğrafya Dergisi 61:63–76, Istanbul (In Turkish)
Shiklomanov LA (1993) World freshwater resources. In: Gleick PH (ed) Water in crisis a guide to world's freshwater resources. Oxford University Press, Oxford
Şilliler Tapan D, Dıvrak BB, Ayas C, İş G, Beton D, Çakıroğlu İ (2008) Wwf Türkiye Ramsar alanları değerlendirme raporu. WWF-TÜRKİYE. (In Turkish)
Somay MA (2006) Hydrogeology of lower Küçük Menderes River coastal wetland. Dokuz Eylul University Graduate School of Natural and Applied Sciences. PhD thesis
Somay MA, Filiz S (2003) Hydrology, hydrogeology and hydrogeochemistry of wetlands: a case study in Izmir Bird Paradise, Turkey. Environ Geol 43:825–835. https://doi.org/10.1007/s00254-002-0697-6
Somay MA, Filiz S (2005) The importance of hydrogeology and hydrogeochemistry in the wetlands. Int Earth Sci Coll on the Aegean Region (IESCA 2005) Abstracts, p 211, Izmir

Somay MA, Gemici U (2009) Assessment of the salinization process at the coastal area with hydrogeochemical tools and geographical information systems (GIS): selçuk plain, Izmir, Turkey. Water Air Soil Pollut 201:55e74. https://doi.org/10.1007/s11270-008-9927-1

Somay MA, Gemici U (2012) Groundwater quality degradation in the buyuk menderes river coastal wetland. Water Air Soil Pollut 223:15–27. https://doi.org/10.1007/s11270-011-0834-5

Somay MA, Gemici U, Filiz S (2008) Hydrogeochemical investigation of kücük menderes river coastal wetland, selçuk-Izmir, Turkey. Environ Geol 55:149–164. https://doi.org/10.1007/s00254-007-0972-7

Somay-Altas MA (2021) Hydrogeochemical Fingerprints of a mixohaline wetland in the Mediterranean: Güllük coastal wetland system-GCWS (Muğla-Turkey). Turk J Earth Sci (in press). https://doi.org/10.3906/yer-2005-11

Somay MA (2016) Importance of hydrogeochemical processes in the coastal wetlands: a case study from Edremit-Dalyan coastal wetland, Balıkesir-Turkey. J Afr Earth Sci 123:29–38. https://doi.org/10.1016/j.jafrearsci.2016.07.003

Somuncu M (1986) Bozkırdaki Doğal Anıtlar: Acıgöl Maarı ve Meke Tuzlası. TÜBiTAK Bilim ve Teknik Dergisi 19(225):42–44. (In Turkish)

Soyaslan İ, Hepdeniz K (2018) Beyşehir Gölü Havzasının Yeraltısuyu Akım Modellemesi Esaslı Hidrojeolojik Özelliklerinin Değerlendirilmesi. Mehmet Akif Ersoy Üniversitesi Fen Bilimleri Enstitüsü Dergisi 9(1):63–74. (In Turkish)

UNESCO-IHP & MedPartnership (2015) Hydrogeological and ecosystem services classification of representative Mediterranean groundwater-related wetlands. Interactive map

Ünsal N, Gurer I, Yildiz E (2010) Determination of water pollution at Sultansazligi wetland Kayseri - Turkey. XXXVIII IAH Congress, Groundwater Quality Sustainability, Krakow

Yalcin MG, Battaloglu R, Ilhan S (2007) Heavy metal sources in sultan marsh and its neighborhood, Kayseri, Turkey. Environ Geol 53:399. https://doi.org/10.1007/s00254-007-0655-4

Yazicigil H, Ekmekci M (2003) Perspectives on Turkish ground water resources. Groundwater Readers Forum 41(3):290–291

Yazicigil H, Ekmekci M (2020) Groundwater. In: Harmancioglu, Altinbilek (eds) Water resources of Turkey. Springer. ISBN 978-3-030-11729-0

Yazicigil H, Karahanoglu N, Yilmaz K, Gundogdu A, Sakiyan J, Yesilnacar E, Tuzcu B (2000) Investigation and management of the Kucuk Menderes Basin groundwater. Hydrogeology, vol. 5. Middle East Technical University, Ankara

Yıldız FE (2007) Kayseri-Sultansazlığı sulak alanında yer altı ve yer üstü suları ilişkisinin belirlenmesi. Gazi Üniversitesi Fen Bilimleri Enstitüsü. PhD thesis (In Turkish)

Yılmaz M (2010) Karapınar Çevresinde Yeraltı Suyu Seviye Değişimlerinin Yaratmış Olduğu Çevre Sorunları. Ankara Üniversitesi Çevre bilimleri Dergisi 2(2): 145–163. (In Turkish)

Yilmaz E, Cicek I (2016) Türkiye Thornthwaite iklim sınıflandırması. J Hum Sci 13(3): 3973–3994. https://doi.org/10.14687/jhs.v13i3.3994

Zal N, Eczacibasi GB, Karauzer ES. (2006) Aşağı Meriç Taşkın Ovasının biyosfer rezervi olarak planlanması. Central Anatolia Forestry Reserach Institute Publications Muhtelif Yayınlar/ Miscellaneous Publications No: 69 (In Turkish)

Chapter 9
Use of Multispectral and Hyperspectral Satellite Imagery for Monitoring Waterbodies and Wetlands

Mahdi Hasanlou and Seyd Teymoor Seydi

Abstract Timely and accurate change detection (CD) of Earth's surface features is important for understanding interactions between human and natural phenomena. Remote sensing (RS) as the most important information resource plays a role key in monitoring and assessment of the environment. One of most applications of hyperspectral imagery is CD. The hyperspectral imagery provides more details from CD compared to multispectral images. Wetlands are one of the most influential ecosystems in the natural environment for which it is very difficult to find an alternative. The monitoring wetland and waterbody areas based on RS imagery need special techniques due to some limitations (existence noise and condition atmospheric, need to high training data and threshold selection, and complexity of water body areas). Based on these problems it is necessary to CD methods to minimize problems so, this research proposed a framework for hyperspectral CD methods on wetland and water body areas. The proposed method is based on incorporating chronochrome, Z-score analysis, Otsu algorithm, SImplex via Split Augmented Lagrangian (SISAL), Harsanyi–Farrand–Chang (HFC), Pearson correlation coefficient (PCC), and support vector machine (SVM) to detect changes using hyperspectral imagery. The proposed method is applied in four main steps: (1) produce a training data for tuning SVM and kernel parameters, (2) predicted change areas based on a chronochrome algorithm and binary change map obtained using SVM classifier, (3) the amplitude of changes is created by Z-Score analysis and binary change mask, and (4) the multiple change map is produced based on the estimation of number and extraction of endmembers and similarity measure. To evaluate the performance of the proposed method, multi-temporal hyperspectral Hyperion images for Shadegan Wetland were used. The results show high accuracy and low false alarms rate of proposed method methods with an overall accuracy of more than 96%, kappa coefficient of more than 0.82. Besides, the proposed method can provide 'multiple changes' as well as the magnitude of the extracted changes.

M. Hasanlou (✉) · S. T. Seydi
School of Surveying and Geospatial Engineering, College of Engineering, University of Tehran, Tehran, Iran
e-mail: hasanlou@ut.ac.ir; seydi.teymoor@ut.ac.ir

L. A. Jawad (ed.), *Southern Iraq's Marshes*, Coastal Research Library 36,
https://doi.org/10.1007/978-3-030-66238-7_9

Keywords Change detection · Hyperspectral · Wetlands · Multiple-change Map · Shadegan · Wetland

9.1 Introduction

Wetlands are one of the most influential ecosystems in the natural environment for which it is very difficult to find an alternative (White et al. 2015). The US Army Corps of Engineers defined wetlands as "Those areas that are inundated or saturated by surface or groundwater at a frequency and duration sufficient to support, and that under normal circumstances do support, a prevalence of vegetation typically adapted for life in saturated soil conditions." Wetlands cover nearby 6–7% of the earth's surface (Keramitsoglou et al. 2015; Mereta et al. 2012). The wetlands and waterbody areas have provided many vital benefits for the environment which are improving the quality of water, controlling the soil erosion, recharging underground water tables, sustaining against flooding, filtering toxic material and sediments, providing a defense mechanism against sandstorms, and providing food and habitat for wildlife (Romshoo and Rashid 2014; Jiang et al. 2014; Whiteside and Bartolo 2015).

The earth's ecosystems are continuously changing due to natural phenomena (flood, drought) and human activities (urban developing) (Gibbes et al. 2009). The wetland changes originated from some events that included dry seasons, alterations in groundwater, and habitat heterogeneity (Taminskasa et al. 2013; Romshoo and Rashid 2014; Rapinel et al. 2015). Fig. 9.1 presented the change of the Shadegan wetland during 1991 through 2015.

Remote sensing (RS) plays a role key in the monitoring of the changes in the environment, especially in wetlands, on different scales (Mabwoga and Thukral 2014; McCarthy et al. 2015). In fact, RS can provide data from the environment on a large scale and real time with minimum cost and time consumption (Bovolo and Bruzzone 2015; Gómez et al. 2016). These properties have made RS a very effective approach in the fields of earth and environment sciences, especially in change detection applications (Liu 2015; Huang et al. 2017; Storey et al. 2017).

Change detection (CD) is a process that aims to measure the difference between two objects at different times (Lu et al. 2011; Singh 1989). One benefit of the CD is to help manage a system more efficiently by using a multi-temporal dataset (Thonfeld et al. 2016). Also, detection of changes can help us create accurate change models based on past information to avoid disastrous events (Hegazy and Kaloop 2015; Thonfeld et al. 2016). With the development of RS systems, it is possible to obtain data from objects in the high spectral resolution which is known as hyperspectral imagery (George et al. 2014). The high spectral resolution of the data helps with distinguishing objects that seem very similar (Seydi and Hasanlou 2018; Barrett 2013; Smith 2012; Yuen and Richardson 2010).

During recent years, the most relevant studies on CD in wetlands have been using remote sensing data. Sica et al. (2016) study on the Paraná River Delta located in Argentina, where change analysis was performed based on post-classification via the

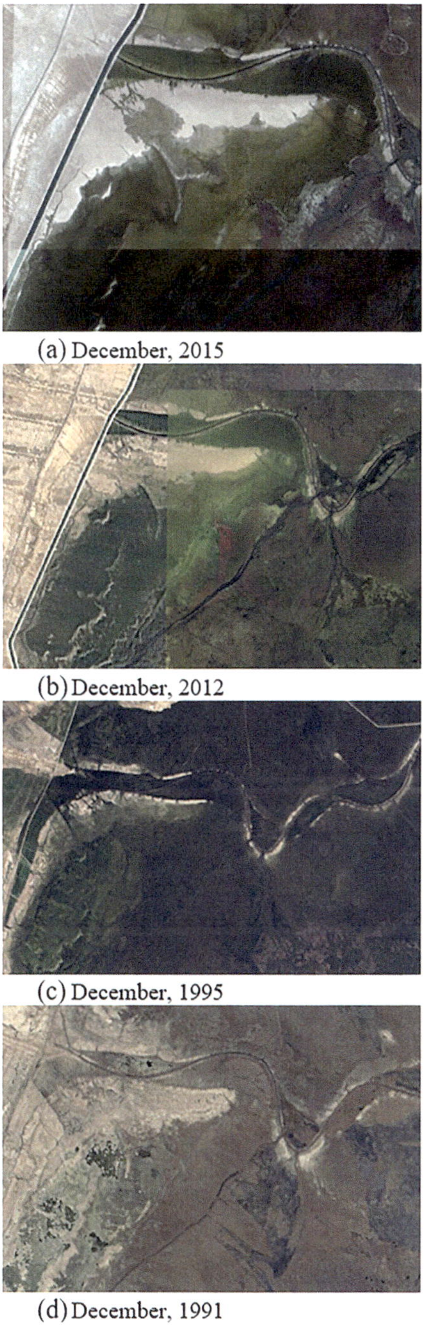

Fig. 9.1 The change of the Shadegan wetland. (**a**) Landsat image in December, 2015, (**b**) Landsat image in December, 2012, (**c**) Landsat image in December, 1995 and (**d**) Landsat image in December, 1991

supervised method and support vector machine (SVM) classifier. Also, Seydi and Hasanlou (2016) studied the Shadegan wetlands located in Iran. This research proposed a new hybrid method for detecting changes that used a semi-supervised method based on iteratively reweighted multivariate alteration detection (IR-MAD) algorithms, Z-score analysis, and Otsu algorithm. They also used hyperspectral image datasets. In an older study, Ghobadi et al. (2015) studied the Al-Hawizeh wetlands located in the southwest region of the Iran-Iraq border. For change analysis, they used maximum likelihood as the post-classification method and classifier on the Landsat datasets (OLI,[1] MSS,[2] and ETM[3]).

Mousazadeh et al. (2015) studied Anzali wetlands in Iran, where their approach integrated supervised classification using maximum likelihood classifier and zonal and object-oriented image analyses. Also, in their study, Landsat 8 and digital topographic maps datasets were used. In another work, Yang and Yan (2016) conducted a change analysis study on Poyang Lake wetland in China where they used supervised classification procedures using error-correcting output code (ECOC) and SVM algorithm. Also in their study, the hydrological data and remotely sensed data contained TM,[4] ETM, OLI, and TIRS.[5] Gunawardena et al. (2014) monitored the eastern river basin region in Sri Lanka using supervised classification methods and Landsat datasets including ETM+,[6] ALOS[7]-AVNIR-2,[8] and ALOS-PALSAR[9] images. Capella Zanotta et al. (2013) studied the central portion of South American areas, specifically the Brazilian Pantanal. They have investigated automatic hybrid methods based on expectation-maximization (EM) and image difference detection. Also, their method showed improvements in CD efficiency by incorporating morphology operators using Landsat dataset for change analysis. Kayastha et al. (2012) analyzed an area in northern Virginia for CD using Z-score and the tasseled cap algorithm. In their paper, threshold selection was performed based on time series analysis of Landsat ETM datasets.

By considering both the CD methods and the employed datasets in related literature, it can be observed that there are several challenges in CD on wetland regions. Firstly, we can see that the most frequently used procedure for CD is the post-classification method. Secondly, the most widely used image datasets for the application of CD in wetland areas are different types of Landsat imagery. Therefore, there is a lack of research based on hyperspectral images for CD applications. On the other hand, hyperspectral imagery has displayed high potential for many

[1]Operational Land Imager.

[2]Multispectral Scanner System.

[3]Enhanced Thematic Mapper.

[4]Thematic Mapper.

[5]Thermal Infrared Sensor.

[6]Enhanced Thematic Mapper Plus.

[7]Advanced Land Observation Satellite.

[8]Advanced Visible and Near Infrared Radiometer type 2.

[9]Phased Array type L-band Synthetic Aperture Radar.

applications such as classification and CD. Also, several studies have been conducted about this type of imagery (Hasanlou et al. 2015; Kumar and Sinha 2014) which can be considered to be applied in monitoring changes in the wetlands and waterbody areas.

The CD methods using RS imagery can be divided into five groups. The first group includes post-classification comparison-based procedures (Castellana et al. 2007).

The first group of methods includes similarity-based methods where the spectral signature of objects is measured (Adar et al. 2012). The advantages of these methods include the simplicity of implementing them and their low computational cost. Nevertheless, they can be affected by noise and atmospheric conditions, need to threshold selection (Liu 2015; Shah-Hosseini et al. 2015).

The second group is the transformation-based methods where the dataset is transformed from image space to another space (Pieper et al. 2015; Shah-Hosseini et al. 2015; Vongsy 2007). These methods have high potential in processing data with high dimensionality and, thus, high capability in CD. The common transformation-based methods include principal component analysis (PCA) (Vongsy et al. 2009), multivariate alteration detection (MAD) (Nielsena and Müllerb 2003), chronochrome (CC) (Eismann et al. 2008), and cross equalization (CE) (Eismann et al. 2008). The main disadvantage of this method, need to threshold selection and finding informative principles for the extraction of changes.

The third group is the post-classification comparisons which are widely used for detecting changes based on comparing classified images in a pixel by pixel class label manner (Dronova et al. 2011; Lee 2011; Zhao et al. 2010). This group provides "multiple-change" or "from-to" information and is not affected by the atmospheric conditions and sensor differences in the acquisition data. However, prior knowledge for the training set is necessary for this group, which is a big challenge for supervised methods due to the fact that acquiring training sets in multi-temporal datasets can be very difficult (Liu 2015; Shah-Hosseini, Homayouni, and Safari 2015). When using hyperspectral imagery, it is inevitable to use dimension reduction procedures due to the Hughes phenomenon (Samadzadegan et al. 2012). Also, for unsupervised methods, it is necessary to label the classes to be able to analyze the change map (Shah-Hosseini et al. 2015). The accuracies of both supervised and unsupervised CD methods depend on the performance of the utilized classifier algorithm (Liu 2015; Pacifici 2007; Shah-Hosseini et al. 2015). The common supervised post-classification comparison-based methods are maximum likelihood (ML) (Lee 2011; Mousazadeh et al. 2015; Yang and Yan 2016), SVM (Sica et al. 2016; Yang and Yan 2016), and random forest (RF) classifiers (Franklin et al. 2015). Also, the common unsupervised methods are ISODATA[10] classifier (Omo-Irabor 2016), fuzzy C-means (FCM) (Ghosh et al. 2011), and K-means (KM) (Fröjse 2011).

[10]Iterative Self-Organizing Data Analysis Techniques.

The fourth group of CD methods that use hyperspectral imagery uses direct multi-date classification (DMC) based on using one classifier algorithm on stacks of multi-date datasets (Ahlqvist 2008). In this group, due to the utilization of one classifier, the computational cost of classification is low. However, this group of method suffers from drawbacks such as providing little knowledge about the "from-to" information and the fact that for supervised methods, it is necessary to have training sets, and also, they need high computational space for process (Shah-Hosseini et al. 2015; Yuan et al. 2005).

The fifth group is hybrid-based procedures that combine the previous methods in order to achieve new automatic or unsupervised methods (Shah-Hosseini et al. 2015; Bovolo et al. 2012).

We described five CD groups and briefly investigated their pros and cons. Generally, there are many challenges in hyperspectral change detection including (1) the outputs of many segment-based threshold selection procedures are not perfect; therefore, these methods require clear histograms of *change* and *no-change* areas. Also, some CD methods require hyper-parameter tuning, which is necessary to be performed based on experimental knowledge (Shah-Hosseini et al. 2015). (2) Many of the automatic methods do not provide information about the nature of changes but only provide the binary change maps, while multiple-change information is important for decision-making. Moreover, these methods do not provide the amplitude of changes (Hussain et al. 2013; Shah-Hosseini et al. 2015). Also, some CD methods need to have knowledge-based threshold that it is hard to set. (3) However, the post-classification and direct classification methods could provide a multiple-change map or "from-to" information, but these methods are supervised; therefore providing training data is inevitable. However, collection of this training data can be very difficult, and (4) as described in the literature review in the previous section, many of the studies used multispectral dataset to monitor the wetland regions; therefore, there is a lack of research that investigates the capabilities of hyperspectral imagery in CD in wetlands and waterbody areas. Nevertheless, a series of spaceborne sensors (e.g., EnMAP,[11] PRISMA,[12] and HyspIRI[13]) will be launched on a schedule that will increase the availability of hyperspectral imagery with improvement in data quality. With this regard, it is necessary to utilize datasets that provide more detail about changes.

Wetlands are very sensitive ecosystems, which implies that monitoring of their changes is necessary for protecting them. In order to provide a monitoring framework to address this issue, we need to focus on informative image datasets and accurate methods. The CD problem could be solved in a simple framework. There are many novel algorithms proposed by researchers for the detection of changes using hyperspectral imagery that solved CD in a complex framework. Therefore, these novel methods improve the performance of CD, but the CD problems become

[11]Environmental Mapping and Analysis Program.

[12]PRecursore IperSpettrale della Missione Applicativa.

[13]Hyperspectral Infrared Imager.

more complex and hard. So, this research proposed a CD method for hyperspectral imagery on wetland and waterbody areas using conventional algorithms. The main novelty proposed method is a simple theme nonetheless preserved accuracy. In addition, the proposed method could be applied in an automatic framework and provides more details of the nature of changes. The main purpose of this chapter is to propose a CD hybrid method that addresses the previously mentioned CD issues. In addition, this study has a number of minor objectives including (1) sensitivity assessment of different kernel functions on hybrid change detection (HCD) and (2) evaluating the effects of normalization steps on input data on the performance of the SVM classifier. In fact, the proposed method is a new HCD method based on the Otsu algorithm, CC, Z-score, PCC,[14] and SVM and has three phases including (1) global predictor phase, (2) analysis phase, and (3) decision phase. More specifically, the global predictor phase uses CC algorithm for highlighting *change* and *no-change* area; the analysis phase uses the SISAL,[15] HFC,[16] and Z-score analysis for data analysis, and the decision phase uses the SVM classifier, Otsu algorithm, and PCC to obtain the binary change map, the amplitude of change map, and the "multiple-change" information map. The criteria selection mentioned methods in the proposed framework are simple for implementation and robust for the analysis of high-dimensional data. In addition, the source codes of these methods are available and can be found online at http://rslab.ut.ac.ir. This hybrid method benefits from several advantages that distinguish it from other HCD methods including (1) sensitivity to subtle changes with high accuracy and low false alarm rates, (2) providing the "multiple-change" information and amplitudes of changes in addition to binary change map, (3) simple implementation compared to common HCD methods, (4) low computational cost and the ability to process high-dimensional data, (5) no need for training set or its unsupervised framework, and (6) incorporating hyperspectral datasets which have high potentials in most applications especially for CD analysis. The rest of this chapter is organized as follows: Section 9.2 describes the general proposed methodology. The details of the proposed method are presented in Sects. 9.3 and 9.4 presents the experimental results of this method.

9.2 Proposed Hybrid Method

This section investigated the detail of the proposed method. The flowchart of the proposed method is illustrated in Fig. 9.2. The proposed method provides three different change maps that are the binary change map, the magnitude of changes, and the multiple-change map.

---

[14]Pearson correlation coefficient.

[15]Simplex identification via split augmented Lagrangian.

[16]Harsanyi–Farrand–Chang.

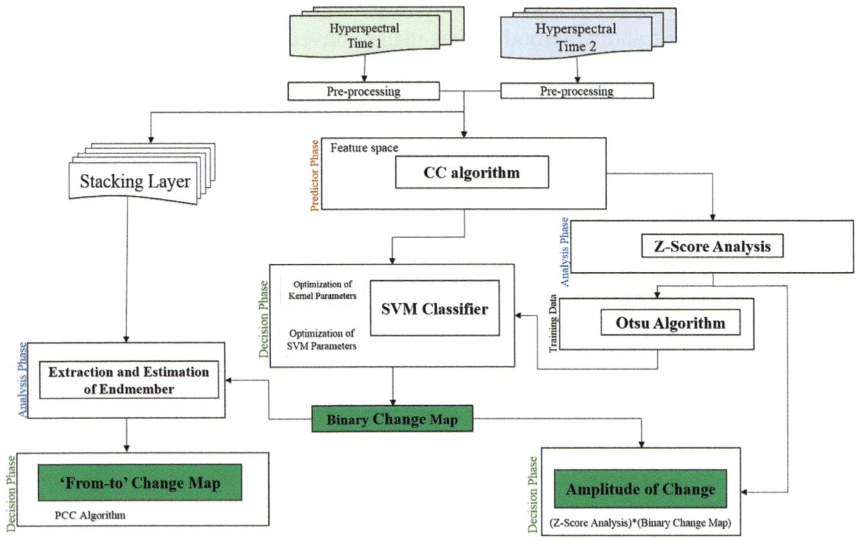

Fig. 9.2 An overview of the proposed method and three blue output boxes

9.2.1 Global Predictor Phase

The main purpose of the predictor phase is to distinguish the *change* pixels from the *no-change* pixels. For this purpose, the CC algorithm was used to highlight the *changing* area from the *no-change* area based on second-order statistics. The result of this phase is a cube data that change areas that differ from *no-change* areas by intensity.

9.2.2 Analysis Phase

The main purpose of this part includes (1) extraction and estimation of endmembers on masked stack data which is performed using HFC and SISAL algorithms, and (2) the output of the CC method is a cubic data; therefore, to aggregate and standardize the output of CC method, the Z-score analysis was applied, and then the single-band data was generated. After extracting and estimating the endmembers, the PCC method is applied to generate the "multiple-change" map. The output of Z-score analysis is used for two purposes: (1) the Z-score analysis is used in combination with the Otsu algorithm in order to generate unsupervised training data, and (2) this method combines binary change maps for extracting the amplitude of change map as a single output.

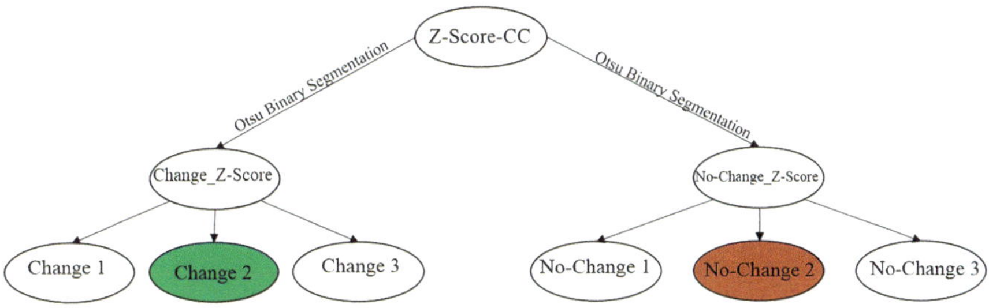

Fig. 9.3 Extraction of training data using iterative Otsu algorithm

9.2.3 Decision Phase

The decision phase is used for (1) locating the *change* and *no-change* pixels (i.e., binary change map), (2) extracting "multiple-change" information, (3) calculating the amplitude of the change map, and (4) automatically generating training data. This phase uses three algorithms: the SVM classifier, the Otsu algorithm, and the PCC algorithm.

9.2.3.1 Training Data Generation

This part explains the automatic production of training data for the SVM classifier by incorporating the Otsu algorithm over the output of CC and the Z-score analysis. After pre-processing of input bitemporal hyperspectral datasets, the first step in the flowchart of the proposed method begins with applying CC transformation and highlighting *change* pixels from *no-change* pixels. The second step of the proposed method is to implement the Otsu algorithm for producing the initial change map. This initial change map contains several *change* pixels mixed with *no-changed* pixel (i.e., unfavorable change pixels) (Fig. 9.3). Therefore, for the initial change map that contains two classes, *change* and *no-change*, the Otsu algorithm is applied once again for each of the previous output classes on the Z-score pixels, and the three classes are divided according to Fig. 9.2.

This process causes more isolation and increases the reliability of *change* and *no-change* pixels. The main reason for dividing the three classes is that the first class for *no-change* class and the third class for change contain many noise pixels because the noise has the minimum value and the maximum value. Therefore, the first class for *no-change* and the third class for *change* are eliminated. The third class for *no-change* and the first class for *change* contain mixed pixels of *change* with *no-change*; therefore, these classes are also removed. In the next step, the pixels whose locations are found via the output of the CC algorithm are selected as the training set for the SVM classifier.

9.2.3.2 Tuning SVM's Kernel Parameters

After producing training data, the parameters of the SVM classifier, including the optimal kernel parameters, are tuned. In this regard, the prepared input datasets are divided into two groups: (1) training data (30% of pixels) and (2) testing data (70% of pixels). The tuning parameters are based on grid search (GS), and the evaluation type is cross-validation that a range is defined for the parameters of kernel and SVM. The SVM classifier is trained using training data based on the defined value of in GS then model made evaluated on training data used criteria such as overall accuracy. The process is repeated until all of the ranges are covered. Finally, the best value of accuracy is equal to tune parameters.

9.2.3.3 Binary Change Map

In the next step, the SVM classifier (based on obtained tune parameters in the previous section) is applied to the output of the CC algorithm. The output of this classifier is a binary change map (i.e., a map with two classes: *change* and *no-change* pixels). The binary change map is determined by assigning each pixel in the image space *change* or *no-change* values. The values of *change* pixels are set to one, and the *no-change* pixels are set to zero.

9.2.3.4 Amplitude of Change

The amplitude of the change map shows the intensity of change. Thus, a high-intensity value represents higher change. The amplitude of the change map is extracted by multiplying the final binary change map by Z-score values, as shown in Fig. 9.4.

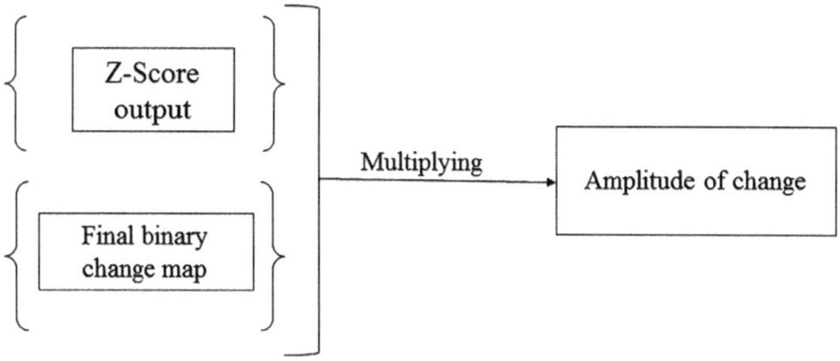

Fig. 9.4 The flowchart of computing amplitude of change

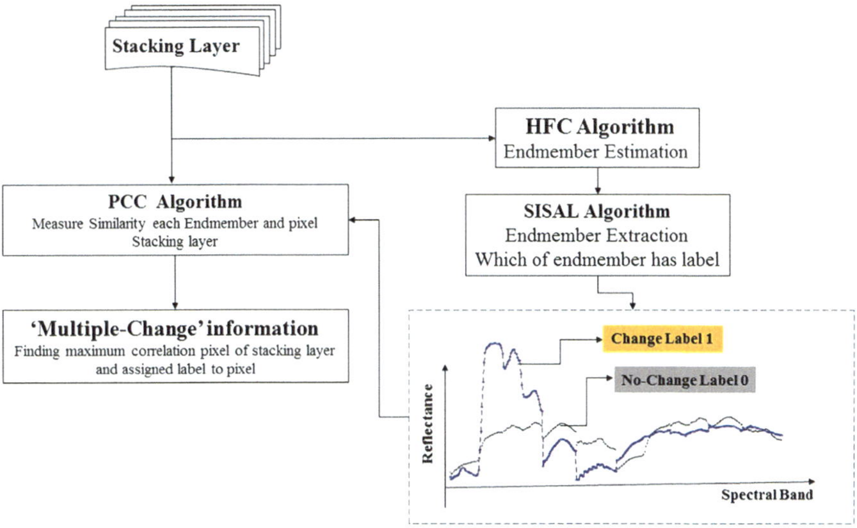

Fig. 9.5 The flowchart retrieving "multiple-change" information

9.2.3.5 Extracting "Multiple-Change" Information

Retrieving accurate "multiple-change" change information is required in many CD analyses (Hussain et al. 2013). In this study, "multiple-change" information is created based on the estimation and extraction of endmembers in the multi-temporal hyperspectral dataset. For this purpose, these four steps are required: (1) masking the *change* area on stack hyperspectral dataset using final binary change map, (2) estimating endmembers by using HFC methods, (3) extracting endmember using SISAL algorithm, and (4) assigning a label for each pixel by finding maximum similarity between pixels of stacking layers and the extracted endmembers by incorporating PCC algorithm. To produce the "multiple-change" map, it is necessary to assign a label to each endmember. Therefore, the PCC algorithm measures the similarity between each endmember and each pixel in stacked hyperspectral data. Usually, assigned labels correspond to endmembers with the highest similarity value. In the next step, pixels with high similarity values related to one of the endmembers are assigned their corresponding labels (Fig. 9.5).

9.3 Methodology

As discussed in the previous section, the proposed hybrid method consists of six main algorithms (Fig. 9.2): (1) chronochrome (CC), (2) Z-score analysis, (3) Otsu algorithm, (4) the SVM classifier, (5) endmember extraction and estimation algorithm, and (6) Pearson correlation coefficient (PCC). Also, the proposed method consists of three main phases including (a) predictor phase, (b) analysis phase, and (c) decision phase. These algorithms and phases are described in more detail in this section.

9.3.1 Chronochrome

The chronochrome approach, proposed by Stocker and Schaum, provides a prediction based on the joint second-order statistics between the reference and test images (Eismann et al. 2008; Schaum and Stocker 1998). The main purpose of this method is to estimate the background in the test image as a linear function of the reference image and detect the changes in the resulting difference image (Schaum and Stocker 1998; Vongsy 2007). For this purpose, given x, a linear predictor is fitted for y. The centered covariance and cross-covariance are computed before fitting a linear estimation to y-data as a function of the x-data (Eq. 9.1):

$$\mathrm{X} = \langle \mathrm{x}|\mathrm{x}^{\mathrm{t}}\rangle, \mathrm{Y} = \langle \mathrm{y}|\mathrm{y}^{\mathrm{t}}\rangle, \mathrm{C} = \langle \mathrm{y}|\mathrm{x}^{\mathrm{t}}\rangle \tag{9.1}$$

A linear estimate of the y-data from the x-data is (Eq. 9.2):

$$\mathrm{y} = \mathrm{Lx} \tag{9.2}$$

where L is the optimal vector wiener filter solution that it is given by (Eq. 9.3):

$$\mathrm{E} = \left\langle (\mathrm{y} - \mathrm{Lx})(\mathrm{y} - \mathrm{Lx})^{\mathrm{t}} \right\rangle \tag{9.3}$$

Also, E is minimized when $L = CX^{-1}$. Therefore, we have:

$$\mathrm{y} = \mathrm{Lx} = \left(\mathrm{CX}^{-1}\right)\mathrm{x} \tag{9.4}$$

And according to Eqs. 9.3 and 9.4:

$$\varepsilon_{\mathrm{cc}} = (\mathrm{y}) - \left(\left(\mathrm{CX}^{-1}\right)\mathrm{x}\right) \tag{9.5}$$

where ε_{cc} is a change residual image. As depicted in the flowchart of the proposed method in Fig. 9.2, chronochrome is incorporated in the predictor phase on the hyperspectral data.

9.3.2 Z-Score Analysis

The Z-score provides the magnitude and directions of deviation from the mean of the distribution which is introduced in the distribution unit of standard deviation. The Z-score is defined in Eq. 9.6 as follows (Cheadle et al. 2003):

$$\text{modified} - \text{Z\_Score} = \sum_{i=1}^{N} ((x_i - \text{mean}_i)/\text{std}_i)) \tag{9.6}$$

In this study, a version of Z-score value is adopted which is, in fact, a normalization that allows us to have the amplitude of change as the output of our proposed method. As stated in Eq . 9.6, the output of this procedure is a single band (Fig. 9.5). The Z-score analysis is then applied to the output of the CC method in the analysis phase as depicted in the flowchart of the proposed method (Figs. 9.2 and 9.6).

9.3.3 Otsu Algorithm

The Otsu algorithm is a group thresholding algorithm that performs image clustering automatically. The idea behind this approach is that the threshold value determines the weight of the variance within the minimum class value. The variance within the class is the variance of the total weight of each defined cluster (Ng 2006; Otsu 1979). In this study, the Otsu algorithm was applied for unsupervised preparation of training data for the SVM classifier according to as shown in the flowchart of the proposed method (Fig. 9.2).

9.3.4 Endmember Extraction

The common method for producing "multiple-change" information is classification, which was discussed in the introduction section. However, this chapter proposes a new procedure for retrieving "from-to" information without applying classification. In this regard, the proposed procedure uses estimation/extraction endmembers as well as the PCC algorithms. On the other hand, many methods are developed for estimating the number of endmembers. We apply the popular HFC method which is based on the distribution of the differences of the eigenvalues of the correlation and the covariance matrices, respectively (Chang and Du 2004). After estimating the

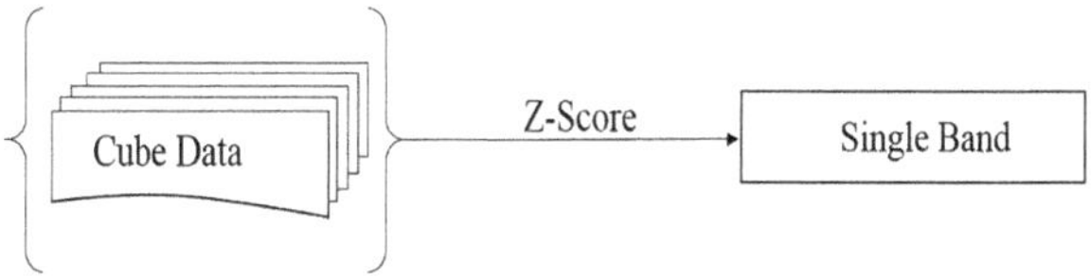

Fig. 9.6 The output of the Z-score analysis has a single band

number of endmembers, endmember extraction begins. Various endmember extraction methods exist in the literature including SISAL (Bioucas-Dias 2009; Keshava 2003; Parente and Plaza 2010). The SISAL algorithm is an unsupervised method for endmember extraction based on fitting a minimum volume simplex to the data subject to a series of constraints. However, it is inevitable to estimate the number of endmembers before using the SISAL algorithm (Bioucas-Dias 2009). In this study, we apply the endmember extraction on the output of the CC algorithm as it can be observed in the flowchart of the proposed method (Fig. 9.2).

9.3.5 Pearson Correlation Coefficient

The PCC is one of the most popular measures for calculating the dependency between two spectral vectors. This measure is widely used in remote sensing applications (Wang 2013). The PCC between spectral random vectors is defined as:

$$\mathrm{PCC} = \frac{\mathrm{cov}(x, y)}{\sigma_x \sigma_y} \tag{9.7}$$

where x and y represent the target and reference spectra and σ_x and σ_y are the standard deviations of x and y spectral vectors, respectively. This study utilizes PCC in the decision phase for stacking layer data to retrieve the "from-to" information, as depicted in the flowchart of the proposed method (Fig. 9.2).

9.3.6 Support Vector Machine

The SVM is a supervised machine learning algorithm that is commonly used for classification purposes and is based on the statistical learning theory (Vapnik 2013). SVM has recently been applied in the classification of multispectral and hyperspectral remote sensing datasets successfully (George et al. 2014; Melgani and Bruzzone 2004). The main idea behind SVM is to find a hyperplane that maximizes the margin between the two classes (Vapnik 2013). This algorithm has several critical parameters including kernel parameters and the penalty coefficient (C). The popular kernels incorporated in SVM include polynomial, radial bias function, and linear kernels (Gaspar et al. 2012; Hasanlou et al. 2015). Different types of kernels and parameters for SVM are presented in Table 9.1.

This study incorporates the SVM algorithm, in the decision phase, on the output of the CC algorithm to make binary changes as illustrated in the flowchart of the proposed method (Fig. 9.2).

Table 9.1 Different types of kernels and parameters in the SVM classifier

Kernel type	Formula	Estimation parameters	Number parameters
Linear	$k(x, y) = x^T y$	C	1
Polynomial	$k(x, y) = (\gamma x^T y + \beta_0)^d$	d, γ, β_0, C	4
Radial basis function	$k(x, y) = e^{\left(-\gamma \|x-y\|^2\right)}$	γ, C	2

9.4 Experiments

In this section, the experimental data and study area are discussed. Also, the results extracted from the proposed method evaluated by qualitative and quantitative criteria are presented. In addition, the change map results of the proposed method are compared with the most common and popular CD algorithms.

9.4.1 Study Area

In this study, three different satellite (i.e., EO-1) hyperspectral image datasets are used for analyzing changes in wetlands and waterbodies as illustrated in Fig. 9.7 and Table 9.2. These datasets have been previously used in many hyperspectral change detection papers (Liu 2015; Seydi and Hasanlou 2016, 2017; Wu et al. 2012), and they can be considered benchmark datasets. The ground truth datasets were developed by the authors through visual analysis and interpretation of the abovementioned researches. Additionally, by using high-resolution image datasets from Google Earth, a detailed visual comparison was carried out. Details and descriptions of each dataset will be presented in the next section. The Hyperion sensor contains 242 spectral bands with wavelengths between 0.4 and 2.5 micrometers and with a spatial resolution of 30 m and a bandwidth of 7.5 km. Hyperion data were obtained at two separate range images using the push broom technology (Jafari and Lewis 2012). One of these spectra was a VNIR range which includes 70 bands between wavelength 356 and 1058 nm and SWIR wavelength consisting of 172 bands between wavelength 852 and 2577 nm ("USGS EO-1" 2017).

9.4.1.1 Poyang Lake (Dataset #1)

The Poyang Lake located in Jiangxi Province is one of largest freshwater resource and biggest floodwater storage wetland areas in China which is located within coordinates 28°24′ to 29°46′N and 115°49′ to 116°46′E (Chan and Xu 2013; Yang and Yan 2016). The extended area of the captured region in the hyperspectral dataset is 232 × 131 pixels. These datasets were acquired from July 16, 2004 and July 27, 2002 (Fig. 9.7a, b).

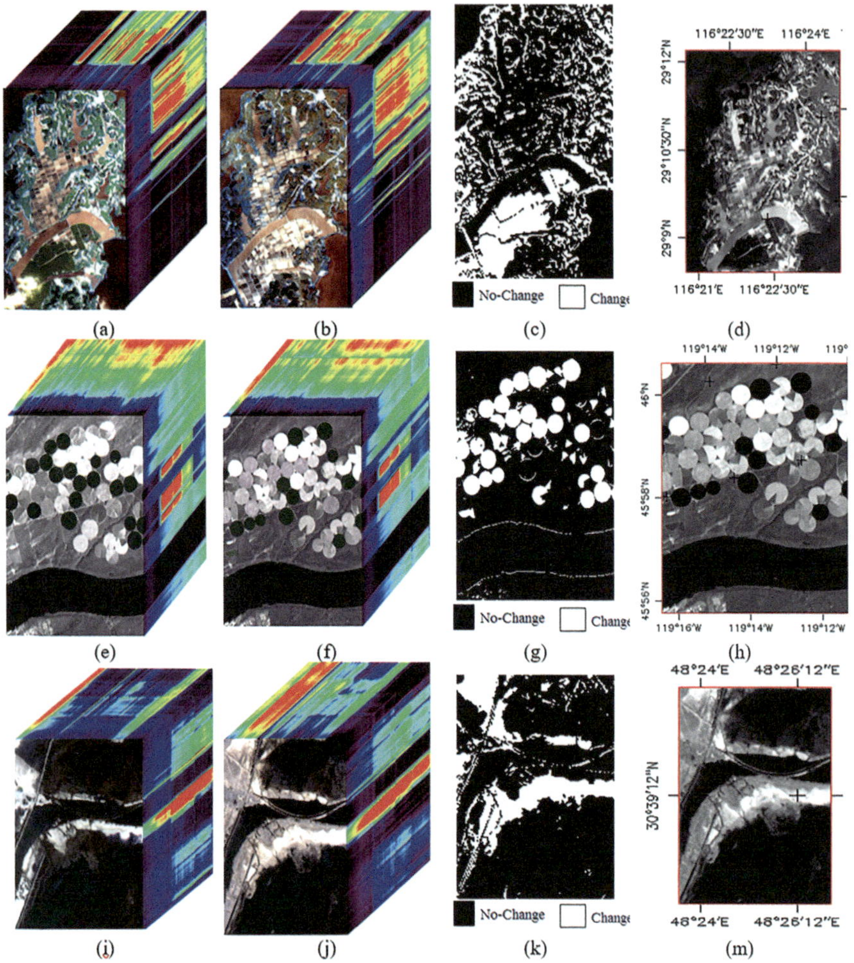

Fig. 9.7 The (**a**) and (**b**) presented false-color composite of the original hyperspectral images acquired in 2002 and 2004 of dataset #1 in China, respectively, (**c**) ground truth, and (**d**) presented geographical location dataset #1. The (**e**) and (**f**) presented false-color composite of the original hyperspectral images acquired in 2004 and 2007 of dataset #2 in the USA, respectively, (**g**) ground truth, and (**h**) presented geographical location dataset #2. The (**i**) and (**j**) present false-color composite of the original hyperspectral images acquired in 2006 and 2006 of dataset #3 in Iran, respectively, (**k**) ground truth, and (**m**) presented geographical location dataset #3

9.4.1.2 Umatilla River (Dataset #2)

The Umatilla River is a gravel-bed river originating in the Blue Mountains of northeastern regions that flows into the Columbia River at Umatilla, OR, USA (Hughes et al. 2006). The extended area of the captured region in the hyperspectral

Table 9.2 The characteristic of datasets in different study areas

Datasets		Acquired date	# of bands	# of pixels	Spatial resolution (m)	Spectral resolution (nm)
Poyang Lake	#1	July 27, 2002	154	232 × 131	30	10
		July 16, 2004				
Umatilla River	#2	May 1, 2004	154	308 × 241	30	10
		May 8, 2007				
Shadegan wetlands	#3	June 6, 2006	154	220 × 123	30	10
		June 29, 2006				

dataset contains 308 × 241 pixels and was acquired on May 1, 2004 and May 8, 2007 (Fig. 9.7c, d).

9.4.1.3 Shadegan Wetlands (Dataset #3)

Shadegan wetland is one of the largest wetlands in Iran. This wetland is created by the downstream part of the river Jarahi and is located at coordinates 30°50′ to 31°00’N and 48°20′ to 49°20′E. The northern section of this wetland includes freshwater, and the salty waterbody is located in the southern part. Also, this wetland is home to different types of plants. The extent of the desired region extracted from EO-1 Hyperion satellite hyperspectral images was 220 × 123 pixels. In this area, we incorporate two multi-temporal datasets acquired on June 29 and June 6, 2006. In Fig. 9.7e, f, a false-color composite of hyperspectral Shadegan wetland images for two different times is illustrated.

9.4.2 *Implementation*

Data pre-processing plays an important role before the beginning of the main process and can be divided into two categories (Jafari and Lewis 2012): spectral and spatial correction. The pre-processing step starts with spectral correction processing; then spatial correction is applied. The first step of pre-processing consists of omitting no-data bands. In this regard, 44 bands (1–7, 58–76, and 225–242) were removed from our imagery (Jafari and Lewis 2012; Scheffler and Karrasch 2013). Also, of the 198 initial bands, two noisy bands including 77 and 78 as well as a number of other bands were removed (Datt et al. 2003; Khurshid et al. 2006). Therefore, 154 bands were selected in total as the input dataset for the proposed change detection method.

Table 9.3 The results obtained from tuning SVM classifier and kernels parameters in different hyperspectral datasets

		Linear		Polynomial					Radial bias function		
Datasets	Normalize	C	# of SV	C	# of SV	γ	d	β_0	C	# of SV	γ
#1	Yes	2–8	210	2–3	16	2–7	3	1	2–6	17	2–7
	No	23	54	2–7	5	21: 220	3	1	2–3	48	2–7
#2	Yes	2–13	408	2–3	13	2–6	3	1	24	125	2–12
	No	25	206	2–6	6	20: 215	3	1	26	125	2–12
#3	Yes	29	5	2–6	608	2–9	3	1	23	65	2–3
	No	25	15	2–9	4	2–9: 29	3	1	25	61	2–5

In the second step, pixels in sample 129 and all lines are shifted to sample 256 in the shortwave infrared (SWIR) spectral bands (Goodenough et al. 2003; Jafari and Lewis 2012; Li et al. 2008). The third step is de-noising, de-striping, and also removing the zero-line by utilizing means and the global approach (Jafari and Lewis 2012; Scheffler and Karrasch 2013). The fourth pre-processing step is a radiometric correction. To achieve this goal, the digital number (DN) values of pixels are converted to physical radiance. The fifth step of the pre-processing is an atmospheric correction, which we used the FLAASH[17] model. The final step of the pre-processing of the hyperspectral dataset is a spatial correction. The accuracy of the geometric correction (RMSE) was less than 0.4 pixel for all three datasets.

As already discussed, the outputs of the proposed method are (1) binary change map, (2) the amplitude of change map, and (3) the "multiple-change" information map. The structure and details of the proposed method are illustrated in Fig. 9.2. This work considered a type of kernel that is widely utilized in the remote sensing community (Liu and Parhi 2016; Ring and Eskofier 2016; Sakthivel et al. 2016; Shah-Hosseini et al. 2015). To tune and select optimized SVM parameters (i.e., gamma (γ) and penalty coefficient (C)), we performed a CV with GS procedures (Gu et al. 2017; Varma and Simon 2006). Also, to have efficient kernel normalization, training and testing data were applied. In the normalization procedure, the data is mapped to values within the *[0,1]* span. The minimum and maximum values were selected based on the minimum and maximum of training data. Table 9.3 presents the results obtained from tuning parameters for kernel and SVM (i.e., the number of support vectors (# of SV), penalty coefficient (C), and gamma (γ) parameter) for three datasets. In addition, this table presented the optimum values for SVM and kernels in two scenarios: normalize and un-normalize.

[17] Fast Line-of-sight Atmospheric Analysis of Spectral Hypercubes.

Table 9.4 Performance of the SVM classifier using a type of kernels in different hyperspectral datasets

Datasets		#1		#2		#3	
Kernel function	Normalize	Overall accuracy (%)	Kappa	Overall accuracy (%)	Kappa	Overall accuracy (%)	Kappa
Linear	Yes	96.77	0.927	97.16	0.907	92.73	0.756
	No	94.68	0.884	97.05	0.906	92.84	0.760
Polynomial	Yes	90.83	0.806	97.10	0.908	96.17	0.885
	No	88.11	0.705	94.75	0.816	89.57	0.628
Radial bias function	Yes	97.40	0.941	97.16	0.908	96.44	0.892
	No	96.65	0.922	97.16	0.908	96.34	0.890

Table 9.5 The number of endmembers and false alarm probability (Pf) for different datasets

Datasets	# of endmembers	P_f
#1	6	10–3
#2	4	10–3
#3	3	10–5

Table 9.4 presented the performance of the SVM classifier using a type of kernels in different hyperspectral datasets. The results show the RBF kernel has the best performance for three datasets. Also, the normalizing dataset improved the accuracy result of the CD.

As mentioned in the above section, to have "multiple-change" information in this study, we used HFC, SISAL, and PCC algorithms. In the HFC algorithm, false alarm probability (P_f) parameters are assigned. A number of endmembers and false alarm probability are listed in Table 9.5.

As we have already discussed in the previous sections, it is essential to compare and check the performance of the proposed method with common and popular CD methods. In this regard, we incorporated ground truth data for all three datasets to compute the validation criteria. In this paper, both quantitative and qualitative criteria were used for comparing the result. The popularly employed CD methods are principal component analysis (PCA) (Adar et al. 2011; Vongsy et al. 2009; Vongsy 2007), cross equalization (CE) (Adar et al. 2011; Eismann et al. 2008), spectral angle mapper (SAM) (Adar et al. 2011), subspace-based (SSB) (Wu et al. 2013), multivariate alteration detection (MAD) (Nielsena and Müllerb 2003), and iterative reweight-MAD (IR-MAD) (Nielsen 2007; Seydi and Hasanlou 2016). All of these methods require assigning suitable thresholds. In this study, unsupervised segmentation by incorporating the Otsu algorithm was used to set these thresholds.

Therefore, by considering the optimum kernel parameters (Table 9.3) for SVM classification for all datasets (#1, #2, and #3), the proposed method begins. Figure 9.8 shows a visual analysis of the proposed method and other CD methods on multi-temporal hyperspectral datasets #1. As it is clear in Fig. 9.8, the proposed method can detect all changes and provide information about the feature changes including

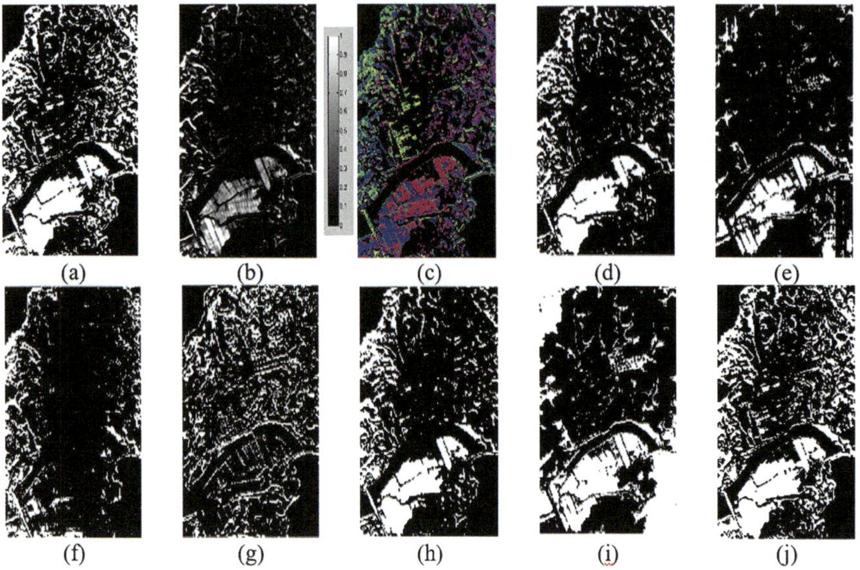

Fig. 9.8 The results of CD methods for dataset #1. (a) Proposed method, binary change map; (b) proposed method, amplitude of change map; (c) proposed method, "from-to" map; (d) CE; (e) SSB; (f) IR-MAD; (g) MAD; (h) PCA; (i) SAM; and (j) ground truth

the "multiple-change" change map as well as the amplitude of changes in the map. This observation empirically proves that the proposed method nearly detects all of the changes compared to the other techniques. In the endmember extraction section, we described that the SISAL and HFC methods were used to obtain the "multiple-change" map. Hence, six classes detected and produced the "multiple-change" change map (Fig. 9.8c) for dataset #1. Figure 9.8b shows the amplitude of changes where the changing intensity is clearly highlighted.

The same computational approach is applied to dataset #2. Figure 9.9 shows changes of Umatilla River where there are many land cover change types in areas that contain different agricultural fields. Also, in this figure, there are low changes in the edges of the river. In this dataset (#2), the "multiple-change" change map has four classes that are detected by the proposed method (Fig. 9.9c). In Fig. 9.9, some algorithms show a certain level of sensitivity to the waterbody area such as the results presented in Fig. 9.9e, f, g. On the other hand, one can clearly observe from the results (Fig. 9.9a–c) that the proposed method has excellent performance compared to other approaches in this area.

Similarly, Fig. 9.10 presents the results of the CD methods on Shadegan wetland (dataset #3). In this region, the main changes are originated from seasonal changes in the water level. The proposed method can find three different classes for the changing area (Fig. 9.10c). As it is clear from the figure, the similarity-based methods such as SAM technique are not suitable for monitoring the changes due to the extraction of false alarm pixels.

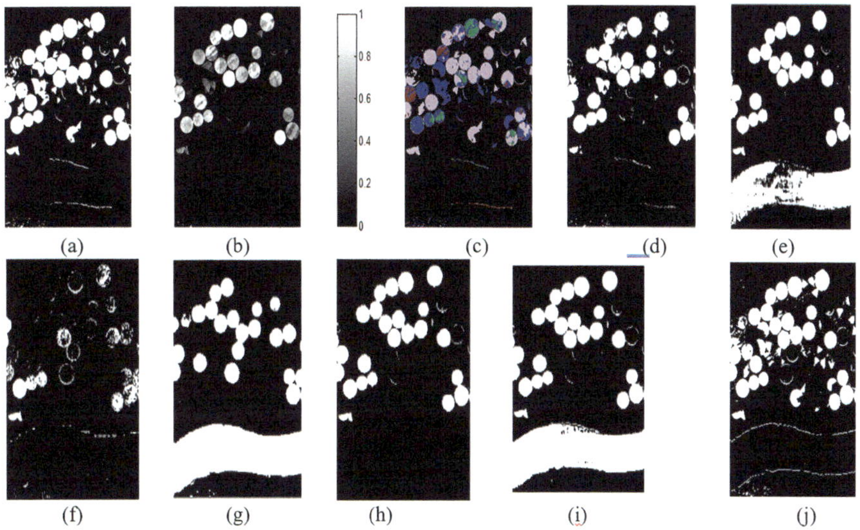

Fig. 9.9 The results of CD methods for dataset #2. (**a**) Proposed method, binary change map; (**b**) proposed method, amplitude of change map; (**c**) proposed method, "from-to" map; (**d**) CE; (**e**) SSB; (**f**) IR-MAD; (**g**) MAD; (**h**) PCA; (**i**) SAM; and (**j**) ground truth

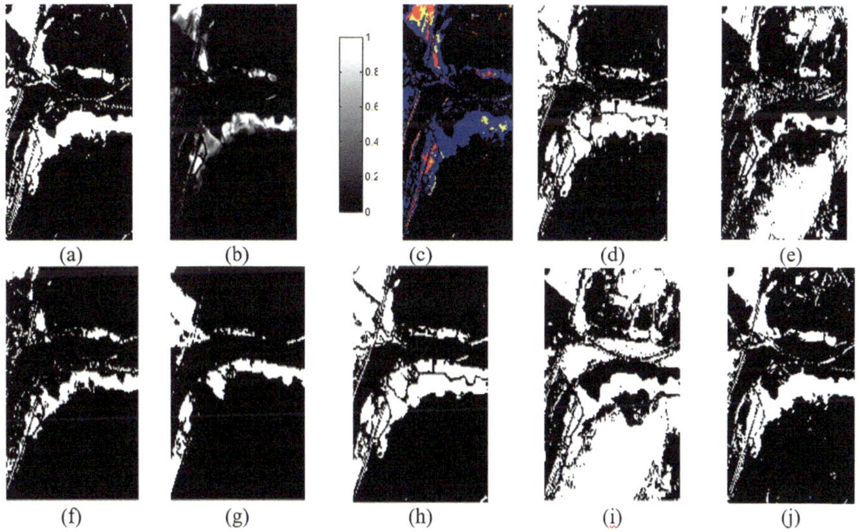

Fig. 9.10 The results of CD methods for dataset #3. (**a**) Proposed method, binary change map; (**b**) proposed method, amplitude of change map; (**c**) proposed method, "from-to" map; (**d**) CE; (**e**) SSB; (**f**) IR-MAD; (**g**) MAD; (**h**) PCA; (**i**) SAM; and (**j**) ground truth

Table 9.6 Performance of the proposed method and other common CD methods for all hyperspectral datasets

Datasets	Indices	SAM	IR-MAD	MAD	CE	SBB	PCA	Proposed method
#1	Overall	66.01	67.07	65.76	88.89	73.11	89.58	97.40
	Kappa	0.223	0.038	0.143	0.727	0.299	0.747	0.941
#2	Overall	75.53	83.85	67.51	93.70	77.855	93.24	97.16
	Kappa	0.371	0.304	0.159	0.778	0.400	0.756	0.907
#3	Overall	52.42	86.06	69.53	83.44	63.86	81.81	96.42
	Kappa	0.111	0.498	0.092	0.527	0.144	0.485	0.890

After the primary observational analysis, we perform a numerical evaluation. In this regard, two common measures are used for evaluating the performance and accuracy of CD methods which include overall accuracy and kappa coefficient. All implemented CD methods are supervised, and their related accuracy is usually computed by examining the best threshold value selection. That means each threshold has a correlation with the accuracy, and maximum accuracy is considered as the final accuracy. Table 9.6 presents the accuracy of the proposed method for the RBF kernel and other CD techniques.

We can clearly observe the superiority of the proposed method compared to other methods in all three different hyperspectral datasets in Table 9.5. Also, as presented in Table 9.5, the PCA and CE methods have efficiencies close to the proposed method, especially for hyperspectral dataset #1 and hyperspectral dataset #2. The IR-MAD algorithm has good performance compared to the MAD algorithm. SAM and SSB have low performance due to sensitivity to noise and atmospheric conditions. These methods utilizing continuous spectral signatures nevertheless, this issue caused to don't suitable for wetland and waterbody change detection using hyperspectral imagery.

This investigation proposed a new change detection method on hyperspectral imagery which included observational and numerical analysis and comparison with other common HCD methods. The proposed method provides three different outputs, which provide more detail about the changes, and thus helps in understanding the changes, while the other CD methods do not give three outputs together. The lecture review in the introduction section and the type of change detection method in five main groups are considered. The challenges and advantages are discussed, and the results of CD certified these issues of CD methods. In addition, more details of visual and numerical analysis show that (1) the hyperspectral imagery has a high capability for CD in waterbodies and wetlands; (2) the CE methods provide better results among common HCD methods; (3) some techniques, such as SSB and SAM, are not suitable for waterbody change detection due to their high sensitivity; (4) the proposed method has the highest accuracy for all employed datasets; therefore, it is efficient for waterbody area; and (5) finally, the proposed methods provide more detail of changes that can help improve the decision-making process.

9.5 Conclusion

Wetlands are critical ecosystems where changes occur frequently and widely. Therefore, creating a framework for monitoring the changes in these ecosystems is essential. In this regard, studying methods that are able to perform accurate change detection in these areas is crucial. This investigation presents a new hybrid method for achieving precise and informative change maps using hyperspectral imagery without requiring prior knowledge of the wetlands and waterbody area. We first discussed all the issues related to CDs using hyperspectral imagery. Therefore, a new change detection method was proposed to address these issues. The proposed hybrid method uses four groups of CD methods to enhance the content and quality of final CD results. The experiments were applied to three real hyperspectral datasets on wetland and waterbody areas from different regions and countries. The output results showed the following: (1) the hyperspectral imagery has high potential to monitoring and assessment of wetland and waterbody areas, however, for this purpose, need special techniques; (2) the visual and numerical analysis proved the excellent performance of the proposed method for hyperspectral change detection compared to other methods; (3) superiority of the proposed unsupervised method without requiring prior knowledge of changes, while some CD methods need training data or setting parameters; (4) the fact that this method can provide binary change map as well as the information about change structure ("multiple-change" map) and also the amplitude map; (5) the use of normalize data and RBF kernel improved the accuracy CD, significantly; and (6) the fact that the implementation of the proposed method is simple and has high efficiency in comparison to other famous and commonly used CD methods like PCA, CE, IR-MAD, SSB, and SAM.

References

Adar S, Notesco G, Brook A, Livne I, Rojik P, Kopackova V et al (2011, October) Change detection over Sokolov open-pit mining area, Czech Republic, using multi-temporal HyMAP data (2009–2010). In: Image and signal processing for remote sensing XVII, vol 8180. International Society for Optics and Photonics, p 81800T

Adar S, Shkolnisky Y, Dor EB (2012, July) New approach for spectral change detection assessment using multistrip airborne hyperspectral data. In: Geoscience and remote sensing symposium (IGARSS), 2012 IEEE International. IEEE, pp 4966–4969

Ahlqvist O (2008) Extending post-classification change detection using semantic similarity metrics to overcome class heterogeneity: a study of 1992 and 2001 US National Land Cover Database changes. Remote Sens Environ 112(3):1226–1241

Barrett EC (2013) Introduction to environmental remote sensing. Routledge

Bioucas-Dias JM (2009, August) A variable splitting augmented Lagrangian approach to linear spectral unmixing. In: Hyperspectral image and signal processing: evolution in remote sensing, 2009. WHISPERS'09. First workshop on. IEEE, pp 1–4

Bovolo F, Bruzzone L (2015) The time variable in data fusion: a change detection perspective. IEEE Geosci Remote Sens Mag 3(3):8–26

Bovolo F, Marchesi S, Bruzzone L (2012) A framework for automatic and unsupervised detection of multiple changes in multitemporal images. IEEE Trans Geosci Remote Sens 50(6):2196–2212
Castellana L, D'Addabbo A, Pasquariello G (2007) A composed supervised/unsupervised approach to improve change detection from remote sensing. Pattern Recogn Lett 28(4):405–413
Chan KKY, Xu B (2013) Perspective on remote sensing change detection of Poyang Lake wetland. Ann GIS 19(4):231–243
Chang CI, Du Q (2004) Estimation of number of spectrally distinct signal sources in hyperspectral imagery. IEEE Trans Geosci Remote Sens 42(3):608–619
Cheadle C, Cho-Chung YS, Becker KG, Vawter MP (2003) Application of z-score transformation to Affymetrix data. Appl Bioinforma 2(4):209–217
Datt B, McVicar TR, Van Niel TG, Jupp DL, Pearlman JS (2003) Preprocessing EO-1 Hyperion hyperspectral data to support the application of agricultural indexes. IEEE Trans Geosci Remote Sens 41(6):1246–1259
Dronova I, Gong P, Wang L (2011) Object-based analysis and change detection of major wetland cover types and their classification uncertainty during the low water period at Poyang Lake, China. Remote Sens Environ 115(12):3220–3236
Eismann MT, Meola J, Hardie RC (2008) Hyperspectral change detection in the presence of diurnal and seasonal variations. IEEE Trans Geosci Remote Sens 46(1):237–249
Franklin SE, Ahmed OS, Wulder MA, White JC, Hermosilla T, Coops NC (2015) Large area mapping of annual land cover dynamics using multitemporal change detection and classification of Landsat time series data. Can J Remote Sens 41(4):293–314
Fröjse L (2011) Multitemporal satellite images for urban change detection
Gaspar P, Carbonell J, Oliveira JL (2012) On the parameter optimization of Support Vector Machines for binary classification. J Integr Bioinform 9(3):33–43
George R, Padalia H, Kushwaha SPS (2014) Forest tree species discrimination in western Himalaya using EO-1 Hyperion. Int J Appl Earth Obs Geoinf 28:140–149
Ghobadi Y, Pradhan B, Shafri HZ, bin Ahmad N, Kabiri K (2015) Spatio-temporal remotely sensed data for analysis of the shrinkage and shifting in the Al Hawizeh wetland. Environ Monit Assess 187(1):1–17. https://doi.org/10.1007/s10661-014-4156-0
Ghosh A, Mishra NS, Ghosh S (2011) Fuzzy clustering algorithms for unsupervised change detection in remote sensing images. Information Sciences 181:699–715. https://doi.org/10.1016/j.ins.2010.10.016
Gibbes C, Southworth J, Keys E (2009) Wetland conservation: change and fragmentation in Trinidad's protected areas. Geoforum 40(1):91–104
Gómez C, White JC, Wulder MA (2016) Optical remotely sensed time series data for land cover classification: a review. ISPRS J Photogramm Remote Sens 116:55–72
Goodenough DG, Dyk A, Niemann KO, Pearlman JS, Chen H, Han T et al (2003) Processing Hyperion and ALI for forest classification. IEEE Trans Geosci Remote Sens 41(6):1321–1331. https://doi.org/10.1109/TGRS.2003.813214
Gu B, Sheng VS, Tay KY, Romano W, Li S (2017) Cross Validation Through Two-Dimensional Solution Surface for Cost-Sensitive SVM. IEEE Trans Pattern Anal Mach Intell 39(6):1103–1121
Gunawardena A, Fernando T, Takeuchi W, Wickramasinghe CH, Samarakoon L (2014) Identification, evaluation and change detection of highly sensitive wetlands in South-Eastern Sri Lanka using ALOS (AVNIR2, PALSAR) and Landsat ETM+ data. In: IOP conference series: earth and environmental science, vol 20. IOP Publishing, p 012050
Hasanlou M, Samadzadegan F, Homayouni S (2015) SVM-based hyperspectral image classification using intrinsic dimension. Arab J Geosci 8(1):477–487
Hegazy IR, Kaloop MR (2015) Monitoring urban growth and land use change detection with GIS and remote sensing techniques in Daqahlia governorate Egypt. Int J Sustain Built Environ 4(1):117–124
Huang S, Ramirez C, Kennedy K, Mallory J, Wang J, Chu C (2017) Updating land cover automatically based on change detection using satellite images: case study of national forests in Southern California. GISci Remote Sens:1–20

Hughes ML, McDowell PF, Marcus WA (2006) Accuracy assessment of georectified aerial photographs: implications for measuring lateral channel movement in a GIS. Geomorphology 74(1–4):1–16

Hussain M, Chen D, Cheng A, Wei H, Stanley D (2013) Change detection from remotely sensed images: From pixel-based to object-based approaches. ISPRS J Photogramm Remote Sens 80:91–106

Jafari R, Lewis MM (2012) Arid land characterisation with EO-1 Hyperion hyperspectral data. Int J Appl Earth Obs Geoinf 19:298–307

Jiang F, Qi S, Liao F, Ding M, Wang Y (2014) Vulnerability of Siberian crane habitat to water level in Poyang Lake wetland, China. GISci Remote Sens 51(6):662–676

Kayastha N, Thomas V, Galbraith J, Banskota A (2012) Monitoring wetland change using inter-annual landsat time-series data. Wetlands 32(6):1149–1162

Keramitsoglou I, Stratoulias D, Fitoka E, Kontoes C, Sifakis N (2015) A transferability study of the kernel-based reclassification algorithm for habitat delineation. Int J Appl Earth Obs Geoinf 37:38–47

Keshava N (2003) A survey of spectral unmixing algorithms. Lincoln Lab J 14(1):55–78

Khurshid KS, Staenz K, Sun L, Neville R, White HP, Bannari A, Champagne CM et al (2006) Preprocessing of EO-1 Hyperion data. Can J Remote Sens 32(2):84–97

Kumar L, Sinha P (2014) Mapping salt-marsh land-cover vegetation using high-spatial and hyperspectral satellite data to assist wetland inventory. GISc Remote Sens 51(5):483–497

Lee S (2011) Detecting wetland change through supervised classification of Landsat Satellite Imagery within the Tunkwa Watershed of British Columbia, Canada. Retrieved January 10, 2017, from http://www.diva-portal.org/smash/record.jsf?pid=diva2:681571

Li H, Zhang D, Zhang Y, Xu Y (2008) Research of image preprocessing methods for EO-1 Hyperion hyperspectral data in tidal flat area. Geoinformatics:71471G–71471G

Liu S (2015) Advanced techniques for automatic change detection in multitemporal hyperspectral images. University of Trento. Retrieved January 10, 2017, from http://eprints-phd.biblio.unitn.it/1393/

Liu Y, Parhi KK (2016) Computing RBF kernel for SVM classification using stochastic logic. In: Signal Processing Systems (SiPS), 2016 IEEE International Workshop on. IEEE, pp 327–332

Lu D, Moran E, Hetrick S, Li G (2011) Land-use and land-cover change detection. In: Advances in environmental remote sensing sensors, algorithms, and applications. CRC Press Taylor & Francis Group, New York, pp 273–290

Mabwoga SO, Thukral AK (2014) Characterization of change in the Harike wetland, a Ramsar site in India, using landsat satellite data. Springerplus 3(1):576

McCarthy MJ, Merton EJ, Muller-Karger FE (2015) Improved coastal wetland mapping using very-high 2-meter spatial resolution imagery. Int J Appl Earth Obs Geoinf 40:11–18

Melgani F, Bruzzone L (2004) Classification of hyperspectral remote sensing images with support vector machines. IEEE Trans Geosci Remote Sens 42(8):1778–1790

Mereta ST, Boets P, Bayih AA, Malu A, Ephrem Z, Sisay A, Endale H et al (2012) Analysis of environmental factors determining the abundance and diversity of macroinvertebrate taxa in natural wetlands of Southwest Ethiopia. Eco Inform 7(1):52–61

Mousazadeh R, Ghaffarzadeh H, Nouri J, Gharagozlou A, Farahpour M (2015) Land use change detection and impact assessment in Anzali international coastal wetland using multi-temporal satellite images. Environ Monit Assess 187(12):1–11

Ng H-F (2006) Automatic thresholding for defect detection. Pattern Recogn Lett 27(14):1644–1649

Nielsen AA (2007) The regularized iteratively reweighted MAD method for change detection in multi-and hyperspectral data. IEEE Trans Image Process 16(2):463–478

Nielsena AA, Müllerb A (2003) Change detection by the MAD method in hyperspectral image data. Retrieved May 9, 2017, from http://citeseerx.ist.psu.edu/viewdoc/summary?doi=10.1.1.608.8367

Omo-Irabor OO (2016) A comparative study of image classification algorithms for landscape assessment of the Niger Delta Region. J Geogr Inf Syst 8(02):163

Otsu N (1979) A threshold selection method from gray-level histograms. IEEE Trans Syst Man Cybern 9(1):62–66
Pacifici F (2007) Change detection algorithms: State of the art. URL: http://www.disp.uniroma2.it/earth_observation/pdf/CD-Algorithms.pdf. Accessed 4 Nov 2011
Parente M, Plaza A (2010) Survey of geometric and statistical unmixing algorithms for hyperspectral images. In: Hyperspectral image and signal processing: evolution in remote sensing (WHISPERS), 2010 2nd workshop on. IEEE, pp 1–4
Pieper M, Manolakis D, Cooley T, Brueggeman M, Weisner A, Jacobson J (2015) New insights and practical considerations in hyperspectral change detection. In: 2015 IEEE International Geoscience and Remote Sensing Symposium (IGARSS). IEEE, pp 4161–4164
Rapinel S, Hubert-Moy L, Clément B (2015) Combined use of LiDAR data and multispectral earth observation imagery for wetland habitat mapping. Int J Appl Earth Obs Geoinf 37:56–64
Ring M, Eskofier BM (2016) An approximation of the Gaussian RBF kernel for efficient classification with SVMs. Pattern Recogn Lett 84:107–113
Romshoo SA, Rashid I (2014) Assessing the impacts of changing land cover and climate on Hokersar wetland in Indian Himalayas. Arab J Geosci 7(1):143–160
Sakthivel NR, Saravanamurugan S, Nair BB, Elangovan M, Sugumaran V (2016) Effect of kernel function in support vector machine for the fault diagnosis of pump. J Eng Sci Technol 11(6):826–838
Samadzadegan F, Hasani H, Schenk T (2012) Simultaneous feature selection and SVM parameter determination in classification of hyperspectral imagery using ant colony optimization. Can J Remote Sens 38(2):139–156
Schaum A, Stocker A (1998) Long-interval chronochrome target detection. In: Proc. 1997 International symposium on spectral sensing research, pp 1760–1770
Scheffler D, Karrasch P (2013) Preprocessing of hyperspectral images: a comparative study of destriping algorithms for EO1-hyperion. In: Image and signal processing for remote sensing XIX, vol 8892. International Society for Optics and Photonics, p 88920H
Seydi ST, Hasanlou M (2016) Novel wetland and water body change detection using multitemporal hyperspectral imagery. In: Presented at the International Water Conference 2016 on Water Resources in Arid Areas. Springer, Oman
Seydi ST, Hasanlou M (2017) A new land-cover match-based change detection for hyperspectral imagery. Eur J Remote Sens 50(1):517–533
Seydi ST, Hasanlou M (2018) Sensitivity analysis of pansharpening in hyperspectral change detection. Applied Geomatics:1–11
Shah-Hosseini R, Homayouni S, Safari A (2015) A hybrid kernel-based change detection method for remotely sensed data in a similarity space. Remote Sens 7(10):12829–12858
Sica YV, Quintana RD, Radeloff VC, Gavier-Pizarro GI (2016) Wetland loss due to land use change in the Lower Paraná River Delta, Argentina. Sci Total Environ 568:967–978
Singh A (1989) Review article digital change detection techniques using remotely-sensed data. Int J Remote Sens 10(6):989–1003
Smith R (2012) Introduction to hyperspectral imaging, MicroImages Inc. Mentor, OH. Retrieved from http://www.microimages.com/documentation/Tutorials/hyprspec.pdf
Storey EA, Stow DA, Coulter LL, Chen C (2017) Detecting shadows in multi-temporal aerial imagery to support near-real-time change detection. GISci Remote Sens:1–18
Taminskasa, J., Petro\liusa R, \Limanauskienėa R, Satkūnasb J, Linkevičienėa R (2013) Prediction of change in wetland habitats by groundwater: case study in Northeast Lithuania. Estonian J Earth Sci 62(2):57ø e72
Thonfeld F, Feilhauer H, Braun M, Menz G (2016) Robust Change Vector Analysis (RCVA) for multi-sensor very high resolution optical satellite data. Int J Appl Earth Obs Geoinf 50:131–140
Varma S, Simon R (2006) Bias in error estimation when using cross-validation for model selection. BMC Bioinform 7(1):91
Vapnik V (2013) The nature of statistical learning theory. Springer, Heidelberg

Vongsy KM (2007). Change detection methods for hyperspectral imagery. Wright State University. Retrieved January 10, 2017, from https://etd.ohiolink.edu/!etd.send_file?accession=wright1184010751&disposition=attachment

Vongsy K, Mendenhall MJ, Hanna PM, Kaufman J (2009) Change detection using synthetic hyperspectral imagery. In: Hyperspectral image and signal processing: evolution in remote sensing, 2009. WHISPERS'09. First workshop on. IEEE, pp 1–4. Retrieved May 21, 2017, from http://ieeexplore.ieee.org/abstract/document/5289016/

Wang J (2013) Pearson correlation coefficient. In: Encyclopedia of systems biology. Springer, pp 1671–1671. Retrieved January 28, 2017, from http://link.springer.com/content/pdf/10.1007/978-1-4419-9863-7_372.pdf

White L, Brisco B, Dabboor M, Schmitt A, Pratt A (2015) A collection of SAR methodologies for monitoring wetlands. Remote Sens 7(6):7615–7645

Whiteside TG, Bartolo RE (2015) Use of WorldView-2 time series to establish a wetland monitoring program for potential offsite impacts of mine site rehabilitation. Int J Appl Earth Obs Geoinf 42:24–37

Wu C, Zhang L, Du B (2012) Targeted change detection for stacked multi-temporal hyperspectral image. In: Hyperspectral image and signal processing (WHISPERS), 2012 4th workshop on. IEEE, pp 1–4. Retrieved July 31, 2017, from http://ieeexplore.ieee.org/abstract/document/6874282/

Wu C, Du B, Zhang L (2013) A subspace-based change detection method for hyperspectral images. IEEE J Sel Top Appl Earth Observ Remote Sens 6(2):815–830

Yang Y, Yan Z (2016) Monitoring and Analyzing of Poyang Lake Wetland Land Use Change Based on RS and GIS. In: Geo-informatics in resource management and sustainable ecosystem. Springer, pp 213–221

Yuan F, Sawaya KE, Loeffelholz BC, Bauer ME (2005) Land cover classification and change analysis of the Twin Cities (Minnesota) Metropolitan Area by multitemporal Landsat remote sensing. Remote Sens Environ 98(2):317–328

Yuen PW, Richardson M (2010) An introduction to hyperspectral imaging and its application for security, surveillance and target acquisition. Imaging Sci J 58(5):241–253

Zanotta DC, Zani H, Shimabukuro YE (2013) Automatic detection of burned areas in wetlands by remote sensing multitemporal images. In: Geoscience and Remote Sensing Symposium (IGARSS), 2013 IEEE International. IEEE, pp 1959–1962

Zhao H, Cui B, Zhang H, Fan X, Zhang Z, Lei X (2010) A landscape approach for wetland change detection (1979-2009) in the Pearl River Estuary. Procedia Environ Sci 2:1265–1278

Chapter 10
Usage of Satellite Technology in Monitoring the Wetlands of Turkey, Tigris, and Euphrates Watershed

Gordana Kaplan, Zehra Yigit Avdan, and Ugur Avdan

Abstract With the increase of anthropogenic activities, the importance of the essence of life, water, proportionally continues to rise. The effective use of water is vital as an important role in socio economic development of nations. Thus, frequent water bodies monitoring is a crucial part of their sustainable management. As remote sensing data is useful in monitoring water and wetland areas, in this chapter we investigate the usage of Earth Observation satellite technology for monitoring wetlands in one of the most significant parts not only in Turkey, but in Western Asia. Along with a brief literature review about wetlands and remote sensing, this chapter investigates the water areas in Tigris and Euphrates Watershed in Turkey using Landsat-8 satellite imagery. The historical investigation showed significant development in the studied area with the several dam constructions in the watershed. The usage of satellite technology is crucial to long-term water planning and management that incorporate principles of sustainability to avoid ecological and environmental catastrophes.

Keywords Remote Sensing · Monitoring · Wetlands · Transboundary waters Turkey

10.1 Background

The importance of water, which has been a vital ingredient since the existence of human beings, continues to increase today. Water cannot be substituted, in other words, it has no alternative and is one of the essential requirements. Because water is the beginning of living life on earth, it is not possible for people to live and continue their activities without water.

G. Kaplan (✉) · U. Avdan
Earth and Space Sciences Institute, Eskisehir Technical University, Eskisehir, Turkey

Z. Y. Avdan
Environmental Engineering Department, Eskisehir Technical University, Eskisehir, Turkey

L. A. Jawad (ed.), *Southern Iraq's Marshes*, Coastal Research Library 36,
https://doi.org/10.1007/978-3-030-66238-7_10

Water also plays an important role in socioeconomic development of nations. Therefore, the effective use of water, not wasting, and the development of water resources are vital issues for the development of countries. The constant increase of the world's population, the deterioration of the ecological balance as a result of environmental pollution, changes in rainfall and flow regimes due to global warming, and the risk of drought have increased the need for water. The situation is extremely serious, especially in arid and semiarid countries.

Both surface and ground waters may flow over several official boundaries between countries, following natural courses before reaching their final destination. As a result, watersheds that may concern more than a single country have emerged. Even now, 280 such watersheds worldwide are being shared between two or more countries, and they are referred to in the literature as "transboundary waters."

In order to protect wetland areas worldwide, the first and most important step would be mapping and monitoring of past and present condition, and setting goals for the future. The condition of many wetland areas is not known due to lack of past and present maps. The only effective platform that can rapidly give information about the past and present wetland area is remote sensing (Maillard et al. 2012).

Remote sensing can be described as the collecting of information about an object or phenomenon without any physical contact. While remote sensing general use includes numerous fields, it has been closely connected with earth science disciplines, and it is used for detecting, monitoring, and classifying objects on earth.

Geographic information system (GIS) is a system constructed to manipulate with geographical data. While, in general, geographical data are collected with remote sensing techniques, their storing, processing, analyzing, and visualizing are conducted through GIS. For the last few decades, the integration of GIS and remote sensing techniques has been investigated in many fields (Baumgartner and Apfl 1996).

With the remote sensing and GIS revolution in the 1980s, several approaches with satellite images in combination with GIS technologies have been used for mapping and monitoring wetlands all over the world.

Satellite remote sensing of wetlands has been used in the early stage of the remote sensing as a science. Thus, Lulla (1983) investigated the capabilities of Landsat Multispectral Scanner (MSS) for wetland monitoring, while Hardisky et al. (1986) investigated the different satellite and aircraft platforms for mapping coastal wetlands. Although it was concluded that the spatial resolution of the National Oceanic and Atmospheric Administration (NOAA) and Advanced Very-High-Resolution Radiometer (AVHRR) is useful for large area mapping, the typical wetland features could not have been resolved. It was also concluded that Landsat MSS data (79 $\times$ 57 meters' spatial resolution) is not enough to provide sufficient information for wetlands, and they point the future studies for this purpose to use Landsat Thematic Mapper (TM) and Satellite Pour l'Observation de la Terre (SPOT).

Using SPOT middle spatial resolution satellite imagery, Rutchey and Vilchek (1994) used both unsupervised and supervised classification techniques for classifying 20 and 12 different wetland classes, separately, where the overall classification accuracy was estimated to be approximately 70 and 80%, respectively. In this research, due to the moderate accuracy results, as a future work, investigation of Landsat satellites was suggested.

In another research, Ozesmi and Bauer (2002) summarized the studies on satellite remote sensing of wetlands, discussing the different classification techniques used for wetland classification such as visual interpretation, unsupervised classification, principal component analysis, pixel-based supervised classification, etc. It was concluded that, in order to improve the wetland classification, ancillary data, such as soil information and elevation data, should be used. It was also concluded that the ideal dates for wetland mapping and monitoring will depend on the type and location of the wetlands. As a future study, the authors have suggested full exploitation of the combined use of optical and radar data for wetland identification.

Li and Chen (2005) developed a rule-based method for mapping wetlands using optical, radar, and digital elevation model (DEM) data. In this paper, data from Landsat-7 Enhanced Thematic Mapper Plus (ETM+) and Radarsat-1/synthetic-aperture radar (SAR) images and DEM data were used. The method was tested in three study sites in eastern Canada. The test results showed that the classification accuracy of the new method ranges from 71% to 92%. In the study, five types of wetlands were defined: fen, bog, swamp, marsh, and shallow water. The DEM data was used for the slope gradient image, the Landsat image was used for retrieving Normalized Difference Vegetation Index (NDVI) and using the near-infrared and shortwave infrared bands, while the radar image was used for an object-based classification. Furthermore, decision rules were applied to wetland mapping.

Xiaobi C. and Xiaoling Ch. (Ji 2007) used an expert system-based image classification for identifying wetland in northern Jiangxi, China. Firstly, they used NDVI and Normalized Difference Water Index (NDWI) for separating water and vegetated areas. Afterward, a supervised classification was performed for extracting other relevant land cover information. Then adding a DEM to the Landsat data, they have set a set of rules for improving the classification. It was concluded that a rule-based expert system approach could improve the classification of wetland-related objects that have similar spectral characteristics.

Dabrowska et al. (2016) applied remote sensing and in situ information for the management of wetlands in Poland. The investigation has been carried out by taking measurements of soil moisture, evapotranspiration, Leaf Area Index (LAI), wet and dry biomass, and the levels of groundwater and meteorological parameters. As a remote sensing data, microwave images from European Remote Sensing-2 SAR (ERS-2SAR) and Environmental Satellite (ENVISAT) were downloaded for modeling soil moisture and humidity changes of the area under investigation. The study also uses optical data from Landsat +ETM, SPOT-Vegetation, ERS-2, the Along Track Scanning Radiometer (ATSR), ENVISAT Medium Resolution Imaging Spectrometer (MERIS), and NOAA/AVHRR for calculation of the biomass and vegetation indices.

Dong et al. (2014) used NDVI and Land Surface Water Index (LSWI) for mapping wetland areas in Northeast China. The algorithm used in this paper is based on the difference between NDVI and LSWI values. NDVI is an index that is calculated from the red and near-infrared (NIR) band, while the LSWI is calculated from NIR and shortwave infrared (SWIR). The results showed an accuracy of 92%.

Cardoso et al. (2014) used spectral analysis of Landsat-5 TM for mapping coastal wetlands in the Amazon River mouth, Brazil. For this purpose, a classification of the 30-meter Landsat image was made using the ENVI Spectral Angle Mapper (SAM) classifier. The results showed that the SAM-supervised classifier is a powerful tool for a spectral analysis of optical sensors with moderate spatial resolution. The study presents a potential for the previously mentioned classifier in discrimination of tropical coastal wetlands, based on the interpretation of spectral curves of soil, water, and vegetation features.

Huang et al. (2014) used Landsat and airborne Light Detection And Ranging (LiDAR) data for wetland mapping and change monitoring. In this paper, a new approach for monitoring wetlands was presented. In this approach, LiDAR data were used to derive highly accurate reference subpixel inundation percentage maps at 30 m resolution. Then the LiDAR-based data were used to model relationships between inundation and the spectral data acquired by Landsat. The results from this paper indicate that the developed approach has a potential for deriving historical inundation changes in areas covered by available Landsat and LiDAR data.

Chen et al. (2015) used Polarimetric SAR (PolSAR) data for classification of coastal wetlands in eastern China. For this purpose, classification over PolSAR image has been made. The classification has been made with a modified algorithm that is a pixel-based method for image segmentation and classification. The results from this study show producer's and user's accuracies of 74.29% and 92.86%, respectively.

Fickas et al. (2016) used Landsat satellite images for monitoring of annual wetland change in the Willamette Valley of Oregon, USA, from 1972 to 2012. Images from Landsat MSS, Landsat TM, and Landsat +ETM were used. The data have been collected in the summer period. Change detection techniques have been applied for the analysis of changes that happened in the wetland areas over the years.

Dronova (2015) investigated the use of object-based image classification techniques in wetland research. Although generally object-based image analyses (OBIA) have been mainly used for very high spatial resolution images generally taken from aircrafts or satellites (Hurd et al. 2005), lately this technique has been also applied to middle and high spatial resolution images for classifying different land covers, where several studies (Kaplan and Avdan 2017; Esetlili et al. 2018) reported improvement of the OBIA classification results compared with pixel-based classification. Similar improvement was reported in several other studies in wetland areas (Esetlili et al. 2018; Kamal and Phinn 2011; Myint et al. 2008; Harken and Sugumaran 2005).

Although wetlands are now recognized worldwide, and their protection is regulated with laws and management plans, the destruction of wetlands has taken a big part in the past and still continues in some parts of the world (Mitsch and Gosselink 2015). Several studies have reported the loss of wetland areas worldwide. The results of the studies can be summarized as follows:

(i) The world is losing wetlands rapidly, particularly in developing countries (Ji 2007).
(ii) The world has lost more than half of the original wetland areas (WWAP 2003).
(iii) The world lost half of its wetlands in the twentieth century (Russi et al. 2013).
(iv) The world lost 87% of its wetlands since 1700.
(v) Wetland loss was 3.7 times faster in the early twenty-first than the twentieth centuries.
(vi) More than 80% of the wetland loss occurs due to human activities.

10.2 Usage of Satellite Technology in Monitoring the Wetlands of Turkey

Turkey is part of the Ramsar agreement that provides the framework for conservation and wise use of wetlands and their resources. Among the high number of wetlands, Turkey has 14 wetland areas that are protected with the Ramsar agreement, 46 wetlands of national, 9 wetlands of local importance, and a number of smaller wetland areas (Kaplan 2019).

Over the years, researchers have used satellite data and technology for monitoring different types of wetlands in Turkey. Thus, satellite data have been used for long-term monitoring of wetlands (three decades surface water bodies change with Landsat images in the Konya basin) (Musaoglu et al. 2018), temporal assessment of natural wetlands and lakes (Dervisoglu et al. 2019) using four Landsat images from 1990 to 2016, land use/land cover change detection in wetlands (Sultan marshes, Turkey) (Kesikoglu et al. 2019), integration of different satellite data for wetland mapping in the central Anatolian region (Balikdami wetlands) (Kaplan and Avdan 2018a; Kaplan et al. 2019), monitoring coastline changes (Burdur wetland and Goksu delta) (FatihTemiz et al. 2017; Ciritci and Türk 2019), evaluation of red-edge bands for wetland classification in the Anatolian region of Turkey (Kaplan and Avdan 2019), etc.

An interesting approach for monthly analysis of wetland dynamics in the Anatolian region of Turkey was presented by Kaplan and Avdan (2018b), where different remote sensing sensors were used such as unmanned aerial vehicle (UAV) and optical, radar, and thermal satellite data, where the relationship between the investigated parameters has been presented.

Several studies conducted in the central Anatolian region in Turkey have recorded a decrease in the water and wetland areas. In the central Anatolian region of Turkey are located few of the most important wetlands in Turkey, such as Eber Lake, Aksehir Lake, Salt Lake, Seyfe Lake, and Sultan wetlands. Several studies have investigated the wetland loss in Turkey in the last past decades. Thus, Ekercin and Ormeci observed Turkey's second biggest lake, Salt Lake or Tuz Golu, and concluded that over 70% of its water surface area decreased from 1987 to 2005 (Ekercin and Örmeci 2010). According to another study's results, where a simple NDVI and NDWI analysis was conducted, it was concluded that the water area of

Aksehir Lake has decreased from 324 km^2 in 1987 to 100 km^2 in 2016 (Kaplan et al. n.d.). In other studies, the NDVI and NDWI were successfully used for water and wetland monitoring and mapping. Thus, the indices were used for estimating the wetland biomass in Sakarbasi wetland, and Eber lake in Turkey. Similarly, with different NDWI threshold values (Kaplan et al. 2019), it was concluded that the index was successful not only in water extraction but also in wetland extraction and can give valuable information about the soil moisture. The conducted studies are valuable for Turkey's wetlands, present the current situation of the wetlands, and have contributed for better understanding of the wetland dynamics, temporal changes, drought, and loss of the wetlands in some of the most important parts of Turkey. However, a very few studies can be found in the literature about the one of the most important wetland parts of Turkey, or even maybe for all Western part of Asia, the Southeastern part of Turkey, the region situated within the Tigris and Euphrates river system, Turkey with the Persian Gulf (Fig. 10.1).

The transboundary stream basins of Euphrates and Tigris cover about one third of Turkey with 256,000 km^2, with 70 km^3/y, and it accounts for 40% of the Turkey's raw water potential, while for Iraq and Syria, it is almost the only water source. Turkey is the upper riparian country of the Euphrates and Tigris, one of the most important waterways in the Middle East along with the Nile River.

The Euphrates and Tigris are the two main branches of the transboundary river basin, which joins 70 km north of the Basra in Iraq and flows another 100 km to form Shatt al-Arab, and thereafter flows into the Persian Gulf.

The Euphrates that is originating in the eastern Turkey is one of the most historically important rivers of Western Asia. Together with the Tigris, it is one of

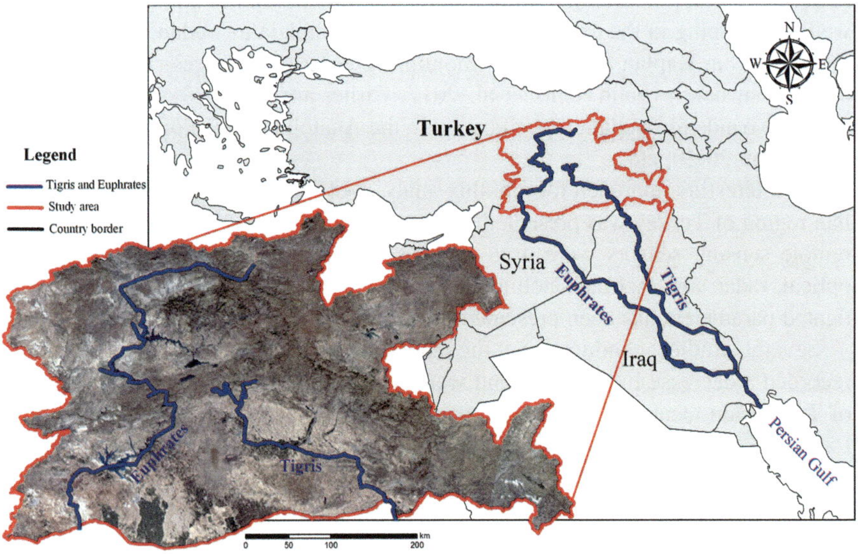

Fig. 10.1 Study area: Tigris and Euphrates watershed in Turkey

the two defining rivers of Mesopotamia (the "Land between the Rivers"). Mesopotamia is the first epoch name of the region between the Euphrates River and the Tigris River, which runs from Southeastern Anatolia region of Turkey to the Persian Gulf.

10.3 Study Area

The study area, the Tigris and Euphrates watershed in Turkey, was extracted from digital elevation model, and it is presented in Fig. 10.2. As it can be seen in Fig. 10.2, the north and the east-south part of the watershed is presented with high mountains, while the south side is presented with low relief with agricultural fields. The Tigris and Euphrates watershed in Turkey covers an area of approximately 177.792 km^2, which is 23% of the total area of Turkey.

Here in this part, data from Landsat-8 satellite was used in order to get a clear perspective about the water and wetland areas in the Tigris and Euphrates watershed in Turkey. Landsat-8, launched in February 2013, is a follow-up mission to the previous Landsat-7, as well as to Landsat-5 that holds the record of a longest living satellite in the space (1984–2013). Landsat missions have given researchers continuous data for over three decades and will continue its mission through the planned Landsat-9, expected to be launched in December 2020 (http n.d.).

Landsat-8 carries two push-broom instruments, the Operational Land Imager (OLI) and the Thermal Infrared Sensor (TIRS). Details about the Landsat-8 bands are given below in Fig. 10.3 and Table 10.1.

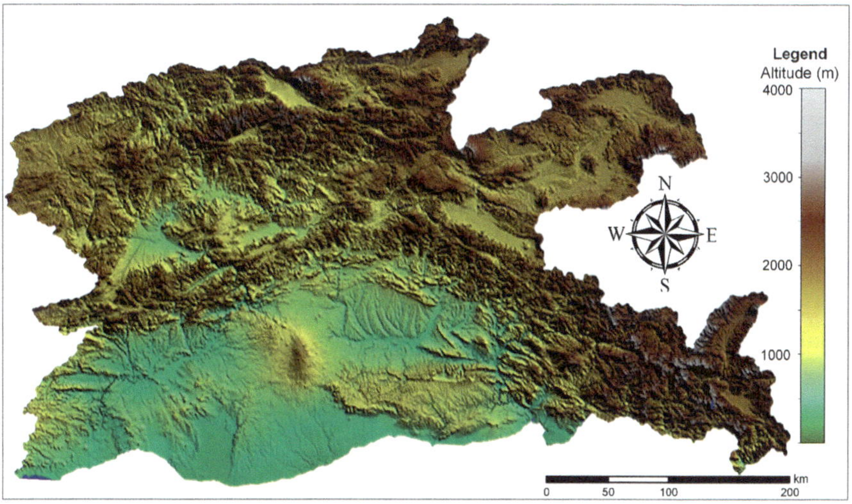

Fig. 10.2 DEM model of the Tigris and Euphrates watershed, Turkey

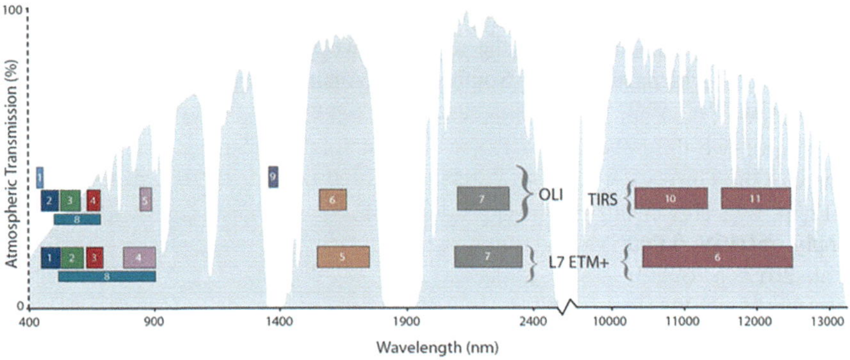

Fig. 10.3 Landsat-7 and Landsat-8 band comparison

Table 10.1 Landsat-8 band designations

Landsat-8 bands	Central wavelength (μm)	Wavelength (μm)	Resolution (m)
Band 1 – coastal aerosol	0.44	0.43–0.45	30
Band 2 – blue	0.48	0.45–0.51	30
Band 3 – green	0.56	0.53–0.59	30
Band 4 – red	0.65	0.64–0.67	30
Band 5 – NIR	0.86	0.85–0.88	30
Band 6 – SWIR-1	1.61	1.57–1.65	30
Band 7 – SWIR-2	2.20	2.11–2.29	30
Band 8 – panchromatic	0.59	0.50–0.68	15
Band 9 – SWIR Cirrus	1.37	1.36–1.38	30
Band 10 – thermal-1	10.89	10.60–11.19	100
Band 11 – thermal-2	12.00	11.50–12.51	100

A total of 14 Landsat satellite images were needed to cover the whole study area. The images were downloaded from the EarthExplorer web page. The images were selected from the summer period in 2019 with time difference of maximum 20 days. The images were selected to be cloud-free and with high quality. Details about the used images are given in Table 10.2. After the careful selection and download of the images, using GIS tools, the images were merged into a single image that was adjusted according to the study areas' border.

Furthermore, the most common index for water and wetland mapping and monitoring, NDWI was applied to the satellite image. The index is calculated from the green and near-infrared bands, which are highly sensitive to water contents:

$$NDWI = \frac{Green - NIR}{Green + NIR}$$

Afterward, in order to get the water area in the watershed, the NDWI image was reclassified, and the results showed that a total of 2.592 km^2 in the study area are

Table 10.2 Details about the used satellite images

	Path	Row	Date
1	169	34	18.07.2019
2	170	33	10.08.2019
3	170	34	10.08.2019
4	171	32	01.08.2019
5	171	33	01.08.2019
6	171	34	01.08.2019
7	172	32	08.08.2019
8	172	33	08.08.2019
9	172	34	08.08.2019
10	173	32	30.07.2019
11	173	33	30.07.2019
12	173	34	30.07.2019
13	174	33	06.08.2019
14	174	34	06.08.2019

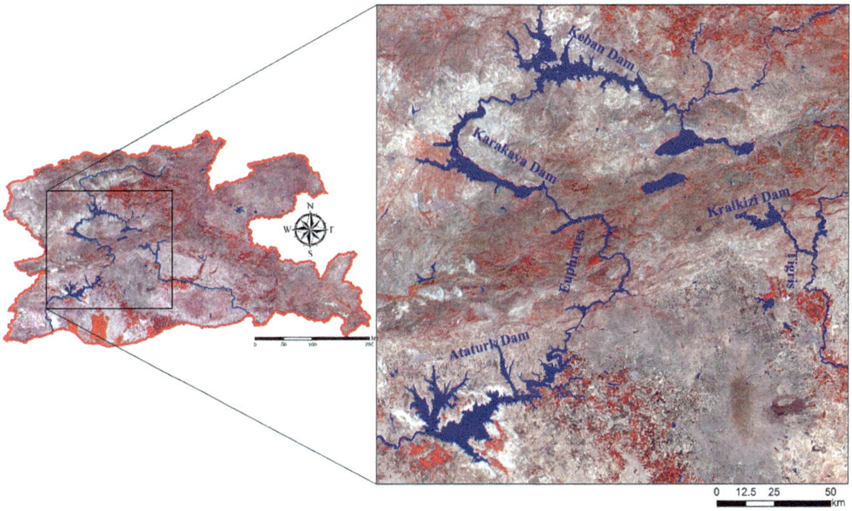

Fig. 10.4 Water areas in the Tigris and Euphrates watershed in Turkey

water bodies. Having in mind the resolution of the Landsat images, it needs to be mentioned that smaller water bodies than 30 m cannot be visible and thus were not taken into consideration.

The results show that there are dozens of big water areas including lakes, dams, and reservoirs, and hundreds of flowing waters including rivers and channels. Three of the largest dams built along the river Euphrates are Ataturk, Karakaya, and Keban dams, while the largest dam along the Tigris River is the Kralkızı dam (Fig. 10.4).

The results show that there are dozens of big water areas including lakes, dams, and reservoirs, and hundreds of flowing waters including rivers and channels. Three of the largest dams built along the river Euphrates are Ataturk, Karakaya, and Keban dams, while the largest dam along the Tigris river is the Kralkizi dam.

While the northern part of the watershed is mostly mountainous, from the satellite images, it can be noticed that big parts of the agricultural areas in the southern part of the watershed are formed along the riverbed of the rivers Euphrates and Tigris which shows the importance of the water bodies in the area. In order to emphasize the importance of the water bodies in this area, analyses were made using satellite images before and after the construction of the Ataturk dam. Thus, one Landsat-5 satellite image from August 1985 and one Landsat-8 satellite image from August 2019 were used in this part. Also, naturally, the presence of the water bodies in the area changes the land cover and land use in the surroundings. Thus, with remote sensing satellite imagery, those changes were analyzed and visually presented. As mentioned before, for the water bodies, the NDWI was used, while for the agricultural areas, NDVI was used. In addition, Land Surface Temperature (LST) maps were used for capturing the changes that occurred after the construction of the Ataturk dam.

LST is one of the curtail parameters to understand the complex wetland areas and its surroundings. However, this parameter has not been a subject of a detailed investigation regarding wetland areas. Kaplan et al. (2019; Kaplan and Avdan 2018b) used LST maps for analyzing the seasonal and monthly temperature changes in wetland areas. Also Jovanovska et al. (2016) used satellite LST maps in order to analyze the changes that happed after the construction of one of the most important dams in the republic of North Macedonia. Similarly, in this part of the study, we analyze the changes in the surroundings of the Ataturk dam. For that purpose, LST algorithm developed by Avdan and Jovanovska (Avdan and Jovanovska 2016) was used. The results of the analyses are shown in Fig. 10.5.

As it can be seen from Fig. 10.5, there is a tremendous land cover and temperature change in the surroundings of the Ataturk dam. Of course, the temperature change is due to the land cover change where local people have made a significant progress in the agricultural area. The construction of the Ataturk dam has opened many opportunities making the region one of the most agricultural productive areas in Turkey. One of the most important progresses is the Harran valley (Fig. 10.6); one of the biggest agricultural areas has been developed after the construction of the dam. Along the Harran valley, the northern part of the watershed has also been developed (Fig. 10.7).

10.4 Conclusion

In many parts of the world, natural wetlands have been replaced with agricultural areas or closed for urban development. Due to agricultural development, there are no natural wetlands left in some parts of China. However, the construction of wetlands

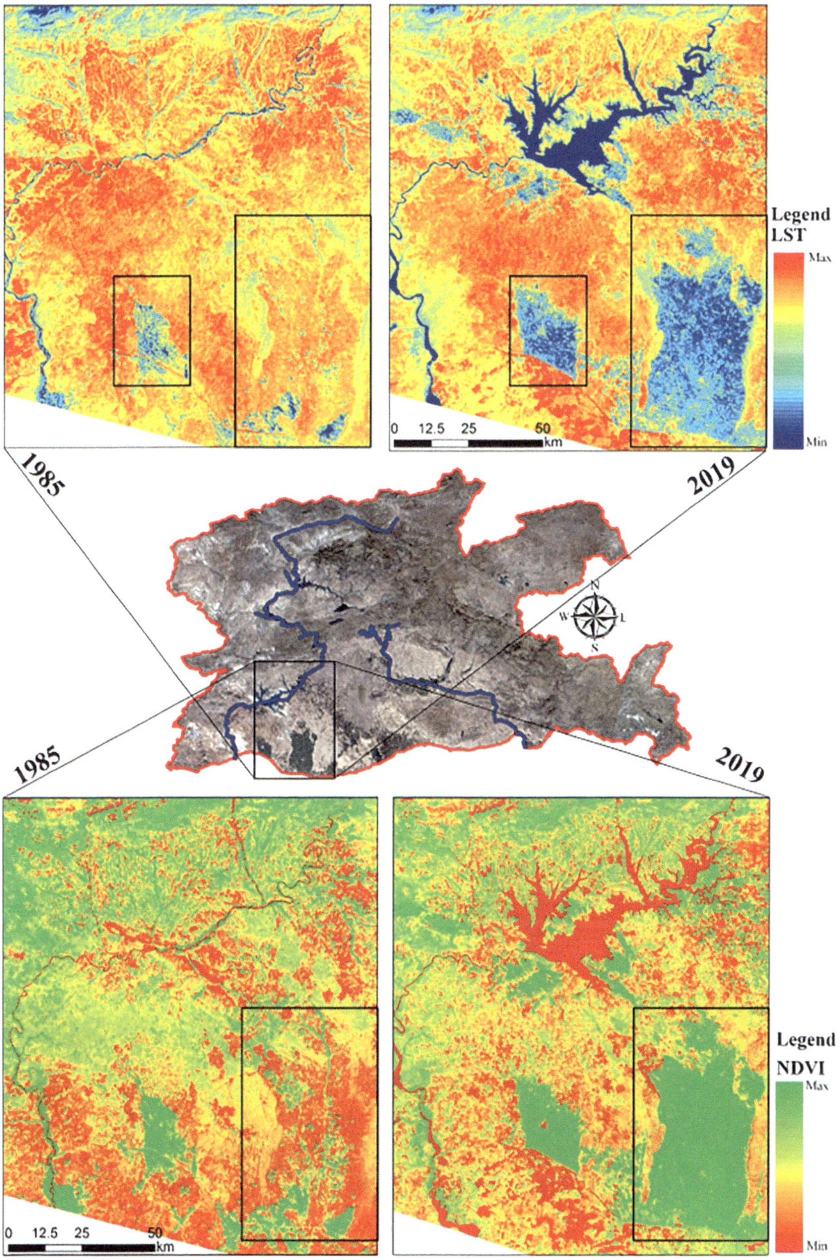

Fig. 10.5 LST and NDVI changes from 1985 to 2019

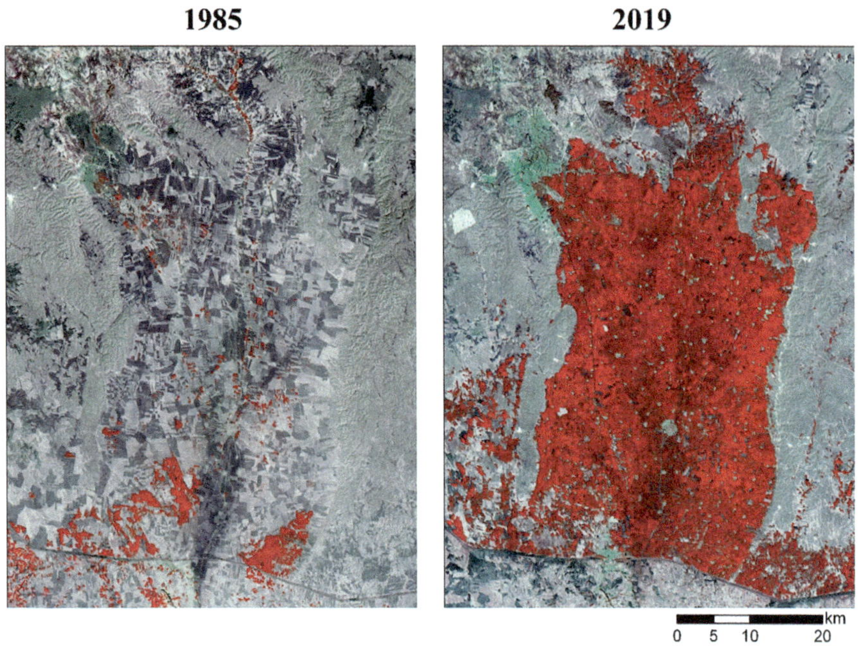

Fig. 10.6 Harran valley before and after the construction of the dam (Landsat, 8; RGB, Band 4, Band 3, Band 2)

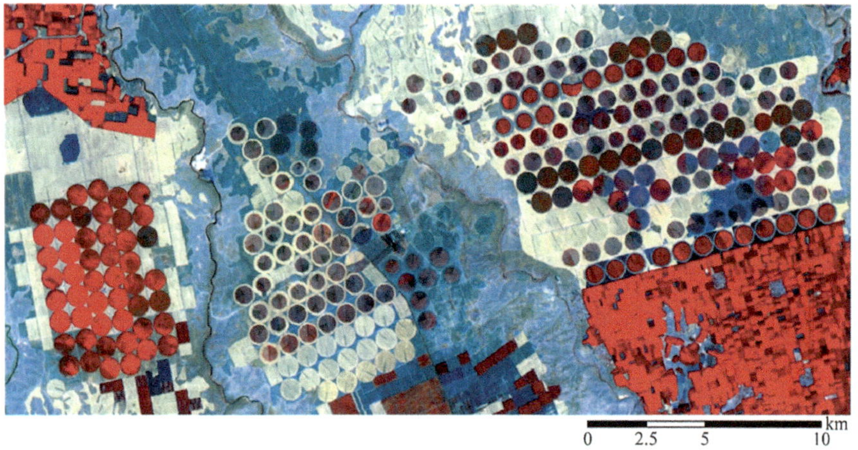

Fig. 10.7 Agricultural areas in the northern part of the Tigris and Euphrates watershed, Turkey (Landsat, 8; RGB, Band 4, Band 3, Band 2)

such as dams has a significant influence over the agricultural, urban, rural, and economic development of a specific area. Such long-term changes in both wetland and agricultural areas can be mapped and monitored only with the use of remote sensing technology. As mentioned before, many researches have already used remote sensing data and techniques for mapping and monitoring wetlands all over the world, and in Turkey. This book chapter not does only cover a brief literature review about wetlands and remote sensing but also covers one of the most significant parts not only in Turkey but in Western Asia. Although there are not many natural wetlands in this region, the construction of wetlands in this region has contributed to a huge development in the area.

The main underlying factor in water-related disputes is water scarcity. The most effective measure that can be taken in this case is undoubtedly to save water by using the existing water at the optimum level. Considering that agriculture is the largest water consumer, water savings should be made in this sector. Consequently, irrigation techniques with minimum water loss should be selected; both surface and groundwater salinity due to agriculture should be kept to a minimum level.

The issue of transboundary waters is not limited to water use only. There is also the environmental dimension of the issue, which has gained importance especially in the recent period. Sustainable use of natural resources in the long term, prevention of pollution, and protection of the environment and natural environment are also the factors to be considered in the policies to be followed. Regulations on transboundary waters cannot be implemented without the good intentioned regional and international cooperation of the concerned countries.

References

Avdan U, Jovanovska G (2016) Algorithm for automated mapping of land surface temperature using LANDSAT 8 satellite data. J Sens 2016

Baumgartner MF, Apfl GM (1996) Remote sensing and geographic information systems. Hydrol Sci J 41(4):593–607

Cardoso GF, Souza C, Souza PWM (2014) Using spectral analysis of Landsat-5 TM images to map coastal wetlands in the Amazon River mouth, Brazil. Wetl Ecol Manag 22(1):79–92

Chen YY, He XF, Wang J (2015) Classification of coastal wetlands in eastern China using polarimetric SAR data. Arab J Geosci 8(12):10203–10211

Ciritci D, Türk T (2019) Automatic Detection of Shoreline Change by Geographical Information System (GIS) and Remote Sensing in the Göksu Delta, Turkey. J Indian Soc Remote Sens 47(2):233–243

Dabrowska-Zielinska K et al (2016) Assessment of carbon flux and soil moisture in wetlands applying sentinel-1 data. Remote Sens 8(9)

Dervisoglu A, et al (2019) Temporal assessment of natural wetlands via remotely sensed data: a case study from Turkey. Fresenius Environ Bull 1005

Dong ZY et al (2014) Mapping wetland areas using landsat-derived NDVI and LSWI: a case study of West Songnen plain, Northeast China. J Indian Soc Remote Sens 42(3):569–576

Dronova I (2015) Object-based image analysis in wetland research: A review. Remote Sens 7(5):6380–6413

Ekercin S, Örmeci C (2010) Evaluating climate change effects on water and salt resources in Salt Lake, Turkey using multitemporal SPOT imagery. Environ Monit Assess 163(1–4):361–368
Esetlili MT et al (2018) Comparison of object and pixel-based classifications for mapping crops using Rapideye imagery: a case study of Menemen Plain, Turkey. Int J Environ Geoinformatics 5(2):231–243
FatihTemiz AB, Durduran SS, Gumus MG (2017) Monitoring coastline change using remote sensing and gis technology: a case study of Burdur Lake, Turkey. Fresen Environ Bullet:7235
Fickas KC, Cohen WB, Yang ZQ (2016) Landsat-based monitoring of annual wetland change in the Willamette Valley of Oregon, USA from 1972 to 2012. Wetl Ecol Manag 24(1):73–92
Hardisky M, Gross M, Klemas V (1986) Remote sensing of coastal wetlands. Bioscience 36(7):453–460
Harken J, Sugumaran R (2005) Classification of Iowa wetlands using an airborne hyperspectral image: a comparison of the spectral angle mapper classifier and an object-oriented approach. Can J Remote Sens 31(2):167–174
Huang CQ et al (2014) Wetland inundation mapping and change monitoring using Landsat and airborne LiDAR data. Remote Sens Environ 141:231–242
Hurd JD et al (2005) Coastal marsh characterization using satellite remote sensing and in situ radiometry data: preliminary results. American Society of Photogrammetry and Remote Sensing, Baltimore
Ji W (2007) Wetland and water resource modeling and assessment: a watershed perspective. CRC Press, Boca Raton
Jovanovska G et al (2016) Land surface temperature change after construction of the Kozjak Dam based on remote sensing data
Kamal M, Phinn S (2011) Hyperspectral data for mangrove species mapping: a comparison of pixel-based and object-based approach. Remote Sens 3(10):2222–2242
Kaplan G (2019) Remote sensing image fusion for mapping and monitoring wetlands in anatolian region, Turkey. PhD thesis
Kaplan G, Avdan U (2017) Object-based water body extraction model using Sentinel-2 satellite imagery. Eur J Remote Sens 50(1):137–143
Kaplan G, Avdan U (2018a) SENTINEL-1 and SENTINEL-2 data fusion for wetlands mapping: Balikdami, Turkey. Int Arch Photogr Remote Sens Spat Inf Sci 42(3)
Kaplan G, Avdan U (2018b) Monthly analysis of wetlands dynamics using remote sensing data. ISPRS Int J Geo Inf 7(10):411
Kaplan G, Avdan U (2019) Evaluating the utilization of the red edge and radar bands from sentinel sensors for wetland classification. Catena 178:109–119
Kaplan G, Avdan, ZY, Avdan U (2019) Mapping and monitoring wetland dynamics using thermal, optical, and SAR remote sensing data. Wetlands management: assessing risk and sustainable solutions. p 87
Kaplan G, et al (n.d.) Drought monitoring using landsat satellite images (Case Study Aksehir Lake) (In Turkish) UZAL-CBS2016 Symposium, Adana, Turkey
Kesikoglu MH et al (2019) Performance of ANN, SVM and MLH techniques for land use/cover change detection at Sultan Marshes wetland, Turkey. Water Sci Technol 80(3):466–477
Li JH, Chen WJ (2005) A rule-based method for mapping Canada's wetlands using optical, radar and DEM data. Int J Remote Sens 26(22):5051–5069
Lulla K (1983) The Landsat satellites and selected aspects of physical geography. Prog Phys Geogr 7(1):1–45
Maillard P, Pivari MO, Luis CHP (2012) Remote sensing for mapping and monitoring wetlands and small lakes in Southeast Brazil. Remote sensing of planet earth. Rijeka, InTech, pp 23–46
Mitsch WJ, Gosselink JG (2015) Wetlands. Wiley, New York, 736 pp. In: ISBN978-1-118-67682-0
Musaoglu N et al (2018) Long-term monitoring of wetlands via remote sensing and GIS: a case study from Turkey. In: Proceedings of The International Conference on Climate Change

Myint SW et al (2008) Identifying mangrove species and their surrounding land use and land cover classes using an object-oriented approach with a lacunarity spatial measure. GISci Remote Sens 45(2):188–208

Ozesmi SL, Bauer ME (2002) Satellite remote sensing of wetlands. Wetl Ecol Manag 10(5):381–402

Russi D et al (2013) The economics of ecosystems and biodiversity for water and wetlands. IEEP, London and Brussels, p 78

Rutchey K, Vilchek L (1994) Development of an Everglades vegetation map using a SPOT image and the Global Positioning System. Photogramm Eng Remote Sens 60(6):767–775

WWAP U (2003) UN World Water Development Report: Water for People. Water for Life

Section III
Geology

Chapter 11
Sedimentology and Mineralogy of Quaternary Sediments of Marshes, South of Iraq

Badir N. Albadran

Abstract The marshes of the southern part of the Mesopotamian Plain lie around the main rivers in the region which are represented by the Tigris and Euphrates. Most surface sediments to a depths exceeding 20 m are sediments dating back to the Quaternary period. These sediments are heterogeneous, with texture and mineral composition from one place to another, which depends on several factors. Sediment colours change between beige, brown, and gray. The marsh sediments vary in texture between areas and depth but mostly are sandy silt, silty sand, and silt, with the mean size from 3.7 to 5.8 phi. The silt fraction is the main component of sediments in the marshes, followed by clay fraction. Since the marshes sediments are variable horizontally and vertically in texture, the minerals that make up these sediments will definitely show a significant change in the quality and quantity of minerals. The percentage of the sand fraction is small in the marshes sediments, but to some extent, it shows a change in heavy and light minerals. The proportion of light minerals is large compared to the proportion of heavy minerals. Light minerals include carbonate, quartz, feldspar, and chert and the heavy minerals are dominant by opaque followed by amphibole, epidote, and pyroxene group's minerals. The proportions of clay minerals also change from one location to another, so it is difficult to note the percentages of clay minerals, but it could conclude that the clay minerals that appear in the marshes sediments are smectite, chlorite, illite, mixed layers illite-smectite, palygorskite, and kaolinite.

Keywords Mesopotamia · Sediments · Minerals · Marshes · Quaternary period · Iraq

B. N. Albadran (✉)
Department of Geology, University of Basrah, Basrah, Iraq

L. A. Jawad (ed.), *Southern Iraq's Marshes*, Coastal Research Library 36,
https://doi.org/10.1007/978-3-030-66238-7_11

11.1 Introduction

The marshes (Ahwar in Arabic) cover the southern part of Mesopotamia; they are distributed around the rivers in the area (Fig. 11.1) as central marshes between Tigris and Euphrates rivers (Baghdad and Zecheri), Al-Hewaza to the east of Tigris River, Al-Hammar south of Euphrates River, etc.; the largest one of these lakes is Al-Hammar. This area is one of the prominent wetlands in the Middle East; it is interesting in many studies of multidisciplinary but still needs more due to its important presence: agricultural, archaeological, touristic, and ecological.

These marshes vary in the salinity of their water; freshwater marshes include the central marshes and the northern part of the Al-Hewaza marsh. The oligohaline marshes include the central of Al-Hewaza and the shallow area of the central marshes and the western part of Al-Hammar marsh (Hussain 2014), whereas the eastern part of Al-Hammar represents a brackish water marsh. The estuarine marshes include the tidal flat creeks of Khor Al-Zubair and the coastal salt marshes which are adjacent to Khor Al-Zubair and Khor Abdullah (Hussain 2014). Some of them are influenced by the tidal phenomenon in the area as the eastern part of Al-Hammar, the creeks of Khor Al-Zubair, the coastal marshes of Khor Abdullah and Khor Al-Zubair, and estuarine marshes of the Shatt Al-Arab River (Fig. 11.2). There are also some marshes that are not affected by the phenomenon of tides such as Al-Hewaza, central marshes, and the western part of Al-Hammar marsh.

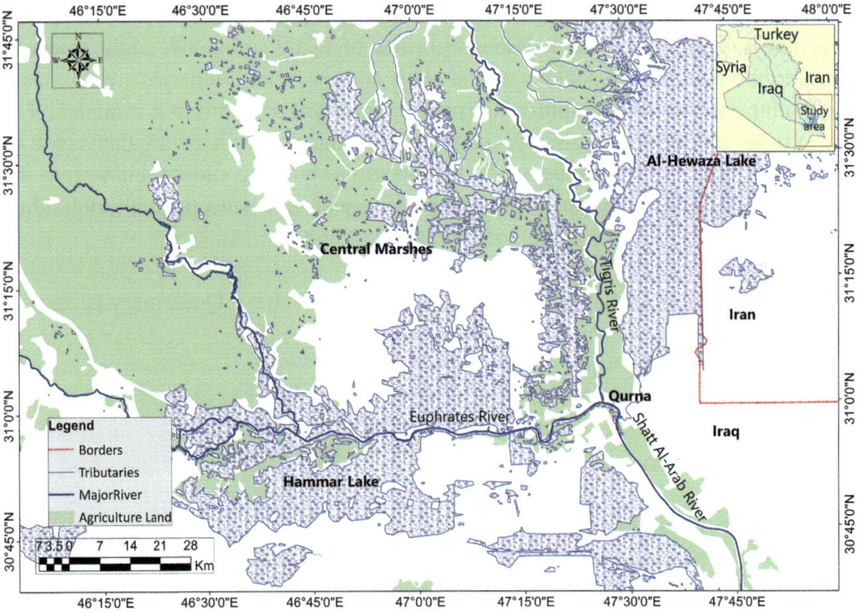

Fig. 11.1 The location of the study area

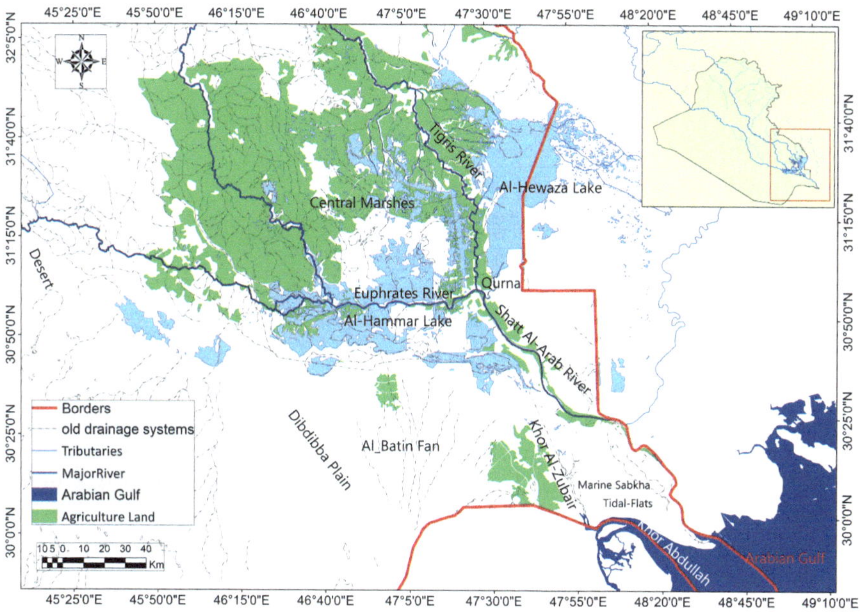

Fig. 11.2 The geomorphology of the study area

The sediments which cover the Ahwar area are mainly sandy clayey silt and clayey silt with a small amount of clay-silt and sandy clay-silt (Abdullah 1982). The main factors controlling the type and distribution of the sediments are climate change and the adjacent sources or transport agents of clastic sediments. The local diagenesis processes play another role in the variation of the sediment type. Ahwar area also was a subject of a large drying in the last two decades by human activity and global climate change.

11.2 Geological Settings

The geological history of the area is a matter of debate from the mid of beginning of the last century. The Quaternary sediments cover the major part of the Mesopotamian plain; the sediments return to the Pleistocene and Holocene. Dibdibba Formation of the Miocene-Pleistocene formation covers the area. This formation is mainly composed of gravels and sand of igneous rocks (Buday 1980). It could deposit in fresh water to the deltaic environment (Buday 1980 and AlSayyab et al. 1982). Dibdibba Formation is covered by the sand of the alluvial fan of Wadi Al-Batin (Jassim and Goff 2006). During the Holocene, Hammar Formation was deposited; this formation is mainly composed of coarse and very coarse sand, with some silt in the lower part, but in the upper part, it contains gray clay and thin washes of shells

(Bellen et al. 1959). The sediments of Holocene comprise sand, silt, clay, or clayey silt that represent a complex alteration of fluvial, deltaic, lacustrine, estuarine, and marine units (Yacoub 2011a).

11.3 Tectonic Settings

The Mesopotamian basin represents a subsidence area that began from the Mesozoic to Recent (Buday 1980). The Ahwar area is located in the lower Mesopotamia basin, and the latter is situated on the Mesopotamian Zone of Buday and Jassim (1987), and it is sited on the unstable shelf of the Arabian Plate (Buday 1980), whereas for Jassim and Goff (2006), the area is located in the stable shelf. The Mesopotamian basin is a structural zone that manifests many surfaces and subsurface structures, like folds, faults, and salt domes (Fouad 2010 and Fouad and Sissakian 2011) due to the collision between the Arabian and Iranian plates. They divided this unstable shelf zone into three subzones; the Tigris, the Euphrates, and the Zubair subzones. The length of the anticlines is from less than 10 to 30 km, and the depth of the basement rocks is more than 9 km (Al-Hawi et al. 2017). The long anticlines have the trend N-S and well developed in the north of the Arabian Gulf Region (Fouad and Sissakian 2011) and also narrow anticlines separated by broader synclines (Buday and Jassim 1987). The major anticlines in the area are Zubair, Rumaila, Nahr Omar, Majnoon, Ruchi, Ratawi, Subba, and Luhais.

The area was affected by many movements; three of them were major movements (Kassler 1973): the first is Pre-Pleistocene movement, and this one was a slow movement which led to the formation of the antiform structures. The second one was Palio-Pleistocene, which is responsible for the formation of the Arabian Gulf, and the third and last one was weak during the Holocene. Lees and Falcon (1952) raised the idea of the tectonic subsidence of the marshes, and evidence of this idea is the presence of the marshes area and even the high rate of sedimentation and the tidal flat existing.

The continuous changing of the watercourses in the area could help to explain the tectonic activity of the area and the uplifting (AL-Sakini 1986 and AL-Sakini 1993). The uplifting movement happens due to the underlying salt domes that introduce subsurface structures (AL-Sakini 1986; Al-Hawi et al. 2017; Darweesh et al. 2018). For Fouad and Sissakian (2011), in some areas there are uplifting processes; they suggested also that the area is under the subsidence process due to the effective neotectonic movements, which are clear through the surface geomorphology. Karim (1991) suggests by the interpretation of the geophysical data that the tectonic subsidence occurred during the Quaternary period along the Euphrates fault. The tectonic events of Mesopotamia effect the river courses as the Tigris and Euphrates, especially during their crossing of the marsh area. This phenomenon is clear through looking to the satellite images; there are many point bars and abounded buried channels.

11.4 Geomorphologic Settings

The Mesopotamian plain is a flat and wide lowland area with emersion of some surface features in some area. The marshes are distributed around the Tigris and Euphrates rivers, Hewaza marsh is to the east of Tigris River, Hammar marsh is to the south of Euphrates River, and the central marshes are between the Tigris and Euphrates rivers. The ground surface of the area is regular with some spreading hills of the ancient archaeological sites (Al-Khafaji 2010). The general slope of the area between Tigris and Euphrates rivers is gentle toward the south, for that the general flow of the streams is toward the south (Al-Ghizi 2005), and the slope of the Mesopotamian plain is 1 m/20 km (Yacoub 2011b). Despite being generally a flat surface, the Mesopotamian plain displays some geomorphological features. In general, these features are linked to the genetic origin (Al-Hmedawy 2008). The main geomorphic features in the area are the lakes (marshes locally called Ahwars), alluvial fans, rivers and their flood plains, deserts, tidal flats, and both sabkhas (marine and inland sabkhas) (Fig. 11.2). The marshes cover an area of about 7500 km2, with depth from few centimeters to 2 m, and the general elevation is about 22 m above sea level (Al-Hilli 1977). The alluvial fans are present in the east at the foot of mountains and in the west near the Al-Batin valley of Dibdibba Plain. The rivers that flow in the area are Tigris in the east and Euphrates in the west; there is new artificial canal that penetrates the area which is called Al-Massab Al-Aam; it is a drainage canal dug for agricultural purposes. There are many branches of the river flowing in the region and fed by water from major rivers. The western side of the marshes is covered by a wide desert of the sandy and gravelly surface. The tidal flats are wide and occupy many kilometer squares. The inland sabkhas are also present near the marsh area and are remarkably clear after the winter season due to the drying of the ephemeral lakes and playas, whereas the coastal sabkhas are distributed around the lagoons and marine tongues in the land.

11.5 Sedimentological and Mineralogical Characters of the Marsh Sediments

11.5.1 Sedimentology

The marsh area is mainly covered by Quaternary sediments; these sediments are from fluvial and dust fallout origin (Aqrawi 1994; Albadran 2006 and Al-Kaaby 2019), where they vary in texture vertically and horizontally. The conditions of oxidation and reduction, as well as the sediment content of organic matter, play an important role in the nature of sediment color; the color varies from greenish gray to dark gray color with depth, as well as light olive; there is also bluish and yellowish gray; the black organic layers are confined to the surface to a depth no more than 10 cm in some locations (Albadran 2006), but most of the marsh areas have sediment color contrasts between brown, beige, and gray. The marshes are covered with mostly sediments of sandy clayey silt and clayey silt with a small amount of clay-silt and sandy mud (Abdullah 1982). The sediment texture in the marshes also varies between areas and depth but is confined between sandy silt, silty sand, and silt, with the mean size from 3.7 to 5.8 phi (Al-Kaaby 2019). According to Al-Kaaby (2019) in the area of Nasiriyah, the sediment components consist of silt, silty sand, and sandy silt, but the silt as a predominant proportion, after which clay and sand come. Thus, the silt forms the predominant percentage of marsh sediments, where the percentage of silt in sediments is from 48 to 82% with an average of 68%. Meanwhile, clay comes second in terms of the percentage of sedimentary components in the marsh sediments, where the sediments consist of 4% sand, 83% silt, and 13% clay. Most sediments showed that they were poor to very poor sorting, which reflected the sedimentary process, and the variation in texture is due to the river and irrigation channel migration with time.

Clay begins to increase with depth especially after about 30 cm depth (Aqrawi 1993). Finally, the sand forms the remaining percentages of sedimentary deposits of marshes and predominantly by fine to very fine grains (Abdullah 1982 and Aqrawi 1993).

The marshes are affected by the aeolian and alluvial sediments, and most of the components of these sediments are calcareous deposits. Aqrawi and Evans (1994) stated that the surficial sediments have been classified into three main layer types:

1. A thin surface layer less than 7 cm in thickness consisting of sandy silt sediments rich in organic matter. This layer contains the remains of the organic matter of the plants, the color of which is from black to dark gray or olive.
2. A shelly clayey silt layer from 7 to 30 cm in depth rich in the upper parts by *Corbicula* sp. and *Bellamya* sp. and by *Melanoids* sp. and *Melanopsis* sp. in the lower parts.
3. The third one more than 30 cm in depth is silty clay and/or clay; for the remaining proportions of coarse silt and fine sand, it is dominated by shells of foraminifera and Ostracoda.

Due to the horizontal and vertical heterogeneity of the sediments, it is difficult to establish a hypothetical succession for the lithology of the area.

11.5.2 Mineralogy

The mineral distribution of the marsh sediments showed significant variation in quality and quantity. In general, light minerals have the highest percentage of heavy minerals. Heavy minerals are important in sediments because of their scarcity and lack of presence and are therefore used to determine the source of sediments. Heavy minerals are also affected by the processes that accompany them during transport and sedimentation (Shehta et al. 2010). Certainly, there is a heterogeneity in the percentage of heavy and light minerals with depth due to some slight heterogeneity in the texture of the sediments and the sources of sediment overlap between them, but it was not a significant variation.

In order to study the minerals of the sediments, they can be divided into minerals of sand size and clay minerals.

11.5.2.1 Minerals of Sand Size

The sediment components of sand are few compared to silt and clay. Therefore, the minerals of this group are of limited proportions and quality; they are divided into two types according to their specific weight, heavy and light minerals. The separation between the two is the relative density as 2.89 g/cm3. The assemblages of the heavy minerals of the surface sediments are similar between the sites of the marshes but in different percentages (Table 11.1).

The percentage of opaque minerals prevails over the rest of the proportions, followed by amphibole group minerals, which consist of hornblende, tremolite, and actinolite and rarely glaucophane. While epidote group minerals consist of epidote, zeosite, and clinozoisite. Pyroxene minerals are composed of augite, diopside, enstatite, and hypersthene. The percentage of these four groups reaches 70%, and it is expected that they are descended from igneous and metamorphic rocks which are outcropped in the north and northeast part of Iraq. These sediments were transported by the Tigris and Euphrates rivers, as well as the sediments that come from the western desert from Iraq through the ephemeral rivers during the rainy seasons (Abdullah 1982).

Table 11.1 Heavy minerals percentages of the surface sediments of the marsh area

Heavy minerals	Range%	Average%
Opaque minerals	14–29	22
Amphiboles group	11–25	18
Epidote group	11–23	17
Pyroxene group	12–25	17
Chlorite	4–10	7
Garnet group	2–7	5
Biotite	1–6	3
Muscovite	1–3	2
Sphene	1–3	2
Zircon	1–7	2
Rutile	1–4	2
Apatite	1–4	2
Tourmaline,		1
Olivine, staurolite, kyanite, spinel, and andalusite		1

After Abdullah (1982)

Table 11.2 Light minerals percentages of the surface sediments of the marsh area

Light minerals	Range%	Average%
Carbonate	35–82	60
Quartz	6–37	16
Feldspars	2–18	10
Chert	2–9	5
Chlorite		4
Muscovite		2
Biotite		1

After Abdullah (1982)

There is a predominant percentage of light minerals in the sand fraction of the surface sediments of the marsh area; the majority of the light minerals are from carbonates (Abdullah 1982) (Table 11.2).

The percentage of carbonates reaches 80% in the sediments of the marshes (Purser 1981; Abdullah 1982; and Aqrawi and Evans 1994). The source of these carbonates is shells of mollusks, detrital carbonate grains originated from outcropped limestone formations transported by Euphrates River and the micrite and microsparite of the carbonate rock fragments. Aqrawi (1995a) concluded the presence of Mg-calcite and calcite in Holocene sediments near-surface, and for High Mg-calcite and low-Mg calcite minerals, they are confined to brackish/marine and lacustrine carbonate environments; he also detected the presence of dolomite in evaporitic and brackish/marine sediments; this was attributed to the sea-level fluctuation and the continuous mixing of fresh and marine waters which is favorable for carbonate precipitation. Low-Mg calcite could also form in the brackish environment of the marshes as an authigenic mineral, whereas high-Mg calcite originates as an authigenic mineral in microcrystalline rhombs. The presence of the dolomite mineral in the form of euhedral rhombs suggests that it is authigenic, and the origin of the aragonite could be biogenic from the shell fragments. The presence of gypsum is in saline surface areas, as well as the presence of pyrite in a euhedral form integrated into the areas that witness an increase in sediments containing organic materials and as an authigenic mineral. In the area between Miassan and Nasiriyah, the percentages of both heavy and light minerals to a depth 8 m are from 3.6 to 9.8% and 90 to 97%, respectively (Table 11.3 and 11.4); there are no significant differences in the percentage of heavy and light minerals with depth in the marshes sediments, and the heavy minerals are dominated by opaque.

Aqrawi (2001) proposed many sedimentary environments due to the sea-level fluctuation, neotectonic activities, differential sedimentation rate, and climatic changes for the last 10,000 years B.C.: the environments of tidal flats, brackish, and marine. Aqrawi (2001) also touched upon the climate of the region for the period from 7000 years B.P. until now, and he said that the period of more than 7000 years B.P. was a semiarid, then from 7000 to 6000 years B.P. was a wet, and from 6000 to 4000 years B.P. was a semiarid and, finally, in the period from 3000 B. P. and until now, the region is experiencing a period of arid climate. After calculating

Table 11.3 Heavy minerals of Thi-Qar area

Depth/m	Opaque	Chlorite	Garnet	Zircon	Orthopyroxene	Clinopyroxene	Amphibole	Epidote	Biotite	Muscovite	Tourmaline	Staurolite	Kyanite	Others
0.5	47.9	9.2	5.6	5.4	2.9	4.3	3.1	4.7	4.5	5.3	2.2	0.6	2.6	1.7
2	43.2	10.4	4.5	5.2	2.6	3	5.8	6.3	4.5	7.2	4.1	–	2.7	0.5
4	43.5	10.3	6.4	5.3	3.2	4.6	6.1	6	4.7	4.3	2.2	–	2.6	0.8
6	49.2	9.6	5.1	4.8	2.3	3.3	4.6	6.4	4.3	5.8	1.9	–	1.3	1.4
8	47	9.1	4.4	4.5	3.9	3.5	6.1	6.4	4.4	5.6	2.6	–	1.4	1.1

After Al-Kaaby (2019)

Table 11.4 Light minerals percentage of Thi-Qar Region

Depth/m	Monocrystalline quartz	Polycrystalline quartz	Orthoclase	Microcline	Plagioclase	Carbonate rock fragments	Chert rock fragments	Igneous rock fragments	Metamorphic rock fragments	Mudstone rock fragments	Evaporites	Coated clay	Light muscovite	Others
0.5	17.6	3.8	2.6	3.3	4.4	33.6	6.5	3.5	3.4	3.9	6.6	2.3	5.7	2.8
2	19.7	3.7	2.7	1.6	2.2	33.8	10.8	2	2.3	3.1	9.1	2	4.5	2.4
4	18.8	4.7	3.9	1.9	3.6	32.7	9.3	3.4	2.7	2.6	7.7	1.7	4.7	2.3
6	17.9	2.6	3.4	2.5	4.7	30.7	7.8	3.9	3.3	3.6	6.8	3.8	7.4	1.6
8	20.6	2.5	2.7	1.9	3.7	29.8	11.2	2.7	2.8	3.9	7.9	3.3	5.3	1.7

After Al-Kaaby (2019)

the compaction of the marshes deposits, Aqrawi (1995b) managed to conclude that the sedimentation rate for the period from 8400 to 3000 years B.P. is between 1.3 and 2.2 mm/year and then it has not exceeded 0.4 mm/year for the most recent period from 3000 years B.P.

11.5.2.2 Clay Minerals

Marsh sediments, like other areas, are characterized by the presence of clay mineral assemblages, in the following descending order of percentage smectite, chlorite, illite, mixed layers of illite-smectite, palygorskite, and kaolinite (Abdullah 1982; and Aqrawi and Evans 1994). The illite and chlorite minerals vary in their percentages from one region to another. For Aqrawi (1993), the clay mineral assemblages of the Holocene sediments of the freshwater lakes of Baghdad and Zechri are the following in descending order: smectite, mixed layer of illite-smectite, illite, palygorskite, kaolinite, and chlorite. In brackish water lake Hammar, the percentage of the clay minerals in descending order is smectite, mixed layer of illite-smectite, and palygorskite, illite, kaolinite, and chlorite. There is a possibility that palygorskite can be detrital and authigenic mineral. The fact that the formation of palygorskite in place is due to the hypersaline, alkaline, and the availability of high magnesium ion in the place (Aqrawi 1995a; and Albadran and Hassen 2003). This means that the marshes are suffering from dryness, which gives an opportunity to form the evaporates, as well as witness the precipitation of carbonates. The percentages of clay minerals in the central marshes with depth showed the predominance of kaolinite minerals as 38–42%, followed by the following clay minerals according to their percentage in descending order, illite 19–28%, palygorskite 15–16%, chlorite 8–13%, mixed layers of montmorillonite-illite 5–7%, and montmorillonite 3%, and the texture of the sediments of the central marshes consist of sand 1–15%, silt 75–86%, and clay 7–21% (Al-Kaaby 2019).

For Abdalrazak et al. (2017), the texture of the Hewaza marsh sediments is from silt, sandy silt, to silty sand, and the mineral composition is calcite 57–64%, quartz 28–35%, and dolomite 5–13% with few and varying percentage of feldspar and gypsum, whereas the clay minerals are kaolinite, montmorillonite, chlorite, palygorskite, and illite and mixed layers of palygorskite-illite and montmorillonite-chlorite.

11.6 Fauna Assemblage

The sediments of the marshes are rich in the presence of mollusks, especially the macrofauna. Most of the mollusks are distributed between three families; these families are Corbiculidae, Thiaridae, and Viviparidae (Purser 1981; Aqrawi and Evans 1994; and Al-Kaeeby 2019). In general, the freshwater mollusks present in the sediments of the southern part of the Mesopotamian belong to Mediterranean

Palearctic domain (Plaziat and Younis 2005). There are several factors that control the distribution of fauna in sediments, including water chemistry, salinity, and temperature. Most of the heterogeneity in fauna distribution is in the rooted vegetation plants in lake-bottom sediments. The most occurrence of fauna is at the bottom of the channels through which the boats move, and this can be due to the turbulence and continuous washing of soft sediments from the bottom of the channels. Ostracoda, foraminifera, and diatoms make up the majority of the microfauna in sediments; the genus of *Ammonia beccarii* of foraminifera is present which reflects the very brackish environment (Murray 1991). The genuses-species *Cyprideis torosa*, *Limnonythere* sp., and *Darwinola* sp. of Ostracoda also present which are typical of the fresh-brackish environment (Murray 1991; and Aqrawi and Evans 1994). For the top sediments of the marshes, some researchers state that the occurrence of *Ammonia beccarii* and *Cyprideis torosa* in the clay-rich horizon in the depth more than 30 cm could reflect the influence of the marine conditions during the deposition of these sediments in the marsh area (Murray 1991; Aqrawi and Evans 1994; and Al-Kaeeby 2019). The radiocarbon results showed that the age of this layer is more than 3000 years B.P. While the layer is between 7 and 30 cm deep, its age is 3000 years B.P., and the surface organic-rich horizon is about 400 years B.P. (Aqrawi and Evans 1994).

Abdalrazak et al. (2017) indicated that the Al-Hewaza marsh sediments showed there are three biofacies and each one returns to a specific sedimentary environment. These biofacies are the following: first, microfacies start from the ground surface of the earth to a depth of 1.5 m, which carries only the fauna *Ammonia tepida*; the presence of these fauna can make this biofacies return to a brackish environment. The second biofacies extend from a depth of 1.5 to 2.45 meters; the following fauna appeared in this biofacies *Ammonia beccarii*, *Elphidium guntri*, *Elphidium advenum*, *Quinqueloculina* sp., *Triloculina trigonula*, *Cypridies torosa*, *Quinqueloculina impressa*, *Littorina*, and *Retusa* with rareness of *Corbicula fluminea*. The environment of this biofacies could belong to a shallow marine, lagoon, or tidal flat. The third one is from depth 2.45 to 3.5 meter; quantities of the following fauna appeared in this biofacies and are represented by *Ammonia tepida*, *Elphidium guntri*, *Elphidium advenum*, *Cypridies torosa*, and *Littorina*. The appearance of these fauna together can suggest that the sedimentary environment is a brackish environment affected significantly by marine water. While four biofacies have appeared for Al-Kaaby (2019) in the north of the central marshes to a depth of 8 m; these biofacies are the following: first biofacies is of freshwater fluvial, the second one is of brackish water environment, the third biofacies are arid and sabkha, and the fourth one is the brackish environment. The same researcher also showed four biofacies in the marsh areas to a depth of 19 m to the west of Missan governorate; these biofacies are the following: biofacies of brackish water environment, second biofacies is a freshwater environment, third biofacies is brackish water of playa environment, and the fourth one is the freshwater environment. The biofacies in the middle marshes are characterized by irregular repetition through the lithological succession.

References

Abdalrazak A, Albadran B, Pournelle J (2017) Sedimentary, mineralogy and biofacies of holocene deposits of Al-Hammar and Al-Hewaiza marshes near Qurna. Int J Eng Appl Sci (IJEAS) 4(8):2394–3661

Abdullah MB (1982) Sedimentology, petrography, geochemistry and hydro chemistry of the recent sediments of Hor AL-Hammar in southern Iraq. Unpubl. MSc thesis, college of science, Univ. of Baghdad, 192p (In Arabic)

Albadran BN (2006) Sedimentology and mineralogy of the Al-Hammar Marsh / Southern Iraq: a review. Marsh Bulletin 1(1):32–39

Albadran BN, Hassen WF (2003) Clay mineral distribution of supratidal area, South of Iraq. Marina Mesopotamica 18(1):25–33

Al-Ghizi HSN (2005) Hydrology of Shatt Al-Garaf and its investments. Thesis, College of Education for human sciences, University of Basrah, 101 p, (In Arabic)

Al- Hawi NA, Albadran BN, Pournelle J (2017) The archaeological sites along the ancient course of Euphrates river. Am Sci Res J Eng Technol Sci 29(1):1–20

Al-Hilli MR (1977) Study of the plant ecology of Ahwar region in southern Iraq. PhD thesis, Univ. Cairo, 477p

Al-Hmedawy H (2008) Geomorphological study of Haur Al-Hammar and adjacent area Southern Iraq using remote sensing data and GIS techniques. MSc thesis University of Baghdad, Iraq

Al-Kaaby LFS (2019) Sedimentological and biofacies study of the quaternary period sediments in selected areas-Southern Iraq. PhD thesis, University of Basrah, Iraq

AL-Khafagi SN (2010) A geomorphologic analysis of the steps of formulating of alluvial plain. Uruk J Human Sci 3(1):107–130

Al-Sakini J (1986) Neotectonic activity in Basrah vicinity and the dryness of western canals of Shatt Al-Arab. Proceedings of the first symposium on Khor Al-Zubair, Marine Science Center, University of Basrah, pp 415–416. (In Arabic)

Al-Sakini J (1993)New window on the Mesopotamia history in the light of geological evidences and archaeology. Dar AL-Shaon AL-Thakafiya, ALAumah, Baghdad, 93p, (In Arabic)

Al-Syyab A, Al-Ansari N, AL-Rawi D, AL-Jassim JA, AL-Omari FS, AL-Saikh ZD (1982) Geology of Iraq. Dar AL-Kutub publishing house, University of Mosul, 277p. (in Arabic)

Aqrawi AAM (1993) Implication of sea level fluctuations, sedimentation and neotectonic for the evolution of the marshlands (Ahwar) of southern Mesopotamia. In: Quaternary proceedings, vol 3. Quaternary Research Association, Cambridge, pp 21–31

Aqrawi AAM (1994) Petrography and mineral content of seafloor sediments of the Tigris–Euphrates delta, NW Arabian Gulf, Iraq. Estuar Coast Shelf Sci 38:569–582

Aqrawi AAM (1995a) Brackish-water and evaporitic Ca-Mg carbonates in the Holocene lacustrine/ deltaic deposits of southern Mesopotamia. J Geol Soc Lond 152:259–268

Aqrawi AAM (1995b) Correction of Holocene sedimentation rates for mechanical compaction: the Tigris–Euphrates Delta, Lower Mesopotamia. Mar Pet Geol 12(4):409–416

Aqrawi AAM (2001) Stratigraphic signature of climate change during the Holocene evolution of the Tigris and Euphrates delta, lower Mesopotamia. Glob Planet Change 28:267–298

Aqrawi AAM, Evan G (1994) Sedimentation in the lakes and marshes (Ahwar) of the Tigris-Euphrates Delta, southern Mesopotamia. Sedimentology 41:755–776

van Bellen RC, Dunnington HV, Wetzel R, Morton DM (1959) Iraq, in Dubertret, Louis, director, Lexiquestratigraphique international, vol 3. Asie, Paris, Internat. Geol. Cong., Comm. Stratig., pt. lOa, p 333

Buday T (1980) The regional geology of Iraq, stratigraphy and palaeogeography. Al-Mosul, State organization of minerals, Dar Al-Kutub Publishing House, 443p

Buday T, Jassim SZ (1987) The regional geology of Iraq tectonism (magmatism and metamorphism) GEOSURV pub, Baghdad, 322 p

Darweesh HA, Obed AM, Albadran BN (2018) Structural study of basins configuration in Mesopotamian area. Int J Eng Appl Scie (IJEAS) 4(9):54–58. ISSN: 2394-3661

Fouad SFA (2010) Tectonic evolution of the Mesopotamia foredeep in Iraq. Iraqi Bull Geol Min 6(2):41–53
Fouad SFA, Sissakian VK (2011) Tectonic and structure evolution of the Mesopotamia plain. Iraqi Bull Geol Min, special Issue, no.4; pp. 33–46
Hussain NA (2014) Biotopes of Iraqi Marshlands. Difaf Publishing, 432p (In Arabic)
Jassim SZ, Goff JC (2006) Geology of Iraq. Dolin, Prague and Moravian museum, Brno, 341p
Karim HH (1991) Structural nature of lower Mesopotamian region from geophysical observations. Proceeding of third symposium, Oceanography of Khor Al-Zubair, Basra, Research, pp 15–25
Kassler P (1973) The structural and geomorphic evaluation of the Persian Gulf. In: Purser BH (ed) The Persian Gulf: Holocene carbonate sedimentation and diagenesis in a shallow epicontinental sea. Springer-Verlag, pp 11–32
Lees GM, Falcon NL (1952) The geographical history of Mesopotamia plain. Geogr J 118:24–39
Murray JW (1991) Ecology and Paleoecology of benthic foraminifera. Longman Scientific and Technical, New York
Plaziat JC, Younis WR (2005) The modern environments of Molluscs in southern Mesopotamia, Iraq: a guide to paleogeographical reconstructions of Quaternary fluvial, palustrine and marine deposits. Notebooks on Geology 1(CG2005-A01):1–18
Purser BH (1981) Preliminary study of the Quaternary Sediments of SE Iraq. A joint research project by the geological survey of Iraq and university of Paris X I (Orsay), 56 p, 111 p, 18 fig., 10 tab
Shehata MG, Abdou AA, El-Hariri TM, Mousa AS (2010) Heavy minerals: a case study on the Gebel Ghorabi member; Bahariya Oasis, Western Desert, Egypt. Int J Acad Res 12(4):159–172
Yacoub SY (2011a) Stratigraphy of Mesopotamia plain. Iraqi Bull Geol Min, Special Issue, no. 4, pp 47–82
Yacoub SY (2011b) Geomorphology of Mesopotamia plain. Iraqi Bull Geol Min, Special Issue 4, pp 7–32

Section IV
Major Biotope

Chapter 12
Phytoplankton and Primary Production in Iraqi Marshes

Bahram K. Maulood and Fikrat M. Hassan

Abstract Mesopotamian marshes were once quite famous by their biodiversity and cultural richness. In fact, the area is an important stopover along the flyway of millions of migratory birds between Europe and Africa. However southern Iraqi marshes had faced a deliberate drainage by previous regime in Iraq, and there are several efforts to restore it again. Theses marshes consist of eastern, central, and southern part that is fed with different sources of water from Tigris, Euphrates, and Shatt al-Arab Rivers. The most obvious feature of algal vegetation in this area is the dominance of filamentous epiphytes with other free floating filamentous algae. Algal habitats include reed stems, lime-encrusted algal felt, and submerged parts of wood and boats. In such habitats, *Chaetophora incrassata, Lyngbya, Calothrix, Aphanocapsa, Tolypothrix*, and *Schizothrix* are found, besides heterocystous blue-green algae which are common epiphytic flora. On the other hand, diatoms are dominant followed by Cyanophyta, Chlorophyta, Cryptophyceae, Pyrrophyceae, and Euglenophyceae. Generally, there are two peaks of phytoplankton bloom that have been recorded: one in autumn and the other in June–July. Chlorophyll *a* concentration had shown different values at different stations; generally it ranges between 0.15 and 21.2 μg/l. The Iraqi marshes are oligotrophic-mesotrophic before 2004, while these marshes are being toward eutrophic according to chlorophyll *a* concentration. Almost all primary productions within marshes are originated from the role of algae and macrophytes, with slight role played by phytoplankton. Primary productivity in Mesopotamian marshes had been estimated to be between 132 and 407 mg/carbon/m3/day estimated by light and dark bottle method. It seems that there are a lot of research to be done in the future regarding the Mesopotamian marshes particularly after the re-flooding process, which the job of the scientists in the near future.

Keywords Diatoms · South of Iraq · Basrah City · Wetlands · Phytoplankton

---

B. K. Maulood (✉)
Department of Biology, College of Science, University of Salahaddin, Erbil, Iraq

F. M. Hassan
Department of Biology, College of Science for Women, University of Baghdad, Baghdad, Iraq

L. A. Jawad (ed.), *Southern Iraq's Marshes*, Coastal Research Library 36,
https://doi.org/10.1007/978-3-030-66238-7_12

12.1 Introduction

Iraqi marshes are an old sustained wetland that had been dwelled originally by ancient Sumerian civilization; however it is known as Mesopotamian marshes or wetland (Thesiger 1964; Douabul et al. 2005; Al-Hilli et al. 2009). In fact such area is an extensive and do cover about 15,000 Km^2 of southern Iraq, around Qurna where Tigris and Euphrates join to form Shatt Al-Arab, locally known as Al-Ahwar (Thesiger 1964). This area comprises the triangle between Amara (Missan), Nasiriyah (Thiqar), and Basra provinces (Al-Hilli et al. 2014). It is shared with Iranian marshes (border) to establish the existing marsh land (Fig. 12.1). Actually it became one of the largest continuous marsh area of the world as its area exceeds 20,000 km^2 that harbor an unspoilt ecosystem; however, this unique ecosystem had been almost completely destroyed in 2000 by previous regime as only about 10% of the marshes of southern Iraq remained functioning as a marsh land (Richardson and Hussain 2006).

Mesopotamian marshes were once considered as the main bread basket of the area as it was the home of hundred species of both terrestrial and aquatic life Zohary (1946) such as birds, mammals, fishes, macrophytes, aquatic plants, ferns, planktons; besides that the area was supported by more than half a million peoples (Fig. 12.1) (Al-Hilli et al. 2014; Richardson and Hussain 2006).

However the military raided settlement killed more than tens of thousands marsh Arabs, burned settlements, and killed livestock. The agriculture and fishing in the area have been shattered so people had to move to the edge of the marshes or deserts (Richardson and Hussain 2006). Nevertheless, decades ago, Maulood et al. (1979) had already pointed out that as the Qurna drainage scheme be implemented which was expected to be finished in the near future, undoubtedly will lead and result in impending huge ecological changes in the area. Furthermore, Maulood et al. (1981) suggested clearly that the main body of the marshes has already been changed recently and pointed out that man-made environmental changes will increasingly effect on the nature of the marshes in the future. Maulood (1991) showed that on the completion of the drainage scheme, disappearance of some marshes will take place. They also pointed out that the water level also will be lowered in the area. Actually, Maulood and Al-Saadi (1990) requested for an essential account for the area in order to preserve the ecological information of the present status of the marshes, particularly literature on such ecosystem in Iraq is quite sparse (Al-Doori and Al-aathami 1990; Maulood and Al-Saadi 1990; Maulood et al. 1981). The establishment of three major dams on upstream tributaries of Tigris and Euphrates rivers, the development of paper industry which use a huge amount of reed, and also the transfer of water to the south of Kuwait all influenced the dryness of the marshes; in fact all these issues have been raised up in order to prevent disasters that could happen in the marshes (The scientific conference for Iraqi biological association Basra, 27th February to 13th March 1990) (Al-Saadi and Maulood 1990).

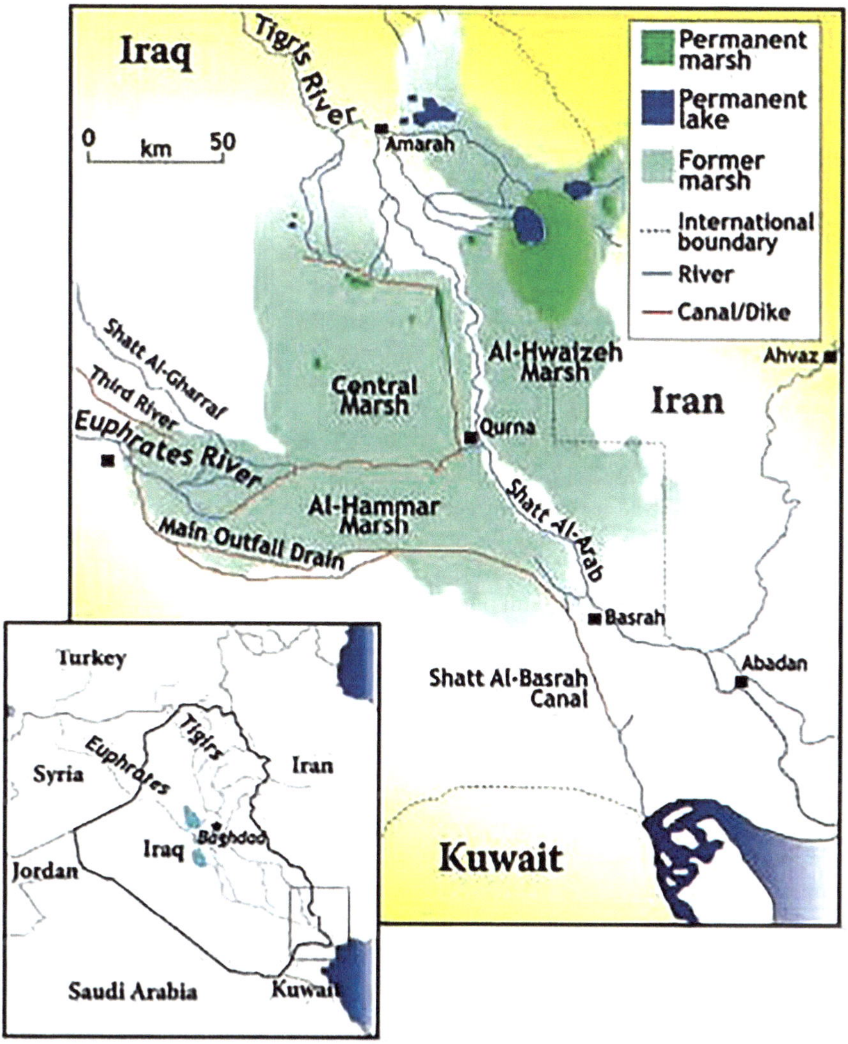

Fig. 12.1 Mesopotamian marshes. (Image courtesy of UNEP 2009)

As a matter of fact, not only marshes but almost all inland water systems in southern Iraq have got almost no attention ecologically, Evans (2002) until the last three decades. Actually investigation on such ecosystems had coincided with establishment of Basrah University particularly with setting up aquatic biology department in Science College, fishery department in Agriculture College, Natural History Museum, Marine Science Centre, and marsh field research station. All these institutions had contributed in one way or another to the existing knowledge on limnology and ecology of the southern inland water system in Iraq including the southern marshes of Iraq (Maulood and Al-Saadi 1990). In fact University of Sulaimani, north Kurdistan part of Iraq, had also a role in investigation of such area at the end of the 1970s of the last century.

Recently, the southern marshes of Iraq have included in some research program since 2003 (Maulood and Douable 2018), but still the whole marsh area was not covered especially the eastern part of the marshes (those marsh areas located between Basrah and Amara cities, south of Iraq). In addition, no ecological investigations were made for those marshes located between Diwaniyah and Nasiriyah (Al-Saadi and Maulood 1990; Richardson and Hussain 2006).

It's a well-known fact that phytoplankton is responsible for the turnover of the carbon dioxide (CO_2) in the aquatic ecosystems and considered as the main primary producer (Carlson 1991; Wetzel 2001; Hassan et al. 2011; Schaum et al. 2017). Such algal growths and outburst are affected by different environmental and non-environment factors, also influencing their dynamics (Falkowski 1998, 1999; Lee 2008; van der Valk 2012).

Many scientists had investigated and recorded various phytoplankton dynamics within different stations (parts) of the marshes. In general, two peaks of phytoplankton growth have been recorded in Mesopotamian marshes, but the date varied with different location and authors; still others have confirmed presence of only one peak of phytoplankton growth (Huq et al. 1977, 1978, and 1981). In general the recorded peaks and outburst of phytoplankton have been already noted in spring and late summer; however the months of March, April, May, July, and October are when the maximum number of phytoplankton has been recorded.

The present study focuses on Al-Hewaizeh, Al-Hammar, and central marshes (Al-Zubaidi 1985; Al-Lami 1986; Al-Saboonchi et al. 1982; Al-Mousawi et al. 1994; Talib 2009). The various patterns of seasonal variation and periodicity of phytoplankton differences might be due to various impacts of hydrology regime and temperature (Kabanova 1968; Jacob et al. 1979; Ibrahim 1980; Abdul-Hussein and Mason 1988; Mitsch and Gosselink 2000); furthermore many other environmental factors may be behind such variation (Rzoska 1980). More comprehensive studies are needed in order to clarify the exact pattern of seasonal variation of phytoplankton within southern marshes. The existing gap of knowledge in this respect hopefully will be fulfilled in years to come.

12.2 Algal Flora of the Marsh Area

In regard to the algal flora before the distraction process of the marshes, Pankow et al. (1979) have recorded 55 non-diatoms taxa of algae from Tigris marshes, whereas Hinton and Maulood (1982) recorded 101 non-diatom taxa in 1978 and 1979 in the region located between Madaina and Gibbaish. However Hinton and Maulood (1980) recorded 76 taxa of diatoms from brackish water habitat in southern marshes, 67 of which were from previously uninvestigated marsh region (Plates 12.1 and 12.2).

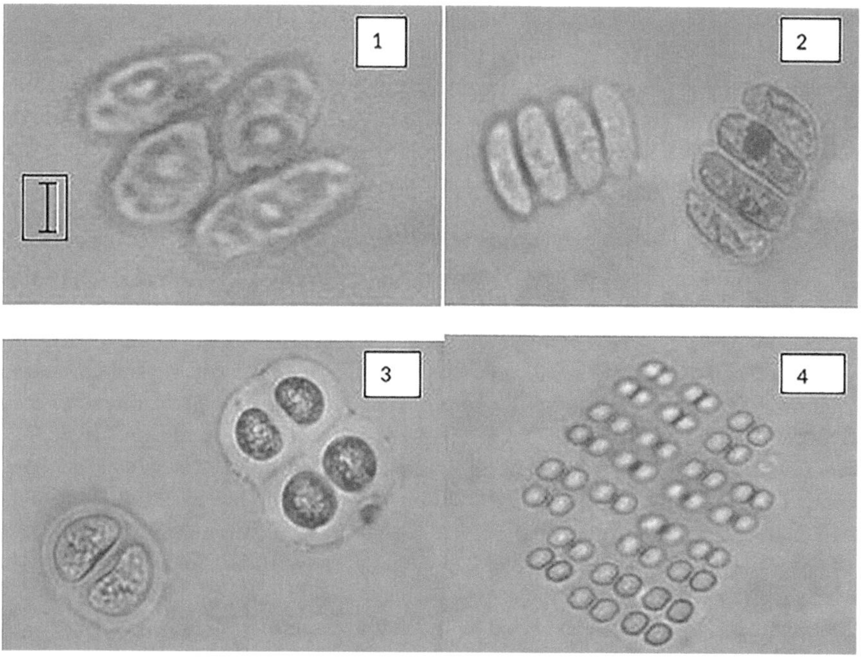

Plate 12.1 Representative's species of algae from the southern marshes of Iraq (non-diatoms species). 1, *Scenedesmus arcuatus*; 2, *Scenedesmus bijuga*; 3, *Chroococcus turgidus*; 4, *Merismopedia tenuissima*. Scale bar = 10 µm

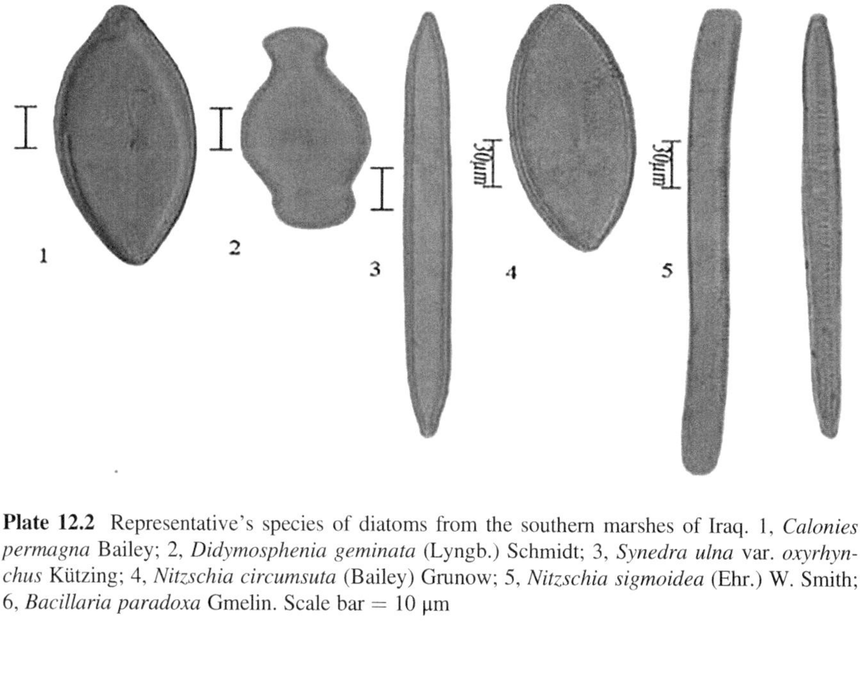

Plate 12.2 Representative's species of diatoms from the southern marshes of Iraq. 1, *Calonies permagna* Bailey; 2, *Didymosphenia geminata* (Lyngb.) Schmidt; 3, *Synedra ulna* var. *oxyrhynchus* Kützing; 4, *Nitzschia circumsuta* (Bailey) Grunow; 5, *Nitzschia sigmoidea* (Ehr.) W. Smith; 6, *Bacillaria paradoxa* Gmelin. Scale bar = 10 µm

The first record of red algae from marshes in particular and for Iraq in general was published by Hinton and Maulood (1980).

One of the best description of algal flora (composition wise) is that of Maulood et al. (1981), who stated that the most obvious feature of the algal vegetation in the Mesopotamian marshes is the abundance of filamentous epiphytes with other free floating filamentous algae. However features of reed stem covered with masses of *Chaetophora incrassate is common*, lime-encrusted algal felts dominated with different *Blue-green algal* species such as *Lyngbya*, *Calothrix*, and *Aphanocapsa is also common*. Presence of *Tolypothrix* and *Schizothrix* was found to be common as well on the submerged parts of the wood, whereas heterocystous blue-green alga were found to be general epiphytic flora.

Many scientists recoded different phytoplankton dynamics in shatt-Al Arab (Schiewer et al. 1982) and in Iraqi marshes. Two peaks of phytoplankton are noticed in spring and autumn (Al-Zubaidi 1985; Al-Lami 1986), while one peak was observed in spring (Al-Saboonchi et al. 1982). Al-Mousawi et al. (1994) recoded two peaks for phytoplankton, one in autumn 1987 and another peak in July 1988 at Al-Hammar marshes. Talib (2009) revealed that phytoplankton has two peaks, one recorded in spring 2006 (March) and the second in May 2006 at Al-Hewaizeh marsh, while in Al-Hammar marsh, two peaks were noticed in April 2006 and October 2006, while in the central marshes, the first peak was in July 2006 and the second in October 2006. These differences might be due to the impact of the hydrology regime and temperature (Abdul-Hussein and Mason 1988; Mitsch and Gosselink 2000).

Al-Obaidi (2006) investigated the Abu-Zirig Marsh and identified a total of 146 algal taxa which included Bacillariophyceae as dominate group and followed by Cyanophyceae, Chlorophyceae, Cryptophyceae, Pyrrophyceae, and Euglenophyceae during the re-flooding process. Another study at Al-Hewaizeh, Suq Shuyukh, and East Hammar Marshes noticed also the dominancy of Bacillariophyceae followed by Chlorophyta and Cyanophyta during the period from summer 2004 to Spring 2005 (Hammadi et al. 2007).

Auda marsh, which is one of important marshes for bird area with code Q072 and belongs to Maysan Province's marshes, is suffered from dryness before 2004 and re-flooding during 2004. The total number of phytoplankton species identified from this marsh ranged from 193 species in 2007 to 246 species in 2013 (Maulood et al. 2013) (Table 12.1), but due to the dry season and the water scarcity in 2018, the number of species reduced to almost 50% (Albueajee et al. 2020).

12.3 Trophic Levels, Chlorophyll *a* and Primary Production

The trophic status and productivity of the marshes have been referred to by Talling (1980a, b) prior to the drainage of the marshes. However, Hinton and Maulood (1982) indicated that Mesopotamian marshes are oligotrophic in nature. This was in accordance to the chemistry and phytoplankton composition; water had possessed

Table 12.1 Most common taxa of phytoplankton in Mesopotamian marshes

Taxa
Phylum: Cyanobacteria
Class: Cyanophyceae
Anabaena sp.
Chroococcus dispersus (Keis.) Lemmermann
C. minutus (Kütz.) Naegeli
C. turgidus (Kütz.) Nägeli
Lyngbya major Meneghinii
Lyngbya sp.
Merismopedia glauca Nägeli
M. minima G.Beck
M. tenuissima Lemmermann
Oscillatoria amoena Gomont
O. limnetica Lemmermann
O. limosa C.Agardh ex Gomont
O. tenuis Agardh
Phormidium tenue (Menegh.) Gomont
Spirulina laxa G.M.Smith
S. major Kützing
Phylum: Chlorophyta
Class: Chlorophyceae
Ankistrodesmus falcatus (Corda) Ralfs
Chlamydomonas sp.
Crucigenia tetrapedia Kirchner West & West
Kirchneriella irregularis (G.M.Smith) Korschikov
Microspora pachyderma (Wil.) Lagerheim
Pediastrum tetras (Ehrenberg) Ralfs
Scenedesmus bijuga (Turp.) Lagerheim
S. dimorphus (Turpin) Kützing
S. quadricauda (Turp.) de Brébisson
Tetraëdron minimum (A. Braun) Hansgirg
T. tringulare Korshikov
Phylum: Ochrophyta
Class: Conjugatophyceae (Zygnematophyceae)
Coelastrum microporum Nägeli
Class: Trebouxiophyceae
Oocystis crassa Wittrock
Phylum: Euglenozoa
Class: Euglenophyceae
Euglena acus Ehrenberg
E. convulata Korshikov
E. elastic Prescott
Euglena gracilis G.A.Klebs

(continued)

Table 12.1 (continued)

Taxa
E. granolata (G.A.Klebs) F.Schmitz
E. oxyuris var. *minor* (Skvortzov) Popova
E. polymorpha Dangeard
E. proxima A.P. Dangeard
E. spirogyra Ehrenberg
E. virdis (O.F.Müller)
Lepocinclis acuta Prescott
L. fusiformis (Carter) Lemmermann
L. ovum (Ehrenberg) Lemmermann
L. glabra Drezepolski
Phacus acutus Pochmann
P. anacoelus Stoke
P. caudatus Swirenko
P. chloroplastes Prescott
P. curvicoda Swirenko
P. longicauda (Ehr) Dujardin
P. nordstedtii Lemmermann
P. orbicularis K.Hübner
P. spirogyra var. *maxima* Prescott
P. totus (Lemm.) skvortzow
Phacus sp.
Trachelomonas sp.
Phylum: Bacillariophyta
Class: Mediophyceae
Cyclotella meneghiniana Kützing
C. striata (Kütz.) Grunow
Cyclotella sp.
Class: Coscinodiscophyceae
Coscinodiscus lacustris Grunow
Melosira sp.
Class: Bacillariophyceae
Achnanthes brevipes var. *intermedia* (Kütz.) Cleve
A. exigua Grunow
A. microcephala (Kütz.) Grunow
A. minutissima Kützing
Amphiprora alata Kützing
A. paludosa W.Smith
Amphora coffeaeformis Agardh
A. mexicana var. *major* Cleve
A. ornata Bailey
A. venata Kützing
Anomoeoneis exilis (Kütz.) Cleve

(continued)

Table 12.1 (continued)

Taxa
Bacillaria paxillifer (Müll.) Hendy
Caloneis permagna (Bail.) Cleve
Campylodiscus clypeus (Ehrenberg) Ehrenberg ex Kützing
Cocconeis pediculus Ehrenberg
C. placentula Ehrenberg
C. placentula var. *euglypta* (Ehr.) Cleve
C. placentula var. *lineata* (Ehr.) Cleve
Cymbella affinis Kützing
C. cistula (Hemp.) Grunow
C. parva (W. Smith)Kitchn
C. tumida (Bréb.)V. Heurck
C. turgida (Greg.) Cleve
C. ventricosa Kützing
Cymbella sp.
Diploneis ovalis (Hisle) Cleve
D. paulla (Schum.) Cleve
Epithemia zebra (Ehr.) Kützing
Eunotia valida Hust.
Eunotia pectinalis (Kützing) Rabenhorst
Eunotia sp.
**Fragilaria construens (*Ehrenberg) Grunow
Fragilaria sp.
Gomphonema constrictum Ehrenberg
G. constrictum var. *capitata* (Ehr.) Cleve
G. gracile Ehrenberg
G. longiceps Ehrenberg
G. olivaceum (Lyng.) Kützing
G. parvulum (Ehr.) Grunow
Gyrosigma acuminatum (Kütz.) Rabenhorst
G. distrodum var. *parkeri* Harrisson
G. spenceri (W. Smith) Cleve
Mastogloia braunii Grunow
M. smithii Thwaites
M. smithii var. *amphicephala* Grunow
M. smithii var. *lacustris* Grunow
Navicula cincta (Ehr.) Kützing
N. cryptocephala Kützing
N. cryptocephala var. *veneta* (Kütz.) Grunow
N. cuspidata Kützing
N. pusilla W. Smith
N. radiosa Kützing
N. radiosa var. *tenella* (Bréb.) Grunow

(continued)

Table 12.1 (continued)

Taxa
N. spicula (Dick.) Cleve
N. viridula var. *rostellata* (Kütz.) Cleve
Navicula sp.
Nitzschia acicularis W.Sm.
N. amphibia Grunow
N. apiculata (Greg.) Grunow
N. filiformis (W. Smith) Hustedt
N. granulata Grunow
N. ignorata Krasske
N. kützingiana Hilse
*N. longissma (*Breb.) Ralfs
N. microcephala Grunow
N. obtusa W. Smith
N. palea (Kütz.) W. Smith
N. romana Grunow
N. sigma (Ehr.) W. Smith
N. tryblionella var. *victoriae Grunow*
Nitzschia sp.
Pleurosigma salinarum Grunow
Pleurosigma sp.
Rhoicosphenia curvata (Kütz.) Grunow
Rhopalodia gibba (Ehr.) Müller
Rhopalodia gibba var. *ventricosa* (Ehr.) Grunow
Surirella ovata Kützing
Surirella sp.
Synedra affinis Kützing
S. amphicephala Kuetzing
S. capitata Ehrenberg
S. fasciculata (Kütz.) Grunow
S. ulna (Nitz.) Ehrenberg
S. ulna var. *biceps* Kützing
S. tenera Gregory
Phylum: Mioza
Class: Dinophyceae
Glenodinium pulvisculus (Ehrenberg) F.Stein

fairly low concentration of inorganic nutrients, Al-Saadi et al. (1989) whereas Maulood et al. (1981) stated that southern marshes in Iraq are characterized as being slightly saline, relatively turbid, and calcareous. Accordingly, diatoms dominate phytoplankton community as the trophic status was mesotrophic and their number, increased Maulood et al. (1981); Al-Saadi et al. (1981). The number were higher in the marshes that connected with Shatt al Arab Antoine (1983); Abdullah (1989); Al-Mousawi et al. (1990); Al-Handal et al. (1992) or the Arabian Gulf and reached 197,000 cells/Lat Burket al Baghdadi station near Gibbaish. The samples were dominated by the colony-forming diatoms *Bacillaria paradoxa* and *Synedra tabulate*. On the other hand, Maulood et al. (1981) suggested that there are no previous data on diatoms from this area for comparison.

Al-Zubaidi (1985) measured the chlorophyll *a* concentration at Qurna marshes and reported concentration of a ranged between 0.21 mg/l and 11.68 μg/l, whereas Al-Lami (1986) studied this ecological factor in Al-Hammar marsh and showed the chlorophyll *a* bimodal pattern of seasonal variations during spring and autumn. Its concentration ranged from 0.15 to 8.46 μg/l during the study period, using chlorophyll *a* as an indicator of primary productivity applied after re-flooding process (after 2004). Hassan et al. (2011) noticed the bimodal pattern of seasonal variations in southern marshes; the chlorophyll *a* concentration was ranged between 1.1 and 21.26 μg/l. In Al-Hewaizeh marsh, the concentration of chlorophyll *a* had ranged between 1.1 and 14.8 μg/l, whereas in two other marshes, the level was 1.2–21.2 and 1.1–14.4 μg/l. These results indicate that the Iraqi marshes are oligotrophic-mesotrophic before 2004, while they tend to be eutrophic according to chlorophyll *a* concentrations.

Al-Saadi (1993) suggested that the primary production in open ocean reaches more than 15×10^9 ton carbon per year. While Ryther and Menzel (1965), Thornton (2012) showed that amount of primary production may reach more than 40 Pg C per year 1Pg = 1gigaton = 1012 g). In contrast, Mitsch and Gosselink (2015) mentioned that primary production in wetland is much higher than that of even terrestrial ecosystem (Platt and Subba Rao 1975; vanan der Valk 2012).

Primary productivity (PP) has been estimated within southern marshes using light and dark bottle and C^{14} method (Al- Zubaidi 1985; Al-Lami 1986). The values ranged between 132 and 407 mg/carbon/m^3/hour and 1.6 to 114 mg/carbon/m^3/hour within stations near Qurna and Alburge within Al-Hammar marsh. It was evident that results were contradictory. The available data on PP using carbon method have been revised and illustrated for various nearby stations in southern Iraq (Table 12.2).

The only studies concerning the PP in the southern Iraqi marshes are those shown in (Table 12.3). Studies prior 2004 Hadi et al. (1989) indicated that the marshes are

Table 12.2 Status of trophic level of Mesopotamia marshes pre and after desiccation

No	Location	Tropic status	Method	Reference
1	Burga Al Baghdadi	Oligohaline	Total count and species composition	Maulood et al. (1979)
2	Abu Suban	Oligotrophic	Chlorophyll and species composition	Maulood et al. (1981)
3	Ashan, Hamma	Mesotrophic	Chlorophyll and species composition	Maulood et al. (1981)
4	Al- Hammar	Mesotrophic	C^{14}	Al-Lami (1986)
5	Al-Hewaizeh	Mesotrophic	Light and dark bottle	Hassan et al. (2011)
6	Al-Hammar	Oligotrophic	Light and dark bottle	Hassan et al. (2011)
7	Central marshes	Mesotrophic	Light and dark bottle	Hassan et al. (2011)
8	Auda marsh	Mesotrophic	Species composition	Albueajee et al. (2020)

Table 12.3 Some published data on primary productivity values in different aquatic systems in Iraq

Marsh	Primary productivity range	Method	References
Near Qurna	12.49–407.09 mg C/m^3. hr.	Light/ dark bottles	Al-Zubaidi (1985)
Al-Hammar	1.6–114.18 mg C /M^3. hr.	C^{-14}	Al-Lami (1986)
Al-Hammar	11.71–119.33 mg C/m3. hr	Light/ dark bottles	Hassan et al. (2011)
Al-Hewaizeh	18.75–256.25 mg C/m3. hr	Light/ dark bottles	Hassan et al. (2011)
Central marshes	15.39–253.13 mg C/m3. hr	Light/ dark bottles	Hassan et al. (2011)
Auda marsh	117.87–175.53 mg C/m^3/ h	Light/ dark bottles	Albueajee (2020)

Table 12.4 Classification of an aquatic system according to Chlorophyll a

Trophic status	Primary productivity mg C/m^2 day
Ultra-oligotrophic	50
Oligotrophic	50–300
Mesotrophic	250–1000
Eutrophic	600–8000
Dystrophic	50–600

Likens (1975)

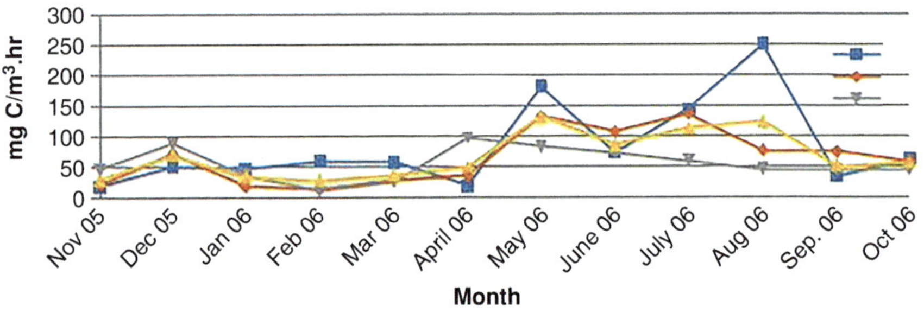

Fig. 12.2 Seasonal variations in primary productivity in the surface for all main marshes during study period. (Image courtesy of Hassan et al. 2011)

most likely oligo-meso-trophic. However, it seems that the re-flooding process had affected the primary production in the marshes as they are tending toward mesotrophic (Maulood et al. 1981) (Table 12.4).

The seasonal variation of PP in the main marshes reflects different patterns of variation when comparing the area before and after the drainage of the marshes (Partow 2001). Bimodal pattern in autumn and spring was recorded by Al-Zubaidi (1985) and Al-Lami (1986), and after the rehabilitation the highest peak was recorded in summer and the lowest peak found in autumn (Fig. 12.2) (Hassan et al. 2011). Recently work on Auda marsh showed unimodal pattern, where the highest value of PP was recorded in summer. These differences are related to the differences in community structure and nutrient availability in addition to hydrology status of the marshes (Schiewer 1984). This disparity might also depend on the seasonal variation of quantitative and quantitative of phytoplankton (Al-Zubaidi

1985; Al-Lami 1986). Temperature might be the most vital factor as it may affect the PP values (Joniak et al. 2003; Teira et al. 2005).

12.4 Conclusion

1. Diatoms are dominating algal flora number-wise, whereas blue-green blooms are mainly represented by *Chroococcus*, *Gomphosphaeria*, and *Microcystis*. On the other hand, diatoms are represented by *Cyclotella, Fragilaria, Synedra*, and *Coscinodiscus*.
2. Marshes are reduced by about 85% or more as a result of drainage through the central government policy at the time in order to diverse Euphrates River water to a newly established drain channel.
3. Trophic status of the marshes indicates that they are oligotrophic-mesotrophic before 2004, whereas they are heading to be eutrophic according to chlorophyll *a*, with the process of eutrophication is in progress.
4. Primary productivity has been estimated within southern marshes and shown to be ranged between 132 and 407 mg/carbon/m^3/day.
5. The seasonal variation of primary productivity in the main marshes reflects different patterns of variation when comparing the pre- and post-drainage settings, where bimodal of variation has been recorded.

References

Abdul-Hussein MM, Mason CF (1988) The phytoplankton community of an eutrophic reservoir. Hydrobiologia 169:265–277

Abdullah DS (1989) Phytoplankton primary productivity and the effecting ecological factors in the Shatt Al-Basrah canal. MSc thesis, College of Agriculture, University of Basrah, Basrah, Iraq

Albueajee AI (2020). Trophic Status of Auda Marsh – Southern of Iraq. PhD thesis, University of Baghdad, College of Science for Women

Albueajee AI, Hassan FM, Douabul AA (2020) Phytoplankton Species composition and Biodivrsity indices in Auda marsh- Southern Iraq. Iraqi J Agric Sci 51(Special Issue):217–228

Al-Doori TY, Al-aathami HA (1990) Status of fish wealth in southern Iraq. Eleventh scientific conference of Iraq. Biology. Soc. Basrah. 27 Feb–2 March 1990

Al-Handal AY; Al-Hakim TM, Saleh TA (1992) Primary production of benthic microflora in Basrah South Iraq. Oceanography of Khor Al-Zubair, 3rd symposium. Marine Science Center. Basrah University, Sc. Publ. No. 16

Al-Hilli MR, Warner A, Asada BG T. Douabul A (2009) An assessment of vegetation and environmental controlsin the 1970s of the Mesopotamian wetlands of southern Iraq. Wet Ecol Manag 17:207–223

Al-Hilli MR, Mayah AA, Hassan FM (2014) Marsh Flora of Southern Iraq before desication. Al-Baser Company. S.A.R.I Printing and Publishing, 195 pp

Al-Lami AA (1986) Ecological study on phytoplankton on some marsh regions Southern Iraq. MSc thesis, College of Science, University of Basrah

Al-Mousawi AH, Al-Saadi HA, Hassan FM (1994) Spatial and seasonal variations of phytoplankton population and related environments in AL-Hammar marsh. Iraq Bas J Sci 12(1):9–20

Al-Mousawi AH, Hadi RA, Kassim TI, Al-lami AA (1990) A study on the algae in the Shatt Al-Arab Estuary, southern Iraq. Marina Mesopotamica 5(2):305–323

Al-Obaidi GSA (2006) A study of phytoplankton community in Abo Zirig Marsh, Southern Iraq. MSc thesis, College of Science, University of Baghdad

Al-Saadi HA, Maulood BK (1990) Aquatic ecology in Southern Iraq (in Arabic). Algal status in southern Iraqi inland waters. Eleventh scientific conference of Iraq. Biology. Soc. Basrah. 27 Feb–2 March 1990

Al-Saadi HA (1993) Primary productivity of phytoplankton in Iraqi aquatic ecosystems. Marina Mesopotamica 8(2):254–276

Al-Saadi HA, Antoine SE, Nurul-Islam AKMC (1981) Limnological investigation in Al Hammar marsh area in Southern Iraq. Nova Hedwigia 35:157–166

Al-Saadi HA, Hadi RAM, Al-Lami AA (1989) Limnological studies on some Marsh areas in Southern Iraq. Limnologica 20(2):26–28

Al-Saboonchi AA, Mohammed ARM, Barak NA (1982) A study of phytoplankton in the Garma Marshes, Iraq. Iraqi J Mar Sci 1(1):67–78

Al-Zubaidi AJM (1985) Ecological Study on the Algae (Phytoplankton) in Some Marshesnear Qurnah Southern Iraq. MSc thesis, College of Science, University of Basrah

Antoine SE (1983) Limnological investigation in the polluted Rabat canal and Shatt Al-Arab River, Basrah, Iraq. NovaHedwigia 38:497–518

Carlson RE (1991) Expanding the trophic state concept to identify non-nutrient limited lakes and reservoirs. In: Enhancing the states's lake management programs, pp 59–71

Douabul AA, Warner Z, Asada BG, Abaychi, T, J K, Alwash AJ (2005) Restoration processes in Abu Zirig-Central Marsh: a case study. Special session 9: restoration of Mesopotamian Marshes of Iraq. The Ecological Society of America (ESA), August 2005 Annual Meeting, Montréal, Canada

Evans MI (2002) The ecosystem. The Iraqi Marshlands: a human and environmental study. London: Politico's pp 201–219

Falkowski PG (1998) Biogeochemical controls and feedbacks on ocean primary production. Science 281:200–206

Falkowski PG (1999) Oceanic sinks for atmospheric CO2. Plant Cell Environ 22:741–755

Hadi RA, Al-Mousawi AH, Al-Zubaidi AM (1989) A study on the primary productivity in Shatt Al-Arab estuary at Basrah. Iraq J Biol Sci Res 20:593–606

Hammadi NS, Jassim AQ, Al-Sodani HM (2007) Occurrence and seasonal variations of phytoplankton in the restored marshes of Southern Iraq. Marsh Bull 2(2):96–109

Hassan FM, Al-Kubaisi AA, Talib AH, Taylor WD, Abdulah DS (2011) Phytoplankton primary production in southern Iraqi marshes after restoration. Baghdad Sci J 8(1):519–530

Hinton GCE, Maulood BK (1980) Some diatoms from brackish water habitats in Southern Iraq. Nova Hedwigia 33:487–497

Hinton GCE, Maulood BK (1982) Contribution to the algal flora of Iraq: The non-diatoms flora of the Southern Marshes. Nova Hedwigia 37:49–63

Huq MF, Hadi RA, Al-saadi HA (1977) Preliminary studies on phytoplankton of north-west Arab Gulf. II. Phytoplankton population dynamics. Bangladesh J Bot 6(1&2):109–121

Huq MF, Al-Saadi HA, Hadi RA (1978) Preliminary studies on the primary production of north-west Arab Gulf during post monsoon period. J Oceangr Soc Japan 34(2):78–80

Huq MF, Al-Saadi HA, Hameed HA (1981) Studies on the primary productivity of the river Shatt al-Arab at Basrah. Iraq Hydrobiologia 77:25–29

Ibrahim MA (1980) Primary productivity of the Coastal waters of Qatar using C^{14}. Raian 4:3–15

Jacob PG, Zarba MA, Anderlini V (1979) Hydrography, Chlorophyll, and plankton of the Kuwait Coastal Waters. Indian Mar Sci 8:150–154

Joniak T, Gołdyn R, Kozak A (2003) The primary production of phytoplankton in the restored Maltański Reservoir in Poland. Hydrobiologia 506–509:311–316

Kabanova JG (1968) Primary production of the northern part of the Indian Ocean. Oceanology 8:214–225

Lee RE (2008) Phycology, 3rd edn. Cambridge University Press, Cambridge, UK. ISBN-13 978-0-521-86408-4

Likens GE (1975) Primary production of inland aquatic ecosystems. In: Lieth H, Whittaker RH (eds) Primary production of the biosphere. Springer-Verlag, New York, pp 185–202
Maulood BK (1991) Contribution to the algal flora of some springs and streams in the Derbendikhan-Iraq, Zanco. Sci J Salahaddin Univ, Arbil, Iraq (2):668–679
Maulood BK, Al-Saadi HA (1990) Aquatic ecology in Southern Iraq. Eleventh scientific conference of Iraq. Biology. Soc. Basrah. 27 Feb–2 March 1990
Maulood BK, Double AA (2018) A study on southern marshes Iraq: past and present. The 4th Conference of Environment and Sustainable Cairo 24–28 November
Maulood BK, Hinton GCF, Kamees HS, Saleh FAK, Shaban AA, Al-Shahawani SMH (1979) An ecologyical survey of some aquatic ecosystem in Southern Iraq. Trop Ecol 20(1):27–40
Maulood BK, Hinton GCF, Whitton BA, Al-Saadi HA (1981) On the algal ecology of the lowland Iraqi marshes. Hydrobiologia 80(3):269–276
Maulood BK, Hassan FM, Al-Lami AA, Toma JJ, Ismail AM (2013) Checklist of algal flora in Iraq. Ministry of Environment, Baghdad
Mitsch WJ, Gosselink JG (2000) Wetlands, 3rd edn. Wiley, New York
Mitsch WJ, Gosselink JG (2015) Wetlands, 5th edn. Wiley, New York
Pankow H, Al-Saadi HA, Huq MF, Hadi RAM (1979) On the algal flora of the marshes near Qurna (southern Iraq). Willdenowia 8:493–506
Partow H (2001) The Mesopotamian marshlands: demise of an ecosystem. UNEP, Nairobi, 56pp
Platt T, Subba Rao DV (1975) Primary production of marine microphytes. In: Cooper JP (ed) Photosynthesis and productivity in different environments. Cambridge University Press, Cambridge, pp 249–280
Richardson CJ, Hussain NA (2006) Restoring the Garden of Eden: an ecological assessment of the marshes of Iraq. Bioscience 56(6):477–489
Ryther JH, Menzel DW (1965) on the production composition and distribution of organic matter in the Western Arabian Sea. Deep Sea Res 12:199–209
Rzoska J (1980) Euphrates and Tigris, Mesopotamian ecology and desting. Junk, W., The Hague, 122 pp.
Schaum C, Barton S, Bestion E et al (2017) Adaptation of phytoplankton to a decade of experimental warming linked to increased photosynthesis. Nat Ecol Evol 1:0094. https://doi.org/10.1038/s41559-017-0094
Schiewer U (1984) In vitro primary productivity (assimilation number) of phytoplankton from the Darsszingst estuary. Limnol (Berlin) 15(2):575–579
Schiewer U, Al-Saadi HA, Hameed HA (1982) On the diel rhythm of phytoplankton productivity in Shatt al-Arab at Basrah. Iraq Arch Hydrobiol 93(2):158–172
Talib AH (2009) Ecological study on the phytoplankton and primary productivity in Southern Iraqi Marshes. PhD dissertation, College of Science, University of Baghdad, Iraq, pp 144
Talling JF (1980a) Water characteristics in Euphrates and Tigris. In: Rzoska J (ed) Mesopotamian ecology and desting. Dr.W.Jund.br. Publishers, The Hyge London
Talling JF (1980b) Phytoplankton. In: Rzoska J (ed) Mesopotamian ecology and desting. Dr. W. Jund.br. Publishers, The Hyge London
Teira E, Mourin B, Maran E, Perez V, Pazo MJ, Serret P, Armas D, Escanez J, Woodward EMS, Fernandez E (2005) Variability of chlorophyll and primary production in the Eastern North Atlantic Subtropical Gyre: potential factors affecting phytoplankton activity. Deep-Sea Res I 52:569–588
Thesiger W (1964). The Marsh Arabs. Penguin, Harmondsworth, 233 pp
Thornton DCO (2012) Primary production in the ocean. Advances in Photosynthesis–Fundamental Aspects. Intech, Rijeca, pp 563–588
UNEP (United Nations Environment Programme) (2009) Support for environmental management of the Iraqi Marshlands: 2004–2009. UNEP, Nairobi, Kenya [Google Scholar]
van der Valk A (2012) The biology of freshwater wetlands. Oxford University Press, Oxford, 280pp
Wetzel RG (2001) Limnology: lake and river ecosystems, 2nd edn. Academic Press, San Diego
Zohary M (1946) The Flora of Iraq and its phytogeographical subdivision. Iraq Dep Agric Bull 31:140–141

Chapter 13
Distribution Patterns, Diversity Centers, and Priorities for Conservation of Aquatic Plants in Iran

Ahmadreza Mehrabian and Farzaneh Khajoi Nasab

Abstract A climatic heterogeneity and geomorphologic complexities that have been shaped during a sophisticated paleo-biogeographical history has led Iran to become a prominent zone of endemism, an endemic center of the Irano-Turanian region, and a global center of diversity for vascular plant taxa in the world. According to the present study, Iranian macrophyte flora includes 68 species of 35 genera of vascular plants belonging to 22 plant families. The Potamogetonaceae family, with 16 species, is the largest family of true aquatic plants in the country. The aquatic plants were distributed between −29 m (*Wolffia arrhiza* (L.) Horkel.) and 3200 m (*Triglochin palustris* L.). The greatest number of species (48 species) occurs at an altitude of 0–500 m. Hydrocharitaceae and Potamogetonaceae comprise the largest number of endangered and critically endangered species of Iranian aquatic plants. The main hotspot of species richness of true aquatic plants occurred in the Hyrcanian phytogeographical region, followed by an area in the Atropatanean region. The highest value of AZEs exists in Gilan Province. Despite the high importance of these habitats from a conservation perspective, numerous direct and indirect man-made threats such as pollution, invasive plants, and climate change have exposed these habitats.

Keywords Geobotany · Ecology · SW Asia · Conservation · Macrophytes

13.1 Introduction

Aquatic ecosystems cover a high amount of biological productivity of the biosphere. They also produce a high proportion of atmospheric oxygen, which ensures survival of life. In addition, archeological studies show that the first large civilizations were established near wetland areas like the Indus, the Nile delta, the Fertile Crescent of

A. Mehrabian (✉) · F. K. Nasab
Department of Plant Sciences and Biotechnology, Faculty of Life Sciences and Biotechnology, Shahid Beheshti University, Tehran, Iran
e-mail: a_mehrabian@sbu.ac.ir

L. A. Jawad (ed.), *Southern Iraq's Marshes*, Coastal Research Library 36,
https://doi.org/10.1007/978-3-030-66238-7_13

the Euphrates and the Tigris rivers, and others (Ramachandra 2001). Aquatic plants such as macrophytes and algae provide many ecological benefits and are essential in promoting the diversity and function of aquatic systems (Carpenter and Lodge 1986). Macrophytes colonize a wide range of hydro-biomes, such as lakes, wetlands, streams, rivers, and marine environments. This has resulted in several evolutionary-ecological adaptations (Wetzel 2001; Kalff 2002). These taxa are the keystone element of hydro-zones (Carpenter and Lodge 1986) and represent an extensive spectrum of ecological roles as well as interact with a wide range of physical factors, such as hydrological, geomorphological, physico-chemical, and biological (e.g. plant, animal, microbe, and algae) (Engelhardt and Ritchie 2001; Wood et al. 2017). Additionally, these taxa, as surrogate elements, can be used to assess biological diversity (Caro 2017). Increases in temperature and CO_2 concentrations and changes in precipitation patterns have had a direct negative effect on the physiology and distribution patterns of aquatic and terrestrial vegetation (Franks et al. 2013; Tokoro et al. 2014). Besides this, fluctuations in chemical and hydrological features have led to changes in the dynamics of aquatic ecosystems, especially in the boreal regions (Knutti and Sedlacek 2013; Alahuhta 2015). Because of their importance, a large range of studies have been conducted over recent decades. Some of the prominent ones include Garrett and Hellquist (2006), Den Hartog (1975), Engelhardt and Ritchie (2001), Fernández-Aláez et al. (2017), Ferreira et al. (2014), Havel et al. (2015), Grutters et al. (2017), Keruzore et al. (2013), O'Hare (2015), Martins et al. (2013), Toivonen et al. (2010), Short et al. (2016), and Wood et al. (2014, 2017). In spite of this, threats from human actions such as climate change, land use change, grazing, and others have been severely imposed upon them (Hughes 2000; Kelly and Goulden 2008; Lacoul and Freedman 2006; Saaa et al. 2010; Akasaka et al. 2010). This underscores the need for conservation studies for this valuable taxa. Measuring priorities is the most important step in conservation programs (Bottrill et al. 2008). Frequently, diversity and distribution have been used to set priorities on small to large scales (Wagnetiz 1986; Barthlott et al. 1996; Olmstead 2013; Mehrabian 2015, 2020a, b). Additionally, the biological patterns mentioned above reveal the ecological niche of taxa (MacArthur 1972; Toledo et al. 2012), and thus, they are considered to be the most important aim of biogeography (Vetaas and Ferrer-Castán 2008).

Iran is considered a prominent example of habitat diversity in the world (Frey and Probst 1986). Climatic heterogeneity and geomorphologic complexities (Stöklin 1968) that have been shaped during a sophisticated paleo-biogeographical history (Zohary 1963) have led Iran to become a prominent zone of endemism (Zohary 1973b). The country is also an endemic center of the Irano-Turanian region (Leonard 1991–1992) and a global center of diversity for vascular plant taxa in the world (Davis et al. 1994; Barthlott et al. 1996, 1999; Kier et al. 2005). Iran has seven catchments (https://frw.ir), including about 77–80 main rivers. Beside this, there are 6 wetlands registered in biosphere reserves and 25 are classified in the Ramsar Convention on Wetlands of International Importance (http://www.wetlandsproject.ir). These hydro-biomes appear to be suitable opportunities to establish high phyto-diversity, especially macrophytes. The first studies on these taxa date back to *Flora Iranica* (Casper 1969; Dandy 1971a-h; Rechinger 1966a-d,

1978, 1990), which published taxonomic literature on aquatic plant families in the scale of the Iranian plateau. Dinarwand (2017) then published *Flora of Aquatic Plants of Iran* within the framework of *Flora of Iran*. The most prominent studies in this scope include "Aquatic plants of Hashilan wetlands" (Karami et al. 2001), "Aquatic plants of Choghakhor wetlands" (Yosefi and Kazemi 2008), "Aquatic plants of Parishan" (Dolatkhahi and Yosefi 2010), "Aquatic plants of western Mazandaran" (Tavakoli et al. 2014), "Ferns and fern allies in Iran" (Akhani et al. 2010), "Wetland flora and diversity of the Western Alborz Mountains, North Iran" (Kamrani et al. 2011), "Plant biodiversity of wetland habitats in the dry steppes of Central Alborz" (Naqinezhad et al. 2010), and "A preliminary checklist of vascular aquatic plants of Iran (Yosefi and Toranj 2015).

Unfortunately, the rate of degradation of Iran's aquatic ecosystems has increased sharply in recent decades. Nevertheless, little attention is paid to distribution patterns, conservation status, or conservation of the vascular aquatic plants. This study aims to provide a general view on distribution of the aforementioned taxa. Conservation status, high diversity zones, and priority zones to conservation are also studied to provide a scientific document for the implementation of aquatic plant management programs in the country. With a new approach, the study also analyzes and introduces Important Plant Areas (IPAs) and Alliance for Zero Extinction (AZEs) as the latest conservation methods for effective conservation management of biodiversity.

13.2 Study Area

13.2.1 Geography

The geographical location of the study area is within the borders of Iran. It covers a total surface area of 1.6 million km^2 located at 24° to 40° N longitude and 44° to 64° E latitude.

13.2.2 Hydrology

This area was divided into 6 catchments including the Caspian Sea (7 sub-catchments), the Persian Gulf and the Sea of Oman (9 sub-catchments), Urmia Lake, the Central Plateau (9 sub-catchments), the Eastern Border (3 sub-catchments) and Qara-Qom. The total surface area is 125 million hectares (https://frw.ir). The diverse geographic structure of Iran has led to the emergence of 41 out of 42 of the known types of wetlands in the world (FAO). The country also has 24 wetlands identified in the Ramsar Convention on Wetlands of International Importance (Ramsar 2016).

13.2.3 Geomorphology and Geology

The orography of Iran is composed of diverse mountainous barriers, such as the Zagros, Alborz, Kopet-Dagh, Central, and Eastern mountain ranges. The Zagros range (Fisher 1968) has a northwest–southeast orientation and shapes a massif between the Iranian plateaus, the Mesopotamian and the Persian Gulf, with the highest peak, Zard Kuh, reaching 4231 m a.s.l. (Homke et al. 2004). The Alborz is an active, arcuate fold-and-thrust belt (Stöklin 1974) that has a soft, sinuous shape that runs east-west in northern Iran. It forms a natural barrier to the Caspian Sea and the central Iranian plateau (Stöklin 1974) and includes Mount Damavand (5670 m a.s.l), the highest peak in the range as well as in Iran. The Kopet-Dagh range is located in the eastern margins of the Caspian Sea in northeastern Iran, Turkmenistan, and northern Afghanistan (Buryakovsky et al. 2001). Scattered interior mountains in central, southern, and eastern Iran comprise the other orographic formations in the country (Fisher 1968). The rest of Iran is comprised of lowlands, coastal and riverine areas, and interior basins enclosed by the geomorphological elements (Fisher 1968), the lowest points of which cover the limited plains of the Caspian Sea (−27 m a.s.l) (Ghorbani 2013).

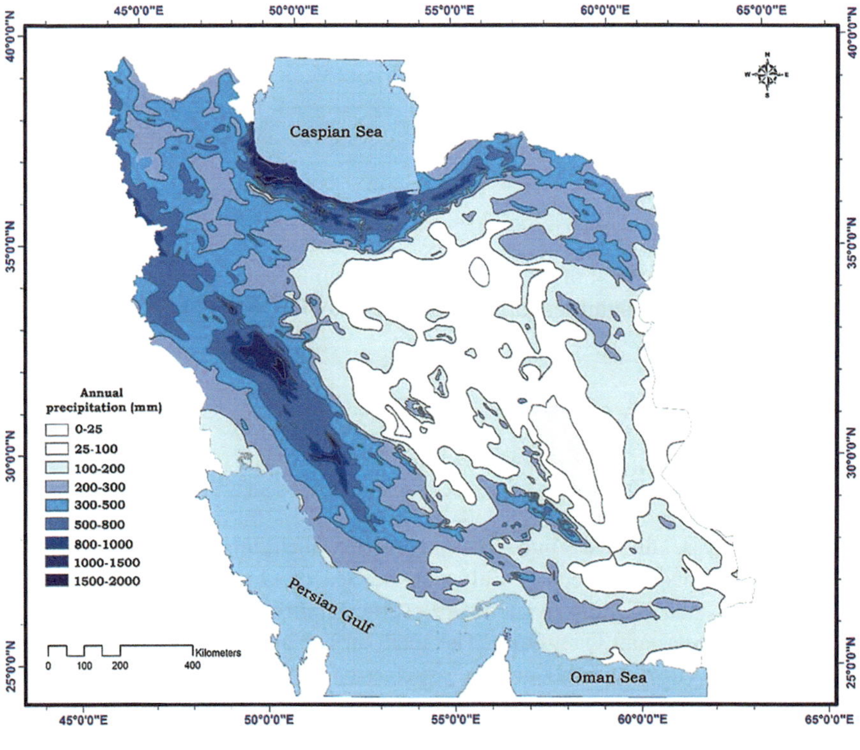

Fig. 13.1 Average rainfall variation (mm) in Iran (1961–2005) (Khalili and Rahimi 2014)

13.2.4 Climatology

Rivas–Martínez et al. (1999) established a new methodology to classify the global macro-bio-climate naturalized by Djamali et al. (2011) for Iran's ecosystems. The classifications include Mediterranean (west, northwest), temperate (north), and tropical (the southern coastal zones of the Persian Gulf and the Gulf of Oman). In addition, zonobiome III covers the southern subtropical arid climates and deserts, while zonobiome VII (rIII) covers the northern parts with arid temperate climates based on World Zonobiomes (Walter and Breckle 2002). Iran, however, is situated in arid zones that receive light rainfall (equivalent to 1/3 of the average global precipitation) (Shakur et al. 2010). The mean annual precipitation is predicted to be 250 mm, ranging from 50 mm in the central zone of Iran to more than 1100 mm in some coastlands of the Caspian Sea (Fig. 13.1) (Khalili and Rahimi 2014). Three different climatic conditions can be identified in its diverse zones. The humid type covers the coastal areas of the Caspian Sea where mean annual rainfall reaches 1600 mm. In the semi-arid type, mean annual rainfall is 450 mm. In addition, the arid type that is established in desert zones receives a mean annual rainfall of 50 mm (Iran Meteorological Organization 2014). Climate conditions show a diverse range from very humid to very arid, and dominant zones of Iran range from arid to very arid. The

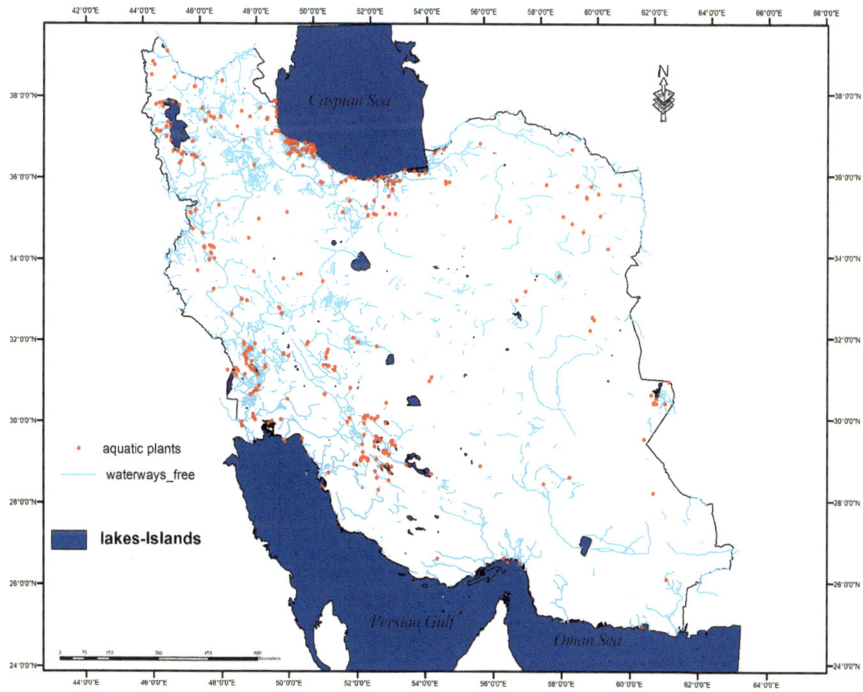

Fig. 13.2 Spatial distribution map of macrophytes in Iran

southern zones have higher winter and summer temperatures with a short rainy season. This is similar to conditions in tropical northeastern African and the hot Sindian deserts that are exposed to extreme fluctuations of maximum and minimum temperatures (Zohary 1973a, b).

13.3 Data Analysis

The macrophyte database was extracted from *Flora Iranica* (Casper 1969; Dandy 1971a-h; Rechinger 1966a-d, 1978, 1990), *Flora of Iran* (Dinarwand 2017), and herbarium data from 662 localities, including 629 specimens, as well as data from several excursions in diverse geologic units (Fig. 13.2). ArcView version 3.2 (ESRI 2000) and DIVA-GIS 7.3 were used to make the spatial data analysis. To evaluate distribution patterns of macrophytes, the distribution points were mapped per $0.25° \times 0.25°$ universal transverse Mercator grid cells (25 km^2 with the exception of boundary areas) to distinguish the pattern of distribution as well as their diversity zones. Subsequently, aquatic species were mapped using the point-to-grid richness analysis tool based on 10×10 km grid cells and the circular neighborhood option with a radius of 25 km. Conservation status of taxa were evaluated by Kew GeoCAT (http://geocat.kew.org/). The extent of occurrence (EOO) and area of occupancy (AOO) were used to classify the threat categories. Some taxa lacked sufficient data of distribution or were data deficient (DD). Some analyses have been used to assess conservation priorities. Index of species rarity in the study area (RI) is based on Williams et al. (1996). The scoring revealed a range from zero (0) to one (1) for these indices. The RI was computed as the inverse of the number of cells with documented species (Ci) in the target area follows as RI = 1/Ci.

13.4 Results and Discussion

13.4.1 Aquatic Plants of Iran

According to the present study, Iranian macrophyte flora includes 68 species of 35 genera of vascular plants belonging to 22 plant families. The Potamogetonaceae family, with 16 species, is the largest family of aquatic plants in the country (Fig. 13.3).

The results of this study show that 69% of the aquatic plants are submerged (Fig. 13.4).

The aquatic plants were distributed between −29 m (*Wolffia arrhiza* (L.) Horkel.) and 3200 m (*Triglochin palustris* L.). The greatest number of species (48 species) occurs at an altitude of 0–500 m (Fig. 13.5).

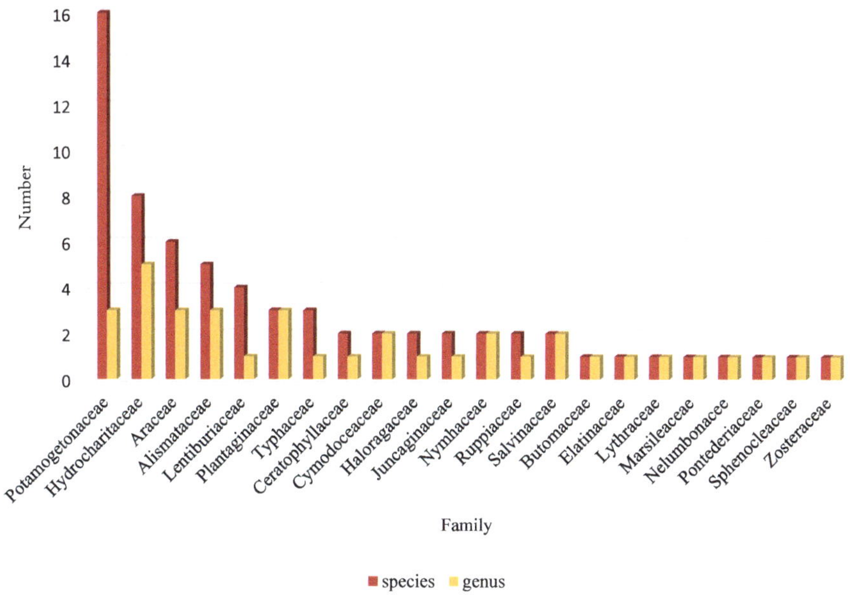

Fig. 13.3 Families of aquatic plants in Iran

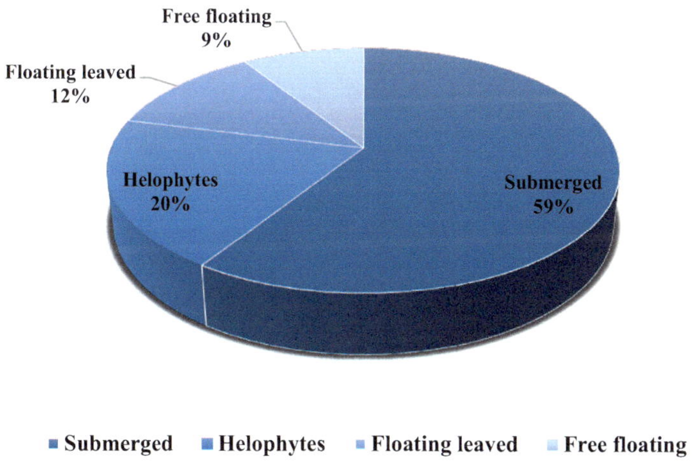

Fig. 13.4 Life-form of aquatic plants in Iran

13.4.2 Distribution Patterns and Species Richness

The species richness map shows that parts of the coastline of the Caspian Sea in Gilan Province have the highest number of species (Fig. 13.6). We categorized the

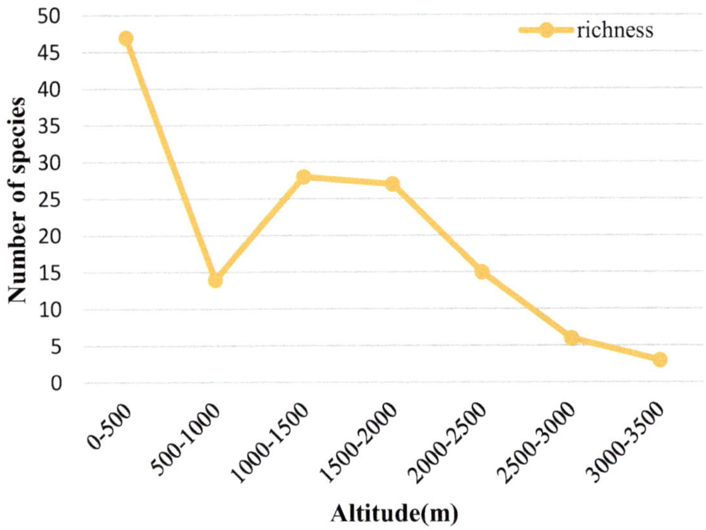

Fig. 13.5 Number of Iranian aquatic species in altitudinal profile

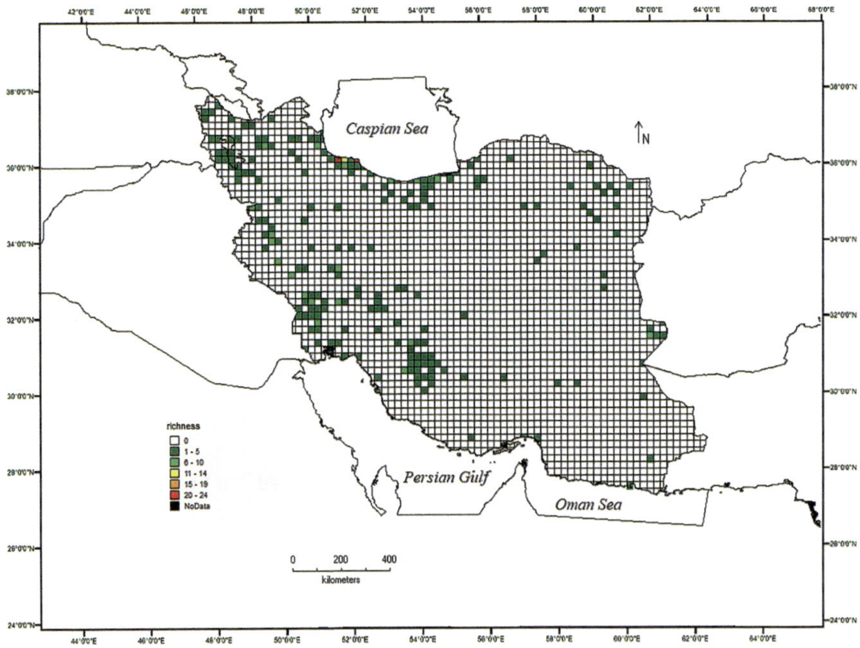

Fig. 13.6 Number of aquatic species per 0.25° × 0.25° grid cell

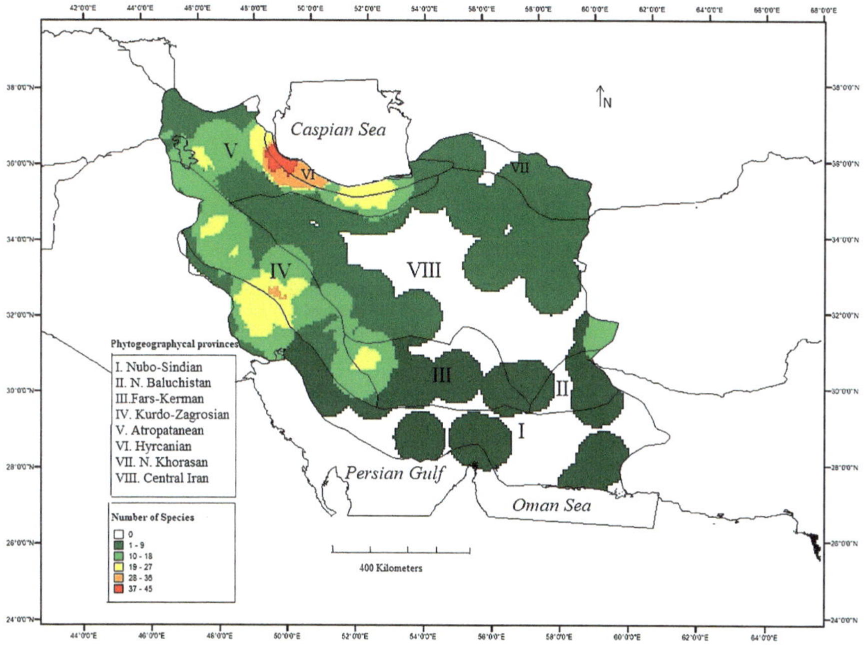

Fig. 13.7 Richness map of the total of aquatic species. A circular neighborhood with a radius of 25 km was used to assign observations to a grid cell

species richness map cells into five groups of 1–24 species. The highest species richness (24 species) of this genus occurred in two grid cells in parts of Gilan, followed by one grid cell with 19 species in Gilan. In addition, 18 grid cells have 6–10 species often concentrated in the north, northwest, and one part of eastern Iran.

Ten species often concentrated in the north, northwest, and one part of eastern Iran.

Figure 13.7 demonstrates the species richness map using 10 × 10 km grid cells, which are highlighted with five different colors. The hotspots (the highest richness values) occur in Gilan Province and then in parts of Ardebil, Mazandaran, Qazvin, and Khuzestan provinces. The main hotspot of species richness of aquatic plants occurred in the Hyrcanian phytogeographical region, followed by an area in the Atropatanean region. A limited area of Kurdo-Zagrosian also has high species richness (with 28–35 species) of aquatic plants. The lowest species richness of aquatic plants is found in central Iran and the Nubo-Sindian regions. More than half of these regions don't have any aquatic plant species at all.

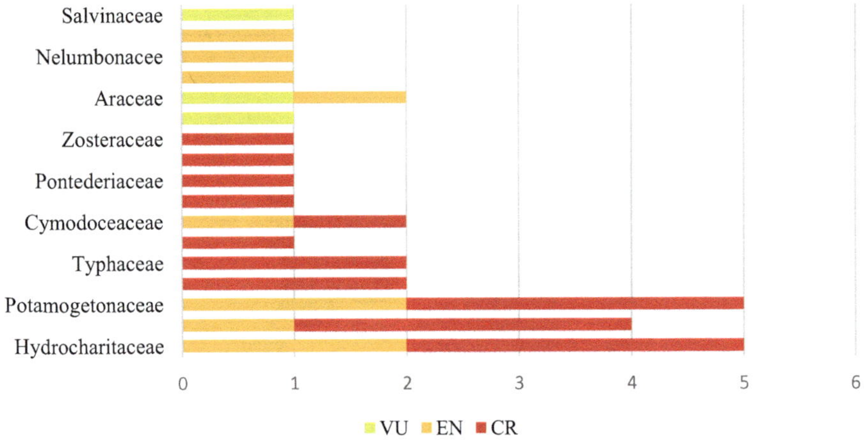

Fig. 13.8 IUCN categories for the families studied

13.4.3 Rare Taxa and Conservation Status (IUCN Red List Category)

Hydrocharitaceae and Potamogetonaceae comprise the largest number of endangered and critically endangered species of Iranian aquatic plants (Fig. 13.8 & Table 13.1). Figure 13.9 demonstrates the IUCN categories for the taxa studied, and about 45% of these species are at high risk of extinction.

According to Table 13.1, 22% of the species, such as *Zostera noltii* Hornem., *Sparganium natans* L., *Sparganium emersum* Rehmann, Verb, *Potamogeton gramineus* L., *Groenlandia densa* (L.) Fourr., *Monochoria vaginalis* (Burm.f.) Presl., *and Callitriche palustis* L., are extremely rare.

We identified the richness of threatened taxa (CR, EN, VU) using the number of endangered and critically endangered species in each grid cell. The highest richness exists in two cells in Gilan Province (7–8 species), followed by another cell in Gilan (6 species) (Fig. 13.10).

Despite the high importance of these habitats from a conservation perspective, numerous direct and indirect man-made threats such as pollution, invasive plants, and climate change have jeopardized these habitats. Severe changes in climate disturb the chemical, physical, and biological functions of freshwater ecosystems (Ejankowski and Lenard 2015; Hossain et al. 2016). Aquatic vegetation, especially macrophytes, are highly vulnerable to these changes. Climate plays an important role in the distribution patterns of plant taxa on a global scale. Therefore, every taxa with specific distribution patterns and limitations are affected by climatic factors. Climatic change slowly alters the distribution patterns of plant taxa and the community composition (Morecroft and Keith 2009). Other menacing factors including heavy metals, agricultural pesticides, changes in land use, dams, sand, and forage

Table 13.1 Aquatic species of Iran

Family	Species	IUCN status	RI
Alismataceae	*Damasonium alisma* Miller	CR	1
Alismataceae	*Sagittaria trifolia* L.	VU	0.2
Alismataceae	*Alisma gramineum* Lej.	LC	0.125
Alismataceae	*Alisma plantago- aquatica* L	LC	0.09
Alismataceae	*Alisma lanceolatum* With.	LC	0.06
Araceae	*Wolffia arrhiza* (L.) Horkel.	VU	0.5
Araceae	*Lemna trisculca* L.	LC	0.25
Araceae	*Spirodela polyrrhiza*(L.) Schlenden.	NT	0.2
Araceae	*Lemna minor* L.	LC	0.16
Araceae	*Lemna perpusilla* Torrey	LC	0.14
Araceae	*Lemna gibba* L.	EN	0.11
Butomaceae	*Butomus umbellatus* L.	LC	0.06
Ceratophyllaceae	*Ceratophyllum submersum* L.	CR	0.5
Ceratophyllaceae	*Ceratophyllum demersum* L.	LC	0.04
Cymodoceaceae	*Thalassodendron cilliatum* (Forssk.) Den Hartog.	CR	1
Cymodoceaceae	*Halodule wrightii* Aschers.	EN	0.33
Elatinaceae	*Bergia capensis* L.	LC	0.33
Haloragaceae	*Myriophyllum spicatum* L.	NT	0.43
Haloragaceae	*Myriophyllum verticillatum* L.	LC	0.25
Hydrocharitaceae	*Najas gracillima* A.Braun ex Magnus.	CR	1
Hydrocharitaceae	*Hydrilla verticillata* (L. f.) Royle, III.	EN	0.5
Hydrocharitaceae	*Hydrocharis morsus - ranae* L.	EN	0.5
Hydrocharitaceae	*Najas gramineae* Delile, Descr.	CR	0.5
Hydrocharitaceae	*Vallisneria spiralis* L.	LC	0.2
Hydrocharitaceae	*Najas minor* All.	LC	0.11
Hydrocharitaceae	*Najas marina* L.	LC	0.1
Hydrocharitaceae	*Halophila ovalis* (R.Br) Hook.	CR	1
Juncaginaceae	*Triglochin palustris* L.	LC	0.5
Juncaginaceae	*Triglochin maritima* L.	LC	0.07
Lentiburiaceae	*Urticularia australis* R.Br.	EN	1
Lentiburiaceae	*Urticularia minor* L.	CR	1
Lentiburiaceae	*Urticularia ochroleuca* R.Hartman	CR	1
Lentiburiaceae	*Urticularia vulgaris* L.	CR	0.5
Lythraceae	*Trapa natans* L.	CR	1
Marsileaceae	*Marsilea quadrifolia* L.	EN	0.5
Nelumbonaceae	*Nelumbo nucifera* Gaertn.	EN	0.5
Nymphaeaceae	*Nuphar luteum* (L.) Sm.	EN	1
Nymphaeaceae	*Nymphaea alba* L.	LC	0.2
Plantaginaceae	*Callitriche brutia* Petagna, Inst.	CR	1
Plantaginaceae	*Callitriche palustis* L.	CR	1
Plantaginaceae	*Hippuris vulgaris* L.	LC	0.25
Pontederiaceae	*Monochoria vaginalis* (Burm.f) Presl.	CR	1

(continued)

Table 13.1 (continued)

Family	Species	IUCN status	RI
Potamogetonaceae	*Groenlandia densa* (L.) Fourr.	EN	1
Potamogetonaceae	*Potamogeton gramineus* L	CR	1
Potamogetonaceae	*Potamogeton trichoides* Cham.& Schleecht.	EN	0.5
Potamogetonaceae	*Potamogeton alpinus* Balb.	CR	0.5
Potamogetonaceae	*Potamogeton natans* L.	CR	0.5
Potamogetonaceae	*Potamogeton filiformis* Pers.	NT	0.33
Potamogetonaceae	*Potamogeton friesii* Rupr.	LC	0.25
Potamogetonaceae	*Potamogeton berchtoldii* Fieb.	LC	0.14
Potamogetonaceae	*Potamogeton pusillus* L.	LC	0.128
Potamogetonaceae	*Potamogeton lucens* L.	LC	0.12
Potamogetonaceae	*Potamogeton amblyphyllus* C.A.	LC	0.08
Potamogetonaceae	*Potamogeton prtfoliatus* L.	LC	0.08
Potamogetonaceae	*Potamogeton crispus* L.	LC	0.06
Potamogetonaceae	*Potamogeton pectinatus* L.	LC	0.05
Potamogetonaceae	*Potamogeton nodosus* Poir.	LC	0.04
Potamogetonaceae	*Zannichellia Palustris* L.	LC	0.04
Ruppiaceae	*Ruppia cirrhosa* (Petagna) Grand, Bull.	LC	0.24
Ruppiaceae	*Ruppia maritima* L.	LC	0.12
Salvinaceae	*Azolla filiculoides* Lam.	VU	0.25
Salvinaceae	*Salvinia natans* (L.) All.	LC	0.2
Sphenocleaceae	*Sphenoclea zeylanica* Gaertn.	CR	0.5
Typhaceae	*Sparganium emersum* Rehmann, Verb.	CR	1
Typhaceae	*Sparganium natans* L.	CR	1
Typhaceae	*Sparganium erectum* L.	LC	0.1
Zosteraceae	*Zostera noltii* Hornem.	CR	1

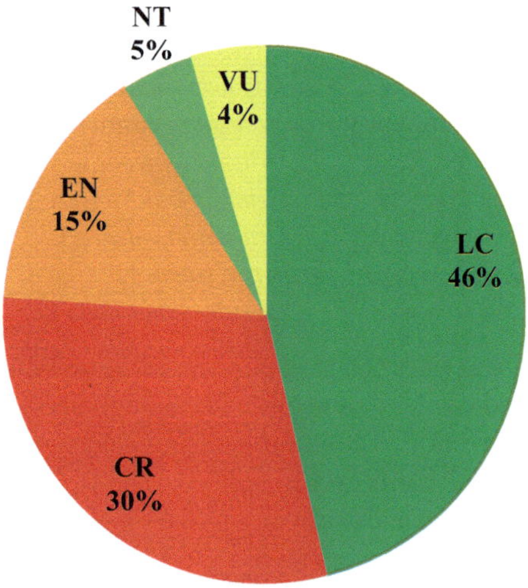

Fig. 13.9 IUCN categories for the studied taxa

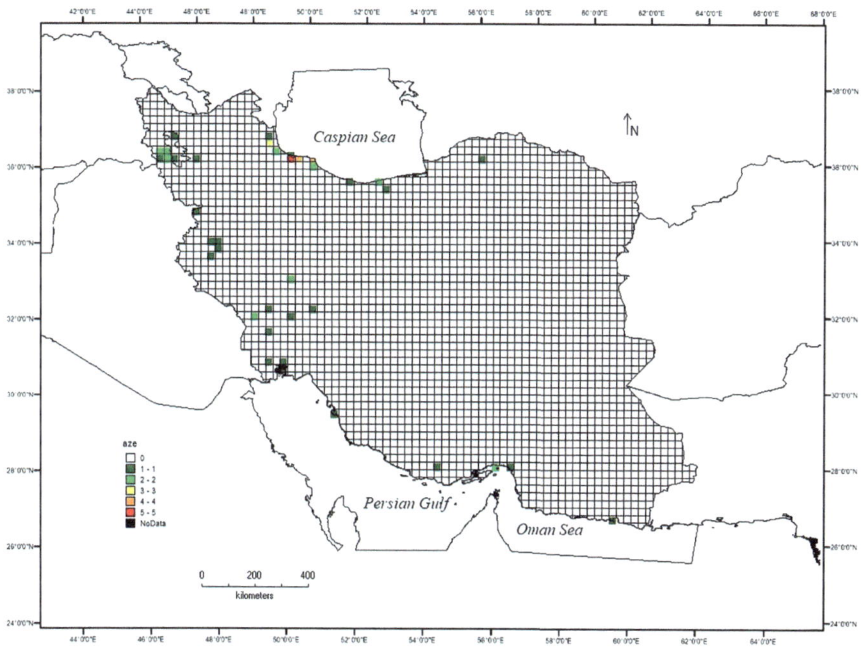

Fig. 13.10 Richness of threatened aquatic plants in Iran

harvesting have disrupted their functioning and led to the destruction of these valuable ecosystems (Hosseini et al. 2010; Ghazban and Khosheghbal 2011; Kamrani et al. 2011). Because of the high levels of transportation, humans have the potential to greatly modify the distribution patterns of plant and animal species (Hodkinson and Thompson 1997). Further, because the anthropogenic threats are severe, wetlands in Iran are facing destruction and a tangible decline of ecosystem performance. Thus, their biological elements (e.g. macrophytes) experience huge alterations. Dam construction, introduction of invasive species, contaminant entry, and changes in land use threaten a wide spectra of natural wetlands in Urmia, Zarivar, Gomishan, Parishan, and others (Babaei and Khodaparast 2009; Hatami et al. 2013; Bayat et al. 2016). Extensive climate change has intensified the threat of these factors (Department of Environment of Iran 2014). Up to now, conservation efforts have not significantly prevented deterioration of the wetlands (National Wetland Conservation Strategy and Action Plan 2014). Our results show that 49% of macrophytes are in endangered categories of IUCN. Unfortunately, a prominent ratio of these species are outside the boundaries of protected areas. Consequently, the threat to these taxa are higher and the need for conservation programs, whether in situ and ex situ, is critical.

References

Akasaka M, Takamura N, Mitsuhash H, Kadono Y (2010) Effects of land use on aquatic macrophyte diversity and water quality of ponds. Freshw Biol 55:909–922
Akhani H, Khoshravesh R, Eskandari M, Greuter W (2010) Ferns and fern allies in Iran. Rostaniha 10(suppl. 1):1–130
Alahuhta J (2015) Geographic patterns of lake macrophyte communities and species richness at regional scale. J Veg Sci 26:564–575
Babaei H, Khodaparast SH (2009) The study of petroleum pollution and heavy metals in Anzali wetlands. The First National conference of Iran Wetlands. Khuzestan, Ahvaz
Barthlott W, Lauer W, Placke A (1996) Global distribution of species diversity in vascular plants: towards a world map of phytodiversity. Erdkunde 50:317–328
Barthlott W, Biedinger N, Braun G, Feig F, Kier G, Mutke J (1999) Terminological and methodological aspects of the mapping and analysis of global biodiversity. Acta Bot Fenn 162(0):103–110
Bayat R, Jafari S, Ghermez Cheshmeh B, Charkhabi AH (2016) Dust effects on vegetation: A case study on Shadegan Wetlands. Remote Sens GIS Nat Resour 7(2):17–32
Bottrill MC, Joseph LN, Carwardine J, Bode M, Cook C, Game ET et al (2008) Is conservation triage just smart decision making? Trends Ecol Evol 23:649–654
Buryakovsky LA, Chilinger GV, Aminzadeh F (2001) Petroleum geology of the South Caspian Basin. Gulf Professional Publishing USA, p 442
Caro T (2017) Conservation by proxy. Island Press
Carpenter SR, Lodge DM (1986) Effects of submersed macrophytes on ecosystem processes. Aquat Bot 26:341–370
Casper SJ (1969) Lentibulariaceae. In: Rechinger KH (ed) Flora Iranica, Lfg. 58. –Akad. Druck- und Verlagsanstalt, Graz
Dandy JE (1971a) Ruppiaceae. In: Rechinger KH (ed) Flora Iranica, Lfg. 84. – Akad. Druckund Verlagsanstalt, Graz
Dandy JE (1971b) Alismataceae. In: Rechinger KH (ed) Flora Iranica, Lfg. 78. – Akad. Druckund Verlagsanstalt, Graz
Dandy JE (1971c) Butomaceae. In: Rechinger KH (ed) Flora Iranica, Lfg. 79. – Akad. Druckund Verlagsanstalt, Graz
Dandy JE (1971d) Hydrocharitaceae. In: Rechinger KH (ed) Flora Iranica, Lfg. 80. –Akad. Druck- und Verlagsanstalt, Graz
Dandy JE (1971e) Juncaginaceae. In: Rechinger KH (ed) Flora Iranica, Lfg. 82. – Akad. Druckund Verlagsanstalt, Graz
Dandy JE (1971f) Najadaceae. In: Rechinger KH (ed) Flora Iranica, Lfg. 86. – Akad. Druckund Verlagsanstalt, Graz
Dandy JE (1971g) Potamogetonaceae. In: Rechinger KH (ed) Flora Iranica, Lfg. 83. –Akad. Druck- und Verlagsanstalt, Graz
Dandy JE (1971h) Zannichelliaceae. In: Rechinger KH (ed) Flora Iranica, Lfg. 85. –Akad. Druck- und Verlagsanstalt, Graz
Davis SD, Heywood VH, Hamilton AC (1994) Centres of plant diversity. Nat Hist 111(1):01–01
Den Hartog C (1975) Aquatic botany—aims and scope of a new journal. Aquat Bot 1:1–2
Dinarwand M (2017) Flora of aquatic plants of Iran. In: Assadi et al (eds) Flora of Iran. Research Institute of Forests and Rangelands
Djamali M, Akhani H, Khoshravesh R, Andrieu-Ponel V, Ponel P, Brewer S (2011) Application of the global bioclimatic classification to Iran: implications for understanding the modern vegetation and biogeography. Ecologia mediterranea 37(1):91–114
Dolatkhahi M, Yosefi M (2010) The study of aquatic and semi-aquatic plants of International Wetlands of Parishan. Fars 1(1):91–104

Ejankowski W, Lenard T (2015) Climate driven changes in the submerged macrophyte and phytoplankton community in a hard water lake. Limnologica - Ecol Manage Inland Waters 52:59–66
Engelhardt KAM, Ritchie ME (2001) Effects of macrophyte species richness on wetland ecosystem functioning and services. Nature 411:687–689
ESRI (2000) ArcView GIS Ver. 3.2a. California: Environmental Systems Research Institute Inc.
Fernández-Aláez C, Fernández-Aláez M, García-Criado F, García Girón J (2017) Environmental drivers of aquatic macrophyte assemblages in ponds along an altitudinal gradient. Hydrobiologia
Ferreira T, O'Hare MT, Szoszkiewicz K, Hellsten S (Guest eds) (2014) Plants in hydrosystems: from functional ecology to weed research. Hydrobiologia 737:1–345
Fisher WB (1968) Physical Geography. In: Fisher WB, editor. The Cambridge History of Iran, vol. 1. Cambridge University Press. pp. 3e110
Franks PJ, Adams MA, Amthor JS, Barbour MM, Berry JA, Ellsworth DS, Farquhar GD, Ghannoum O, Lloyd J, McDowell N et al (2013) Sensitivity of plants to changing atmospheric CO2 concentration: from the geological past to the next century. New Phytol 197:1077–1094
Frey W, Probst W (1986) A synopsis of the vegetation of Iran. In: Kürschner H (ed) Contributions to the vegetation of Southwest Asia. TAVO, Naturwissenshaften, Beih, pp 61–73
Garrett EC, Hellquist CB (2006) Aquatic and wetland plants of Northeastern North America, vol 2. The University of Wisconsin Press, 537pp
Ghazban F, Khosheghbal M (2011) The origin of heavy metals in Anzali wetland sediments. J Ecol 57:45–56. (In Persian)
Ghorbani M (2013) The economic geology of Iran. Springer, p 581
Grutters MC, Haury J, Van Valkenburg JLCH, Brundu G, Newman J, Clayton JS, Anderson LWJ, Hofstra D (2017) Management and control methods of invasive alien freshwater aquatic plants: a review. Aquat Bot 136:112–137
Hatami M, Mirzayi M, Bandegani M, Sadeghi M, Nazsabet F (2013) Determination of mercury, lead, arsenic, cadmium and chromium in salt and water of Maharloo Lake, Iran, in different seasons. J Univ Med Sci Mazandaran 23(108):91–98
Havel JE, Kovalenko KE, Thomaz SM, Amalfitano S, Kats LB (2015) Aquatic invasive species: challenges for the future. Hydrobiologia 750:147–170
Hodkinson DJ, Thompson K (1997) Plant dispersal: the role of man. J Appl Ecol 34:1484–1496
Homke S, Verges J, Emami H, Garces M, Karpuz R (2004) Magnetostratigraphy of Miocene Pliocene Zagros foreland deposits in the front of the push–e Kush arc (Lurestan Province Iran). Earth Planet Sci Lett 225(3–4):397–410
Hossain K, Yadav S, Quaik S, Pant G, Maruthi AY, Ismail N (2016) Vulnerabilities of macrophytes distribution due to climate change. Theor Appl Climatol 129:1123–1132
Hosseini M, Nabavi MN, Rajabzadeh A, Omidvar V (2010) The alternation comparison of conservative valuations of Shadegan Wetlands during 1980–2000. Wetland (Shahid Chamran-Ahvaz) 1(4):21–37. (In Persian)
Hughes L (2000) Biological consequences of global warming: is the signal already apparent? Trends Ecol Evol 15:56–61
Iran Meteorological organization (2014) Annual report, Tehran
Kalff J (2002) Limnology. Printice Hall, New Jersey, 592 p
Kamrani A, Naqinezhad A, Attar F, Jalili A, Charlet D (2011) Wetland flora and diversity of the Western Alborz Mountains, North Iran. Phytologia Balcanica 17(1):53–66
Karami M, Kasmani ME, Almaseh AA (2001) Plants of Hashilan Wetland, Kermanshah, Iran. J Sci Iran 1(3):2–7
Kelly AE, Goulden ML (2008) Rapid shifts in plant distribution with recent climate change. Proc Natl Acad Sci USA 105:11823–11826
Keruzore AA, Willby NJ, Gilvear DJ (2013) The role of lateral connectivity in the maintenance of macrophyte diversity and production in large rivers. Aquat Conserv Mar Freshwat Ecosyst 23:301–315

Khalili A, Rahimi J (2014) High-resolution spatiotemporal distribution of precipitation in Iran: a comparative study with three global precipitation data sets. Theor Appl Climatol 118(1–2): 211–221

Kier G, Mutke J, Dinerstein E, Ricketts TH, Küper W, Kreft H, Barthlott W (2005) Global patterns of plant diversity and floristic knowledge. J Biogeogr 32(7):1107–1116

Knutti R, Sedlacek J (2013) Robustness and uncertainties in the new CMIP5 climate model projections. Nat Clim Chang 3(4):369–373

Lacoul P, Freedman B (2006) Environmental influences on aquatic plants in freshwater ecosystems. Environ Rev 14:89–136

Leonard J (1991–1992) Contribution a l'etude de la flora et de la vegetation des deserts d'Iran Fascicule 10 Etude des aires de distribution les phytochories les chorotypes Meise: Jardin botanique national de Belgique

MacArthur RH (1972) Geographical ecology. Harper & Row, New York

Martins SV, Milne J, Thomaz SM, McWaters S, Mormul RP, Kennedy M, Murphy K (2013) Human and natural drivers of changing macrophyte community dynamics over 12 years in a Neotropical riverine floodplain system. Aquat Conserv Mar Freshwat Ecosyst 23:678–697

Mehrabian AR (2015) Distribution patterns and diversity of Onosma in Iran: with emphasis on endemism conservation and distribution pattern in SW Asia. Rostaniha 16(1):36–60. (in Persian)

Mehrabian AR, Khajoei Nasab F, Fraser-Jenkins CR, Tajik F (2020a) Distribution patterns and priorities for conservation of Iranian pteridophytes. Fern gaz 21(4):141–160

Mehrabian AR, Sayadi S, Majidi Kuhbenani M, Hashemi Yeganeh V, Abdoljabari M (2020b) Priorities for conservation of endemic trees and shrubs of Iran: important plant areas (IPAs) and Alliance for zero extinction (AZE) in SW Asia. Journal of Asia-Pacific Biodiversity 13(2):295–305

Morecroft MD, Keith SA (2009) Plant ecology as an indicator of climate and global change. In: Climate change: observed impacts on planet earth, 1st edn. Elsevier B.V, pp 297–305

Naqinezhad AR, Attar F, Jalili A, Mehdigholi K (2010) Plant biodiversity of wetland habitats in dry steppes of Central Alborz Mountains, N. Iran. Aust J Basic Appl Sci 4:321–333

O'Hare MT (2015) Aquatic vegetation—a primer for hydrodynamic specialists. J Hydraul Res 53:687–698

Olmstead RG (2013) Phylogeny and biogeography in *Solanaceae*, *Verbenaceae* and *Bignoniaceae*: a comparison of continental and intercontinental diversification patterns. Bot J Linn Soc 171:80–102

Ramachandra TV (2001) Restoration and management strategies of wetlands in developing countries. Indian Inst Sci Electron Green J 4(3):126–138

Ramsar International Convention of Wetlands. Available from http://www.ramsar.org/. Accessed 10 May 2016

Rechinger KH (1966a) Ceratophyllaceae. In: Rechinger KH (ed) Flora Iranica, Lfg. 28. –Akad. Druck- und Verlagsanstalt, Graz

Rechinger KH (1966b) Haloragaceae. In: Rechinger KH (ed) Flora Iranica, Lfg. 18. –Akad. Druck- und Verlagsanstalt, Graz

Rechinger KH (1966c) Hippuridaceae. In: Rechinger KH (ed) Flora Iranica, Lfg. 31. – Akad. Druck- und Verlagsanstalt, Graz

Rechinger KH (1966d) Nymphaeaceae. In: Rechinger KH (ed) Flora Iranica, Lfg. 33. –Akad. Druck- und Verlagsanstalt, Graz

Rechinger KH (1978) Trapaceae. In: Rechinger KH (ed) Flora Iranica, Lfg. 127. –Akad. Druck- und Verlagsanstalt, Graz

Rechinger KH (1990) Pontederiaceae. In: Rechinger KH (ed) Flora Iranica, Lfg. 167. –Akad. Druck- und Verlagsanstalt, Graz

Rivas–Martínez S, Sanchez-Mata D, Costa M (1999) Boreal and western temperate forest vegetation (syntaxonomical synopsis of the potential natural plant communities of North America II). Itinera Geobotanica 12:3–311

Saaa LL, Bozek M, Hauxwell J, Wagner K (2010) Response of aquatic macrophytes to human land use perturbations in the watersheds of Wisconsin lakes, USA. Aquat Bot 93(1):1–8
Shakur A, Roshan Gh, Najafe R et al (2010) Evaluating climatic potential for palm cultivation in Iran with emphasize on degree-day index. Afr. J. Agric. Res. 13:99e118
Short FT, Kosten S, Morgan PA, Malone S, Moore GE (2016) Impacts of climate change on submerged and emergent wetland plants. Aquat Bot 135(3):17
Stöklin J (1968) Structural history and tectonics of Iran: a review. AAPG Bull 52:1229–1258
Stöklin J (1974) Northern Iran: Alborz Mountains. Geological Society, London, Special Publications, 4: 213–234
Tavakoli S, Ejtehadi H, Amini T, Zarre H, Razavi S (2014) Pajooheshhaye Giahi 26(4):423–433
Toivonen L, Triest P, Uotila, Willby N (Guest eds) (2010) Aquatic invasions and relation to environmental changes: proceedings of the 12th international symposium on aquatic weeds, European weed research society. Hydrobiologia 656:1–267
Tokoro T, Hosokawa S, Miyoshi E, Tada K, Watanabe K, Montani S et al (2014) Net uptake of atmospheric CO2 by coastal submerged aquatic vegetation. Glob Change Biol 20:1873–1884
Toledo JA, Kaminski J, Santanna MA, Santos DR (2012) Tampão Santa Maria (TSM) como alternativa ao tampão SMP para medição da acidez potencial de solos ácidos. Braz Soil Sci Soc 36:427–435
Vetaas OR, Ferrer-Castán D (2008) Patterns of woody plant species richness in the Iberian Peninsula: environmental range and spatial scale. J Biogeogr 35:1863–1878
Wagnetiz G (1986) *Centaurea* in Southwest Asia: patterns of distribution and diversity. Proc R Soc Edinburgh. Sect B Biol Sci 89B:11–21
Walter H, Breckle S-W (2002) Zonobiome of the evergreen tropical rain forest (Zonobiome of the equatorial humid diurnal climate). — In Walter's vegetation of the earth: the ecological systems of the geo-biosphere, Springer, Berlin: 115–162
Wetzel RG (2001) Limnology: lake and river ecosystems. Academic Press, San Diego, 998 p
Williams PH, Gibbons D, Margules C, Rebelo A, Humphries C, Pressey R (1996) A comparison of richness hotspots rarity hotspots and complementary areas for conserving diversity of British birds. Conserv Biol 10(1):155–174
Wood KA, O'Hare MT, Daunt F, Stillman RA (2014) Can sacrificial feeding areas protect aquatic plants from herbivore grazing? Using behavioral ecology to inform wildlife management. PLoS One 9
Wood KA, O'Hare MT, McDonald C, Searle KR, Daunt F, Stillman RA (2017) Herbivore regulation of plant abundance in aquatic ecosystems. Biol Rev
Yosefi M, Kazemi R (2008) Aquatic and semi -aquatic plants of Choghakhor Wetlands. First National Conference on Iranian Wetlands, Ahvaz, pp 1–8
Yosefi M, Toranj S (2015) A preliminary checklist of vascular aquatic plants of Iran. Res J Recent Sci 4(1):1–8
Zohary M (1963) On the geobotanical structure of Iran. The Bulletin of the Research Council of Israel, Section D, Botany, Supplement 11D: 1–114
Zohary M (1973a) On the geobotanical structure of Iran. Bull Res Council Israel 11D (supplementary):1–113
Zohary M (1973b) Geobotanical foundations of the Middle East. vol. 1 and 2. Stuttgart, Gustav Fischer Verlag and Amsterdam, Swets and Zeitlinger

Chapter 14
Compiled Checklist of Aquatic Invertebrates of the Southern Marshes of Iraq

Laith A. Jawad

Abstract Freshwater habitats are recognized for their significance in regard to biodiversity. In the present chapter, information available in the international database related to the invertebrate fauna of the southern marshes of Iraq were reviewed. Such revision will assist the evaluation of the present status of this group animal in order to locate which invertebrate taxa need to be investigated further. The results of the revision performed in this study showed that only five invertebrate taxa were examined from the southern marshes of Iraq; these are Rotifera, Annelida, Crustacea, aquatic insects and Mollusca. The most specious group was aquatic insects with 105 species belonging to 18 families and the group with the least number of species was Annelida with 44 species belonging to 4 families. The results also showed that there are still other invertebrate taxa that were not yet investigated in comparison to what is happening in the neighbouring countries. As recommendations, continuous comprehensive investigations should be continued in order to cover all the southern marsh areas and investigate all the invertebrate groups inhabiting these areas in the future.

Keywords Insects · Crustacean · Worms · Mollusca · South of Iraq · Annelida · Rotifers

14.1 Introduction

Aquatic invertebrates are intricate in numerous diverse routes in riverine environment (e.g. Wallace and Webster 1996). For instance, they subsidize considerably to nutrient circulation and the output of organic matters, whether manufactured within the system or inflowing from the wet zone (Malmqvist 2002). The invertebrates can feed on a variety of food resources such as the biofilm that consists of algae, fungi,

L. A. Jawad (✉)
School of Environmental and Animal Sciences, Unitec Institute of Technology, Mt Albert, Auckland, New Zealand

L. A. Jawad (ed.), *Southern Iraq's Marshes*, Coastal Research Library 36,
https://doi.org/10.1007/978-3-030-66238-7_14

bacteria and protozoans together with dissolved and particulate organic matter (Malmqvist 2002). These invertebrates are important feed items for predators such as fish (Huryn and Wallace 2000).

The geographical dispersal of invertebrates in the freshwater system is governed not only by ecosystem settings but also to a large degree by vigorous activities of the invertebrates themselves (Malmqvist 2002). Dispersal of these animals supports the communication of their groups along their geographical distribution, and separate populations get linked together (e.g. Nakano and Murakami 2001).

In Iraq, invertebrate groups are not fully investigated, and there are still a large part of this group that needs to be covered in different landscapes of Iraq including the southern marshes (Garstecki and Amr 2011).

Among the important aquatic invertebrates are the dragonflies, which is considered one of the liveliest constituents of the freshwater system (Schneider 1982, Van Straalen 1997). further studies about 40 species of this aquatic insects that were recorded from Iraq, but this number is not confirmed and expect the number will rise in case further studies take place in the inland waters of Iraq (Morton 1919, 1920, 1921, Sage 1960a, b, c, St. Quentin 1964, Asahina 1973, 1974). Among the important species found in the lower reaches of Mesopotamia are *Orthetrum sabina*, *Crocothemis servilia*, *Crocothemis erythraea*, *Diplacodes lefebvrii*, *Brachythemis fuscopalliata*, *Trithemis annulata* and *Selysiothemis nigra* (Asahina 1974).

The aquatic insects and in particular members of the order Coleoptera represents a major part in the environment of the southern marshes of Iraq. This group of invertebrates has not received enough attention. They were mainly studied by Ali (1978a, b). Recently, a few studies were published on the aquatic insects of the southern marshes, but they are way behind the full coverage of the diversity of this group in southern Iraq (Brown 1953; Asahina 1974; Ali 1978a, b; Hassan et al. 2000; Ali et al. 2002; Kareem and Al-Edani 2015; Augul et al. 2016; Porter 2016; Abd and Al-Asady 2017; Al-Hashmi et al. 2018).

The other groups of invertebrates were not in a better taxonomic status than the aquatic insects. Until this moment, the investigations of the aquatic invertebrates in Iraq and in the marsh areas are still lacking, and they have been performed on personal capacities.

Among those groups of invertebrates that attracted some attention are the malacofauna (molluscs, which includes gastropods and bivalves) and some crustacean groups. The freshwater mollusca of Iraq was first described by Mousson (1874). Later on, several investigators took parts in identifying this group, but although their works were comprehensive, they remain in complete as no further workers have continued investing the molluscs of Iraq (Annandale 1918, 1920; Annandale and Prashad 1919; Germain 1924; Haas 1969). In recent years, some fragmentary studies mostly ecological have been published, but no profound and comprehensive taxonomic works have accomplished. Therefore, this group remains to be identified properly (Al-Qarooni 2005; Plaziat and Younis 2005; Ali et al. 2007; Glöer and Naser 2007, 2008; Lucy and Graczyk 2008; Gmais et al. 2008; Abdul-Sahib and Abdul-Sahib 2009; Al-Waaly et al. 2014; Marrone et al. 2014; Qazar et al. 2015).

Similarly, the different groups of crustaceans are still suffering from the lack in their identification and full taxonomic coverage due to the severe shortage in the field expeditions. There were some new taxa that were described from the southern part of Iraq such as *Atyaephyra desmarestii mesopotamica* and *Caridinia baboulti basrensis* (Al-Adhub 1987); *Elamenopsis kempi* and *Sesarma boulengeri*, recorded from Al-Hawizeh and Al- Hammar marshes (Ali et al. 2007); and *Potamon mesopotamicum* from the Al-Hawizeh marshes (Naser 2009).

Among the other groups of crustaceans found in the lower reaches of Mesopotamia is the calanoid copepod that also shared the bad taxonomic status with the other aquatic invertebrates. Certain species were reported early in the last century such as *Canthocamptus staphylinus*, *Diaptomus vulgaris*, *Diaptomus blanci* and *Diaptomus chevreuxi*, while Mohamed and Salman (2009) reported on the presence of three species of calanoid copepods in the marshes, *Phyllodiaptomus irakiensis*, *Acanthodiaptomus denticornis*, *A. salinus* and *Eudiaptomus vulgaris*. On the other hand, the amphipod *Parhyale basrensis* was recorded from the marsh area by Ali et al. (2007). Several other species of crustacean were reported from the southern marshes of Iraq in the last few decades (Al-Khafaji et al. 2016, 2017; Al-Adhub and Hamzah 1987; Al-Adhub 1987; Ajeel and Abaas 2013; Saoud 2005; Naser et al. 2011, 2012a, b; Ali and Al-Maliky 2017; Salman et al. 1990, 2006; Ali and Salman 1986, 1998; Mohammed and Ali 2010; Mohammed et al. 2008, 2014; Al-Saboonchi et al. 1986; Abbas et al. 2015).

The Annelida might have a better taxonomic standing as the species of this group have seen slightly better chance of investigations (Al-Mayah and Al-Abbad 2010; Almukhtar et al. 2012; Jaweir 2014; Al-Abbad 2010, 2012, 2014).

The rotifers are among the invertebrate groups that has no chance to receive much scientific attention in whole Iraq in general and in the marsh areas in particular. No studies were published on this group earlier than 1989. This makes the information gathered about this group very limited in species and geographical aspects (Abdul-Hussein et al. 1989; Ahmed and Mohammed 2006; Ghazi and Ahmed 2008; Ghazi et al. 2011; Ghazi and Ali 2012; Ahmed and Ghazi 2009, 2014; Hammadi et al. 2015; Hammadi 2016; Al-Ameen et al. 2019).

Some of the invertebrates species living in the marsh areas are endemic and are nearly threatened, vulnerable or data deficient (Garstecki and Amr 2011). Among these are four species of dragonflies, *Gomphus kinzelbachi* DD Endemic, *Brachythemis fuscopalliata* VU Endemic, *Libellula pontica* NT Occurrence unclear and *Anormogomphus kiritshenkoi* NT Historically common. Boudot et al. (2009) assessed the conservation status of the dragonflies of Iraq and suggested that *Sympecma paedisca* as endangered, *Lindenia tetraphylla* as near-threatened and *Onychogomphus flexuosus* and *Brachythemis fuscopalliata* as vulnerable (Garstecki and Amr 2011).

Some other invertebrate species might also be at the verge of different levels of threatened, but their conservation status was not yet assessed. In addition, the ecocide that happened to the marsh areas might put more invertebrate species at the threatened stage. Therefore, a comprehensive investigation should be accomplished in order to reveal the true picture of the state of the invertebrate animals in Iraq in general and in the marsh areas in particular (Evans 2002).

In the present chapter, a compilation of the reported invertebrates species from the southern marshes of Iraq were arranged in a checklist. This checklist is far from being a complete as large numbers of references were not available to the author at the date of writing this paper. It is also far from being comprehensive in the sense of number of species as there is a severe lack of investigations on this group of animals. Although this checklist is preliminary, it is considered useful as it provides a base line for future attempts to build up more comprehensive list for this important group of animals.

14.2 The Checklist

The present literature review investigation has shown that only four invertebrate taxa, i.e. Rotifera, Annelida, Crustacea, aquatic insects and Mollusca, have been studied and explored so far in Iraq (Table 14.1, Lists 14.1, 14.2, 14.3, 14.4 and 14.5). The most diverse group of these was the aquatic insects as it showed the presence of 105 species belonging to 18 families, with Dytiscidae as the spacious insect family with 43 species. The second group in the order of diversity is Crustacea, which showed to contain 64 species belonging to 23 families, with Chydoridae as the spacious family with 19 species. The phylum Mollusca comes next with 60 species of 20 families. The gastropods represent 49 species belonging to 14 families, with Lymnaeidae as the spacious family of 10 species. On the other hand, the bivalves were characterized by 11 species belonging to 6 family, with both Cyrenidae and Unionidae as spacious families of 3 species for each. The Rotifera group comes with 55 species and 12 families and with Brachionidae as a spacious family with 21 species. The last group is the Annelida, with 44 species belonging to 4 families reported so far, with family Naididae as a spacious species of 40 species.

Comparing the spacious families of the 5 invertebrate taxa examined, the insect family Dytiscidae showed the highest number of species, 43, followed by annelid

Table 14.1 Number of species, families and spacious families of invertebrate taxa reported to be present in the southern marshes of Iraq. Information retrieved from the literature

Taxa	Number of species	Number of families	Spacious family
Rotifera	55	12	Brachionidae (21)
Annelida	44	4	Naididae (40)
Crustacea	23	62	Chydoridae (19)
Aquatic insects	105	18	Dytiscidae (43)
Mollusca	49	14	Lymnaeidae (10)
Gastropoda	11	6	Cyrenidae (3)
Bivalvia			Unionidae (3)

family with 40 species. Crustacea was the group that showed having the highest number of families (23) followed by Mollusca with 20 families.

List 14.1: Species of the phylum Rotifera compiled from literature published on the southern marshes of Iraq

Phylum Rotifera

Family: Asplanchnidae

fissa (Gosse 1850)

Family: *Brachionidae*

Anuraeopsis fissa (Gosse 1851)

Brachionus angularis (Gosse 1851)

B. calyciflorus (Pallas 1766)

B. calyciflorus f. *amphiceros* (Ehrenberg 1838)

B. calyciflorus f. *dorcas* (Gosse 1851)

B. leydigii (Cohn 1862)

B. plicatilis (Müller 1786)

B. quadridentatus (Hermann 1738)

B. rubens (Ehrenberg 1838)

B. urceolaris (Müller 1773)

Keratella cochlearis (Gosse 1851)

K. quadrata (Müller 1786)

Keratella tropica (Apstein 1907)

K. valga (Ehrenberg 1834)

Lepadella ovalis (Müller 1786)

Lepadella patella (Müller 1773)

Notholca acuminata (Ehrenberg 1832)

N. squamula (Müller 1786)

N. striata (Müller 1786)

Platyias patulus (Muller 1786)

P. quadricornis (Ehrenberg 1832)

Family: Euchlanidae

Euchlanis dilatata (Ehrenberg 1832)

Family: Gastropodidae

Gastropus hyptopus (Ehrenberg 1838)

Family: Hexarthridae

Hexarthra mira (Hudson 1871)

Family: Lecanidae

Lecane closterocerca (Schmarda 1859)

L. hamata (Stokes 1896)

L. luna (Müller 1776)

L. lunaris (Ehrenberg 1832)

Lecane scutata (Harring & Myers 1926)

(continued)

List 14.1 (continued)

L. stenroosi (Meissner 1908)
L. thalera (Harring & Myers 1926)
L. ungulata (Gosse 1887)
Monostyla bulla (Gosse 1851)
Family: Lepadellidae
Colurella adriatica (Ehrenberg 1831)
Lepadella patella (Müller 1773)
Squatinella mutica (Ehrenberg 1832)
Family: Notommatidae
Cephalodella gibba (Ehrenberg 1830)
Family: Synchaetidae
Polyarthra dolichoptera (Idelson 1925)
P. vulgaris (Carlin 1943)
Synchaeta lakowitziana (Lucks 1930)
Synchaeta oblonga (Ehrenberg 1832)
Family: Testudinellidae
Pompholyx complanata (Gosse 1851)
Testudinella patina (Hermann 1783)
Family: Trichocercidae
Trichocerca cylindrica (Imhof 1891)
T. elongata (Gosse 1886)
T. flagellata (Hauer 1937)
T. porcellus (Gosse 1851)
Family: Trichotriidae
Macrochaetus subquadratus (Perty 1850)
Trichotria pocillum (Müller 1776)
T. tetractis (Ehrenberg 1830)
T. truncata (Whitelegge 1889)

List 14.2: Species of the phylum Annelida compiled from literature published on the southern marshes of Iraq

Phylum Annelida
Class Polychaeta
Class Aelosomata
Family: Aeolosomatidae
Aeolosoma leidyi (Cragin 1887)
Class Clitellata
Family: Naididae

(continued)

List 14.2 (continued)

Allonais gwaliorensis
A. pectinata (Stephenson 1910)
Chaetogaster limnaei (von Baer 1827)
Dero cooperi (Stephenson 1932)
Dero (Dero) digitata (Müller 1773)
D. dorsalis (Forroniere 1899)
Dero (Aulophorus) furcatus (Müller 1773)
Dero (Dero) nivea (Aiyer 1929)
D. obtusa (d`Udekem 1855)
D. nivea (Aiyer 1929)
Dero (Aulophorus) furcata (Müller 1773)
Dero sawayai (Marcus 1943)
Nais communis (Piguet 1906)
N. elinguis (Müller 1773)
N. pardalis (Piguet 1906)
N. variabilis (Piguet 1906)
Paranais litoralis (Müller 1784)
P. frici (Hrabe 1941)
Slavina appendiculata (d`Udekem 1885)
Stephensoniana trivandrana (Aiyer 1926)
Stylaria lacustris (Linnaeus 1767)
S. fossularis (Leidy 1852)
Subfamily: Pristininae
Pristina idrensis (Sperber 1948)
P. longiseta (Ehrenberg 1828)
P. macrochaeta (Stephenson 1931)
P. notopora (Cernosvitov 1937)
P. osborni (Walton 1906)
P. proboscidea (Baddard 1896)
P. sima (Marcus 1944)
Pristinella aequiseta (Bourne 1891)
Subfamily: Rhyacodrilinae
Branchiura sowerbyi (Beddard 1892)
Subfamily: Tubicinae
Aulodrilus pigueti (Kowalewski 1914)
Branchiura sowerbyi (Beddard 1892)
Embolocephalus velutinus (Grube 1879)
Limnodrilus claparadeanus (Ratzel 1868)
L. hoffmeisteri (Claparéde 1862)
L. maumeensis (Brinkhurst and Cook 1966)
L. profundicola (Verrill 1871)

(continued)

List 14.2 (continued)

Slavina appendiculata (d'Udekem 1855)
Tubifex tubifex (Müller 1774)
Family: Lumbriculidae
Lumbriculus variegatus
Stylodrilus herringianus
Family: Lumbricidae
Eiseniella tetraedra (Savigny 1826)

List 14.3: Species of the class Crustacea compiled from literature published on the southern marshes of Iraq

Class Crustacea
Order: Anostraca
Family: Artemiidae
Artemia franciscana (Kellogg 1906)
Order: Harpacticoida
Elaphoidella grandidieri (Guerne & Richard 1893)
Elaphoidella sewelli (Chappuis 1928)
Order: Cyclopoida
Family: Cyclopidae
Apocyclops dengizicus (Lepeshkin 1900)
Cyclops vernalis (Fischer 1853)
Mesocyclops isabellae (Dussart & Fernando 1988)
Microcyclops varicans (Sars 1863)
Thermocyclops crassus (Fischer 1853)
Order: Cladocera
Family: Bosminidae
Bosmina (Bosmina) longirostris (Müller 1785)
Bosmina (Liederobosmina) meridionalis (Sars 1904)
Family: Chydoridae
Alona circumfimbriata (Megard 1967)
A. costata (Sars 1862)
Alona rectangula (Sars 1862)
Alona rustica rustica (Scott 1895)
A. cambouei (Guerne & Richard 1893)
A. karua (King 1853)
Alonella diaphana (King 1853)
Camptocercus rectirostris (Schoedler in 1862)
C. uncinatus (Smirnov 1971)

(continued)

List 14.3 (continued)

Chydorus sphaericus sphaericus (Mueller hadton 1785)
Dunhevedia crassa (King in 1853)
Eurycercus glacialis (Lilljeborg 1887)
Indialona macronyx (Daday 1898)
Kurzia longirostris (Daday 1898)
Leydigia acanthocercoides (Fischer 1854)
Leydigia macrodonta macrodonta (Sars 1916)
Pleuroxus aduncus (Jurine 1820)
P. paraplesius (Frey 1993)
P. similis (Vavra 1900)
Family: Daphniidae
Ceriodaphnia rigaudi (Richard in 1894)
Daphnia exilis (Herrick 1895)
D. hyalina (Leydig 1860)
D. lumholtzi (Sars 1885)
D. magna (Straus 1820)
D. pulex (Leydig 1860)
Scapholeberis kingi (Sars in 1888)
Simocephalus expinosus (De Geer 1778)
Simocephalus vetulus (Müller 1776)
Family: Ilyocryptidae
Ilyocryptus agilis (Kurz 1878)
Ilyocryptus spinifer (Herrick in 1882)
Family: Macrothricidae
Macrothrix spinosa (King 1853)
Family: Sididae
Diaphanosoma brachyurum (Liévin in 1848)
D. orghidani (Nigrea 1982)
Latonopsis fasciculata (Daday 1905)
Family: Macrothricidae
Macrothrix spinosa (King 1853)
Family: Moinidae
Moina affinis (Birge 1893
M. brachiata (Jurnie 1820)
M. micrura (Kurz in 1874)
Order: Decapoda
Family: Atyidae
Atyaephyra desmarestii mesopotamica (Al-Adhub 1987)
Caridina babaulti basrensis (Al-Adhub 1987)
Family: Cyclopettidae
Limnoithona tetraspina (Zhang & Li 1976)

(continued)

List 14.3 (continued)

Family: Hymenosomatidae

Elamenopsis kempi (Chopra & Das 1930)

Family: Matutidae

Matuta planipes (Fabricius 1798)

Family: Palaemonidae

Macrobrachium nipponense (De Haan 1849)

Palaemon elegans (Rathke 1837)

Family: Penaeidae

Metapenaeus affinis (H. Milne Edwards 1837 [in Milne Edwards 1834-1840])

Family: Potamidae

Potamon mesopotamicum (Pretzmann 1962)

Family: Sesarmidae

Sesarma boulengeri (Calman 1920)

Family: Varunidae

Eriocheir hepuensis (Dai 1991)

Eriocheir sinensis (H. Milne Edwards 1853)

Order: Isopoda

Family: Cymothooidea

Annina mesopotamica (Ahmed 1971)

Family: Sphaeromatidae

Sphaeroma annandalei annandalei (Stebbing 1910)

Order: Amphipoda

Family: Hyalidae

Parhyale basrensis (Salman 1986)

Order: Mysida

Family: Mysidae

Indomysis nybini (Biju & Panampunnayil 2010)

List 14.4: Species of the aquatic insects compiled from literature published on the southern marshes of Iraq

Class Insecta

Order: Coleoptera

Family: Culicidae

Anopheles multicolour (Cambouliu 1902)

Culex pipiens (Linnaeus 1758)

Family: Dytiscidae

Acilius sulcatus (Linnaeus 1758)

(continued)

List 14.4 (continued)

Agabus biguttulus (Thomson 1867)
A. conspersus (Marsham 1802)
A. solieri (Aubé 1837)
Anthracinus procerus (Régimbart 1883: 229)
Bidessus confusus (Klug 1833)
Bidessus major (Sharp 1882)
Brychius elevatus (Panzer 1793)
Canthydrus diophthalmus (Reiche & Saulcy 1855)
C. luctuosus (Aubé 1838)
Copelatus discoideus (Sharp 1882)
Copelatus pulchellus (Klug 1834)
Cybister lateralimarginalis (De Geer 1774)
Cybister tripunctatus asiaticus (Olivier 1795)
Deronectes turca (Seidlitz 1887)
Donacia simplex (Fabricius 1775)
Eretes sticticus (Linnaeus 1776)
Graptodytes sedilloti phrygius (Guignot 1942)
Herophydrus musicus (Klug 1834)
Hydaticus (*Prodaticus*) *histrio* (Clark 1864)
H. (*P.*) *leander* (Rossi 1790)
***H.* (*P.*) *pictus* (Sharp 1882)**
***H.* (*P.*) *ponticus* (Sharp 1882)**
Hydrocoptus subvittulus (Motschulsky 1859)
Hydroporus inscitus (Sharp 1882)
Hydroporus thermalis (Germar 1838)
H. notabilis (LeConte 1850)
Hydrovatus sordidus (Sharp 1882)
Hyphydrus pictus (Klug 1834)
Hygrotus inscriptus (Sharp 1882)
H. lernaeus (Schaum 1857)
H. saginatus (Schaum 1857)
Ilybius chalconatus (Panzer 1796)
Laccophilus hyalinus (De Geer 1774)
L. maculosus maculosus (Say 357)
L. sharpi (Régimbart 1889)
L. sublineatus (Sharp 1882)
Methles rectus (Sharp 1882)
Noterus clavicornis (De Geer 1774)
N. ponticus (Sharp 1882)
Orthochaetes insignis (Aubé 1863)
Platambus lunulatus (Fischer von Waldheim 1829)

(continued)

List 14.4 (continued)

Rhantus suturalis (Macleay 1825)

Family: Gyrinidae

Aulonogyrus striatus (Fabricius 1792)

A. concinnus (Klug in Ehrenberg 1834)

Dineutus arabicus (Régimbart 1907)

D. grandis (Klug in Ehrenberg 1834)

D. subspinosus (Klug in Ehrenberg 1834)

Gyrinus caspius (Ménétries 1832)

G. colymbus (Erichson 1837)

G. dejeani (Brullé 1832)

G. fairmairei (Régimbart 1883)

G. libanus (Aubé 1838)

G. luctuosus (Régimbart 1883)

G. natator (Linnaeus 1758)

G. suffriani (Scriba 1855)

G. urinator (Illiger 1807)

Orectochilus villosus (Müller 1776)

Family: Hydrophilidae

Berosus luridus (Linnaeus 1761)

Order: Hemiptera

Family: Chironomidae

Chironomus piger (Strenzke 1959)

C. lacunarius (Wulker & Klotzi 1973)

C. plumosus (Linnaeus 1758)

Cricotopus bicinctus (Meigen 1818)

C. sylvestris (Fabricius 1794)

Family: Corixidae

Helicorisa vermiculata (Puton 1874)

Sigara lateralis (Leach 1817)

Family: Macrovellidae

Macrovelia hornii (Uhler 1872)

Family: Mesoveliidae

Mesovelia vittigera (Horvath 1895)

Family: Pleidae

Plea leachi (McGregor and Kirkaldy 1899)

Family: Veliidae

Velia saulii saulii (Tamanini 1947)

Order: Odonata

Family: Aeshnidae

Aeshna mixta (Latreille 1805)

Anax ephippiger (Burmeister 1839)

(continued)

List 14.4 (continued)

Anax parthenope (Selys 1839)

Family: Calopterygidae

Calopteryx splendens (Harris 1780)

Family: Coenagrionidae

Erythromma lindenii (Selys 1840)

Ischnura evansi (Morton 1919)

Ischnura fountaineae (Morton 1905)

Ischnura pumilio (Charpentier 1825)

Ischnura senegalensis (Rambur 1842)

Family: Euphaeidae

Epallage fatime (Charpentier 1840)

Family: Gomphidae

Onychogomphus flexuosus (Schneider 1845)

Onychogomphus lefebvrii (Rambur 1842)

Lindenia tetraphylla (Vander Linden 1825)

Family: Lestidae

Chalcolestes parvidens (Artobelevski 1929)

Lestes barbarus (Fabricius 1798)

Sympecma paedisca (Brauer 1877)

Family: Libellulidae

Brachythemis fuscopalliata (Selys 1887)

Crocothemis erythraea (Brullé 1832)

C. servilia (Drury 1773)

Diplacodes lefebvrii (Ramber 1842)

D. nebulosa (Fabricius 1793)

D. trivialis (Rambur 1842)

Orthetrum chrysostigma (Burmeister 1839)

O. sabina (Drury 1770)

O. taeniolatum (Schneider 1845)

O. trinacria (Selys 1841)

Pantala flavescens (Fabricius 1798)

Selysiothemis nigra (Vander Linden 1825)

Sympetrum arenicolor (Jödicke 1994)

Sympetrum fonscolombii (Selys 1840)

Sympetrum striolatum (Charpentier 1840)

Trithemis annulata (Beauvois 1807)

Trithemis festiva (Rambur 1842)

Family: Platycnemididae

Platycnemis kervillei (Martin 1909)

List 14.5: Species of the Gastropoda and Bivalvia compiled from literature published on the southern marshes of Iraq

Phylum Mollusca

Class Gastropoda

Family: Ampullariidae

Pila ovata (Olivier 1804)

Family: Bithyniidae

Bithynia badiella (Parreyss 1849

B. hareerensis (Glöer & Naser 2008

B. iraqensis (Pallary 1939)

Family: Hydrobiidae

Amnicola (*Alocinna*) *ejecta* (Mousson 1874)

Amnicola limosus (Say 1817)

Tryonia clathrata (Stimpson 1865)

Family: Lymnaeidae

Bulimnea megasoma (Say 1824)

Lymnaea (*R.*) *canalifera* (Mousson 1874)

L. cor Annandale and Prashad (1919)

L. gedrosiana (Annandale and Prashad 1919)

L. (*R.*) *tenera* (Parreyss 1849)

Radix auricularia (Linnaeus 1758)

R. lagotis (Schrank 1803)

Radix natalensis (Krauss 1848)

Radix peregra (Müller, 1774)

Stagnicola palustris (Müller 1774)

Family: Neritidae

Neritina reclivata (Say 1822)

Neritina (*D.*) *schlaeflii* (Mousson 1874)

Neritina (*Dostia*) *violacea* (Gmelin 1790)

Theodoxus (*N.*) *euphraticus* (Mousson 1874)

T. jordani (Sowerby 1832)

T. (*N.*) *macrii* (Recluz 1841)

T. (*N.*) *mesopotamicus* (Mousson 1874)

Family: Paludomidae

Cleopatra bulimoides (Olivier 1804)

Family: Physidae

Physella acuta (Draparnaud 1805)

Physella gyrina (Say 1821)

Family: Planorbidae

Bulinus truncatus contortus (Michaud 1829)

Ferrissia fragilis (Tryon 1863)

Gyraulus albus (Müller 1774)

G. convexiusculus (Hutton 1849)

(continued)

List 14.5 (continued)

G. ehrenbergi (Beck 1837)
G. euphraticus (Mousson 1874)
G. huwaizahensis (Glöer & Naser 2007)
Planorbis intermixtus (Mousson 1874)
Family: Pomatiopsidae
Tricula palmyrae (Dautzenberg 1894)
***Family:* Potamididae**
Pirenella cingulata (Gmelin 1791)
Family: Pyramidellidae
Odostomia insculpta var. *laevissima* (Marshall 1893)
Family: Stenothyridae
Stenothyra iraqensis (Dance & Eames 1966)
Family: Tateidae
Potamopyrgus antipodarum (Gray 1843)
Family: Thiaridae
Melanopsis buccinoidea (Olivier 1801)
Melanoides costata (Olivier 1804)
M. tuberculata (Müller 1774)
Melanopsis nodosa (Férussac 1823)
Melanopsis (Melanopsis) praemorsum (Linnaeus 1758)
M. subtingitana (Annandale 1918)
Family: Viviparidae
Bellamya bengalensis (Lamarck 1822)
Bellamya unicolor (Olivier 1804)
Viviparus intertextus (Say 1829)
Class Bivalvia
Family: Cyrenidae
Corbicula fluminalis (Müller 1774)
Corbicula fluminea (Müller 1774)
Corbicula tigridis (Monsoon 1874)
Family: Dreissenidae
Dreissena polymorpha (Pallas 1771)
Mytilopsis leucophaeata (Conrad 1831)
Family: Mycetopodidae
Anodonta vescoiana (Bourguignat 1857)
Family: Pisidiidae
Pisidium dubium (Say 1817)
Family: Semelidae
Theora mesopotamica (Annandale 1918)
Family: Unionidae
Gonidea angulata (Lea 1838)
Unio tigridis (Bourguignat 1852)
Pseudodontopsis euphraticus (Bourguignat 1852)

14.3 Remarks

The five invertebrate taxa investigated were not fully studied through the literature as there are large number of families belonging to these groups that were never considered. As to the remaining invertebrate taxa other than the five groups investigated in the present chapter the literature review showed no evidence of any study has been done so far on them. For example, the protozoa, worms, cnidarians, leeches and other minor phyla. Therefore, the present compiled list is far from being complete as it does not reflect the actual and present biodiversity of aquatic invertebrates in the southern marshes. Two main causes might be behind such shortage: (1) field studies are concentrated on the main invertebrate groups and on certain families of these main groups; and (2) publication of the results of the studies performed by Iraqi scientists are mainly published locally in journals that are not covered by international database making it very difficult to reach them.

The examination of the literature available on the international database has shown that the aquatic insect contains the highest number of both species and families among the five invertebrate taxa studied in the southern marshes of Iraq (105 species and 18 families). Although the number of species looks high in comparison with number of species and families obtained for the other invertebrate groups investigated, it is very low in comparison to the other regions in the area (Pourang 1996; Grosser and Pešić 2006; Sharifinia 2015; Askari et al. 2009; Zamanpoore et al. 2011; Shayeghi et al. 2015).

The space prided in this chapter is too small to accommodate a full comparison with the invertebrate biodiversity in the wetlands in the neighbouring Iran, but it is possible to give examples of the main groups that did not appear in the literature review of the invertebrates inhabiting the southern marshes of Iraq. The following taxa were absent from literature on the invertebrates of the marsh areas in Iraq, while they are represented in the neighbouring Iran (Shayeghi et al. 2015; Zamanpoore et al. 2011; Askari et al. 2009; Sharifinia 2015): insects (Ephemeroptera, Plecoptera, Megaloptera, Neuroptera, Diptera, Trichoptera, Lepidoptera, Heteroptera and Hymenoptera); Crustacea (*Gammarus*); and macroinvertebrates.

As a recommendation, further comprehensive field expeditions in the southern marshes are needed in order to give a complete cover to all the areas of the marshes of Iraq. Once such expeditions achieved, two main things need to be established, first, a deposition of all materials collected in a national collection centre that allows scientists from all over the world to examine and study these specimens for further studies, and second, to publish a series of monographs about each group of invertebrate to represent a main reference to be used for generations to come.

References

Abbas MF, Salman SD, Al Mayahy SH (2015) Diversity and seasonal changes of zooplankton communities in the Shatt Al-Arab River, Basrah, Iraq, with a special reference to Cladocera. Int J Marine Sci 5(24):1–14

Abd IF, Al-Asady HS (2017) External morphological study of Diplacodes Triviales Rambure (Odonata: Libellulidae) new record to Iraq. Ibn AL-Haitham J Pure and Appl Sci 25(2)

Abdul-Hussein MM, Al-Saboonchi AA, Ghani AA (1989) Brachionid rotifers from Shatt Al-Arab River, Iraq. Marina Mesopotamica 4(1):1–17

Abdul-Sahib EM, Abdul-Sahib IM (2009) A new record of the freshwater clam, Anodonta vescoiana Bourguignat, 1857 (Mollusca: Bivalvia) from Al-Ezz River, Iraqi marshes. Mesopotamian J Marine Sci 24(1):7–12

Ahmed HK, Ghazi AH (2009) A taxonomical and environmental study of the genus Brachionus (Rotifera: Monogononta) (Pallas, 1776) in Al-Hammer marsh, south of Iraq. Iraqi J Aquacul 6(2):105–112

Ahmed HK, Ghazi AHH (2014) Rotifers diversity at Shatt Al-Arab River and Shatt Al-Basrah canal, South of Iraq, during the abnormal rising of water salinity. Mesopotamian J Marine Sci 29(2):145–154

Ahmed HK, Mohammed HH (2006) The Rotifera community in the south Marshes of Iraq. Marsh Bulletin 1(1):54–58

Ajeel SG, Abaas MF (2013) Abundance and diversity of Cladocera in South Al-Hammar Marshes-Southern Iraq. Tishreen Univ J Res Sci Stud-Biol Sci Series 35:253–266

Al-Abbad MY (2010) New records of Pristina proboscidea and P. aequiseta. Marsh Bulletin 5(2):131–140

Al-Abbad MY (2012) New records of twelve species of Oligochaeta (Naididae and Aeolosomatidae) from the Southern Iraqi Marshes, Iraq. Jordan J Biol Sci 147(618):1–7

Al-Abbad MYM (2014) Addition of four new records of Naidid (Oligochaeta) to the Iraqi fauna. Mesopotamian J Marine Sci 29(2):137–144

Al-Adhub AHY (1987) On a new subspecies of a freshwater shrimp (Decapoda, Atyidae) from the Shatt Al-Arab River, Iraq. Crustaceana 53:1–4

Al-Adhub AHY, Hamzah HA (1987) Caridina babaulti basrensis Subsp. nov., from the Shatt al-Arab region, Iraq (Decapoda, Caridea, Atyidae). Crustaceana 52(3):225–228

Al-Ameen FAM, Kadhim ZK, Thamir AJ, (2019, September) Review of Rotifers in Iraqi waters. In Journal of Physics: Conference Series (Vol. 1294, No. 7, p. 072005). IOP Publishing

Al-Hashmi AH, Al-Safar HH, Augul RS (2018) Key to the species of the Orthetrum Newman, 1833 (Odonata, Libellulidae) with a New Record Species in Iraq. Bulletin of the Iraq Natural History Museum (P-ISSN: 1017-8678, E-ISSN: 2311-9799), 15(1), pp. 15–29

Ali HA (1978a) A list of some aquatic beetles of Iraq (Coleoptera: Dytiscidae). Bull Nat Hist Res Centre Univ Baghdad 7(2):11–14

Ali HA (1978b) Some taxonomic studies on the aquatic beetles of Iraq (Coleoptera: Gyrinidae). Univ Baghdad Bull Nat Hist Res Centre 7:15–20

Ali MH, Al-Maliky TH (2017) Fecundity of the crab, Potamon mesopotamicum Brandis, Storch & Turkay, 1998 from the Mesopotamian Marshlands, Iraq. J Fisher & Environ 41(3):6–11

Ali MH, Salman SD (1986) The reproductive biology of Parhyale basrensis Salman (Crustacea, Amphipoda) in the Shatt al-Arab river. Estuar Coast Shelf Sci 23(3):339–351

Ali M, Salman SD (1998) Production of the crab Elamenopsis kempi (Hymenosomatidae) in the Garmat- Ali region, Basrah, Iraq. Mar Ecol 19:67–75

Ali MH, Anon MR, Mohammed HH (2002) The seasonal variations of abundance and biomass of the two odonate naiads Ischnura evansi Morton (Odonata: Coenagrionidae) and Brachytemis fuscocapillata Selys (Odonata: Libellulideae) in the Qarmat Ali region, Basrah. Marina Mesopotamica 17(2):405–415

Ali AH, Aziz NM, Hamza HA (2007) Abundance, occurrence, seasonal changes and species composition of Macroinvertebrates in the restored Iraqi southern marshes. Marsh Bulletin 2(1):80–95

Al-Khafaji KK, Al Qarooni IH, Al Abbad MY, Al-Lateef NMA (2016) Study of the growth, reproductive biology and abundance for invasive shrimps Palaemon elegans Rathke from Garmat Ali River Basrah, Southern Iraq. J Coast Life Med 4(7):536–540

Al-Khafaji KK, Mutlak FM, Yaseen AT, Al-Sharaa AS (2017) First finding of Matutid crabs: Matuta planipes Fabricius, 1798 (Crustacea: Decapoda) from Iraqi coast and Garmat Ali River, Basrah area, Iraq. Int J Aquacul 7

Al-Mayah SH, Al-Abbad MYM (2010) New records of two species of oligochaetes (Naididae): Pristina longiseta and P. macrochaeta from Iraq, with notes on their morphology and reproduction. Mesopotamian J Marine Sci 25(2):166–175

Almukhtar EA, Sabtie HA, Jaweir HJ (2012) Aquatic oligochaetes of Iraq's Southern Marshes. Baghdad Sci J 9(3):472–480

Al-Qarooni IHM (2005) Abundance and occurrence studies on some of zooplankton and aquatic snails in Al-Chabaish, Al-Hammar and Al-Fuhud marshes southern Iraq (Doctoral dissertation, M. Sc. Thesis-college of Education–Uni. of Basrah. 2005.97 pp. (in Arabic))

Al-Saboonchi AA, Barak NA, Mohamed AR (1986) Zooplankton of Garma Marshes, Iraq. J Biol Sci Res 17(1):33–40

Al-Waaly ABM, Mohammad MK, Al-Miali HM (2014) Freshwater snail's diversity in the middle and south regions of Iraq. Adv Bio Res 5(3):166–171

Annandale TN (1918) Freshwater shells from Mesopotamia.– Records of the Indian Museum, Calcutta, t. 15, pp. 159–170, 1 fig., 1 pl.

Annandale TN (1920) Report on the freshwater Gastropod Molluscs of lower Mesopotamia. Part II – the family Planorbidae.- Records of the Indian Museum, Calcutta, t. 18, pp. 147–149, 1 fig

Annandale TN, Prashad B (1919) Report on the freshwater Gastropod molluscs of lower Mesopotamia. Part 1: the genus Limnaea.– Records of the Indian Museum, Calcutta, t. 18, pp. 103–115

Asahina S (1973) Odonata of Iraq. Japan J Zool 17(1):17–36

Asahina S (1974) An additional note to the Odonata of Iraq. Kontyû, Tokyo 42:107–109

Askari O, Farshbaf Pourabad R, Khaganinia S (2009) Faunistic study of Heteroptera of Zanjanroud region in Zanjan province of Iran. Munis Entomol & Zool 4(2):560–563

Augul RS, Al-Hashmi AH, Al-Saffar HH (2016) New record of the Epaulet Skimmer: Orthetrum chrysostigma (Burmeister) (Odonata: Libellulidae) from Iraq. J Biodivers Environ Sci 8(5):233–238

Boudot JP, Kalkman VJ, Azpilicuetaamorin M, Bogdanovic T, Cordero RA, Degabriele G, Dommanget JL, Ferreira S, Garrigos B, Jovic M, Kotarac M, Lopau W, Marinov M, Mihokovic N, Riservato E, Samraoui B, Schneider W (2009) Atlas of the Odonata of the Mediterranean and North Africa. Libellula Suppl 9:1–256

Brown ES (1953) LII.—notes on aquatic Hemiptera from Syria and Iraq. Ann Mag Nat Hist 6(68):579–600

Evans MI (2002) The ecosystem. In: The Iraqi Marshlands: a human and environmental study. Politico's, London, pp 201–219

Garstecki T, Amr Z (2011) Biodiversity and ecosystem management in the Iraqi Marshlands–screening study on potential World Heritage Nomination. IUCN, Amman, 189p

Germain L (1924) Catalogue of the Planorbidae in the Indian museum (natural history), Calcutta.– Records of the Indian Museum, Calcutta, t. 21, 210 p., 4 pl.

Ghazi AH, Ahmed HK (2008) Abundance and diversity of rotifera in the Garmat Ali region ponds, Basrah-Iraq. Iraqi J Aquacul 5(1):33–40

Ghazi AH, Ali MH (2012) Rotifers community structure and abundance along Shatt Al-Arab River from Garmmat Ali to Al-Fao, Southern Iraq. Marsh Bulletin 7(2):150–161

Ghazi AH, Hassan HF, Khathem SA (2011) Effect of water level declinein Saffia restoration site on the density and diversity of Rotifer, Al-Hawaizah Marshes, Iraq. Marsh Bulletin 6(2):159–166

Glöer P, Naser MD (2007) Gyraulus huwaizahensis n. sp.–a new species from Mesopotamia, Iraq (Mollusca: Gastropoda: Planorbidae). Mollusca 25(2):147–152

Glöer P, Naser MD (2008) Bithynia hareerensis n. sp., a new Bithynia from Mesopotamia (Iraq) (Gastropoda: Bithyniidae). Mollusca 26(2):159–162

Gmais SA, Aziz NM, Al-Khafaji KK, Ama'al GY, Naser MD (2008) The genus Lymnaea (Lamarck, 1799) from Southern Mesopotamia: are the morphological and anatomical studies enough to solve its complexity? Mesopotamian J Marine Sci 23(2):349–362

Grosser C, Pešić V (2006) On the diversity of Iranian leeches (Annelida: Hirudinea). Archiv Biol Sci 58(1):21–24

Haas F (1969) Superfamilia Unionacea.– Das Tierreich, 88, Walter de Gruyter & Co, Berlin, pp. 1–663

Hammadi NS (2016) An ecological survey of Littoral Rotifera from some selected areas of Iraq: II. South of Iraq. Res J Sci & Technol 8(3):125–128

Hammadi NS, Salman SD, Abbas MF (2015) Rotifera of the Southern Iraqi Marshes, with a reference to the major zooplankton groups in the region. Marsh Bulletin 10(1):46–55

Hassan KS, Habeeb MA, Al-Mousawi NJ (2000) Occurrence of aquatic insects with algae in Basrah province. Marina Mesopotamica 15(1):137–143

Huryn AD, Wallace JB (2000) Life history and production of stream insects. Annu Rev Entomol 45:83–110

Jaweir HJ (2014) Checklist of aquatic oligochaetes species in Tigris–Euphrates river basin. Baghdad Sci J 11(3):1397–1404

Kareem DK, Al-Edani A (2015) Diagnosis and ecological distribution of aquatic (Hemiptera: Heteroptera) in Sullein marsh in Basrah, South of Iraq. Mesopotamian J Marine Sci 30(1):33–46

Lucy FE, Graczyk TK (2008) First record of the New Zealand mud snail Potamopyrgus antipodarum JE Gray, 1843 (Mollusca: Hydrobiidae) in Greece–notes on its population structure and associated microalgae. Aquat Invasion 3(3):341–344

Malmqvist B (2002) Aquatic invertebrates in riverine landscapes. Freshw Biol 47(4):679–694

Marrone F, Naser MD, Amaal GY, Sacco F, Arculeo M (2014) First record of the North American cryptic invader Ferrissia fragilis (Tryon, 1863) (Mollusca: Gastropoda: Planorbidae) in the Middle East. Zoology in the Middle East 60(1):39–45

Mohamed ARM, Hussain NA, Al-Noor SS, Mutlak FM, Al-Sudani IM, Mojer AM, Toman AJ, Abdad MA (2008) Fish assemblage of restored Al-Hawizeh marsh, Southern Iraq. Ecohydrol & Hydrobiol 8(2–4):375–384

Mohamed HH, Salman SD, Abdullah AAM (2009) Some aspects of the biology of two copepods: Apocyclops dengizicus and Mesocyclops isabellae from a Pool in Garmat-Ali, Basrah, Iraq. Turk J Fish Aquat Sci 8(2):239–247

Mohammed DS, ALI MH (2010) A morphological and molecular study on Artemia franciscana (Branchiopoda, Anostraca) from Basrah, Iraq. Crustaceana 83:941–956

Mohammed HH, Ferrari FD, Abbas MF (2014) Introduction of the copepod Limnoithona tetraspina Zhang & Li, 1976 (Cyclopoida, Cyclopettidae) into the Shatt Al-Arab, Iraq. Crustaceana 87(3):376–382

Morton KJ (1919) Odonata from Mesopotamia. Entomologist's Monthly Magazine 55:143–151

Morton KJ (1920) Odonata collected in North-Western Persia and Mesopotamia by Captain P. X. Buxton - R. A. M. C. Entomologist's Monthly Magazine 56:82–87

Morton KJ (1921) Neuroptera, Mecoptera and Odonata from Mesopotamia and Persia. Entomologist Monthly Magaz 7:213–225

Mousson A (1874) Coquilles terrestres et fluviatiles recueillies par M. le Dr Alex. Schlæfli en Orient. Journal de Conchyliologie 22:5–60

Nakano S, Murakami M (2001) Reciprocal subsidies: dynamic interdependence between terrestrial and aquatic food webs. Proc Natl Acad Sci U S A 98:166–170

Naser MD (2009) First record of the freshwater crab, Potamon mesopotamicum Brandis, Storch & Turkay, 1998 (Decapoda, Brachyura, Potamidae) from the Al-Huwaizah Marshes, Iraq. Crustaceana 82(12):1599–1602

Naser MD, Son MO, Yasser AG (2011) Assessing the risks of invasions of aquatic invertebrates in the Shatt Al-Arab River. Russian J Biol Invasion 2(2-3):120

Naser MD, Page TJ, Ng NK, Apel M, Yasser AG, Bishop JM, Ng PK, Clark PF (2012a) Invasive records of Eriocheir hepuensis Dai, 1991 (Crustacea: Brachyura: Grapsoidea: Varunidae): implications and taxonomic considerations. BioInvasions Record 1(1)

Naser MD, Khalaf TA, Yasser AG, Darweesh HS (2012b) New records of Indomysis nybini Biju & Pananpunnayil, 2010 from the South of Iraq. Crustaceana 85(3):379–383

Plaziat JC, Younis RW (2005) The modern environments of Molluscs in southern Mesopotamia, Iraq: A guide to palaeogeographical reconstructions of Quaternary fluvial, palustrine and marine deposits. Carnets de géologie

Porter RF (2016) Odonata from Iraq, with three new records. Notulae Odonatologicae 8(8):247–318

Pourang N (1996) Heavy metal concentrations in surficial sediments and benthic macroinvertebrates from Anzali wetland, Iran. Hydrobiologia 331(1–3):53–61

Qazar IA, Azeez NM, Marina BA (2015) Freshwater snails of east Hammar Marsh and Shatt Al-Arab during 2008–2009. Marsh Bulletin 10(2):149–159

Sage BL (1960a) Field notes on some birds of Eastern Iraq. Ardea 48:160–178

Sage BL (1960b) Notes on the Odonata of Iraq. Iraq nat Hist Mus Publ 18:1–11

Sage BL (1960c) Notes on the Odonata of Iraq. The Entomologist 93:118–125

Salman SD, Ali MH, Al-Adhub AH (1990) Abundance and seasonal migrations of the penaeid shrimp Metapenaeus affinis (H. Milne-Edwards) within Iraqi waters. Hydrobiologia 196(1):79–90

Salman SD, Page TJ, Naser MD, Yasser AAG (2006) The invasion of Macrobrachium nipponense (De Haan, 1849)(Caridea: Palaemonidae) into the southern Iraqi marshes. Aquat Invasion 1(3):109–115

Saoud KD (2005) Distribution and abundance of some crustacean in Shatt al- Arab River. Marina Mesopotamica 21:131–147

Schneider W (1982) Man-induced changes in the dragonfly fauna of the Jordan Valley. Adv Odonatol 1:243–249

Sharifinia M (2015) Macroinvertebrates of the Iranian running waters: a review. Acta Limnol Bras 27(4):356–369

Shayeghi M, Nejati J, Shirani-Bidabadi L, Koosha M, Badakhshan M, Bavani MM, Arzamani K, Choubdar N, Bagheri F, Saghafipour A, Veysi A (2015) Assessing the fauna of aquatic insects for possible use for malaria vector control in large river, Central Iran. Acta Med Iran 53(9):523–532

St. Quentin D (1964) Odonaten aus Anatolien und dem Irak. Entomologische Mitteilungen des Zoologischen Museums Hamburg 3(50):49–51

Van Straalen NM (1997) In: Pankhurst CE, Doube BM, Gupta VVSR (eds) Community structure of soil arthropods as a bioindicator of soil health. Biological indicators of soil health. Centre for Agriculture and Biosciences International, Wallingford

Wallace BJ, Webster JR (1996) The role of macroinvertebrates in stream ecosystem function. Annu Rev Entomol 41:115–139

Zamanpoore M, Grabowski M, Poeckl M, Schiemer F (2011) Taxonomic review of freshwater Gammarus (Crustacea: Amphipoda) from Iran. Zootaxa 3140(1):1–14

Chapter 15
The Indian Shad *Tenualosa ilisha* (Hamilton 1822) in the Southern Marshes of Iraq: A Revision and Evaluation of a Compiled Data

Laith A. Jawad

Abstract The Indian shad *Tenualosa ilisha* (Hamilton 1822) is one of the important commercial fish species along its geographical distribution from the Arabian Gulf to the East Asia at the vicinity of Indonesia. Once this species reaches the northern part of the Arabian Gulf, it ascends the large rivers such as Shatt al-Arab River in Iraq and Karun River in Iran for spawning. Juveniles of this species were found in the southern marshes of Iraq, where they were considered as nursery grounds. Although this species has been heavily investigated in all the countries that it is found, such studies in Iraq are scarce in comparison with those countries in the Indian subcontinent and Southeast Asia. The present chapter revised the literature on *T. ilisha* inhabiting the southern marshes of Iraq in order to provide an up-to-date data base for future studies on this species. The results of this review showed that the number of studies achieved so far is very few and they cover academic issues rather applied. In addition, the subjects of these studies were not diverse. Evaluations of these studies were given, and recommendations were provided at the end of the chapter.

Keywords Shatt al-Arab River · Juvenile fishes · Clupeidae · Migration · Commercial fishes

15.1 Introduction

Among the commercial species belonging to the fish family Clupeidae is *Tenualosa ilisha* (Hamilton 1822), which is also known as Hilsa shad. This species is anadromous, and its geographical distribution line covers several countries in Southeast and South Asia extending from the Arabian Gulf to the China Sea (Sahoo et al. 2018; Mandal et al. 2018; Mohindra et al. 2019). This species is categorised in ascending rivers and other freshwater systems (Chattopadhyay et al. 2019).

L. A. Jawad (✉)
School of Environmental and Animal Sciences, Unitec Institute of Technology, Mt Albert, Auckland, New Zealand

L. A. Jawad (ed.), *Southern Iraq's Marshes*, Coastal Research Library 36,
https://doi.org/10.1007/978-3-030-66238-7_15

Along its geographical distribution, *T. ilisha* is considered one of the most commercial species, and millions of Asian inhabitants use it for their daily food due to the high content of omega fatty acid and other therapeutic applications (Dutta and Hazra 2017).

The capability of *T. ilisha* in standing variable degrees of salinity makes this species a perfect ascender of the rivers. Such range can reach 22.4–3.4 ppt (Barat et al. 1996; Ali et al. 2014a, b), and during spawning dispersal, individuals of this species can stand low level of salinity down to 0.05 ppt (Barat et al. 1996). With such physiological extreme, individuals showed wide range of adaptation in the freshwater-sea water environment (Mohindra et al. 2019).

In Iraq and especially in the southern part, *T. ilisha* is considered as one of the main commercial fish species. In spring time and during ascending the individuals of this species Shatt al-Arab River, people in south of Iraq usually refrain from eating any kind of fish other than *T. ilisha*, which is locally known as "Subour" (Al-Hassan 1999).

As an important commercial species, the biology and ecology of *T. ilisha* were fully investigated especially around the Indian subcontinent, but unfortunately, the information on the different aspects of this species in Iraq have been poorly studied (Al-Hassan 1993a, b).

In order to advance the research on studies on *T. ilisha* in the southern marshes of Iraq where this species usually visits during its fluvial migration, a review to the previous works performed on this species in the marsh area is needed to form a baseline study for further studies on this important species. Therefore, the aim of the present chapter is to review the previous studies in different aspects of the biology and ecology of this species so far being published. The present review also evaluates the available data and gives suggestions and recommendations for the scientist interested in pursuing further studies on *T. ilisha* in the southern marshes of Iraq.

15.2 The Identity of the Indian Shad *T. ilisha*

In 1803, Russel described the Indian shad for the first time and gave it the name *palash*. The specimens were taken from the vicinity of Vizagapatam, India. After a thorough examination, Hamilton (1822) transferred the species to another taxa, so it came to known as *Clupanodon ilisha*. In 1917, Regan created a new genus of clupeid fish and gave it a name of *Hilsa*. This genus was deigned to comprise the *Hilsa*-like clupeoids of the Indo-Pacific, and in doing so, Regan (1917) designates the full identity of the Indian shad and separated it from the other *Hilsa* species such as *H. toli* and *H. kanagurta* (Coad 2017). Later, the species has been assigned to the genus *Tenualosa* by Whitehead (1985).

The Indian shad has the following set of morphological characters: presence of a notch at the middle of the upper jaw, the anal fin with less than 30 rays, mouth that is usually terminal, no pores on the posterior part of the scales, top of head with fine frontoparietal striae, straight gill arches, presence of 30–33 ventral scutes making a keel alongside the abdomen (Al-Nasiri and Al-Mukhtar 1988a, b; Marammazi et al. 1995), presence of 44–51 scales on the lateral line and presence of nearly 275 fine gill rakers.

15.3 Revision of Some Aspects of the Indian Shad Studied from the Southern Marshes of Iraq

In this section, the previous published literature on the Indian shad from the southern marshes of Iraq will form the base for the discussion and revision. These literatures will be grouped according to their subject, revised and evaluated accordingly. Unlike other countries where Indian shad is found, the marine waters of Iraq in particular and the southern marshes of Iraq are lacking technical data on this important commercial fish species. Therefore, the subjects that researchers dealt with about this species are very limited and concentrated around certain subjects of fish biology and ecology. Several other important issues are lacking, and evaluation and recommendations will be given for each subject.

15.3.1 Morphology

Under this heading, there are only four papers published fairly recently in the period 2007–2017. These works talk about some characters of the gills and the role of their surface area in the respiratory function in addition to examination of the red and white muscle proportions (Abdul Kareem 2007), on some osteological characters (Mohamed et al. 2011), on the morphometric and meristic characters (Abed et al. 2012) and on the percentage of the red and white muscles (Hamza 2017).

In his work, Abdul Kareem (2007) has examined the gill surface of three clupeid fish species including *T. ilisha*. The main results can be obtained from the work of Abdul Kareem (2007) that the Indian shad showed larger gill surface area than the other two species studied. As to the red-white muscles proportions, there is a trend of increase in the amount of red muscle towards the caudal region. Also, the results of this work showed that the percentage of the small diameter muscle fibres is more towards the caudal region. The main body of this work is located in the results, but these results were not fully discussed. Although the author attempted to relate his results to the type and mode of swimming of the three clupeid fish species he examined, he did not give the readers the ability to comprehend the goal of the work.

Mohamed et al. (2011) studied some osteological features of the head and vertebral column of the Indian shad and another clupeid species *Nematalosa nasus*. The results of their work reach to not unexpected finding as the characters chosen for their study showed to be different between the two species. This work lack the goal as the authors have chosen two distinct clupeid species that their characters are completely different. For such a study, they should choose species that belong to the same genus so their similarities and differences can be revealed and compared. In addition, the selection of the bones was not comprehensive as the osteological characters were few. The results of this work should be connected to phylogenetic conclusion; otherwise they have no meaning as the two species are already distinct.

Abed et al. (2012) investigated the morphometric and meristic characters of the Indian shad taken from three localities along Shatt al-Arab River. The results of this work have the potential for a rich discussion, but the authors failed to do so. The discussion section is a repetition of the result section. The authors should relate the differences they obtain in some body proportions of fish specimens taken from three localities to environmental, biological or ecological factors. Also, they should give an explanation for the similarity in the morphometric and meristic characters of specimens sampled from the three localities they examined. Instead, the results were left to the readers to give a conclusion.

The work of Hamza (2017) looks dubious. They stated that the fish specimens were collected from City of Karbala about 200 Km southwest of the capital Baghdad. No previous reports were on record about the presence of the Indian shad in the vicinity of this city on the freshwater system of Iraq. Therefore, the results of this work cannot be evaluated and compared with that of Abdul Kareem (2007) as the fish species might come from anywhere and not particularly from the southern marshes of Iraq.

There are general trends in all morphometric works mentioned above; these are short results and lack of real discussion, where the authors need to explain their findings.

15.3.2 Occurrence

The percentage of occurrence of *T. ilisha* has been studied in the southern marshes of Iraq and for several years after the reflooding process (Abdullah 2017; Mohamed and Hussain 2012; Mohamed et al. 2008a, b, 2009, 2011, 2012a, b, 2013, 2014, 2015). The percentage of the occurrence of the total catch of the specimens of the Indian shad showed wide variation. The minimum value (0.53%) was observed in 2013 (Mohamed et al. 2013) and the maximum value (27.4%) in 2015 (Mohamed et al. 2015). The data obtained for marsh areas directly after reflooding were 10.4% and 10.1% in year 2008 and 2009, respectively (Mohamed et al. 2008a, b, 2009). Within this range, the percentage of occurrence has dropped considerably to 0.53% in year 2013 (Mohamed et al. 2013), and in 2017 the value was low (2.88%) (Abdullah 2017). No reason was given for such discrepancies.

This group of research looks fine, but still the readers get lost once they reach to the discussion section because it is a repetition of the results. Several parts of the results were not explained, and those clarified were given unrealistic reasons.

15.3.3 Fisheries and Stock Assessment

Studies on the fisheries and stock assessment of the Indian shad are very few in spite of the commercial importance of this species and their unique presence in the marsh areas during the spawning season (Mohamed et al. 2008a, b, 2016; Sultan and Nasir 2018). The status of the artisanal fisheries in the lower reaches of the Euphrates-Tigris Rivers was studied by Mohamed et al. (2008a, b) in the year 2005. The relative abundance of *T. ilisha* in the marsh areas is shown to be 0.1%. This value is similar to that of the lower reaches of Euphrates and smaller than that of the lower reaches of Tigris River (1.2%). Although the study of Mohamed et al. (2008a, b) was designed to study the artisanal fisheries status in the lower reaches of Mesopotamia, the authors concentrate in their work on certain freshwater fish species and left the remaining species without giving any comments on their results of catch. They relate the low value of the relative abundance of *T. ilisha* present in the marsh area to the factors of pollution, but if this is the case, then individuals of the Indian shad should not ascend Euphrates River as the marsh area is the corridor for the lower reaches of Euphrates River. On the contrary, they reported high relative abundance value (2.1%) for Euphrates River.

In the period 2009–2020, Mohamed et al. (2016) investigated the stock assessment of *T. ilisha* and other fish species in Al-Hammar marsh area. Their results showed that the percentage of abundance for *T. ilisha* was 6.1%. On the other hand, the growth and mortality parameters $L\infty$, K, Z, M and F were 57.1, 0.33, 1.55, 0.68 and 0.87 cm, while the exploitation rate was 0.56. They suggested that all the species investigated including *T. ilisha* were overexploited. They recommend an enforcement for fishing regulation measures on the fish stock in the marsh area. In this work, there are plenty of results, but the authors failed to explain them. They kept comparing these results between the species they studied and those have been investigated before without giving reasons for any similarities or differences. Moreover, they did not use their results to come up with a strategic solution to conserve the fishery of this important commercial species.

Sultan and Nasir (2018) studied the availability of the Indian shad in the fish landing at the edge of the marsh area at the city of Thigar for the period 2005–2016. Their results reveal that the percentage of landing of *T. ilisha* was 36% of the total catch from 2005 to 2016. With such high percentage, the authors have suggested for building a small factory to process the large amount of fish specimens obtained. This suggestion is quite good, but first the regulation for fishing should be put in order to maintain the catch that will form the base for any fish industries in the marsh area.

15.3.4 Food of **T. ilisha**

In 2008, two papers were published on the food of the juveniles of *T. ilisha* from the marsh area (Mohamed et al. 2008a, b, Mutlak et al. 2008). Mohamed et al. (2008a, b) investigated the food of the juveniles of *T. ilisha* inhabiting Al-Hammar marsh in 2006, and their results showed that algae were the most imperative food items, with a percentage > 33%, diatoms ranked second >28%, and zooplankton (Copepoda, Cladocera, Rotifera and Ostracoda) came third. The general percentage configuration of dietary constituents of *T. ilisha* was filamentous algae (40%), diatoms (34%), organic materials (8%), Copepoda (7%), Cladocera (5%), Rotifera (4%) and Ostracoda (2%). The results also showed that East Hammar marsh serves as setting ground. The main result this study obtained is that no adult fish of *T. ilisha* was caught at the eastern part of Al-Hammar marsh, but the north most extent was the northern banks of Shatt al-Arab River. Mohamed et al. (2008a, b) suggested that the adults *T. ilisha* do not spawn in Al-Hammar marshes, but the juveniles are present there. They gave further evidence for their suggestion, and that is represented by catching eggs and larvae of *T. ilisha* from the northern banks of Shatt al-Arab River. The other result this paper showed is that the juveniles of *T. ilisha* feed on algae and diatoms in the autumn season and then change to zooplankton later in the fall season. This paper showed an interesting results, but again the discussion of these results did not continue further to suggest a general theory or did not give any feedback that help the future researches. The work by Mutlak et al. (2008) and Younis et al. (2013) showed the same results of Mohamed et al. (2008a, b) in regard to food habits of the juveniles of *T. ilisha*, but on different part of the marsh area.

15.3.5 Reproduction

Information on the reproduction of *T. ilisha* in the marsh areas of Iraq are very infrequent. Although the reported data showed that these information appeared in 1989 (Jabir and Faris 1989), no work was published since then until 2016 (Al-Mukhtar et al. 2016). No other works were published on the fecundity of *T. ilisha* in Iraq.

The fecundity of the Indian shad in Shatt al-Arab River was investigated by Jabir and Faris (1989). Their study was based on 270 specimens, with 54 mature individuals that used to estimate the fecundity of this species. Their results showed that the fecundity range between 444,960 and 1,616,560 eggs per female, with average of 879.989 eggs. There were some variations in fecundity of individuals of different sizes. The authors end up their work without giving any suggestions, and the discussion section was mainly a comparison with results on Indian shad from the Indian subcontinent.

Almukhtar et al. (2016) studied the reproduction of *T. ilisha* from both Shatt al-Arab River and Al-Hammar marsh during the period 2013–2014. In their work, they used 1456 individuals, with overall sex ratio of 1:1.7 observed in this study. The main finding of this work is that the Indian shad spawn in Shatt al-Arab River and not in the marsh area. The absolute fecundity ranged between 223,750 and 1,477,780 eggs for fishes with total length of 200–450 mm and mean weight range of 246.3–891.7 g. Comparing this result with that of Jabir and Faris (1989), it is clearly shown that there is a drop in the fecundity of this species occurred within 26 years, but the authors stated differently. They suggested that their result of fecundity was higher than that of Jabir and Faris (1989). This work looks fine in giving satisfactory reasons for the different aspects of the reproduction of the Indian shad, but still the authors failed to explain their results of fecundity and its comparison with the other published works.

15.3.6 Other Biological Aspects of Tenualosa ilisha *in the Southern Marshes of Iraq*

Several biological issues of the Indian shad in the southern marshes of Iraq were investigated (Al-Mukhtar 2013; Al-Mukhtar et al. 2015, 2016; Mohamed et al. 2017a). These include the validation of the fish scales for estimation of age, length-weight relationship, sex ratio and age of male and female individuals.

Al-Mukhtar (2013) reviewed the previous studies that used scales to estimate the age of the fish and concluded that the scales need patience and precision during the measurements of the distances between the annuli. In 2015, Almukhtar studied some biological aspects of *T. ilisha* collected from Shatt al-Arab River. Among the important findings that this work provided, the first appearance of the Indian shad in Shatt al-Arab River was during February of the year. This information is useful as there were no exact date for the entrance of the Indian shad into Shatt al-Arab River provided. This work also showed that large adult fish continue entering Shatt al-Arab River during June of the year, with age composition of 0-V^{+} year. The results of Al-Mukhtar et al. (2015) have also showed that the Indian shad population was mostly medium-sized fishes, with RSD-T value of 3.8.

Al-Mukhtar et al. (2016) investigated the possibility of using fish scales to estimate the age of the fish. They successfully estimated the age of fish individuals collected from Shatt al-Arab River and the southern marshes and also evaluated the growth of these fishes. They recommended at the end of their work, the scales are useful tools for estimation of age and calculating the growth rate in the Indian shad.

Mohamed et al. (2017a) examined different populations of *T. ilisha*. They sampled fish individuals inhabiting the marshes areas in order to report on some of their biological features. Their results showed that the sex ratio was 1:1 in favour of females.

Overall, the works that have been achieved on the different biological aspects of the Indian shad in both Shatt al-Arab River and the southern marshes are useful as a baseline for further studies in the future in spite of their scarcity.

15.3.7 *Ecological Studies Related to the Indian Shad* T. ilisha *Inhabiting the Southern Marshes of Iraq*

The ecological aspects related to the Indian shad living in the southern marshes of Iraq have been poorly investigated. The work of Mohamed et al. (2017b) represents the only published work on this issue. In this work, the authors evaluated the water quality index of Garmat Ali River, one of the existing branches of the Euphrates River from the southern marshes. Their results showed that the water quality of this river lies on the margin of the international standard (WQI). The main finding of this work is the clear effect of the state of the water of the river on the presence of fish assemblage. In this study, the following fish groups were observed and examined: 8 native, 8 exotic and 18 marine fish species. This result differs significantly from the previous assemblage studies done in the area such as Younis et al. (2010) and Mohamed et al. (2013). According to the study of Younis et al. (2010), there were 13 native, 10 marine and 5 exotic fish species, while Mohamed et al. (2013) showed 10 native, 10 marine and 6 exotic species. It is clear from the comparison of the results of Mohamed et al. (2017b) with those of Younis et al. (2010) and Mohamed et al. (2013) that there was a drop down in the number of both native and marine species from 13 and 10 according to Younis et al. (2010) and Mohamed et al. (2013), respectively, to 8 species for both categories according to Mohamed et al. (2017b). In the same time, the number of the exotic species upsurged to become 18 according to Mohamed et al. (2017b), while it was 5 and 6 species according to Younis et al. (2010) and Mohamed et al. (2013). The drop in the native and marine fish species categories was related to the deteriorating status of the water quality of the river to below the international standard and to the increase in water salinity.

The work of Mohamed et al. (2017b) represents step forward towards achieving researches that can propose ecological solutions for an existing aquatic problems. The authors of this work have dealt with the problem in a scientific way and provided the readers with the scope of the problem, showing the method of solving it and giving the solution afterwards.

15.3.8 *Parasites and Diseases of Indian Shad in the Southern Marshes of Iraq*

This category of studies are no not well investigated in the southern marshes of Iraq (Al-Shatti et al. 2009; Jawad et al. 2014; Bannai and Muhammad 2016; Mhaisen and Al-Niaeem 2019). There are several diseases that may affect the Indian shad in the southern marshes of Iraq. These are parasite-, bacterial- and fungal-related diseases in addition to diseases that are related to non-organism such as certain morphological abnormalities. The published works so far on the Indian shad are related to parasites, bacteria, fungi and morphological aberration.

Regarding the studies related to the parasites, Bannai and Muhammad 2016 studied some parasites that are considered as a biological indicators for immigration. The authors examined 1456 specimens of *T. ilisha*. The results showed the presence of three parasites, *Nothobomolochus* sp. (copepod), *Faustula* sp. and *Ectenurus papillatus* (Digenea). This work is quite confusing as the reader easily gets lost through reading the results and complete his/her lost during reading the discussion, which is a repeat of the results and bombarded with references that seem not in their right place. Half of the result of this work deals with the size and number of the fish species which is considered a secondary target.

Mhaisen et al. (2013) studied the trematodes in some freshwater and marine fish species from Iraq including *T. ilisha*. This study includes 54 trematode taxa distributed in 44 fish species. There were only two trematode species found in the Indian shad, *Ascocotyle coleostoma* and *Faustula rahemii*.

The nematodes were reviewed in 8 elasmobranchs and 38 teleosts fish species by Ali et al. (2014a, b). Only *Contracaecum* larva was reported from *T. ilisha*. This species was reported by Al-Janae'e (2010) originally.

In their checklist, Mhaisen and Abdul-Ameer (2013) listed monogenean species taken for a number of fish species in the waters of Basrah Province. There was only *Gyrodactylus* sp. reported to be present in *T. ilisha*.

For the crustacean's external parasites, several works have been published on this issue. Adday (2013) reported the presence of *Nothobomolochus* sp. and *Ergasilus* sp. on the gills of *T. ilisha*. AL-Niaeem et al. (2015) studied the parasitic copepods of some fish species from Basrah Province, which include *T. ilisha*, but the authors do not state what parasite species are found in each of the fish species examined. This is a drawback to the scientific value of the published work. Khamees et al. (2015) revised the literature that dealt with the parasitic crustaceans on fishes inhabiting waters of Basrah Province and reported the following crustacean's species parasitised on *T. ilisha*: *Acanthocolax* sp. which was reported as *Bomolochus* sp., *Anchistrotos tangi*, *Clavella adunca*, *Ergasilus ogawai*, *E. rostralis*, *E. sieboldi*, *Gnathia* sp., *Nerocila phaiopleura* and *Nothobomolochus ilhoikimi*.

Recently, Mhaisen et al. (2018) published a comprehensive review of the parasite fauna of marine fishes of Iraq. This review comprises 253 parasite species taken from 86 fish species (13 Elasmobranchs and 73 Teleosts). Seventeen different parasite species were reported from *T. ilisha*. The parasites are as follows: Platyhelminthes includes *Ascocotyle coleostoma*, *Ectenurus papillatus*, *Leptomazocraes indica*, *Acanthocolax* sp., *Anchistrotos tangi*, *Ergasilus rostralis*, *Ergasilus sieboldi sieboldi*, *Gnathia* sp., *Nerocila phaiopleura* and *Nothobomolochus ilhoikimi*.

The work of Mhaisen et al. (2018) represents a major contribution to the marine parasitology science in Iraq and in the Arabian Gulf area. Such work will be considered as a baseline data base for the future investigations on fish parasites in marine fish species in the Arabian Gulf area.

Bacteriological studies related to the Indian shad were also scarce, and there is only one published work on the southern marshes of Iraq. Al-Shatti et al. (2009) examined specimens of *T. ilisha* and isolated the bacterium *Staphylococcus aureus* and *Escherichia coli* present on the surface of the gills. Except for the record of the two bacteria, this work adds nothing to the information about the bacterial diseases of *T. ilisha*.

Mhaisen and Al-Niaeem (2019) reviewed the literature on the fungi and fungal-like organisms infecting fish species including *T. ilisha* at Basrah Province, Iraq. The results in general showed the presence of three species of the phylum Ascomycota, one species each of the phylum Zygomycota and the phylum Microsporidia, two species of the phylum Choanozoa and five species of the phylum Oomycota. The result showed that *T. ilisha* is infected with only one fungal species, *Ichthyophonus hoferi*.

The work of Mhaisen and Al-Niaeem (2019) is comprehensive and considered a baseline for any future study on the fungi or fungal-related organism that infect freshwater fish species of Iraq or even the neighbouring countries. Further studies are needed to know why *T. ilisha* is infected with only one species, while other freshwater species like *Oreochromis aureus* can get infected with up to six species.

The report on the incidences of morphological aberrations in fishes is a new subject to the fish fauna of Iraq. It has been started when Al-Hassan (1982) published his first observations on cases of skeletal anomalies in some freshwater fish species. No reports on abnormal fishes appeared on *T. ilisha* until Jawad et al. (2014) reported on the incidence of severe lordosis involving two flexions shown by specimen of *Tenualosa ilisha* collected from the marsh area. The absence of reports on the abnormal fishes from the marsh area is related to the lack of expertise in this field. Thousands of fish specimens are usually examined every year by researchers, but they have no knowledge and ability to recognise the deformity and reporting it. Therefore, as a recommendation a short course should be given for all researchers that are dealing with fresh fish specimens to educate them about the abnormalities that may appear on fish specimens. The attendees of this course should recognise the incidence of deformity and know how to deal with it.

15.4 Remarks and Recommendations

The above revision showed the following: (1) a limitation in the choice of subjects that dealt with the Indian shad in the southern marshes of Iraq, where a concentration of researches on certain subjects was observed; (2) subjects chosen for the different study reviews which have an academic trend rather than applied; and (3) studies that were published in either local or grey journals that are unavailable or difficult to obtain by the international audience.

As a recommendations, researchers should start to introduce applied studies in their research plans on the Indian shad inhabiting the southern marshes of Iraq. *Tenualosa ilisha* is an important commercial fish species, and management of its stock and fisheries while its individuals are in Iraq is a vital need for the conservation of the fishery of this species. The policy holders in Iraq should think of building aquaculture facilities to cultivate this species in order to assist its natural stocks in both Shatt al-Arab River and the southern marshes of Iraq from depletion on one hand and to create a national industry that support the economy of Iraq.

References

Abdul Kareem HM (2007) Relationship between respiratory area of gills and skeletal muscles in three species of fishes (Clupeiformes). J Meaisan Res 3:1–23

Abdullah AHJ (2017) Diversity, abundance and community structure of fishes in the lower part of the Euphrates River Southern Iraq. Mesopotamian J Marine Sci 32(2):64–77

Abed JM, Ismaeel AK, Kahtan AA (2012) Morphometric and Meristic of Young Hilsa Shad Tenualosa ilisha (Hamlton-Buchanan, 1822) in Shatt Al-Arab River. Basrah J Agric Sci 25(Special Issue):74–80

Adday TK (2013) Parasitic crustaceans of some marine fishes of Basrah province, Iraq (Doctoral dissertation, Ph. D. Thesis, Coll. Agric., Univ. Basrah: 302 pp)

Al-Hassan LAJ (1982) Vertebral abnormalities in fishes from Iraq and the United Arab Emirates, Arabian Gulf. Iraqi J Marine Sci 1:13–23

Al-Hassan LAJ (1993a) Additional synopsis of biological data on Tenualosa ilisha. Marina Mesopotamica, Supplement 2:1–23

Al-Hassan LA (1993b) Additional synopsis of biological data on Tenualosa ilisha. Marina Mesopotamica 2:1–23

Al-Hassan LAJ (1999) Shad of the Shatt Al-Arab River in Iraq. Shad J Seattle 4(2):1–4

Ali AD, Naser MN, Bhaumik U, Hazra S, Bhattacharya SB (2014a) Migration, Spawning Patterns and Conservation of Hilsa Shad (Tenualosa ilisha) in Bangladesh and India. Published by Academic Foundation India, New Delhi and International Union for Conservation of Nature and Natural Resources (IUCN), 95

Ali AH, Mhaisen FT, Khamees NR (2014b) Checklists of nematodes of freshwater and marine fishes of Basrah Province, Iraq. Mesopot J Mar Sci 29(2):71–96

Al-Janae'e AMS (2010) Parasites of some Iraqi fishes in two localities varied in their trophic levels in inland water of Basrah. M. Sc. Thesis, College of Agriculture, University of Basrah: 228 pp. (In Arabic)

Al-Mukhtar MA (2013) Review of the validity of scales to estimate age, growth, life history and maturity schedule of Hilsa shad Tenualosa ilisha (Hamilton, 1822). Iraqi J Aquaculture 10:155–169

Al-Mukhtar MA, Mutlak FM, Mahdi AA (2015) Population biology of the Hilsa Shad Tenualosa ilisha (Hamilton, 1822) during migration to Shatt Al-arab River and southern Al-Hammar marsh, Basra-Iraq. Iraqi J Aquaculture 12:1–16

Al-Mukhtar MA, Jasim W, Mutlak F (2016) Reproductive biology of Hilsa Shad Tenualosa ilisha (Teleostei: Clupeidae) during spawning migration in the Shatt Al Arab River and Southern Al Hammar Marsh, Basra, Iraq. J Fish Aquat Sci 11(1):43–55

Al-Nasiri SK, Al-Mukhtar MA (1988a) On the biology of sbour Hilsa ilisha (Hamilton) (Pisces, Clupeidae) from Ashar Canal, Basrah. Iraqi J Agricul Sci "Zanco" 6(1):97–106

Al-Nasiri SK, Al-Mukhtar MA (1988b) On the biology of sbour Hilsa ilisha (Hamilton) (Pisces, Clupeidae) from Ashar Canal, Basrah. Pak J Zool 20(4):321–328

AL-Niaeem, K. S., Al-Saboonchi, A. A. and Ahmed, R.A. (2015) Effect of water quality on fishes infected with copepods from three stations in Basrah province, Iraq. J Int Academ Res Multidiscip 3(4):428–436

Al-Shatti SMH, Al-Sehlani STJ, Al-Khuzai AKN (2009) Susceptibility of Staphylococcus aureus and Escherichia coli isolated from fish to some antibiotics. Scientific J Karbala Univ 4:221–227

Bannai MA, Muhammad ET (2016) Parasites as biological tags of Tenualosa ilisha (Hamilton-Buchanan, 1822) as an indicator of fish immigration in Iraqi waters. Int J Fish Aquat Stud 4(3):665–669

Barat A, Punia P, Ponniah AG (1996) Karyotype and localization of NOR in threatened species, Tenualosa ilisha (Ham.) (Clupeidae: Pisces). Chromosome Sci 82(II):2828–2832

Chattopadhyay D, Chakraborty A, Ray PK, Mandal R, Banik SK, Suresh VR, Ghosh K (2019) Larval rearing of Hilsa shad, Tenualosa ilisha (Hamilton 1822). Aquacult Res 50(3):778–785

Coad BW (2017) Review of the herrings of Iran (family Clupeidae). Int J Aquatic Biol 5(3):128–192

Dutta S, Hazra S (2017) From biology to management: a critical review of Hilsa Shad (Tenualosa ilisha). Indian J Geo-Mar Sci 46:975–1033

Hamza NM (2017) Study of percentage of red and white muscles in two species of bony fishes Tenualosa ilisha (Hamilton, 1822) and Cyprinus carpio (Linnaeus, 1758). Ibn AL-Haitham J Pure & Appl Sci 30(1):288–296

Jabir MK, Faris AA (1989) Fecundity of sbour, Tenualosa ilisha (Hamilton-Buchanan, 1822) in the Shatt al-Arab River, Basrah. Marina Mesopotamica 4(2):281–296

Jawad LA, Al-Faisal AJ, Al-Mutlak FM (2014) Incidence of lordosis in the cyprinid fish, Carasobarbus luteus and the Shad, Tenualosa ilisha collected from Barash waters, Iraq. Int J Marine Sci 4

Khamees NR, Mhaisen FT, Ali AH (2015) Checklists of crustaceans of freshwater and marine fishes of Basrah Province, Iraq. Mesopot J Mar Sci 30(1):1–32

Mandal S, Lal KK, Singh RK, Sah RS, Jena JK, Singh A, Mohindra V (2018) Comparative length-weight relationship and condition factor of Hilsa Shad, Tenualosa ilisha (Hamilton, 1822) from freshwater, estuarine and marine environments in India. Indian J Fish 65(2):33–41

Marammazi J Gh (1995) A study of identification and some biological aspects of sobour Tenualosa ilisha (Ham. Buch 1822) in Bahmanshir River, Khuzestan, Iran. M.Sc. Thesis, Tarbiat Modarres University. 78 p. (In Farsi)

Mhaisen FT, Abdul-Ameer KN (2013) Checklists of Gyrodactylus species (Monogenea) from fishes of Iraq. Basrah J Agric Sci 1:8–25

Mhaisen FT, Al-Niaeem KS (2019) Checklists of Fungi and fungal-like organisms infecting fishes of Basrah Province, Iraq. Biol & Appl Environ Res 3(1):42–56

Mhaisen FT, Khamees NR, Ali AH (2013) Checklists of trematodes of freshwater and marine fishes of Basrah Province, Iraq. Basrah J Agric Sci 1:50–77

Mhaisen FT, Ali AH, Khamees NR (2018) Marine fish parasitology of Iraq: a review and checklists. Biol Appl Environ Res 2(2):231–297

Mohamed ARM, Hussain NA (2012) Evaluation of fish assemblage environment in east Hammar using integrated biological index. Basrah J Sci 30. (2B Environmental Conference):87–105

Mohamed ARM, Al-Noor SS, Faris RA (2008a) The status of artisanal fisheries in the lower reaches of Mesopotamian rivers, North Basrah, Iraq. Proc 5th Int Con Biol Sci(Zool) 5:126–132

Mohamed ARM, Hussain NA, Al-Noor SS, Mutlak FM (2008b) Occurrence, abundance, growth and food habits of sbour Tenualosa ilisha juveniles in three restoresd marshes Southern Iraq. Basrah J Agric Sci 21:528–538

Mohamed ARM, Hussain NA, Al-Noor SS, Coad BW, Mutlak FM (2009) Status of diadromous fish species in the restored East Hammar marsh in southern Iraq. Am Fisher Soc Sympos 69:577–588

Mohamed ARM, Al-Noor SS, Habbeb FS (2011) Comparative study of some skeletal features of two Clupeid species, Hilsa shad Tenualosa ilisha and Gizzard shad Nematalosa nasus in Iraq waters. Basrah J Agric Sci 24(3):29–40

Mohamed ARM, Ahmed SM, Al-Okailee MT (2012a) Variations in occurrence, abundance and diet of Hilsa, Tenualosa ilisha larvae in the north of Shatt Al-Arab River, Basrah. Iraq Basrah J Agri Sci 25

Mohamed ARM, Resen AK, Taher MM (2012b) Longitudinal patterns of fish community structure in the Shatt Al-Arab River, Iraq. Basrah J Sci 30(2B Environmental Conference):65–86

Mohamed ARM, Hussein SA, Lazem LF (2013) Fish assemblage of Garmat Ali River, north of Basrah. Iraq Agric Sci 26:150–166

Mohamed ARM, Hussein SA, Mutlak FM (2014) Composition of fish assemblage in the East Hammar marsh, southern Iraq. Baghdad J Sci 11:1373–1381

Mohamed ARM, Hussein SA, Lazem LF (2015) Spatiotemporal variability of fish assemblage in the Shatt Al-Arab River, Iraq. J Coastal Life Med 3(1):27–34

Mohamed ARM, Hussein SA, Mutlak FM (2016) Stock assessment of four fish species in the East Hammar Marsh, Iraq. Asian J Appl Sci 4(3) June 2016

Mohamed ARM, Hussein SA, Mutlak FM (2017a) Some biological aspects of four fish species in East Hammar Marsh. Iraq J of Scientific & Eng Res 4(8):278–287

Mohamed ARM, Younis KH, Hameed EK (2017b) The ecological condition of the Garmat Ali River, Iraq. Global J Biol Agricul & Health Sci 6:2319–5584

Mohindra V, Dangi T, Tripathi RK, Kumar R, Singh RK, Jena JK, Mohapatra T (2019) Draft genome assembly of Tenualosa ilisha, Hilsa shad, provides resource for osmoregulation studies. Sci Rep 9(1):1–14

Mutlak FM, AL-Okailee MT, Younis KH, Yasin AT (2008) Al-Hammar marsh as nursery and feeding ground for some marine fish species. Marina Mesopotamica 23(1):201–214

Regan CT (1917) A revision of the Clupeoid fishes of the genera Pomobbus, Brevoortia, and Dorosoma, and their allies. Ann Mag Nat Hist 19(18):304

Sahoo AK, Wahab MA, Phillips M, Rahman A, Padiyar A, Puvanendran V, Bangera R, Belton B, De DK, Meena DK, Behera BK (2018) Breeding and culture status of Hilsa (Tenualosa ilisha, Ham. 1822) in South Asia: a review. Rev Aquac 10(1):96–110

Sultan QM, Nasir NA (2018) A survey of the availability of fish species in the landing centres of DhiQar Province, Iraq, during the period from 2005 to 2016. Mesopotamian J Marine Sci 33(2):83–90

Whitehead PJP (1985) FAO species catalog. Clupeoid fishes of the world (suborder Clupeoidei). Part 1 – Chirocentridae, Clupeidae and Pristigasteridae. FAO (Food and Agriculture Organization of the United Nations) Fisheries Synopsis No. 125, v. 7 (pt 1): i-x + 1-303

Younis KH, Hussain NA, Mohamed ARM (2010) Ecological assessment of fish assemblage in the Shatt Al-Arab River-Karmat Ali, Basrah using integrated biological index (IBI). J Univ Karbala (special issued), pp. 22–31

Younis KH, Al-Shamary AJ, Al-Zawar JK (2013) Trophic pyramids and feeding habits of fish assemblage in East Hammar marsh. Proceedings of the 1st Conference of College of Science, University of Karbala, Iraq, pp 92–97

Chapter 16
The Presence of *Gambusia* Fish in the Southern Marshes of Iraq: Bad or Good

Laith A. Jawad

Abstract The mosquitofish is a very important species in human health aspects. This species is larvivorous and found very efficient in controlling the population of the larvae of the mosquitos that transmit malaria parasites. Therefore, it has been transferred to many countries around the world in an aim to control malaria with the assistance of the World Health Organization. The genus *Gambusia* contains 44 species, but only *G. affinis* and *G. holbrooki* were known for controlling the larvae of the mosquitos. Besides this positive side of this species, *Gambusia* fishes, several negative factors were observed after their introduction to the new environment. This chapter deals with the positive and negative aspects of the mosquitofish, and at the end of the chapter, a set of recommendations are given in a hope that the policy makers in Iraq would consider when they review the status of the freshwater ecosystem in Iraq.

Keywords Mosquitofish · South of Iraq · Larvae · Aquatic insects · *Gambusia affinis* · *Holbrooki*

16.1 Introduction

The genus *Gambusia* (Poey 1854) (family: Poeciliidae) contains 44 valid species (Fricke et al. 2020) distributed almost in every freshwater system on earth. These are as follows: *Gambusia affinis* (Baird and Girard 1853), *Gambusia alvarezi* (Hubbs and Springer 1957), *Gambusia amistadensis* (Peden 1973), *Gambusia atrora* (Rosen and Bailey 1963), *Gambusia aurata* (Miller and Minckley 1970), *Gambusia baracoana* (Rivas 1944), *Gambusia beebei* (Myers 1935), *Gambusia bucheri* (Rivas 1944), *Gambusia puncticulata* (Poey 1854), *Gambusia clarkhubbsi* (Garrett and Edwards 2003), *Gambusia dominicensis* (Regan 1913), *Gambusia eurystoma*

L. A. Jawad (✉)
School of Environmental and Animal Sciences, Unitec Institute of Technology, Mt Albert, Auckland, New Zealand

L. A. Jawad (ed.), *Southern Iraq's Marshes*, Coastal Research Library 36,
https://doi.org/10.1007/978-3-030-66238-7_16

(Miller 1975), *Gambusia gaigei* (Hubbs 1929), *Gambusia geiseri* (Hubbs and Hubbs 1957), *Gambusia georgei* (Hubbs and Peden 1969), *Gambusia heterochir* (Hubbs 1957), *Gambusia hispaniolae* (Fink 1971), *Gambusia holbrooki* (Girard 1859), *Gambusia hurtadoi* (Hubbs and Springer 1957), *Gambusia krumholzi* (Minckley 1963), *Gambusia lemaitrei* (Fowler 1950), *Gambusia longispinis* (Minckley 1962), *Gambusia luma* (Rosen and Bailey 1963), *Gambusia manni* (Hubbs 1927), *Gambusia marshi* (Minckley and Craddock 1962), *Gambusia melapleura* (Gosse 1851), *Gambusia monticola* (Rivas 1971), *Gambusia nicaraguensis* (Günther 1866), *Gambusia nobilis* (Baird and Girard 1853), *Gambusia panuco* (Hubbs 1926), *Gambusia pseudopunctata* (Rivas 1969), *Gambusia punctata* (Poey 1854), *Gambusia puncticulata* (Poey 1854), *Gambusia quadruncus* (Langerhans 2012), *Gambusia regani* (Hubbs 1926), *Gambusia rhizophorae* (Rivas 1969), *Gambusia senilis* (Girard 1859), *Gambusia sexradiata* (Hubbs 1936), *Gambusia speciosa* (Girard 1859), *Gambusia vittata* (Hubbs 1926), *Gambusia wrayi* (Regan 1913), *Gambusia, xanthosoma* (Greenfield 1983), *Gambusia yucatana* (Regan 1914) and *Gambusia zarskei* (Meyer, Schories and Schartl 2010). It is not yet clear whether all these species are useful in combating malaria parasite transferring factor, the mosquito by eating their larvae. But what is sure is the two species *G. affinis* and *G. holbrooki* were tested on an international scale about their validity to control the larvae of the mosquitos that transferring malaria parasite as they feed on them. Therefore, *G. affinis* and *G. holbrooki* were comprehensively investigated, and the global ichthyological data bases are full of literature on these two species.

After the discovery of the ability of *G. affinis* and *G. holbrooki* to be used as biological control, they were introduced in over 60 countries, and they remain the best larvivorous fishes so far (Gerberich and Laird 1985; Walton 2007). Besides the importance of *Gambusia* fish in controlling mosquito larvae, this fish showed some adverse effects on the environment. Some investigations were published on the influence of *Gambusia* fish on the fauna and flora in the habitats they live from different localities around the world (Stockwell and Henkanaththegedara 2011). Such effects can be seen as struggle for food, consuming eggs, larvae and adults, spread of parasites and disease, hostile behaviour causing injury and death and hybridization (Cabrera et al. 2017) as the process of introducing a species to a new area will cause extermination for the native species and such extreme effect is one of the problems facing the conservation of biodiversity worldwide (Cabrera et al. 2017).

The present chapter explores the negative effects of the presence of the *Gambusia* species in the freshwater environment and to give a recommendation for the policy makers in Iraq to take action against this important conservation point.

16.1.1 Mosquitofish in Iraq

There are two uncertainties about *Gambusia* populations in Iraq: first, there is confirmation about what species has been introduced in Iraq; and second, there is no confirmed date of introduction of mosquitofish in Iraq (Jawad 2003). This mosquitofish was introduced into Iraq to combat mosquito larvae that is a vector of malaria especially in southern part of Iraq.

16.2 The Identity of *Gambusia* Fish

In general members of the family Poeciliidae were introduced into many countries through the ornamental fish trade (Stockwell and Henkanaththegedara 2011). Among these species are the two species *G. affinis* and *G. holbrooki* that attract great attention due to their ability to control mosquito larvae that transmit the parasite of malaria. The original habitat of these two species is the southern United States (Van Dine 1907; Krumholz 1948; Welcomme 1988; Pyke 2008). The factors that enable the mosquitofish to have success in their new environment are broad diet, broad physiological tolerance, fast population growth rates, high genetic differences, high levels of hostility and high spreading abilities (Arthington and Mitchell 1986; Ehrlich 1986; Arthington 1989; Courtenay and Meffe 1989; Leberg 1990). Among the species of the genus *Gambusia*, the studies have shown that both *G. affinis* and *G. holbrooki* have more ability for dispersal in the new environment than other species of the genus (Rehage and Sih 2004). The readers may refer to Swanson et al. (1996) for detailed discussion of mosquitofish biology, culture systems and use of *Gambusia* spp. as a biological control factor for mosquitoes and to Gerberich and Laird (1985), Ahmed et al. (1988), Legner (1995) and Hurst (2004) for bibliographies of larvivorous fishes used for mosquito control.

The two species *G. affinis* and *G. holbrooki* were separated by a set of characters such as distributions (Rosen and Bailey 1963), morphology (Rivas 1963), chromosome studies (Black and Howell 1979) and biochemical variations (Wooten et al. 1988).

Female and male of *Gambusia* fish can be differentiated morphologically. Males reach maturity at 35 mm in total length, while females become mature at 60 mm in total length. Males are usually smaller than females, have moderately slender body shape and have a gonopodium, which is an altered anal fin used as an intromitting organ (Walton 2007). Among the important morphological characters that separate the two species *G. affinis* and *G. holbrooki* is the shape of the male sexual organ (Rivas 1963; Rauchenberger 1989). The other characters that these two species can be differentiated from each other are the structure of hypobranchial basket.

Mosquitofish prefers shallow, calm, productive temperate and tropical lowland lakes, ponds and streams (Meffe and Snelson 1989). These fishes are frequently linked with submerged vegetation and found near the shore (Hubbs 1971; Arthington et al. 1986). In streams and rivers, mosquitofish is usually found in quiet backwaters and pools (Walton 2007).

As a member of the family Poeciliidae, mosquitofish is livebearers, which means internal fertilization of the eggs and the female has gestation period after which a free swimming young will be released. Each broad of young can hold between 1 and

300 individuals (Krumholz 1948; Gerberich and Laird 1985; Reznick and Miles 1989), with mean number of young survivors of 5.4 having total length ranging between 14 and 16 mm (Gerberich and Laird 1985). Growth of the young is fast, and young reach maturity in 3–10 weeks at total length ranging between 11 and 15 mm (Daniels and Felley 1992; Haynes and Cashner 1995).

16.3 The Impact of *Gambusia* Fish on the Environment

During the last century, research of the effect of *Gambusia* on mosquito populations have developed. Preliminary findings were subjective, containing non-substantial remarks of ponds with *Gambusia* added (Howard 1910; Moore 1922; Pyke 2008; Seale 1917). These unreliable notes have persisted until recently (Duryea et al. 1996). The subjective notes have indicated that abundance of mosquito larvae increases after *Gambusia* deletion (Duryea et al. 1996; Krumholz 1944; Murdoch and Bence 1987).

During the period of 1930 and 1960, *Gambusia* turned to be the only species to be used in control of mosquito larvae (Homski et al. 1994; Pyke 2008). Since then, *Gambusia* have been attributed with the ability to reduce significantly mosquito-borne diseases in several parts of the world (Ghosh and Dash 2007; Howard 1920), but other investigators have determined that there is no clear indication that *Gambusia* have ever decreased mosquito density adequately to control mosquito-transmitted disease (Lounibos and Frank 1994).

In addition to their benefit in combating mosquito larvae through their feeding process, the mosquitofishes have shown negative signs through their introduction in different parts of the world in order to control malaria disease. These unfavourable effects were clear on the ecosystem constituents, and they are summarized in this section.

16.3.1 Effects of Gambusia *Fish on Invertebrates*

Hurlbert et al. (1972) were among the pioneered investigators in offering an experiment showing the effects of *Gambusia* on the ecosystem. In their study, they showed that these fishes can create a decrease in zooplankton abundance and aquatic-insect populations and a succeeding upsurge in phytoplankton populations (Hurlbert et al. 1972). Therefore, mosquitofish deceptively can perform in a top-down manner (Carpenter and Kitchell 1988) by instigating a decrease in zooplankton abundance, which in turn removes the pressure on phytoplankton production in the area they are inhabiting due to the absence of the main plankton feeders, the zooplankton.

On the other hand, mosquitofish can also influence the biodiversity of the ecosystem through eliminating some crustacean populations and considerably decreasing several other zooplankton and macroinvertebrate taxa (Hurlbert and Mulla 1981). In additional controlled study, different species of aquatic insects appeared to be eliminated as a result of introducing mosquitofish (Farley and Younce 1977).

16.3.2 Gambusia *Effects on Fish*

The influence of mosquitofish on native fish has attracted much consideration. In some instances, the effects have been well recognized, but in many others, influences have been indirect and obtained from the concurrent establishment of mosquitofish and decline of native fish(es). In reality, many investigators have accredited the deterioration and extermination of native fishes to the concurrent introduction of non-native mosquitofish (Minckley and Deacon 1968; Pister 1974; Soltz and Naiman 1978).

Several studies have reported on the loss of several populations of different fish species such as White River springfish (*Crenichthys baileyi*) and Moapa dace (*Moapa coriacea*) (Deacon et al. 1964; Scoppettone 1993). Similarly, Hubbs and Brodrick (1963) stated on the loss of many populations of the rare Big Bend *Gambusia* (*G. gaigei*) subsequent the establishing of western mosquitofish (*G. affinis*).

Other studies have indicated that *G. holbrooki* had negative effects via size-selective predation on experimental populations of the least killifish (*Heterandria formosa*) (Lydeard and Belk 1993; Belk and Lydeard 1994). On the other hand, Mills et al. (2004) stated that *G. affinis* is able to decrease body growth and severely reduced endurance of young of the year in least chub (*Lotichthys phlegethontis*).

Perfunctorily, the influence of mosquitofish is mainly owed to hybridization, resource competition and predation on fish eggs, larvae, young or even adults (Courtenay & Meffe 1989). Predation of *Gambusia* on juveniles of Australian black mudfish (*Neochanna diversus*) was demonstrated by Barrier and Hicks (1994). On the other hand, Rincon et al. (2002) gave signs for heavy predation by *G. holbrooki* on juveniles of two endangered Spanish toothcarps (*Aphanius iberus* and *Valencia hispanica*). Towards adult fish, *Gambusia* has shown this behaviour to the pygmy perch (*Edelia vittata*) (Gill et al. 1999). Such study revealed that the damage of the aggressive action of *Gambusia* towards this species is clearly seen in the caudal fin damage and finally in the rate of mortality of western pygmy perch (*Edelia vittata*). Also as an effect of the presence of *Gambusia* fish, retarded growth was stated for the Australian fish, *Pseudomugil signifer*, due to the occurrence of *G. holbrooki* (Howe et al. 1997).

The other possibility of *Gambusia* fish in influencing other fish species inhabiting their environment is through transmitting various foreign parasites and diseases (Arthington and Lloyd 1989; Eldredge 2000). Archdeacon (2007) showed that the Asian tapeworm (*Bothriocephalus acheilognathi*) has apparently became available when mosquitofish (*G. affinis*) are introduced and now infects endangered Mohave tui chub (*Siphateles bicolor mohavensis*).

16.3.3 Gambusia *Influence on Amphibians*

Impacts of *Gambusia* fish on amphibians were reported by Hamer, Lane and Mahony (2002). Such influence includes preying on the eggs and larvae of the amphibians such in the case of Grub California red-legged frog (*Rana aurora draytonii*), which eventually decrease population of this species. Experimental and field works showed an adverse effect on the amphibians through the observation on the survivals of the tadpoles and larvae, postponed metamorphosis and reduced growth (Gamradt and Kats 1996; Lawler et al. 1999; Hamer et al. 2002). On the other hand, Lawler et al. (1999) presented a case of injury of the tadpoles of the endangered California red-legged frog (*Rana aurora draytonii*) caused by *G. affinis*.

Interactions between different species of *Gambusia* usually happened when two or more species are sharing the same niche as they will compete for several constituents of ecosystem they are living in (Pyke 2008), but such co-existence is very rare in nature, and if it happens then each species will occupy different habitats (Hubbs and Peden 1969; Meffe and Snelson 1989; Wooten and Lydeard 1990; Hubbs and Peden 1969).

16.4 Some Remarks on the Control of Mosquitofish

Lydeard and Belk (1993) suggested two management policies to get rid of an unfavourable introduced fish species. The first step was a partial but continuous removal of a substantial number of individuals regularly in order to reduce the size of the population. The second policy is to complete removal by poisoning. The programme of removing major part of the population of the invasive species was favoured by Ruiz-Navarro et al. (2013), where they perform their study in the Iberian Peninsula.

On the other hand, Willis and Ling (2000) suggested to use rotenone applications in order to remove *Gambusia* fish. Rotenone can be added to small water bodies when water level is in at its lowest level in summer time. This procedure depends on the idea that *Gambusia* fish can stand rotenone more than several other freshwater fish species. Therefore, many fish species will be affected first, which then can be removed and kept alive in freshwater confinement until the eradication of *Gambusia* takes place in the second round of adding more concentrated rotenone to affect the *Gambusia* fish. In this round, large number of *Gambusia* individuals will be affected and can be removed easily. Using rotenone in any freshwater body in order to clear mosquitofish would therefore need to ensure that those rates were between 100 and

200 μg. L^{-1}. Rotenone suggested to be used to perform this suggestion is in the form of powdered derris, which is cheap and readily available but slowly reduces on exposure to air and light (Willis and Ling 2000). Therefore, it has been suggested to use always fresh product. Liquid rotenone is not recommended for this choice as fish can detect the chemical and avoid entering the area where the rotenone is applied (Willis and Ling 2000). Also, it can be an additional contaminant to the environment (Bettoli and Maceina 1996). The drawbacks of this method are as follows: (1) individuals of fish that are found hiding or in an area away from the area of rotenone application are safe and can re-enter the treated area; and (2) resistance to rotenone may develop in *Gambusia* after several applications of this chemical (Fabacher and Chambers 1972).

References

Ahmed SS, Linden AL, Cech JJ Jr (1988) A rating system and annotated bibliography for the selection of appropriate, indigenous fish species for mosquito and weed control. Bull Soc Vector Ecol 13:1–59

Archdeacon TP (2007) Effects of Asian tapeworm, mosquitofish, and food ration on Mohave tui chub growth and survival. Master's thesis, University of Arizona

Arthington AH. 1989. Impacts of introduced and translocated freshwater fishes in Australia. In *Proceedings of Workshop on Introduction of Exotic Aquatic Organisms in Asia. Asian Fisheries Special Publication 3*, ed. SS De Silva, pp. 7–20. Manilla, Philipp.: Asian Fish. Soc..

Arthington AH, Lloyd LN (1989) Introduced Poeciliidae in Australia and New Zealand. In: Meffe GK, Snelson FF (eds) Evolution and ecology of livebearing fishes (Poeciliidae). Prentice-Hall, New York, pp 333–348

Arthington AH, Mitchell DS (1986) Aquatic invading species. In: Groves RH, Burdon JJ (eds) Ecology of biological invasions: an Australian perspective. Aust. Acad. Sci, Canberra, pp 34–53

Arthington AH, McKay RJ, Milton DA (1986) The ecology and management of exotic and endemic freshwater fishes in Queensland. In: Hundloe TJA (ed) Fisheries management: theory and practice in Queensland. Griffith Univ Press, Brisbane, pp 224–245

Barrier RFG, Hicks BJ (1994) Behavioural interactions between black mudfish (*Neochanna diversus* Stokell, 1949 Galaxiidae) and mosquitofish (*Gambusia affinis*). Ecol Freshw Fish 3:93–99

Belk MC, Lydeard C (1994) Effect of *Gambusia holbrooki* on a similar-sized, syntopic poeciliid, *Heterandria formosa*: competitor or predator? Copeia 1994:296–302

Bettoli PW, Maceina MJ (1996) Sampling with toxicants. In: Murphy BR, Willis DW (eds) Fisheries techniques, 2nd edn. American Fisheries Society, Bethesda, pp 303–333

Black DA, Howell WM (1979) The North American mosquitofish, Gambusia affinis: a unique case in sex chromosome evolution. Copeia 1979:5069–5173

Cabrera MB, Bogan S, Posadas P, Somoza GM, Montoya-Burgos JI, Cardoso YP (2017) Risks associated with introduction of poeciliids for control of mosquito larvae: first record of the non-native Gambusia holbrooki in Argentina. J Fish Biol 91(2):704–710

Carpenter SR, Kitchell JF (1988) The trophic cascade in lakes. Cambridge University Press, Cambridge

Courtenay WR Jr, Meffe GK (1989) Small fishes in strange places: a review of introduced poeciliids. In: Meffe GK, Snelson FF Jr (eds) Ecology and evolution of livebearing fishes (Poeciliidae). Englewood Cliffs, Prentice Hall, pp 301–331

Daniels GL, Felley JD (1992) Life history and foods of Gambusia affinis in two waterways of southwestern Louisiana. Southwest Nat 37:157–165
Deacon JE, Hubbs C, Zahuranec BJ (1964) Some effects of introduced fishes on the native fish fauna of southern Nevada. Copeia 1964:384–388
Duryea R, Donnelly J, Guthrie D et al (1996) Gambusia affinis effectiveness in New Jersey mosquito control. Proceedings of the eighty-third annual meeting of the New Jersey Mosquito Control Association, Inc., pp. 95–102
Ehrlich PR (1986) Which animal will invade? In: Mooney HA, Drake JA (eds) Ecology of biological invasions of north American and Hawaii. Springer, New York, pp 79–95
Eldredge LG (2000) Non-indigenous freshwater fishes, amphibians and crustaceans of the Pacific and Hawaiian islands. In: Sherley G (ed) Invasive species in the Pacific: a technical review and draft regional strategy. South Pacific Regional Environment Program, Samoa, pp 173–190
Fabacher DL, Chambers H (1972) Rotenone tolerance in mosquitofish. Environ Pollut 3:139–141
Farley DG, Younce LC (1977) Stocking date versus efficacy of Gambusia affinis in Fresno County rice fields. Proc Calif Mosq Vector Control Assoc 45:83–86
Fricke R, Eschmeyer WN, Van der Laan R (eds.) (2020) Eschmeyer's catalog of fishes: genera, species, references (http://researcharchiveCalacademyorg/research/ichthyology/catalog/fishcatmainasp) Electronic version accessed 26 May 2020
Gamradt SC, Kats LB (1996) Effect of introduced crayfish and mosquitofish on California newts. Conserv Biol 10:1155–1162
Gerberich JB, Laird M (1985) Larvivorous fish in the biocontrol of mosquitoes, with a selected bibliography of recent literature. In: Laird M, Miles JW (eds) Integrated mosquito control methodologies. Vol. 2. Biocontrol and other innovative components and future directions. Academic Press, London, pp 47–76
Ghosh SK, Dash AP (2007) Larvivorous fish against malaria vectors: a new outlook. Trans R Soc Trop Med Hyg 101:1063–1064
Gill HS, Hambleton SJ, Morgan DL (1999) Is the mosquitofish, *Gambusia holbrooki* (Poeciliidae), a major threat to the native freshwater fishes of South-Western Australia? In: Seret B, Sire JY (eds) Proceedings of the 5th Indo-Pacific Fish Conference, Noumea. Society of French Ichthyology, Paris, pp 393–403
Hamer AJ, Lane SJ, Mahony MJ (2002) The role of introduced mosquitofish (Gambusia holbrooki) in excluding the native green and golden bell frog (*Litoria aurea*) from original habitats in South-Eastern Australia. Oecologia (Berl) 132:445–452
Haynes JL, Cashner RC (1995) Life history and population dynamics of the western mosquitofish: a comparison of natural and introduced populations. J Fish Biol 46:1026–1041
Homski D, Goren M, Gasith A (1994) Comparative evaluation of the larvivorous fish *Gambusia affinis* and *Aphanius dispar* as mosquito control agents. Hydrobiologia 284:137–146
Howard LO (1910) Preventative and remedial work against mosquitoes. US Dept Agric Div Entomol Bull 88:1–126
Howard HH (1920) Malaria control in rural communities by anti-mosquito measures. South Med J 13:260–266
Howe E, Howe C, Lim R, Burchett M (1997) Impact of the introduced poeciliid Gambusia holbrooki (Girard, 1859) on the growth and reproduction of Pseudomugil signifer (Kner, 1865) in Australia. Mar Freshw Res 48:425–433
Hubbs C (1971) Competition and isolation mechanisms in the Gambusia affinis 3 G. heterochir hybrid swarm. Bull Texas Memorial Museum 19:1–46
Hubbs C, Brodrick HJ (1963) Current abundance of *Gambusia gaigei,* an endangered fish species. Southwest Nat 8:46–48
Hubbs C, Peden E (1969) *Gambusia georgei* sp. nov. from San Marcos, Texas. Copeia 1969:357–364
Hurlbert SH, Mulla HS (1981) Impacts of mosquitofish (*Gambusia affinis*) predation on plankton communities. Hydrobiologia 83:125–151

Hurlbert SH, Zedler J, Fairbanks D (1972) Ecosystem alteration by mosquitofish (*Gambusia affinis*) predation. Science 175:639–641

Hurst TP (2004) Evaluation of Australian native fish and larvicides for the integrated control of freshwater mosquito vectors. Ph.D. Thesis. Queensland Institute of Medical Research and School of Population Health, University of Queensland, Brisbane, Australia

Jawad L (2003) Impact of environmental change on the freshwater fish fauna of Iraq. Int J Environ Stud 60(6):581–593

Krumholz LA (1944) Northward acclimatization of the western mosquitofish, *Gambusia affinis affinis*. Copeia 1944:82–85

Krumholz LA (1948) Reproduction in the western mosquitofish, *Gambusia affinis affinis* (Baird & Girard), and its use in mosquito control. Ecol Monogr 18:1–43

Lawler SP, Dritz D, Strange T, Holyoak M (1999) Effects of introduced mosquitofish and bullfrogs on the threatened California red-legged frog. Conserv Biol 13:613–622

Leberg PL (1990) Influence of genetic variability on population growth: implications for conservation. J Fish Biol 37:193–195

Legner EF (1995) Biological control of Diptera of medical and veterinary importance. J Vector Ecol 20:59–120

Lounibos LP, Frank JH (1994) Biological control of mosquitoes. In: Rosen D, Bennett FD, Capinera JL (eds) Pest Management in the Subtropics. Intercept, Andover, pp 395–407

Lydeard C, Belk MC (1993) Management of indigenous fish species impacted by introduced mosquitofish: an experimental approach. The Southwestern Naturalist 38:370–373

Meffe GK, Snelson FF (1989) In: Meffe GK, Snelson FF (eds) An ecological overview of Poeciliid fishes. Prentice Hall, Englewood Cliffs, pp 13–31

Mills MD, Rader RB, Belk MC (2004) Complex interactions between native and invasive fish: the simultaneous effects of multiple negative interactions. Oecologia 141:713–721

Minckley WL, Deacon JE (1968) Southwestern fishes and the enigma of "endangered species". Science 159:1424–1432

Moore JP (1922) Use of fishes for control of mosquitos in northern fresh waters of the United States. Rep US Comm Fish Append 4:1–60

Murdoch WW, Bence J (1987) General predators and unstable prey populations. In: Kerfoot WC, Sih A (eds) Predation: direct and indirect impacts on aquatic communities. Univ. Press N. Engl, Hanover/London, pp 17–30

Pister EP (1974) Desert fishes and their habitats. Trans Am Fish Soc 103:531–540

Pyke GH (2008) Plague minnow or mosquito fish? A review of the biology and impacts of introduced Gambusia species. Annu Rev Ecol Evol Syst 39:171–191

Rauchenberger M (1989) Systematics and biogeography of the genus Gambusia (Cyprinodontiformes: Poecilidae). Am Mus Novit 2951:1–74

Rehage JS, Sih A (2004) Dispersal behavior, boldness, and the link to invasiveness: a comparison of four Gambusia species. Biol Invasions 6:379–391

Reznick DN, Miles DB (1989) Review of life history patterns in poeciliid fishes. Chap. 7. In: Meffe GK, Snelson FF Jr (eds) Ecology and evolution of livebearing fishes (Poeciliidae). Englewood Cliffs, Prentice Hall, pp 125–148

Rincon PA, Correas AM, Morcillo F, Risueno P, Lobon-Cervia J (2002) Interaction between the introduced eastern mosquitofish and two autochthonous Spanish toothcarps. J Fish Biol 61:1560–1585

Rivas R (1963) Subgenera and species groups in the Poeciliid fish genus Gambusia Poey. Copeia 1963:331–347

Rosen DE, Bailey RM (1963) The poeciliid fishes (Cyprinodontiformes), their structure, zoogeography and systematics. Bull Am Mus Nat Hist 126:1–176

Ruiz-Navarro A, Verdiell-Cubedo D, Torralva M, Oliva-Paterna FJ (2013) Removal control of the highly invasive fish Gambusia holbrooki and effects on its population biology: learning by doing. Wildl Res 40(1):82–89

Scoppettone GG (1993) Interactions between native and non-native fishes of the Upper Muddy River, Nevada. Trans Am Fish Soc 122:599–608

Seale A (1917) The mosquitofish *Gambusia affinis* in the Philippines. Philipp J Sci 12:177–187

Soltz DL, Naiman RJ (1978) The natural history of native fishes in the Death Valley system. Nat Hist Mus Los Angeles Co Sci Ser 30:1–76

Stockwell CA, Henkanaththegedara SM (2011) Evolutionary conservation biology. In: *Ecology and evolution of poeciliid fishes*. University of Chicago Press, Chicago, p 424

Swanson C, Cech JJ Jr, Piedrahita RH (1996) Mosquitofish: biology, culture, and use in mosquito control. Mosq. Vector Control Assoc Calif and Univ Calif Mosq Research Program, Sacramento

Van Dine DL (1907) The introduction of top-minnows (natural enemies of mosquitoes) into the Hawaiian Islands. Hawaii Agric Exper Station Press Bull 20:1–10

Walton WE (2007) Larvivorous fish including Gambusia. J Am Mosq Control Assoc 23(sp2):184–220

Welcomme RL (1988) International introductions of inland aquatic species. FAO Fisheries Technical Papers

Willis K, Ling N (2000) Sensitivities of mosquitofish and black mudfish to a pesticide: could rotenone be used to control mosquitofish in New Zealand wetlands? NZ J Zool 27(2):85–91

Wooten MC, Lydeard C (1990) Allozyme variation in a natural contact zone between *Gambusia affinis* and *G. holbrooki*. Biochem Syst Ecol 18:169–173

Wooten MC, Scribner KT, Smith MH (1988) Genetic variability and systematics of Gambusia in the southeastern United States. Copeia 1988:283–289

Chapter 17
Checklist of Freshwater Fishes of Southwestern Wetlands of Iran

Hamid Reza Esmaeili

Abstract In our time, freshwater biological diversity is under different heavy anthropogenic and natural pressures, and therefore, obtaining information and monitoring of ecosystems and freshwater species under the threats should be considered as a routine work for conservation management planning. The regularly obtained information on species diversity and distribution has a great potential to act as a tool for the prioritization of conservation and management efforts in highly international important water bodies such as wetlands especially in the dried area. Study of two wetlands in southwestern Iran (Shadegan and Haur Al-Azim Wetlands) reveals presence of 37 species belonging to 30 genera, 13 families, and 8 orders. The ichthyofaunas of both wetlands are dominated by cyprinid species. They have also experienced accidentally or intentionally introductions of alien /non-native fish species. Presently, there are 11 (34.4% of the fish fauna) and 10 (31.3% of the fish fauna) introduced species in the Shadegan and Haur Al-Azim Wetlands, respectively. Xenocyprididae with four exotic species (36.4% of non-native fishes) is ranked the first and is followed by Cichlidae, Cyprinidae, and Poeciliidae, each with two species (18.2%) and Heteropneustidae with one species (9.1%). The alien species, *Carassius auratus*, *Hemiculter leucisculus*, *Heteropneustes fossilis*, *Coptodon zillii*, *Oreochromis aureus*, and *G. holbrooki* have established breeding populations. Currently, the region is facing oil and other developmental projects, modernization, dam construction, hydropower plant developments, overfishing, channelization, and intentional or accidental introduction of exotic elements. These resulted in strong impacts on the population size, species diversity, genetic diversity, and natural function of freshwater ecosystems of the whole region.

Keywords Fish diversity · Taxonomy · Native species · Exotic species · Threats · Conservation

---

H. R. Esmaeili (✉)
Ichthyology and Molecular Systematics Research Lab., Zoology Section, Department of Biology, College of Sciences, Shiraz University, Shiraz, Iran
e-mail: hresmaeili@shirazu.ac.ir

L. A. Jawad (ed.), *Southern Iraq's Marshes*, Coastal Research Library 36,
https://doi.org/10.1007/978-3-030-66238-7_17

17.1 Introduction

In our time, freshwater biological diversity is under heavy anthropogenic and natural pressures including habitat destruction/loss, overexploitation, climate change, environmental pollution, and introduction of exotic species, and therefore, obtaining information and monitoring of ecosystems and freshwater species (particularly current diversity and distribution) under the threats should be considered as a routine work for conservation management planning (Dudgeon et al. 2006; Kuljanishvili et al. 2020). The regularly obtained information on species diversity and distribution has a great potential to act as a tool for the prioritization of conservation and management efforts in highly international important water bodies such as wetlands especially in the dried area.

Wetlands are fundamental characteristics of coastal marine and freshwater systems around the world that provide highly diverse services and sustain highly diverse and productive aquatic communities including fishes (e.g., Elliott et al. 2007; Abell et al. 2008; Kaller et al. 2013). According to the Ramsar Convention, the wetlands are considered “as areas of marsh, fen, peatland or water, whether natural or artificial, permanent or temporary, with water that is static or flowing, fresh, brackish or salt, including areas of marine water the depth of which at low tide does not exceed six metres” (see the Ramsar Convention). This Convention is an international treaty accepted on 2 February 1971 in Ramsar city in northern Iran (thus the common name “Ramsar Convention”). One main tool of the Convention is its List of Wetlands of International Importance (the “Ramsar List”) which presently comprises 2210 Ramsar Sites, covering more than 210 million hectares (Ramsar Site 2020).

Wetlands offer a wide range of ecological, social, economic, cultural, and educational benefits known as ecosystem services (Costanza et al. 1997; Clarkson et al. 2013), including sustaining exclusive native biodiversity; providing suitable habitat for threatened species; fisheries support; production of food, water, wood, fiber, and genetic resources; regulation of air quality, climate, water purification, disease, pest, pollination, and natural hazard regulation; reduction of river flooding; protection of coastlines; landscape aesthetics; sites for recreation and transport; ecology education; sustenance of human cultures; ecotourism; bird-watching; fish-watching; bloom-watching; hydric soil development; primary productivity; serving as chemical sources, sinks, and transformers; and water storage (see Millennium Ecosystem Assessment 2005; Dise 2009; Clarkson et al. 2013; Mitsch et al. 2015). Despite covering only 1.5% of the Earth’s surface, wetlands provide a disproportionately high 40% of global ecosystem services (Zedler and Kercher 2005; Clarkson et al. 2013). Due to their specific function in the downstream receivers of water and waste from both natural and human sources, stabilizing water supplies, cleaning up polluted waters, and recharge groundwater aquifers, the wetlands are sometimes called kidneys of the landscape (Mitsch et al. 2015). Nature’s supermarket is another names for wetlands due to the extensive food chain that they supply and rich biodiversity which they maintain (Mitsch et al. 2015).

Iran currently has 25 sites designated as Wetlands of International Importance (Ramsar Sites), with a surface area of 1,488,624 hectares including the Shadegan and Haur Al-Azim Wetlands located in southwestern Iran in border with Iraq (Ramsar Site 2020). However, despite the ecological, social, economic, cultural, and educational benefits of these two unique ecosystems, they are currently under severe threats due to strong human-induced environmental changes which started during different wars in the region and will continue to occur because of modernization and industrial development. Such threats change the whole function of ecosystems and thus decrease their biodiversity especially ichthyodiversity. Hence, it is necessary to assemble most up-to-date list of fishes in the Shadegan and Haur Al-Azim Wetlands, which is aimed to act as a baseline information for future investigation and conservation of fish diversity in these regions.

17.2 Material and Methods

17.2.1 Study Area

The study area (Figs. 17.1, 17.2, and 17.3) covers two international important wetlands located in the southwestern Iran in border with Iraq, which either is considered as a part of the Mesopotamian marshland complex (Haur Al-Azim Wetland) or had been connected to the marshland complex in the past (Shadegan Wetland).

Fig. 17.1 Shadegan Wetland and three Mesopotamian marshland complex

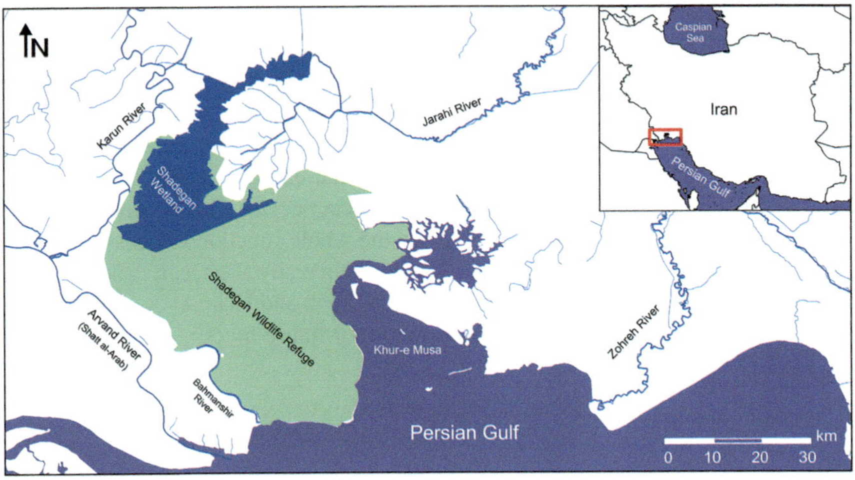

Fig. 17.2 Shadegan Wetland, southwestern Iran

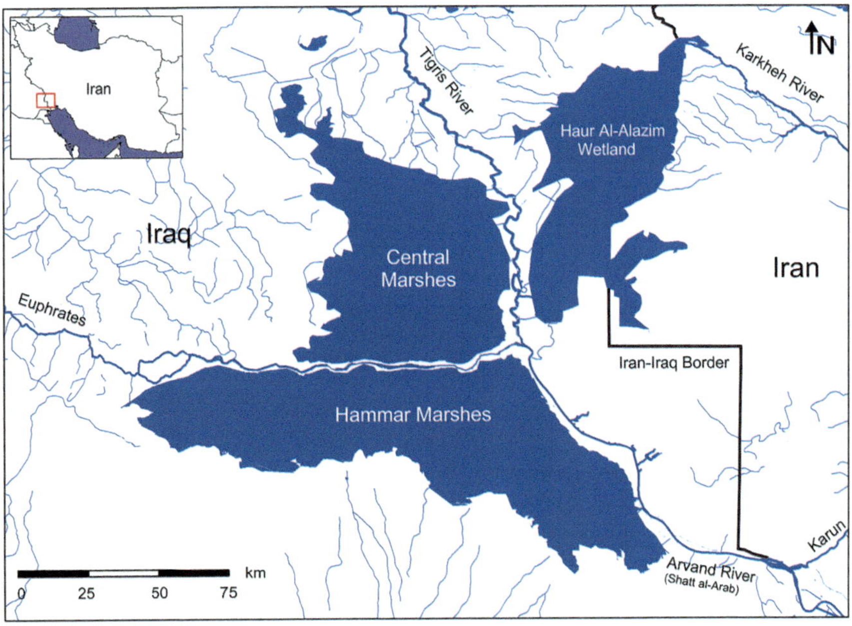

Fig. 17.3 Haur Al-Azim Wetland, southwestern Iran

17.2.2 The Shadegan Wetland

The Shadegan Wetland (Shadegan Marshes, Mudflats of Khor-al Amaya & Khor Musa; 30°30′N, 48°45′E) with about 400,000 to 537,700 ha was designated in Ramsar Site on 23 June1975 with site number 41, and it is the largest wetland in Iran and among the largest designated Ramsar sites in the world. Due to chemical pollution, the Shadegan site was placed on the Montreux Record in 16 June 1993. The site includes fresh and brackish sedge marshes, tidal flats, creeks, sandbanks, and a low island. The Shadegan Wetland lies in the southwest of Iran in the most downstream reaches of the Jarahi/Jarrahi River drainage at the head of the Persian Gulf (Lotfi 2016). It has been considered to be remnant of a much larger marsh, which earlier stretched to the Haur Al-Azim marsh on the Iraqi border and is thus a far-off component of the Mesopotamian marshlands complex (Lotfi 2016). The Karoun River flows southwards just a few kilometers from the western boundary of the intertidal flats. The Shadegan Wetland can be broadly divided into i) a large freshwater marsh, ii) extensive intertidal flats, iii) the Khur Musa Bay and its associated islands, and iv) the coastal fringes of the Persian Gulf (Figs. 17.1 and 17.2).

Due to habitat heterogeneity and unique biodiversity of the Shadegan Wetland, this water body has been considered as one of the magnificent natural landscapes of the world (Hashemi and Ansary 2012; Lotfi 2016). The wetland supports a diverse flora, with about 110 plant species in 17 plant communities, 40 species of mammals, 174 species of birds, 9 species of reptiles (including breeding marine turtles), 3 species of amphibians, 4 species of shrimp, and over 90 fresh and marine fish species (Lotfi 2016). A large number of rural indigenous peoples around the Shadegan Wetland are dependent on the services and resources provided by this wetland. Fisheries activities along the shorelines of Persian Gulf and in the Khur Musa Bay are an important source of income to the rural families. Buffalo and cow farming is also a common commerce in all the rural areas around the wetland (Lotfi 2016).

The Shadegan Wetland is usually supplied by i) freshwater from the Jarahi River (about 90%), ii) tidal influxes of seawater from the Persian Gulf, and iii) floodwater spills from the Karun/Karoun River.

However, construction of storage dams in the upstream, development of several irrigation projects, and reduced rainfall have changed the water flow regime into the wetland.

17.2.3 The Haur Al-Azim Wetland (Hawizeh Marsh)

The Haur Al-Azim Wetland (Hawizeh Marsh; 31°25′N 47°37′59″E) with about 137,700 ha is a transboundary wetland and a part of the Mesopotamian marshland complex located in the southwestern Iran (about 20–25%) and the southern Iraq. It was designated as the first Ramsar Site of Iraq in 2007 with site number 1718 (Ramsar Site 2020). This international important wetland was placed on the Montreux Record in April 2010 (Ramsar Site 2020).

This wetland is fed by the Tigris and Euphrates rivers in the Iraqi side and the Karkheh River in the Iranian side (Figs. 17.1 and 17.3).

The Mesopotamian marshland complex are wetland areas composed of three separate but neighboring water bodies, namely, Central, Haur Al-Azim (Hawizeh), and Hammar marshes located in southwestern Iran, southern Iraq, and Kuwait (Partow 2001).

The Haur Al-Azim Wetland is internationally important from ecological, economic, historical, social, and cultural point of views. This water body is a biodiversity reservoir of priority importance for conservation, and it is home to some threatened species of reptiles, birds, mammals, and fishes (Ramsar Site 2020). Despite its international importance, Haur Al-Azim (Hawizeh) continues to face internal and external large-scale threats because of a decrease in the water flows due to the construction of dams on the feeder rivers in Turkey and Iran, establishment of a 90 km soil embankment on the Iranian side and decreasing rainfall. The oil production in the areas adjacent to the marshes and overhunting of birds are other threats.

The Haur Al-Azim marsh is characterized by thickets of aquatic vegetation, generally comprising of common reed, *Phragmites australis*, and rushes, *Typha domingensis* (Al-Hilli et al. 2009; Al-Thahaibawi et al. 2019), and therefore constitutes a natural shelter for several bird, fish, buffalo, and other aquatic life (Partow 2001), and it is one of the sources of inland fisheries (Al-Thahaibawi et al. 2019).

17.2.4 Preparation of the Checklist

To pile up this fish regional checklist, i) the previously published articles (see the references cited in the discussion section) were reviewed; ii) ichthyological collections in Iran, e.g., ZM-CBSU, Zoological Museum of Shiraz University, Collection of Biology Department, Shiraz and Canada, CMNFI, Canadian Museum of Nature, Ottawa were checked; and iii) specimens collected from extensive field expeditions up to 2020 from the regions were examined. The checklist is presented in the form of a table (Table 17.1) which is ordered by taxonomic ranks including orders, families, and genera according to the classification given by Nelson et al. (2016) and the Eschmeyer's Catalogue of Fishes (Fricke et al. 2020). Alien (exotic) and native species and fish origin (e.g., marine or freshwater sources) are shown in Table 17.1. Additional notes on taxonomy of certain recorded species, hybridization, and habitat changes are also given in the discussion section. Photographs of several fish species are provided.

Table 17.1 Checklist of freshwater fishes of the Shadegan (SH) and Haur Al-Azim (HA) Wetlands

Order	Family	Genus	Species	Oc.	SH	HA
Clupeiformes	Clupeidae	*Tenualosa*	*T. ilisha* (Hamilton, 1822)	N	*	
Cypriniformes	Cyprinidae	*Arabibarbus*	*A. grypus* (Heckel, 1843)	N	*	*
		Capoeta	*C. trutta* (Heckel, 1843)	N		*
		Carasobarbus	*C. luteus* (Heckel, 1843)	N	*	*
			C. sublimus (Coad & Najafpour, 1997)	N		*
		Carassius	*Carassius auratus* (Linnaeus, 1758)	I	*	*
		Cyprinion	*C. kais* Heckel, 1843	N	*	*
			C. macrostomum Heckel, 1843	N	*	*
		Cyprinus	*C. carpio* Linnaeus, 1758	I	*	*
		Garra	*G. rufa* (Heckel, 1843)	N		*
		Luciobarbus	*L. barbulus* (Heckel, 1849)	N	*	*
			L. esocinus Heckel, 1843	N		*
			L. kersin (Heckel, 1843)	N	*	*
			L. xanthopterus Heckel, 1843	N	*	*
	Leuciscidae	*Acanthobrama*	*A. marmid* Heckel, 1843	N	*	*
		Alburnoides	*A. idignensis* Bogutskaya & Coad, 2009	N	*	*
		Alburnus	*A. sellal* Heckel, 1843	N	*	*
		Chondrostoma	*C. regium* (Heckel, 1843)	N	*	*
		Leuciscus	*L. vorax* (Heckel, 1843)	N	*	*
	Xenocyprididae	*Ctenopharyngodon*	*C. idella* (Valenciennes, 1844)	I	*	*
		Hemiculter	*H. leucisculus* (Basilewsky, 1855)	I	*	*
		Hypophthalmichthys	*H. molitrix* (Valenciennes, 1844)	I	*	*
		Hypophthalmichthys	*H. nobilis* (Richardson, 1844)	I	*	*
Siluriformes	Bagridae	*Mystus*	*M. pelusius* (Solander, 1794)	N	*	*
	Heteropneustidae	*Heteropneustes*	*H. fossilis* (Bloch, 1794)	I	*	*

(continued)

Table 17.1 (continued)

Order	Family	Genus	Species	Oc.	SH	HA
	Siluridae	*Silurus*	*S. triostegus* Heckel, 1843	N	*	*
Synbranchiformes	Mastacembelidae	*Mastacembelus*	*M. mastacembelus* (Banks & Solander, 1794)	N	*	*
Cichliformes	Cichlidae	*Coptodon*	*C. zillii* (Gervais, 1848)	I	*	*
		Oreochromis	*O. aureus* (Steindachner, 1864)	I	*	*
Cyprinodontiformes	Aphaniidae	*Aphanius*	*A. mesopotamicus* Coad, 2009	N	*	*
		Aphaniops	*A. stoliczkanus* (Day, 1872)	N	*	*
	Poeciliidae	*Gambusia*	*G. holbrooki* Girard, 1859	I	*	*
		Poecilia	*P. latipinna* (Lesueur, 1821)	I	*	
Mugiliformes	Mugilidae	*Planiliza*	*P. abu* (Heckel, 1843)	N	*	*
			P. subviridis (Valenciennes, 1836)	N	*m	
		Ellochelon	*E. vaigiensis* (Quoy & Gaimard, 1825)	N	*m	
Perciformes	Sparidae	*Acanthopagrus*	*A. arabicus* Iwatsuki, 2013	N	*m	

Abbreviations: *N* native, *I* introduced, *m* marine origin, * present

17.3 Results

In total, 37 species, belonging to 30 genera, 13 families, and 8 orders, were listed for both the Shadegan and Haur Al-Azim Wetlands (Table 17.1). From these, 33 species are recorded from the Shadegan Wetland and also 32 from the Haur Al-Azim Wetland. The ichthyofaunas of both wetlands are dominated by cyprinid species. These two studied wetlands have also experienced accidentally or intentionally introductions of alien/non-native fish species. Presently, there are 11 (34.4% of the fish fauna) and 10 (31.3% of the fish fauna) introduced species in the Shadegan and Haur Al-Azim Wetlands, respectively (Table 17.1). Recently introduced *P. latipinna* (Poeciliidae) has been recorded from the Shadegan Wetland only. Xenocyprididae with four species (36.4% of non-native fishes) is ranked the first and is followed by Cichlidae, Cyprinidae, and Poeciliidae, each with two species (18.2%) and Heteropneustidae with one species (9.1%). The alien species, *Carassius auratus*, *Hemiculter leucisculus*, *Heteropneustes fossilis*, *Coptodon zillii*, *Oreochromis aureus*, and *G. holbrooki* have established breeding populations. Interestingly, there is no record of *Pseudorasbora parva*, a widely introduced species in Iran, in both water bodies.

17.4 Discussion

In the following, additional notes on several fish taxa are provided for further argumentations with regard to taxonomy, distribution, habitat changes, and hybridization.

17.4.1 Clupeidae

There is only one clupeid species, the Hilsa shad, *Tenualosa ilisha* (Fig. 17.4), which enters into freshwater part of the Shadegan Wetland. This species is also found in the coastal areas, estuaries, and brackish water parts of this wetland (Fig. 17.1). This anadromous clupeid fish migrates from the Persian Gulf to the Arvand and Bahmanshir River for spawning (Roomiani et al. 2014). *Tenualosa ilisha* supports a commercial fishery in the region and is a very important food fish in southwest of Iran (Roomiani et al. 2014). There are records of two more clupeid form fishes, *Sardinella sindensis* (Day, 1878) and *Thryssa hamiltonii* (Gray, 1835) (Engraulidae), from the Shadegan Wetland (see Esmaeili et al. 2014a; Hashemi et al. 2015), but they are only found in the coastal area, estuaries, and the Khur-e Musa Bay of this wetland.

Fig. 17.4 The Hilsa shad, *Tenualosa ilisha*, from the Persian Gulf. (Photo by Dr. J.E. Randal)

Fig. 17.5 The gold fish, *Carassius auratus*, from the Tigris River drainage

17.4.2 Cyprinidae

There are 13 species of cyprinids in the studied wetlands. Members of this family were previously classified under the subfamily Cyprininae, but it is now elevated to the family rank (Fricke et al. 2020; Kuljanishvili et al. 2020). *Carassius* populations from the both wetlands have been considered as *Carassius auratus* (Linnaeus, 1758) by several authors (Mohamed et al. 2008; Rezai and Papahn 2013; Esmaeili et al. 2014a; Hashemi et al. 2015) based on morphological characteristics including color pattern (Fig. 17.5). Recently, morphologically identified population of the Shadegan Wetland and also the Karun River was confirmed by molecular approaches using the mitochondrial gene, cytochrome *b* (Cyt *b*). According to Khosravi et al. (2020), *Carassius* individuals collected from the Karun River, Shadegan Wetland, and Hamun-e-Saberi Wetland corresponded to the *C. auratus* clade, those from the Siah Palas Stream (Caspian Sea basin) corresponded to *C. langsdorfii*, and individuals from the Anzali Wetland (Caspian Sea basin) belonged to *C. gibelio*. Although *C. auratus* and *C. gibelio* can be distinguished by molecular characteristics (Rylková

Fig. 17.6 Putative hybrid and parents. a, *Luciobarbus barbulus* ZM-CBSU G1158, 214 mm SL; b, putative hybrid between *L. barbulus* and *L. kersin*, ZM-CBSU G1151, 328 mm SL; c, *L. kersin* ZM-CBSU G1150, 306 mm SL; all from Iran: Khuzestan Prov.: Hor al-Azim wetland

et al. 2013; Khosravi et al. 2020), however, there is no agreement on the use of morphological characters to separate these congeneric species (Kuljanishvili et al. 2020). Record of the Crucian carp, *Carassius carassius* (Linnaeus, 1758) from the Shadegan Wetland (Hashemi et al. 2015), might be a misidentification.

Luciobarbus pectoralis (Heckel, 1843) has been frequently reported from the Shadegan and Haur Al-Azim Wetlands (e.g., Hashemi et al. 2015). However, based on the mitochondrial (COI) gene sequences of individuals collected from the upper reaches of Jarahi River which drains into the Shadegan Wetland and also from the Haur Al-Azim Wetland, all of them belong to the other species *L. barbulus* (Khaefi et al. 2018). It seems that *L. pectoralis* is restricted to the Orontes River system in Turkey and Syria and Mediterranean River drainages of Turkey (Esmaeili et al. 2017a, 2018).

Presence of two hybrids of *Luciobarbus barbulus* x *Luciobarbus kersin* and *Luciobarbus barbulus* x *Luciobarbus xanthopterus* (Figs. 17.6 and 17.7.) has been reported from the Haur Al-Azim Wetland (see Khaefi et al. 2018). The hybrid individuals were identified based on their intermediate morphological characteristics with the parent species, their position in the clade, and their close genetic distance with the maternal gene sequences (Khaefi et al. 2018). Presence of the Kersin barbel, *Luciobarbus kersin* (Heckel, 1843) in the Haur Al-Azim Wetland (Fig. 17.6), has been confirmed applying morphological characteristics and molecular markers (Khaefi et al. 2018).

Fig. 17.7 Putative hybrid and parents. a. *L. barbulus* ZM-CBSU G1158, 214 mm SL; b, putative hybrid between *L. barbulus* and *L. xanthopterus* ZM-CBSU G1169, 294 mm SL; c. *L. xanthopterus* ZM-CBSU G1160, 191 mm SL. All from Iran: Khuzestan Prov.: Hor al-Azim Wetland

Fig. 17.8 The Kersin barbel, *Luciobarbus kersin*, from the Shadegan Wetland

Luciobarbus kersin has also been recorded from the Shadegan Wetland (Fig. 17.8). *Luciobarbus esocinus* Heckel, 1843, is another cyprinid fish in the region (Fig. 17.9).

Fig. 17.9 *Luciobarbus esocinus* from the Tigris River drainage

17.4.3 Leuciscidae

Members of this family were previously classified under the family Cyprinidae, but it is now considered as a distinct family (Fricke et al. 2020; Kuljanishvili et al. 2020). There are five native species of leucisids in the studied wetlands. Taxonomic position of *Alburnoides* and *Alburnus* populations from the region are changed, and the populations are now designated as *Alburnoides idignensis* Bogutskaya & Coad, 2009 and *Alburnus sellal* Heckel, 1843 (Fig. 17.10) (see Bogutskaya and Coad 2009; Mohammadian-Kalat et al. 2017; Esmaeili et al. 2017a, 2018).

Fig. 17.10 *Alburnus sellal* from the Tigris River drainage

Fig. 17.11 *Hemiculter leucisculus* from the Tigris River drainage

17.4.4 Xenocyprididae

There are four exotic species of xenocypridids in the studied wetlands. Members of this family were previously classified under the family Cyprinidae, but it is now considered as distinct family (Fricke et al. 2020; Kuljanishvili et al. 2020). From these, *H. leucisculus* (Basilewsky, 1855) (Fig. 17.11) is the only fish that has been established in both wetlands and now produces breeding populations.

17.4.5 Siluriformes

There are three families of siluriforms including Bagridae, Heteropneustidae, and Siluridae comprising one species each. There are two native species, *Mystus pelusius* (Solander, 1794) and *Silurus triostegus* Heckel, 1843 (Fig. 17.12), and one exotic species, *Heteropneustes fossilis* (Bloch, 1794).

17.4.6 Cichlidae

Members of this family were previously classified under the order Perciformes, but it is now considered to be a distinct order (Cichliformes) (see Nelson et al. 2016; Betancur et al. 2017). There are two endemic species, *Iranocichla hormuzensis*

Fig. 17.12 *Silurus triostegus* from the Shadegan Wetland

Fig. 17.13 An exotic cichlid fish, *Coptodon zillii*, from the Shadegan Wetland

Coad, 1982 and *I. persa* Esmaeili, Sayyadzadeh & Seehausen, 2016, which are distributed in the lower river drainages of the Persian Gulf and Oman Sea, and two exotic cichlid species, *Coptodon zillii* (Gervais, 1848) and *Oreochromis aureus* (Steindachner, 1864) in Iran (Coad 1982; Esmaeili et al. 2016, 2017a, 2018). From these, *C. zillii* and *O. aureus* are found in the Shadegan (Figs. 17.13 and 17.14) and Haur Al-Azim Wetlands (Khaefi et al. 2014). The first record of the redbelly tilapia, *C. zillii*, in the freshwaters of Iran (Shadegan Wetland) was given by Khaefi et al. (2014). The second species, the blue tilapia, *O. aureus*, was reported from the Arvand, Karun, and Bahmanshir River drainages and also from the Korramshahr Cane Canals in Khuzestan Province by Valikhani et al. (2016). Further investigations reveal range expansion of exotic cichlids in southwestern Iran and presence of *C. zillii* in the regular catches of the Shadegan Wetland (Valikhani et al. 2018). Both of these exotic cichlids are also recorded from Iraq by Mutlak and Al-Faisal (2009) at the main outfall drain in the city of Basrah. Tilapias have high

Fig. 17.14 An exotic cichlid fish, the blue tilapia, *Oreochromis aureus*, from the Karun River. (Photo by H. Valikhani)

Fig. 17.15 The exotic cichlid fish, Nile tilapia *Oreochromis niloticus* from the Arvand River. (Photo by Abbas J. Al-Faisal)

resistance for various ecological conditions and can easily reproduce in new habitats. This ability may contribute to the successful breeding and establishment of this exotic cichlid fish in other inland waters of the region in the near future. Monitoring of these species is highly recommended. Presence of another exotic cichlid, the Nile tilapia *Oreochromis niloticus* (Linnaeus, 1758) (Fig. 17.15) in the Arvand River (Shatt Al-Arab River), southern Iraq, has been confirmed by Al-Faisal and Mutlak (2014), and like the other tilapias, this fish also may become established in Iran through connected water bodies. Reports of the Nile tilapia in the Euphrates River at Al-Hindia Barrage, middle of Iraq (Abulheni and Abbas 2017); the Tigris River, south of Baghdad (Attee et al. 2017); and the Garmat Ali River, east of the Hammar Marsh to the Arvand River/Shatt Al-Arab, Basreh (Mohamed and Al-Wan 2020a), and also presence of the redbelly tilapia in the AL-Delmj Marsh, Iraq (Al-Zaidy 2013), and the Garmat Ali River, East of the Hammar Marsh to the Arvand River/ Shatt Al-Arab, Basreh (Mohamed and Al-Wan 2020b), all reveal establishment and range expansion of these exotic fishes in the Tigris-Euphrates River system and need of conservation plans and management programs.

17.4.7 Aphaniidae

Members of this family were previously classified under the family Cyprinodontidae, but it is now considered as distinct family (Esmaeili et al. 2020; Fricke et al. 2020). *Aphanius mesopotamicus* Coad, 2009, and *Aphaniops stoliczkanus* (Day, 1872) (=*Aphanius dispar*) are two native toothcarps (Figs. 17.16 and 17.17) of the Shadegan and Haur Al-Azim Wetlands (Keivany et al. 2012; Esmaeili et al. 2014a, 2020; Masoudi et al. 2016). All aphaniid fishes were previously classified in one genus, *Aphanius*, but now they are considered in three distinct genera of *Aphanius*, *Aphaniops*, and *Paraphanius* (Esmaeili et al. 2020). There is possibility of presence of *Paraphanius mento* (Heckel, 1843) in the region.

Sympatricity of the Mesopotamian toothcarp, *Aphanius mesopotamicus*, with *Aphaniops stoliczkanus* in the Shadegan Wetland and their possible hybridization have been reported and discussed by Masoudi et al. (2016). They reported

Fig. 17.16 A male specimen of *Aphanius mesopotamicus* from the Shadegan Wetland

Fig. 17.17 Female (above) and male (below) *Aphaniops stoliczkanus* from the Shadegan Wetland

intermediate characters in otolith morphology, pigmentation, and some morphometric values in several individuals recovered from the Shadegan Wetland and considered them as putative hybrids (Masoudi et al. 2016).

17.4.8 Poeciliidae

There are two species of poeciliids distributed in the region (Figs. 17.18 and 17.19), namely, *Gambusia holbrooki* Girard, 1859, and *Poecilia latipinna* (Lesueur, 1821) (Al-Faisal and Mutlak 2014; Esmaeili et al. 2014a; Esmaeili et al. 2017b) in the

Fig. 17.18 Female and male specimens of the eastern mosquitofish, *Gambusia holbrooki*

Fig. 17.19 Female and male specimens of the sailfin molly, *Poecilia latipinna*, collected from the Choibdeh canal (Arvand River)

Fig. 17.20 The abu mullet, *Planiliza abu*

studied area. The eastern mosquitofish, *G. holbrooki*, is the most widespread invasive species in Iran (Esmaeili et al. 2014a).

17.4.9 Mugilidae

Planiliza abu (Heckel, 1843) (Fig. 17.20), *P. subviridis* (Valenciennes, 1836), and *Ellochelon vaigiensis* (Quoy & Gaimard, 1825) are three mugilid species reported from the area (Table 17.1). *Planiliza abu* is found in both wetlands (Mohamed et al. 2008; Rezai and Papahn 2013; Esmaeili et al. 2014a; Hashemi et al. 2015); but *P. subviridis* (Fig. 17.21) and *E. vaigiensis* (Fig. 17.22) with marine origin are found in the southern parts of the Shadegan Wetland (Esmaeili et al. 2014a; Hashemi et al. 2015). *Ellochelon vaigiensis* enters fresh waters (Al-Nasiri and Hoda 1976; Coad 2017), and has been reported from the Arvand River, Iraq near Iranian Khuzestan.

17.4.10 Sparidae

Sparidae presents only one species, the Arabian yellowfin seabream, *Acanthopagrus arabicus* Iwatsuki, 2013 (Fig. 17.23), which was already considered as *A. latus* (see Esmaeili et al. 2014a, b). *Acanthopagrus arabicus* is widely distributed in estuaries and rivers of coastal area of the Persian Gulf and Strait of Hormuz including the Arvand River (Shatt al Arab) in Iraq on the border with Khuzestan in Iran, the Bahmanshir River, Mahshahr creeks, and the Shadegan Wetland (Hashemi and Ansary 2012; Khaefi et al. 2014, Esmaeili et al. 2014a, b).

Fig. 17.21 The greenback mullet, *Planiliza subviridis* from the Persian Gulf. (Photo by Dr. J.E. Randal)

Fig. 17.22 The squaretail mullet, *Ellochelon vaigiensis*. (Photo by Dr. J.E. Randal)

Fig. 17.23 The Arabian yellowfin seabream, *Acanthopagrus arabicus*

17.5 Conclusion

During the period of different wars in the region, both wetlands were under severe pressure. Currently, the region is facing oil and other developmental projects, modernization, dam construction, hydropower plant developments, overfishing, channelization, and intentional or accidental introduction of exotic elements. These resulted in strong impacts on the population size, species diversity, genetic diversity, and natural function of freshwater ecosystems of the whole region. As a result, a significant part of the spawning areas for many species has been lost, exotics compete with native elements for food and space and transmit diseases and parasites, total catch of native fishes decreases, and the whole landscape of wetlands is affected. All of these will also change the social and cultural activities of the people in the region.

Acknowledgments F. Zarei is acknowledged for preparation of some maps. This study was financially supported by Shiraz University.

References

Abell R, Thieme ML, Revenga C, Bryer M, Kottelat M, Bogutskaya N et al (2008) Freshwater ecoregions of the world: A new map of biogeographic units for freshwater biodiversity conservation. Bioscience 58(5):403–414

Abulheni AJ, Abbas LM (2017) First record of the tilapia *Oreochromis niloticus* (Linnaeus, 1758) in Euphrates River at Al-Hindia barrage–middle of Iraq. J Kerbala Univ 2017:18–21

Al-Faisal AJ, Mutlak FM (2014) First record of the Nile tilapia *Oreochromis niloticus* (Linnaeus, 1758), from the Shatt Al-Arab River, South Iraq. Mesop J Mar Sci 29(1):45–50

Al-Hilli MRA, Warner BG, Asada T, Douabul A (2009) An assessment of vegetation and environmental controls in the 1970s of the Mesopotamian wetlands of southern Iraq. Wetlands Ecol Manage 17(3):207–223

Al-Nasiri SK, Hoda SMS (1976) A guide to the freshwater fishes of Iraq. Basrah Natural History Museum Publication, Basrah, 124p

Al-Thahaibawi BMH, Younis KH, Al-Mayaly IK (2019) Fish assemblage structure in Al-Huwaizah marsh southern of Iraq after inscribed on the World Heritage List. Iraqi J Sci 60:1430–1441

Al-Zaidy KJ (2013) First record of *Tilapia zillii* (Gewais, 1848) in AL-Delmj Marsh West AL-Diwania city middle of Iraq. Diyala Agric Sci J 5(1):9–16

Attee RS, Jassem AK, Al Heni A, Khalifa SZ (2017) Description of the growth of Nile tilapia (Linnaeus, 1758) *Oreochromis niloticus* in the Tigris River south of Baghdad. International J Chem Tech Res 10(9):1093–1102

Betancur RR, Wiley EO, Arratia G, Acero A, Bailly N, Miya M et al (2017) Phylogenetic classification of bony fishes. BMC Evol Biol 17(1):162

Bogutskaya NG, Coad BW (2009) A review of vertebral and fin-ray counts in the genus *Alburnoides* (Teleostei: Cyprinidae) with a description of six new species. Zoosystematica Rossica 18(1):126–173

Clarkson BR, Ausseil AGE, Gerbeaux P (2013) Wetland ecosystem services. In: Ecosystem services in New Zealand: Conditions and trends. Manaaki Whenua Press, Lincoln, pp. 192–202

Coad BW (1982) A new genus and species of cichlid endemic to southern Iran. Copeia 1982 (1):28–37

Coad B (2009) A new species of tooth-carp, *Aphanius mesopotamicus*, from Iran and Iraq (Actinopterygii, Cyprinodontidae). ZooKeys 31:149–163

Coad BW (2017) Review of the freshwater mullets of Iran (Family Mugilidae). Iran J Ichthyol 4 (2):75–130

Costanza R, d'Arge R, De Groot R, Farber S, Grasso M, Hannon B et al (1997) The value of the world's ecosystem services and natural capital. Nature 387(6630):253–260

Dise NB (2009) Peatland response to global change. Science 326:810–811

Dudgeon D, Arthington AH, Gessner MO, Kawabata ZI, Knowler DJ, Lévêque C, Naiman RJ, Prieur-Richard AH, Soto D, Stiassny ML, Sullivan CA (2006) Freshwater biodiversity: importance, threats, status and conservation challenges. Biol Rev 81(2):163–182

Elliott M, Whitfield AK, Potter IC, Blaber SJ, Cyrus DP, Nordlie FG, Harrison TD (2007) The guild approach to categorizing estuarine fish assemblages: A global review. Fish and Fisheries 8 (3):241–268

Esmaeili HR, Teimori A, Feridon OWFI, Abbasi K, Brian WC (2014a) Alien and invasive freshwater fish species in Iran: Diversity, environmental impacts and management. Iran J Ichthyol 1(2):61–72

Esmaeili HR, Masoudi M, Mehraban HR (2014b) Assignment of *Acanthopagrus* populations in the Persian Gulf drainage system of Iran to *Acanthopagrus arabicus* Iwatsuki, 2013 (Perciformes: Sparidae). Iran J Ichthyol 1(1):23–28

Esmaeili HR, Sayyadzadeh G, Seehausen O (2016) *Iranocichla persa*, a new cichlid species from southern Iran (Teleostei, Cichlidae). ZooKeys 636:141–161

Esmaeili HR, Mehraban H, Abbasi K, Keivany Y, Brian WC (2017a) Review and updated checklist of freshwater fishes of Iran: Taxonomy, distribution and conservation status. Iran J Ichthyol 4 (Suppl. 1):1–114

Esmaeili HR, Masoudi M, Chermahini MA, Esmaeili AH, Zarei F, Ebrahimi M (2017b) Invasion of the Neotropical and Nearctic fishes to Iran. FishTaxa 2(3):126–133

Esmaeili HR, Sayyadzadeh G, Eagderi S, Abbasi K (2018) Checklist of freshwater fishes of Iran. FishTaxa 3(3):1–95

Esmaeili HR, Teimori A, Zarei F, Sayyadzadeh G (2020) DNA barcoding and species delimitation of the Old World tooth-carps, family Aphaniidae Hoedeman, 1949 (Teleostei: Cyprinodontiformes). PLoS One 15(4):e0231717

Fricke R, Eschmeyer WN, Van der Laan R (Eds) (2020) Eschmeyer's Catalog of fishes: Genera, species, references. Retrieved from http://researcharchive.calacademy.org/research/ichthyology/catalog/fishcatmain.asp

Hashemi SAR, Ansary H (2012) Biomass and production of fish species in the Shadegan wetland, Iran. Global Veterinaria 9:123–128

Hashemi SA, Ghorbani R, Kymaram F, Hossini SA, Eskandari G, Hedayati A (2015) Fish species composition, distribution and abundance in Shadegan Wetland. Fish Aquac J 6(2):1000128

Kaller MD, Kelso WE, Trexler JC (2013) Wetland fish monitoring and assessment. In: Wetland techniques. Springer, Dordrecht, pp. 197–263

Keivany Y, Alavi-Yeganeh MS, Seyfabadi J (2012) A new record confirms the occurrence of *Aphanius mesopotamicus* Coad, 2009, in southwestern Iran (Actinopterygii: Cyprinodontidae). Check List 8(2):283–285

Khaefi R, Esmaeili HR, Chermahini MA (2018) Natural hybridization of *Luciobarbus barbulus* x *Luciobarbus kersin* and *Luciobarbus barbulus* x *Luciobarbus xanthopterus* in the Persian Gulf Basin. Turk J Fish Aquat Sci 18(12):1399–1407

Khaefi R, Esmaeili HR, Zareian H, Babaei S (2014) The first record of the redbelly tilapia, *Tilapia zillii* Gervais, 1848 in freshwaters of Iran. Turk J Zool 38:96–98

Khosravi M, Abdoli A, Ahmadzadeh F, Saberi-Pirooz R, Rylková K, Kiabi BH (2020) Toward a preliminary assessment of the diversity and origin of cyprinid fish genus *Carassius* in Iran. J Appl Ichthyol 36(4):422–430

Kuljanishvili T, Epitashvili G, Freyhof J, Japoshvili B, Kalous L, Levin B et al (2020) Checklist of the freshwater fishes of Armenia, Azerbaijan and Georgia. J Appl Ichthyol 36(4):501–514

Lotfi A (2016) Shadegan wetland (Islamic Republic of Iran). In: Finlayson C, Milton G, Prentice R, Davidson N (eds) The wetland book. Springer, Dordrecht

Masoudi M, Esmaeili HR, Teimori A, Gholami Z, Gholamhosseini A, Sayyadzadeh G et al (2016) Sympatry and possible hybridization among species of the killifish genus *Aphanius* Nardo, 1827 (Teleostei: Cyprinodontidae) in Southwestern Iran. Limnologica 59:10–20

Millennium Ecosystem Assessment (2005) Ecosystems and human well-being: Synthesis. Island Press, Washington, DC

Mitsch WJ, Bernal B, Hernandez ME (2015) Ecosystem services of wetlands. International Journal of Biodiversity Science, Ecosystem Services & Management 11:1–4

Mohamed ARM, Al-Wan SM (2020a) Biological aspects of an invasive species of *Oreochromis niloticus* in the Garmat Ali River, Basrah, Iraq. J Agric Vet Sci 13(2):15–26

Mohamed ARM, Al-Wan SM (2020b) Evaluation of biological characters of the invasive species, *Coptodon zillii* in the Garmat Ali River, Basrah, Iraq. Int J Fish Aquat Stud 8(2):176–185

Mohamed ARM, Hussain NA, Al-Noor SS, Mutlak FM, Al-Sudani IM, Mojer AM et al (2008) Fish assemblage of restored Al-Hawizeh marsh, southern Iraq. Ecohydrol Hydrobiol 8(2–4):375–384

Mohammadian-Kalat T, Esmaeili HR, Aliabadian M, Freyhof J (2017) Re-description of *Alburnus doriae*, with comments on the taxonomic status of *A. amirkabiri*, *A. mossulensis*, *A. sellal* and *Petroleuciscus esfahani* (Teleostei: Cyprinidae). Zootaxa 4323(4):487–502

Mutlak FM, Al-Faisal AJ (2009) A new record of two exotic cichlids fish *Oreochromis aureus* Steindacher, 1864 and *Tilapia zillii* Gervais, 1848 from south of the main outfall drain in Basrah city. 24(2):160–170

Nelson JS, Grande TC, Wilson MV (2016) Fishes of the world, 5th edn. John Wiley & Sons, Hoboken

Partow H (2001) The Mesopotamian marshlands: Demise of an ecosystem. Publication UNEP/DEWA, Nairobi (Kenya), 103p

Ramsar Site (2020). www.ramsar.org

Rezai M, Papahn F (2013) A Study of the fish fauna in Hoor-Al azim Wetland. J Appl Ichthyol Res 1(2):53–59. In Persian

Roomiani L, Sotudeh AM, Hakimi Mofrad R (2014) Reproductive biology of Hilsa shad (*Tenualosa ilisha*) in coastal waters of the Northwest of Persian Gulf. Iran J Fish Sci 13 (1):201–215

Rylková K, Kalous L, Bohlen J, Lamatsch DK, Petrtýl M (2013) Phylogeny and biogeographic history of the cyprinid fish genus *Carassius* (Teleostei: Cyprinidae) with focus on natural and anthropogenic arrivals in Europe. Aquaculture 380:13–20

Valikhani H, Abdoli A, Kiabi BH, Nejat F, Sadeghsaba M, Khosravi M (2018) A study on the status of invasive tilapia species (*Coptodon zillii* Gervais, 1848 and *Oreochromis aureus* Steindachner, 1864) in the aquatic ecosystems of Khuzestan Province, Iran. Environ Sci 15 (4):29–44

Valikhani H, Abdoli A, Kiabi BH, Nejat F (2016) First record and distribution of the blue tilapia, *Oreochromis aureus* (Steindachner, 1864)(Perciformes: Cichlidae) in inland waters of Iran. Iran J Ichthyol 3(1):19–24

Zedler JB, Kercher S (2005) Wetland resources: Status, trends, ecosystem services and restorability. Annu Rev Environ Resour 30:39–74

Chapter 18
Checklist of Fishes of the Caspian Sea Basin: Land of Wetlands

Hamid Reza Esmaeili and Keyvan Abbasi

Abstract Research works on the biodiversity of the Iranian plateau including terrestrial, marine, brackish, and freshwater ecosystems have been steadily increasing over the last few decades, mainly as a result of a growing scientific interest, presence of unique species, introduction of exotic elements, and an increased awareness of the importance of conservation and sustainable use of biological resources. Being located entirely in the southwest Palaearctic, Iran receives elements from both of Afrotropical and Oriental regions. In addition, due to recent anthropological activities, it receives other faunal elements from the Nearctic and the Neotropical regions. The Caspian Sea which is a part of a highly diverse area of Paratethys basin is the largest lake or inland water body in the world presenting both brackish and fresh water habitats, rivers, lakes, lagoons, marshes, and marine environments presenting a high biodiversity especially fish species. This paper presents a list of fishes in the southern Caspian Sea basin including two wetlands within it and compares some of its fish elements with the Tigris-Euphrates River system. In total, 116 species, belonging to 65 genera and 29 families, are listed here for the southern Caspian Sea basin of Iran. From these, 5 species have not recently been collected. There are 8 endemic and 11 exotic species. The Anzali and Chamkhale Wetlands present 75 (63% of the whole basin species) and 25 (21.5%) species. The Caspian Sea basin is characterized by presence of high diverse fishes of the Ponto-Caspian families, Gobiidae and Leuciscidae, with 20 species each. It is followed by Cyprinidae (12); Clupeidae (8); Acipenseridae and Nemacheilidae (each with 7 species); Xenocyprididae (5 species); Cobitidae, Oxudercidae, and Salmonidae (each with 4 species); and Percidae (3 species). The rest of families present two or one species. *Arabibarbus*, *Carasobarbus*, *Cyprinion*, *Garra*, and *Mesopotamichthys* genera which are the main cyprinid fish elements of the Tigris-

H. R. Esmaeili (✉)
Ichthyology and Molecular Systematics Research Lab., Zoology Section, Department of Biology, College of Sciences, Shiraz University, Shiraz, Iran
e-mail: hresmaeili@shirazu.ac.ir

K. Abbasi
Inland Waters Aquaculture Research Centre, Iranian Fisheries Sciences Research Institute, Agricultural Research, Education and Extension Organization, Bandar Anzali, Iran

L. A. Jawad (ed.), *Southern Iraq's Marshes*, Coastal Research Library 36,
https://doi.org/10.1007/978-3-030-66238-7_18

Euphrates River system, and also aphaniids, cichlids, and mastacembelids, are absent in the Caspian Sea basin.

Keywords Fishes · List · Endemic species · Exotic species · Marshes · Tigris-Euphrates Rivers

18.1 Introduction

Research works on the biodiversity of the Iranian plateau including terrestrial, marine, brackish, and freshwater ecosystems have been steadily increasing over the last few decades, mainly as a result of a growing scientific interest, presence of unique species, introduction of exotic elements, and an increased awareness of the importance of conservation and sustainable use of biological resources (e.g., Karami et al. 2008; Esmaeili et al. 2010, 2017a, b, 2020; Safaei-Mahroo et al. 2015; Khaleghizadeh et al. 2017; Abbasi et al. 2019).

Although Iran is being located entirely in the southwest Palaearctic, however, this country is geographically close to the Afrotropical and the Oriental regions, to receive elements from both of them as well, presenting high biodiversity. Moreover, due to recent anthropological activities, it receives other faunal elements from the Nearctic and the Neotropical regions (Esmaeili et al. 2017a, b). The country is usually divided into 14 bioclimatic groups based on Emberger's method (Sabeti 1997; Safaei-Mahroo et al. 2015). Iran also presents six main watershed areas comprised of Central (52%), Persian Gulf and Gulf of Oman (24%), Caspian Sea (11%), Urmia Lake (3.4%), Eastern Iran (6.8%), and Qareh Qom (2.8%) (Safaei-Mahroo et al. 2015).

The Caspian Sea (=Khazar, Mazandaran), which is a part of a highly diverse area of Paratethys basin (Naseka and Bogutskaya 2009; Esmaeili et al. 2014a, b), is the largest lake or inland water body in the world covering 18% of the total area of all lakes in the world. The Caspian Sea basin includes both brackish and fresh water habitats, rivers, lakes, lagoons, marshes, and marine environments presenting a good example of a long history of connection and isolation from fresh and marine waters promoting high biodiversity.

The southern Caspian Sea basin, herein, includes the rivers draining to the sea, the wetlands, and the sea itself within Iranian territorial waters (Fig. 18.1). Some major river systems of the southern Caspian basin of Iran from west to east are Aras, Safidrud, Chalus, Haraz, Talar, Tajan, Gorgan, and Atrak/Atrek rivers. Anzali, Miankaleh, Amirkelayeh, and Chamkhale are some of the major wetland systems within this basin. It is worth to mention that Iran shares 250 wetlands (Khorami Pour et al. 2015), among which 25 wetlands including the Anzali Wetland (with an approximate area of 1.5 million ha) have been recorded in the Ramsar Convention as the Wetlands of International Importance (Khorami Pour et al. 2015; Ramsar Site 2020).

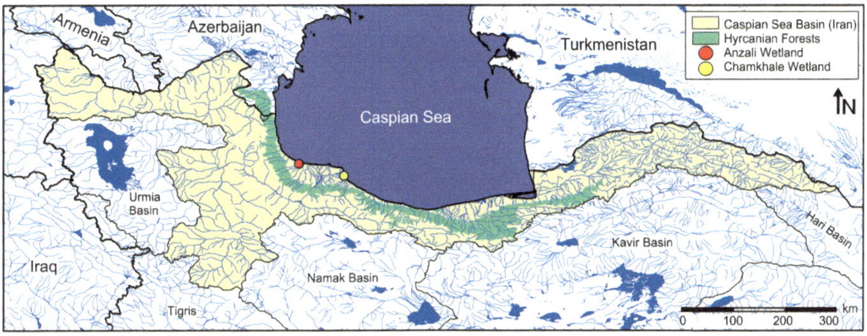

Fig. 18.1 The Caspian Sea basin and two selected wetlands, Anzali and Chamkhale

However, despite ecological, social, economic, cultural, and educational benefits of the southern Caspian Sea ecosystems especially wetlands, they are currently under severe human-induced threats (modernization and industrial development). Such threats change the whole function of ecosystems and thus decrease their biodiversity especially ichthyodiversity. Hence, it is necessary to assemble most up-to-date list of fishes in the southern Caspian Sea basin, in general, and two important wetlands within this basin, the Anzali and Chamkhale Wetlands, in specific concern. The herein presented data can act as a baseline information for future investigation and conservation of fish diversity in these regions.

18.2 Material and Methods

18.2.1 Study Area

The study area (Figs. 18.1 and 18.2) covers the southern Caspian Sea basin in the Iranian territory and two wetlands within this basin including the Anzali and Chamkhale:

The Anzali Wetland (Talab) complex with about 15,000 ha is located in the western part of the southern Caspian Sea basin, northern Iran, Gilan/Guilan Province, north of Iran. It was designated as a Ramsar site on 23 June 1975 with site number 40 (Ramsar Site 2020). This internationally important wetland is a large, freshwater lagoon fed by several rivers (Chafroud, Bahambar, Morghak, Masal, Palangvar, Masolehrodkhan, Pasikhan, Siahdarvishan, Lakanroud, and Siahroud) and separated from the sea by a dune system, supporting extensive reedbeds, abundant submerged, and floating vegetation (Ramsar Site 2020), with very diverse flora and fauna including fishes. This Ramsar site was placed on the Montreux Record in 1993 due to change in water levels and increased nutrient enrichment, resulting in the quick spread of the reed species *Phragmites australis* (Ramsar Site 2020). The Chamkhale/Chamkhaleh Wetland and River is much smaller than Anzali

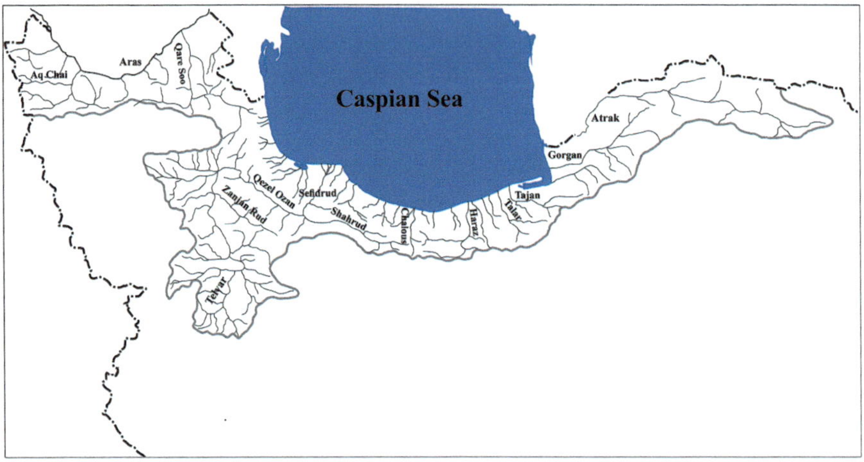

Fig. 18.2 The southern Caspian Sea basin and its major rivers

Wetland located at 37°12′49.75″N, 50°16′16.97″E (Fig. 18.1), East of Anzali, Guilan province.

18.2.2 Preparation of the Checklist

To compile this fish regional checklist, (1) the previously published articles (see the references cited in the discussion section) were reviewed, (2) ichthyological collections in Iran, e.g., ZM-CBSU, Zoological Museum of Shiraz University, Collection of Biology Department, Shiraz and Canada, CMNFI, Canadian Museum of Nature, Ottawa, were checked, and (3) specimens collected from extensive field expeditions up to 2020 from the regions were examined. Here a list of fish species adopting all changes from the other comprehensive checklists (e.g., Abbasi 2006a, b, 2017; Esmaeili et al. 2010, 2014a, b, 2017a, b, 2018; Jouladeh et al. 2015; Abbasi et al. 2019) is provided.

The checklist is presented in the form of a table (Table 18.1) which is ordered by alphabetic families and genera names according to the classification given by Nelson et al. (2016) and the Eschmeyer's Catalog of Fishes (Fricke et al. 2020). Alien (introduced/exotic), native, and endemic species are shown in Table 18.1. Additional notes on taxonomy of certain recorded species are also given in the discussion section. Photographs of several habitats (Appendix 1, Figs. 18.4–18.9) and fish species (Appendix 2) are provided.

Table 18.1 Checklist of freshwater fishes of the Caspian Sea basin and their presence in the Anzali Wetland (A) and Chamkhaleh Wetland (C)

Taxa	ST	A	C
Acheilognathinae			
Rhodeus amarus (Bloch, 1782) (Fig. 18.10, upper image form 1, lower image form 2)	N	*	*
Acipenseridae			
Acipenser baerii Brandt, 1869	I		
Acipenser gueldenstaedtii Brandt & Ratzeburg, 1833	N		
Acipenser nudiventris Lovetsky, 1828	N		
Acipenser persicus Borodin, 1897 (Fig. 18.11)	N	*	
Acipenser ruthenus Linnaeus, 1758	N		
Acipenser stellatus Pallas, 1771	N	*	
Huso huso (Linnaeus, 1758)	N		
Anguillidae			
Anguilla anguilla (Linnaeus, 1758) (Fig. 18.12)	I	*	
Atherinidae			
Atherina caspia Eichwald, 1831 (Fig. 18.13)	N	*	*
Channidae			
Channa micropeltes (Cuvier, 1831)	I	*	
Clupeidae			
Alosa braschnikowi (Borodin, 1904) (Fig. 18.14)	N	*	
Alosa caspia (Eichwald, 1838) (Fig. 18.15)	N	*	
Alosa kessleri (Grimm, 1887)	N	*	
Alosa saposchnikowii Grimm, 1885	N		
Alosa sphaerocephala (Berg, 1913)	N		
Clupeonella caspia Svetovidov, 1941	N	*	
Clupeonella engrauliformis (Borodin, 1904) (Fig. 18.16)	N		
Clupeonella grimmi Kessler, 1877 (Fig. 18.17)	N		
Cobitidae			
Cobitis faridpaki Mousavi-Sabet, Vasil'eva, Vatandoust & Vasil'ev, 2011 (Fig. 18.18)	NE		
Cobitis saniae Eagderi, Jouladeh-Roudbar, Jalili, Sayyadzadeh & Esmaeili, 2017	NE	*	*
Sabanejewia aurata (De Filippi, 1863) (Fig. 18.19)	N		
Sabanejewia caspia (Eichwald, 1838) (Fig. 18.20)	N	*	
Coregonidae			
Stenodus leucichthys (Güldenstaedt, 1772) (Fig. 18.21)	N?		
Cyprinidae			
Barbus cyri De Filippi, 1865 (Fig. 18.22)	N	*	*
Capoeta capoeta (Güldenstaedt, 1773)	N		
Capoeta kaput Levin, Prokofiev & Roubenyan, 2019 (Fig. 18.23)	N		
Capoeta razii Jouladeh-Roudbar, Eagderi, Ghanavi & Doadrio, 2017 (Fig. 18.24)	N	*	*
Carassius auratus (Linnaeus, 1758)	I	*	

(continued)

Table 18.1 (continued)

Taxa	ST	A	C
Carassius gibelio (Bloch, 1782) (Fig. 18.25)	I	*	*
Carassius langsdorfii Temminck & Schlegel, 1846	I	*	
Cyprinus carpio Linnaeus, 1758 (Fig. 18.26)	N/I	*	
Luciobarbus brachycephalus (Kessler, 1872)	N		
Luciobarbus capito (Güldenstaedt, 1773) (Fig. 18.27)	N	*	*
Luciobarbus caspius (Berg, 1914)	N	*	
Luciobarbus mursa (Güldenstaedt, 1773) (Fig. 18.28)	N		
Esocidae			
Esox lucius Linnaeus, 1758 (Fig. 18.29)	N	*	*
Gasterosteidae			
Gasterosteus aculeatus Linnaeus, 1758 (Fig. 18.30)	I	*	*
Pungitius platygaster (Kessler, 1859) (Fig. 18.31)	N	*	
Gobiidae			
Anatirostrum profundorum (Berg, 1927)	N		
Benthophilus baeri Kessler, 1877	N		
Benthophilus ctenolepidus Kessler, 1877	N	*	
Benthophilus granulosus Kessler, 1877	N		
Benthophilus leobergius Berg, 1949	N	*	
Benthophilus leptocephalus Kessler, 1877	N		
Benthophilus macrocephalus (Pallas, 1787)	N		
Benthophilus pinchuki Ragimov, 1982	N		
Benthophilus ragimovi Boldyrev & Bogutskaya, 2004	N		
Mesogobius nonultimus (Iljin, 1936)	N		
Neogobius caspius (Eichwald, 1831) (Fig. 18.32)	N	*	
Neogobius melanostomus (Pallas, 1814)	N	*	
Neogobius pallasi (Berg, 1916) (Fig. 18.33)	N	*	
Ponticola bathybius (Kessler, 1877)	N	*	
Ponticola cyrius (Kessler, 1874)	N		
Ponticola gorlap (Iljin, 1949) (Fig. 18.34)	N	*	*
Ponticola iranicus Vasil′eva, Mousavi-Sabet & Vasil′ev, 2015 (Fig. 18.35)	NE	*	*
Ponticola ratan (Nordmann, 1840)	N	*	
Ponticola syrman (Nordmann, 1840)	N	*	
Proterorhinus nasalis (De Filippi, 1863) (Fig. 18.36)	N	*	*
Gobionidae			
Pseudorasbora parva (Temminck & Schlegel, 1846) (Fig. 18.37)	I	*	*
Romanogobio macropterus (Kamensky, 1901)	N		
Leuciscidae			
Abramis brama (Linnaeus, 1758) (Fig. 18.38)	N	*	
Acanthobrama microlepis (De Filippi, 1863)	N		
Alburnoides eichwaldii (De Filippi, 1863)	N		
Alburnoides parhami Mousavi-Sabet, Vatandoust & Doadrio 2015	NE		*
Alburnoides samiii Mousavi-Sabet, Vatandoust & Doadrio, 2015	NE	*	

(continued)

Table 18.1 (continued)

Taxa	ST	A	C
Alburnoides tabarestanensis Mousavi-Sabet, Anvarifar & Azizi, 2015 (Fig. 18.39)	NE		
Alburnus chalcoides (Güldenstaedt, 1772)	N	*	*
Alburnus filippii Kessler, 1877	N	*	
Alburnus hohenackeri Kessler, 1877 (Fig. 18.40)	N	*	*
Ballerus sapa (Pallas, 1814)	N?		
Blicca bjoerkna (Linnaeus, 1758) (Fig. 18.41)	N	*	*
Chondrostoma cyri Kessler, 1877	N		
Leucaspius delineatus (Heckel, 1843)	N	*	
Leuciscus aspius (Linnaeus, 1758)	N	*	
Pelecus cultratus (Linnaeus, 1758)	N	*	
Rutilus frisii (Nordmann, 1840)	N	*	*
Rutilus lacustris (Pallas, 1814) (Fig. 18.42)	N	*	
Scardinius erythrophthalmus (Linnaeus, 1758) (Fig. 18.43)	N	*	
Squalius turcicus De Filippi, 1865	N	*	*
Vimba persa (Pallas, 1814)	N	*	*
Loricariidae			
Hypostomus plecostomus (Linnaeus, 1758)	I	*	
Lotidae			
Lota lota (Linnaeus, 1758)	N?		
Mugilidae			
Chelon auratus (Risso, 1810)	I	*	*
Chelon saliens (Risso, 1810)	I	*	*
Nemacheilidae			
Oxynoemacheilus bergianus (Derjavin, 1934)	N	*	
Oxynoemacheilus brandtii (Kessler, 1877)	N		
Oxynoemacheilus lenkoranensis (Abdurakhmanov, 1962)	N		
Oxynoemacheilus veyselorum Cicek, Eagderi & Sungur, 2018	N		
Paracobitis abrishamchiani Mousavi-Sabet, Vatandoust, Geiger& Freyhof, 2019	NE		
Paracobitis atrakensis Esmaeili, Mousavi-Sabet, Sayyadzadeh, Vatandoust & Freyhof, 2014 (Fig. 18.44)	N		
Paracobitis hircanica Mousavi-Sabet, Sayyadzadeh, Esmaeili, Eagderi, Patimar & Freyhof, 2015 (Fig. 18.45)	NE		
Oxudercidae			
Hyrcanogobius bergi Iljin, 1928	N		
Knipowitschia caucasica (Berg, 1916)	N	*	
Knipowitschia longecaudata (Kessler, 1877)	N		
Rhinogobius lindbergi Berg, 1933 (Fig. 18.46)	I	*	
Percidae			
Perca fluviatilis Linnaeus, 1758 (Fig. 18.47)	N	*	
Sander lucioperca (Linnaeus, 1758)	N	*	
Sander marinus (Cuvier, 1828)	N	*	

(continued)

Table 18.1 (continued)

Taxa	ST	A	C
Pangasiidae			
Pangasius sanitwongsei Smith, 1931	I	*	
Petromyzontidae			
Caspiomyzon wagneri (Kessler, 1870) (Fig. 18.48)	N	*	
Poeciliidae			
Gambusia holbrooki Girard, 1859	I	*	*
Poecilia reticulata Peters, 1859	I	*	
Salmonidae			
Oncorhynchus keta (Walbaum, 1792)	I?		
Oncorhynchus mykiss (Walbaum, 1792)	I	*	
Salmo caspius Kessler, 1877	N	*	
Salmo trutta Linnaeus, 1758 (Fig. 18.49)	N	*	
Serrasalmidae			
Piaractus brachypomus (Cuvier, 1818)	I	*	
Siluridae		*	
Silurus glanis Linnaeus, 1758 (Fig. 18.50)	N	*	
Syngnathidae			
Syngnathus caspius Eichwald, 1831	N	*	*
Tincidae			
Tinca tinca (Linnaeus, 1758) (Fig. 18.51)	N	*	
Xenocyprididae			
Ctenopharyngodon idella (Valenciennes, 1844)	I	*	
Hemiculter leucisculus (Basilewsky, 1855) (Fig. 18.52)	I	*	*
Hypophthalmichthys molitrix (Valenciennes, 1844)	I	*	
Hypophthalmichthys nobilis (Richardson, 1844)	I	*	
Mylopharyngodon piceus (Richardson, 1846)	I?	*?	
Total	116	75	25

N native, *I* introduced, *E* endemic, *?* record, * present

18.3 Results

In total, 116 species, belonging to 65 genera and 29 families, are listed here for the southern Caspian Sea basin of Iran. From these, 5 species have not recently been collected. There are 8 endemic and 11 exotic species. The Anzali and Chamkhale Wetlands present 75 (63% of the whole basin species) and 25 (21.5%) species. The Caspian Sea basin is characterized by presence of high diverse fishes of the Ponto-Caspian families, Gobiidae and Leuciscidae, with 20 species each. It is followed by Cyprinidae (12 species); Clupeidae (8 species); Acipenseridae and Nemacheilidae (each with 7 species); Xenocyprididae (5 species); Cobitidae, Oxudercidae, and Salmonidae (each with 4 species); and Percidae (3 species). The rest of families present two or one species (Fig. 18.3). The Caspian Sea basin has also experienced accidentally or intentionally introductions of alien/non-native fish species. The alien

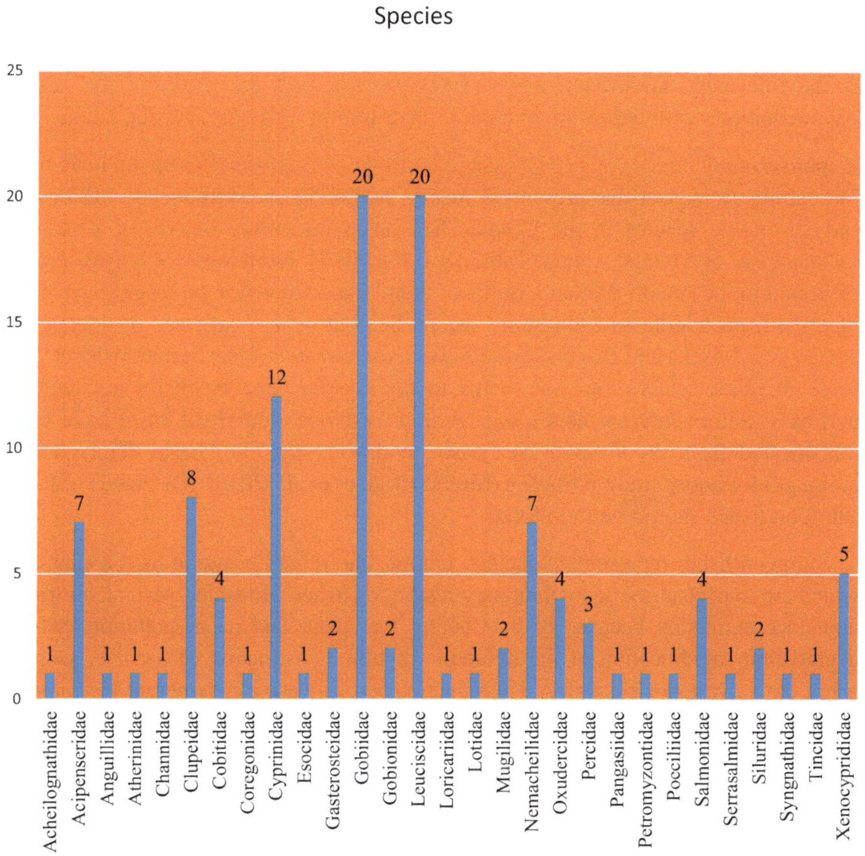

Fig. 18.3 Number of species in each family recorded from the southern Caspian Sea basin

species, *Anguilla anguilla*, *Carassius gibelio*, *Chelon auratus*, *Chelon saliens*, *Gasterosteus aculeatus*, *Hemiculter leucisculus*, *Gambusia holbrooki*, *Pseudorasbora parva*, and *Rhinogobius lindbergi*, have already established breeding populations. *Channa micropeltes* (Channidae), *Carassius langsdorfii* (Cyprinidae), *Hypostomus plecostomus* (Loricariidae), *Pangasius sanitwongsei* (Pangasiidae), *Poecilia reticulata* (Poeciliidae), and *Piaractus brachypomus* (Serrasalmidae) have been recently recorded from this basin (the Anzali Wetland). There is no record of alien cichlids (e.g., *Coptodon zillii*, *Oreochromis aureus*, and *O. niloticus*) which have been established in the southern water bodies of Iran and Iraq, in the Shadegan Wetland or in the Mesopotamian marshland complex and Arvand River (see Esmaeili et al. 2020 in press).

18.4 Discussion

In the following, additional notes on several fish taxa are provided for further argumentations with regard to taxonomy, distribution, and conservation status.

Acipenseridae
The family Acipenseridae presents 7 species including one introduced/translocated and six native species in the Caspian Sea basin inhabiting the rivers, estuaries, coastal areas, and inland waters (Table 18.1, Fig. 18.3). From these, only two species are found in the Anzali Wetland. Some of them (e.g., *Acipenser ruthenus*) have been reported from middle and southern Caspian Sea but never confirmed by specimens of Iran (see Naseka and Bogutskaya 2009). *Acipenser baerii* has been translocated to some research centers and fish farms in the Caspian Sea basin for aquaculture purposes. Some sturgeon fishes, e.g., *Acipenser baerii* and *Huso huso*, have been translocated from the Caspian Sea basin to the Persian Gulf basin (Tigris River drainage) for aquaculture purposes (Mousavi-Sabet et al. (2019). The status of these sturgeon fishes should be monitored.

Sturgeon fishes represent an ancient, unique, and relict lineage of the Caspian Sea which are important in the evolutionary history of fishes. However, several sturgeons now face extinction because of overfishing for caviar and meat, pollution, habitat degradation (modification of the hydraulic regime, construction of barriers, such as weirs, dams, irrigation canals), and introduction of the invasive ctenophore, *Mnemiopsis leidyi*. They are now considered as critically endangered or vulnerable species (see Esmaeili et al. 2017a, b, 2018; Abbasi et al. 2019; Jouladeh-Roudbar et al. 2020).

Coregonidae
There is only one coregonid species, *Stenodus leucichthys*, in the southern Caspian Sea basin. Members of this family were previously classified under the family Salmonidae, but it is now elevated to the family rank (Fricke et al. 2020; Kuljanishvili et al. 2020). This species is rare in Iranian waters. There is report of presence of a single specimen of *Stenodus leucichthys* (about 55 cm and 3500 gr) in the fish market at Fereydunkenar city in 2018 (Jouladeh-Roudbar et al. 2020). This native coregonid fish is listed as Extinct in the Wild in IUCN's Red Data Book due to illegal fishing, barrier construction, restricted access to spawning grounds, and environmental pollution (Freyhof and Kottelat 2008; Poursaeid and Falahatkar 2012).

Cyprinidae
There are 12 species of cyprinids in the southern Caspian Sea basin. *Barbus*, *Capoeta*, *Carassius*, *Cyprinus*, and *Luciobarbus* are the only genera present in the southern Caspian Sea basin. *Arabibarbus*, *Carasobarbus*, *Cyprinion*, *Garra*, and *Mesopotamichthys* genera, which are the main cyprinid fish elements of the Tigris-Euphrates River system, are absent in the Caspian Sea basin (Esmaeili et al. 2010; 2017a, b, 2018). The native *Barbus cyri* of the Caspian Sea basin is taxonomically replaced by *B. lacerta*, and *Luciobarbus brachycephalus*, *L. capito*, *L. caspius*, and *L. mursa* are replaced by diverse congeneric taxa, e.g., *L. barbulus*, *L. esocinus*,

L. kersin, and *L. xanthopterus* (Esmaeili et al. 2010, 2017a, b, 2018). In addition, *Capoeta capoeta*, *C. kaput*, and *C. razii* which are native elements of the Caspian Sea basin are replaced with other congeneric species, e.g., *Capoeta birunii*, *C. coadi*, *C. mandica*, *C. macrolepis*, and *C. trutta* in the Tigris-Euphrates River system and some endorheic basins (Alwan et al. 2016a, b; Zareian and Esmaeili 2017; Zareian et al. 2016, 2017).

The exotic genus *Carassius* presents three species, *Carassius auratus*, *C. gibelio*, and *C. langsdorfii* (Abbasi et al. 2019; Khosravi et al. 2020). Although *C. auratus* and *C. gibelio* can be distinguished by molecular characteristics (Rylková et al. 2013; Khosravi et al. 2020), however, there is no agreement on the use of morphological characters to separate these congeneric species (Kuljanishvili et al. 2020). It should be noted that several records of *C. auratus* may actually refer to *C. gibelio* and *C. langsdorfii*. However, it seems that ornamental aquarium *C. auratus* has low distribution range and population size.

Gobiidae

Neilson and Stepien (2009) included some *Ponticola ratan* specimens from the Black Sea basin in their phylogenetic analysis, but taxonomic status of *Ponticola goebelii* from the Caspian Sea basin based on molecular data still remains unexplored. Today, Eschmeyer's Catalog of Fishes (Fricke et al. 2020) recognizes both taxa, *Ponticola ratan* (distributed in the Black Sea basin and Sea of Azov) and *Ponticola goebelii* (distributed in the Caspian Sea basin), as valid species. However, molecular study is required to validate *Ponticola goebelii*.

Leuciscidae

Members of this family were previously classified under the family Cyprinidae, but it is now considered as a distinct family (Fricke et al. 2020; Kuljanishvili et al. 2020). There are 20 native/endemic species of leuciscids in the studied area. Taxonomic position of *Alburnoides* populations from the Caspian Sea basin is changed, and the populations are now designated as *A. eichwaldii*, *A. parhami*, *A. samiii*, and *A. tabarestanensis* (Mousavi-Sabet et al. 2015). The congeneric native taxa, *A. idignensis* and *A. nicolausi*, inhabit in the Tigris-Euphrates River system (see Bogutskaya and Coad 2009). *Alburnus chalcoides*, *A. filippii*, and *A. hohenackeri* are three native bleak fishes of the Caspian Sea basin. From these, *A. hohenackeri* has been translocated to the other basins/river drainages, and it is now found in sympatricity with the widely distributed bleak, *A. sellal*, in some water bodies of southern Iran (Mohammadian-Kalat et al. 2017; Esmaeili et al. 2017a, b, 2018).

The white-eye bream, *Ballerus sapa*, is a Ponto-Caspian element distributed in the basins of the Black, Caspian, and Aral seas. It is reported from Iran by Derzhavin (1934), Berg (1949), and Holčík and Oláh (1992) but is not found in recent years in the southern Caspian Sea basin.

The native species, *Acanthobrama macrolepis*, *Chondrostoma cyri*, and *Leuciscus aspius*, occur in the Caspian Sea basin, while *A. marmid*, *C. esmaeilii*, *C. regium*, *C. zagrosensis*, and *L. vorax* occur in the Tigris-Euphrates River system.

Rutilus caspicus from the Caspian Sea basin was treated as a synonym of *R. lacustris* based on the mtDNA sequences (Levin et al. 2017). This view is followed here too.

For a long time, the Caspian Sea *Rutilus kutum* populations and the Black Sea *Rutilus frisii* populations were treated as a valid taxonomic unit, e.g., species (see Bogutskaya and Iliadou 2006). However, based on (1) the molecular data presented by Kotlík et al. (2008), which proposed an extensive gene flow between the Caspian and Black Sea populations, and (2) unpublished COI molecular data, Kuljanishvili et al. (2020) synonymized *R. kutum* with *R. frisii*.

Lotidae

A single specimen of *Lota lota* was recorded from the lower reach of Safid River (Safidrud) in 1921, but apparently it is extremely rare in Iran. It has no recent record (Coad 2016).

Oxudercidae

Members of this family were previously classified under the family Gobiidae, but it is now considered as a distinct family (Fricke et al. 2020; Kuljanishvili et al. 2020). The family includes three native and one introduced species. The taxonomic status of the introduced freshwater goby of the genus *Rhinogobius* to Iran was reviewed by Sadeghi et al. (2019), and the introduced populations were considered as *R. lindbergi*.

Siluridae

The family Siluridae presents only one native species, *Silurus glanis*, in the Caspian Sea basin. The order Siluriformes comprises Bagridae, Heteropneustidae, Siluridae, and Sisoridae in the Persian Gulf basin. *Mystus pelusius*, *Silurus triostegus*, *Glyptothorax kurdistanicus*, and *G. silviae* are native siluriforms, and *Heteropneustes fossilis* is an alien species in the Tigris-Euphrates River system (Esmaeili et al. 2010, 2017a, b, 2018).

Xenocyprididae

There are five exotic species of xenocypridids in the Caspian Sea basin/Anzali Wetland. Members of this family were previously classified under the family Cyprinidae, but it is now considered as a distinct family (Fricke et al. 2020; Kuljanishvili et al. 2020). From these, *H. leucisculus* is the only fish that has been established in both wetlands and now produces breeding populations. There is no recent record of *Mylopharyngodon piceus* in Iran.

Absence of Aphaniids, Cichlids, and Mastacembelids

Members of the family Aphaniidae were previously classified under the family Cyprinodontidae, but it is now considered as a distinct family (Esmaeili et al. 2020; Fricke et al. 2020). Although 15 aphaniid species in three distinct genera, *Aphanius*, *Aphaniops*, and *Paraphanius* (Esmaeili et al. 2020), occur in the inland waters of Iran including the neighboring basins (Urmia, Namak, Kavir), however, there is no record of any *aphaniids* in the Caspian Sea basin.

Members of Cichlidae family were previously classified under the order Perciformes, but it is now considered to be a distinct order (Cichliformes) (see Nelson et al. 2016; Betancur-R et al. 2017). There are two endemic species, *Iranocichla hormuzensis* and *I. persa*, which are distributed in the lower river drainages of the Persian Gulf and Oman Sea in Iran and three exotic cichlid species, *Coptodon zillii*, *Oreochromis aureus*, and *O. niloticus*, in Iran and Iraq (Coad 1982; Mutlak and Al-Faisal 2009; Al-Faisal and Mutlak 2014; Khaefi et al. 2014; Esmaeili et al. 2016, 2017a, 2018; Valikhani et al. 2018; Mohamed and Al-Wan 2020a, b).

Presence of the mastacembelid fish, *Mastacembelus mastacembelus*, is one of the characteristic features of the biodiversity of Tigris-Euphrates River system which is absent in several inland basins (e.g., Caspian Sea, Esfahan, Namak, Kavir, and Urmia Lake).

18.5 Conclusion

The ichthyodiversity of the Caspian Sea basin with 116 species (no recent record for 5 species) (65 genera and 29 families) is significantly richer than any other inland basins of Iran (see Esmaeili et al. 2017a, b, 2018; Jouladeh-Roudbar et al. 2020). The high diversity is observed in the Anzali Wetland as compared to other wetlands located in the southern Caspian Sea basin. It harbors 75 native and exotic species although there is no record for some species. While the Chamkhale Wetland comprises 25 species (present study), the Amirkelayeh Wetland has 15 species (Nezami Baluchie and Khara 2004), Kiashahr National Park includes 25 species (Khara and Nezami Baluchie 2005), and Gomishan Wetland harbors 15 species (Patimar et al. 2009). The fish species richness of the southern Caspian Sea basin can be linked to the permanent connection with the Caspian Sea, variable habitat structure, and the larger area in comparison with the other basins. The total number of recorded species is changing due to the newly described species, changing the taxonomic status due to synonymy, species concept (phylogenetic vs. biological), new introductions, using new techniques and tools, and number of ichthyologists involved.

The high sedimentation in various rivers and wetlands, pollution, barrier construction, channelization, industrial developmental projects, modernization, and intentionally or accidentally introduction of exotic elements have affected the population size of several species resulting in (1) possible extinction of some in the wild (*Stenodus leucichthys*) and (2) assignment of several others in the list of threatened species of IUCN (sturgeon fishes) and low catch of the native roaches (*Rutilus* spp.). Hence, regular monitoring and conducting conservation and management programs/plans are needed to be considered.

Acknowledgments I would like to thank F. Zarei for preparing the map and G. Sayyadzadeh, H. Mehraban, F. Zarei, M. Masoudi, Zammaniannejad, R. Khaefi, and Z. Ganjali for helping with fish collections. This study was financially supported by Shiraz University.

Appendices

Appendix 1: Representative Habitats of the Southern Caspian Sea Basin

Fig. 18.4 Western parts of the Anzali Wetland

Fig. 18.5 Anzali Wetland

Fig. 18.6 Outlet region of the Anzali Wetland

Fig. 18.7 Chamkhale Wetland

Fig. 18.8 Chamkhale River

Fig. 18.9 Safid River (Safidrud)

Appendix 2: Representative Fish Species in the Southern Caspian Sea Basin. The Order of Appearance of Species Is in Accordance with Table 18.1

Fig. 18.10 *Rhodeus amarus* (Acheilognathinae)

Fig. 18.11 *Acipenser persicus* (Acipenseridae)

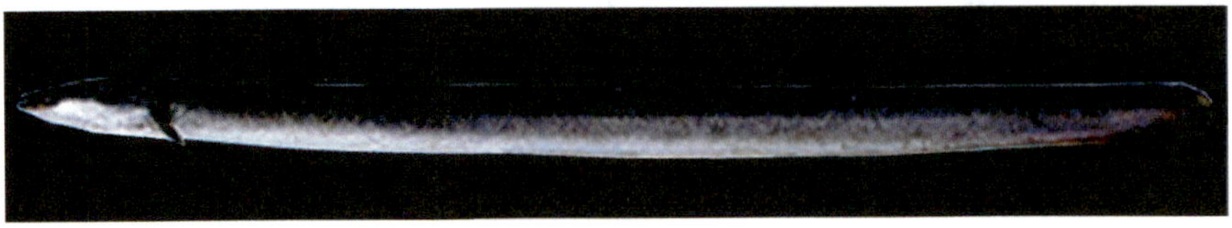

Fig. 18.12 *Anguilla anguilla* (Anguillidae)

Fig. 18.13 *Atherina caspia* (Atherinidae)

Fig. 18.14 *Alosa braschnikowi* (Clupeidae)

Fig. 18.15 *Alosa caspia* (Clupeidae)

Fig. 18.16 *Clupeonella engrauliformis* (Clupeidae)

Fig. 18.17 *Clupeonella grimmi* (Clupeidae)

Fig. 18.18 *Cobitis faridpaki* (Cobitidae)

Fig. 18.19 *Sabanejewia aurata* (Cobitidae)

Fig. 18.20 *Sabanejewia caspia* (Cobitidae)

Fig. 18.21 *Stenodus leucichthys* (Coregonidae)

Fig. 18.22 *Barbus cyri* (Cyprinidae)

Fig. 18.23 *Capoeta kaput* (Cyprinidae)

Fig. 18.24 *Capoeta razii* (Cyprinidae)

Fig. 18.25 *Carassius gibelio* (Cyprinidae)

Fig. 18.26 *Cyprinus carpio* (Cyprinidae)

Fig. 18.27 *Luciobarbus capito* (Cyprinidae)

Fig. 18.28 *Luciobarbus mursa* (Cyprinidae)

Fig. 18.29 *Esox lucius* (Esocidae)

Fig. 18.30 *Gasterosteus aculeatus* (Gasterosteidae)

Fig. 18.31 *Pungitius platygaster* (Gasterosteidae)

Fig. 18.32 *Neogobius caspius* (Gobiidae)

Fig. 18.33 *Neogobius pallasi* (Gobiidae)

Fig. 18.34 *Ponticola gorlap* (Gobiidae)

Fig. 18.35 *Ponticola iranicus* (Gobiidae)

Fig. 18.36 *Proterorhinus nasalis* (Gobiidae)

Fig. 18.37 *Pseudorasbora parva* (Gobionidae)

Fig. 18.38 *Abramis brama* (Leuciscidae)

Fig. 18.39 *Alburnoides tabarestanensis* (Leuciscidae)

Fig. 18.40 *Alburnus hohenackeri* (Leuciscidae)

Fig. 18.41 *Blicca bjoerkna* (Leuciscidae)

Fig. 18.42 *Rutilus lacustris* (Leuciscidae)

Fig. 18.43 *Scardinius erythrophthalmus* (Leuciscidae)

Fig. 18.44 *Paracobitis atrakensis* (Nemacheilidae)

Fig. 18.45 *Paracobitis hircanica* (Nemacheilidae)

Fig. 18.46 *Rhinogobius lindbergi* (Oxudercidae)

Fig. 18.47 *Perca fluviatilis* (Persidae)

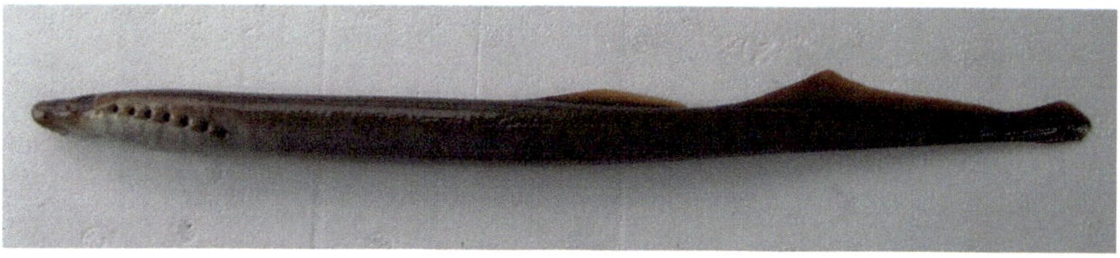

Fig. 18.48 *Caspiomyzon wagneri* (Petromyzontidae)

Fig. 18.49 *Salmo trutta* (Salmonidae)

Fig. 18.50 *Silurus glanis* (Siluridae)

Fig. 18.51 *Tinca tinca* (Tincidae)

Fig. 18.52 *Hemiculter leucisculus* (Xenocyprididae)

References

Abbasi K (2006a) Identification and distribution of fish fauna in Shafarud River, Guilan Province. Iran Sci Fish J 15(2):73–86. [in Persian]

Abbasi K (2006b) Identification and distribution of fish fauna in Hevigh River (Guilan Province). Iran J Biol 18(4):370–382. [in Persian]

Abbasi K (2017) Fishes of Guilan. Encycl Guilan Culture Civil 66:205 p. [in Persian]

Abbasi K, Moradi M, Mirzajani A, Nikpour M, Zahmatkesh Y, Abdoli A, Mousavi-Sabet H (2019) Ichthyo-diversity in the Anzali Wetland and its related rivers in the southern Caspian Sea basin, Iran. J Anim Divers 1(2):90–135

Al-Faisal AJ, Mutlak FM (2014) First record of the Nile tilapia *Oreochromis niloticus* (Linnaeus, 1758), from the Shatt Al-Arab River, South Iraq. Mesopotamian J Mar Sci 29(1):45–50

Alwan NH, Zareian H, Esmaeili HR (2016a) *Capoeta coadi*, a new species of cyprinid fish from the Karun River drainage, Iran based on morphological and molecular evidences (Teleostei, Cyprinidae). ZooKeys 572:155–180. https://doi.org/10.3897/zookeys.572.7377

Alwan N, Esmaeili HR, Krupp F (2016b) Molecular phylogeny and zoogeography of the *Capoeta damascina* species complex (Pisces: Teleostei: Cyprinidae). PLoS One 11(6):e0156434. https://doi.org/10.1371/journal.pone.0156434

Berg LS (1949) Freshwater fishes of the USSR and adjacent countries. Program for Scientific Translations, Jerusalem. 496 p

Betancur-R R, Wiley EO, Arratia G, Acero A, Bailly N, Miya M et al (2017) Phylogenetic classification of bony fishes. BMC Evol Biol 17(1):162

Bogutskaya NG, Coad BW (2009) A review of vertebral and fin-ray counts in the genus *Alburnoides* (Teleostei: Cyprinidae) with a description of six new species. Zoosyst Rossica 18(1):126–173

Bogutskaya NG, Iliadou K (2006) *Rutilus panosi*, a new roach from Western Greece (Teleostei: Cyprinidae). Zoosyst Rossica 14(2):293–298

Coad BW (1982) A new genus and species of cichlid endemic to southern Iran. Copeia 1982(1):28–37

Coad BW (2016) Review of the freshwater Cuskfishes of Iran (Family Lotidae). Iran J Ichthyol 3 (4):229–235

Derzhavin AN (1934) Presnovodnye ryby yuzhnogo poberezh'ya Kaspiya. Vstuplenie [Freshwater fishes of the southern shore of the Caspian Sea. Introduction]. Trudy Azerbaidzhanskogo Otdeleniya Zakavkazskogo Filiala Akademii Nauk SSSR, Sektor Zoologii, Baku 7:91–126

Esmaeili HR, Coad BW, Gholamifard A, Teimory A (2010) Annotated checklist of the freshwater fishes of Iran. Zoosyst Rossica 19:361–386

Esmaeili HR, Coad BW, Mehraban HR, Masoudi M, Khaefi R, Abbasi K, Mostafavi H, Vatandoust S (2014a) An updated checklist of fishes of the Caspian Sea basin of Iran with a note on their zoogeography. Iran J Ichthyol 1(3):152–184

Esmaeili HR, Teimori A, Feridon OWFI, Abbasi K, Brian WC (2014b) Alien and invasive freshwater fish species in Iran: Diversity, environmental impacts and management. Iran J Ichthyol 1(2):61–72

Esmaeili HR, Sayyadzadeh G, Seehausen O (2016) *Iranocichla persa*, a new cichlid species from southern Iran (Teleostei, Cichlidae). ZooKeys 636:141–161

Esmaeili HR, Mehraban H, Abbasi K, Keivany Y, Brian WC (2017a) Review and updated checklist of freshwater fishes of Iran: Taxonomy, distribution and conservation status. Iran J Ichthyol 4 (Suppl. 1):1–114

Esmaeili HR, Masoudi M, Chermahini MA, Esmaeili AH, Zarei F, Ebrahimi M (2017b) Invasion of the Neotropical and Nearctic fishes to Iran. FishTaxa 2(3):126–133

Esmaeili HR, Sayyadzadeh G, Eagderi S, Abbasi K (2018) Checklist of freshwater fishes of Iran. FishTaxa 3(3):1–95

Esmaeili HR, Teimori A, Zarei F, Sayyadzadeh G (2020) DNA barcoding and species delimitation of the Old World tooth-carps, family Aphaniidae Hoedeman, 1949 (Teleostei: Cyprinodontiformes). PLoS One 15(4):e0231717

Freyhof J, Kottelat M (2008) *Stenodus leucichthys*. The IUCN red list of threatened species 2008: e. T20745A9229071

Fricke R, Eschmeyer WN, Van der Laan R (eds) (2020) Eschmeyer's catalog of fishes: Genera, species, references. Retrieved from http://researcharchive.calacademy.org/research/ichthyology/catalog/fishcatmain.asp

Holcik J, Oláh J (1992) Fish, fisheries and water quality in Anzali Lagoon and its watershed. Report prepared for the project – Anzali Lagoon productivity and fish stock investigations. Food and Agriculture Organization, Rome, FI: UNDP/IRA/88/001 Field Document 2: x + 109 pp

Jouladeh-Roudbar A, Vatandoust S, Eagderi S, Jafari-Kenari S, Mousavi-Sabet H (2015) Freshwater fishes of Iran; an updated checklist. Aquacult Aquar Conserv Legis 8(6):855–909

Jouladeh-Roudbar A, Ghanavi HR, Doadrio I (2020) Ichthyofauna from Iranian freshwater: Annotated checklist, diagnosis, taxonomy, distribution and conservation assessment. Zool Stud 59:21

Karami M, Hutterer R, Benda P, Siahsarvie R, Kryštufek B (2008) Annotated check-list of the mammals of Iran. Lynx, series nova 39(1):63–102

Khaefi R, Esmaeili HR, Zareian H, Babaei S (2014) The first record of the redbelly tilapia, *Tilapia zillii* Gervais, 1848 in freshwaters of Iran. Turk J Zool 38:96–98

Khaleghizadeh A, Roselaar K, Scott DA, Tohidifar M, Mlíkovský J, Blair M, Kvartalnov P (2017) Birds of Iran: Annotated checklist of the species and subspecies. Iranian Research Institute of Plant Protection, 350 p

Khara H, Nezami Baluchie S (2005) Studying fish biodiversity and abundance in Boujagh Wetland of Kiashahr, South-Western Caspian Sea. Iran Sci Fish J 13(4):41–54. [in Persian]

Khorami Pour SK, Monavari SM, Riazi B, Khorasani N (2015) Caspian rapid assessment method: A localized procedure for assessment of wetlands at southern fringe of the Caspian Sea. Environ Monit Assess 187(7):420

Khosravi M, Abdoli A, Ahmadzadeh F, Saberi-Pirooz R, Rylková K, Kiabi BH (2020) Toward a preliminary assessment of the diversity and origin of cyprinid fish genus *Carassius* in Iran. J Appl Ichthyol 36(4):422–430. https://doi.org/10.1111/jai.14039

Kotlík P, Marková S, Choleva L, Bogutskaya NG, Ekmekci FG, Ivanova PP (2008) Divergence with gene flow between Ponto-Caspian refugia in an anadromous cyprinid *Rutilus frisii* revealed by multiple gene phylogeography. Mol Ecol 17:1076–1088. https://doi.org/10.1111/j.1365-294X.2007.03638.x

Kuljanishvili T, Epitashvili G, Freyhof J, Japoshvili B, Kalous L, Levin B et al (2020) Checklist of the freshwater fishes of Armenia, Azerbaijan and Georgia. J Appl Ichthyol 36(4):501–514. https://doi.org/10.1111/jai.14038

Levin BA, Simonov EP, Ermakov OA, Levina MA, Interesova EA, Kovalchuk OM, Malinina YA, Mamilov NS, Mustafayev NJ, Pilin DV, Pozdeev IV (2017) Phylogeny and phylogeography of the roaches, genus *Rutilus* (Cyprinidae), at the eastern part of its range as inferred from mtDNA analysis. Hydrobiologia 788(1):33–46

Mohamed ARM, Al-Wan SM (2020a) Biological aspects of an invasive species of *Oreochromis niloticus* in the Garmat Ali River, Basrah, Iraq. J Agric Vet Sci 13(2):15–26

Mohamed ARM, Al-Wan SM (2020b) Evaluation of biological characters of the invasive species, *Coptodon zillii* in the Garmat Ali River, Basrah, Iraq. Int J Fish Aquat Stud 8(2):176–185

Mohammadian-Kalat T, Esmaeili HR, Aliabadian M, Freyhof J (2017) Re-description of *Alburnus doriae*, with comments on the taxonomic status of *A. amirkabiri*, *A. mossulensis*, *A. sellal* and *Petroleuciscus esfahani* (Teleostei: Cyprinidae). Zootaxa 4323(4):487–502

Mousavi-Sabet H, Vatandoustm S, Doadrio I (2015) Review of the genus *Alburnoides* Jeitteles, 1861 (Actinopterygii, Cyprinidae) from Iran with description of three new species from the Caspian Sea and Kavir basins. Casp J Environ Sci 13(4):293–331

Mousavi-Sabet H, Salehi M, Sarpanah A, Pirali Kheirabadi E (2019) First records of *Acipenser baerii* and *Huso huso* (Actinopterygii: Acipenseriformes: Acipenseridae) from the Tigris–Euphrates basin, Iran. Acta Ichthyol Piscat 49(3):265–267

Mutlak FM, Al-Faisal AJ (2009) A new record of two exotic cichlids fish *Oreochromis aureus* Steindacher, 1864 and *Tilapia zilli* Gervais, 1848 from south of the main outfall drain in Basrah city. Mesop J Mar Sci 24(2):160–170

Naseka NM, Bogutskaya NG (2009) Fishes of the Caspian Sea: Zoogeography and updated checklist. Zoosyst Rossica 18:295–317

Neilson ME, Stepien CA (2009) Escape from the Ponto-Caspian: Evolution and biogeography of an endemic goby species flock (Benthophilinae: Gobiidae: Teleostei). Mol Phylogenet Evol 52 (1):84–102

Nelson JS, Grande TC, Wilson MV (2016) Fishes of the world, 5th edn. Wiley, Hoboken

Nezami Baluchie S, Khara H (2004) Species composition and abundance of fishes in Amirkelayeh Wetland. Iran Sci Fish J 12(4):193–206. [in Persian]. https://doi.org/10.22092/isfj.2004.113709

Patimar R, Abdoli A, Hasanzade Kiabi B, Allahyari S, Naderi Jelowdar M (2009) Fish species diversity of the coastal areas in Gomishan Wetland. J Agric Sci Nat Resour 1-A):10 pp. [in Persian]

Poursaeid S, Falahatkar B (2012) Threatened fishes of the world: *Stenodus leucichthys leucichthys* Guldenstadt, 1772 (Salmonidae). Aqua, Int J Ichthyol 18:31–35

Ramsar Site (2020). www.ramsar.org

Rylková, Kalous L, Bohlen J, Lamatsch DK, Petrtýl M (2013) Phylogeny and biogeographic history of the cyprinid fish genus Carassius (Teleostei: Cyprinidae) with focus on natural and anthropogenic arrivals in Europe. Aquaculture 380:13–20

Sabeti H (1997) Forest, trees and shrubs of Iran. Yazd, University Pub, Yazd. 806 p

Sadeghi R, Esmaeili HR, Zarei F, Esmaeili A, Abbasi K (2019) The taxonomic status of an introduced freshwater goby of the genus *Rhinogobius* to Iran (Teleostei: Gobiidae). Zool Middle East 65(1):51–58

Safaei-Mahroo B, Ghaffari H, Fahimi H, Broomand S, Yazdanian M, Najafi-Majd E et al (2015) The herpetofauna of Iran: Checklist of taxonomy, distribution and conservation status. Asiatic Asian Herpetol Res 6(4):257–290

Valikhani H, Abdoli A, Kiabi BH, Nejat F, Sadeghsaba M, Khosravi M (2018) A study on the status of invasive tilapia species (*Coptodon zillii* Gervais, 1848 and *Oreochromis aureus* Steindachner, 1864) in the aquatic ecosystems of Khuzestan Province, Iran. Environ Sci 15 (4):29–44

Zareian H, Esmaeili HR (2017) Mitochondrial phylogeny and taxonomic status of the *Capoeta damascina* species group (Actinopterygii: Cyprinidae) in Iran with description of a new species. Iran J Ichthyol 4:231–269

Zareian H, Esmaeili HR, Freyhof J (2016) *Capoeta anamisensis*, a new species from the Minab and Hasan Langhi River drainages in Iran (Teleostei: Cyprinidae). Zootaxa 4083:126–142. https://doi.org/10.11646/zootaxa.4083.1.7

Zareian H, Esmaeili HR, Gholamhosseini A, Japoshvili B, Özuluğ M, Mayden RL (2017) Diversity, mitochondrial phylogeny, and ichthyogeography of the *Capoeta capoeta* complex (Teleostei: Cyprinidae). Hydrobiologia 806:363–409. https://doi.org/10.1007/s10750-017-3375-0

Chapter 19
The Ornithological Importance of the Southern Marshes of Iraq

Mudhafar A. Salim, Salwan Ali Abed, and R. F. Porter

Abstract The Key Biodiversity Areas (KBA) surveys conducted by Nature Iraq and the Iraqi Ministry of Environment from 2005 to 2011, followed by surveys by the Iraqi Organization for Conservation of Nature (IOCN), have greatly increased our knowledge of the ornithological diversity of the Mesopotamian Marshes in southern Iraq.

Keywords Ornithology · Red List species · Endemic species · Key Biodiversity Areas (KBAs) · Conservation · Threats

Together with observations by ornithologists in the last century, there is now a good understanding of the bird populations of this, the largest wetland complex in the Middle East. A total of 264 species have been recorded in the marshes and their environs. Of these 77 have been found breeding; a further 11 may breed. Fifty-four of the breeding species are resident in the marshes though some also have migratory populations. Some 197 species are regular winter visitors or passage migrants, and a further 20 species are rare visitors or vagrants having been recorded on less than 3 occasions. Some species that were regular winter visitors or passage migrants 50 or more years ago no longer occur. Despite the drainage faced by the marshes that started in the 1980s, no breeding bird species has been lost. In terms of conservation priority, 16 species that regularly occur are red listed by IUCN/BirdLife

M. A. Salim
Natural Heritage Division, Arab Regional Centre for World Heritage (ARC-WH), Manama, Kingdom of Bahrain
e-mail: m.salim@arcwh.org

S. A. Abed (✉)
College of Science, University of Al-Qadisiyah, Al Diwaniyah, Iraq
e-mail: salwan.abed@qu.edu.iq

R. F. Porter
Adviser to Middle East Programme, BirdLife International, Cambridge, UK
e-mail: RFPorter@talktalk.net

L. A. Jawad (ed.), *Southern Iraq's Marshes*, Coastal Research Library 36,
https://doi.org/10.1007/978-3-030-66238-7_19

International as globally threatened or near threatened. Also, an additional four breeding species are regionally threatened, and a further six breeding species are regionally near threatened. Furthermore the marshes hold globally or regionally important populations of 30 species, 2 of which, the Basra reed warbler *Acrocephalus griseldis* and Iraq babbler *Turdoides altirostris*, are endemic to Iraq. The economic value of birds, their threats and conservation are discussed as well as opportunities and recommendations for future actions.

19.1 Introduction

The marshes of southern Iraq (Fig. 19.1a, b) are the largest and most important wetland complex for wildlife in the Middle East (Evans 1994, Scott 1995, Salim et al. 2009). In this paper these marshes are defined as the permanent wetlands of the Central Marshes, East Hammar, West Hammar and Hawizeh. Maps and satellite images show that the maximum area of the marshes was >10,000 km^2 in the 1970s, but was reduced to 14% of the original size at the peak of drainage in 2002 (New Eden 2006). Currently, based on the recordings of Centre for Restoration of Iraqi Marshlands (CRIM), the area of the marshes ranges between 3500 and 4000 km^2 (CRIM 2013).

Historically and currently they regularly support over 250 species of breeding, wintering or migrating birds, most importantly waterbird species (see, e.g. Allouse 1953, 1960, 1961, 1962; Scott and Carp 1982; Salim et al. 2012). Extensive surveys undertaken by Nature Iraq from 2004 to 2011(Nature Iraq 2017) followed by those of the IOCN have shown that, despite the drainage undertaken in the 1980s and 1990s, no species has become extinct as a breeding bird in the marshes of southern Iraq.

A total of 264 species have been recorded in the marshes and their immediate environs (Table 19.1). Of these 77 have been found breeding and a further 11 may breed. Fifty-four of the breeding species are resident in the marshes though some also have populations that migrate. A total of 197 are regular winter visitors or passage migrants from Europe and Asia. A further 20 species are rare visitors or vagrants, the latter having been recorded on less than 3 occasions (see Porter et al. 2010; Salim et al. 2006, 2012). Some species that were regular winter visitors or passage migrants 50 or more years ago no longer occur; these include Pallas's fish eagle *Haliaeetus leucoryphus*, white-tailed eagle *Haliaeetus albicilla* and slender-billed curlew *Numenius tenuirostris*. The latter is probably now globally extinct (Kirwan et al. 2015).

In terms of conservation priority, 25 species that regularly occur are red listed as globally threatened or near threatened (Table 19.2). Of these the marshes are considered to be of the greatest importance for marbled duck *Marmaronetta angustirostris*, ferruginous duck *Aythya nyroca*, common pochard *Aythya ferina*, greater spotted eagle *Clanga clanga*, Eastern imperial eagle *Aquila heliaca*, Basra reed warbler and probably black-tailed godwit *Limosa limosa*. An additional four breeding species have been recently designated as regionally threatened (Symes

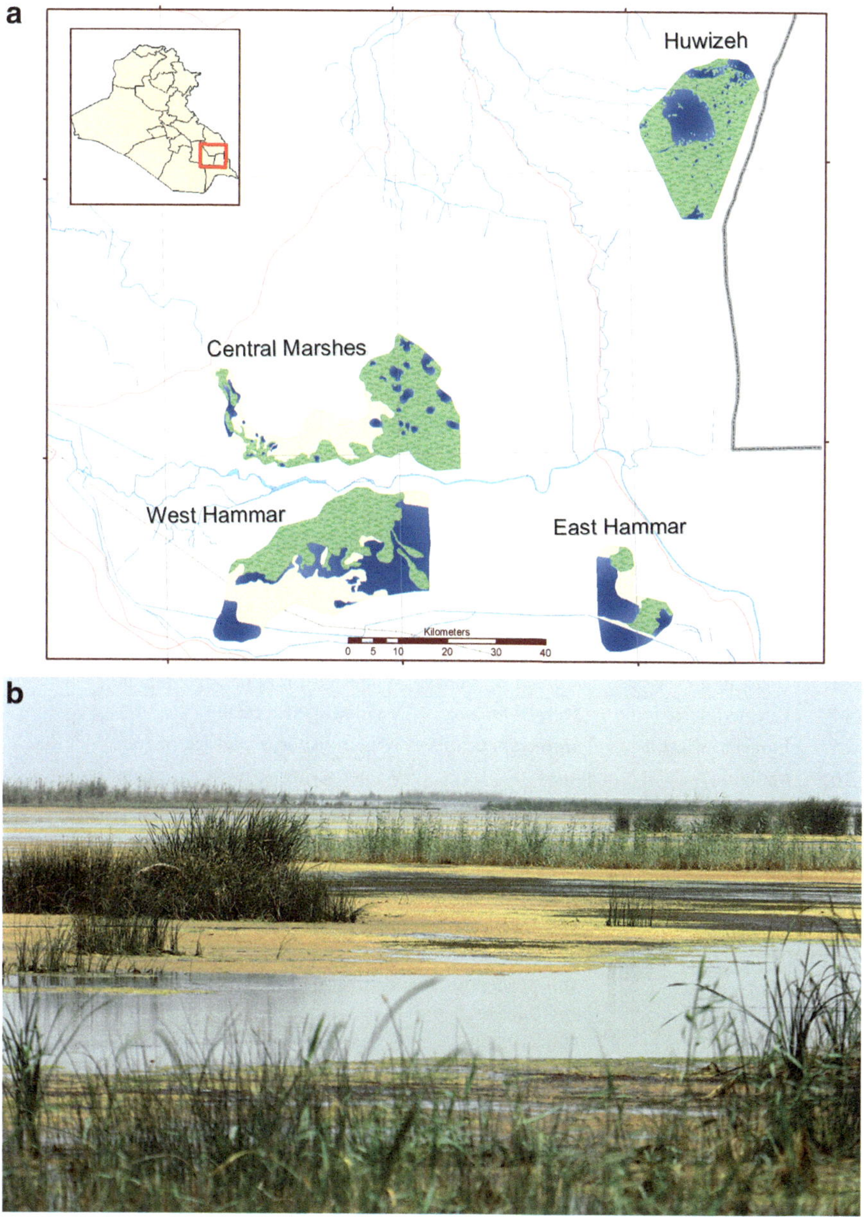

Fig. 19.1 (**a**), map showing the location of the southern marshes of Iraq (Ahwar). (Image courtesy of IMOE, 2014). (**b**), the predominant landscape of the environment of the marshes of Iraq. (Image courtesy of Mudhafar A. Salim)

Table 19.1 The status of birds recorded in the Mesopotamian Marshes of southern Iraq

#	English name	Scientific name	Status in the marshes
1.	Black francolin	*Francolinus (f.) francolinus*	Resident breeder
2.	Common quail	*Coturnix coturnix*	Passage migrant; may breed
3.	Eastern greylag goose	*Anser anser rubrirostris*	Resident breeder, historically; winter visitor
4.	Greater white-fronted goose	*Anser albifrons*	Uncommon winter visitor
5.	Lesser white-fronted goose	*Anser erythropus*	Uncommon winter visitor
6.	Red-breasted goose	*Branta ruficollis*	Rare winter visitor historically
7.	Mute swan	*Cygnus olor*	Very rare winter visitor
8.	Bewick's swan	*Cygnus columbianus bewickii*	Rare winter visitor
9.	Whooper swan	*Cygnus cygnus*	Very rare winter visitor
10.	Common shelduck	*Tadorna tadorna*	Winter visitor
11.	Ruddy shelduck	*Tadorna ferruginea*	Winter visitor & passage migrant
12.	Cotton pygmy goose	*Nettapus coromandelianus*	Vagrant – historically
13.	Gadwall	*Mareca strepera*	Winter visitor & passage migrant
14.	Falcated duck	*Mareca falcata*	Vagrant – historically
15.	Eurasian wigeon	*Mareca penelope*	Winter visitor & passage migrant
16.	Mallard	*Anas platyrhynchos*	Winter visitor & passage migrant
17.	Northern pintail	*Anas acuta*	Winter visitor & passage migrant
18.	Eurasian teal	*Anas crecca*	Winter visitor & passage migrant
19.	Northern shoveler	*Spatula clypeata*	Winter visitor & passage migrant; may breed
20.	Garganey	*Spatula querquedula*	Passage migrant; may breed
21.	Marbled duck	*Marmaronetta angustirostris*	Resident breeder; winter visitor
22.	Red-crested pochard	*Netta rufina*	Resident breeder; winter visitor & passage migrant
23.	Common pochard	*Aythya ferina*	Winter visitor & passage migrant
24.	Ferruginous duck	*Aythya nyroca*	Resident breeder; winter visitor & passage migrant
25.	Tufted duck	*Aythya fuligula*	Winter visitor & passage migrant
26.	Greater scaup	*Aythya marila*	Rare winter visitor
27.	Smew	*Mergellus albellus*	Rare winter visitor; commoner historically
28.	Goosander	*Mergus serrator*	Vagrant, historically
29.	Red-breasted merganser	*Mergus merganser*	Vagrant, historically
30.	White-headed duck	*Oxyura leucocephala*	Uncommon winter visitor

(continued)

Table 19.1 (continued)

#	English name	Scientific name	Status in the marshes
31.	Little grebe	*Tachybaptus ruficollis*	Resident breeder (endemic race *iraquensis*); winter visitor
32.	Great crested grebe	*Podiceps cristatus*	Winter visitor
33.	Black-necked grebe	*Podiceps nigricollis*	Winter visitor
34.	Greater flamingo	*Phoenicopterus roseus*	Winter visitor & passage migrant
35.	Black stork	*Ciconia nigra*	Rare passage migrant
36.	Western white stork	*Ciconia c. ciconia*	Passage migrant & winter visitor
37.	African sacred ibis	*Threskiornis aethiopicus*	Rare resident breeder
38.	Glossy ibis	*Plegadis falcinellus*	Rare resident breeder; winter visitor & passage migrant
39.	Eurasian spoonbill	*Platalea leucorodia*	Rare resident breeder; passage migrant
40.	Eurasian bittern	*Botaurus stellaris*	Winter visitor & passage migrant; may breed
41.	Little bittern	*Ixobrychus minutus*	Breeding summer visitor; passage migrant; some winter
42.	Black-crowned night heron	*Nycticorax nycticorax*	Resident or breeding summer visitor; passage migrant
43.	Squacco heron	*Ardeola ralloides*	Breeding summer visitor; passage migrant; some may winter
44.	Western cattle egret	*Bubulcus ibis*	Resident and breeding summer visitor; passage migrant
45.	Grey heron	*Ardea cinerea*	Rare resident breeder; winter visitor & passage migrant
46.	Goliath heron	*Ardea goliath*	Very rare resident breeder
47.	Purple heron	*Ardea purpurea*	Breeding summer visitor; passage migrant; a few winter
48.	Western great egret	*Ardea a. alba*	Winter visitor & passage migrant; some remain in summer and may breed
49.	Little egret	*Egretta garzetta*	Resident or breeding summer visitor; winter visitor & passage migrant
50.	Great white pelican	*Pelecanus onocrotalus*	Winter visitor & passage migrant
51.	Dalmatian pelican	*Pelecanus crispus*	Rare winter visitor
52.	Pygmy cormorant	*Microcarbo pygmaeus*	Resident breeder; winter visitor
53.	Continental great cormorant	*Phalacrocorax sinensis*	Winter visitor & passage migrant

(continued)

Table 19.1 (continued)

#	English name	Scientific name	Status in the marshes
54.	African darter	*Anhinga rufa chantrei*	Rare resident breeder
55.	Western osprey	*Pandion haliaetus*	Uncommon winter visitor & passage migrant
56.	European honey buzzard	*Pernis apivorus*	Passage migrant or vagrant
57.	Black-winged kite	*Elanus caeruleus*	Uncommon resident; winter visitor
58.	Eurasian black kite	*Milvus migrans*	Winter visitor & passage migrant
59.	Black-eared kite	*Milvus migrans lineatus*	Status uncertain; has been recorded in winter
60.	Pallas's fish eagle	*Haliaeetus leucoryphus*	Former winter visitor, not recorded since the 1940s
61.	White-tailed eagle	*Haliaeetus albicilla*	Former winter visitor, not recorded since the 1940s
62.	Short-toed snake eagle	*Circaetus gallicus*	Passage migrant
63.	Western marsh harrier	*Circus aeruginosus*	Local breeding resident; winter visitor & passage migrant
64.	Hen harrier	*Circus cyaneus*	Winter visitor & passage migrant
65.	Pallid harrier	*Circus macrourus*	Winter visitor & passage migrant
66.	Montagu's harrier	*Circus pygargus*	Passage migrant; some winter
67.	Eurasian sparrowhawk	*Accipiter nisus*	Winter visitor & passage migrant
68.	Steppe buzzard	*Buteo buteo vulpinus*	Winter visitor & passage migrant
69.	Long-legged buzzard	*Buteo rufinus*	Winter visitor & passage migrant
70.	Greater spotted eagle	*Clanga clanga*	Winter visitor & passage migrant
71.	Steppe eagle	*Aquila nipalensis*	Winter visitor & passage migrant
72.	Eastern imperial eagle	*Aquila heliaca*	Uncommon winter visitor & passage migrant
73.	Booted eagle	*Hieraaetus pennatus*	Passage migrant; some winter
74.	Lesser kestrel	*Falco naumanni*	Passage migrant
75.	Common kestrel	*Falco tinnunculus*	Resident breeder; winter visitor
76.	Merlin	*Falco columbarius*	Uncommon winter visitor & passage migrant
77.	Eurasian hobby	*Falco subbuteo*	Passage migrant
78.	Saker falcon	*Falco cherrug*	Very rare winter visitor; formerly much commoner
79.	Peregrine falcon	*Falco peregrinus*	Uncommon winter visitor
80.	Macqueen's bustard	*Chlamydotis macqueenii*	Winter visitor in a very few numbers around the marshes; may breed (bred formerly)

(continued)

Table 19.1 (continued)

#	English name	Scientific name	Status in the marshes
81.	Water rail	*Rallus aquaticus*	Winter visitor & passage migrant; probably breeds
82.	Corncrake	*Crex crex*	Passage migrant
83.	Little crake	*Zapornia parva*	Winter visitor & passage migrant; may breed
84.	Western Baillon's crake	*Zapornia intermedia*	Winter visitor & passage migrant
85.	Spotted crake	*Porzana porzana*	Passage migrant
86.	Grey-headed swamphen	*Porphyrio poliocephalus*	Resident breeder
87.	Common moorhen	*Gallinula chloropus*	Resident breeder; winter visitor & passage migrant
88.	Eurasian coot	*Fulica atra*	Uncommon resident breeder; winter visitor & passage migrant
89.	Demoiselle crane	*Grus virgo*	Status uncertain; common winter visitor in the 1910s
90.	Common crane	*Grus grus*	Passage migrant; some winter
91.	Eurasian stone curlew	*Burhinus oedicnemus*	Resident and breeding summer visitor around the marshes; winter visitor & passage migrant
92.	Black-winged stilt	*Himantopus himantopus*	Resident breeder; winter visitor & passage migrant
93.	Pied avocet	*Recurvirostra avosetta*	Resident breeder; winter visitor
94.	Northern lapwing	*Vanellus vanellus*	Uncommon winter visitor
95.	Spur-winged lapwing	*Vanellus spinosus*	Resident breeder; passage migrant
96.	Red-wattled lapwing	*Vanellus indicus*	Resident breeder
97.	White-tailed lapwing	*Vanellus leucurus*	Resident breeder; winter visitor
98.	Pacific golden plover	*Pluvialis fulva*	Status uncertain; probably a rare winter visitor
99.	Grey plover	*Pluvialis squatarola*	Uncommon winter visitor & passage migrant
100.	Common ringed plover	*Charadrius hiaticula*	Winter visitor & passage migrant
101.	Little ringed plover	*Charadrius dubius*	Passage migrant; may breed
102.	Eurasian dotterel	*Charadrius morinellus*	Uncommon winter visitor & passage migrant
103.	Kentish plover	*Anarhyncus alexandrinus*	Resident breeder; winter visitor & passage migrant
104.	Lesser sand plover	*Anarhyncus atrifrons*	Uncommon winter visitor & passage migrant
105.	Greater sand plover	*Anarhyncus leschenaultii*	Winter visitor & passage migrant

(continued)

Table 19.1 (continued)

#	English name	Scientific name	Status in the marshes
106.	Caspian plover	*Anarhyncus asiaticus*	Status uncertain; probably a rare passage migrant
107.	Jack snipe	*Lymnocryptes minimus*	Rare winter visitor & passage migrant
108.	Great snipe	*Gallinago media*	Rare passage migrant
109.	Common snipe	*Gallinago gallinago*	Winter visitor & passage migrant
110.	Black-tailed godwit	*Limosa limosa*	Winter visitor & passage migrant
111.	Bar-tailed godwit	*Limosa lapponica*	Winter visitor & passage migrant
112.	Eurasian whimbrel	*Numenius phaeopus*	Winter visitor & passage migrant
113.	Slender-billed curlew	*Numenius tenuirostris*	Last recorded in 1979; may be globally extinct
114.	Eurasian curlew	*Numenius arquata*	Winter visitor & passage migrant
115.	Spotted redshank	*Tringa erythropus*	Winter visitor & passage migrant
116.	Common redshank	*Tringa totanus*	Winter visitor & passage migrant
117.	Marsh sandpiper	*Tringa stagnatilis*	Winter visitor & passage migrant
118.	Common greenshank	*Tringa nebularia*	Winter visitor & passage migrant
119.	Green sandpiper	*Tringa ochropus*	Winter visitor & passage migrant
120.	Wood sandpiper	*Tringa glareola*	Winter visitor & passage migrant
121.	Terek sandpiper	*Xenus cinereus*	Winter visitor & passage migrant
122.	Common sandpiper	*Actitis hypoleucos*	Winter visitor & passage migrant
123.	Ruddy turnstone	*Arenaria interpres*	Winter visitor & passage migrant
124.	Red knot	*Calidris canutus*	Vagrant
125.	Sanderling	*Calidris alba*	Winter visitor & passage migrant
126.	Little stint	*Calidris minuta*	Winter visitor & passage migrant
127.	Temminck's stint	*Calidris temminckii*	Winter visitor & passage migrant
128.	Curlew sandpiper	*Calidris ferruginea*	Winter visitor & passage migrant
129.	Dunlin	*Calidris alpina*	Winter visitor & passage migrant
130.	Broad-billed sandpiper	*Calidris falcinellus*	Uncommon winter visitor & passage migrant
131.	Ruff	*Calidris pugnax*	Winter visitor & passage migrant
132.	Red-necked phalarope	*Phalaropus lobatus*	Passage migrant
133.	Cream-coloured courser	*Cursorius cursor*	Resident or migrant breeder around the marshes
134.	Collared pratincole	*Glareola pratincola*	Breeding summer visitor; passage migrant

(continued)

Table 19.1 (continued)

#	English name	Scientific name	Status in the marshes
135.	Black-winged pratincole	*Glareola nordmanni*	Reported breeding in 1920s; not recorded since
136.	Slender-billed gull	*Chroicocephalus genei*	Resident breeder; winter visitor & passage migrant.
137.	Common black-headed gull	*Chroicocephalus ridibundus*	Winter visitor & passage migrant
138.	Little gull	*Hydrocoloeus minutus*	Vagrant
139.	Great black-headed gull	*Ichthyaetus ichthyaetus*	Winter visitor
140.	Common gull	*Larus canus*	Rare winter visitor
141.	Caspian gull	*Larus cachinnans*	Winter visitor & passage migrant
142.	Armenian gull	*Larus armenicus*	Winter visitor & passage migrant
143.	Baltic (lesser black-backed) gull	*Larus fuscus fuscus*	Uncommon winter visitor & passage migrant
144.	Gull-billed tern	*Gelochelidon nilotica*	Breeding summer visitor; passage migrant; few winter
145.	Caspian tern	*Hydroprogne caspia*	Uncommon breeding summer visitor; passage migrant
146.	Little tern	*Sternula albifrons*	Breeding summer visitor; passage migrant; few winter
147.	Common tern	*Sterna hirundo*	Breeding summer visitor; passage migrant
148.	Whiskered tern	*Chlidonias hybrida*	Resident and breeding summer visitor; winter visitor & passage migrant
149.	White-winged tern	*Chlidonias leucopterus*	Uncommon breeding summer visitor; passage migrant
150.	Black tern	*Chlidonias niger*	Vagrant
151.	Pin-tailed sandgrouse	*Pterocles alchata*	Resident breeder around the marshes
152.	Spotted sandgrouse	*Syrrhaptes senegallus*	Uncommon resident breeder around the marshes
153.	Black-bellied sandgrouse	*Syrrhaptes orientalis*	Rare winter visitor around the marshes
154.	Common wood pigeon	*Columba palumbus*	Resident breeder; winter visitor
155.	European turtle dove	*Streptopelia turtur*	Breeding summer visitor; passage migrant
156.	Eurasian collared dove	*Streptopelia decaocto*	Resident breeder
157.	Laughing dove	*Spilopelia senegalensis*	Resident breeder
158.	Namaqua dove	*Oena capensis*	Rare resident breeder
159.	Common cuckoo	*Cuculus canorus*	Passage migrant

(continued)

Table 19.1 (continued)

#	English name	Scientific name	Status in the marshes
160.	Western barn owl	*Tyto alba*	Rare resident breeder
161.	Pallid scops owl	*Otus brucei*	Resident or breeding summer visitor
162.	Eurasian scops owl	*Otus scops*	Passage migrant; may breed
163.	Little owl	*Athene noctua*	Uncommon resident breeder
164.	European nightjar	*Caprimulgus europaeus*	Passage migrant
165.	Egyptian nightjar	*Caprimulgus aegyptius*	Breeding summer visitor; passage migrant.
166.	Common swift	*Apus apus*	Passage migrant
167.	Pallid swift	*Apus pallidus*	Status uncertain; passage migrant
168.	European roller	*Coracias garrulus*	Passage migrant
169.	Indian roller	*Coracias benghalensis*	Resident and breeding summer visitor
170.	White-throated kingfisher	*Halcyon smyrnensis*	Resident breeder
171.	Common kingfisher	*Alcedo atthis*	Resident breeder; winter visitor & passage migrant
172.	Pied kingfisher	*Ceryle rudis*	Resident breeder
173.	Blue-cheeked bee-eater	*Merops persicus*	Breeding summer visitor; passage migrant
174.	European bee-eater	*Merops apiaster*	Passage migrant
175.	Eurasian hoopoe	*Upupa epops*	Passage migrant
176.	Eurasian wryneck	*Jynx torquilla*	Passage migrant
177.	Red-backed shrike	*Lanius collurio*	Passage migrant
178.	Daurian shrike	*Lanius isabellinus*	Winter visitor & passage migrant
179.	Turkestan shrike	*Lanius phoenicuroides*	Uncommon passage migrant
180.	Lesser grey shrike	*Lanius minor*	Passage migrant
181.	Steppe grey shrike	*Lanius excubitor pallidirostris*	Uncommon winter visitor & passage migrant
182.	Great grey shrike	*Lanius [e.] excubitor*	Status uncertain; probably only a passage migrant
183.	Woodchat shrike	*Lanius senator*	Passage migrant
184.	Masked shrike	*Lanius nubicus*	Passage migrant
185.	Eurasian golden oriole	*Oriolus oriolus*	Passage migrant
186.	Eurasian magpie	*Pica pica*	Uncommon resident breeder
187.	Rook	*Corvus frugilegus*	Winter visitor
188.	Mesopotamian crow	*Corvus cornix capellanus*	Resident breeder
189.	Brown-necked raven	*Corvus ruficollis*	Uncommon resident breeder around the marshes

(continued)

Table 19.1 (continued)

#	English name	Scientific name	Status in the marshes
190.	Hypocolius	*Hypocolius ampelinus*	Breeding summer visitor
191.	Eurasian penduline tit	*Remiz pendulinus*	Rare winter visitor
192.	Greater hoopoe-lark	*Alaemon alaudipes*	Resident breeder around the marshes
193.	Calandra lark	*Melanocorypha calandra*	Winter visitor
194.	Greater short-toed lark	*Calandrella brachydactyla*	Winter visitor & passage migrant
195.	Lesser short-toed lark	*Alaudala rufescens*	Status uncertain; probably a winter visitor
196.	Crested lark	*Galerida cristata*	Resident breeder
197.	Eurasian skylark	*Alauda arvensis*	Winter visitor & passage migrant
198.	White-eared bulbul	*Pycnonotus leucotis*	Resident breeder
199.	Sand martin	*Riparia riparia*	Breeding summer visitor; passage migrant
200.	Eurasian barn swallow	*Hirundo rustica*	Breeding summer visitor; passage migrant
201.	Red-rumped swallow	*Cecropis rufula*	Passage migrant
202.	Common house martin	*Delichon urbicum*	Passage migrant
203.	Cetti's warbler	*Cettia cetti*	Winter visitor
204.	Willow warbler	*Phylloscopus trochilus*	Passage migrant
205.	Common chiffchaff	*Phylloscopus collybita*	Winter visitor & passage migrant
206.	Basra reed warbler	*Acrocephalus griseldis*	Breeding summer visitor
207.	Great reed warbler	*Acrocephalus arundinaceus*	Breeding summer visitor; passage migrant; may winter
208.	Clamorous reed warbler	*Acrocephalus stentoreus*	Uncommon winter visitor & passage migrant; possibly rare resident breeder
209.	Moustached warbler	*Acrocephalus melanopogon*	Probably bred in the 1920s; not recorded since
210.	Sedge warbler	*Acrocephalus schoenobaenus*	Passage migrant
211.	Eurasian reed warbler	*Acrocephalus scirpaceus*	Passage migrant
212.	Marsh warbler	*Acrocephalus palustris*	Passage migrant
213.	Eastern olivaceous warbler	*Iduna pallida*	Breeding summer visitor; passage migrant
214.	Upcher's warbler	*Hippolais languida*	Passage migrant

(continued)

Table 19.1 (continued)

#	English name	Scientific name	Status in the marshes
215.	River warbler	*Locustella fluviatilis*	Vagrant
216.	Zitting cisticola	*Cisticola juncidis*	Uncommon resident breeder
217.	Graceful prinia	*Prinia gracilis*	Resident breeder
218.	Iraq babbler	*Turdoides altirostris*	Resident breeder
219.	Afghan babbler	*Turdoides huttoni*	Resident breeder
220.	Eurasian blackcap	*Sylvia atricapilla*	Passage migrant
221.	Garden warbler	*Sylvia borin*	Passage migrant
222.	Barred warbler	*Curruca nisoria*	Passage migrant
223.	Lesser whitethroat	*Curruca curruca*	Passage migrant
224.	Common whitethroat	*Curruca communis*	Passage migrant
225.	Ménétries's warbler	*Curruca mystacea*	Winter visitor & passage migrant
226.	Common starling	*Sturnus vulgaris*	Winter visitor
227.	Common blackbird	*Turdus merula*	Winter visitor
228.	European robin	*Erithacus rubecula*	Winter visitor & passage migrant
229.	Bluethroat	*Luscinia svecica*	Winter visitor & passage migrant
230.	Thrush nightingale	*Luscinia luscinia*	Passage migrant
231.	Common nightingale	*Luscinia megarhynchos*	Passage migrant
232.	White-throated robin	*Irania gutturalis*	Passage migrant
233.	Rufous-tailed scrub robin	*Cercotrichas galactotes*	Breeding summer visitor; passage migrant
234.	Black redstart	*Phoenicurus ochruros*	Winter visitor; races *ochruros* & *phoenicuroides* recorded
235.	Common redstart	*Phoenicurus phoenicurus*	Passage migrant
236.	Whinchat	*Saxicola rubetra*	Passage migrant
237.	European stonechat	*Saxicola rubicola*	Winter visitor & passage migrant
238.	Siberian stonechat	*Saxicola maurus*	Winter visitor & passage migrant
239.	Isabelline wheatear	*Oenanthe isabellina*	Winter visitor & passage migrant
240.	Northern wheatear	*Oenanthe oenanthe*	Passage migrant
241.	Pied wheatear	*Oenanthe pleschanka*	Passage migrant
242.	Eastern black-eared wheatear	*Oenanthe melanoleuca*	Passage migrant
243.	Desert wheatear	*Oenanthe deserti*	Winter visitor and passage migrant
244.	Finsch's wheatear	*Oenanthe finschii*	Winter visitor
245.	Spotted flycatcher	*Muscicapa striata*	Passage migrant

(continued)

Table 19.1 (continued)

#	English name	Scientific name	Status in the marshes
246.	Semi-collared flycatcher	*Ficedula semitorquata*	Passage migrant
247.	Red-breasted flycatcher	*Ficedula parva*	Rare passage migrant
248.	House sparrow	*Passer domesticus*	Resident breeder
249.	Spanish sparrow	*Passer hispaniolensis*	Winter visitor
250.	Dead Sea sparrow	*Passer moabiticus*	Resident and breeding summer visitor
251.	Yellow-throated sparrow	*Gymnoris xanthocollis*	Breeding summer visitor; passage migrant
252.	Western yellow wagtail (includes all races)	*Motacilla flava*	Passage migrant
253.	Black-headed wagtail	*Motacilla flava feldegg*	Passage migrant; may breed
254.	Citrine wagtail	*Motacilla citreola*	Winter visitor & passage migrant
255.	Grey wagtail	*Motacilla cinerea*	Winter visitor & passage migrant
256.	White wagtail	*Motacilla alba*	Winter visitor & passage migrant
257.	Tawny pipit	*Anthus campestris*	Passage migrant
258.	Meadow pipit	*Anthus pratensis*	Winter visitor & passage migrant
259.	Tree pipit	*Anthus trivialis*	Passage migrant
260.	Red-throated pipit	*Anthus cervinus*	Passage migrant
261.	Water pipit	*Anthus spinoletta*	Winter visitor & passage migrant
262.	Desert finch	*Rhodospiza obsoleta*	Winter visitor around the marshes
263.	Corn bunting	*Emberiza calandra*	Winter visitor
264.	Common reed bunting	*Emberiza schoeniclus*	Uncommon winter visitor

et al. 2015), namely, African sacred ibis *Threskiornis aethiopicus*, Goliath heron *Ardea goliath*, African darter *Anhinga rufa* and pygmy cormorant *Microcarbo pygmaeus* with a further six breeding species designated as regionally near threatened. See ‘Conservation Priority Species’ below.

Furthermore the marshes hold globally or regionally important populations of 30 species, 2 of which, the Basra reed warbler and Iraq babbler, are endemic to Iraq and 3 have highly restricted ranges in the Middle East: African darter, Goliath heron and African sacred ibis.

Table 19.2 Globally threatened birds in the southern marshes of Iraq, based on IUCN/BirdLife International Red List at 2020

Critically endangered
Sociable lapwing *Vanellus gregarius***.** Status in the lower Mesopotamian Marshes uncertain. May occur. Historically a regular winter visitor to Iraq, however in small numbers. No recent records but known to pass through on migration (from satellite tracking)
Slender-billed curlew *Numenius tenuirostris***.** Now probably globally extinct; historically has occurred in small numbers on migration, last in 1979
Endangered
White-headed duck *Oxyura leucocephala***.** Probably a regular winter visitor in small numbers
Egyptian vulture *Neophron percnopterus***.** Passage migrant
Steppe eagle *Aquila nipalensis.* Winter visitor and passage migrant
Saker falcon *Falco cherrug***.** Very rare winter visitor; formerly much commoner
Basra reed warbler *Acrocephalus griseldis***.** Endemic breeding summer visitor; recent NI surveys have shown the population to be c4,000 pairs
Vulnerable
Red-breasted goose *Anser ruficollis***.** Occasional in winter
Lesser white-fronted goose *Anser erythropus***.** Uncommon winter visitor
Marbled duck *Marmaronetta angustirostris*. Breeding resident and winter visitor. NI winter counts in the southern marshes have exceeded 20,000 – the highest concentration in the world
Common pochard *Aythya ferina*. Passage migrant and winter visitor
Greater spotted eagle *Clanga clanga***.** Passage migrant and winter visitor
Eastern imperial eagle *Aquila heliaca*. Passage migrant and winter visitor
Macqueen's bustard *Chlamydotis macqueenii.* Winter visitor to areas near marshes
Turtle dove *Streptopelia turtur*. Passage migrant
Near threatened
Ferruginous duck *Aythya nyroca*. Breeding resident, passage migrant and winter visitor
Dalmatian pelican *Pelecanus crispus*. Rare winter visitor
Pallid harrier *Circus macrourus*. Passage migrant
Northern lapwing *Vanellus vanellus*. Uncommon winter visitor
Great snipe *Gallinago media*. Passage migrant but no recent records
Black-tailed godwit *Limosa limosa*. Passage migrant and winter visitor
Bar-tailed godwit *Limosa lapponica*. Passage migrant and winter visitor
Eurasian curlew *Numenius arquata*. Passage migrant and winter visitor
Curlew sandpiper *Calidris ferruginea*. Passage migrant and winter visitor
Armenian gull *Larus armenicus*. Passage migrant and winter visitor
Meadow pipit *Anthus pratensis*. Passage migrant and winter visitor

19.2 Important Breeding Birds

About 77 species have been recorded breeding in the southern marshes, with a further 11 probably breeding (see Table 19.1).

The marshes provide an important breeding habitat for many wetland birds (Ararat et al. 2011) and, in the case of some species, hold the highest population in the world. In terms of global significance, the endangered Basra reed warbler

Fig. 19.2 Adult Basra reed warbler *Acrocephalus griseldis*, an endangered species for which the southern marshes of Iraq hold the highest population in the world. June 2008. (Image courtesy of Mudhafar A. Salim)

Fig. 19.3 Nest with chicks of the endangered Basra reed warbler *Acrocephalus griseldis* in the southern marshes of Iraq. May 2009. (Image courtesy of Mudhafar A. Salim)

(Figs. 19.2 and 19.3) is the most important with Iraq probably holding over 90% of the world breeding population, with breeding also in the Mesopotamian Marshes in Iran (Keramat Hafezi Birgani in litt) and isolated populations Kuwait and Israel (Porter

Fig. 19.4 Marbled duck *Marmaronetta angustirostris*, a threatened species for which the southern marshes of Iraq hold a globally important breeding population. October 2019. (Image courtesy of Mudhafar A. Salim)

Fig. 19.5 The Middle East race of the African darter *Anhinga rufa chantrei* in breeding plumage. The Hawizeh Marshes of the southern marshes of Iraq hold most of the world population. February 2010. (Image courtesy of Mudhafar A. Salim)

and Aspinall 2010). Based on KBA survey counts from 2008 to 2010, the breeding population of Basra reed warbler is estimated at over 4000 pairs in the southern marshes (Nature Iraq 2017); however, a detailed census is essential to provide a more accurate figure.

The only other globally threatened breeding species is the marbled duck (see Table 19.2 and Fig. 19.4), which is the most widespread and commonest duck in the southern marshes in summer. Although no population estimates have been attempted, breeding season counts during the KBA surveys suggest that there are over 1000 pairs (Nature Iraq 2017). The breeding population of the near-threatened ferruginous duck (Table 19.2) has not been calculated, but it is fairly common and widespread. The only other duck to breed is the red-crested pochard *Netta rufina*, which is globally of least concern.

Twelve species of herons and their allies and one species of cormorant breed (see Table 19.1) of which the commonest and most widespread are the squacco heron *Ardeola ralloides*, purple heron *Ardea purpurea* and pygmy cormorant. Three have highly restricted ranges in the Middle East: African darter, Goliath heron and African sacred ibis. The darter (Fig. 19.5) is assigned to the race *chantrei*, and the Mesopotamian Marshes (of Iraq and Iran) now hold the only breeding colonies in the Middle East for this taxon (Schodde et al. 2012 and Keramat Hafezi Birgani in litt). The

Goliath heron also has a small, non-feral breeding population in Yemen, whilst the ibis only breeds regularly in a non-feral state in Iraq (Porter and Aspinall 2010).

The southern marshes hold the highest population of grey-headed swamphen *Porphyrio poliocephalus* in the Middle East. Whilst amongst the breeding waders, gulls and terns, the KBA surveys show that the marshes hold important populations of black-winged stilt *Himantopus himantopus*, white-tailed lapwing *Vanellus leucurus* and collared pratincole *Glareola pratincola*, slender-billed gull *Chroicocephalus genei* and whiskered tern *Chlidonias hybrida*, all of which exceed 1% of the biogeographic population (see Table 19.3).

In the case of the lapwing, provisional calculations based on the KBA surveys and sample surveys (Salim and Porter in prep) suggest the population may be as high as 2000 pairs in the southern marshes and their environs, which is over 8% of the global population.

Three species of kingfisher breed, by far the commonest of which is the pied kingfisher *Ceryle rudis*, and from personal experience in the region and beyond, RFP believes the marshes hold the highest concentration he has ever witnessed. White-throated kingfisher *Halcyon smyrnensis* and common kingfisher *Alcedo atthis* are less common, but nevertheless occur regularly.

Finally the population of the endemic Iraq babbler (Fig. 19.6) as suggested by the KBA surveys is probably in excess of 4000 pairs. Its conservation status is least concern.

19.3 Endemic Species and Subspecies

These have already been mentioned under 'Important Breeding Birds', but we feel it is worth highlighting the importance of the southern marshes of Iraq for endemic species and subspecies. The two internationally acclaimed endemics are Basra reed warbler and Iraq babbler (see Porter and Aspinall 2010). In addition the hypocolius *Hypocolius ampelinus*, a Middle East endemic, has a large breeding population, though its size is unknown.

Four endemic subspecies are also worthy of specific mention: black francolin *Francolinus francolinus arabstanicus*, little grebe *Tachybaptus ruficollis iraquensis*, African darter *Anhinga rufa chantrei* and Mesopotamian crow *Corvus cornix capellanus*.

19.4 Winter Visitors and Passage Migrants

The marshes of southern Iraq are crucial as a stop-over site for migrating and wintering birds. Table 19.1 shows that 197 species are regular winter visitors or passage migrants, breeding in Eurasia and dependant on these vast wetlands for resting and refuelling on their long migrations. Such wetlands are crucial for the survival of many species that have to face the long migration over the inhospitable

Table 19.3 Species with globally or regionally important populations (>1%) in the southern marshes of Iraq

Common name	Scientific name	Status in Iraq marshes	Population where known in southern marshes
Ruddy shelduck	*Tadorna ferruginea*	Winter visitor	
Gadwall	*Mareca strepera*	Winter visitor	
Northern shoveler	*Spatula clypeata*	Winter visitor	
Common teal	*Anas crecca*	Winter visitor	
Marbled duck	*Marmaronetta angustirostris*	Resident; winter visitor	Probably & gt;1000 pairs breed; 20,000+ winter, probably & gt;50% of global population
Ferruginous duck	*Aythya nyroca*	Resident; winter visitor & passage migrant	Population unknown, but probably >5% of global population
Tufted duck	*Aythya fuligula*	Winter visitor	
Little grebe (Iraq race)	*Tachybaptus ruficollis iraquensis*	Endemic; resident	Population unknown, but 100% of global population
African sacred ibis	*Threskiornis aethiopicus*	Resident	50+ pairs (probably & gt;90% of Middle East population)
Squacco heron	*Ardeola ralloides*	Resident; winter visitor & passage migrant	
Goliath heron	*Ardea goliath*	Rare resident	Probably <5 pairs, probably >20% of Middle East population
Purple heron	*Ardea purpurea*	Resident; winter visitor & passage migrant	
Little egret	*Egretta garzetta*	Resident; winter visitor & passage migrant	
Great white pelican	*Pelecanus onocrotalus*	Winter visitor & passage migrant	
Pygmy cormorant	*Microcarbo pygmaeus*	Resident; winter visitor	
African darter	*Anhinga rufa*	Resident; endemic race *chantrei*	Population probably <50 pairs; probably >80% of Middle East population
Grey-headed swamphen	*Porphyrio poliocephalus*		Population not known but probably <50% of the Middle East population
Common coot	*Fulica atra*	Resident; winter visitor	
Black-winged stilt	*Himantopus himantopus*	Resident; winter visitor & passage migrant	
White-tailed lapwing	*Vanellus leucurus*	Resident; probably winter visitor	Probably >2000 pairs (>8% of global population)
Black-tailed godwit	*Limosa limosa*	Winter visitor & passage migrant	

(continued)

Table 19.3 (continued)

Common name	Scientific name	Status in Iraq marshes	Population where known in southern marshes
Collared pratincole	*Glareola pratincola*	Breeding summer visitor; passage migrant	Probably >4000 pairs (probably >5% of regional population)
Slender-billed gull	*Chroicocephalus genei*	Resident; breeding summer visitor; winter visitor & passage migrant.	Breeding population c7,000 pairs (>5% of global population)
Whiskered tern	*Chlidonias hybrida*	Resident; winter visitor & passage migrant	
Pied kingfisher	*Ceryle rudis*	Resident	Population not known, but probably >20% of Middle East population
Mesopotamian crow	*Corvus cornix capellanus*	Endemic; resident	Population not known, but probably >80% of global population
Hypocolius	*Hypocolius ampelinus*	Breeding summer visitor	Population not known, but probably >5% of global population
Basra reed warbler	*Acrocephalus griseldis*	Endemic; breeding summer visitor	Population estimated at >4000 pairs, probably >80% of global population
Iraq babbler	*Turdoides altirostris*	Endemic; resident	Population estimated at >4000 pairs, probably >50% of global population
Dead Sea sparrow	*Passer moabiticus*	Resident; breeding summer visitor	Population estimated at >5000 pairs, probably >25% of global population

deserts of Arabia and north-east Africa. This is especially so during spring migrations when arrival at a vast wetland complex will, for many species, provide the first large resting and feeding area for probably over 3000 km. Those species for which the marshes are known to be most important for migrants and winter visitors are shown in Table 19.3; however the true value of the marshes for many other migrant birds probably cannot be overestimated.

19.5 Conservation Priority Species

In terms of species conservation priority, 25 species that regularly occur in the marshes are red listed as globally threatened or near threatened. They are listed in Table 19.2. Of these the marshes are considered to be of the greatest importance for marbled duck (vulnerable; Fig. 19.7), ferruginous duck (near threatened), greater spotted eagle (vulnerable; Fig. 19.8), Eastern imperial eagle (vulnerable) and Basra

Fig. 19.6 The endemic Iraq babbler *Turdoides altirostris* in the southern marshes of Iraq, April 2013. (Image courtesy of R F Porter)

Fig. 19.7 Marbled ducks *Marmaronetta angustirostris* in winter. A threatened species for which the southern marshes of Iraq hold a globally important wintering population. February 2010. (Image courtesy of Mudhafar A. Salim)

reed warbler (endangered). As has already been stated, a recent study (Symes et al. 2015) considered four species breeding in the Iraq marshes to be regionally threatened in the Arabian Peninsula, namely, pygmy cormorant, African darter, Goliath heron and sacred ibis, and six species to be regionally near threatened: red-crested

Fig. 19.8 The globally threatened Greater Spotted Eagle *Clanga clanga* is one of the wintering raptors in the southern marshes of Iraq. March 2010. (Image courtesy of Mudhafar A. Salim)

pochard, grey heron *Ardea cinerea*, purple heron, Western marsh harrier *Circus aeruginosus*, pied avocet *Recurvirostra avosetta* and common kingfisher.

19.6 The Economic Value of Birds in the Marshes

Birds are considered as one of the main natural protein resources for most of the Ma'dan (the marsh dwellers). Garstecki and Amr (2011) gave this as one of the key threats to the birds in the southern marshes. They produced a list of bird species and families that are traditionally hunted in the marshes, though recent surveys suggest this is higher (Salim personal observations).

Along with fish, birds are the favourite dish for the Marsh Arabs. The most desired species is the mallard *Anas platyrhynchos*, which is hunted in considerable numbers over the entire marshes of southern Iraq. Other favourite duck species in winter are Eurasian teal *Anas crecca*, shoveler *Spatula clypeata* and pintail *Anas acuta*, as well as coot *Fulica atra*, whilst in spring, summer and autumn, it is marbled duck, garganey *Spatula querquedula*, ferruginous duck and red-crested pochard, as well as some other duck species.

19.7 Important Bird and Biodiversity Areas in the Southern Marshes

The four marshlands that comprise the southern marshes of Iraq (Central Marshes, East Hammar, West Hammar and Hawizeh) have all been identified as Important Bird and Biodiversity Areas and KBAs by Nature Iraq surveys (Nature Iraq 2017). IBAs are recognised by strict global criteria (BirdLife International 2013, 2020).

The key bird species recorded at each of the four marshes during the period 2005–2011 are given below. The breeding populations are, in most cases, very approximate estimates based on sample area counts. These are the first attempts at giving population figures for some of the important bird species in the marshes, but we cannot overemphasise the need for a comprehensive survey to estimate the populations more accurately.

19.8 The Southern Marshes of Iraq

19.8.1 Central Marshes

Marbled duck, >100 pairs breeding, up to 12,000 wintering; Basra reed warbler, >2000 pairs breeding; Iraq babbler, >1000 pairs breeding; white-tailed lapwing, >50 pairs breeding. In addition to greater spotted eagle and Eastern imperial eagle winter in the Central Marshes, ferruginous duck occurs in summer and winter, and pallid harrier *Circus cyaneus*, Eurasian curlew *Numenius arquata* and black-tailed godwit are found on passage. The endemic race of little grebe breeds widely. Over 70,000 waterbirds have been recorded in winter (Salim et al. 2010).

19.8.2 East Hammar

Marbled duck, breeds and up to 800 winter; slender-billed gull, >3000 pairs breed; Basra reed warbler, >300 pairs; Iraq babbler, >400 pairs. In addition greater spotted eagle and Eastern imperial eagle winter and the endemic race of little grebe and endemic race of hooded crow (Mesopotamian crow) breed. The site regularly holds >30,000 wintering waterbirds (Salim et al. 2010).

19.8.3 West Hammar

Marbled duck, >500 pairs breeding, >4000 wintering; Basra reed warbler, >1000 pairs breeding; Iraq babbler, >2000 pairs breeding; white-tailed lapwing, >1000 pairs breeding; slender-billed gull >3000 pairs; whiskered tern >1000 pairs breeding; and hypocolius >500 pairs breeding (Salim et al. 2010).

In addition Eastern imperial eagles winter, Eurasian curlew *Numenius arquata* occurs on passage, and ferruginous duck is a widespread breeding species as well as winter visitor. The endemic race of little grebe and hooded crow (Mesopotamian crow) breeds. The site regularly holds >50,000 wintering waterbirds (Salim et al. 2010).

19.8.4 Hawizeh

Marbled duck, >200 pairs breeding and >1000 wintering; white-headed duck, 38 in winter 2005, Basra reed warbler, >500 pairs breeding, Iraq babbler >800 pairs breeding. The Hawizeh marshland is the only known wetland in Iraq that holds a breeding population of African darter and African sacred ibis. According to frequent reports of locals and hunters, the Goliath heron occurs in the northern part of this marsh, but in small numbers, probably <20 pairs.

In addition Hawizeh holds wintering Eastern imperial and greater spotted eagles; ferruginous ducks breed and winter, and pallid harriers and black-tailed godwits occur on passage and in winter. The Iraqi resident race of little grebe breeds as does the Mesopotamian crow. The site holds over 35,000 waterbirds in winter (Salim et al. 2010).

19.9 Threats, Conservation and Opportunities

The highest threat to the marshes of southern Iraq is posed by shortage of water in the Tigris and Euphrates that is the life blood on which the marshes are dependant. Any activity that reduces the flow of water of these rivers, inside or outside Iraq, will have a serious impact on the biodiversity of the marshes. The continuing traditional practice of reed cutting is considered to have a low impact on wildlife; indeed it often encourages healthy reed growth which can be good for biodiversity.

Reeds, channels and areas of open water are the main habitats on which breeding, migrating and wintering birds are dependant, and it is encouraging to witness just how quickly reeds became re-established once the marshes were reflooded (Salim, personal observation). After reflooding in 2003, the marshes naturally restored themselves, plant cover increased, and birds and other fauna quickly returned and now have healthy populations once again.

The hunting of waterfowl must be put on a sustainable level, and globally and regionally threatened and near-threatened species should receive full protection. It is encouraging to hear that a hunters' awareness programme is now in an advanced planning stage.

A potential threat to the fauna and flora of the marshes is posed by expanding oil exploration and its associated infrastructure in a small number of areas. Conservationists and developers must work closely to ensure that any impact on the biodiversity of these globally important marshes is minimal.

The expanding practice of electrofishing is considered to be a serious threat to fish populations and thus on fish-eating bird and mammal populations. It might prove to be unsustainable, thus impacting a human food resource. The impact of electrofishing should be urgently addressed.

The marshes hold several large colonies of nesting herons and their allies that are regionally important in the Middle East. These colonies are highly vulnerable to disturbance, and consideration should be given to their protection by wardening during the breeding season. Finally, a multi-organisational education programme should be launched to show the global importance of the marshes for wildlife, people and culture.

The designation of the Hawizeh Marshes as a Ramsar site and the recent designation of the Central Marshes as a national park provide an excellent focus for this. Also now that the southern marshes have become a World Heritage Site, it would be appropriate to have education and scientific facilities for visitors – both national and international – that shows the global importance of the marshes for birds, other wildlife and people.

19.10 Recommendations

The following require active encouragement:

- Greater co-operation internationally and especially between Iraq, Iran and Turkey to understand the global importance of the Mesopotamian Marshland complex and its dependence on the water flows in the Tigris and Euphrates.
- Greater awareness of the globally importance of the Southern Marshes, especially amongst local people.
- A co-ordinated multidisciplinary scientific and technical programme to better understand the biodiversity and threats of the marshes, this should especially include dedicated surveys to determine the populations of the species for which the Mesopotamian Marshes are most important; this should include a special study of the significance of the marshes for the survival of migratory birds.
- The establishment of a monitoring programme and database for birds and other fauna.
- The preparation of a management plan that encapsulates conservation action plans for species of conservation concern, especially endemic and globally and regionally threatened species.
- The activation of laws and education to control hunting and electrofishing.

Acknowledgements We would like to pay special thanks to our colleagues in Nature Iraq, past and present, who have contributed extensively to the KBA field surveys, namely, Laith Ali Al Obeidi, Omar Fadhel, Mohammad Turki and Zaman Salman. Support and advice have been received from many colleagues especially in the IOCN and BirdLife International. We also thank Mike Blair who was very helpful in discussing taxonomy and nomenclature.

The following institutions and organisations have supported the KBA programme, and hence this paper, in a variety of ways: Nature Iraq, the Ministry of Environment, the Canadian-Iraqi Marshlands Initiative, BirdLife International and the Italian Ministry of Environment, Land and Sea, and College of Science/University of Al-Qadisiyah. We thank them all.

References

Allouse B (1953) The Avifauna of Iraq. Iraq Natural History Museum, Baghdad

Allouse B (1960, 1961, 1962) [*Birds of Iraq*], vols I, II, III. Al-Rabita Press, Baghdad. [In Arabic]

Ararat K, Fadhil O, Porter RF, Salim M (2011) Breeding birds in Iraq: important new discoveries. Sandgrouse 33:12–33

BirdLife International (2013) Global IBA Criteria. Downloaded from http://www.birdlife.org/datazone/info/ibacritglob

BirdLife International (2020) IUCN Red List for birds. www.birdlife.org

Centre for Restoration of the Iraqi Marshlands (2013) The Iraqi Ministry of Water Resources. www.crim.mowr.gov.iq

Evans MI (1994) Important bird areas in the Middle East. BirdLife International, Cambridge

Garstecki T, Amr Z (2011) Biodiversity and ecosystem management in the Iraqi Marshlands: screening study on potential world heritage nomination. IUCN – ROWA, Jordan

Nature Iraq (2017) Key biodiversity areas of Iraq: priority sites for conservation & protection. Tablet House Publishing, Sulimaniyah

Kirwan GM, Porter RF, Scott DA (2015) Chronicle of an extinction? A review of Slender-billed Curlew records in the Middle East. Br Birds 108:669–682

New Eden Group (2006) The New Eden master plan for integrated water resources management in the Marshlands Area, vol 1–4. Iraqi Ministries of Environment, Water Resources, and Municipalities and Public Works, Baghdad

Porter R, Aspinall S (2010) Birds of the Middle East. Christopher Helm, London

Porter RF, Salim M, Ararat K, Fadhel O (2010) A provisional checklist of the birds of Iraq. Marsh Bull 5(1):56–95

Salim, MA & RF Porter (in prep). Towards a population assessment of the breeding population of White-tailed Lapwing in Iraq

Salim MA, Porter R, Schiermacker-Hansen P, Christensen S, Al- Jbour S (2006). [Field guide to the birds of Iraq]. Nature Iraq/BirdLife International, Baghdad. [In Arabic]

Salim MA, Porter R, Rubec C (2009) A summary of birds recorded in the marshes of southern Iraq, 2005–2008. BioRisk 3:205–219

Salim MA, Abd IM, Abdlhasan NA, Minjal MSh (2010) Key biodiversity survey of Southern Iraq: 2009 site review. Nature Iraq report. Sulaimaniya, Iraq. Nature Iraq Bub. No. NI-1209-01

Salim MA, Al-Sheikhly OF, Majeed KA, Porter RF (2012) An annotated checklist of the birds of Iraq. Sandgrouse 34:4–43

Schodde R, Kirwan GM, Porter RF (2012) Morphological differentiations and speciation among darters (*Anhinga*). Bull Br. Ornithol Club 132:283–294

Scott DA (1995) A directory of wetlands in the Middle East. IUCN/International Waterfowl and Wetlands Research Bureau, Gland/Slimbridge

Scott DA, Carp E (1982) A midwinter survey of wetlands in Mesopotamia, Iraq: 1979. Sandgrouse 4:6–76

Symes A, Taylor J, Mallon D, Porter R, Simms C, Budd K (2015) The conservation status and distribution of the breeding birds of the Arabian Peninsula. IUCN/Environment and Protected Areas Authority, Cambridge/Gland/Sharjah

Section V
Biodiversity Aspects

Chapter 20
Northern Gulf Marine Biodiversity in Relevance to the River Discharge

Faiza Al-Yamani, Igor Polikarpov, and Maria Saburova

Abstract This chapter summarizes the results of research studies, which addressed the peculiarity of the estuarine and northern Arabian Gulf marine environment and the influence of the Shatt Al-Arab River and its associated marshes on the oceanography and biodiversity of the northern Gulf. The findings of the studies indicated that the northern waters of Kuwait, which are impacted by the Shatt Al-Arab River discharge, displayed lower salinity, higher nitrate concentration, higher chlorophyll-*a*, and higher sedimentation. The biodiversity of the northern waters of the Gulf also was significantly distinguished from the adjacent area away from the influence of the river flow. Discharge of the Shatt Al-Arab is considered to be a dominating driver of the northern Gulf's ecology and largely responsible for the primary and secondary productivity of Kuwait's waters. Long-term reduction in river discharge due to man-made alterations (damming and diversion of rivers) impacted the oceanographic characteristics of Kuwait's waters with implication to fisheries resources. Management of Shatt Al-Arab River discharge into the Gulf requires cooperation among the riparian countries and downstream countries like Kuwait to preserve the productivity, biodiversity, and uniqueness of the northern Gulf ecosystems. Long-term monitoring and assessment as well as joint research programs involving scientists from the three concerned countries, Kuwait, Iraq, and Iran, are needed.

Keywords Aquatic biodiversity · Salinity changes · Kuwait · Mesopotamian marshes · Plankton · Benthos · Shatt Al-Arab delta

---

F. Al-Yamani · I. Polikarpov (✉) · M. Saburova
Ecosystem-based Management of Marine Resources Program, Kuwait Institute for Scientific Research, Salmiya, Kuwait

L. A. Jawad (ed.), *Southern Iraq's Marshes*, Coastal Research Library 36,
https://doi.org/10.1007/978-3-030-66238-7_20

20.1 Introduction

The Arabian Gulf, also known as the Persian Gulf, is a remarkable West Asian Mediterranean-type marginal sea of the Indian Ocean (Tomczak and Godfrey 2003), located in a region of the Middle East with subtropical hyperarid and arid climate due to the influence of the surrounding arid land masses (Johns et al. 1999; Hamza and Munawar 2009; Sheppard et al. 2010). Sheppard et al. (2010) mentioned that 14 historical variants of the name of the Arabian/Persian Gulf are known. Here the name "Gulf" is used, as is the case in several preceding scientific publications. Circulation in the Gulf is in an anticlockwise motion, driven primarily by density gradients creating a reverse estuarine flow (Reynolds 1993, 2002). The narrow Strait of Hormuz restricts water exchange between the Gulf and the northwestern Indian Ocean.

Salinity of the Gulf is generally determined by the balance between evaporation, freshwater inflow, and water exchange with the Sea of Oman (Reynolds 1993; Sheppard et al. 2010). As a result of strong evaporation particularly during long and hot summer coupled with very limited freshwater river runoff and negligible rainfall in the arid to hyperarid climate, the Gulf's waters are among the most saline in the world's oceans. Salinities generally range from about 38 to 41 (hereafter practical salinity units, psu), increasing to more than 50–70 in some embayments (e.g., in the shallow, enclosed Gulf of Salwah in the western Gulf) but falling to around 37 toward the Sea of Oman (John et al. 1990; Reynolds 1993; Sheppard et al. 2010). The Gulf is a shallow body of water with an average depth of 36 m (Emery 1956), which allows it to absorb more heat than typical oceanic waters. Its shallowness along with climatic conditions makes this region one of the hottest places worldwide with notable seasonal amplitude of seawater temperatures. Sea surface temperature ranges between 17 °C in winter and 36 °C in summer (Sheppard et al. 1992, 2010), and in some shallow areas the temperature range is even more (Al-Yamani et al. 2004). Thus, the Gulf is an exceptional marine system of special interest because of the naturally high levels of environmental stresses it experiences, forcing marine inhabitants to survive here at tolerance thresholds along a broad range of salinity and temperature (Sheppard et al. 2010).

The Shatt Al-Arab River system with its main tributaries Tigris, Euphrates, and Karun rivers is the main source of freshwater input to the northern Gulf and plays a vital role in support of the Mesopotamian marshes, the largest wetlands in southwest Asia (Richardson and Hussain 2006; UN-ESCWA and BGR 2013). Until recently, the northern region of the Gulf was clearly distinguished from the adjacent hypersaline waters by a unique hydrology due to significant freshwater inflow and salinity offset toward the lower levels (e.g., Reynolds 1993; Al-Yamani et al. 2004, 2017; Polikarpov et al. 2016). The anticlockwise-flowing currents in the north of the Gulf deflect the Shatt Al-Arab River plume (Reynolds 1993; Fig. 20.1a, f) making waters along southwestern Gulf's coast the main receiving basin for nutrient-rich and turbid freshwater discharge of the river (e.g., Al-Abdul-Razzaq et al. 1982;

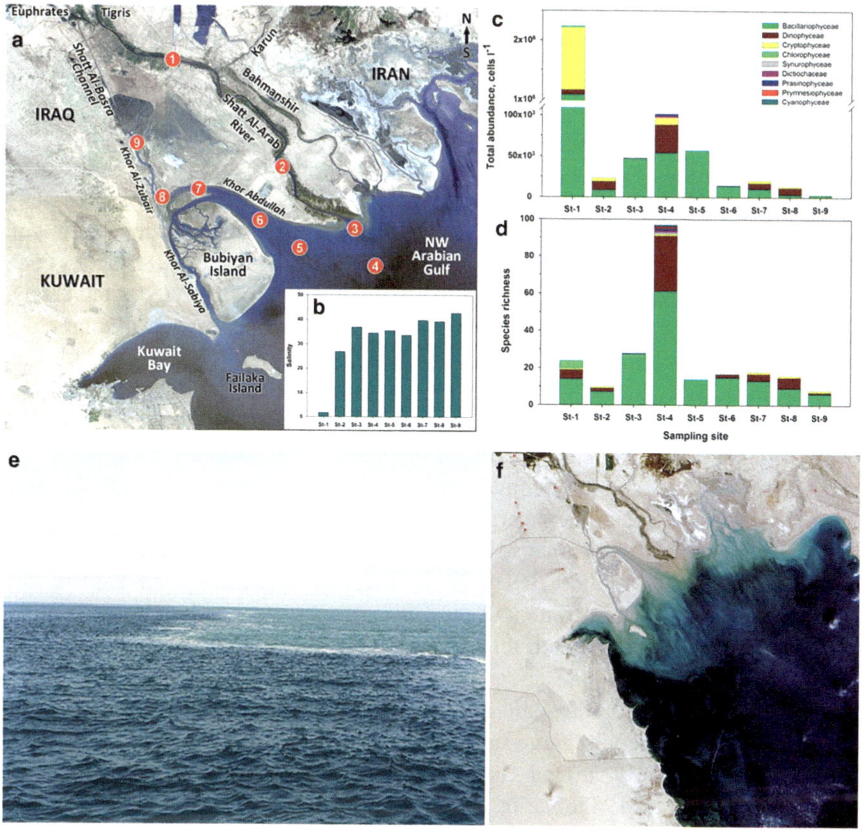

Fig. 20.1 Shatt Al-Arab deltaic system. (**a**) Satellite map of the Shatt Al-Arab River basin and northwestern Arabian Gulf; sampling site locations for "Basra Pearl" cruise (Marine Science Centre, University of Basra, Iraq, 21–27 June 2012) are indicated by red circles; (**b**) salinity records (expressed as psu) along the Shatt Al-Arab deltaic system, June 21–27, 2012; (**c** and **d**) total phytoplankton abundance and composition (**c**) and species richness (**d**, number of species) along the Shatt Al-Arab deltaic system (phytoplankton groups are represented by colored bars); (**e**) interaction between discharge of Shatt Al-Arab River and saline waters of the Gulf (observed in April 2006 in the vicinity of Failaka Island in Kuwait's northern waters, photo by Igor Polikarpov). (**f**) Satellite map of the northern Gulf with river plume is clearly visible. (Source: ORNL DAAC 2018. MODIS and VIIRS Land Products Global Subsetting and Visualization Tool. ORNL DAAC, Oak Ridge, Tennessee, USA. Accessed October 11, 2005. Subset obtained for Terra 0.250 km)

Abaychi et al. 1988; Al-Saadi et al. 1989; Al-Yamani 1989, 2008; Jones et al. 2002; Al-Yamani et al. 2004, 2007).

The influence of the river discharge is particularly evident at the northernmost area of the Gulf near Iraq and Kuwait where Shatt Al-Arab meets the hypersaline Gulf's waters (Fig. 20.1a, e). The reduction of salinity is traced across Kuwait's waters up to offshore area about 100 km southward particularly during maximum river discharge in spring (Al-Yamani et al. 2004, 2017; Polikarpov et al. 2009;

Devlin et al. 2015a). High negative correlation between salinity of Kuwait's waters and quantity of the Shatt Al-Arab discharge indicated primary contribution of freshwater to salinity balance among other factors including effect of evaporation and desalination effluent (Al-Yamani et al. 2017).

The biological productivity of the Gulf is highly variable and closely limited to the nutrient distribution throughout the basin (Brewer and Dyrssen 1985; Jones et al. 2002; Al-Yamani et al. 2004; Polikarpov et al. 2016; Al-Yamani and Naqvi 2019). Unlike the Sea of Oman, the lack of the favorable upwelling conditions results in nutrient limitation in most of the offshore Gulf's waters (El Samra 1988; Jones et al. 2002), which, in turn, determines the low to moderate productivity. Conversely, some coastal areas, especially influenced by river discharge, are rich in total chlorophyll and are much more productive (e.g., Nezlin et al. 2007, 2010; Sheppard et al. 2010; Polikarpov et al. 2016, 2019).

The discharge of the Shatt Al-Arab is probably the main driving force of the northern Gulf and responsible for much of productivity, strongly influencing the pelagic communities and sustaining high biomass of the phytoplankton in this area compared to the adjacent marine waters (Al-Abdul-Razzaq et al. 1982; Al-Yamani 2008; Subba-Rao and Al-Yamani 1998; Al-Yamani et al. 2007; Nezlin et al. 2007, 2010; Polikarpov et al. 2009, 2016, 2019; Sheppard et al. 2010; Al-Yamani and Naqvi 2019). The northern waters of the Gulf are characterized by elevated phytoplankton biomass based on both remotely sensed and field measured chlorophyll-*a* concentrations (e.g., Subba-Rao and Al-Yamani 1998; Al-Yamani et al. 2004; Nezlin et al. 2007, 2010; Sheppard et al. 2010; Polikarpov et al. 2019). In addition, the Shatt Al-Arab River discharge supplies significant amounts of fine sediments that accumulate specifically around the northern islands of Kuwait (Kassler 1973; Douabul and Al-Saad 1985; Karim and Salman 1987; Alosairi et al. 2011).

The interaction of freshwater inflow with hypersaline water masses in the northern Gulf results in distinct environmental gradients along the southwestern coast. A north to south increasing trend in salinity exists across Kuwait's waters from sites in close proximity to Shatt Al-Arab River mouth up to distant southern offshore waters. Salinity, in turn, is inversely correlated with turbidity and nutrient concentrations (Al-Yamani et al. 2007, 2019b; Al-Ghadban et al. 2008; Polikarpov et al. 2009). Opposing gradients of key physical and chemical variables diversify the environment of the northern Gulf and provide a range of ecological niches for the development of diverse pelagic and benthic biota there (Al-Yamani et al. 2004, 2019b; Polikarpov et al. 2009).

Over the past decades, the upper drainage basin of the Shatt Al-Arab River has been highly impacted by a series of dams, reservoirs, and artificial channels built by Turkey, Syria, Iran, and Iraq. Large-scale modifications in the upper basins of the Tigris, Euphrates, and Karun rivers as well as the destruction of the Mesopotamian marshes and their further partial restoration have also impacted the quantity and quality of the Shatt Al-Arab discharge. In addition, climate variability in recent years (lower precipitation, occasional droughts, temperature increase, and higher evaporation rates) also led to the decline of the river runoff. Over time, these factors jointly have altered geomorphological and hydrodynamic patterns and sediment transport within the basin and have strongly affected its ecosystem (Al-Yamani and Khan 2002; Al-Yamani et al. 2007; Al-Ghadban et al. 2008; Sheppard et al. 2010, 2012). For the past three decades, the northern Gulf has experienced an incessant raise in salinity due to a decline in freshwater inflow from the Shatt Al-Arab River system (Al-Yamani et al. 2017). Currently, Kuwait's marine environment is affected by higher salinity since the 1980s (Al-Yamani et al. 2017, 2019b, c). Seawater intrudes for up to about 100 km upward along the Shatt Al-Arab River course (Moyel and Hussain 2015; Lawler 2016).

The recorded sharp drop of the Shatt Al-Arab River discharge in recent decades due to climatic changes and upstream regulation of the river flow has posed a serious threat to the basic components of the northern Gulf ecosystem in terms of productivity and diversity. Reduced river runoff affected the abundance, size, and composition of the phytoplankton and impoverished these previously productive waters, diminishing the food supply for zooplankton, which, in turn, have an adverse effect on landings of commercial fish and shrimps (Sheppard et al. 2010; Al-Husaini et al. 2015).

The salinity crisis and associated ecological degradation currently facing the northern Gulf have been well documented by numerous publications in the international and regional journals and scientific reports of Kuwait Institute for Scientific Research (KISR). However, most available data relevant to long-term impact of river discharge to the northern Gulf's ecosystem are scattered over numerous published sources and often restricted to a particular biological group and vary greatly by timeframe of surveys. It remains difficult to obtain up-to-date information on the status of the Shatt Al-Arab River discharge, whereas current data on abundance and diversity of marine biota are yet to be very fragmental and often weakly comparable in long-term scale.

The overall objective of this study is to assess the changes in structure and diversity of marine biota in the northern Gulf, with special reference to Kuwait's waters, which may have been caused by long-term alteration of the Shatt Al-Arab River discharge. In this study, we outlined the up-to-date status of biodiversity in the northern Gulf ranging from prokaryotes and unicellular protists to marine vertebrates based on available published records. A set of uniformly collected and processed data with broad coverage across Kuwait's waters was obtained from a range of KISR's research projects during 2005–2016 to explore the local diversity trends of main plankton groups including phytoplankton, loricate ciliates, and copepods and to describe long-term changes in their composition. To assess long-term trends in basic oceanographic variables, dataset from systematic monitoring of Kuwait's waters conducted by KISR since 1995 onward was used. Remote sensing parameters were retrieved using the NASA GIOVANNI online tool (Acker and Leptoukh 2007).

20.2 Biodiversity of the Northern Gulf

The northernmost part has been identified to be among the most important ecological and biological areas of the Gulf due to its biodiverse nature, harboring endemic, rare, and threatened species and supporting high abundance and productivity of marine biota (Khan 2002; Al-Yamani et al. 2004; Sheppard et al. 2010; UNEP/CBD 2016). This area provides spawning and nursery grounds, productive habitat, food, and protection for many species, including fish and crustaceans of regional economic value and resident and migratory waterfowl and shorebirds (e.g., Price 1993; Al-Yamani and Khan 2002; Sheppard et al. 2010; Bishop et al. 2013; Al-Husaini et al. 2015). Due to the unique oceanographical conditions, this area is characterized by diverse and endemic marine biota adapted to extreme conditions and is considered a hotspot of marine biodiversity (UNEP/CBD 2016). Projection of the global and regional climatic changes on the biodiversity patterns showed an expectable role of this area as a climatic refuge for many marine species from other areas of the Gulf (Wabnitz et al. 2018).

The early taxonomic surveys of Kuwait's marine biota, going back to the 1960–70s, have been focused mostly on several groups of marine organisms, including fish and crustaceans, due to their commercial importance, while a large number of pelagic and benthic inhabitants have been poorly considered or neglected. Over the past decade, great taxonomic efforts applied to organisms inhabiting Kuwait's marine environment have resulted in the publication of numerous identification and illustrated guides and checklists for a wide range of taxonomic groups from prokaryotes and protists to marine vertebrates, making this area to be among the most well-documented parts of the Gulf in terms of marine biodiversity (Jones 1986; Al-Yamani and Prusova 2003; Bishop 2003; Al-Yamani and Skryabin 2006; Al-Yamani and Khvorov 2010; Al-Yamani and Saburova 2010, 2011, 2019a, b; Al-Yamani et al. 2011, 2012a, b, 2014, 2019a; Al-Kandari et al. 2009, 2019, 2020a, b; Al-Enezi et al. 2020). The important scientific outcomes of these research activities include the comprehensive taxonomic account of marine biota. The descriptions of new species and new records of rare and exotic species for the Gulf extend their global distribution and known tolerance limits. The revealed taxonomic diversity of marine biota in the northern part of the Gulf provides baseline information on the species composition and constitutes an important tool for further taxonomical and ecological assessments, including studies on the long-term trends and variability in regional diversity and associated ecosystem services, detection of invasive species, the identification of endangered and threatened species, and other related investigations.

On the basis of the assembled data from the most recent published records, this study infers a total of over 2700 taxa inhabiting Kuwait's marine environment. This great taxonomic diversity is represented by both pelagic and benthic organisms belonging to 34 phyla of 5 kingdoms from the prokaryotic and eukaryotic domains.

The diversity of planktonic prokaryotes belonging to the phylum Cyanobacteria in Kuwait' waters is restricted to 3 taxa including filamentous *Trichodesmium erythraeum*, ectosymbiont *Synechococcus carcerarius* associated with heterotrophic dinophysoid dinoflagellates (Al-Yamani and Saburova 2019a), and *Synechococcus* sp. which is abundant in nanoplankton (Al-Yamani et al. 2004), whereas the benthic cyanobacteria are more diverse with 6 genera (Al-Zaidan et al. 2006) and 12 common species (Jones 1986) identified along Kuwait's coast. In contrast, the composition of bacterioplankton of Kuwait's waters has been found highly diverse based on 16S rRNA denaturing gradient gel electrophoresis fingerprinting and phylogenetic analysis (Almutairi 2015). The obtained sequences across Kuwait's waters were assigned to the Proteobacteria phylum, and five bacterial classes were identified including Alphaproteobacteria, Betaproteobacteria, Gammaproteobacteria, Deltaproteobacteria, and Epsilonproteobacteria, among which Alphaproteobacteria and Gammaproteobacteria were the two most predominant classes.

Highly diverse assemblages of diatom algae (phylum Bacillariophyta) are represented by a total of 449 taxa, of which 158 species are planktonic, 90 species can be considered as tychoplanktonic that occasionally are washed out from the bottom sediments and carried into the water column (Al-Yamani and Saburova 2019b), and 201 species occur in various benthic habitats along Kuwait's coast (Al-Yamani and Saburova 2011). A total of 281 dinoflagellate species (phylum Miozoa) largely contribute to the diversity of phytoplankton (221 taxa) and microphytobenthos (59 taxa), and 1 taxon was recognized as an intracellular parasite of loricate ciliates (Al-Yamani and Saburova 2010, 2019a; Chomérat et al. 2012; Saburova et al. 2012a, b, 2013).

Unicellular flagellates comprise a diverse assemblage of autotrophic and heterotrophic protists with more than 50 species identified, forming an important component of both pelagic (31 taxa) and benthic (22 taxa) communities of Kuwait's marine environment, and are defined by their ecological niches rather than taxonomic classification (Al-Yamani and Saburova 2010, 2019a). This taxonomically diverse group is represented by a broad range of taxa belonging to eight phyla of three kingdoms including Chromista (phyla Cryptophyta, Haptophyta, Ochrophyta, Bigyra, Katablepharidophyta, and Cercozoa), Plantae (phylum Chlorophyta), and Protozoa (phylum Euglenozoa).

Taxonomical account of ciliates (phylum Ciliophora) of Kuwait is limited to the pelagic habitat, where 1 photosynthetic ciliate, 63 loricated taxa, and 2 sessile species attached to plankton diatoms were recognized (Al-Yamani and Skryabin 2006; Al-Yamani et al. 2011; Al-Yamani and Saburova 2019a, b), whereas a highly diverse naked ciliates in both pelagic and benthic habitats (unpublished observations of the third author) are totally overlooked. The most recent comprehensive overview of foraminiferal diversity (phylum Foraminifera) listed a total of 451 species, which largely contribute to the diversity of benthic microfaunal assemblages of Kuwait (Al-Enezi et al. 2020). One protozoan taxon (phylum Myzozoa) is identified to be parasitic on calanoid copepods (Al-Yamani et al. 2011).

An assemblage of metazoan zooplankton of Kuwait's waters comprises a wide range of organisms belonging to 9 invertebrate phyla, among which crustaceans (phylum Arthropoda) represents the most diverse group with 54 species of copepods, 3 species each of branchiopods and decapods, 2 species of ostracods, and 1 species each of mysids and cumaceans. Other representatives belong to the phyla Cnidaria (14 species), Ctenophora (2 species), Platyhelminthes (1 species), Rotifera (2 species), Annelida (2 species), Mollusca (1 species), Chaetognatha (4 species), and Chordata (8 species) (Al-Yamani et al. 2011). Meroplankton encompasses pelagic larval stages of a number of benthic invertebrates belonging to 8 phyla, of which the young life stages of decapods (phylum Arthropoda, 29 species identified) and gastropods (phylum Mollusca, 4 species) are the most diverse groups. Other representatives of Kuwait's meroplankton include larval stages of Platyhelminthes, Nemertea, Brachiopoda, Phoronida, Bryozoa, and Chordata (Al-Yamani and Khvorov 2007, 2010; Al-Yamani et al. 2011).

A total of 80 marine macroalgae were recognized in the composition of Kuwait's benthic flora belonging to phyla Chlorophyta (23 species), Rhodophyta (27 species), and Heterokontophyta (27 species) (Al-Yamani et al. 2014). Vegetation of Kuwait's salt marshes is dominated by 18 halophylic flowering vascular plants belonging to phylum Tracheophyta (Jones 1986), and two species of seagrass occur on sheltered sandy beaches and extend into the subtidal zone forming seagrass meadows (Jones 1986; Al-Yamani et al. 2014).

A total of 472 species, which represent 11 phyla of macrozoobenthos, were identified during recent faunistic survey along Kuwait's coast (Al-Yamani et al. 2012b). The most diverse groups were represented by phyla Annelida with 120 polychaeta species, Mollusca (233 species including 130 gastropods, 99 bivalves, 3 scaphopods, and 1 polyplacophoran), and Arthropoda (85 species including 34 decapods, 25 amphipods, 10 cumaceans, 5 cirripedians, 4 isopods, 3 tanaidaceans, 2 pycnogonids, 1 copepod, and 1 mysid). Lower number of species was encountered for phyla Echinodermata (16 species), Cnidaria (5 species), Sipuncula (5 species), Chordata (3 species), and Bryozoa (2 species). Brachiopoda, Phoronida, and Hemichordata were represented with one species each. Organisms attributed to Nemertini, Turbellaria, and Oligochaeta were not identified to species level (Al-Yamani et al. 2012b). The diversity of Cnidaria is largely complemented by the 35 recorded species of scleractinian corals in Kuwait's southern waters (Hodgson and Carpenter 1995; Carpenter et al. 1997).

A comprehensive faunistic 2-year survey was conducted during 2013–2015 to document intertidal macrofauna throughout Kuwait entire coastal zone from the Khor Al-Subbiya channel in the north to Saudi Arabian border in the south, including the major offshore islands. This most recent taxonomic survey has resulted in significant enlargement of the recorded macrofaunal diversity up to 737 invertebrate species (Al-Kandari et al. 2017a, 2018) mostly due to more species revealed in the composition of polychaetes (241) and crustaceans (213, mainly in decapods, amphipods, and isopods) and in previously neglected marine sponges Porifera (20). The results of this research project supplemented the known diversity of mollusks and amphipods of the northern Gulf waters. The checklist of molluskan species of Kuwait's marine environment has enlarged the diversity of this group up to 271, of which more than 60 taxa were recognized as new records for Kuwait and many taxa, mostly micro-mollusks, were tentatively identified to be new for science (Al-Kandari et al. 2020a). Moreover, the diversity of this group has been supplemented by records of 12 rare nudibranch species (Nithyanandan 2012). The latest assessment of amphipods species diversity has revealed 86 species in Kuwait's benthic habitats (Al-Yamani et al. 2019a; Al-Kandari et al. 2020b).

At present, Kuwait's fish list includes 345 species belonging to two classes. Cartilaginous fishes (class Elasmobranchii) are represented by 39 species, and the most diverse is group of ray-finned bony fishes (class Actinopterygii) with 306 species recorded (Bishop 2003). Further, the list of marine vertebrates recorded in Kuwait's waters includes five species of sea snakes, two species of sea turtles (green turtle *Chelonia mydas* and hawksbill turtle *Eretmochelys imbricata*), and three species of marine mammals including Indo-Pacific humpback dolphin (*Sousa chinensis*), bottlenose dolphin (*Tursiops truncatus*), and finless porpoise (*Neophocaena phocaenoides*) (Bishop et al. 2013).

Intensive long-term taxonomic surveys of Kuwait's marine environment have resulted in numerous new records of previously overlooked species in local and regional scales and description of new taxa mostly endemic for the northern Gulf. A new large and peculiar dinoflagellate *Prorocentrum bimaculatum* Chomérat and Saburova has been described from Kuwait's intertidal sediments (Chomérat et al. 2012), and marine benthic dinoflagellate *Durinskia agilis* (Kofoid and Swezy) Saburova, Chomérat, and Hoppenrath has been redescribed and reclassified (Saburova et al. 2012a). New record of tropical potentially toxigenic dinoflagellate *Fukuyoa* (=*Gambierdiscus*) *yasumotoi* from Kuwait's southern shores has extended its known geographical distribution and temperature tolerance limits (Saburova et al. 2013), and observations of dinoflagellate *Kryptoperidinium foliaceum* from the hypersaline tidal creek in Kuwait's northern area have extended the known range of salinity tolerance for this brackish water microalga to salinity of more than 100 psu (Saburova et al. 2012b).

Seven new tintinnid species have been described from the composition of loricate ciliates inhabiting Kuwait's waters, namely, *Leprotintinnus bubiyanicus*, *Leprotintinnus elongatus*, *Tintinnopsis failakkaensis*, *Luminella kuwaitensis*, *Metacylis pithos*, *Eutintinnus conicus*, and *Eutintinnus contractus* Skryabin and Al-Yamani (2006, 2007a). New copepod species have been described from Kuwait's waters including *Subeucalanus flemingeri* Prusova, Al-Yamani, and Al-Mutairi (2001), *Bestiolina arabica* Ali, Al-Yamani, and Prusova (2007), and *Labidocera kuwaitiana* Prusova and Al-Yamani (2014).

Until today, the faunistic surveys of Kuwait's coast have added numerous new species to science from a range of macrofaunal groups including corals, crabs, snapping shrimps, amphipods, and polychaetes. Hodgson and Carpenter (1995) have described new species of scleractinian coral *Acropora arabensis* from the coral reef surrounding Kubbar Island in the southern waters of Kuwait. To date, the species is only known from the Arabian Gulf, Southern Red Sea, and Madagascar. Recently, new coral species *Psammocora albopicta* has been described with Indo-West Pacific distribution (Benzoni 2006) that was previously identified as *Psammocora superficialis* by Carpenter et al. (1997).

Jones and Clayton (1983) have described two new camptandriid crabs from Kuwait's muddy shores, namely, *Cleistostoma kuwaitense* (currently accepted as

Leptochryseus kuwaitense) and *Paracleistostoma arabicum* (*Manningis arabicum*) (Jones and Clayton 1983). Being originally described from Kuwait, *M. arabicum* is now known to be widely distributed throughout the Gulf and the Sea of Oman including the entire coast of Iran, Iraq, Kuwait, Saudi Arabia, and Qatar, while the geographic distribution of *L. kuwaitense* is mainly restricted to the northern Gulf (Naderloo 2017). A new snapping shrimp, *Alpheus lutosus*, has been described from the intertidal mudflats of Bubiyan Island in the northern Kuwait, south of the Shatt Al-Arab delta (Anker and De Grave 2009). Two other unidentified *Alpheus* species were observed off Kuwait Bay in Ras Al-Ardh and Al-Julaia'h (Al-Yamani et al. 2012b).

Faunistic surveys along Kuwait's coast and islands by Mohammad (1970a, b, 1971, 1972, 1973, 1980) documented 20 new polychaete species to what previously known as the following: *Prionospio rotalis* (Mohammad 1970a), *Nereis* (*Nereis*) *neogracilis*, *Perinereis kuwaitensis* (Mohammad 1970b), *Nereis* (*Neanthes*) *deplanata* (currently accepted as *Neanthes deplanata*), *Perinereis arabica*, *Isolda albula* (Mohammad 1971), *Polydora vulgaris* (Mohammad 1972), *Alluaudella longicirrata* (currently accepted as *Odontosyllis freycinetensis*), *Autolytus zonatus* (currently accepted as *Imajimaea zonata*), *Bookhoutia oligognatha*, *Marphysa gemmata* (currently accepted as *Paucibranchia gemmata*), *Kuwaita magna*, *Polydora spondylana*, *Magelona heteropoda* (currently accepted as *Magelona obockensis*), *Streblosoma longa* (currently accepted as *Pseudostreblosoma longum*), *Neoleprea clavata* (Mohammad 1973), *Phylo kubbarensis*, *Scyphoproctus aciculatus*, *Clymenura annulata* (currently accepted as *Leiochone annulata*), and *Telothelepus macrothoracicus* (Mohammad 1980). Further morphological and molecular studies of polychaetes collected from Kuwait's coast have revealed the identity of *Hydroides operculata* from Kuwait and distinguished it from morphologically similar taxa from India, Hong Kong, and Australia (Sun et al. 2017).

Seven new for science amphipod species have been described from the Sea City waterway at Kuwait's southern shore by Myers and Nithyanandan (2016) including *Protohyale arabica*, *Bemlos acuticoxa*, *Laticorophium bifurcatum*, *Podocerus mamlahensis*, *Latigammaropsis pseudojassa*, *Ceradocus* (*Denticeradocus*) *alama*, and *Maera irregularis*. The latest assessment of amphipods species diversity from the entire Kuwait's coast has recognized 18 taxa new for science that currently being described (Al-Yamani et al. 2019a; Al-Kandari et al. 2020b).

20.3 Overall Biodiversity Pattern Across the Northern Gulf: Habitats

The Shatt Al-Arab River contributes to the cyclonic circulation, drives the salinity balance through the dilution of freshwater outflow, promotes the biological productivity by the discharge of nutrients, and affects the sedimentary influx in the northern part of the Gulf (e.g., Karim and Salman 1987; Al-Bakri and El-Sayed 1991;

Reynolds 1993; Al-Yamani et al. 2004, 2017, 2019b; Riegl et al. 2010; Sheppard et al. 2010). Strongly influenced by the Shatt Al-Arab River discharge, the northern part of the Gulf is known for its unique oceanographic conditions supporting endemic biota adapted to extreme and highly dynamic environment and constitutes one of the most biologically productive areas of the Gulf (e.g., Price 1993; Al-Yamani et al. 2004; Nezlin et al. 2010; Sheppard et al. 2010). Different combinations of physical, chemical, and geological variables discriminate a range of environments, among which the Shatt Al-Arab delta, northern waters surrounding Bubiyan and Failaka islands in close proximity to Shatt Al-Arab mouth, Kuwait Bay, and offshore open area southward are clearly delineated. Contrasting gradients of salinity, water temperature, turbidity, nutrient concentrations, and textural variability of sediments (Fig. 20.3) diversify Kuwait's marine environment and support characteristic associations of pelagic and benthic species (Jones 1986; Jones et al. 2002; Al-Yamani et al. 2004, 2007, 2012b, 2019b; Polikarpov et al. 2009; Al-Zaidan et al. 2013; Bishop et al. 2013).

20.3.1 Shatt Al-Arab Delta

The Shatt Al-Arab delta includes the Shatt Al-Arab estuary, Khor Abdullah, and Khor Al-Zubair channels (Fig. 20.1a) and contains numerous unique marine, coastal, and tidal habitats, including muddy intertidal areas. This area harbors both freshwater and marine species with a higher level of species richness and endemism than the adjacent Gulf waters. Due to high nutrient concentrations, the delta is one of the most productive coastal ecosystems of the Gulf. The most of the delta region is located in shallow muddy areas where the depths range from 0.5 to 5 m. Most of the coastal regions bordering this area are tidal flats and salt flats, devoid of extensive vegetation (Isaev and Mikhailova 2009). This area supports a growth of salt-tolerant vegetation (mainly cyanobacterial mats). There are freshwater wetlands just north of the active delta, which support freshwater vegetation and biota (Richardson and Hussain 2006; UN-ESCWA and BGR 2013).

The various deltaic habitats support 415 species of microalgae, more than 120 marine and brackish water species of free swimming copepods, diverse non-copepod zooplankton assemblage, and 380 species of fishes (32 species of freshwater fishes and 348 species of marine and brackish fishes) (Al-Saroonchi and Al-Saad 1988; Khalaf et al. 2014; UNEP/CBD 2016). Four species form seagrass meadows, of which seagrass *Halophila decipiens* constitutes the first record for the Gulf and can be considered endemic species for Iraqi marine coastal area (Ibrahim 2018). The delta area is a critical habitat for several indigenous species, including three copepods, *Acartia* (*Acartiella*) *faoensis*, *Bestiolina arabica*, and *Phyllodiaptomus irakiensis* (Khalaf 1991, 2008a, b), which are currently heavily affected by natural and anthropogenic changes of their environment (Khalaf et al. 2015).

The relatively healthy complex of coral reefs has been recently discovered in Iraqi marine coastal waters at the mouth of the Shatt Al-Arab River (Pohl et al. 2014). These newly found coral reefs cover an area of 28 km^2 at depths of 7 to 20 m, being composed of at least eight coral species with associated echinodermates, sponges, and mollusks.

An expedition survey "Basra Pearl" was conducted to study the Shatt Al-Arab deltaic system during June 21–27, 2012, by the Marine Science Centre (University of Basra, Iraq). This research cruise provided an opportunity to assess the phytoplankton abundance and composition along a wide range of salinity from brackish waters (salinity of 1.71 psu) at the downstream part of the Shatt Al-Arab River through marine area off the river mouth with salinity of 34.5–36.9 psu up to the hypersaline waters in Khor Al-Zubair channel (stations 8 and 9) with salinity up to 42.7 (Fig. 20.1a, b).

Phytoplankton community was characterized by high variability along the salinity gradient in terms of both diversity and abundance (Fig. 20.1c, d). A wide variety of phytoplankton (116 taxa) belonging to 9 algal classes were identified from the Shatt Al-Arab River to Khor Al-Zubair channel. Diatoms (Bacillariophyceae) were the richest algal group with 73 taxa, followed by dinoflagellates (Dinophyceae, 33 taxa) and Chlorophyceae with 4 taxa. Other algal groups (Dictyochophyceae, Cryptophyceae, Synurophyceae, Prasinophyceae, Prymnesiophyceae, and Cyanophyceae) were represented by one taxon each.

The diatom assemblage was dominated by centric species. The largest taxonomic diversity was associated with the genus *Chaetoceros* (16 taxa). A pronounced prevalence of diatoms was typical for taxonomic structure of the phytoplankton throughout the deltaic area. Their prevalence was greatest (95.8–100%) within inshore Gulf waters and reduced down to 56–58% in both brackish and hypersaline waters. Dinoflagellates accounted for 17.3% of the total species richness on the average. High contribution of this group was associated with Gulf's waters off Shatt Al-Arab mouth and with Bubiyan's coastal waters near Warba Island and Khor Al-Zubair channel.

In contrast to river and channel waters, the Gulf waters in close proximity to the river discharge strongly differed from the other sampled area in its highest species richness and diversity. Among 116 encountered phytoplankton taxa, 97 species were found off the Shatt Al-Arab mouth that constituted 84% of the total phytoplankton species richness. Overall, phytoplankton composition was mostly composed of typical marine species; however, freshwater taxa constituted a noticeable component of both phytoplankton abundance and diversity in the brackish water environment. At the Shatt Al-Arab River downstream, the genus *Peridiniopsis* was found in an appreciable abundance ($5.6 \cdot 10^4$ cells$\cdot l^{-1}$), and chlorophytes *Actinastrum*, *Eudorina*, and *Scenedesmus* constituted a noticeable component of the phytoplankton, accounting for 16.7% of the total species richness. Furthermore, the brackish waters of the Shatt Al-Arab River were strongly distinguished from the other sampled sites by having appreciable high phytoplankton abundance ($2.2 \cdot 10^6$ cells$\cdot l^{-1}$), which is contributed mainly by bloomed small-sized diatoms (*Thalassiosira* and *Skeletonema*) and cryptophycean flagellates. In contrast, the lowest phytoplankton abundance ($3.2 \cdot 10^3$ cells$\cdot l^{-1}$) and diversity (8 taxa) were recorded in the hypersaline waters of Khor Al-Zubair channel.

Salinity was identified as the main factor structuring the distribution of phytoplankton. The total abundances of the main phytoplankton groups were strongly negatively correlated with water salinity (Pearson correlation coefficient with values from −0.89 to −0.93, $p < 0.001$), whereas high positive correlation was detected between salinity and evenness of the phytoplankton community ($r = 0.8$, $p < 0.001$).

The occurrence of dinoflagellate belonging to *Peridiniopsis* genus in bloom proportions has been observed previously in 2009–2010 in the Shatt Al-Arab River in vicinity of Basra City at salinity range from 2 to 7.2 psu (Hameed et al. 2016). Thorough taxonomic survey has assigned this bloom-forming dinoflagellate to *Peridiniopsis minima* that was originally described from the freshwater phytoplankton assemblage of the Jiulongjiang River in China (Zhang et al. 2014). Observations of freshwater *P. minima* in the brackish waters of the Shatt Al-Arab River system in appreciable concentrations extend the known range of salinity tolerance for this dinoflagellate and expand its geographic distribution (Hameed et al. 2016). This species can be considered endemic to Shatt Al-Arab River, and its occasional records in northern Gulf's waters around Bubiyan Island (Al-Yamani and Saburova 2019a) most probably reflect anomalous intrusion of freshwaters from the Shatt Al-Arab River system. The outbreaks of this dinoflagellate in the Shatt Al-Arab River signify a real ecological threat to river system as microalgal blooms could be a major problem for aquatic biota and for human living in Basra City because the river serves as a drinking water supply.

20.3.2 Bubiyan Island

The largest island of Kuwait, Bubiyan Island, is located in the western side of the Shatt Al-Arab delta in the extreme northwest corner of the Gulf, surrounded by the Khor Abdullah channel in the east and by the Khor Al-Sabbiya channel on its western side (Fig. 20.1a). Most of its coastline consists of broad intertidal areas, which are largely inundated by spring tides (Al-Yamani et al. 2004). Bubiyan's waters occupy a transition zone between the Gulf's high salinity waters and the freshwaters received from Shatt Al-Arab and the Third River (connected to Shatt Al-Basra) through Khor Al-Zubair and provide some of the most interesting and likely productive marine habitats in Kuwait. Nearby freshwater discharge reduces usual high salinities of the Gulf, discriminating the marine life around Bubiyan from that elsewhere in Kuwait, and supports richness of threatened, rare, and taxonomically uncommon species (Bishop 2002; Bishop et al. 2013; Alsaffar and Chen 2019).

A marked prevalence of diatoms in both composition and abundance of the phytoplankton is evident everywhere around Bubiyan Island. A total of 200 microalgal taxa were identified during two phytoplankton surveys in 2004–2005 and 2015–2016. The diverse diatom assemblage with 144 taxa was dominated by centric species. The largest taxonomic diversity was associated with the genera *Chaetoceros* (20 taxa), *Nitzschia* (15), *Rhizosolenia* (11), and *Coscinodiscus* (9). Dinoflagellates contributed 48 taxa to the total phytoplankton composition with most diverse genera being *Protoperidinium* (15 taxa) and *Tripos* (6). The shallow waters around Bubiyan Island are affected by intense tidal currents resulting in frequent occurrence of pennate diatoms (species of *Pleurosigma*, *Gyrosigma*, *Amphora*, *Diploneis*, *Amphiprora*, *Surirella*, *Nitzschia*, and *Trachyneis*) in the phytoplankton composition due to resuspension of these benthic species from the sediments to the surface waters via turbulent mixing (Al-Yamani and Saburova 2011, 2019b; Bishop et al. 2013).

The taxonomic composition of benthic microalgae inhabiting mudflats around Bubiyan Island was mainly restricted to epipelic pennate diatoms in contrast to sandy sediments toward the south, where sand-dwelling dinoflagellates and phototrophic flagellates were quite diverse (Al-Yamani and Saburova 2010, 2011). The abundant diatom populations stabilize the mud surface against resuspension and erosion by secreting mucilaginous films and often forming thin brownish carpets. Multispecific colonial growth of diatom populations (e.g., *Nitzschia*, *Amphora*, *Berkeleya*, *Navicula*) within mucilaginous films on the mud surface was typically observed in extensive mudflats of Bubiyan Island (Al-Yamani and Saburova 2011).

Floristic surveys of Khor Al-Sabbiya coast revealed the occurrences of rare exotic microalgae representing the first records of these taxa in Kuwait's marine environment often from nontypical habitats. Recently, the dinoflagellate *Kryptoperidinium foliaceum* has been recorded in small hypersaline tidal creek flowing into the Khor Al-Sabbiya channel in Kuwait's northern shore with salinity ranging over tidal cycle from 63.7 up to 178.5 psu (Saburova et al. 2012b). These observations of *K. foliaceum* in hypersaline conditions in bloom proportions extend significantly the known range of salinity tolerance for this species. Whereas the bottom sediments of the creek were occupied by diatom-dominated algal mats (Fig. 20.2c), high productivity of the water column was contributed mainly by abundant population of this dinoflagellate providing a key food source for the diverse and dense protozoan community of the tidal creek. High content of carotenoid-rich lipid globules within the cytoplasm in Kuwait's strain of *K. foliaceum* has been attributed to adaptation to a combination of extremely high salinity with low nutrient status within its environment in Kuwait similar to other halotolerant microalgae, such as the chlorophycean flagellate *Dunaliella*, which thrives in hypersaline environment and can produce great quantities of β-carotene in lipid globules within the chloroplasts. One more exotic record is based on the observation of the chlorophycean colonial flagellate *Oltmannsiella lineata* in the small rock pool in highly saline conditions (73 psu) on the shore of Khor Al-Sabbiya channel, where the plentiful population of this species resulted in greenish water discoloration (Al-Yamani and Saburova 2019a).

The composition of foraminiferal assemblages around Bubiyan Island has been found to be impoverished (46 species recognized) compared to the adjacent areas due to the absence of some characteristic families that commonly dominated intertidal and subtidal sediments of warm seas (Al-Zamel and Cherif 1998). Based on the cluster analysis, three assemblages related to salinity conditions and physiographic

Fig. 20.2 Different biota and habitats in Kuwait's marine environment. (**a**) A colony of slender-billed gulls (*Larus genei*) at the shore of Bubiyan Island, April 2006; (**b**) cyanobacteria-dominated algal mat at the surface of muddy sediments in Sulaibikhat Bay, February 2007; (**c**) mucilaginous brown film of diatom-dominated algal mat on the mud surface at the bottom of the hypersaline tidal creek at Khor Al-Sabbiya coast, February 2012; (**d**) mudskipper (*Periophthalmus waltoni*) near its burrow and glasswort *Salicornia herbacea* at the intertidal mudflat of Sulaibikhat Bay, April 2019; (**e**) ocypodid crab (*Macrophthalmus depressus*) near its burrow at the intertidal mudflat of Sulaibikhat Bay, April 2019; (**f** and **g**) coral reefs around Qaruh (**f**) and Umm Al-Maradim (**g**) islands in Kuwait's southern offshore area, November 2014 (**a**–**e** – photos by Igor Polikarpov; **f** and **g** – courtesy of Vladimir Grintsov)

settings were identified with distinct faunal elements discriminating the highly stressed habitats of tidal creeks with abnormal salinity from the low stressed south Bubiyan offshore area.

The coastal waters around Bubiyan Island are inhabited by abundant zooplankton population with density ranging over three orders of magnitude from $4.6{\cdot}10^3$ to $5.3{\cdot}10^6$ ind.·m^{-3} during the 2004–2005 survey. Zooplankton composition is comprised of diverse range of protozoan and metazoan representatives from 14 phyla including Ciliophora, Cnidaria, Ctenophora, Phoronida, Rotifera, Arthropoda, Chaetognatha, and Chordata and pelagic larvae of Annelida, Mollusca, Echinodermata, and Decapoda with high prevalence of loricate ciliates (tintinnids, 25 species) and copepods (47 species) (Bishop et al. 2013).

Composition of tintinnids is largely dominated by the genera *Tintinnopsis* and *Luminella*. Furthermore, gigantic tintinnids *Leprotintinnus bubiyanicus*, endemic species of the northern waters, and *Tintinnopsis ampla* are the markers of Bubiyan waters (Skryabin and Al-Yamani 2006; Al-Osairi et al. 2016). Copepods overwhelmingly dominated zooplankton in terms of both abundance (up to $2.5{\cdot}10^5$ ind. ·m^{-3}) and diversity (47 species representing 22 genera and 17 families). Families Paracalanidae and Oithonidae most largely dominate the copepod composition. In addition, the waters around the island host some indigenous species such as the copepods *Bestiolina similis*, *Pseudodiaptomus arabicus*, and *P. ardjuna* (Al-Yamani et al. 2007).

Meroplankton assemblage in Bubiyan's coastal waters is dominated by zoeae of brachyuran crabs and larvae of commercially important shrimps *Parapenaeopsis stylifera* and *Metapenaeus affinis*. In addition, fish eggs and larvae of clupeid, sciaenid, and gobiid fish accounted by nearly 100% of the ichthyoplankton in this area in spring and summer, indicating this area as an essential spawning and nursery ground for commercially important penaeid shrimps and fish including zobaidy *Pampus argenteus* (Al-Attar 1984; Bishop and Khan 1991, 1999; Al-Yamani and Khvorov 2007; Bishop et al. 2011, 2013).

The coastal waters around Bubiyan Island are richer in fish, crustaceans, and mollusks than nearby Kuwait Bay. Species caught during the trawl surveys in 2004–2005 included 90 fishes, 32 crustaceans, 9 mollusks, 5 echinoderms, 1 tunicate, and 1 sea snake. Two of Kuwait's important shrimp species, *M. affinis* and *P. stylifera*, dominated numerically in trawl surveys during spring and autumn seasons. Important commercial fish species such as zobaidy (pomfret, *Pampus argenteus*) and saboor (hilsa shad, *Tenualosa ilisha*) frequently occurred in the gill net survey around Bubiyan Island (Bishop et al. 2013; Alsaffar and Chen 2019). Northern Kuwait's waters are one of only two locations where the smoothtooth blacktip shark (*Carcharhinus leiodon*) is known to occur (Moore et al. 2014).

The extensive creek systems and mudflats around Bubiyan Island are of great importance for migratory waterfowl and numerous shorebirds (Fig. 20.2a). Both resident and migratory birds use this area as their wintering habitats. Out of Kuwait's total 335 bird species known, 40% (143 species) were recorded in Bubiyan Island area during 2003–2005. This area serves as a nesting and foraging habitats for over 30 species of waterfowl including high densities of birds of national, regional, and international importance. Crab-plover colonies in Bubiyan Island are considered to be the largest nesting populations along the Gulf and possibly the world (Delima et al. 2013). Rare Indo-Pacific humpback dolphin (*Sousa chinensis*) uses Bubiyan's waters as a resident home. It is believed to be the largest and possibly seeding population of this species along the Gulf (Bishop et al. 2013).

20.3.3 Failaka Island

Failaka Island is located southward of Bubiyan Island and southeastward of Khor Al-Sabbiya channel (Fig. 20.1a). The coastline of the island includes a belt of low rocky outcrops, sandy and muddy intertidal flats, tidal pools, and sargassum beds (Al-Sarawi et al. 1995). The recent faunistic surveys have revealed a total of 316 species of marine benthic invertebrates and 17 vertebrates. Crustacea was the most diverse group accounting for 37% of the total zoobenthic species richness, followed by mollusks (29%), polychaetes (26%), and Echinodermata (3%). Other invertebrate groups (Cnidaria, Porifera, Sipuncula, Bryozoa, and Nemertea) and Chordata were poorly represented (Al-Kandari et al. 2017b). Limited *Sargassum* beds dominated by *S. boveanum* have been mapped along the southeastern coast of Failaka Island representing the northernmost site of *Sargassum* occurrence in Kuwait's waters. Seasonally, these *Sargassum* beds offer nursery habitat for economically important shrimps and support the assemblages of epiphytic diatoms, macroalgae, and a range of marine invertebrates (mostly amphipods) (Jones and Al-Attar 1982; Bishop and Khan 1991; Ali et al. 2018; Alghunaim et al. 2019, 2020).

20.3.4 Kuwait Bay

One of the most prominent habitats of Kuwait's marine environment is the semi-enclosed Kuwait Bay, which indents the coastline for approximately 40 km inland and accounts for nearly one-half the country's shoreline. It is a shallow, non-estuarine, tide-dominated water body with average depth of 5.2 m and maximum depth at its mouth of 23 m experiencing a high tidal range of 4 m. Textural characteristics of the bottom sediments indicate a low-energy zone that includes most of the Bay with primarily muddy sediments and a moderate-energy zone restricted to the southern offshore area with primarily sandy deposits. The Bay is characterized by gently sloping mudflats, which resulted from the deposition of river silt originating mostly from the Shatt Al-Arab River discharge. The extensive intertidal mudflats alongside the coastline of Kuwait Bay are considered to be the most extensive found anywhere in the Gulf region. During low tide, large areas of mudflats reaching widths up to 5 km are exposed, particularly at the inner part of Kuwait Bay and in the small embayment of Sulaibikhat Bay at the southern coast

(Emery 1956; Khalaf et al. 1982; Al-Abdul-Razzaq et al. 1983a; Al-Sarawi et al. 1985; Abou-Seida and Al-Sarawi 1990; Al-Yamani et al. 2004; Pokavanich et al. 2013).

Kuwait Bay is influenced by nutrient-rich Shatt Al-Arab River discharge. It supports the abundant and diverse assemblages of primary producers in both pelagic and benthic habitats (Al-Zaidan et al. 2003, 2006; Al-Yamani et al. 2004; Polikarpov et al. 2009). Extensive mudflats of Kuwait Bay covered by dense microbial mats have been recognized among the most fertile habitats contributing far more to intertidal productivity than other resources including phytoplankton, salt marsh halophyte vegetation, and mangroves. Apart from supporting high productivity, mudflats contribute to cycling nutrients, treating water to prevent strong eutrophication, controlling biological diversity, and regulating water quality in the northern Gulf and worldwide (Al-Zaidan et al. 2006; Almojil 2017; Al-Said et al. 2019; Hale 2020).

In the intertidal zone, the surface of the muddy sediments is covered with tight algal mats (Fig. 20.2b) composed of a thin layer of detritus reinforced densely by filaments of cyanobacteria and inhabited by numerous small- and medium-sized diatoms mainly from *Halamphora*, *Seminavis*, *Navicula*, and *Nitzschia* genera and euglenoid flagellates (Al-Zaidan et al. 2006; Al-Yamani and Saburova 2010, 2011). Cyanobacteria dominate the coarser grained top shore sediments which are subject to desiccation but are increasingly replaced by diatoms at lower intertidal zone where fine sediments retain water at low tide. Numerically, diatoms show a high mean contribution of 61% followed by cyanobacteria with 30.6% (Al-Zaidan et al. 2006).

Diatom assemblages of organically enriched sands in the eastern part of Kuwait Bay are predominantly comprised of pennate forms living closely attached to the sediment particles including *Achnanthes*, *Opephora*, small-sized *Navicula*, and *Amphora*. Apart from these attached forms, chain-forming diatoms, e.g., *Dimeregramma*, *Plagiogramma*, *Biddulphia*, *Grammatophora*, and *Melosira*, also occur in considerable abundances. The presence of such chained forms changes the community structure considerably and results in a shift from a two- to a three-dimensional community (Al-Yamani and Saburova 2011).

The species diversity of the phytoplankton community in Kuwait Bay with 211 identified taxa is largely comprised of diatoms (107 taxa) and dinoflagellates (93 taxa). The largest species richness was associated with centric diatoms belonging to the genera *Chaetoceros*, *Rhizosolenia*, and *Coscinodiscus* as well as the pennate diatoms from *Nitzschia* and *Pleurosigma* genera. A large contribution of typically benthic, large and small diatoms from periphytone, epipelon, and epipsammon associations (e.g., *Pleurosigma*, *Diploneis*, *Surirella*, *Trachyneis*, *Nitzschia*, *Entomoneis*, *Plagiotropis*) indicates a strong mixing of the water column by tidal currents in this shallow embayment. Phytoplankton abundance in Kuwait Bay during 2007–2009 was composed mainly of diatoms, small-sized cryptophycean flagellates, and gymnodinioid dinoflagellates. The development of the phytoplankton community was supported mainly by medium- to small-sized forms throughout the studied period. High positive correlations of this dimensional fraction with total inorganic nitrogen and silicates indicated strong relationship between nutrient concentrations and phytoplankton abundance.

The composition of macroalgae assemblage in Kuwait Bay is largely dominated by the green algae including *Ulva*, *Enteromorpha*, *Cladophora*, and *Bryopsis*, with a few species of brown and red algae (Al-Yamani et al. 2014; UNEP/CBD 2016). These macroalgae occur as a thick layer on the surface of intertidal mudflats for a short period of time, particularly at the upper intertidal zone. This layer is often grazed upon by herbivorous invertebrates (UNEP/CBD 2016). Vegetation of *Sargassum* beds is observed seasonally along the southeastern coast of the Bay (Bishop and Khan 1991; Bishop et al. 1994; Ali et al. 2018; Alghunaim et al. 2019, 2020).

The cycling of microalgae, both within the intertidal mudflats and between the intertidal sediments and overlying water column, and the presence of dense macroalgal assemblage have considerable importance both for benthic and pelagic herbivores. For this reason, mudflats provide fertile feeding grounds for a number of species of wading birds and are breeding and nursery grounds of several commercially important fish and shrimp species. These mudflats maintain a constant supply of juveniles to fisheries stocks that grow rapidly due to high food supply and relatively high water temperatures and then emigrate as subadults to deeper waters (Bishop 1989; Al-Yamani et al. 2004; Al-Zaidan et al. 2006; Chen et al. 2009; Almojil 2017). Recent study has revealed a diverse and abundant ichthyoplankton in the inshore waters of Kuwait Bay, represented by a large number of fish eggs and larvae of a total of nine fish families along with other zooplankton dominated by copepods, radiolarians, and larvae of mollusks. Abundant ichthyoplankton has been favored by food availability and low abundance of ichthyoplankton predators (Ali et al. 2019).

A total of 70 fish species have been captured in Sulaibikhat Bay between 1986 and 1990, of which the long-term dominant species, *Leiognathus decorus*, was present in all life history stages and in consistent numbers every year, whereas the two numerical dominants, *Liza carinata* and *Pomadasys stridens*, showed marked between-year variability (Wright et al. 1996). Later data acquired through monthly otter trawl sampling in Kuwait Bay during 2002–2004 have consisted of 80 fish species representing 41 families with *Leiognathus* (*Photopectoralis*) *bindus* and *Plicofollis tenuispinis* dominating numerically (Chen et al. 2009).

The extensive mudflats of Kuwait Bay, through the productivity of microbial mats, form a key food link supporting abundant and diverse benthic fauna. Foraminifera have been found to be the most abundant and diverse group among microfauna in Sulaibikhat Bay. A number of surveys evaluating the environmental quality of this embayment resulted in the recognition from 45 to 59 taxa from the total benthic foraminiferal assemblages (Al-Zamel et al. 2009; Al-Enezi and Frontalini 2015). A total of 49 species of benthic ostracods were identified in Kuwait Bay and were mostly found living on clayey silt sediments and sandy mud (Al-Abdul-Razzaq et al. 1983b).

A total of 200 species of macrozoobenthos and representatives of higher taxonomic groups belonging to 10 phyla have been identified from the subtidal sediments of Kuwait Bay during faunistic survey in 2009–2010. Mollusca (85 species),

Polychaeta (60), and Crustacea (41) were the most diverse groups, while Echinodermata (7), Sipunculida (3), Tentaculata (2), and Hemichordata (1) were poorly represented. Macrofauna was found to be abundant (1010–9140 ind.·m^{-2}) with numerical prevalence of polychaetes, gastropods, and crustaceans (Al-Rifaie et al. 2012).

Intertidal mudflats of Sulaibikhat Bay are inhabited by 14 crustaceans, 2 mollusks, 1 sipunculoid, and 8 fish species, in addition to the 33 fish and shellfish species occurring in the subtidal area. Camptandriid crabs *Leptochryseus kuwaitense*, *Manningis arabicum*, and *Nasima dotilliformis* are the most indicative species. Ocypodid crabs *Opusia* (*Tylodiplax*) *indica*, *Ilyoplax stevensi*, *Macrophthalmus* (*Mareotis*) *depressus* (Fig. 20.2e), and *Macrophthalmus* (*Venitus*) *dentipes* and the piluminid crab *Eurycarcinus orientalis* commonly occur. Intertidal mudflats are characterized by dense populations of four mudskipper species (*Periophthalmus waltoni* (Fig. 20.2d), *Periophthalmus koelreuteri*, *Boleophthalmus boddarti*, and *Acentrogobius ornatus*), among which *B. boddarti* is commonly observed to display a site-specific behavior by building walls of mud around its polygonal territory, for defense against its competitors and is large enough to provide adequate food (Clayton and Vaughan 1982; Jones and Clayton 1983; Clayton 1987; Al-Zaidan et al. 2003).

Since 2018, the diversity of marine biota inhabiting Sulaibikhat Bay is a subject of an ongoing comprehensive study by the Kuwait Institute for Scientific Research within a marine protected area (MPA), which was recently designated by the Kuwait Government. Large collection of marine invertebrates are currently being identified using morphological and molecular approaches, and knowledge on biodiversity of marine fauna in this ecologically important area are expected to be of high scientific significance.

The shore of Sulaibikhat Bay is considered the last area in Kuwait Bay, which is left almost undamaged, especially in relevance to salt marshes bordering the mudflats. This habitat provides fertile feeding grounds for a large number of seabirds using this area as permanent habitat throughout the year or visiting the mudflats of Sulaibikhat Bay as temporary stations for refueling and resting during their migratory route to their final destination. The greater pink flamingo *Phoenicopterus ruber* and the common seagull *Larus canus* are the most dominant seabirds in the Bay. The little stint *Calidris minuta*, sandpiper *Actitis hypoleucos*, gray heron *Ardea cinerea*, and the little egret *Egretta garzetta* are frequently observed along the Bay's shore (Al-Zaidan et al. 2003; UNEP/CBD 2016).

20.3.5 Kuwait's Southern Marine Waters

The marine environment southward from Kuwait Bay, away from the direct influence of the Shatt Al-Arab River discharge, is characterized by relatively clear, nutrient-poor, and more saline waters compared to the northern Kuwaiti waters (Al-Yamani et al. 2004, 2017, 2019b, c; Polikarpov et al. 2009; Sheppard et al.

2010; Alghunaim et al. 2020). Phytoplankton composition of Kuwait's southern waters is distinguished by larger contribution of dinoflagellate species (Polikarpov et al. 2009; Sheppard et al. 2010). While the bulk of phytoplankton composition in Kuwait's waters is composed of neritic species, the occurrence of true oceanic species is mainly restricted to the southern waters indicating the influence of the Gulf's water mass and adding to species richness mostly by dinoflagellate taxa (Saburova et al. 2013; Al-Yamani and Saburova 2019a).

The distinctive feature of Kuwait's waters is the presence of one of the most northerly coral reefs in the world. A range of coral platforms and smaller patch reefs and fringing reefs are distributed across the southern offshore waters and along the southern coastline (Fig. 20.2f, g) but most pronounced around three small offshore islands, Kubbar, Qaruh, and Umm Al-Maradim (e.g., Downing 1988; Hodgson and Carpenter 1995; Carpenter et al. 1997; Al-Yamani et al. 2004; Sheppard et al. 2010, 2012). Kuwait's coral reefs are smaller and less diverse compared to other Arabian Gulf coral reef systems occurring further south. Kuwait's southern waters host 35 recorded species of scleractinian corals, of which massive *Porites* corals are the most dominant, particularly on the reef flats, with patches of other corals, including small colonies of branching *Acropora* and *Stylophora*.

A wide range of invertebrate and vertebrate animals are associated with Kuwait's coral reefs contributing to the total biological diversity of the southern waters of Kuwait. The coral reef system harbors numerous macroalgae, sponges, bryozoans, hydroids, zoanthids, medusas, polychaetes, sea snails, sea slugs, bivalves, cephalopods, sea urchins, starfish, sea cucumbers, brittle stars, crinoids, crustaceans, and tunicates, whose diversity is still poorly explored. Echinoderms such as the black reef-boring sea urchin *Echinometra mathaei* and long-spined sea urchin *Diadema setosum* are the most abundant and commonly recorded species (Downing and Roberts 1993; Carpenter et al. 1997; Alsaffar and Lone 2000; Pilcher et al. 2000; Papathanasopoulou and Zogaris 2015; UNEP/CBD 2016).

A total of 124 fish species have been recorded on Kuwait's coral reefs including sharks and rays, eels, and bony fish (Pilcher et al. 2000; UNEP/CBD 2016) representing more than one-third of the total Kuwait's fish list (Bishop 2003). Damselfish *Chromis xanthopterygia* and *Neopomacentrus sindensis* are among the most abundant fish on Kuwait's reefs and form large schools near corals. A range of fish species use the coral reefs on a seasonal basis as a breeding ground including species of commercial importance, and many pelagic fish are attracted to the coral reefs for feeding and shelter.

Qaruh and Umm Al-Maradim are coral islands regularly used as the most northern area in the Gulf for breeding of the green turtle *Chelonia mydas* and hawksbill turtle *Eretmochelys imbricata* (Downing 1988; Pilcher et al. 2014; Papathanasopoulou and Zogaris 2015; UNEP/CBD 2016). Four species of sea tern breed each year on Kubbar Island and feed on the abundant fish resources in the vicinity of coral reefs (Downing 1988; Papathanasopoulou and Zogaris 2015).

20.4 Overall Biodiversity Pattern: North to South Gradient

The discharge of the Shatt Al-Arab River is considered an important factor strongly affecting the environment of the northern waters of the Gulf and largely governing the marine biota of these waters (e.g., Reynolds 1993; Al-Yamani et al. 2004, 2007, 2017, 2019b; Polikarpov et al. 2009, 2016, 2019; Sheppard et al. 2010). Strong latitudinal gradients in water quality across Kuwait's marine environment have been studied over the period from 2004 to 2006 and well documented (Fig. 20.3). The

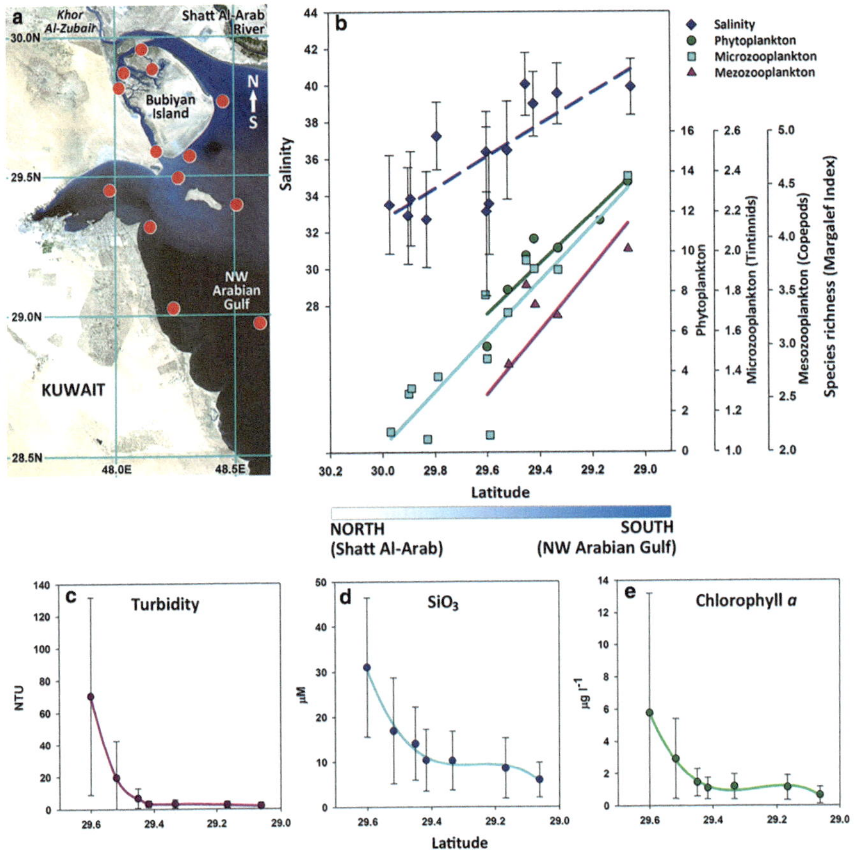

Fig. 20.3 Key oceanographic variables and biodiversity pattern of Kuwait's waters (2004–2006 dataset). (**a**) Satellite map of Kuwait's marine area; sampling sites are indicated by red circles; (**b**) change of salinity and species richness (Margalef index) in the phytoplankton, microzooplankton, and mesozooplankton communities along a north-south latitudinal gradient (blue rhombs represent the average values of salinity ± SD; green circles, cyan squares, and pink triangles represent Margalef index values for phyto-, microzoo-, and mesozooplankton, respectively; lines represent the linear regressions); (**c**–**e**) north-south latitudinal gradients in turbidity, silicate, and chlorophyll-*a* concentrations (circles represent the average values ± SD)

effect of the river discharge is most pronounced in the northern waters around Bubiyan Island and felt in Kuwait Bay and on most of the Kuwait's coast, up to southern waters near the Saudi Arabian border. The salinity increases by moving away from the river mouth (Fig. 20.3b) indicating lower impact of river discharge, whereas the turbidity and nutrient concentrations, particularly silicate, display decreasing trends (Fig. 20.3c, d). In general, a decline of abundance for both pelagic and benthic populations in Kuwait's marine environment from north to south is mainly driven by a decrease in the influence of Shatt Al-Arab discharge, whereas distinct upward trends in species richness of marine biota are evident across Kuwait's waters southward. This pattern is illustrated by north to south gradients in species richness of the main plankton groups including phytoplankton, tintinnids, and copepods (Fig. 20.3b) over the period of 2004–2007. Similar trends in distribution and diversity of a wide range of marine organisms have been reported.

20.4.1 *Bacterioplankton*

The phylogenetic diversity of bacterioplankton exhibited an increasing gradient from north to south along Kuwait's coast. Although the northern sampling site in Khor Al-Sabbiya channel received a nutrient-rich river discharge expecting to be favored to eutrophication-associated bacteria, the bacterial community of Khor Al-Sabbiya was composed of few phylotypes, whereas the higher richness of bacterial composition was recorded in Kuwait Bay, and most diverse bacterioplankton was associated with the southern coastal waters (Almutairi 2015).

20.4.2 *Phytoplankton*

In terms of phytoplankton, the most plentiful diatom-dominated assemblage is associated with low-saline and nutrient-rich Kuwait's northern waters and characterized by high chlorophyll level (as a proxy of phytoplankton biomass, Fig. 20.3e) but impoverished in species composition. An increase in phytoplankton species richness was observed moving away from the river mouth (Fig. 20.3b) and was characterized by increase in dinoflagellate species (non-silicious species), mainly due to depletion in silicates in Kuwait's southern waters (Fig. 20.3d) due to lower influence of the Shatt Al-Arab River discharge (Al-Yamani et al. 2004, 2012c; Polikarpov et al. 2009; Sheppard et al. 2010). Nutrient-rich northern waters favor intensive diatom-dominated phytoplankton blooms (more than 10^6–10^7 cells·l^{-1}), whereas extensive blooms of diazotrophic cyanobacterium *Trichodesmium erythraeum* which has the capacity to fix atmospheric nitrogen occasionally occur in nutrient-depleted southern waters (Al-Yamani et al. 2012c; Al-Yamani and Saburova 2019a; Polikarpov et al. 2020).

20.4.3 Microzooplankton

Between 2004 and 2008, the abundance and diversity of tintinnids exhibited reversing trends along Kuwait's coast, with high abundance and low diversity in the northern waters under the influence of the nutrient-rich Shatt Al-Arab River and Shatt Al-Basra discharge (Skryabin and Al-Yamani 2007b; Al-Yamani et al. 2019b). The northern waters surrounding Bubiyan Island were distinguished from the adjacent area in terms of tintinnid abundance and composition, being characterized by impoverished species composition with 25 species out of 62 tintinnid taxa reported in Kuwait's waters and elevated abundance (up to $3.9 \cdot 10^3$ ind.$\cdot l^{-1}$), whereas the southern waters were ranked by the highest species richness of tintinnids (Fig. 20.3b).

20.4.4 Mesozooplankton

A similar ascending trend in a southerly direction is characterized for species richness of copepods (Fig. 20.3b), and species composition of this zooplankton group varied across Kuwait's waters. In general, copepod composition is dominated by calanoid species, among which *Parvocalanus crassirostris* is the most abundant and ubiquitous (Michel et al. 1986; Al-Yamani and Prusova 2003; Al-Yamani et al. 2004). One more calanoid copepod, *Acartia pacifica*, codominates with *P. crassirostris* in the northern waters in Khor Al-Sabbiya channel, but its abundance decreases in the southern offshore waters (Al-Yamani et al. 2004). Further, the distribution of some copepods (*Acartia* (*Acartiella*) *faoensis*, *Bestiolina arabica*, *B. similis*, *Pseudodiaptomus ardjuna*, *P. arabicus*) is restricted to the northern waters when salinity is lower than 36 psu (Al-Yamani et al. 2007).

20.4.5 Microphytobenthos

A wide range of benthic organisms are also affected by the river runoff, as their distribution and abundance depend upon the sediment texture. In the north, suspended material from the Shatt Al-Arab River settles to form extensive soft sediment areas of intertidal flats around Bubiyan Island and within Kuwait Bay. On the more exposed open coast southward of Kuwait Bay, medium and coarse sandy beaches extend down to the Saudi Arabian border and beyond (Al-Abdul-Razzaq et al. 1982; Jones 1986; Al-Yamani et al. 2004).

There is a large distinction in the abundance and composition of microflora between muddy sediments on the northern coast and sandy beaches along Kuwait's southern shore. High biomass of cyanobacteria-diatom dominating algal mats in the north is replaced by diverse complex of sand-dwelling microalgae in southerly direction. Distribution of sand-dwelling dinoflagellates is limited to intertidal sandflats and shallow bottom sediments south of Kuwait Bay where sandy beaches dominate the inshore habitats. Similarly, a diverse complex of benthic dinoflagellates occurs epiphytically on macroalgae in the southern areas. High diversity of the sand-dwelling dinoflagellates is supported mainly by gymnodinioid and peridinioid taxa, whereas the representatives of *Prorocentrum*, *Coolia*, *Ostreopsis*, and *Fukuyoa* (*Gambierdiscus*) are among the most abundant and diverse epiphytic dinoflagellates and discriminating taxa for Kuwait's southern microphytobenthic assemblages (Saburova et al. 2009, 2013; Al-Yamani and Saburova 2010). Along with numerous species of benthic diatoms (Al-Yamani and Saburova 2011), both sand-dwelling and epiphytic dinoflagellates contribute to the high diversity of microphytobenthic community along Kuwait's southern coast.

20.4.6 Macrophytobenthos

In contrast to Chlorophyta-dominated macroalgal assemblages in Kuwait's northern area where a great amount of *Ulva* and *Enteromorpha* are occasionally washed upon shores after winter storms (Sheppard et al. 2010), both brown (Phaeophyceae) and red (Rhodophyta) algae essentially contribute to the composition of macroalgae along the southern cost of Kuwait (Al-Yamani et al. 2014). *Sargassum* beds extend along Kuwait's coast seasonally from December to March varying in coverage area and composition (Bishop et al. 1994; Alghunaim et al. 2019, 2020).

Apart from the availability of hard substratum to attach, the overall trend in distribution and composition of seaweeds most likely reflects the bathymetric and turbidity gradients across Kuwait's waters and depth-related underwater light spectrum and its intensity and is determined by the differences in the pigment compositions among macroalgal groups (e.g., Saffo 1987). Green algae prevail in the shallowest and turbid northern waters, while brown and red algae with green wavelength-harvesting accessory pigments (carotenoids and phycobiliproteins) occur in deeper southern waters.

Discharge of highly turbid waters from the Shatt Al-Arab River in the north results in gradient of turbidity between northern and southern areas (Fig. 20.3c) that, in turn, alters the underwater light availability across Kuwait's waters (Alghunaim et al. 2020). The turbidity and underwater light conditions were found to be prime drivers on the distribution and composition of the brown algae *Sargassum* in Kuwait's waters (Alghunaim et al. 2019, 2020). Northern very turbid waters around Failaka Island offer the harshest environment for *Sargassum* growth, whereas clear and better illuminated southern inshore waters are more favorable for this seaweed. Further, along with widespread but low abundant *Sargassum angustifolium*, the *Sargassum* bed along Failaka Island in the north is largely dominated by shade-tolerant *S. boveanum*, whereas the distribution of susceptible to light availability *S. asperifolium* was mostly restricted to the southern waters (Alghunaim et al. 2019).

20.4.7 *Microzoobenthos*

Foraminiferal fauna of Kuwait completely lacks representatives of the two families, Soritiidae and Peneropliidae, that commonly occur in sediments of subtropical and tropical areas in the Gulf and worldwide, including the United Arab Emirate's coast, Red Sea, and the Mediterranean. This gap in foraminiferal species composition has been ascribed to the influence of the Shatt Al-Arab River discharge on Kuwait's marine environment. Further, composition of foraminifers in Kuwait is poorly represented by characteristic and diverse microalgal symbiont-bearing genus *Amphistegina*. Species of this genus are widespread in warm, clear, nutrient-depleted shallow marine habitats. However, in Kuwait's waters, the distribution of a few representatives of this genus is largely restricted to southern areas, especially in the shallow sediments around coral islands of Qaruh and Um Al-Maradim. The river plume most probably limits the proliferation of these species northward by the reduction of available light due to high turbidity of the northern waters (Al-Enezi and Frontalini 2015).

20.4.8 *Macrozoobenthos*

The Shatt Al-Arab River plume and associated environmental factors influence the distribution of both invertebrates and vertebrates. Al-Bakri et al. (1985) considered Kuwait's macrobenthic community impoverished when compared to the other intertidal communities and believed the harsh environmental conditions, especially high summer temperatures and salinities, were responsible. On the other hand, Apel and Turkay (1999) revealed that the maximum grapsid and ocypodid crab diversity in the Gulf is found in Kuwait. High biodiversity, presence of endemic species, and abundance of biota on the intertidal mudflats in the northern area of Kuwait have been ascribed to lower salinity originating from the nearby Shatt Al-Arab River discharge (Sheppard et al. 1992; Apel and Turkay 1999; Al-Zaidan et al. 2003). More recently, Al-Yamani et al. (2012b) showed that macrozoobenthos of Kuwait's marine environment have high species diversity and low dominance level though biodiversity estimates may markedly differ depending upon the locality, with less diverse benthic community occurring in Kuwait Bay.

The distribution and abundance patterns of amphipods associated with *Sargassum* beds were significantly affected by turbidity, being found much lower in highly turbid waters around Failaka Island compared to habitats located southward (Ali et al. 2018). Wider comparisons with the molluskan faunas of the Gulf and the Arabian region suggested that the fauna of Kuwait has some unique elements not found elsewhere. Molluskan diversity was found to be higher in rocky-sandy areas than intertidal mudflats, with the highest species richness occurring to the south of Kuwait Bay and in the coastal area around Failaka Island (Al-Kandari et al. 2020a).

For shrimp populations, there are substantial data showing that species, numbers, and sizes in Kuwait are different from those of the other Gulf countries south of Kuwait due to the Shatt Al-Arab River discharge influence (Bishop et al. 2001, 2011). Kuwait is the only country in the northern Gulf to have commercial populations of shrimps *Parapenaeopsis stylifera* and *Metapenaeus affinis*. The Mesopotamian marshes of the lower Tigris and Euphrates River system serve as a nursery habitat for *M. affinis*, and the subadults egressed to the northern Gulf where they mature and spawn (Mathews 1986; Salman et al. 1990). The green tiger prawn (*Penaeus semisulcatus*), the primary shrimp species throughout the Gulf and Kuwait's main species, grows considerably larger in Kuwait than its counterparts further south due to food availability, which results from proximity of Kuwait's waters to the Shatt Al-Arab River mouth (Bishop et al. 2001). Moreover, two of Kuwait's most popular fishes, the silver pomfret (*Pampus argenteus*) and the hilsa shad (*Tenualosa ilisha*), are also restricted to the northern area of the Gulf (Al-Yamani et al. 2004). Hilsa shad is anadromous and migrates up the Tigris and Euphrates rivers to spawn (Al-Hassan 1999). Throughout their range, silver pomfret are also closely associated with river mouths and adjacent estuaries. A second population of silver pomfret is present near Qeshm Island (eastern Gulf coast) in the southern part of the Gulf, but there are no commercial populations between this stock and that of the northern Gulf (Almatar et al. 2006). Moreover, a number of commercially important fish have been considered to have Shatt Al-Arab-related life stages, including javelin grunter (*Pomadasys kaakan*), fourfinger threadfin (*Eleutheronema tetradactylum*), tigertooth croaker (*Otolithes ruber*), klunzinger's mullet (*Liza klunzingeri*), and greenback mullet (*Liza subviridis*) (Bishop et al. 2011; Al-Husaini et al. 2015; Ben-Hasan et al. 2018).

20.5 Shatt Al-Arab River System

The northern Gulf receives the waters of only one large perennial river system that of the Tigris, Euphrates, Karkheh, and Karun – which, with their tributaries, combine in bringing down the snow waters from the Zagros ranges and from the Armenian Highlands (Wilson 1928). The Tigris and Euphrates have their sources within 80 km of each other in eastern Turkey and run mostly southeast through northern Syria and Iraq to the head of the Gulf. The total length of the Euphrates is about 2800 km and of Tigris is about 1900 km. In their upper reaches, rivers flow far from one another (~400 km, near the Turkey-Syria border), while in the middle reaches they gradually move closer. These two rivers confluence at Qurna City to form the Shatt Al-Arab River with a total length of nearly 200 km before it drains into the Gulf (Coleman and Huh 2004; Isaev and Mikhailova 2009).

Several tributaries join the Shatt Al-Arab during its course to the Gulf, most importantly the Karkheh and the Karun rivers (UN-ESCWA and BGR 2013). Karkheh River discharges in the Al-Hawizeh marshes east of the Tigris at the Iran-Iraq border. The silt moving by Karkheh is entirely deposited in the marshes (Wilson

1928). Al-Hawizeh is the best remaining natural marsh in the region (Richardson and Hussain 2006) supporting high diversity of birds and freshwater fish and being important area for spawning migrations of marine fish stocks of the Gulf (Lawler 2005). The river continues to flow from this wetland for about 15 kilometers until it enters the Shatt Al-Arab at approximately 6 kilometers southeastward from Qurna City (Abdullah et al. 2015).

One more significant contributor of Shatt Al-Arab is Karun River, the longest river in Iran, which drains a mountainous area with relatively high level of precipitation (Cressey 1958). The Karun River diverges into two main outlets: the Haffar, the main channel of the Karun River uniting the Shatt Al-Arab at Khorramshahr City, and the Bahmanshir channel that is a secondary estuary of the Karun River and going in parallel the Shatt Al-Arab waterway for 70 km before discharging into the Gulf (Borjian 2012). In ancient times, the Karun had several channels, none of which connected with the Shatt Al-Arab, because the Karun emptied directly into the Gulf. The Haffar channel, according to the medieval Arab geographer Al-Muqaddasi, was excavated in the tenth century CE with the aim to provide waterway transportation between the cities of Basra on the Shatt Al-Arab and Ahvaz on the Karun. Since 1765, it has been the main channel of the Karun River (Isaev and Mikhailova 2009; Afary et al. 2020).

The Shatt Al-Arab Delta area is classified as estuarine-deltaic because the river's sediment seeps into a shallow, narrow part of the Gulf. The Shatt Al-Arab Delta is 140 km wide and splits into more than ten branches. The network of irrigation ditches in the delta region appears to be responsible for a nearly 64% water loss after contributing sources reach the main channel (Coleman and Huh 2004). The landscape is characterized by green marshy areas, lakes, lagoons, and estuaries, bordered by irrigated lands and date palm plantations and surrounded by desert (Isaev and Mikhailova 2009).

The regional climatic characteristics of the major suppliers of the Shatt Al-Arab River, Tigris and Euphrates basins, are increasing of aridity toward downstream, with gradual reduction in precipitation and air humidity due to a higher evaporation tax together with decreases in flow and increasing salinity as a natural tendency. However, the increasing requirements for irrigation water due to rising aridity toward downstream can also intensify these natural processes. The annual average precipitation upstream in Turkey can reach 1000 mm and, at Turkish-Syrian border, is 250 mm but significantly decreases downstream to less than 100 mm reaching only 50 mm per year at the Syria-Iraq border (UN-ESCWA and BGR 2013; Venturi and Capozzoli 2017).

Historically, runoff in both the Euphrates and Tigris rivers had been characterized by pronounced spring floods. These floods had both positive and negative implications: they fertilized cultivated lands with sediments, but they also threatened farms and human livelihood whenever flood seasons produced extreme events. The average annual discharge of the Euphrates and Tigris rivers together is difficult to determine due to the large yearly fluctuation. According to the records for 1938–1980, there have been years when 68 km^3 were observed in both rivers in the mid-1960s and years when the amount was over 84 km^3 in the mid-1970s.

However, there was the critical drought year with less than 30 km^3 at the beginning of the 1960s (UNDG 2005; FAO 2009).

Around 7500–6000 BCE, the oldest hydraulic civilization of the world started in Mesopotamia from first systematic irrigation and flood control when the Sumerians dug canals to irrigate wheat and barley in present-day southern Iraq (e.g., Wittfogel 1956; Adams 1965). In modern times, Iraq was the first country to begin engineering work along the Euphrates River, starting in 1914 with the Hindiya Barrage Dam. This dam was built to transfer floods to a lower land area, forming Lake Razzaza. Large dam projects in the Tigris and Euphrates basin began in the 1950s in Iraq and were predominantly focused on flood control and irrigation. Large irrigation projects, such as the North Al-Jazeera and East Al-Jazeera, involved the installation of network of canals to bring the waters of the basin into areas to serve thousands of hectares of newly irrigated farmland (FAO 2009).

Syria and Turkey began large hydro-engineering projects on the Euphrates in the 1960s. These countries completed their first dams on the Euphrates within a year of one another: the Tabqa Dam (better known as the Euphrates Dam) by Syria in 1973 and the Keban Dam by Turkey in 1974. After the first large embankment dams have been operated in Turkey and Syria, hydrologic dynamics changed drastically. The floods that occurred until the early 1970s did not repeat after the construction of the Keban and Tabqa dams. From the late 1970s until the early 1990s, the flow regime of the Euphrates was clearly smoothed by reduction of peak spring floods and compensation of low flows during dry season due to beneficial reservoir operations (Cullmann 2013).

In 1977, Turkey's Southeast Anatolia Project (GAP, or Guneydogu Anadolu Projesi in the Turkish language) was initiated on the upper parts of the Euphrates and Tigris basins (Bilgen 2019). Turkey's Ataturk Dam on the Euphrates, the centerpiece of GAP, was completed in 1992. The filling of the Ataturk Dam has increased water regulation and resulted in drastic reduction of the Euphrates flows since the mid-1990s. The difference in mean monthly flows before and after the construction of the big dams was about 50%. This share of the water was diverted in Turkey, Syria, and Iraq and no longer made it to the lower reaches of the Euphrates. This hydrological condition was among the causes of serious water shortage and water quality problems in Iraq. Development of hydrodynamics in the Tigris exhibits parallels to the situation of the Euphrates (Cullmann 2013).

The most recent part of this huge project, the Ilisu Dam on the Tigris River, has been completed recently. On May 19, 2020, Turkey commissioned the first turbine of the power plant at this dam (Hürriyet Daily News 19/05/2020). The Ilisu Dam is intended for hydroelectric power generation only but will be followed by another regulator dam at Cizre. The latter is planned to use for irrigation purposes. Cizre Dam is to be constructed 45 km downstream of Ilisu Dam and about 20 km upstream of the border line between Turkey and Iraq (Al-Madhhachi et al. 2020).

Iran has only in the last few decades started large dam projects on Karun, Dez, and Karkheh rivers (e.g., Chomani and Bijnens 2016). The Karkheh and Karun supply an estimated 41% of the Shatt-al-Arab's historical mean annual discharge of 73.6 km^3. Iran's upstream exploitation increasingly utilizes nearly all of the tributaries' waters. On the Karkheh River, dams and irrigation have drastically curtailed downstream flows, shrinking the Al-Hawizeh marshes at the Iran-Iraq border by two-thirds from 1991 to 2009. Dams and irrigation projects have also pared mean annual discharge on the Karun. The effect of this has already been felt, as small rivers and private wells in northern Iraq have begun to dry up. In 2004, Iraq lodged formal complaints against Iran's development activities, leading to the creation of a joint technical committee for shared water issues, but as yet no common management measures have resulted (Michel 2017).

At the present time, a basin-scale river flow regulation in the region has resulted in construction of more than 60 dams and reservoirs along Tigris, Euphrates, Karun, and Karkheh river basins in Turkey, Syria, Iraq, and Iran for purposes of irrigation and hydroelectric power generation. Long-term and large-scale river flow manipulations ranked the Shatt Al-Arab River basin among the most highly regulated river systems worldwide along with Volta River in West Africa, Manicouagan in Canada, Colorado in the USA, Rio Negro in Argentina, and Mae Khlong in Thailand. The total storage capacity of largest existing 26 dams along the Tigris and Euphrates basin is large enough to hold more than entire annual average discharge (124%) of both rivers (Nilsson et al. 2005), and the remaining smaller dams definitely contribute to additional flow regulation.

The assessments of the potential long-term mean annual discharge of the Shatt Al-Arab into the Gulf calculated from the summarized flow estimates of its main tributaries ranged widely between 46 $km^3 \cdot year^{-1}$ (Reynolds 1993), approximately 74 $km^3 \cdot year^{-1}$ (UN-ESCWA and BGR 2013), and 35–133 $km^3 \cdot year^{-1}$ (Sheppard et al. 2010) and require adjustments for large runoff losses through evaporation from lakes and swamps mostly in the Mesopotamian marshlands (Reynolds 1993; Isaev and Mikhailova 2009). Long-term anthropogenic development along the course of the Tigris, Euphrates, and Karun rivers including upstream dam and reservoir construction in Turkey, Syria, Iraq, and Iran and the drainage of the Mesopotamian marshlands in Iraq has severely impacted freshwater input to the northern Gulf through the Shatt Al-Arab River system (Isaev and Mikhailova 2009; UN-ESCWA and BGR 2013; Abdullah et al. 2015; Lawler 2016). Significant declining trends have been revealed over the period between 1933 and 2011 in the annual flows of the main tributaries of the Shatt Al-Arab, Tigris (-0.14 $km^3 \cdot year^{-1}$), and Euphrates (-0.19 $km^3 \cdot year^{-1}$) (Abdullah et al. 2015). Most recent assessment has indicated drastic reduction of the Shatt Al-Arab River outflow by 95% compared to historical values. The mean annual river discharge at Faw City close to river mouth dropped down from 1189 $m^3 \cdot s^{-1}$ during 1948–1960 to only 58 $m^3 \cdot s^{-1}$ in 2017–2018 with seasonal variation from 42 $m^3 \cdot s^{-1}$ in October to 90 $m^3 \cdot s^{-1}$ in May (Al-Asadi and Alhello 2019).

20.6 Long-Term River-Related Salinity Changes in the Northern Gulf

Kuwait's northern marine environment is most influenced by river discharge due to its closest proximity to Shatt Al-Arab River mouth. Since the salinity balance of Kuwait's waters has been shown to be primarily governed by freshwater input from the Shatt Al-Arab River (Al-Yamani et al. 2017), this easily measured variable can be served as a reliable tracer for the magnitude of riverine outflow into the northern Gulf. To show long-term variability in salinity related to Shatt Al-Arab River outflow since mid-1990s onward, we used the northernmost systematically monitored sampling site at Khor Al-Sabbiya channel (Fig. 20.4a, b). Based on the long-term dataset, a general increasing trend in salinity was identified; however, several periods of lower and higher salinity alternated over time (Fig. 20.4a), reflecting variability of weather conditions and water regulation activities upstream of the Shatt Al-Arab River system. Similar trends have been shown across Kuwait's waters from Kuwait Bay to the southern offshore area over the period from 1982 to 2015 (Al-Yamani et al. 2017).

Historical data on salinity in Khor Al-Sabbiya are restricted to a scarce number of measurements in 1981–1982, when an average salinity of 36.3–36.6 psu with a range from 32.7 in April to 40.5 psu in June was recorded (Dames and Moore 1983; Grabe et al. 1992). A significant decrease of salinity occurred during 1995–1998, when mean values dropped down to 24.2–28.2 psu with a range from 18.7 to 34.9 psu. This drastic drop in salinity has been attributed to high discharge rates of the man-made Third River (Al-Yamani et al. 2007) due to the diversion of the Euphrates River into Third River (connected to Shatt Al-Basra canal) in Iraq in 1992 to provide irrigation water for the land between the Euphrates and the Tigris rivers (FAO 2008). The northern Gulf received drainage waters from more than 1.5 million ha of agricultural land through the Khor Al-Zubair channel outflow nearby Warba Island (UN-ESCWA and BGR 2013; Al-Ansari et al. 2014). Moreover, this period of low salinity in Kuwait's northern waters coincided with elevated precipitation over the interannual average level in Iraq (Fig. 20.4a).

The Middle East region has been subject to an almost continuous drought since 1998, which is probably the most intense of the past 900 years (e.g., Cook et al. 2016). Large-scale decrease in the amount of rainfall within the watershed of the Euphrates and Tigris rivers in Turkey, Syria, Iran, and Iraq (UNESCO 2014; Al-Faraj et al. 2015; Amini et al. 2016; Hameed et al. 2018; Bachmann et al. 2019) affected the riverine flow and resulted in reduction of the freshwater inflow into the northern Gulf. The period between 2000 and 2003 was characterized by higher salinity of Khor Al-Sabbiya waters, which ranged from 33.3 to 45.0 psu with mean salinity values of 37.7–40.6 psu (Fig. 20.4a).

After 2003, waters of the Tigris and Euphrates were released uncontrolled in order to restore Iraqi southern marshes, and about 20% of the marshlands were inundated in the spring of 2004 (Richardson et al. 2005) and later up to 58% by 2006 (Al-Handal and Hu 2015). The restoration of the marshes was facilitated by the onset

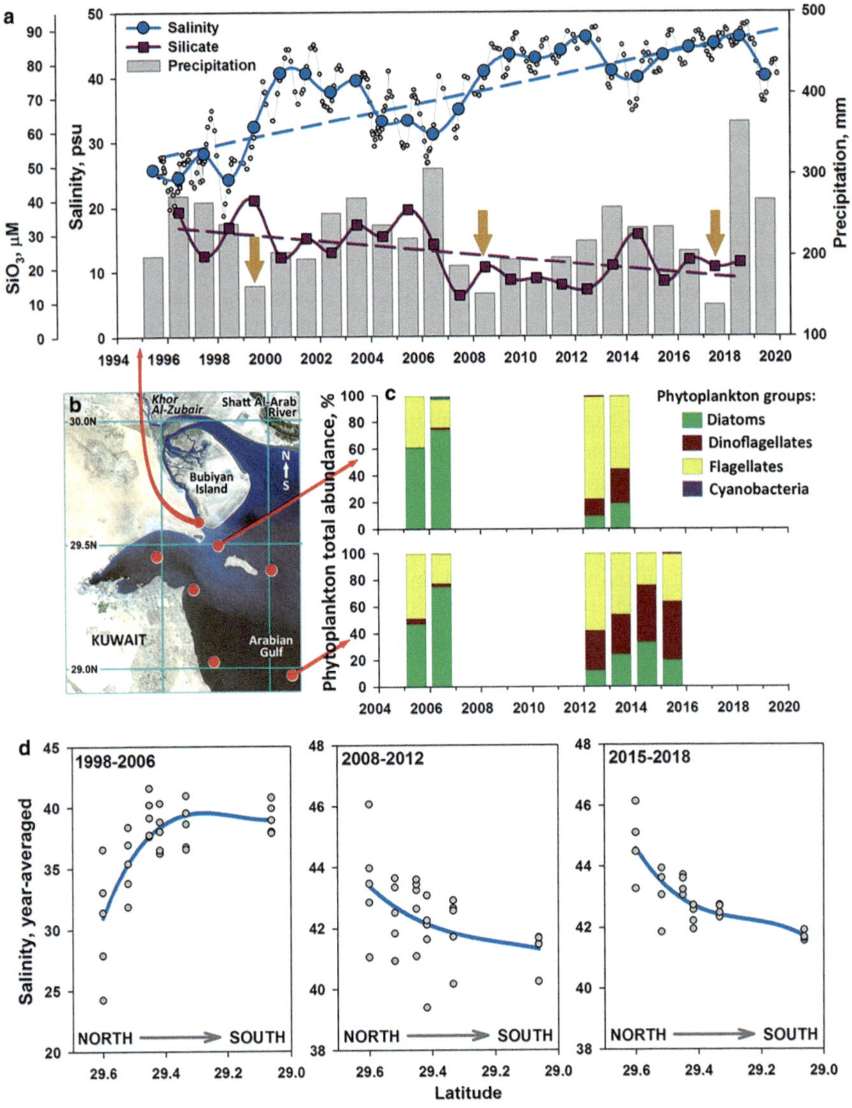

Fig. 20.4 Long-term trends in oceanographic variables and change in phytoplankton community in Kuwait's waters. (**a**) Long-term trends in salinity and silicate concentrations at Khor Al-Sabbiya superimposed on the annual precipitation trend over Iraq; small empty dots represent monthly records of salinity, large blue circles represent annual average salinity, dark-pink squares represent annual average silicate concentrations, gray bars represent annual precipitation (source: average combined satellite-gauge precipitation monthly 0.5 deg. [AIRS, SSMI GPCPMON v3.0] mm/month for 1995–2019, Shape Iraq), dash lines represent linear regression trends for salinity ($R^2 = 0.613$) and silicates ($R^2 = 0.346$), and orange arrows point to extremely low precipitation rates; (**b**) satellite map of Kuwait's marine area; sampling sites are indicated by red circles; (**c**) changes between 2005–2006 and 2012–2016 periods in percentage abundance of different phytoplankton groups represented by colored bars off Kuwait Bay (upper plot) and at Kuwait's southern offshore waters (bottom plot); (**d**) changes in north-south latitudinal gradients of salinity between 1998–2006, 2008–2012, and 2015–2018 periods

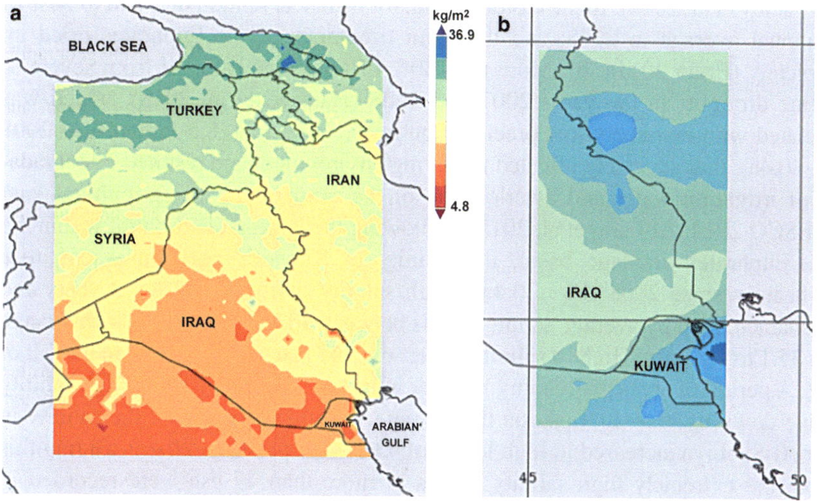

Fig. 20.5 Weather anomalies over the Shatt Al-Arab River basin. (**a**) Spatial distribution of satellite-derived surface soil moisture during 2008 over the Middle East (kg/m^2), Global Land Data Assimilation System (GLDAS NOAH) dataset (Rodell et al. 2004); (**b**) satellite-derived average precipitation rate monthly (rainfall, mm/month), November 2018, Tropical Rainfall Measuring Mission (TRMM) dataset, 3B43 v.7. (Hong et al. 2007). Datasets accessed using NASA GIOVANNI v. 4.34 https://giovanni.gsfc.nasa.gov/giovanni/

of the wet period across the Tigris-Euphrates drainage basin replacing drought. Several years around the mid-2000s were characterized by higher than average rainfall in Turkey, Iran, and Iraq (Karabulut 2015; Amini et al. 2016; Awchi and Kalyana 2017; Al-Nassar 2018; Hameed et al. 2018; see Fig. 20.4a for Iraq), which led to a significant rise in flow rates of the rivers and the collection of water in catchment reservoirs. For instance, the mean combined satellite-gauge-derived precipitation rate was 267.7 mm over Iraq in 2006 compared to the mean annual precipitation of 214.9 $\pm$ 54.8 mm/year for the entire period between 1995 and 2018 (Fig. 20.4a). Moreover, as based on a satellite radar altimetry, Karakaya reservoir on the Euphrates River and Thartar Lake in between the Tigris and Euphrates rivers in Iraq experienced water depletion since 1996–1997, but in 2002 the water level in these reservoirs started to rise and was enough to be normally operated by 2004 (Zakharova et al. 2007). Consequently, during the period from 2004 to 2007, Kuwait's northern waters were influenced by higher flow rates from Shatt Al-Arab, and mean salinity dropped down to 31.2–35.0 psu with a salinity range between 26.7 and 40.9 psu (Fig. 20.4a).

In 2008–2009, large-scale and long-lasting drought spread across much of the Middle East and Central Asia, with drastic restrictions on water usage in Turkey, Syria, Iran, and Iraq due to decline of reservoir levels along courses of the Tigris and Euphrates rivers (e.g., Sheffield and Wood 2011). Across the Fertile Crescent region as a whole, the winter of 2007/2008 proved to be the driest climate record (Fig. 20.5a), with precipitation 35% below the 1961–1990 average (Kelley et al.

2015; Selby et al. 2017). In the eastern Syria, for instance, rainfall dropped to 30% of the annual average in 2008, and the main tributaries of the Euphrates dried up completely (Erian 2011a, b). More than 80% of Iraqi area suffered from severe to extreme droughts in the years 2007 and 2008 (Hameed et al. 2018), which was associated with extremely low precipitation rates, down to 128.6 mm/year in 2008 (Fig. 20.4a), that adversely affected the living environment and resulted in degradation of irrigation water and deterioration of the productivity of agricultural land (UNESCO 2014; Al-Faraj et al. 2015). Following the large-scale drought within the Tigris-Euphrates drainage basin, the salinity in Kuwait's northern waters rose significantly since 2008 (Fig. 20.4a). High salinity period lasted for 5 years and was characterized by average salinity levels between 40.9 and 46.1 psu with a range from 37.1 to 47.6 psu. Highest salinity values over 47 psu were recorded in the fall of 2012. A period of abnormally rainy weather in Iraq in 2013 resulted in slight salinity decline down to 39.9–40.9 psu on the average. In subsequent years, the salinity in Khor Al-Sabbiya increased to high levels of 43.3–46.1 psu (Fig. 20.4a). During this period, the extremely high salinity values of more than 47 psu were recorded in November and December 2017 and July, August, and September 2018. The highest salinity over the entire analyzed period (48.2 psu) was documented in Khor Al-Sabbiya waters on October 10, 2018.

Heavy rain episodes in part of the Middle East caused flooding in Syria, northern Iran, and Jordan during late October 2018. Later on, Kuwait, Iran, and Iraq have suffered from unusually heavy rainfall and experienced flash flooding in November 2018 (Fig. 20.5b). During heavy rainfall and flash flooding in Kuwait, meteorological stations have recorded 49.2 mm of rain on November 9, 2018, that is proportionated to the amount normally observed for the most rainy period of October, November, and December combined. Subsequently, during extremely heavy rainfall episode on November 18–19, Kuwait received nearly half of a year's rate of rainfall. In northern Kuwait, up to 74–94 mm of rain had fallen during these 2 days (FloodList 2018a, b) compared to long-term annual average estimation of 114 mm by the World Meteorological Organization (WMO) based on rain gauge data from Kuwait's International Airport during 1961–2002 (Marcella and Eltahir 2008). A total, in November 2018, Kuwait received roughly double the annual rainfall rate (Fig. 20.5b) comparable with historic maximum records of 300 mm in 1872 and 1954 (Mrs. Janan Behzad, Secretary General of the Kuwaiti Society for the Protection of the Environment; KUNA 2018). A great amount of rainwater was directly discharged into the Gulf through drainage systems causing local lowered salinity of the marine waters. Salinity dropped down from 48.2 psu in October to 45.6 psu in November and December 2018 in Khor Al-Sabbiya (Fig. 20.4a), and 2 weeks later, the salinity records decreased from 42 to 38 psu in Kuwait Bay.

The wettest period occurred over Iraq in 2018–2019 with extremely high annual precipitation of 364.7 mm in 2018 and 268.3 mm in 2019 above the 90 percentile. The salinity of Kuwait's northern waters declined following regional heavy rainfall episodes at the end of 2018 (Fig. 20.4a). In 2019, lowered average salinity value of 40.06 psu was recorded, and from April to July, the lowest salinity range from 36.81

to 37.32 psu was documented comparable to historical records of salinity in Khor Al-Sabbiya during 1981–1982 (Dames and Moore 1983; Grabe et al. 1992).

Long-term reduction in the Shatt Al-Arab River discharge affected the seasonality and spatial distribution of salinity ranges across Kuwait's waters. Over an annual cycle in Kuwait's waters, the lowest salinity generally occurs in April to May (Fig. 20.4a) due to the spring increase of freshwater inflow from the Shatt Al-Arab during the period of snow melt and seasonal rainfall in the Euphrates and Tigris basins (Al-Yamani et al. 2004, 2017; UN-ESCWA and BGR 2013; Abdullah et al. 2015). However, during the high salinity periods since 2008, amplitude of seasonal variations was less (Fig. 20.4d) compared to the previous years throughout Kuwait's waters from north to south (Al-Yamani et al. 2017). After 2008, the steady north to south increase in salinity across Kuwait's waters showed a reverse salinity pattern (Fig. 20.4d). Recent data showed that after 2008, salinity of the northern waters, which are of close proximity to the river mouth, became almost always higher than that at the southern offshore waters indicating very low freshwater input from the river, which did not compensate for the strong evaporation in the branched network of channels in shallow waters around Bubiyan Island. As a tracer of the river plume, the salinity is highly correlated with spatial distribution of nutrient concentrations and turbidity across Kuwait's waters (Al-Yamani et al. 2004, 2007, 2019b; Polikarpov et al. 2009, 2020; Fig. 20.4a). During the past decade, the overall structure of the marine environment was altered by the reduction of freshwater input into the northern Gulf from the Shatt Al-Arab River, possibly impacting the spatial and seasonal patterns of the marine biota.

20.7 River Discharge Impact on Marine Biota in the Northern Gulf

In the present review, we provide enough evidence to conclude that the marine environment of the northern Gulf is highly dependent on the nutrient-rich turbid Shatt Al-Arab River discharge, affecting water quality, biological productivity, species composition, species richness, and endemism. A number of studies have documented the impact of the reduction of the freshwater flow from the Shatt Al-Arab River and the loss of the important filtering function of the marshes due to the drying up of the marshlands on the marine environment of the northern Gulf. Reduction in freshwater river runoff resulted in increased salinity both in the Shatt Al-Arab River for about 100 km upward and along Kuwait's waters southward up to more than 100 km away from the river mouth.

The Shatt Al-Arab mouth area is influenced by the tides of the Gulf. Historically, tidal impact was limited to the mouth of the river, and the freshwaters of the Shatt Al-Arab in the lower basin only became brackish near the towns of Seeba, Iraq, and Abadan, Iran (Rahi 2018). Over the past decades, however, the flow of freshwater is no longer strong enough to hold back the saline waters of the Gulf, and tidal effects have been observed as far as north of Basra, 100 km inland, turning river and ground waters brackish (Isaev and Mikhailova 2009; Lawler 2016). Calculated over the period of 1977–2016, the average salinity of river waters at Basra, approximately 50 km upstream, is 2.55 ppt with high interannual variability. However, since then the waters of Tigris and Euphrates were diverted to restore the Mesopotamian marshes, and up to now, the salinity at the upstream of the Shatt Al-Arab has not dropped below 2 ppt, which is above the standard limits for human consumption (drinking) and of limited use for agricultural applications. Moreover, high salinity range of 16–18 psu has been recorded in 2009 (Hameed et al. 2013; Rahi 2018).

20.7.1 Lower Reach of the Shatt Al-Arab

Apart from the decline of water quality, long-term salinization has strongly affected the river ecosystem, altering composition and abundance of native species and facilitating the biological invasion of marine and euryhaline species in the Shatt Al-Arab River. High salinity in Shatt Al-Arab River and Mesopotamian marshes during recent years seems to be the major factor behind remarkable change in biodiversity where several native species disappeared and more alien species appeared (e.g., Naser et al. 2011; Khalaf et al. 2015).

Recently, dense populations of dinoflagellate species have been recorded occasionally in the composition of generally diatom-dominated phytoplankton of the Shatt Al-Arab River. The dinoflagellate *Peridiniopsis minima* was found to constitute a major proportion of the overall phytoplankton assemblage in the river waters during November and December 2009 about 30 km south of Qurna City and downstream up to Basra City, within the salinity range from 2 to 7.2 psu (Hameed et al. 2016). The presence of this species in the composition of phytoplankton in the upper reach of Shatt Al-Arab in 2012 (Fig. 20.1c, d) points out the successful introduction of this dinoflagellate in the riverine phytoplankton. The occurrence of typically marine dinoflagellate *Scrippsiella trochoidea* in the Shatt Al-Arab waters was limited within salinities ranging from 2.3 to 16.7 psu during 2009–2010 (Hameed and Saburova 2015). Both these dinoflagellates have never been reported previously from the Iraqi aquatic system. The presence of these bloom-forming invasive dinoflagellate species in the Shatt Al-Arab waters can pose a risk to water quality, as microalgal blooms could be a major problem for aquatic biota and for human living along the river and using the river as a water supply.

Most of alien invertebrates in the inland Iraqi waters belong to marine and euryhaline species, which facilitates the success of their expansion in an ecosystem subjected to artificial changes in the hydrological regime due to competitive advantages with respect to native species and formation of free ecological niches in degraded native communities (Naser et al. 2011). The occurrence of the marine harpacticoid copepod, *Delavalia longifurca*, in brackish water areas of Shatt Al-Arab River up to Qurna City and in the southern Iraqi marshes was reported during 2006–2009, although the distribution of this species has previously been restricted to the lower reaches of the river at Faw and Seeba (Mohammed 2018). Native invertebrate communities of the Shatt Al-Arab River and Mesopotamian marshes were strongly affected by the invasion of a range of alien crustacean and molluskan species, of which widespread species are the Chinese mitten migrating crab *Eriocheir sinensis*, the oriental river shrimp *Macrobrachium nipponense*, the rockpool shrimp *Palaemon elegans*, the New Zealand mud snail *Potamorpyrgus antipodarum*, the zebra mussel *Dreissena polymorpha*, the freshwater gastropod *Physa acuta*, and the barnacles *Balanus amphitrite*, *Amphibalanus subaldidus*, and *Amphibalanus improvisus* (Clark et al. 2006; Salman et al. 2006; Naser and Son 2009; Naser et al. 2011, 2015; Shahdadi et al. 2014).

Very high and uniform dispersal of alien invertebrates along the Shatt Al-Arab River waterway results in biotic homogenization exerted by alien species on native fauna of invertebrates (Naser et al. 2011). Invaded invertebrates have a direct and indirect impact on native species through predation, competition for space and food resources both in benthic and pelagic habitats. Apart from fouling of hydraulic structures by mature barnacles, their pelagic larvae contribute to alteration of zooplankton composition. The prevailed cirripede larvae (53–59%) replaced the previously dominant Cladocera and reduced the contribution of copepods in the composition of freshwater zooplankton of the Shatt Al-Arab River between Qurna and Basra in 2008–2009 (Abbas et al. 2014).

The diversity of fish has declined in the Shatt Al-Arab River system from 68 to 26 species over the past decades. The high salinity has led to the death of fish in fish farms in Seeba area and in the Iraqi province of Maheala. Many of the fish left the Shatt Al-Arab waters because of the high level of salinity and the entry of other exotic fish. Some of the valuable fish in the Shatt Al-Arab River disappeared and migrated northward of Basra toward the marshes (Yaseen et al. 2016). An increase in the number of exotic and marine fish was recorded for the inland Iraqi waters in the southern marshes recently compared to historical observations (Hussain et al. 2009). The deterioration of water quality of the southern Iraqi marshes has resulted in several native cyprinid species to disappear (e.g., *Barbus subquincunciatus* and *B. scheich*) due to an increase in salinity in the early 1990s. After inundation of the marshes in 2003, populations of a few native species (e.g., *B. xanthopterus* and *B. grypus*) largely decreased due to scarcity of benthic food resources, competition with the alien species *Cyprinus carpio*, and increased salinity. Other native species (e.g., *B. sharpeyi* and *B. luteus*) became rare due to loss of their habitat which was occupied by the invasive species *Ctenopharyngodon idella* and *Carassius auratus* (Hussain et al. 2009).

20.7.2 Shatt Al-Arab Delta

Recent zooplankton surveys have revealed a decline in abundances of endemic estuarine copepods *Acartia* (*Acartiella*) *faoensis*, *Bestiolina arabica*, and *Phyllodiaptomus irakiensis* in Shatt Al-Arab deltaic waters since these copepods cannot tolerate high salinity (Khalaf et al. 2015). *Acartia* (*Acartiella*) *faoensis* was considered one of the most dominant species in the Iraqi brackish water of Khor Al-Zubair, playing a great nutritional role in the food chain of the region (Khalaf 2007). However, since salinity in the river delta began to increase due to drainage of the Mesopotamian marshes in 1990s, a decline in copepod abundances has been recorded. In Khor Al-Zubair, for instance, the salinity has drastically increased from a range of 25.4–33.5 psu in 1990 up to 48–50 psu in 2008 (Khalaf et al. 2015), and extremely high salinity (52.76 psu) was recorded there in the autumn of 2010 (Khalaf et al. 2014).

For a long time, Kuwait's southern waters have been thought to be the northern limit of the coral distribution in the Gulf. The northernmost Gulf area encompassing Kuwait's northern waters as well as Iraqi waters adjacent to the freshwater discharge from the Shatt Al-Arab River was considered coral-free. This was primarily ascribed to the river plume, which carries large amounts of sediment and causes large variations in salinity (e.g., Hodgson and Carpenter 1995; Carpenter et al. 1997; Sheppard et al. 2010, 2012). The recent discovery of a healthy living coral reef in Iraqi marine water, 26 km away from the river mouth, has been attributed to declined river runoff to acceptable salinity and transparency of the adjacent waters allowing coral growth (Pohl et al. 2014). Species-specific tolerances to highly variable environment could be considered a deciding factor in the coral community differentiation all over the Gulf (Sheppard et al. 2012). Accordingly, the abundance of silica-containing demo-sponges in the discovered Iraqi coral reef is a distinguished character most probably caused by silicate-rich waters from the Shatt Al-Arab discharge (Pohl et al. 2014).

20.7.3 Kuwait's Northern Waters

The modifications of river flow and drainage of the Mesopotamian marshes caused long-lasting adverse effect on the ecosystem of the northern Gulf (e.g., Al-Yamani et al. 1997b, 2007; Al-Ghadban et al. 1999, 2008; Saeed et al. 1999; Al-Yamani and Khan 2002). The Mesopotamian marshes are an important integral part of the Tigris and Euphrates basin, forming an aquatic landscape with a unique wildlife populations and being a buffer zone for sediment and nutrient transport through the river basin (Partow 2001; Al-Yamani and Khan 2002; Al-Handal and Hu 2015). Draining of limited portions of the marshes to reclaim land for agriculture and oil exploration began in the 1950s, but human activities were expanded greatly during 1980s–1990s. In 1983, a 150-km-long artificial Shatt Al-Basra canal linked the Euphrates River with oligohaline Al-Hammar marsh and further down with Khor Al-Zubair to find an outlet for the disposal of flood water to the sea, turning the environment of the Khor Al-Zubair from a hypersaline lagoon to an estuarine one. Further, to provide irrigation water for the land between the Euphrates and the Tigris rivers, Iraq began constructing in the 1960s a 565-km-long canal, the Third River, which was completed in 1992 and captured almost the entire flow of the Euphrates that was once filling the Al-Hammar marsh (Partow 2001; Al-Yamani and Khan 2002).

The complex hydro-engineering projects appeared to be very effective, and water diversion and agricultural reclamation in the marshlands of the southern Iraq in 1992 alone have reduced the marsh area by 80% compared with the figures in 1984–1985 and have destroyed the integrity of the extensive wetland complex of the Qurna marshes (Maltby 1994; Partow 2001; Al-Mudaffar Fawzi and Mahdi 2014). Much of the suspension load of the Tigris and Euphrates rivers is normally deposited in the marshes and the delta of Shatt Al-Arab River prior to the discharge into the Gulf. After large-scale drainage, the water filtering role of the marshes has ceased, and the Third River became a canal to discharge polluted agricultural, industrial, and urban drainage directly into the Shatt Al-Basra canal, before emptying into the northern Gulf through the Khor Al-Zubair channel. Kuwait's northern waters were influenced by higher flow rates from the Third River following the year 1992, and hence, its waters became less saline and more turbid and polluted, with adverse impact on hydrodynamic regime and marine environment (Al-Yamani et al. 1997b, 2007; Al-Ghadban et al. 1999, 2008; Saeed et al. 1999; Al-Yamani and Khan 2002). Variability in salinity, turbidity, sediment deposition, and nutrient levels as well as plankton community compositions in Kuwait's waters served as good tracers for monitoring changes occurring as a result of fluctuating river discharge volume and marsh destruction/restoration (Al-Yamani et al. 1997b, 2007; Al-Ghadban et al. 2008).

The drainage of the marshes and water discharge through the Shatt Al-Basra canal into the northern Gulf resulted in the increase of sedimentation and altered hydrodynamics of Kuwait's northern area. The draining process of the Iraqi marshes has enhanced the deposition of finer material (less than 63μm) in Kuwait's northern area in the late 1990s when compared with the pattern of grain size distribution in the bottom sediments in the 1980s. At the same period, the water depths of the northern marine area became shallower when compared to the bathymetric records in the 1956 and 1986 Admiralty Charts, suggesting an increase in sediment deposition (Al-Ghadban et al. 1999, 2008). During the marsh drainage period, mean turbidity in the waters of Khor Al-Sabbiya channel was more than tenfold higher than the rest of Kuwait's waters and much higher compared to historic levels in 1979/1980 prior to the drainage period and after 2003 when drainage period ended (Al-Yamani et al. 2007).

During the period of high discharge of the Third River into the Gulf through the Khor Al-Zubair from 1992 to 2003, Kuwait's northern waters received surplus nutrient influx attributed to the discharge of agricultural drainage water. The highest concentration of nitrate (23.4μM) was recorded in 1996 when the lowest salinity occurred for that year (Al-Yamani et al. 2007; Fig. 20.4a for silicate). The 1995–1998 survey showed much higher nitrate concentrations (mean 10μM) in Khor Al-Sabbiya waters and north of Failaka Island compared to mean concentration of 0.8μM toward the south. Similar declining trends were observed in distribution of phosphate and silicate from the Khor Al-Sabbiya in the north to the southern waters (Al-Yamani and Khan 2002; Al-Ghadban et al. 2008).

The changes in nutrient fluxes and increase of stoichiometric N:Si and P:Si ratios (Al-Yamani et al. 1997a, b) due to the influence of the Third River outflow altered the abundance, composition, and size spectrum of the phytoplankton. In turbid and nutrient-rich waters of Khor Al-Sabbiya channel, smaller-sized phytoplankton proliferated (Al-Yamani et al. 1997a, b, 2007) being more tolerated to reduced light penetration. Further, there has been an increase in the frequency of occurrence of monospecific blooms in Kuwait's waters since 1995 (Al-Yamani et al. 2012c; Al-Yamani and Saburova 2019a; Polikarpov et al. 2020). Sporadic blooms of a wide range of microalgae were becoming a common occurrence, among which harmful bloom of toxic dinoflagellate *Karenia selliformis* implicated in massive fish kill in Kuwait Bay in 1999 (Heil et al. 2001) was of special importance.

The highest mean density of tintinnids (471,156 per m^3) was observed in 1997–1998 in Kuwait's northern waters near Failaka Island, and foraminiferans were most abundant (5378 per m^3) in the Khor Al-Sabbiya waters when compared to the adjacent area (Al-Ghadban et al. 2008). During the period of lowered salinity due to freshwater discharge from the Third River, five estuarine copepods (*Acartia (Acartiella) faoensis*, *Bestiolina arabica*, *Bestiolina similis*, *Pseudodiaptomus ardjuna*, and *Pseudodiaptomus arabicus*) were found for the first time in Kuwait's northern waters when salinity was less than 36 psu (Al-Yamani et al. 2007). The estuarine copepod species could serve as a bioindicator for the magnitude of river discharge into the northern Gulf environment. The copepod species *Acartia (Acartiella) faoensis* has been originally described from the Khor Al-Zubair and Khor Abdullah waters (Khalaf 1991) and has appeared in northern Kuwait's waters during lower salinity period, pointing out the salinity fluctuations as a main invasion vector.

20.7.4 Bubiyan Island

The plankton structure can be used as a water quality indicator across Kuwait's waters because the local species composition changes strongly with environmental gradients (Al-Yamani et al. 2007, 2019b; Polikarpov et al. 2009, 2016). The only two comparable surveys in duration and sampling design, which have included quantitative dataset on phytoplankton and microzooplankton, are those that were conducted around Kuwait's northern Bubiyan Island in 2004–2005 (Bishop et al. 2006, 2013) and 2015–2016 (Al-Osairi et al. 2016) (Fig. 20.6b). This allows to assess the alterations of the basic links of the food chain in a long-term scale in Kuwait's northern waters in close proximity to freshwater discharge. During the above two sampling periods, the phytoplankton and microzooplankton populations were identified and counted.

Changes in habitat conditions alter the phytoplankton community. Salinity may be considered as an important factor governing the phytoplankton structure in the coastal waters around Bubiyan Island. The comparison between the two datasets in 2004–2005 and 2015–2016 revealed the dramatic rise in the salinity level of the

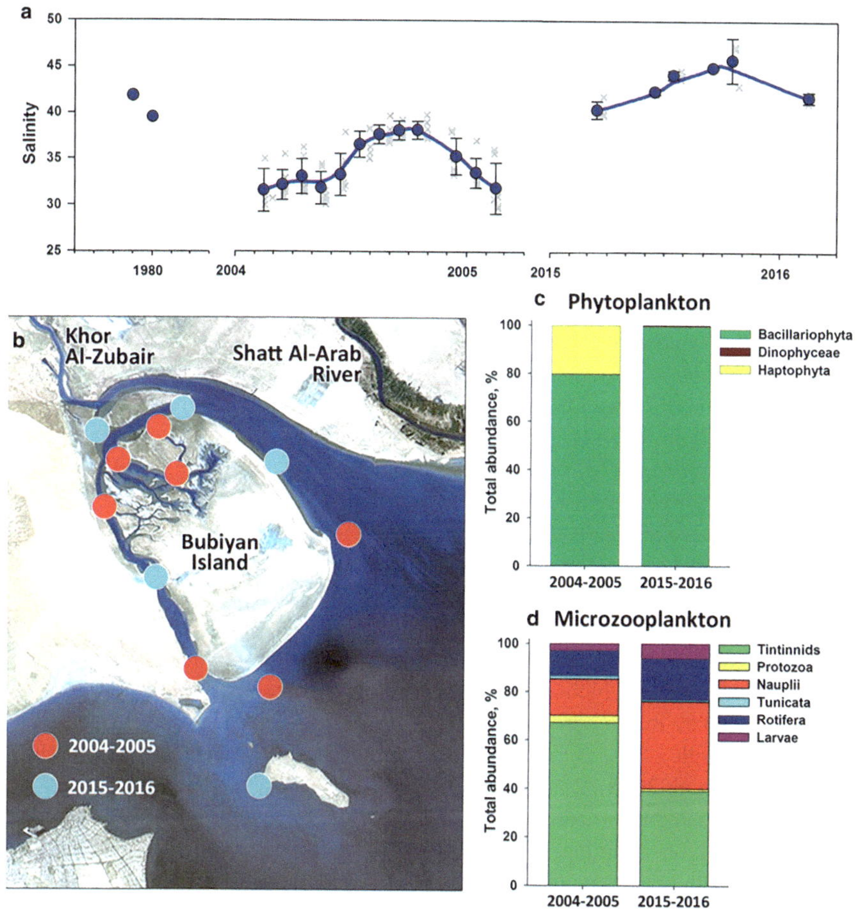

Fig. 20.6 Long-term changes in salinity and microplankton in Kuwait's northern waters around Bubiyan Island between 2004–2005 and 2015–2016. (**a**) Annual cycles of salinity during 2004–2005 and 2015–2016 compared to reference measurements in 1979/1980 (source: Dames and Moore 1983); circles represent monthly averaged values of salinity ± SD; (**b**) satellite map of Kuwait's northern waters; sampling sites are indicated by red circles for 2004–2005 and by cyan circles for 2015–2016 (due to some differences in sampling design applied in 2004–2005 compared to 2015–2016 studied period (seven sampling sites and monthly sampling in 2004–2005 versus five sampling sites and quarterly sampling in 2015–2016), the appropriate dataset for microplankton was extracted from closely located sampling sites at similar sampling dates); (**c** and **d**) changes between 2004–2005 and 2015–2016 periods in percentage abundance of different phytoplankton (**c**) and microzooplankton (**d**) groups represented by colored bars

coastal waters around Bubiyan Island from 34.44 ± 3.57 psu in 2004–2005 to 43.02 ± 2.29 psu in 2015–2016, higher than the historic level of 39.5–41.8 psu in winter of 1979/1980 (Jacob et al. 1981) (Fig. 20.6a). Apart from the high diatom abundance during the two periods, dinoflagellate contribution to the phytoplankton species composition was greater during the high salinity phase of 2015–2016

(25.2%) than in the low salinity phase in 2004–2005 (19.7%). The two sampling periods share 49.5% of their total species pool. The phytoplankton inhabiting the coastal waters around Bubiyan Island differed during low and high salinity periods. In 2015–2016, the phytoplankton community was found to be more diverse and equilibrated than in the period of 2004–2005 (Shannon-Wiener's index 3.20 vs. 1.75; Pielou's index 0.65 vs. 0.35).

The composition of the dominant species and their contribution to the total phytoplankton abundance significantly differed between the two periods (Fig. 20.6c). In 2004–2005, the phytoplankton around the island was characterized by winter blooms of small-sized diatom *Chaetoceros socialis* (49% of the total phytoplankton abundance) and haptophyte algae *Phaeocystis globosa* (20.2%). In 2015–2016, *P. globosa* was completely absent from the phytoplankton composition, and contribution of *C. socialis* to the total abundance was negligible (0.2%). These small-sized bloom-forming species were replaced by medium- and large-sized diatoms *Thalassiosira* spp. (24.5%), *Chaetoceros decipiens* (11%), *Thalassionema nitzschioides* (8.4%), and *Guinardia flaccida* (5%).

Blooms of colonial haptophyte algae *Phaeocystis globosa* are recurrent events in Kuwait's waters (Al-Hassan et al. 1990; Al-Kandari et al. 2009; Al-Yamani et al. 2012c; Al-Yamani and Saburova 2019a; Polikarpov et al. 2020). Seasonal blooms of this species occurred yearly in autumn, winter, and spring primarily in the northern waters of Kuwait during 2004–2006. The main effect of *Phaeocystis* blooms is formation of foam, which appears when blooms are washed ashore. During gillnetting survey in the coastal waters of Bubiyan in December 2004, the gill net was clogged with a huge amount of mucilage colonies due to a heavy bloom of this alga (Bishop et al. 2013).

When compared to the earlier observation period, the contribution of tintinnids to the microplankton abundance was reduced from 34 to 20% due to the greater prevalence of copepod nauplii, rotifers, and larval plankton with the increase of salinity in 2015–2016 (Fig. 20.6d). The species richness of tintinnids dropped from 25 to 16 taxa over the decade. Similarly, a reduction in tintinnid species richness was observed in waters around Bubiyan Island during 2007–2008 compared to the lower salinity period of 2004–2006 (Al-Yamani et al. 2019b).

While *Tintinnopsis parvula*, *Luminella kuwaitensis*, and *Codonellopsis lusitanica* were the most numerically important tintinnid species in 2004–2005, *Tintinnopsis gracilis* became the dominant species in 2015–2016, accounting for 46% of the total tintinnid abundance compared to 2% in the 2004–2005 period. Nine species of tintinnids seem to disappear from the Bubiyan fauna, and the endemic species *Leprotintinnus bubiyanicus* disappeared from its type locality. In 2004–2005, composition of tintinnids of Bubiyan coastal waters distinguished this area from the adjacent waters; however, this distinction became negligible in subsequent years due to loss of habitat-specific taxa.

20.7.5 *Off Bubiyan and Southwards*

Close proximity to river mouth makes Kuwait's northern waters the most indicative area for monitoring long-term changes in freshwater discharge and its related impacts on the northern Gulf marine environment. However, the influence of the river plume extends throughout the entire marine area of Kuwait (Al-Yamani et al. 1997a, b, 2004, 2007, 2017, 2019c; Al-Ghadban et al. 1999, 2008; Polikarpov et al. 2009).

The temporal patterns in nutrient variability across Kuwait's waters over three decades (1983–2013), particularly for nitrogen and silicate, have been shown to be influenced by the long-term changes of the nutrient-rich Shatt Al-Arab River discharge (Devlin et al. 2015a). The fluvial outflow from Shatt Al-Arab into the northern Gulf is a main source of silicate for Kuwait's waters (Brewer and Dyrssen 1985). A long-term decline in silicate concentrations in Khor Al-Sabbiya waters over the past two decades was observed with concurrent reduction of the river outflow as indicated by the increase in salinity in the Khor waters (Fig. 20.4a). Even over far shorter time frame covering sharp shift in salinity between 2004 and 2008, the reduction in freshwater availability has affected the nutrient levels, with phosphate and nitrate concentrations dropping down in 2007–2008 by 54–57% of the corresponding levels in 2004–2006 (Al-Yamani et al. 2019b).

The revealed significant relationships between salinity and phytoplankton abundance and composition, nutrient concentrations, and their ratios indicate the significant role of the Shatt Al-Arab River runoff (as indexed by salinity) in governing phytoplankton composition and development in the northern Gulf (Polikarpov et al. 2009, 2016, 2020; Al-Said et al. 2017; Al-Yamani et al. 2019b; Devlin et al. 2019). Long-term analysis of phytoplankton samples from the Environment Public Authority (EPA) and Kuwait Institute for Scientific Research (KISR) monitoring programs has revealed the shift in phytoplankton functional groups with dominance of non-siliceous phytoplankton (dinoflagellates and flagellates) over siliceous fraction (diatoms) during the past decade (Al-Said et al. 2017; Al-Yamani et al. 2019b; Devlin et al. 2019; Polikarpov et al. 2020). Changes in phytoplankton composition in Kuwait Bay and in the southern offshore waters are clearly illustrated in Fig. 20.4c.

According to Margalef (1978), who related turbulence and nutrient concentrations as key factors determining the size and morphology of the phytoplankton species in oceanic waters, the small-sized forms tend to dominate the phytoplankton in low-mixing oligotrophic conditions, whereas nutrient-rich and high-mixing waters favor the prevalence of larger species. The different phylogenetic groups or life-forms in the phytoplankton exhibit distinct and often contrasting adaptive ecologies, supporting their niche partitioning along the turbulence-nutrient gradient (Margalef 1978; Smayda 2002). Accordingly, the period of lower salinity of Kuwait's waters corresponding to surplus of river discharge with large inputs of nutrients and turbidity was characterized by high dominance of chain-forming diatoms that constituted the medium to large phytoplankton size fraction.

Conversely, in the period of drastic river runoff decline since 2007–2008 onward, the highest contribution of the small-sized dinoflagellates and flagellates was observed with a low proportion of large-sized phytoplankton (Fig. 20.4c). Recent outcomes from flow cytometry studies supplementing the microscopy-based observations revealed an overwhelming dominance (up to 80%) of particles with equivalent spherical diameter below 20μm in the autotrophic plankton community in Kuwait's waters (Al-Yamani et al. 2019b).

Change in composition driven by increased proportion of small-sized dinoflagellates and flagellates could be a reason for the decrease in phytoplankton biomass during the past decade as inferred from the declined trends in chlorophyll-*a* concentrations in Kuwait's waters (Al-Said et al. 2017; Al-Yamani et al. 2019b, c; Devlin et al. 2019). The recent changes in nutrient flux driven by the decline of Shatt Al-Arab River runoff and the rising coastal urbanization of Kuwait Bay shifted nutrient proportions toward favorable conditions for phytoplankton bloom development in Kuwait's waters. Inorganic nitrogen to phosphorus stoichiometric ratio was found to be governing the frequency and composition of phytoplankton blooms in Kuwait's waters, with low N:P ratio favoring frequent algal blooms over the past decade, particularly in Kuwait Bay (Al-Yamani and Saburova 2019a; Polikarpov et al. 2020).

The long-term changes in phytoplankton abundance, biomass, size spectrum, and composition induced by river runoff have the potential, in turn, to impact the pelagic trophodynamics by cascading effect on the food web. A decline in phytoplankton stock and change in dominant size fraction since 2007 have affected the microzooplankton community in terms of its abundance, biomass, diversity, and size spectrum (Al-Yamani et al. 2019b). High correlations between tintinnid abundance, phytoplankton biomass, and nutrients pointed out the microplankton as an important link in the pelagic food web. Species composition with predominance of neritic tintinnids with loricae agglutinated by biogenic or mineral particles was indicative to highly turbulent environment during the period of elevated river runoff in 2004–2006, whereas the runoff reduction increased the relative abundance of tintinnids with hyaline loricae, which are better adapted to open water conditions. Moreover, the difference in the dominance of tintinnid species between periods could be related to prevailed prey size (Dolan and Pierce 2013). Predominance of small-sized phytoplankton fraction during the low river runoff since 2007 has favored the development of tintinnids with a small oral diameter of lorica. Thus, the contribution of *Luminella kuwaitensis* with a small oral lorica diameter (< 30μm) ingesting nano-sized phytoplankton increased after 2007 compared with the previous period when larger diatom phytoplankton predominated. At the same time, some gigantic tintinnids with large oral lorica diameters (40–70μm) disappeared from the community, as in the case of *Leprotintinnus bubiyanicus* (Al-Yamani et al. 2019b).

Copepod abundance estimates in 2008–2009 revealed higher cyclopoid (omnivores) to calanoid (herbivores) ratio in the composition of zooplankton compared to low saline periods of 1997–1998. Recent survey in 2014–2016 showed the prevalence of small-sized copepods belonging to the genera *Oithona*, *Parvocalanus*, and *Acartia* in Kuwait's waters. Calanoid copepods like *Parvocalanus* are mostly herbivorous, and cyclopoid copepods (mainly *Oithona*), which are omnivorous, could exert strong top-down control on microplankton by grazing (Al-Yamani and Madhusoodhanan 2018; Al-Yamani et al. 2019b).

20.8 Kuwait's Fishery

Fishing has a long cultural and economical importance in Kuwait (e.g., Al-Yamani et al. 2004; Al-Husaini et al. 2015) yet is one of the main pressures impacting biodiversity (Devlin et al. 2015b). Kuwait's fisheries are multispecies and multi-gear, with 23 commercial species considered as most important. The most recent overview of Kuwait's fisheries status (Al-Husaini et al. 2015) has reported a significant drop of total fisheries landings by over half in the 10-year period since the mid-1990s including zobaidy (*Pampus argenteus*), suboor or hilsa shad (*Tenualosa ilisha*), hamoor (*Epinephelus coioides*), newaiby (*Otolithes ruber*), hamra (*Lutjanus malabaricus*), and green tiger prawn (*Penaeus semisulcatus*). A decline of the Shatt Al-Arab River discharge, along with overfishing, has been identified among the major reasons for drastic reduction of fish and shrimp stock in Kuwait's waters due to alteration of salinity, nutrient cycles, and subsequent reduction of primary and secondary productivity of the Gulf's northern waters (Al-Yamani et al. 2004, 2007; Sheppard et al. 2010; Bishop et al. 2011; Al-Husaini et al. 2015; Ben-Hasan et al. 2018).

Particularly hard hit are fish species directly related to the Shatt Al-Arab River basin in their life cycles, such as the anadromous hilsa shad and estuarine-dependent zobaidy. These two main migratory fish stocks in the northern area of the Gulf are shared among Kuwait, Iraq, and Iran. Both species are considered economically valuable and important for the fisheries in the region. In Kuwait, catch of both species has decreased by more than 90% between 1995 and 2007. These two species accounted for 32% of Kuwait's total finfish landings in 1995 but only 6% in 2007 and 5–12% in 2012 (Sheppard et al. 2012; Al-Husaini et al. 2015; Ben-Hasan et al. 2018). A similar steady decline of hilsa shad in Iraqi marine catches from 52.9% in 1991–1994 to 18.9% over the 2007–2011 has been reported (Al-Dubakel 2011; Almukhtar et al. 2016). The catch rate of zobaidy in Iranian Khouzestan Province was also decreased between 1993 and 1999 (Al-Husaini 2003).

The hilsa shad migrates from the Gulf toward freshwater up the tributaries of the Shatt Al-Arab and associated marshes for spawning and juvenile development. During its upstream migrations in breeding season, this fish ascends the Tigris, Euphrates, and Karun rivers and may reach up to Qal'at Saleh on the Tigris River and to Al-Fahod on the Euphrates River, about 120–180 km north of Basra City. The rise of salinity in Shatt Al-Arab River during the past years drives adult individuals of hilsa shad to enter the East Al-Hammar marsh for spawning (Al-Dubakel 2011 and references therein). Juveniles, about 15 cm total length, migrate downstream to the Gulf for feeding and maturation. Primary spawning grounds for zobaidy are Khor Musa in Iran and Kuwait Bay, while the main nursery areas include the extensive estuarine mudflats in Iran and Iraq and Khor Abdullah east of Bubiyan Island in Kuwait (Al-Husaini et al. 2015).

Among commercially catched shrimps, the life cycle of the most important green tiger shrimp *Penaeus semisulcatus* (accounting for about 60% of the landings in Kuwait) is weakly dependent on the Shatt Al-Arab River system. The nursery areas of this shrimp occur in Kuwait Bay and the adjacent southern parts of the Gulf, away from the estuary. In contrast, the kiddi shrimp *Metapenaeus affinis*, the second most important species accounting for about 30% of Kuwait's shrimp landings, is a euryhaline species using the Iraqi marshes as a nursery habitat. *M. affinis* juveniles immigrate from Kuwait's waters to the Iraqi marshes, where they grow to the size of sexual maturation, and then migrate back to the marine spawning grounds (Bishop et al. 2011; Al-Husaini et al. 2015; Ben-Hasan et al. 2018).

Migratory and estuarine-dependent fish and shrimp species of the northern Gulf were defined to be particularly vulnerable to ever-reduced riverine flux and degradation of the Iraqi marshes. Therefore, large-scale upstream river flow regulation through the Shatt Al-Arab basin have been ranked one of the main causes for catch declines of these species due to possible alterations of their spawning, feeding, and migration patterns (Bishop et al. 2011; Sheppard et al. 2012; Al-Husaini et al. 2015; Ben-Hasan et al. 2018).

Inversely, the shift in salinity balance toward hypersaline conditions by reduced river discharge has been favorable for some fish species whose distribution was formerly limited mostly to the southern waters of Kuwait with low catching rates. For instance, commercial catches of spangled emperor (*Lethrinus nebulosus*) increased by 441% from 17 to 92 t between 1994 and 2007 (Sheppard et al. 2012), and its percentage contribution to the total fish landing in 2012 increased by 912% compared to 1995 (Al-Husaini et al. 2015). Other species whose catches have increased drastically since the mid-1990s include Japanese threadfin bream *Nemipterus japonicus* (Sheppard et al. 2012), spotted croaker *Protonibea diacantha*, and king soldier bream *Argyrops spinifer* (Al-Husaini et al. 2015). However, despite the increase in landing, the contribution of these species to the total fish landing is negligible.

20.9 Conclusion

Although Kuwait contributes only 0.4% to the combined Tigris-Euphrates River watershed, it benefits more than any other drainage-basin country from the Shatt Al-Arab discharge entering to Kuwait's marine area (Al-Yamani and Khan 2002). In the present review, the emphasis was placed on the consequences of the Shatt Al-Arab freshwater flow reduction on the marine environment of the northern Gulf. The results of the different studies indicated that large-scale flow regulation activities in the river basin have far-reaching impacts downstream, resulting in changes in the hydrology of the northern part of the Gulf and affecting the marine ecosystem and some of its components.

Discharge of the Shatt Al-Arab is considered to be a dominating driver of the northern Gulf's ecology and largely responsible for primary and secondary productivity of Kuwait's waters. The sharp drop in the river runoff in recent decades due to global climatic changes and regulation of river flows in the upper reaches of the river has increased the water salinity and decreased the nutrient supply, altering the spatial and seasonal hydrographic settings in the northern Gulf. Such changes have adversely affected the structure and composition of pelagic biota and shifted the balance toward less diverse communities dominated by small-sized forms. The long-term variability in phytoplankton abundance and species composition contributed to the decline in the abundance, biomass, and size structure of zooplankton communities, probably causing a significant drop in fishery stocks coupled with overfishing impact, ultimately affecting the ecosystem function, and threatening the marine food security in the northern Arabian Gulf. Management of Shatt Al-Arab River discharge into the Gulf requires cooperation among the riparian countries and downstream countries like Kuwait to preserve the productivity, biodiversity, and uniqueness of the northern Gulf ecosystems. Long-term monitoring and assessment as well as joint research programs involving scientists from the three concerned countries Kuwait, Iraq, and Iran are needed.

References

Abaychi JK, Darmoian SA, DouAbul AAZ (1988) The Shatt Al-Arab River: a nutrient salt and organic matter source to the Arabian Gulf. Hydrobiologia 166:217–224

Abbas MF, Salman SD, Al-Mayahy SH (2014) Diversity and seasonal changes of zooplankton communities in the Shatt Al-Arab River, Basrah, Iraq, with a special reference to Cladocera. Mesopot J Mar Sci 29:51–70

Abdullah AD, Masih I, van der Zaag P, Karim UF, Popescu I, Al Suhail Q (2015) Shatt al Arab River system under escalating pressure: a preliminary exploration of the issues and options for mitigation. Int J River Basin Manag 13:215–227

Abou-Seida MM, Al-Sarawi MA (1990) Utilisation and management of coastal areas in Kuwait. Coast Manag 18:385–401

Acker JG, Leptoukh G (2007) Online analysis enhances use of NASA Earth science data. Eos Trans AGU 88(2):14–17

Adams RM (1965) Heartland of cities, surveys of ancient settlement and land use on the central floodplain of the Euphrates. University of Chicago Press, Chicago

Afary J, Avery PW, Mostofi K (2020) Iran. In: Encyclopædia Britannica. https://www.britannica.com/place/Karun-River. Downloaded on 6 May 2020

Al-Abdul-Razzaq S, Khalaf F, Al-Bakri D, Shublaq W, Al-Sheikh Z, Kittaneh W, Al-Ghadban AN, Al-Saleh S (1982) Marine sedimentology and benthic ecology of Kuwait marine environment. EES-31C. Report KISR 694. Kuwait Institute for Scientific Research, Kuwait

Al-Abdul-Razzaq S, Shublaq W, Al-Sheikh Z (1983a) The marine benthic microfauna of the tidal flats of Kuwait. J University Kuwait (Sci) 10:101–108

Al-Abdul-Razzaq S, Shublaq W, Al-Sheikh Z, Kittaneh W (1983b) Ecology and distribution of ostracods in Kuwait Bay. J Micropaleontol 2:39–45

Al-Ansari N, Ali AA, Knutsson S (2014) Present conditions and future challenges of water resources problems in Iraq. J Water Resour Prot 6:1066–1098

Al-Asadi SAR, Alhello AA (2019) General assessment of Shatt Al-Arab River, Iraq. Int J Water 13:360–375
Al-Attar MH (1984) Kuwait Bay: a nursery area for penaeid shrimps. I. *Penaeus semisulcatus*. In: Mathews CP (ed) Proceedings of 3rd shrimp and fin fish. Management workshop, Report KISR 1156, vol 1. Kuwait Institute for Scientific Research, Kuwait, pp 85–105
Al-Bakri D, El-Sayed MI (1991) Mineralogy and provenance of the elastic deposits of the modern intertidal environment of the northern Arabian Gulf. Mar Geol 97:121–135
Al-Bakri D, Foda M, Behbehani M, Khalaf F, Shublaq W, El-Sayed M, Al-Sheikh Z, Kittaheh W, Khuribet A, Al-Kadi A (1985) The environmental assessment of the intertidal zone of Kuwait. Report KISR 1687. Kuwait Institute for Scientific Research, Kuwait
Al-Dubakel AY (2011) Commercial fishing and marketing of hilsa shad *Tenualosa ilisha* (Hamilton-Buchanon, 1822) in Basrah-Southern Iraq. Emir J Food Agric 23(2):178–186
Al-Enezi E, Frontalini F (2015) Benthic foraminifera and environmental quality: the case study of Sulaibikhat Bay (Kuwait). Arab J Geosci 8:8527–8538
Al-Enezi E, Khader S, Balassi E, Frontalini F (2020) Modern benthic foraminiferal diversity: an initial insight into the total foraminiferal diversity along the Kuwait coastal water. Diversity 12:142
Al-Faraj FAM, Scholz M, Tigkas D, Boni M (2015) Drought indices supporting drought management in transboundary watersheds subject to climate alterations. Water Policy 17:865–886
Al-Ghadban AN, Saeed T, Al-Dousari AM, Al-Shemmari H, Al-Mutairi M (1999) Preliminary assessment of the impact of draining of Iraqi marshes on Kuwait's northern marine environment. Part I. Physical manipulations. Wat Sci Tech 40:75–87
Al-Ghadban A, Uddin S, Beg M, Al-Dousari A, Gevao B, Al-Yamani F (2008) Ecological consequences of river manipulations and drainage of Mesopotamian marshes on the Arabian Gulf ecosystem: investigations on changes in sedimentology and environmental quality, with special reference to Kuwait Bay. Technical report EM013C. Kuwait Institute for Scientific Research, Kuwait
Alghunaim A, Taqi A, Al-Kandari M, Al-Said T (2019) Distribution and nature of *Sargassum* species in the Kuwait waters. J Geosci Environ Prot 7:53–59
Alghunaim A, Yamamoto T, Madhusoodhanan R, Polikarpov I, Chen W, Al-Said T, Al-Zekri W, Alenezi Y, Al-Yamani F (2020) Evaluating the potential area and environmental key factors for *Sargassum* bed restoration in highly turbid waters of the northwestern Arabian Gulf. Mar Pollut Bull 153:110949
Al-Handal A, Hu C (2015) MODIS observations of human-induced changes in the Mesopotamian Marshes in Iraq. Wetlands 35:31–40
Al-Hassan LAJ (1999) Shad of the Shatt Al-Arab River in Iraq. Shad J 4(2):1–4
Al-Hassan RH, Ali AM, Radwan SS (1990) Lipids and their constituent fatty acids of *Phaeocystis* sp. from the Arabian Gulf. Mar Biol 105:9–14
Al-Husaini M (2003) Fishery of shared stock of the silver pomfret, *Pampus argenteus*, in the Northern Gulf: a case study. In: FAO papers presented at the Norway-FAO expert consultation on the Management of Shared Fish Stocks Bergen, Norway, 7–10 October 2002. FAO fisheries report No. 695 Supplement, Rome, pp 44–56
Al-Husaini M, Bishop JM, Al-Foudari HM, Al-Baz AF (2015) A review of the status and development of Kuwait's fisheries. Mar Pollut Bull 100:597–606
Ali M, Al-Yamani F, Prusova I (2007) *Bestiolina arabica* sp. nov. (Copepoda, Calanoida, Paracalanidae), a new species from the northwestern Arabian Gulf. Crustaceana 80:195–205
Ali M, Al-Ghunaim A, Subrahmanyam MNV, Al-Enezi Y, Al-Said T, Al-Zakri W, Al-Adila H, Grintsov VA (2018) On the diversity of amphipods inhabiting *Sargassum* as well as clear areas in Kuwait coastal waters, with an assessment of the effect of turbidity and notes on their abundance, composition, and distribution: a preliminary study. Crustaceana 91:767–819
Ali M, Al-Mutairi M, Subrahmanyam MNV, Isath S, AlAwadi MA, Kumar PN, Al-Hebini K, Omar SAS (2019) Temporal variations in abundance and species richness of zooplankton with

emphasis on ichthyoplankton in the subtidal waters of Umm Al-Namil Island, Northwestern Arabian Gulf of the ROPME Sea Area. Am J Environ Sci 15(6):188–204

Al-Kandari M, Al-Yamani FY, Al-Rifaie K (2009) Marine phytoplankton atlas of Kuwait's waters. Kuwait Institute for Scientific Research, Kuwait. ISBN:99906-41-24-2

Al-Kandari M, Bishop J, Chen W, Skryabin V, Polikarpov I, Sattari Z, Taqi A, Hussain S (2017a) Biodiversity, distribution and abundance of intertidal macrofauna in Kuwait. Final report FM075C. Report KISR 14461. Kuwait Institute for Scientific Research, Kuwait

Al-Kandari M, Sattari Z, Al-Mansouri H, Hussain S (2017b) Intertidal macrofauna of Failaka Island. Presented in: Workshop on Intertidal Macrofauna Biodiversity of Kuwait (IMBK), 1–2 May 2017, Kuwait

Al-Kandari M, Bishop J, Chen W, Skryabin V, Satari Z, Hussain S (2018) Treasures of the intertidal area of Kuwait. In: Managing and health of the Gulf ecosystem: dealing with climate change, invasion species, and coastal alterations, 5–7 November 2018, Al Ain, United Arab Emirates University, UAE, abstracts, p 25

Al-Kandari M, Sattari Z, Hussain S, Radashevsky VI, Zhadan A (2019) Checklist of intertidal polychaetes (Annelida) of Kuwait, Northern part of the Arabian Gulf. Reg Stud Mar Sci 32:100872

Al-Kandari M, Oliver PG, Chen W, Skryabin V, Raghu M, Yousif A, Al-Jazzaf S, Taqi A, AlHamad A (2020a) Diversity and distribution of the intertidal Mollusca of the State of Kuwait, Arabian Gulf. Reg Stud Mar Sci 33:100905

Al-Kandari M, Al-Yamani F, Nithyanandan M, Polikarpov I, Grintsov V (2020b) An annotated checklist of the Amphipoda of Kuwait, Northern Arabian Gulf. Crustaceana. (in press)

Al-Madhhachi A-S, Rahi K, Leabi W (2020) Hydrological impact of Ilisu Dam on Mosul Dam; the river Tigris. Geosciences 10:120

Almatar S, Abu Rezaq TS, Al-Baz AF, RahmanYosuf AR (2006) The zobaidy: a study of its life history and aquaculture trials in Kuwaiti waters. Safat, Kuwait: Kuwait Centre for Research and Studies (in Arabic)

Almojil D (2017) The hidden treasures of Kuwait's mudflats. https://www.envearth.com/the-hidden-treasures-of-kuwait%E2%80%99s-mudflats-historical-cultural-and-environmental/. Downloaded on 14 March 2020

Al-Mudaffar Fawzi N, Mahdi BA (2014) Iraq's inland water quality and their impact on the north-western Arabian Gulf. Marsh Bull 9(1):1–22

Almukhtar MA, Jasim W, Mutlak F (2016) Reproductive biology of Hilsa shad *Tenualosa ilisha* (Teleostei: Clupeidae) during spawning migration in the Shatt Al-Arab River and Southern Al-Hammar Marsh, Basra, Iraq. J Fish Aquat Sci 11(1):43–55

Almutairi A (2015) Spatial-temporal variations and diversity of the bacterioplankton communities in the coastal waters of Kuwait. Mar Pollut Bull 100:699–709

Al-Nassar ART (2018) Dynamics of cyclones and precipitation over the Middle East. PhD thesis. Universitat Politècnica de Catalunya, Barcelona, 111 pp

Alosairi Y, Imberger J, Falconer RA (2011) Mixing and flushing in the Persian Gulf (Arabian Gulf). J Geophys Res 116:184–193

Al-Osairi Y, Al Sudairawi M, Neelamani S, Polikarpov I, Kwarteng A, Small C, Al-Handal AY (2016) The effects of developments in Mesopotamian marshland on Boubyan and Warba islands. KISR report EC060K. Kuwait Institute for Scientific Research, Kuwait

Al-Rifaie K, Al-Yamani F, Lennox A, Boltachova N, Revkov N, Grintsov V, Kolesnikova E, Murina V (2012) Macrozoobenthos community structure during four seasons in Kuwait Bay. Int J Oceans Oceanogr 6:45–67

Al-Saadi HA, Hadi RA, Schiewer MU, Al-Mousawi AH (1989) On the influence of the sewage drainage from Basrah City on the phytoplankton and related nutrients in the Shatt Al-Arab estuary. Iraq Arch Hydrobiol 114:443–452

Alsaffar A, Chen W (2019) Spatial and temporal diversity of marine fauna of Boubyan Island of Kuwait. Aquat Ecosyst Health Manag 22:141–148

Alsaffar AH, Lone KP (2000) Reproductive cycles of *Diadema setosum* and *Echinometra mathaei* (Echinoidea: Echinodermata) from Kuwait (northern Arabian Gulf). Bull Mar Sci 67:845–856

Al-Said T, Al-Ghunaim A, Rao DS, Al-Yamani F, Al-Rifaie K, Al-Baz A (2017) Salinity-driven decadal changes in phytoplankton community in the NW Arabian Gulf of Kuwait. Environ Monit Assess 189:268

Al-Said T, Naqvi SWA, Ahmed A, Madhusoodhanan R, Fernandes L, Kedila R, Almansouri H, Al-Rifaie K, Al-Yamani F (2019) Heterotrophic consumption may mask increasing primary production fuelled by anthropogenic nutrient loading in the northern Arabian/Persian Gulf. Mar Pollut Bull 148:30–46

Al-Sarawi M, Eundlach ER, Baca BJ (1985) An atlas of shoreline types and resources, Kuwait. Kuwait University Press, Kuwait, 60 pp

Al-Sarawi M, Marmoush AYM, Jen-Men LO, Al-Hulail F (1995) Coastal morphology of Failaka Island, Kuwait. J Univ Kuwait (Sci) 22:263–282

Al-Saroonchi AA, Al-Saad HT (1988) Check list of the algae from Shatt Al-Arab River, Iraq. J Univ Kuwait (Sci) 15:79–96

Al-Yamani FY (1989) Impact of river diversion in Southern Iraq on the hydrobiology of the northern Arabian Gulf. PW1/E/07-A6 1500. Kuwait Institute for Scientific Research, Kuwait

Al-Yamani F (2008) Importance of the freshwater influx from the Shatt Al-Arab River on the Gulf marine environment. In: Abuzinada AH, Barth H-J, Krupp F, Böer B, Al-Rifaie K (eds) Protecting the Gulf's marine ecosystems from pollution. Springer, Berlin, pp 207–222

Al-Yamani F, Khan NY (2002) Chapter 9: Changes in riverine input and loss of wetlands. In: Khan NY, Al-Ajmi D, Munawar M (eds) The Gulf ecosystem: health and sustainability. Backhuys Publishers, Leiden, pp 235–251

Al-Yamani F, Khvorov SA (2007) Spatial and temporal variability in larval decapod abundance in Kuwait's waters off Bubiyan Island. Int J Oceans Oceanogr 2(1):69–84

Al-Yamani FY, Khvorov SA (2010) Interactive guide of planktonic decapod larvae of Kuwait's waters. Kuwait Institute for Scientific Research, Kuwait. Registry number: 2010/009–4 (Digital-CD)

Al-Yamani F, Madhusoodhanan R (2018) Long-term changes in the planktonic ecosystem of the NW Arabian Gulf: impact of coastal alterations. In: International Conference Gulf 3: Managing the Health of the Gulf Ecosystem: Dealing with Climate Change, Invasive Species, and Coastal Alterations, 5–7 November 2018, Al Ain, UAE. Abstracts, p 26

Al-Yamani F, Naqvi SWA (2019) Chemical oceanography of the Arabian Gulf. Deep Sea Res Part II: Top Stud Oceanogr 161:72–80

Al-Yamani FY, Prusova I (2003) Common copepods of the northwestern Arabian Gulf: identification guide. Kuwait Institute for Scientific Research, Kuwait, 162 pp. ISBN:99906-41-18-8

Al-Yamani FY, Saburova MA (2010) Illustrated guide on the flagellates of Kuwait's intertidal soft sediments. Kuwait Institute for Scientific Research, Kuwait. ISBN:978-99906-95-04-5

Al-Yamani FY, Saburova MA (2011) Illustrated guide on the benthic diatoms of Kuwait's marine environment. Kuwait Institute for Scientific Research, Kuwait. ISBN:978-99966-37-00-1

Al-Yamani F, Saburova M (2019a) Marine phytoplankton of Kuwait's waters. Vol. 1. Cyanobacteria, dinoflagellates, flagellates. Kuwait Institute for Scientific Research, Kuwait, 467 pp. ISBN:978-99966-37-20-9

Al-Yamani F, Saburova M (2019b) Marine phytoplankton of Kuwait's waters. Vol. 2. Diatoms. Kuwait Institute for Scientific Research, Kuwait, 336 pp. ISBN:978-99966-37-20-9

Al-Yamani FY, Skryabin VA (2006) Identification guide for protozoans from Kuwait's waters. Coastal planktonic ciliates: I. Tintinnids. Kuwait Institute for Scientific Research, Kuwait. ISBN:99906-41-25-0

Al-Yamani F, Durvasula R, Ismail W, Al-Rifaie K, Al-Saeed T, Al-Yaqout A, Al-Omran L (1997a) Dynamic oceanography of the northwestern waters of the Arabian Gulf: ecological significance of the marine food web. Final report KISR 5173. Kuwait Institute for Scientific Research, Kuwait

Al-Yamani FY, Bishop J, Al-Rifaei K, Ismail W, Al-Yaqout A, Al-Omran L, Kwarteng A, Al-Ghadban A, Sheppard C (1997b) Assessment of the effects of the Shatt Al-Arab's altered discharge regimes on the ecology of the northern Arabian Gulf. Final report KISR 5174. Kuwait Institute for Scientific Research, Kuwait

Al-Yamani FY, Bishop J, Ramadhan E, Al-Husaini M, Al-Ghadban AN (2004) Oceanographic atlas of Kuwait's waters. Kuwait Institute for Scientific Research, Kuwait, 202 pp. ISBN:99906-41-19-6

Al-Yamani FY, Bishop JM, Al-Rifaie K, Ismail W (2007) The effects of the river diversion, Mesopotamian Marsh drainage and restoration, and river damming on the marine environment of the northwestern Arabian Gulf. Aquat Ecosyst Health Manag 10:277–289

Al-Yamani FY, Skryabin V, Gubanova A, Khvorov S, Prusova I (2011) Marine zooplankton – practical guide. Volume 1 & 2. Kuwait Institute for Scientific Research, Kuwait. ISBN:978-99966-95-07-0

Al-Yamani F, Saburova M, Polikarpov I, Skryabin V (2012a) Marine biodiversity in the arid environment of Kuwait. In: BIT's 1st annual world congress of Ocean-2012. New wave of world marine economy, 20–23 September 2012, Dalian, China, Abstracts

Al-Yamani FY, Skryabin V, Boltachova N, Revkov N, Makarov M, Grinstov V, Kolesnikova E (2012b) Illustrated Atlas on the zoobenthos of Kuwait. Kuwait Institute for Scientific Research, Kuwait

Al-Yamani F, Saburova M, Polikarpov I (2012c) A preliminary assessment of harmful algal blooms in Kuwait's marine environment. Aquat Ecosyst Health Manag 15(sup1):64–72

Al-Yamani FY, Polikarpov I, Al-Ghunaim A, Mikhaylova T (2014) Field guide for the marine macroalgae (Chlorophyta, Rhodophyta, Phaeophyceae) of Kuwait. Kuwait Institute for Scientific Research, Kuwait, 190 pp. ISBN:978-99966-37-02-5

Al-Yamani F, Yamamoto T, Al-Said T, Alghunaim A (2017) Dynamic hydrographic variations in northwestern Arabian Gulf over the past three decades: temporal shifts and trends derived from long-term monitoring data. Mar Pollut Bull 122:488–499

Al-Yamani FY, Al-Kandari M, Polikarpov I, Grintsov V (2019a) Field guide of order Amphipoda (Malacostraca, Crustacea) of Kuwait. Kuwait Institute for Scientific Research, Kuwait, 390 pp. ISBN:978-99966-95-07-0

Al-Yamani F, Madhusoodhanan R, Skryabin V, Al-Said T (2019b) The response of microzooplankton (tintinnid) community to salinity related environmental changes in a hypersaline marine system in the northwestern Arabian Gulf. Deep Sea Res Part II: Top Stud Oceanogr 166:151–170

Al-Yamani F, Polikarpov I, Al-Said T, Ahmed A, Madhusoodhanan R, Sarkar A, Fernandes L, Yamamoto T, Al-Zakri W, Al-Ghunaim A, Al-Fakhraldeen S, Thuslim F (2019c) Sustained oceanographic investigations for Kuwait's waters. Final report KISR FM001O. Kuwait Institute for Scientific Research, Kuwait

Al-Zaidan ASY, Jones DA, Al-Mohanna SY, Meakins R (2003) Endemic macrofauna of the sulaibikhat bay salt marsh and mudflat habitats, Kuwait: status and need for conservation. J Arid Environ 54(1):115–124

Al-Zaidan ASY, Kennedy H, Jones DA, Al-Mohanna SY (2006) Role of microbial mats in Sulaibikhat Bay (Kuwait) mudflat food webs: evidence from δ13 C analysis. Mar Ecol Prog Ser 308:27–36

Al-Zaidan ASY, Al-Mohanna SY, George P (2013) Status of Kuwait's fishery resources: assessment and perspective. Mar Policy 38:1–7

Al-Zamel Z, Cherif O (1998) Subtidal foraminiferal assemblages of the western part of the Shatt Al-Arab Delta, Kuwait, Arabian Gulf. J Foraminifer Res 28:327–344

Al-Zamel A, Al-Sarawi M, Khader S, Al-Rifaiy I (2009) Benthic foraminifera from polluted marine environment of Sulaibikhat Bay (Kuwait). Environ Monit Assess 149:395–409

Amini A, Zareie S, Taheri P, Wan KBY, Mustafa MR (2016) Drought analysis and water resources management inspection in Euphrates-Tigris Basin. In: Bucur D (ed) River Basin Management. InTech Publication, Romania, pp 97–111

Anker A, De Grave S (2009) A new snapping shrimp (Crustacea Decapoda, Alpheidae, Alpheus) from the estuarine mudflats of Kuwait. Zoologisch Mededelingen 83:811–817

Apel M, Türkay M (1999) Taxonomic composition, distribution and zoogeographic of the Grapsid and Ocypodid crab fauna of intertidal soft bottoms in the Arabian Gulf. Estuar Coast Shelf Sci 49:131–142

Awchi TA, Kalyana MM (2017) Meteorological drought analysis in northern Iraq using SPI and GIS. Sustain Water Resourc Manag 3:451–463

Bachmann A, Tice V, Al-Obeidi LA, Kılıç DT, Challenge C (2019) Ecosystem. Tigris-Euphrates river ecosystem: a status report. In: Paper for the Mesopotamia water forum 2019. Kurdistan Region of Iraq, Sulaymaniyah

Ben-Hasan A, Walters C, Christensen V, Al-Husaini M, Al-Foudari H (2018) Is reduced freshwater flow in Tigris-Euphrates rivers driving fish recruitment changes in the Northwestern Arabian Gulf? Mar Pollut Bull 129:1–7

Benzoni F (2006) *Psammocora albopicta* sp. nov., a new species of scleractinian coral from the Indo-West Pacific (Scleractinia; Siderastreidae). Zootaxa 1358:49–57

Bilgen A (2019) The Southeastern Anatolia Project (GAP) in Turkey: an alternative perspective on the major rationales of GAP. J Balk Near East Stud 21:532–552

Bishop JM (1989) Review of shrimp nursery ground studies in Kuwait Bay 1984–1986. Kuwait Bull Mar Sci 10:37–50

Bishop JM (2002) Fishing and mariculture. In: Khan NY, Munwar M, Price ARG (eds) The Gulf ecosystem, health and sustainability. Backhuys Pub, Leiden, pp 253–278

Bishop JM (2003) History and current checklist of Kuwait's ichthyofauna. J Arid Environ 54 (1):237–256

Bishop JM, Khan MH (1991) Depth as a factor in abundance and size of juvenile penaeid shrimps in the absence of estuaries and marshes. Mar Biol 109:103–114

Bishop JM, Khan MH (1999) Use of intertidal and adjacent mudflats by juvenile penaeid shrimps during 24-h tidal cycles. J Exp Mar Biol Ecol 232:39–60

Bishop JM, Xu X, Kwarteng AY, Alsaffar A (1994) Estimating the subtidal extent of Kuwait's benthic sargassum beds. Papers presented at the technical consultation on Shrimp Management in the Arabian Gulf, Al-Khobar. Marine Fisheries Directorate, Ministry of Agriculture and Water, Riyadh, Saudi Arabia

Bishop JM, Ye Y, Alsaffar AH, Fetta N, Abdulqader E, Liu Q, Al-Mohammadi J, Al-Qasserer J, Al-Mulla AA, Sudiqe AR, El-Abdul Had ARM, Al-Jazzaf ST, Al-Dhubaib AH, Al-Juwaied KH, Al-Foudary H, Almatar S, El-Ramadan AHM, El-Salah AA, El-Rabeigh AA, El-Askary SA, Al-Ibrahim EH (2001) Shrimp stock assessment in the western Arabian Gulf by countries of the Gulf Cooperation Council. Final report FM021C. Report KISR 6291. Kuwait Institute for Scientific Research, Kuwait, 149 pp

Bishop J, Al-Yamani F, Alsaffar AH, Lennox A, Yousef A, Al-Rifaie K, Abdul-Razzaq HA, Ismail W, Saburova M, Altukhov D, Khvorov S, Klimova TN (2006) Bubiyan Island and environmental assessment and preparation of a master plan: fisheries and oceanography. Final report FM040C. Report KISR 8407. Kuwait Institute for Scientific Research, Kuwait

Bishop JM, Chen W, Alsaffar AH, Al-Foudari HM (2011) Indirect effects of salinity and temperature on Kuwait's shrimp stocks. Estuar Coasts 34:1246–1254

Bishop J, Ismail W, Al-Yamani F, Saburova M, Alsaffar A, Lennox A, Khvorov S, Yousef A, Klimova T, Al-Rifiae K, Al-Mansouri H (2013) Chapter 11: Marine life. In: Omar SA, Roy WY (eds) Ecology and environment of Boubyan Island in Kuwait. Kuwait Institute for Scientific Research, Kuwait, pp 167–239. ISBN:978-99966-37-08-7

Borjian H (2012) Karun River I. Geography and hydrology II. Encyclopædia Iranica 2012 XV (Fasc. 6):629–633. http://www.iranicaonline.org/articles/karun_1_2. Downloaded on 18 May 2020

Brewer PG, Dyrssen D (1985) Chemical oceanography of the Persian Gulf. Prog Oceanogr 14:41–55

Carpenter KE, Harrison PL, Hodgson G, Alsaffar AH, Alhazeem SH (1997) The corals and reef fishes of Kuwait. Kuwait Institute for Scientific Research, Environmental Public Authority, Fahed Al-Marouk Press Est, Kuwait, 166 pp

Chen W, Almatar S, Bishop JM (2009) Spatial and temporal variability of fish assemblage in Kuwait Bay. Mar Biol 156:415–424

Chomani K, Bijnens T (2016) The impact of Daryan Dam on Kurdistan Region of Iraq. Save the Tigris and Iraq Marshes Campaign. http://www.iraqicivilsociety.org/wp-content/uploads/2016/10/Daryan-Dam-Report.pdf. Downloaded on 4 June 2020

Chomérat N, Saburova M, Bilien G, Al-Yamani F (2012) *Prorocentrum bimaculatum* sp. nov. (Dinophyceae, Prorocentrales), a new benthic dinoflagellate species from Kuwait (Arabian Gulf). J Phycol 48(1):211–221

Clark PF, Abdul-Sahib IM, Al-Asadi MS (2006) The first record of *Eriocheir sinensis* H. Milne Edwards, 1853 (Crustacea: Brachyura: Varunidae) from the Basrah area of Southern Iraq. Aquat Invasions 1:51–54

Clayton DA (1987) Why mudskippers build walls. Behaviour 102:185–195

Clayton DA, Vaughan TC (1982) Pentagonal territories of the mudskipper *Boleophthalmus boddarti* (Pisces: Gobiidae). Copeia 1:233–235

Cook BI, Anchukaitis KJ, Touchan R, Meko DM, Cook ER (2016) Spatiotemporal drought variability in the Mediterranean over the last 900 years. J Geophys Res Atmos 121 (5):2060–2074

Coleman JM, Huh OK (2004) Major world deltas: a perspective from space. Coastal Studies Institute, Louisiana State University. http://128.138.136.5/science/groups/wessman/projects/wdn/papers/colemanManuscript.pdf. Downloaded on 14 May 2020

Cressey GB (1958) The Shatt al-Arab Basin. Middle East J 12:448–460

Cullmann J (2013) Hydrology. In: Kibaroglu A, Kirschner A, Mehring S, Wolfrum R (eds) Water law and cooperation in the Euphrates-Tigris region. A comparative and interdisciplinary approach. Brill, Leiden, pp 177–189

Dames A, Moore S (1983) Aquatic biology investigations: studies for Subiya area, Kuwait Bay and development of electrical networks contract N. MEW/CP/PGP-1113-80/81. Final report, vol I. Government of Kuwait, Ministry of Electricity and Water, Kuwait, Safat

Delima M, Al-Mutairi M, Dasgti J, Al-Dosseri S, Loughland R, Gregory G, Khalil F, Siddiqui K (2013) Chapter 9: Wildlife. In: Omar SA, Roy WY (eds) Ecology and environment of Boubyan Island in Kuwait. Kuwait Institute for Scientific Research, Kuwait, pp 144–157. ISBN:978-99966-37-08-7

Devlin MJ, Massoud MS, Hamid SA, Al-Zaidan A, Al-Sarawi H, Al-Enezi M, Al-Ghofran L, Smith AJ, Barry J, Stentiford GD, Morris S, da Silva ET, Lyon BP (2015a) Changes in the water quality conditions of Kuwait's marine waters: long term impacts of nutrient enrichment. Mar Pollut Bull 100:607–620

Devlin MJ, Le Quesne WJF, Lyons BP (2015b) The marine environment of Kuwait – emerging issues in a rapidly changing environment. Mar Pollut Bull 100:593–596

Devlin MJ, Breckels M, Graves CA, Barry J, Capuzzo E, Huerta FP, Al Ajmi F, Al-Hussain MM, LeQuesne WJF, Lyons BP (2019) Seasonal and temporal drivers influencing phytoplankton community in Kuwait marine waters: documenting a changing landscape in the Gulf. Front Mar Sci 6:141

Dolan JR, Pierce RW (2013) Diversity and distributions of tintinnid ciliates. In: Dolan JR, Agatha S, Coats DW, Montagnes DJS, Stocker DK (eds) Biology and ecology of tintinnid ciliates: models for marine plankton. Wiley Blackwell, Oxford, pp 214–243

Douabul AAZ, Al-Saad HT (1985) Seasonal variation of oil residues in water of Shatt Al-Arab river, Iraq. Water Air Soil Pollut 24:237–246

Downing N (1988) The coral reefs and coral islands of Kuwait. In: Proceeding of the ROPME workshop on coastal area development, UNEP regional seas report and Strudies #90. ROPME Publication # GC-5/006, pp 73–77

Downing N, Roberts C (1993) Has the Gulf War affected coral reefs of the Northwestern Gulf? Mar Pollut Bull 27:149–156

El Samra MI (1988) Chemical observations in the Arabian Gulf and the Gulf of Oman. Arab Gulf Sci Res A6(2):205–215

Emery KO (1956) Sediments and water of the Persian Gulf. Bull Am Assoc Petrol Geol 40:2354–2383

Erian W (2011a) Agriculture drought and land degradation risk in Arab region. In: The first Arab regional conference for disaster risk reduction, 19–21 March 2011, Aqaba, Jordan

Erian W (2011b) Drought vulnerability in the Arab region: case study – drought in Syria. Ten years of scarce water (2000–2010). Syria, Damascus: Arab Center for the Studies of Arid Zones and Dry Lands (ACSAD)/International Strategy for Disaster Reduction (ISDR), 74 pp

FAO (Food and Agriculture Organization) (2008) AQUASTAT Country Profile – Iraq. FAO of the United Nations, Rome

FAO (Food and Agriculture Organization) (2009) Irrigation in the Middle East region in figures: AQUASTAT survey – 2008. In: Frenken K (ed). FAO Water Reports, Rome, p 34

FloodList (2018a) Floods Across the Gulf Region. http://floodlist.com/asia/floods-across-the-gulf-region. Downloaded on 28 June 2020

FloodList (2018b) Middle East – Heavy Rain and Flooding in Iran, Iraq and Kuwait. http://floodlist.com/asia/flooding-iran-iraq-and-kuwait-november-2018. Downloaded on 28 June 2020

Grabe SA, Lees DC, Allaire HP (1992) Macrozooplankton studies in Kuwait Bay (Arabian Gulf). I: distribution and composition of the ichthyoplankton. J Plankton Res 14(4):607–623

Hale SS (2020) In praise of mud. Bull Ecol Soc Am:e01709

Hameed AH, Saburova M (2015) First record of *Scrippsiella trochoidea* (Dinophyceae) in Shatt Al-Arab River (Southern Iraq). Mar Biodiver Rec 8:e150

Hameed AH, Ali MH, Aljorany YS, Hassan WF, Al-Hello AAZN (2013) Assessing changes in seawater intrusion and water quality of the Shatt Al-Arab River, Iraq. Ann Limnol Int J Lim 49:199–206

Hameed AH, Saburova M, Chomérat N (2016) Taxonomic characterization of a brackish water, bloom-forming *Peridiniopsis* species (Peridiniales, Dinophyceae) from Iraq. Mar Biodiver Rec 9:e40

Hameed M, Ahmadalipour A, Moradkhani H (2018) Apprehensive drought characteristics over Iraq: results of a multidecadal spatiotemporal assessment. Geosciences 8:58

Hamza W, Munawar M (2009) Protecting and managing the Arabian Gulf: past, present and future. Aquat Ecosyst Health Manag 12:429–439

Heil CA, Glibert PM, Al-Sarawi MA, Faraj M, Behbehani M, Husain M (2001) First record of a fish-killing *Gymnodinium* sp. bloom in Kuwait Bay, Arabian Sea: chronology and potential causes. Mar Ecol Prog Ser 214:15–23

Hodgson G, Carpenter K (1995) Scleractinian corals of Kuwait. Pac Sci 49:227–246

Hong Y, Adler RF, Hossain F, Curtis S, Huff GJ (2007) A first approach to global runoff simulation using satellite rainfall estimation. Water Resour Res 43:W08502

Hürriyet Daily News. May 19, 2020. Turkey inaugurates Ilisu Dam. https://www.hurriyetdailynews.com/turkey-inaugurates-ilisu-dam-154904. Downloaded on 10 June 2020

Hussain NA, Mohamed ARM, Al Noo SS, Mutlak FM, Abed IM, Coad BW (2009) Structure and ecological indices of fish assemblages in the recently restored Al-Hammar Marsh, Southern Iraq. BioRisk 3:173–186

Ibrahim FH (2018) Mapping the distribution and occurrence of four newly recorded species of seagrasses in the North-West Arabian Gulf. Mesopot J Mar Sci 33(1):37–48

Isaev VA, Mikhailova MV (2009) The hydrography, evolution, and hydrological regime of the mouth area of the Shatt Al-Arab River. Water Resour 36(4):380–395

Jacob PG, Anderlini V, Mohammad OS, Zarba MA (1981) Physical and chemical oceanography of Kuwait's coastal waters, December 1979–May 1980. Report KISR 168. Kuwait Institute for Scientific Research, Kuwait

John VC, Coles SL, Abozed AI (1990) Seasonal cycle of temperature, salinity and water masses of the Western Arabian Gulf. Oceanol Acta 13:273–281

Johns WE, Jacobs GA, Kindle JC, Murray SP, Carron M (1999) Arabian marginal seas and gulfs: report of a workshop held at Stennis space center, Mississippi, 11–13 May, 1999. Technical report 2000–01. University of Miami RSMAS

Jones DA (1986) A field guide to the sea shores of Kuwait and the Arabian Gulf. University of Kuwait, Kuwait

Jones DA, Al-Attar M (1982) Observations on the postlarval and juvenile habitats of *Penaeus semisulcatus* in Kuwait Bay and adjacent waters. In: Mathews CP (ed) Revised proceedings. Shrimp fisheries management workshop, report KISR 670. Kuwait Institute for Scientific Research, Kuwait, pp 112–129

Jones DA, Clayton D (1983) The systematic and ecology of crabs belonging to the genera *Cleistostomade* Haan and *Paracleistostomade* Man on Kuwait mudflats. Crustaceana 45:183–199

Jones DA, Price ARG, Al-Yamani F, Al-Zaidan A (2002) Coastal and marine ecology. In: Khan NY, Munawar M, Price ARG (eds) The Gulf ecosystem: health and sustainability. Backhuys Publishers, Leiden, pp 65–104

Karabulut M (2015) Drought analysis in Antakya-Kahramanmaraş Graben, Turkey. J Arid Land 7 (6):741–754

Karim HH, Salman H (1987) Estimation of sediment discharge, sedimentation rate and the fate of hydrocarbon residues of Shatt Al-Arab river sediments, Northwest Arabian Gulf. Mar Mesop 2:103–115

Kassler P (1973) The structural and geomorphic evolution of the Persian Gulf. In: Purser BH (ed) The Persian Gulf. Holocene carbonate sedimentation and diagenesis in a shallow epicontinental sea, 1st edn. Springer, Berlin, pp 11–32

Kelley CP, Mohtadi S, Cane MA, Seager R, Kushnir Y (2015) Climate change in the fertile crescent and implications of the recent Syrian drought. Proc Natl Acad Sci 112(11):3241–3246

Khalaf TA (1991) New calanoid copepods of the genus *Acartia* from Khor Abdulla and Khor Al-Zubair waters, Iraq. Mar Mesop 6(1):80–91

Khalaf TA (2007) Post-naupliar stages of *Acartia* (*Acartiella*) *faoensis*, Khalaf (Copepoda: Calanoida), from Khor Al-Zubair south of Iraq. Int J Oceans Oceanogr 2(1):179–186

Khalaf TA (2008a) A new record of *Bestiolina arabica* Ali et al., 2007 (Calanoida, Copepoda) from Khor Al-Zubair canal and Shatt Al-Arab River, Southern Iraq. Mar Mesop 23(2):377–386

Khalaf TA (2008b) A new species of *Phyllodiaptomus* Kiefer (Copepoda, Calanoida) from the Shatt Al-Arab River, Southern Iraq. Cructaceana 81(3):257–269

Khalaf FI, Al-Ghadban AN, Al-Saleh S, Al-Omran L (1982) Sedimentology and mineralogy of Kuwait Bay bottom sediments, Kuwait, Arabian Gulf. Mar Geol 46:71–99

Khalaf TA, Morad M, Awad AH (2014) Zooplankton in Iraqi marine and brackish water: diversity and abundance of zooplankton and its infection with protozoan parasites in Iraqi marine and brackish water. LAP Lambert Academic Publishing, Saarbrücken

Khalaf TA, Ali M, Ajeel SG (2015) Is the Northwestern Arabian/Persian Gulf a biodiversity hotspot? In: CBD Regional workshop to facilitate the description of ecologically or biologically significant marine Areas (EBSAs) in the North-West Indian Ocean and Adjacent Gulf Areas, and Training Session on EBSAs, 19–25 April 2015, United Arab Emirates, Dubai. https://www.cbd.int/meetings/EBSAWS-2015-02/iraq-submission1-en.pdf. Downloaded on 17 March 2020

Khan NY (2002) Physical and human geography. In: Khan NY, Munawar M, Price ARG (eds) The Gulf ecosystem: health and sustainability. 1st ed. Backhuys Publishers, Leiden, pp 3–21

KUNA (Kuwait News Agency) (2018) Experts: climate changes behind weather fluctuations, heavy rains in Kuwait. 17 November 2018. https://www.kuna.net.kw/ArticleDetails.aspx?id=2759884&language=en. Downloaded on 5 June 2020

Lawler A (2005) Ecology. Reviving Iraq's wetlands. Science 307:1186–1189

Lawler A (2016) Iraq confronts a new enemy – the Persian Gulf. Science 351(6269):111–112

Maltby E (1994) An environmental and ecological study of the marshlands of Mesopotamia. AMAR Appeal Trust, London

Marcella MP, Eltahir EAB (2008) The hydroclimatology of Kuwait: explaining the variability of rainfall at seasonal and interannual time scales. Am Meteorol Soc 9:1095–1105

Margalef R (1978) Life-forms of phytoplankton as survival alternatives in an unstable environment. Oceanol Acta 1:493–509

Mathews CP (1986) Shrimp fisheries management project. Phase III. Final report (MB-47). Report KISR 2180. Kuwait Institute for Scientific Research, Kuwait

Michel D (2017) Iran's impending water crisis. In: Reed D (ed) Water, security and U.S. foreign policy. Routledge, New York, pp 168–188

Michel HB, Behbehani M, Herring D, Arar M, Shoushani M, Brakoniecki T (1986) Zooplankton diversity, distribution and abundance in Kuwait waters. Kuwait Bull Mar Sci 8:37–105

Mohammad M-BM (1970a) A new species of the genus *Prionospio* (Annelida: Polychaeta). Hydrobiologia 36(1):23–26

Mohammad M-BM (1970b) Description of two new species of Nereidae (Annelida: Polychaeta). Zool J Linn Soc Lond 49(3):183–186

Mohammad M-BM (1971) Intertidal polychaetes from Kuwait, Arabian Gulf, with descriptions of three new species. J Zool London 163(3):285–303

Mohammad M-BM (1972) Infestation of the Pearl Oyster *Pinctada margaritifera* (Linné) by a new species of *Polydora* in Kuwait, Arabian Gulf. Hydrobiologia 39(4):463–477

Mohammad M-BM (1973) New species and records of polychaete annelids from Kuwait, Arabian Gulf. Zool J Linn Soc Lond 52(1):23–44

Mohammad M-BM (1980) Polychaete annelids from Kuwaitian islands, Arabian Gulf, with descriptions of four new species. Zool J Linn Soc Lond 69(1):31–42

Mohammed HH (2018) *Delavalia longifurca* (Sewell, 1934) (Copepoda: Harpacticoida) from the Southern Iraqi Marshes and Shatt Al-Arab River, Basrah, Iraq. Biol Appl Environ Res 2:34–43

Moore ABM, Almojil D, Harris M, Jabado RW, White WT (2014) New biological data on the rare, threatened shark *Carcharhinus leiodon* (Carcharhinidae) from the Persian Gulf and Arabian Sea. Mar Freshw Res 65(4):327–332

Moyel MS, Hussain N (2015) Water quality assessment of the Shatt Al-Arab River, Southern Iraq. J Coast Life Med 3(6):459–465

Myers AA, Nithyanandan M (2016) The Amphipoda of Sea City, Kuwait. – The Senticaudata (Crustacea). Zootaxa 4072(4):401–429

Naderloo R (2017) Atlas of crabs of the Persian Gulf. Springer eBooks, 444 pp

Naser MD, Son MO (2009) First record of the New Zealand mud snail *Potamopyrgus antipodarum* (Gray 1843) from Iraq: the start of expansion to Western Asia? Aquat Invasions 4:369–372

Naser MD, Son MO, Yasser AG (2011) Assessing the risks of invasions of aquatic invertebrates in the Shatt Al-Arab River. Rus J Biol Invasions 2:120–125

Naser MD, Rainbow PS, Clark PF, Yasser AG, Jones DS (2015) The barnacle *Amphibalanus improvisus* (Darwin, 1854), and the mitten crab *Eriocheir*: one invasive species getting off on another! BioInvasions Rec 4:205–209

Nezlin NP, Polikarpov IG, Al-Yamani F (2007) Satellite-measured chlorophyll distribution in the RSA: spatial, seasonal and inter-annual variability. Int J Oceans Oceanogr 2(1):139–156

Nezlin NP, Polikarpov IG, Al-Yamani FY, Rao DVS, Ignatov AM (2010) Satellite monitoring of climatic factors regulating phytoplankton variability in the Arabian (Persian) Gulf. J Mar Syst 82:47–60

Nilsson C, Reidy CA, Dynesius M, Revenga C (2005) Fragmentation and flow regulation of the world's large river systems. Science 308(5720):405–408

Nithyanandan M (2012) New and rare nudibranch records from Kuwait, Arabian Gulf (Mollusca, Opisthobranchia). Mar Biodiver Rec 5:e115

Papathanasopoulou N, Zogaris S (2015) Coral reefs of Kuwait. KUFPEC, Biodiversity East, Cyprus. ISBN:978-9963-2811-2-1

Partow H (2001) The Mesopotamian marshlands: demise of an ecosystem. UNEP early warning and assessment technical report, UNEP/DEWA/TR.01-3 Rev. 1, Nariboi, Kenya, 47 pp

Pilcher NJ, Wilson S, Alhazeem SH, Shokri MR (2000) Status of coral reefs in the Arabian/Persian Gulf and Arabian Sea region (Middle East). In: Wilkinson C (ed) Status of coral reefs of the world: 2000. Australian Institute of Marine Science, Australia, pp 55–64

Pilcher NJ, Antonopoulou M, Perry L, Abdel-Moati MA, Al Abdessalaam TZ, Albeldawi M, Al Ansi M, Al-Mohannadi SF, Al Zahlawi N, Baldwin R, Chikhi A, Das HS, Hamza S, Kerr OJ, Al Kiyumi A, Mobaraki A, Al Suwaidi HS, Al Suweidi AS, Sawaf M, Tourenq C, Williams J, Willson A (2014) Identification of important sea turtle areas (ITAs) for hawksbill turtles in the Arabian region. J Exp Mar Biol Ecol 460:89–99

Pohl T, Al-Muqdadi SW, Ali MH, Al-Mudaffar Fawzi N, Ehrlich H, Merkel B (2014) Discovery of a living coral reef in the coastal waters of Iraq. Sci Rep 4:4250

Pokavanich T, Polikarpov I, Lennox A, Al-Hulail F, Al-Said T, Al-Enezi E, Al-Yamani F, Stokozov N, Shuhaibar B (2013) Comprehensive investigation of summer hydrodynamic and water quality characteristics of desertic shallow water body: Kuwait Bay. J Coast Dyn 12 (2):1253–1264

Polikarpov I, Al-Yamani FY, Saburova M (2009) Space-time variability of phytoplankton structure and diversity in the north-western part of the RSA (Kuwait's waters). In: Krupp F, Musselman LJ, Kotb MMA, Weidig I (eds) Environment, biodiversity and conservation in the Middle East. Proceedings of the first Middle Eastern Biodiversity Congress, Aqaba, Jordan, 20–23 October 2008. BioRisk 3:83–96

Polikarpov I, Saburova M, Al-Yamani F (2016) Diversity and distribution of winter phytoplankton in the Arabian Gulf and the Sea of Oman. Cont Shelf Res 199:85–99

Polikarpov I, Al-Yamani F, Saburova M (2019) Remote sensing of phytoplankton variability in the Arabian/Persian Gulf. In: Barale V, Gade M (eds) Remote sensing of the Asian seas. Springer, Cham, pp 485-501

Polikarpov I, Saburova M, Al-Yamani F (2020) Decadal changes in diversity and occurrence of microalgal blooms in the NW Arabian/Persian Gulf. Deep Sea Res Part II: Top Stud Oceanogr 179:104810

Price ARG (1993) The Gulf: impacts and management initiatives. Mar Pollut Bull 27:17–27

Prusova I, Al-Yamani F (2014) A new species of *Labidocera* (Copepoda, Calanoida, Pontellidae) from the Arabian Gulf. Crustaceana 87(10):1149–1167

Prusova I, Al-Yamani F, Al-Mutairi H (2001) *Subeucalanus flemingeri* sp. n. from the Arabian Gulf (Copepoda: Eucalanidae). Zoosystematica Rossica 10(1):47–54

Rahi KA (2018) Salinity management in the Shatt Al-Arab River. Int J Eng Technol 7 (4.20):128–133

Reynolds RM (1993) Physical oceanography of the Gulf, Strait of Hormuz, and the Gulf of Oman: results from the Mt. Mitchell expedition. Mar Pollut Bull 27:35–59

Reynolds RM (2002) Oceanography. In: Khan NY, Munawar M, Price ARG (eds) The Gulf ecosystem: health and sustainability. Backhuys Publishers, Leiden, pp 53–64

Richardson CJ, Hussain NA (2006) Restoring the Garden of Eden: an ecological assessment of the marshes of Iraq. Bioscience 56:477–489

Richardson CJ, Reiss P, Hussain NA, Alwash AJ, Pool DJ (2005) The restoration potential of the Mesopotamian marshes of Iraq. Science 307:1307–1311

Riegl B, Poiriez A, Janson X, Bergman KL (2010) The Gulf: Facies belts, physical, chemical, and biological parameters of sedimentation on a carbonate ramp. In: Westphal H, Riegl B, Eberli GP (eds) Carbonate depositional systems: assessing dimensions and controlling parameters, 1st edn. Springer, New York, pp 145–213

Rodell M, Houser PR, Jambor U, Gottschalck J, Mitchell K, Meng C-J, Arsenault K, Cosgrove B, Radakovich J, Bosilovich M, Entin JK, Walker JP, Lohmann D, Toll D (2004) The global land data assimilation system. Bull Am Meteorol Soc 85(3):381–394

Saburova M, Al-Yamani F, Polikarpov I (2009) Biodiversity of free-living flagellates in Kuwait's intertidal sediments. In: Krupp F, Musselman LJ, Kotb MMA, Weidig I (Eds) Environment, biodiversity and conservation in the Middle East. Proceedings of the First Middle Eastern Biodiversity Congress, Aqaba, Jordan, 20–23 October 2008. BioRisk 3:97–110

Saburova M, Chomérat N, Hoppenrath M (2012a) Morphology and SSU rDNA phylogeny of *Durinskia agilis* (Kofoid & Swezy) comb. nov. (Peridiniales, Dinophyceae), a thecate, marine, sand-dwelling dinoflagellate formerly classified within *Gymnodinium*. Phycologia 51 (3):287–302

Saburova M, Polikarpov I, Al-Yamani F (2012b) First record of *Kryptoperidinium foliaceum* (Dinophyceae: Peridiniales) from a hypersaline environment in Kuwait, Northwestern Arabian Gulf. Mar Biodiver Rec 5:e104

Saburova M, Polikarpov I, Al-Yamani F (2013) New records of the genus *Gambierdiscus* in marginal seas of the Indian Ocean. Mar Biodiver Rec 6:e91

Saeed T, Al-Ghadban AN, Al-Shemmari H, Al-Mutairi M, Al-Hashash H (1999) Preliminary assessment of the impact of draining of Iraqi marshes on Kuwait's northern marine environment. Part II. Sediment associated pollutants. Water Sci Technol 40(7):89–98

Saffo MB (1987) New light on seaweeds source. Bioscience 37(9):654–664

Salman SD, Ali MH, Al-Adhub AHY (1990) Abundance and seasonal migrations of the penaeid shrimp *Metapenaeus affinis* (H. Milne-Edwards) within Iraqi waters. Hydrobiologica 96:79–90

Salman SD, Rage TJ, Naser MD, Yasser AG (2006) The invasion of *Macrobrachium nipponense* (De Haan, 1849) (Caridea: Palaemonidae) into the Southern Iraqi Marshes. Aquat Invasions 1:109–115

Selby J, Dahi OS, Fröhlich C, Hulme M (2017) Climate change and the Syrian civil war revisited. Polit Geogr 60:232–244

Shahdadi A, Sari A, Naderloo R (2014) A checklist of the barnacles (Crustacea: Cirripedia: Thoracica) of the Persian Gulf and Gulf of Oman with nine new records. Zootaxa 3784:201–223

Sheffield J, Wood EF (2011) Drought: past problems and future scenarios. Earthscan Ltd., Washington, DC. 210 pp

Sheppard C, Price A, Roberts C (1992) Marine ecology of the Arabian Region: patterns and processes in extreme tropical environments. Academic, London

Sheppard C, Al-Husiani M, Al-Jamali F, Al-Yamani F, Baldwin R, Bishop J, Benzoni F, Dutrieux E, Dulvy NK, Subba Rao DVSR, Jones DA, Loughland R, Medio D, Nithyanandan M, Pilling GM, Polikarpov I, Price AR, Purkis S, Riegl B, Saburova M, Namin KS, Taylor O, Wilson S, Zainal K (2010) The Gulf: a young sea in decline. Mar Pollut Bull 60:13–38

Sheppard C, Al-Husiani M, Al-Jamali F, Al-Yamani F, Baldwin R, Bishop J, Benzoni F, Dutrieux E, Dulvy NK, Durvasula SRV, Jones DA, Loughland R, Medio D, Nithyanandan M, Pilling GM, Polikarpov I, Price ARG, Purkis SJ, Riegl BM, Saburova M, Samimi-Namin K, Taylor O, Wilson S, Zainal K (2012) Environmental concerns for the future of Gulf coral reefs. In: Riegl B, Purkis S (eds) Coral reefs of the Gulf. Coral reefs of the world, vol 3. Springer, Dordrecht, pp 349–373

Skryabin VA, Al-Yamani FY (2006) New species of the genera *Tintinnopsis*, *Metacylis* and *Eutintinnus* (Ciliophora: Spirotrichea: Tintinnida) from waters off Kuwait (Arabian Gulf). Biologiya Morya (Vladivostok) 32(5):358–362

Skryabin VA, Al-Yamani FY (2007a) New species of genera *Leprotintinnus* and *Luminella* (Ciliophora, Spirotrichea, Tintinnida) from Kuwait's Waters of the Arabian Gulf. Kuwait J Sci Eng 34(1A):79–90

Skryabin V, Al-Yamani F (2007b) Composition, distribution and seasonal changes of tintinnids in the northwestern Arabian Gulf around the Bubiyan Island. Int J Oceans Oceanogr 2(1):99–106

Smayda TT (2002) Adaptive ecology, growth strategies and the global bloom expansion of dinoflagellates. J Oceanogr 58:281–294

Subba-Rao DV, Al-Yamani FY (1998) Phytoplankton ecology in the waters between Shatt Al-Arab and Straits of Hormuz, RSA: a review. Plankton Biol Ecol 45(2):101–116

Sun Y, Al-Kandari M, Kubal P, Walmiki N, Kupriyanova EK (2017) Cutting a Gordian knot of tubeworms with DNA data: the story of the *Hydroides operculata*-complex (Annelida, Serpulidae). Zootaxa 4323(1):039–048

Tomczak M, Godfrey JS (2003) Regional oceanography: an introduction, 2nd edn. Daya Publishing House, New Delhi

UNDG (United Nations Development Group) (2005) The national water master plan – Phase 1 Water Resources Assessment, 26 pp. http://www.fao.org/nr/water/aquastat/basins/euphrates-tigris/print1.stm. Downloaded on 15 June 2020

UNEP/CBD (United Nations Environment Programme/Convention on Biological Diversity) (2016) Subsidiary Body on Scientific, Technical and Technological Advice. Twentieth meeting, Montreal, Canada, 25–30 April 2016, Item 4.4 of the provisional agenda

UNESCO (United Nations Educational, Scientific and Cultural Organization) (2014) Integrated Drought Risk Management (DRM). National Framework for Iraq. Executive Summary. http://reliefweb.int/sites/reliefweb.int/files/resources/Drought%20executive%20summary.pdf. Downloaded on 12 February 2020

UN-ESCWA and BGR (United Nations Economic and Social Commission for Western Asia and Bundesanstalt für Geowissenschaften und Rohstoffe) (2013) Chapter 5: Shatt al Arab, Karkheh, and Karun Rivers. In: Inventory of shared water resources in Western Asia. Beirut, Salim Dabbous Printing Co.. 626 pp

Venturi L, Capozzoli C (2017) Changes in the water quantity and quality of the Euphrates river are associated with natural aspects of the landscape. Water Policy 19:233–256

Wabnitz CCC, Lam VWY, Reygondeau G, Teh LCL, Al-Abdulrazzak D, Khalfallah M, Pauly D, Deng Palomares ML, Zeller D, Cheung WWL (2018) Climate change impacts on marine biodiversity, fisheries and society in the Arabian Gulf. PLoS One 13(5):e0194537

Wilson AT (1928) The Persian Gulf. An historical sketch from the earliest times to the beginning of the twentieth century. Oxford University Press, Clarendon

Wittfogel KA (1956) The hydraulic civilization: Man's role in changing the Earth. University of Chicago Press, Chicago

Wright JM, Abou-Seed F, Clayton DA (1996) Long term changes in the fish assemblage of Sulaibikhat Bay, Kuwait. Kuwait J Sci Eng 23:47–60

Yaseen BR, Al Asaady KA, Kazem AA, Chaichan MT (2016) Environmental impacts of salt tide in Shatt Al-Arab Basra/Iraq. IOSR J Environ Sci Toxicol Food Technol 10(1-II):35–43

Zakharova EA, Kouraev AV, Crétaux J-F, Al-Yamani F, Polikarpov I (2007) Radar altimetry for studies of large river basins: hydrological regime of the Euphrates-Tigris rivers. In: Proceedings of 'Envisat symposium 2007', Montreux, Switzerland, 23–27 April 2007, ESA SP-636, July 2007, 6 pp

Zhang Q, Liu G, Hu Z (2014) Description of a new freshwater bloom-forming dinoflagellate with a diatom endosymbiont, *Peridiniopsis minima* sp. nov. (Peridiniales, Dinophyceae) from China. Algol Stud 145(/146):119–122

Chapter 21
A Possible Threat to the Fish Biodiversity in the Southern Marshes of Iraq: A Mini-Review

Laith A. Jawad

Abstract The diversity of the freshwater biota is not investigated, in particular the invertebrate groups that support all the aquatic vertebrates and mainly the fish groups that depend on invertebrates in their daily food and growth. At the present time, thousands of taxa are in the process of depletion due to the global environment changes as a result of human activities, and at the same time, the efforts to stop such a loss in the freshwater biodiversity is very slow and does not rise to the required attention in order to restore this productive and important environment. There are several continuing pressures that affect the freshwater ecosystem, which cause a loss for the sensitive species and ultimately reduce the biodiversity. This chapter gives a short review for the main threats that the fish group can face through the changes in the environment. Some of them might not appear as direct, but they are important for the existing different species of fishes. The threats given in this chapter are baseline for the future studies on the biodiversity conservation in Iraq and the neighbouring countries.

Keywords Aquatic biota · Diversity · South of Iraq · Impact · Environment · Fishes

21.1 Introduction

In the freshwater environment and worldwide, there are over 10,000 fish species inhabiting these habitats (Lundberg et al. 2000), which represents nearly 40% of global fish variety and a quarter of the world's vertebrate species (Cazzolla Gatti 2016a, b). Besides this large number of fish species, there are other freshwater animals such amphibians, aquatic reptiles and mammals that share the freshwater environment with the fish species (Cazzolla Gatti 2016a, b). In spite of the fact that 0.01% of the water covering the surface of earth is freshwater, a third of the vertebrates are inhabiting this habitat (Gleick 1996).

L. A. Jawad (✉)
School of Environmental and Animal Sciences, Unitec Institute of Technology, Mt Albert, Auckland, New Zealand

L. A. Jawad (ed.), *Southern Iraq's Marshes*, Coastal Research Library 36,
https://doi.org/10.1007/978-3-030-66238-7_21

The diversity present in the biota inhabiting the freshwater ecosystem is a priceless natural reserve in economic, cultural, aesthetic, scientific and educational aspects (Cazzolla Gatti 2016a, b). Maintenance and supervision are decisive, but freshwater diversity is suffering a deterioration extremely greater than that in the most impacted land environments (Strayer and Dudgeon 2010). If condition of human demands and activities continue, with present losses of species, the likelihood of conserving much of the enduring biodiversity in freshwater will be very little (Cazzolla Gatti 2016a, b).

Generally speaking, the information on the total freshwater biodiversity is sadly inadequate, specifically among the invertebrate and microbe groups that are inhabiting the tropical latitudes that maintain most of the species of the world (Cazzolla Gatti 2016a, b). Similarly, vertebrates also are not entirely identified, containing well-studied taxa, such as the fish (Stiassny 2002). Freshwater fishes, which have a tendency to be more or less restricted to drainage systems, offer a moderately conventional system for exploratory designs of dispersal that may reveal the symbol of past continental and climate changes (Levêque et al. 2007).

The aim of the chapter at hand is to present a concise idea about the threat that can affect the freshwater fish fauna inhabiting the southern marshes of Iraq. Stating such threats will be the cornerstone for any future conservation projects which might be performed in the southern marshes region of Iraq.

21.2 The Possible Threats to the Freshwater Fish Fauna of the Southern Marshes of Iraq

There are two different types of threats that might impact the fish fauna of the southern marshes of Iraq; these are local threats and global threats. In this section, a discussion of these two types of threats will be given.

21.2.1 Local Threats

Freshwater habitats are progressively affected by numerous pressures that cause a loss of subtle species and an overall decrease in the biodiversity (Meyer et al. 1999). Pressures to world freshwater diversity biota can come together under seven interrelating classes: overexploitation; water pollution; flow modification; destruction or degradation of habitat; invasion by exotic species; and hydropower (Arthington et al. 2010). Any changes in the global climate status in conjoint with nitrogen deposition and runoff patterns (Poff et al. 2002) are overlaid pressures upon all of the preceding groups.

21.2.1.1 Over-Misuse and Niche Deprivation

The overutilization disturbs mostly vertebrates, in particular fish, reptiles and some amphibians, whereas other pressures have their outcome for the entire diversity of biota living in the freshwater from microbes to megafauna (Cazzolla Gatti 2016a, b).

The pollution harms arise as a plague (Dudgeon et al. 2006), and while some developed countries have made a substantial advancement in dropping water pollution from local and manufacturing sources, the pressures from extreme nutrient supplementation (Smith 2003) are mounting (Colburn et al. 1996). Ecosystem deprivation is produced by a number of interrelating issues (Dudgeon et al. 2006). It might contain direct influences on the aquatic habitat (such as the excavation of river sand) or indirect pressures that can be an outcome of deviations within the drainage basin (Cazzolla Gatti 2016a, b). For instance, dissertation is frequently linked with variations in surface runoff and augmented river sediment loads that can cause habitat modifications, such as blockage of river bottoms and alluvial aggradation (Dudgeon et al. 2006).

21.2.1.2 Flow Changes and Water Pollution

The changes in the drainages of water occurred everywhere on this globe (Dynesius and Nilsson 1994; Nilsson et al. 2005). Such changes may differ in harshness and kind, but incline to be further hostile in areas with extremely inconstant flow systems (Cazzolla Gatti 2016a, b). The divergence of waterways is due to human activities in order to protect communities from flooding or conservation in the amount of water for drinking and other usages. The results of the human interference in the natural waterways cause partial dryness of rivers in different parts of the world (Richter et al. 2010). In Iraq, such events are present as the dams built in Turkey, Syria and Iran have obstructed a huge amount of water that supposed to feed the Euphrates and Tigris Rivers, which in turn affect extensively the water level in the southern marshes located in south of Iraq. Freshwater flow fluctuations are expected to be worsened by global climate variation due to the augmented occurrence of floods and drought, and, therefore, an upsurge in water-engineering reactions can be expected (Vorosmarty et al. 2000; Kirpotin 2014). As a consequence of such changes in the water flow direction, the pressure on the freshwater biodiversity will be austere (Dudgeon et al. 2006).

Physical pollutants in rivers, lakes and other freshwater bodies include the presence of particles of soil originated from the landscape or washed from surface areas by flowing water (Free et al. 2014). On the other hand, chemical pollution originated from human activities can cause complicated conditions to the inhabitants of the freshwater habitats (Jürgens et al. 2013). For instance, any excess in the level of nutrients will lead to an ecological catastrophic event as it leads to huge plant growth. In addition, artificial organic compounds may affect the physiology of the aquatic animals living in the freshwater ecosystem, or if the concentration is high enough, morbidity of the fauna might be expected (Fu et al. 2013). Consequently, after the contaminants are taken by animals and plants, they pass on in the food chain to higher-level organisms such as predators and consumers. Numerous aquatic taxa were proposed to be used as bioindicators of pollution in the freshwater ecosystem (Bustos-Baez and Frid 2003; Lundberg et al. 2000; MacNeil et al. 2002). Such studies have recommended the success of using macrobenthic invertebrates as bioindicators of pollution (Azrina et al. 2006).

21.2.1.3 Susceptibility of Freshwater Biodiversity

Susceptibility is among the characteristic features of the freshwater system, and such nature is responsible for being as a settlement of the organisms in (Cazzolla Gatti 2016a, b; Davis et al. 2013). In some instances, the changes that occurred for a freshwater system can last for a long time; as an example for such event is what happened in China, where changes in some major river continued for 4000 years (Dudgeon 2000). In some cases, the changes that human activities were the causative agent can stay forever (Cazzolla Gatti 2016a, b).

Usually, freshwater system is exposed to strong human population's struggle that might lead in many occasions to armed conflicts especially those countries separated by waterways such as Iraq, Iran, Turkey and Syria (Poff et al. 2010). In water agreements between such countries, the biodiversity issues are usually overlooked (Poff et al. 2002). Such case can be seen in the area between China and India, where there are 55% of large dams in the world and no considerations were given for diversity of biota downstreams (Tharme 2003). The collective impacts and interfaces of the chief groups of pressures have caused a failure in population and a decrease in the variety of biota in the world (Hershkovitz et al. 2015). Looking at the numbers of taxa that have been affected due to such changes, we can see many freshwater vertebrates (for instance a loss of 19 mammals, 92 birds, 72 reptiles and 44 species of fish) (Cazzolla Gatti 2016a, b). Moreover, 32% of amphibian species – reliant on wet and water ecosystems – in the world are now endangered by death. A considerable percentage of vulnerable birds (12%) or mammals (23%) and 168 species may already be vanished (Amphibia Web 2005).

21.2.2 Global Threats to Freshwater Biodiversity

21.2.2.1 Invasions and Land-Use Change

In addition to the anthropogenic pressures in the form of physical and chemical effects, deliberate and undeliberate invasion of species has added difficulties to those already present against the biodiversity in the freshwater system as these species are more expected to occupy the newly altered or ruined freshwater (Bunn and

Arthington 2002; Koehn 2004). Several instances were reported on such events, e.g. Nile perch in Lake Victoria, the crayfish plague in Europe and salmonids in the southern hemisphere lakes and streams (Rahel 2002).

The urbanization of soils, as a changeover from natural, seminatural or agricultural usage to an urban one, is a factual 'consumption' of the soil, which yields a modification of the landscape. The development of urban zones and associated substructure is augmenting the impairment to the diversity of biota in the freshwater system. Arienzo et al. (2001) indicate that the pressure of land use and urban runoff on the pollution of the Sarno River basin in southern Italy has ruined the river water quality, particularly near the river mouth.

Among the other threats to the freshwater fauna and flora are the changes in the normal catchment land use for agriculture, industry or urban zones and the linked eutrophication of the water (Cazzolla Gatti 2016a, b). To stop freshwater biodiversity damage, it is vital to enumerate the associations between the diversity of freshwater biota and catchment land use and water nutrient concentrations (Weijters et al. 2009). Aquatic science desires to exchange rapidly into a controlling or investigational stage, related with renovation and measuring ecosystem reaction (Cazzolla Gatti 2016a, b).

21.2.2.2 Eutrophication and Acidification

Eutrophication is a natural event that occurs in lakes around the world (in lakes it occurs naturally with the increase in the available nutrients (Dodds and Smith 2016) and forms one of the chief problems for lake globally (Jeppesen et al. 2014)). This effect can be seen in the overgrown of algae, which disturb the natural equilibrium in the aquatic ecosystem (Zhang et al. 2016). Such changes in the lake ecosystem will lead to a decrease in the amount of the dissolved oxygen in the water, which in turn has lethal effects on the biota inhabiting the lake and upsurging the release of phosphates and nitrates from the sediments (Dunck et al. 2015).

The other important threat is the acidification of water, which resulted as an atmospheric precipitation with nitrogen oxides (from motor vehicles) and sulphur oxides (mainly from the combustion of petroleum products and coal) (Manca et al. 2016). The acid rain forms when these products combine with water vapour in the air and turn to rain affecting the freshwater system significantly (Cazzolla Gatti 2016a, b).

21.2.2.3 Climate Change

What is called 'the global centre of biodiversity' that includes unusual accumulation of endemic species, this spot can cause a drastic disaster if the habitats of this area are changed, due to the restricted movement of the species outside this area (Fialkowski et al. 2003). Among the changes that might affect these endemic species are the variation in the amount of rainfall and temperature (Davies 2010) as sometimes water temperatures can exceed the thermal tolerances of aquatic fauna (Portner and Knust 2007). Thus, the fauna cannot alter their dispersal as a reaction to cumulative temperatures. So, all alleviation reactions need be in situ to yield a biophysical cover appropriate for refining the flexibility of species (Davies 2010).

Climate changes have a significant effect on the freshwater ecosystem. For example, higher temperature can downgrade the quality of water and turn it unfavourable for the aquatic biota to live in (IPCC 2007). Due to the change in the water temperature, groups of phytoplankton may also change their composition, and in return some of the taxa will disappear, and accordingly some zooplankton taxa and other micro-invertebrates that feed on the phytoplankton also disappear. Such absence of some taxa from the freshwater ecosystem will create an imbalance in the aquatic equilibrium (Cazzolla Gatti 2016a, b).

Climate variation is likely to influence the lotic ecosystems in two chief ways: first, overdeviations in the water cycle and, second, via the linked modifications in terrestrial ecosystems within a given basin (Cazzolla Gatti 2016a, b). For inland wetlands, variations in rainfall configurations and floods in enormous zones of arid land will unfavourably disturb bird species that depend on a network of wetlands and lakes that are otherwise, or even sporadically, humid and fresh, and more dry and saline (Roshier and Rumbachs 2004).

Such effects might be augmented by habitat disintegration or disruption or damage of migration pathways, or even make alterations in other biota, such as bigger contact with the predatory birds (Butler and Vennesland 2000).

21.3 Conclusions

Freshwater environment maintains thousands of species, though their conservation and management are serious, and the biota diversity of this ecosystem is a valuable natural reserve. These habitats are suffering deteriorations in their biota more than those of the most impacted terrestrial environments, and the probability of preserving much of the enduring biodiversity in these environments is shown to be needed (Cazzolla Gatti 2016a, b).

These ecosystems are endangered by several human-made impacts like overfishing, water pollution, flow changes, destruction or degradation of habitat, invasion of exotic species, hydroelectric power stations and climate change (Cazzolla Gatti 2016a, b).

Climate variation is already affecting the capability of ecosystems to control water flows. Controlling the quality and quantity of water is a main ecosystem need globally. Rising temperatures, altering radiation and cloud coverage and the deprivation of ecosystem structure can cause the appearance of new and higher flow points, and also hamper the capacity of ecosystems to regulate the flow of water. This has significant results both for ecosystems, with linked species groupings, and people, in the scale of whole catchment zones (Cazzolla Gatti 2016a, b). The damage of the marshes owing to over-extraction of groundwater, drainage for human use, decreased overflow and reduced biodiversity is a result (Cazzolla Gatti 2016a, b).

Finally, the diversity of biota of the freshwater environment may be considered a vital constituent in the process of adaptation strategies for both droughts and floods through the management of river basins, wetlands, forests and agricultural systems (Cazzolla Gatti 2016a, b).

The short review given in this chapter is only a part of the major threats that the aquatic biota can face, in particular fish groups as they are the most vulnerable aquatic animals towards the changes in their environment. Therefore, these threats are put forward for those who will perform any conservation programme(s) on the freshwater biota in Iraq in the future as this study is considered only a baseline that the future studies can follow.

References

Amphibia Web (2005) Amphibia Web species numbers. Amphibia Web: information on amphibian biodiversity and conservation, Berkeley, U.S.A. Available online at: http://amphibiaweb.org/. Accessed 20 Mar 2016

Arienzo M, Adamo P, Rosaria Bianco MR, Violante P (2001) Impact of land use and urban runoff on the contamination of the Sarno river basin in southwestern Italy. Water Air Soil Pollut 131(1/4):349–366

Arthington AH, Naiman RJ, Mcclain ME, Nilsson C (2010) Preserving the biodiversity and ecological services of rivers: new challenges and research opportunities. Freshw Biol 55(1):1–16

Azrina MZ, Yap CK, Rahim Ismail AR, Ismail A, Tan SG (2006) Anthropogenic impacts on the distribution and biodiversity of benthic macroinvertebrates and water quality of the Langat River, Peninsular Malaysia. Ecotoxicol Environ Saf 64(3):337–347

Bunn SE, Arthington AH (2002) Basic principles and ecological consequences of altered flow regimes for aquatic biodiversity. Environ Manage 30(4):492–507

Bustos-Baez S, Frid C (2003) Using indicator species to assess the state of macrobenthic communities. Hydrobiologia 496(1–3):299–309

Butler RW, Vennesland RG (2000) Integrating climate change and predation risk with wading bird conservation research in North America. Waterbirds Int J Waterbird Biol 23:535–540

Cazzolla Gatti R (2016a) Biological diversity and current threats of lotic ecosystems. In: Pokrovsky OS (ed) Riparian zones: characteristics, management practices and ecological impacts. Nova Publisher, New York, pp 5–22

Cazzolla Gatti R (2016b) Freshwater biodiversity: a review of local and global threats. Int J Environ Stud *73*(6):887–904

Colburn T, Myers JP, Dumanoski P (1996) Hormonal sabotage. Nat Hist 81(3):45–46

Davies PM (2010) Climate change implications for river restoration in global biodiversity hotspots. Restor Ecol 18(3):261–268

Davis J, Pavlova A, Thompson R, Sunnucks P (2013) Evolutionary refugia and ecological refuges: key concepts for conserving Australian arid zone freshwater biodiversity under climate change. Glob Change Biol 19(7):1970–1984

Dodds W, Smith VH (2016) Nitrogen, phosphorus, and eutrophication in streams. Inland Waters 6(2):155–164

Dudgeon D (2000) The ecology of tropical Asian rivers and streams in relation to biodiversity conservation. Annu Rev Ecol Syst 31:239–263

Dudgeon D, Arthington AH, Gessner MO, Kawabata ZI, Knowler DJ, Lévêque C, Naiman RJ (2006) Freshwater biodiversity: importance, threats, status and conservation challenges. Biol Rev 81(02):163–182

Dunck B, Lima-Fernandes E, Cássio F, Cunha A, Rodrigues L, Pascoal C (2015) Responses of primary production, leaf litter decomposition and associated communities to stream eutrophication. Environ Pollut 202:32–40
Dynesius M, Nilsson C (1994) Fragmentation and flow regulation of river systems in the northern third of the world. Ecol Appl 1:182–195
Fialkowski W, Rainbow PS, Smith BD, Zmudzinski L (2003) Seasonal variation in trace metal concentrations in three talitrid amphipods from the Gulf of Gdansk, Poland. J Exp Mar Biol Ecol 288:81–93
Free CM, Jensen OP, Mason SA, Eriksen M, Williamson NJ, Boldgiv B (2014) High-levels of microplastic pollution in a large, remote, mountain lake. Mar Pollut Bull 85(1):156–163
Fu J, Hu X, Tao X, Yu H, Zhang X (2013) Risk and toxicity assessments of heavy metals in sediments and fishes from the Yangtze River and Taihu Lake, China. Chemosphere 93(9):1887–1895
Gleick PH (1996) Water resources. In: Schneider SH (ed) Encyclopaedia of climate and weather. Oxford University Press, New York, pp 817–823
Hershkovitz Y, Dahm V, Lorenz AW, Hering D (2015) A multi-trait approach for the identification and protection of European freshwater species that are potentially vulnerable to the impacts of climate change. Ecol Indic 50:150–160
Intergovernmental Panel on Climate Change (IPCC) (2007) Climate change 2007: synthesis report
Jeppesen E, Meerhoff M, Davidson TA, Trolle D, Sondergaar M, Lauridsen TL, Beklioglu M, Brucet S, Volta P, González-Bergonzoni I, Nielsen A (2014) Climate change impacts on lakes: an integrated ecological perspective based on a multi-faceted approach, with special focus on shallow lakes. J Limnol 73(1):88–111
Jürgens MD, Johnson AC, Jones KC, Hughes D, Lawlor AJ (2013) The presence of EU priority substances mercury, hexachlorobenzene, hexachlorobutadiene and PBDEs in wild fish from four English rivers. Sci Total Environ 461–462:441–452
Kirpotin SN (2014) Western Siberia in a changing climate. Int J Environ Stud 71(5):591–594
Koehn JD (2004) Carp (Cyprinus carpio) as a powerful invader in Australian waterways. Freshw Biol 49(7):882–894
Levêque C, Oberdorff T, Paugy D, Stiassny MLJ, Tedesco PA (2007) Global diversity of fish (Pisces) in freshwater. In: Freshwater animal diversity assessment. Springer, Dordrecht, pp 545–567
Lundberg JG, Kottelat M, Smith GR, Stiassny MLJ, Gill AC (2000) So many fishes, so little time: an overview of recent ichthyological discovery in continental waters. Ann MO Bot Gard 87:26–62
MacNeil C, Dick JT, Bigsby E, Elwood RW, Ian Montgomery WI, Gibbins CN, Kelly DW (2002) The validity of the Gammarus: Asellus ratio as an index of organic pollution: abiotic and biotic influences. Water Res 36(1):75–84
Manca MM, Bonacina C, Yan ND (2016) What have we learned about ecological recovery from liming interventions of acid lakes in Canada and Italy? J Limnol 75(s2):1–3
Meyer JL, Sale MJ, Mulholland PJ, Poff N (1999) Impacts of climate change on aquatic ecosystem functioning and health. J Am Water Resourc Assoc 35(6):1373–1386
Nilsson C, Reidy CA, Dynesius M, Revenga C (2005) Fragmentation and flow regulation of the world's large river systems. Science 308(5720):405–408
Poff NL, Brinson MM, Day JW (2002) Aquatic ecosystems and global climate change. Pew Centre on Global Climate Change, Arlington
Poff NL, Richter BD, Arthington AH, Bunn SE, Naiman RJ, Kendy E, Acreman M (2010) The ecological limits of hydrologic alteration (ELOHA): a new framework for developing regional environmental flow standards. Freshw Biol 55(1):147–170
Portner HO, Knust R (2007) Climate change affects marine fishes through the oxygen limitation of thermal tolerance. Science 315(5808):95–97
Rahel FJ (2002) Homogenization of freshwater faunas. Annu Rev Ecol Syst 33:291–315
Richter BD, Postel S, Revenga C, Scudder T, Lehner B, Churchill A, Chow M (2010) Lost in development's shadow: the downstream human consequences of dams. Water Altern 3(2):14–42

Roshier DA, Rumbachs RM (2004) Broad-scale mapping of temporary wetlands in arid Australia. J Arid Environ 56(2):249–263

Smith VH (2003) Eutrophication of freshwater and coastal marine ecosystems a global problem. Environ Sci Pollut Res 10(2):126–139

Stiassny MLJ (2002) Conservation of freshwater fish biodiversity: the knowledge impediment. Verhandlungen der Gesellschaft für Ichthyologie 3:7–18

Strayer DL, Dudgeon D (2010) Freshwater biodiversity conservation: recent progress and future challenges. J North Am Benthol Soc 29(1):344–358

Tharme RE (2003) A global perspective on environmental flow assessment: emerging trends in the development and application of environmental flow methodologies for rivers. River Res Appl 19(5–6):397–441

Vorosmarty CJ, Green P, Salisbury J, Lammers RB (2000) Global water resources: vulnerability from climate change and population growth. Science 289(5477):284–288

Weijters MJ, Janse JH, Alkemade R, Verhoeven JTA (2009) Quantifying the effect of catchment land use and water nutrient concentrations on freshwater river and stream biodiversity. Aquat Conserv Mar Freshw Ecosyst 19(1):104–112

Zhang Y, Liu X, Qin B, Shi K, Deng J, Zhou Y (2016) Aquatic vegetation in response to increased eutrophication and degraded light climate in Eastern Lake Taihu: implications for lake ecological restoration Sci Rep 6, Article number: 23867

Chapter 22
Biotic Homogenization: A Process That Is Happening in the Southern Marshes of Iraq

Laith A. Jawad

Abstract Biotic homogenization designates the course by which species incursions and exterminations upsurge the genetic, taxonomic, or functional similarity of two or more taxa over a definite time interlude. The subject of biotic homogenization is a young and speedily developing research area in the promising area of conservation biogeography, and this chapter aims to give an idea about the basic concepts of biotic or taxonomic homogenization. In the same time, this study sheds lights on the fish biotic homogenization that is occurring in the environment of the southern marshes of Iraq due to two factors, the effects of the invasive species and the environment changes that implement huge ecological changes in the studied area.

Keywords Biodiversity · South of Iraq · Taxonomy · Distribution · Environment · Impact

22.1 Introduction

Biotic homogenization is the steady upsurge of resemblance between regions which is a prevalent course that form the structure and function of biotic groups and is chiefly determined by the joint influences of species conquests and exterminations (Olden et al. 2004; Olden 2006; Winter et al. 2009; Villeger et al. 2011). This biotic homogenization resulted in an interface between several species to determine the settings for the endurance of a specific species and has turned out to be one of the serious concerns in conservation biology (McKinney and Lockwood 1999; Traveset and Richardson 2006; Berg et al. 2010). The failure or extermination of species in any ecosystem has an effect on other species. Such an influence is through the trophic levels, and in the end the local biodiversity will be impacted (Petchey et al. 2008; Dunn et al. 2009).

L. A. Jawad (✉)
School of Environmental and Animal Sciences, Unitec Institute of Technology, Mt Albert, Auckland, New Zealand

L. A. Jawad (ed.), *Southern Iraq's Marshes*, Coastal Research Library 36,
https://doi.org/10.1007/978-3-030-66238-7_22

In certain cases, some species are dependent on another species to continue their life; such species are particularly vulnerable by biotic homogenization (Douda et al. 2013). The capability of those species that need another species to assist in their lives to survive hangs mainly on the aptitude of the affiliate species to replace their former hosts with settlers (Moir et al. 2010). Therefore, the most endangered of these species are deliberated to be the species that have a restricted and specialized way of life and rely on a limited host species. Several instances on such a case can be drawn from the entomology-parasitology world (e.g., Dunn 2005).

Considerable information available on biotic homogenization has concentrated on enumerating how the species composition of separated zones has become more similar, though the course of homogenization encompasses through all levels of biological organization (Rahel 2002). For instance, environment homogenization has ensued in similar habitats across North America such as urban business districts, golf courses, canals, and warmwater reservoirs. In these man-made environment habitats, endemic species are substituted by diverse globalized species with the result that whole ecosystems similar to each other now happen in unlike parts of the country (Rahel 2002). Blair (2001) stated that bird and butterfly groups from urban areas in California and Ohio were more alike to each other than they were to the native populations they substituted. The same occurrence happens in urban lakes through North America that incline to be subjugated by a set of organisms that can face levels of degraded water quality, such as aquatic oligochaetes, common carp *Cyprinus carpio*, and goldfish *Carassius auratus*.

In this chapter and for the sake of introducing the subject of bio homogenization, certain topics will be discussed such as measurement of homogenization in the nature, what causes biotic homogenization, types of homogenization, ecological mechanisms of biotic homogenization, ecological and evolutionary values of biotic homogenization, and the magnitude and spatial level of biotic homogenization. At the end of the chapter, a section is allocated to discuss the possibility of the presence of biotic homogenization in southern marshes of Iraq.

22.2 The Measurement of Biotic Homogenization

Homogenization is known as an upsurge in the similarity of biotas over time, and a direct way to measure this rise is done by applying similarity indices (Rahel 2002). A generally used index is Jaccard's coefficient of similarity, calculated as percent similarity D [a = (a + b + c)] × 100 where a is the number of species present in both biotas, b the number of species present only in the first biota, and c the number of species present only in the second biota (Radomski and Goeman 1995; Marchetti and Moyle 2001). Values can range from 0%, which represents no similarity or taxa have no species in common, to 100%, which denote to complete similarity, i.e., taxa have indistinguishable species contents. A typical method for evaluating changes through time in homogenization is to estimate the similarity between a pair of biotas at two times, typically pre- and post-alteration by humans (Rahel 2002). If similarity is augmented, then the biotas have become more homogeneous. Rahel (2000) used this method to enumerate the modification in similarity of fish faunas among the 48 adjoining United States from pre-European settlement to the present.

Ecologists differentiate between α-diversity (the number of species in a specific habitat) and β-diversity (the turnover of species across habitats). A reduction in diversity indicates that homogenization has occurred. Spatial access or β-diversity can be measured by counting the number of species lost or gained as one moves from one site to another site along a spatial continuum (Russell 1999). The resultant value is then scaled to the size of the combined species pool at both sites. One formula for measuring β-diversity is $T = (G + L)/\alpha$ where T is the spatial turnover, G is the number of species found in the first site but not in the second, L is the number of species found in the second site but not the first, and α is the total number of species found within both. Duncan and Lockwood (2001) used this method to inspect the modification in spatial turnover of fish, amphibian, and mussel species through zoogeographic regions in Tennessee.

Cluster analysis and ordinations are multivariate methods exploited to evaluate similarity between groups of sampling sites. With the cluster analysis, sites with similar faunal groups are arranged into hierarchical groups displaying a gradual upsurge in their similarity. This method permits one to see interactions among all sites, not just one pair at a time. Blair (2001) utilized cluster analysis to show that the physical habitat and biotic groups of urban areas in California and Ohio were more alike than the original ecosystems present in these zones. Therefore, these areas have practiced homogenization of both habitats and biotas.

22.3 What Causes Biotic Homogenization

There are three routes that can lead to the biotic homogenization: introductions of non-native species, extirpation of native species, and habitat alterations that facilitate these two processes (Rahel 2002). While incursions have continuously been a natural phenomenon, they nowadays happen at a faster pace owing to man-made events (Vermeij 1991), grouping of a local zone due to the filters that decrease the regional species pooling to a subgroup of species that have had the chance to settle in a new habitat, are physiologically acclaimed to the abiotic conditions, and have the ecological features required to interrelate efficaciously with the other species present. With this model of community gathering, the first filter is embodied by glacial actions and biogeographic barriers that stop numerous species from settling a region (Rahel 2002). A few of these species are physiologically and ecologically matched to the region but have not had the chance to comprehend this possibility (Rahel 2002). A chief impact of humans is introducing species across obstructions for establishment and therefore to eradicate the biogeographic strainer as an issue in defining the species configuration of local populations (Brown and Moyle 1997; Rahel 2002).

Destructions also can upsurge or decline the similarity of biotas. An upsurge in similarity happens when each biota misses its exclusive species but holds prevalent species (Rahel 2002). Duncan and Lockwood (2001) claim that this situation will lead to homogenization later in the future for amphibian, fish, and mussel faunas in Tennessee. Their justification is based on that regions differ ecologically have a large number of exclusive species that are restricted with small population sizes and therefore extremely susceptible to extermination. Once these species become vanished in the future, only comparatively prevalent and abundant species will endure (Rahel 2002). Therefore the ecoregions will become more alike, even without introductions of new, cosmopolitan species. On the other hand, reduction in similarity owing to destruction could occur in case two taxa lose the species they have in common and hold their exclusive species (Rahel 2002).

Habitat modifications may cause an establishment of an alien species or destruction of native species (Marchetti and Moyle 2001). Under natural flow systems, high flows in winter and spring avert the establishing of alien species. But when flows are alleviated by dams, aliens dominate and relocate natives through competition and predation. The part of habitat in the process of alteration and causing destruction is obviously recognized for freshwater mussels. The elimination of many populations in the southeastern United States is accredited to reservoir construction that eradicated the riffle habitats mandatory for many species (Williams et al. 1993; Bogan 1993).

22.4 Types of Biotic Homogenization

Biotic homogenization is deliberated as a predominant ecological practice that includes the loss of genetic, taxonomic, or functional uniqueness (Olden et al. 2004). Genetic homogenization is known as an upsurge in the genetic resemblance of gene pools over time due to intra- and interspecific hybridization mainly with alien species. Such phenomenon can be demarcated in terms of the allelic configuration of a specific locus or set of loci (i.e., identity of genotypes) or their frequencies (i.e., relative abundance of genotypes) (Olden 2006). On the other hand, taxonomic homogenization denotes to an upsurge in the phylogenetic likeness of biota over time due to the formation of diverse species and extermination of endemic species (Olden 2006). Taxonomic homogenization has been the chief focus of preceding studies and is used conversely with the expression biotic homogenization through the literature. The functional homogenization is denoted as a rise in the functional likeness of biotas over time due to the instituting of species with similar "roles" in the ecosystem (i.e., species with several functional counterparts) and the diminish of species owning exclusive functional "roles." One method to evaluating functional resemblance is based on the occurrence or lack of species features or the frequency distribution of characters in the population (i.e., similarity in species' ecological niches) (Olden 2006). Functional homogenization has likenesses to the idea of ecological homogenization described by McKinney and Lockwood (1999).

22.5 The Ecological Mechanisms of Biotic Homogenization

Anthropogenic activities are the main source of biotic homogenization due to their role in upsurging the rates of alien species introduction and destruction of the native species through changing the habitat that they live in (Rahel 2002). It is important to understand the several ecological methods that are fundamental for the creation of the biotic homogenization procedure. Lately, Olden and Poff (2003) suggested a theoretical design showing a number of discrete mechanisms functioning in both aquatic and terrestrial habitats by which biotic homogenization or differentiation might happen in reaction to the explicit series of incursions and losses. They showed graphically how the incursion of a group of similar species and the extermination of exclusive species can be an outcome in varying levels of biotic homogenization (Rahel 2002). This representation stresses that the introduction of an international species will, by itself, upsurge homogenization, but this consequence is expanded if endemic species that subsidize to the exclusivity of the system are vanished (Olden 2006). This instance also shows that biotic homogenization or differentiation can happen with or without variations in species richness (Olden 2006).

22.6 Biotic Homogenization in the Southern Marshes of Iraq

There is a clear indication for the presence of biotic homogenization in the environment of the southern marshes of Iraq. These indications were built on the observations made on the fish fauna. Similar observations might also present for other animal groups such invertebrates, but they need to be recovered.

The biotic homogenization in fish fauna of the southern marshes is due to two factors, first, the invasive species and, second, the changes in the environment. For the first factor, several invasive species have been introduced in the lower reaches of Euphrates-Tigris Rivers; among these are the carp species, common carp, *Cyprinus carpio*, grass carp, *Ctenopharyngodon idella*, and the silver carp, *Hypophthalmichthys molitrix*. There are other invasive species, but the carp species are the most successful species that are established in the freshwater system of Iraq and brought about a wide range of changes to the native fish fauna more than any other invasive fish species. Such changes were through predation on eggs and larvae of native species, competition on food and niches, and disturbing the environment and turn it unfavorable to live by young of the native species. In such a way, all native fish species such as several cyprinid species like *Mesopotamichthys sharpeyi*, *Luciobarbus xanthopterus*, and *Arabibarbus grypus* will be affected. The abundance of these species has dropped down significantly in both the marsh area and Shatt al-Arab River due to the competition with the invasive carp species. In the early 1970s when there was no invasive carp species present in the freshwater system of

Iraq, the environment of both Shatt al-Arab River and the marsh areas was not taxonomically homogenized, i.e., the species composition is quite different, but 30 years later, biotic homogenization became clear, and the environment of both water bodies became homogenized in regard to their species contents. Such homogenization will continue in its effect until an equilibrium is reached when what remain of the native species can bear the mode of life that the invasive species follow.

The second factor, the changes in the environment, has also created biotic homogenization. The environment alterations through the rising of the salinity in the Shatt al-Arab River have caused more marine species to move upstream and reach the southern parts of the marsh area. With such changes, the marine species composition has changed in both Shatt al-Arab River and the southern marshes of Iraq, with more species have invaded these water bodies. At the moment, the marine species composition of both Shatt al-Arab River and the marsh area is taxonomically homogenized, but this homogenization is only in the southern part of the marsh that is close to the Shatt al-Arab River. In further northern part of the marshes, the presence of marine species is negligible.

The abovementioned homogenization created by the invasive species and changes in the environment can bring a multiplicity of ecological problems. With the absence of the native and the dominance of the invasive species, the invertebrate and flora of the ecosystem of the marsh area will be significantly affected, which in turn several species of both plants and animals will vanish from the area. On the other hand, the presence of more marine species in the marsh area might cause extra predation and competition added to that already present and created by the carp species. In addition, the marine species might bring with them new parasite's species that may be acclimatized in the marsh area and bring further impact to this ecosystem.

As an author, I aimed in this chapter to give a quick picture of one of the ecological problems that the ecosystem of the marsh areas in Iraq is experiencing in the present time. Therefore, the attention of those who plan for a future development in this area needs to be directed toward the information given in this chapter.

References

Berg MP, Kiers ET, Driessen GERA, Van Der Heijden MARC, Kooi BW, Kuenen FRAN, Liefting MAAR, Verhoef HA, Ellers JACI (2010) Adapt or disperse: understanding species persistence in a changing world. Glob Change Biol 16:587–598

Blair BB. 2001. Birds and butterflies along urban gradients in two ecoregions of the United States: is urbanization creating a homogeneous fauna? See Lockwood and McKinney 2001, pp 33–56

Bogan AE (1993) Freshwater bivalve extinctions (Mollusca: Unionoida): a search for causes. Am Zool 33:599–609

Brown LR, Moyle PB (1997) Invading species in the Eel River, California: successes, failures, and relationships with resident species. Environ Biol Fish 49:271–291

Douda K, Lopes-Lima M, Hinzmann M, Machado J, Varandas S, Teixeira A, Sousa R (2013) Biotic homogenization as a threat to native affiliate species: fish introductions dilute freshwater mussel's host resources. Divers Distrib *19*(8):933–942

Duncan JR, Lockwood JL (2001) Spatial homogenization of the aquatic fauna of Tennessee: extinction and invasion following land use change and habitat alteration. In: Biotic homogenization. Springer, Boston, MA, pp. 245–257
Dunn RR (2005) Modern insect extinctions, the neglected majority. Conserv Biol 19:1030–1036
Dunn RR, Harris NC, Colwell RK, Koh LP, Sodhi NS (2009) The sixth mass coextinction: are most endangered species parasites and mutualists? Proc R Soc B Biol Sci 276:3037–3045
Marchetti MP, Moyle PB (2001) Effects of flow regime on fish assemblages in a regulated California stream. Ecol Appl 11:530–539
McKinney ML, Lockwood JL (1999) Biotic homogenization: a few winners replacing many losers in the next mass extinction. Trends Ecol Evol 14:450–453
Moir ML, Vesk PA, Brennan KEC, Keith DA, Hughes L, McCarthy MA (2010) Current constraints and future directions in estimating coextinction. Conserv Biol 24:682–690
Olden JD (2006) Biotic homogenization: a new research agenda for conservation biogeography. J Biogeogr *33*(12):2027–2039
Olden JD, Poff NL (2003) Toward a mechanistic understanding and prediction of biotic homogenization. Am Nat **162**:442–460
Olden JD, Poff NL, Douglas MR, Douglas ME, Fausch KD (2004) Ecological and evolutionary consequences of biotic homogenization. Trends Ecol Evol 19:18–24
Petchey OL, Eklof A, Borrvall C, Ebenman B (2008) Trophically unique species are vulnerable to cascading extinction. Am Nat 171:568–579
Radomski PJ, Goeman TJ (1995) The homogenizing of Minnesota lake fish assemblages. Fisheries 20(7):20–23
Rahel FJ (2000) Homogenization of fish faunas across the United States. Science 288:854–856
Rahel FJ (2002) Homogenization of freshwater faunas. Annu Rev Ecolo Syst *33*(1):291–315
Russell GJ (1999) Turnover dynamics across ecological and geological scales. In: McKinney ML, Drake JA (eds) Biodiversity dynamics: turnover of populations, taxa, and communities. Columbia University Press, New York, pp 377–404
Traveset A, Richardson D (2006) Biological invasions as disruptors of plant reproductive mutualisms. Trends Ecol Evol 21:208–216
Vermeij GJ (1991) When biotas meet: understanding biotic interchange. Science 253:1099–1104
Villeger S, Blanchet S, Beauchard O, Oberdorff T, Brosse S (2011) Homogenization patterns of the world's freshwater fish faunas. Proc Natl Acad Sci USA 108:18003–18008
Williams JD, Warren ML Jr, Cummings KS, Harris JL, Neves RJ (1993) Conservation status of freshwater mussels of the United States and Canada. Fisheries 18(9):6–22
Winter M, Schweiger O, Klotz S, Nentwig W, Andriopoulos P, Arianoutsou M, Basnou C, Delipetrou P, Didziulis V, Hejda M, Hulme PE, Lambdon PW, Pergl J, Pysek P, Roy DB, Kuhn I (2009) Plant extinctions and introductions lead to phylogenetic and taxonomic homogenization of the European flora. Proc Natl Acad Sci USA 106:21721–21725

Section VI
Natural Resources

Chapter 23
Fish, Fishing Methods and Fisheries of the Southern Marshes of Iraq

Laith A. Jawad

Abstract Iraq is one of the fortunate countries in having fisheries activities in three different environments, rivers, wetlands and coastal areas. In each of these habitats, the fishing methods and gear are different owing to the fish species targeted, the ethnicity of the fishers and the topography of the area. In the southern marshes of Iraq, both fishing methods and gear are different from those used in rivers and coastal areas. Some of these gears and methods are ancient and could be used by the ancestors of the Marsh Arabs. In this chapter, a short review about the fishing methods and gear used in the southern Mesopotamia is given. Such review will give an idea about how fish catching is going on in this part of the and for the researchers to compare with the other areas in the future.

Keywords Nets · Traps · Marsh Arabs · Fishing gear · Spears

23.1 Introduction

The southern Iraq perceived the development of one of the first urban civilization of mankind, Sumer. This civilization has developed the first trading system with the other parts of the world at that time, the ancient Meluhha (India), Magan (Oman) and Dilmun (Bahrain). The Sumerians were clever enough to take advantage of their most important asset, their strategic position, and try to trade with others outside their area. During the third millennia b.c., the Gulf coast was much closer to the city of Ur, one of the most famous archaeological sites of that period, which enabled its development as a major trading harbour. In exchange for the raw materials, the Sumerians exported manufactured high-quality product always attractive for Sumerians' trading partners.

Iraq's southern marshlands provide important bird habitat and a flyway between Siberia and Africa, served as spawning grounds for different fish species including

L. A. Jawad (✉)
School of Environmental and Animal Sciences, Unitec Institute of Technology, Mt Albert, Auckland, New Zealand

L. A. Jawad (ed.), *Southern Iraq's Marshes*, Coastal Research Library 36,
https://doi.org/10.1007/978-3-030-66238-7_23

Fig. 23.1 Icons of the southern marshes of Iraq. (**a**) Al-Mudhif, (guest house), external view; (**b**) Al-Mudhif (guest house), internal view; (**c**) one of the water channels in the marshes, with water buffalo swimming near the bank; (**d**) floating reed island; (**e**) old fisherman in his boat; (**f**) familiar view of swimming herd of water buffaloes. (Image courtesy of Laith Jawad)

some marine species such as the clupeid *Tenualosa ilisha* and a nursery for panaeid shrimp (*Metapenaeus affinis*). They are also an important natural filter for polluted water from the Tigris and Euphrates Rivers before they entered the Arabian Gulf.

The marshlands of southern Iraq have served humans for more than 6000 years in being the land of agriculture, economics and livelihoods. The people who live in those marshes are known as "The Marsh Arabs". They are living in arched reed houses and depend on water buffalo with rice, barley, fish and birds to keep living. In doing so, the Marsh Arabs have preserved for centuries a unique lifestyle that separate them from any human populations in the Middle East (Fig. 23.1).

The marshes situated mainly within southern Iraq and a portion of southwestern Iran. Originally covering an area of 20,000 km^2 (7700 sq mi) and divided into three major areas, the Central Marshes lie between the Tigris and Euphrates, while the Hammar Marshes lie south of the Euphrates, and the Hawizeh Marshes are bound east of the Tigris. The principal Marshes are divided from each other by the Euphrates and Tigris Rivers: The Hammar Marshes are located south of the Euphrates River, and to the west of its confluence with the Tigris. The Central Marshes are

Fig. 23.2 Map showing the locations of the marshes in Iraq both before and after draining

situated north of the Euphrates River and west of the Tigris River, and the Al-Hawizeh Marshes lie east of the Tigris River, overlapping the border with Iran. The marshlands are formed by permanent and seasonal shallow and deep water lakes, as well as mudflats that are regularly inundated during periods of seasonal flooding (Fig. 23.2).

23.2 The Geological History of the Southern Marshes of Iraq

The huge marshes of southern Iraq are a depression formed tectonically as a result of the Arabian plate being subducted under the Iranian or Eurasian plate (Baeteman et al. 2005). The original process involved natural and physical changes in the climate and sea level, before and during the Holocene period, which prepared for wetland formation in several stages. Initially, during the Last Glacial Maximum (16,000 b.c.e.), the marshes were absent because the groundwater was too low and the Arabian Gulf was quite shallow (Sanlaville 2002). Then, during the Postglacial (12,000 b.c.e.) the sea level rose, bringing the coastline further inland. Elevation range is considered an additional important physical factor in the process of marsh formation. Marshes are formed in the deepest and lowest areas of the large and flat lower Mesopotamian plain. The studies have shown that by the sixth to the eighth centuries c.e., the sedimentation from the rivers was greater than the rate of sea-level rise, and thus the coasts started to prograde farther, the coastal *sabkhas* were replaced by floodplain, and freshwater marshes are formed in the river estuaries (Sanlaville 2002). By the seventh century c.e., the process of marsh formation had alleviated into an enormous wetland (370 × 90 km).

The southern reaches of Mesopotamia were always inclined to forming wetlands. This process of marsh formation has been considered as a complex natural process involving changes to hydrology and sedimentation beyond simply tectonic subsidence and marine invasion (Eger 2011).

23.3 The Dilemma of the Marshes: The Drainage

The most vital threat to the marshes has been the drainage and diversion of water supplies for agriculture and oil exploration and production. This threat was worsened after the first Gulf War (1980–1988). The failed uprising of Shiite Muslims, who used the abundant reeds and maze of waterways as a way to hide out, led the former regime (regime of Saddam Hussein) to drain the marshes intentionally through construction of dykes to remove the threat of further rebellion. The rebellion was crushed, and the drainage further accelerated to cover the Central and Hammar marshes in order to evict Shias that have taken refuge in the marshes. The daily life of fishing and rice production was substituted with dry agriculture (Lawler 2002). This intentional obliteration had an overwhelming impact on the ecosystem, threatening species and the Marsh Arabs, who were forced to leave their culture as their environment was destroyed. The full level of the devastation became clear when the UNEP released satellite images in 2001 showing that 90% of the marshlands had been lost (UNEP 2009) (Fig. 23.2). The indigenous Marsh Arabs became either internally displaced persons or environmental refugees expatriate to other countries (UNEP 2001). Accordingly, the population of the Marshes was reduced from nearly 500,000 people in the 1950s to as few as 20,000 by 2003 (Nicholson and Clark 2002) (Fig. 23.3).

With the fall of Saddam Hussein regime in 2003, the initiative of re-flooding the marshes was taken by the locals, and some Marsh Arab tribes returned to retrieve

Fig. 23.3 Two devastating locations in the marshes due to the draining crime. (Image courtesy of Laith Jawad)

their lives. The substantial restoration efforts in 2004–2008 resulted in successful rehydration of approximately 58% of the Marshes (UNEP 2005; Al-Handel and Hu 2015). Some of the marsh areas were very difficult to restore due to the high soil and water salinity (Richardson et al. 2005; Richardson and Hussain 2006).

23.4 Ichthyofauna of the Marshes

The marshes, in general, play a key role not only for ichthyofauna of the Tigris-Euphrates basin but also for numerous diadromous species of the Arabian Gulf. The ichthyofauna of the marshes supports local fisheries, and hence forms part of the resource base of the unique lifestyle of the marsh inhabitants. It is therefore a requirement for the maintenance of the outstanding universal value of the property under World Heritage.

In general, the ichthyofauna of the marsh areas is similar in species composition to that of the Tigris-Euphrates river system. The marine fish fauna of Iraq has been investigated from the taxonomic point of view over the last 50 years (Khalaf 1961; Mahdi 1962; Banister 1980; Al-Hassan and Miller 1987; Coad 1991, 2010; Ziyadi et al. 2018). Coad (2010) studied the freshwater of Iraq and showed it comprises 52 species in 11 families, dominated by the Cyprinidae with 37 species. He also showed that there are 44 native and 13 exotic freshwater species. Recently, several freshwater fish species were recorded from the Tigris-Euphrates river system (Jawad 2012); this demonstrates that the exploration of Iraq's ichthyofauna is still ongoing and may yield further species in the future.

There are 14 species considered endemic to the Tigris-Euphrates basin (Garstecki and Amr 2011); these are *Luciobarbus esocinus*, *Luciobarbus kosswigi*, *Mesopotamichthys sharpeyi*, *Luciobarbus subquincunciatus*, *Luciobarbus xanthopterus*, *Caecocypris basimi*, *Cyprinion kais*, *Hemigrammocapoeta elegans*, *Typhlogarra widdowsoni* (family, Cyprinidae), *Barbatula frenata* (family, Balitoridae), *Glyptothorax kurdistanicus* (family, Sisoridae), *Glyptothorax steindachneri*, *Silurus triostegus* (family, Siluridae) and *Aphanius mesopotamicus* (family, Cyprinodontidae) (Fig. 23.4a). It is of interest that the majority of the freshwater fish species belong to the family Cyprinidae and in particular to the genera *Barbus* and *Luciobarbus*, which are economically important. The current distribution, trends, conservation status and threats of the endemic fish species of the Marshes need to be studied for conservation and sustainable fisheries management planning. The current status of endemic ichthyofauna is also crucial for the assessment of the integrity of the marsh ecosystem as a whole.

Among the important freshwater fish species living in the marshes are *Carasobarbus luteus*, *Cyprinus carpio*, *Ctenopharyngodon idella*, *Luciobarbus pectoralis*, *L. xanthopterus*, *Mesopotamichthys sharpeyi* and *Arabibarbus grypus* (Fig. 23.4b, c).

Fish species of the marine origin ascend Shatt al-Arab River and have been recorded as far inland as Hammar Marsh. Some used to travel up the Tigris and Euphrates Rivers to varying degrees, but dams and water diversion schemes now prevent more extensive movements (Coad 2010). A total of marine species have been listed for the Marshes such as *Tenualosa ilisha* (Clupeidae); *Thryssa hamiltonii*; *Thryssa whiteheadi* (Engraulidae); *Carcharhinus leucas* (Carcharhinidae); *Netuma bilineatus* (Ariidae); *Liza klunzingeri*; *Liza subviridis* (Mugilidae); *Acanthopagrus berda*; *Acanthopagrus latus*; *Sparidentex hasta*

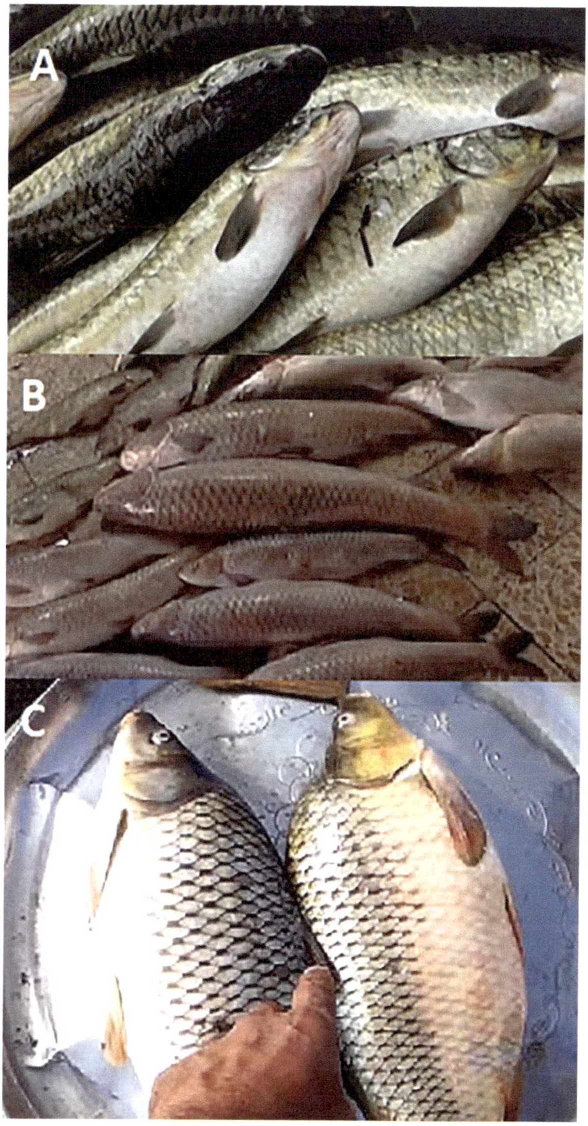

Fig. 23.4 (**a**) Endemic fish *Mesopotamichthys sharpeyi* to the Tigris-Euphrates system; (**b**, **c**) Common fish species, *Arabibarbus grypus* and *Cyprinus carpio*, respectively. (Image courtesy of Laith Jawad)

(Sparidae); *Johnius belangerii*; *Otolithes ruber* (Sciaenidae); *Pampus argenteus*; *Pampus chinensis* (Stromateidae); and *Brachirus orientalis* (Soleidae). Of these species, only the bull shark *Carcharhinus leucas* is listed as near-threatened on the IUCN Red List of Threatened Species (IUCN 2010), but the importance of the Marshes for diadromous species nevertheless adds to their overall biodiversity importance.

23.5 Fishing Gears and Methods of the Southern Marshes of Iraq

Fishing gear is conveniently divided into two broad categories based on the method of capture: active gear that is propelled or towed in chase of the target species or passive gear, which targets species that move into or towards (Jennings et al. 2001). There is a variety of different net models, each developed to catch particular species or for fishing in a particular environment (Von Brandt 1984; Sainsbury 1996). More specifically, Botros (1968) divided fishing gear into three main groups, i.e. entangling, luring and pursuit. Both active and passive primitive gears and methods are used to catch fish in the lower plain of Mesopotamia, where the native people of the marshlands follow their own fishing methods that they invent (Al-Khait 1978). Differences in the fishing areas within the lower Mesopotamian plain will directly affect the type of fishing gears and methods to be used. Sometimes different gears and methods were used at the same place due to such factors as the behaviour of target fish and the nature of the bottom. People in lower Mesopotamia fish in all seasons, but fishing becomes very heavy from March to May when there is internal migration of a member of the carp family (Cyprinidae) towards the marsh area for reproduction (Al-Hamed 1966). This phenomenon is locally known as "Zara" (Arabic local dialect for schooling).

The Marsh Arabs used the usual fishing gear and methods in addition to some methods that are unique to them and never being even used in other parts of Iraq. The following are the main types of fishing gear and methods used by those indigenous people.

I. Nets

1. Surface gill net (Fig. 23.5a, b): The simplicity of construction and operation makes them one of the most basic and widespread methods for fishing in the inland waters and the preferred method in most lakes. In addition, gill nets are vulnerable to poaching, and operators frequently spend the night watching over their gear (Welcomme 2001). Gill nets are among the most selective fishing gear with respect to the size range of target species captured.

   Fishermen in the marsh areas use gill net, and the usual net size used in the marshland is 180 × 270 ft. Headline floats are made of pieces of date palm stems, while the footrope is weighted with small stones rather than lead. Different construction methods are used for nets with different mesh sizes. The common legal gill net is called "Suba'e" (Suba'a: Arabic for seven), where there are seven mesh holes to the foot. The illegal gill net is called "Hedash" (Heda'ash: Arabic for 11) where there are 11 mesh holes to the foot. The latter are considered illegal due to their small mesh size, which can capture the undersized fish. The locals in the shallow marsh areas characteristically direct fish towards the net by using noise generated by hitting empty tins to lead fish towards the nets.

Fig. 23.5 Surface gill net, (**a**) net lifted with fish catch; (**b**) releasing fish from the net; (**c**) fish entangled with net; (**d**) pulling gill net. (Image courtesy of Laith Jawad)

2. Seine net (Fig. 23.5c, d): Seine nets are encircling nets of various types. The usual size of this net used in the marshlands ranges between 400 and 1000 metres long and 6 and 7 metres high. The full efficiency of this net can yield 500–2500 kg of fish (Al-Khait 1978). There are usually 20–50 pockets connected to the main panel of the net. The local people of the marshlands anchor one end of the net to the shore while the other end is taken away by small boat, paying out the net in a circle, eventually returning to pick up the anchored end before hauling. The fish captured using this technique are usually landed in excellent condition, because they spend little time in the purse, and command a high price at market. The effectiveness of fishing is partly due to the movement of the warps across the bottom, which

disturbs and guides the fish within the area being worked. Several species of fish are energetic jumpers and may escape by leaping over the net as it is closed. Sometimes illegal small-meshed nets are used in the area.

There are several types of seine net used in the inland water bodies in southern Mesopotamia, all depending on the size of the mesh.

3. Cast net (Fig. 23.6a, b): The cast net is circular, measuring from 5 to 6 metres in diameter, with bars of lead attached to the edge, and is used by a single fisherman. The cheapness and transportability make cast nets one of the most common gears in inland water fisheries. There are over 20 brail lines attached to a 9-metre hand line made of sash cord, which causes the net to pouch or bag around the trapped fish with the help of the lead at the edge of the net. One end of the hand line is tied to the wrist to leave both hands free to throw the net. The fisherman stands either in shallow water or in a boat and throws the net forcefully out onto the water, where it lands like a parachute and sinks to the bottom. Cast nets are selective for lower size ranges, and larger, faster-moving fish can escape the falling net but may become entangled in the process (Welcomme 2001). The catching efficiency may be improved by the addition of pockets at the circumference. The cast net for big fish has a larger mesh and heavier sinkers. The marshlands' native people usually retrieve the catch by taking the net up into the boat with the catch inside. There are some illegal versions of this net where a small mesh size is used.

II. Pot Trap (Fig. 23.7a)

Pot traps are among the most primitive of fishing implements and have remained little changed. Generally, traps take advantage of the movements of fishes along a tidal gradient or migration route. This type of trap is basically designed to catch crustaceans, but the native people at the marshlands found it good for catching fish. Most pots are similar in design, comprising a ridged frame usually made of palm leaves, with a mesh covering in which there is a single entrance. The entrance is fitted with baffles to prevent animals from escaping. Pot traps are mainly used at the time of breeding migration. Pots are usually deployed randomly with both ends anchored and marked by a surface float of date palm bark. Most local fishers use "dough" or a piece of unwanted fish as bait. Pots tend to be set for longer than other gear, as it takes time for the bait within the pot to begin to attract the target species. Catch rate increases over several days as the feeding activities of animals consuming the bait increase the dispersion of attractant odours. Pot buoys characteristically bear the owner's mark, and individuals from different Marsh Arab tribes generally respect pots of the other tribes and do not poach them. This behaviour arose from their religious and cultural origin. Sometimes, fishermen write some religious verses for those people who can read and signs for those who cannot tell the poacher that God will punish any poacher trying to steal the catch.

III. Spears (Fig. 23.7b)

The use of spears to catch fish is among the most ancient forms of hunting, and they are still used to the present day. Sumerians and Babylonians used spears of various designs to catch species of barbel in the rivers and marshlands of southern

Fig. 23.6 (**a**) fisherman preparing to through cast net; (**b**) fisherman throwing cast net; (**c**) cast landing on the surface of the water; (**d**) cast net pulled full of fish. (Image courtesy of Laith Jawad)

Mesopotamia (Saggs 1987). Similar spears are in use by modern Ma'dan tribes (Marsh Arabs) (Hadid and Al-Mahdawi 1977).

The type of spear usually used in the marshlands is generally comprised of a 2-metre handle, usually of hard palm tree wood, tipped with single- or multiple-pointed heads. The spear is used from a standing position in a small canoe. Most of

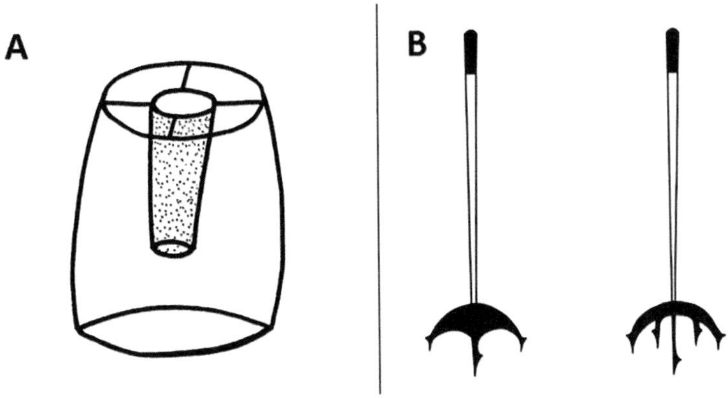

Fig. 23.7 (**a**) Pot trap; (**b**) two types of spears. (Image courtesy of Jawad 2006)

the time, the fishermen throw some food items in the water to attract fish. This fishing method is used mainly at night where fishermen take with them an oil lantern to attract fish to the surface of the water. Spearfishing is a quick fishing method, but it is dangerous from a hygiene standpoint, as the spearhead is seldom clean and may contaminate the catch. This becomes evident when the longer the catch remains before eating or selling at the local market. Marshlands fishermen often sell their spearfishing catch at the nearest city fish markets, which are often overseen by regional health and safety officials.

IV. Hook and Line (Fig. 23.8a)

The ancient Mesopotamian people used this type of fishing method; different types of hooks have been discovered during the excavations in both Sumer and Babylonia (Sahrhage 1999). They are also mentioned in the writings of the ancient Mesopotamians (Contenau 1954). At the present time, Mad'an tribes living in the marshlands of southern Iraq use similar types of hooks to catch fish. Instead of lures, the local people use various baits such as meat, small fish and dough in their hook and line fishing gear. Movement of the line when fish are eating the bait is indicated by a piece of date palm bark connected to the line and floating on the surface of the water.

Primitive methods of catching fish are usually practised by Marsh Arabs at the centre of the marsh area and away from the cities (Al-Khait 1978). These are:

A. Al-Shiah (mud dams) (Fig. 23.8b): In this very primitive fishing method, the fisherman makes a small mud dam across an enclosed water area and covers its entrance with a net made of cotton threads. The hole in the net is designed to let the fish in only. On the top of the net, near the surface, the fisherman usually fastens some empty tins so that they make noise when the net moves as the fish enter. Then the fisherman jumps in the enclosed area and catches fish by hand or encircles them with a piece of cloth.
B. Al-Suwaise (burning method) (Fig. 23.9a, b): A small reed island is usually selected by the fisherman, encircles it with a net and then sets fire to the reeds.

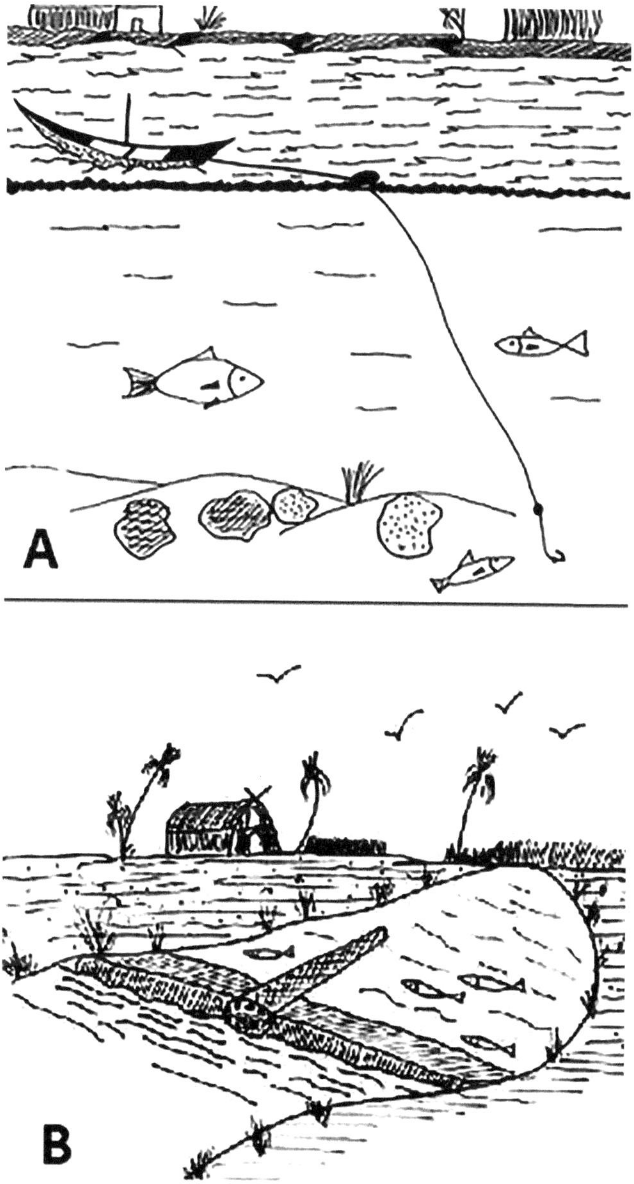

Fig. 23.8 (**a**) Line and hook; (**b**) Al-Shiah (mud dams) fishing method. (Image courtesy of Jawad 2006)

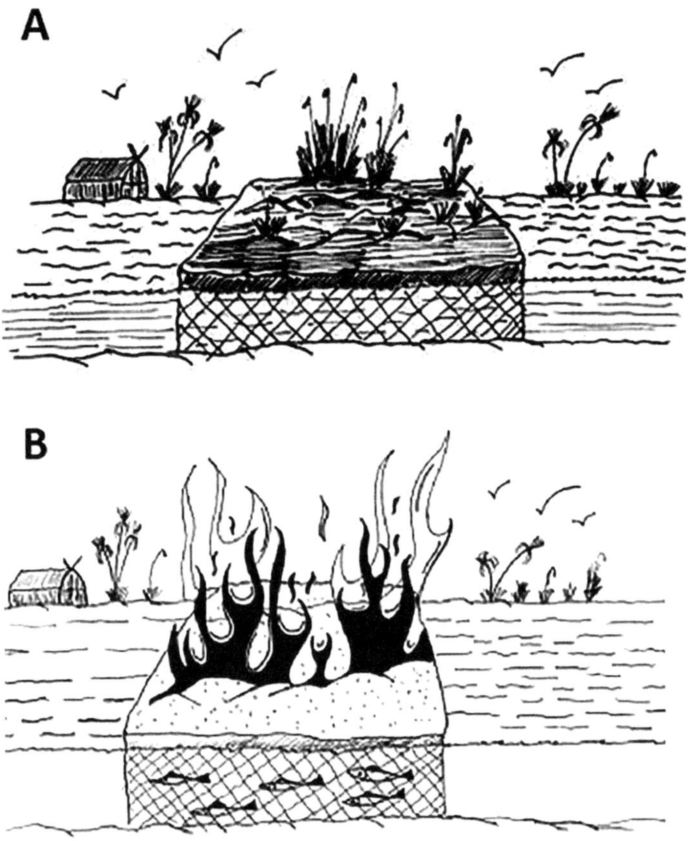

Fig. 23.9 Al-Suwaise (burning fishing method); (**a**) encircling a floating reed island with net; (**b**) burning the reed to push fish toward the net. (Image courtesy of Jawad 2006)

Fish will move away from the shore of the island towards the net. In this way they got entangled and they will be easily caught and removed.

C. Al-Tawamees (fishing by diving) (Fig. 23.10a): Small reed floating islands are selected for this type of fishing. More than one fisherman is involved in this operation. A hole is dug in the floor of the reed island by one of the fishermen while the other ties a rope to his ankle. He then enters the water through the hole, leaving the other fisherman to hold the rope. This is used as a guide to the diver to surface. The fish are caught by hand one at a time. The diver paralyses the fish by bending, breaking its backbone.

D. Al-Zahar (fishing with poison): The marshland Arabs have a very thorough knowledge of the nature of the aquatic plants living in their areas. They knew them by local names given by their ancestors over the long history of the marshlands, the names relating to the physiological action of the individual plant species. They also knew what could be extracted from each plant and for

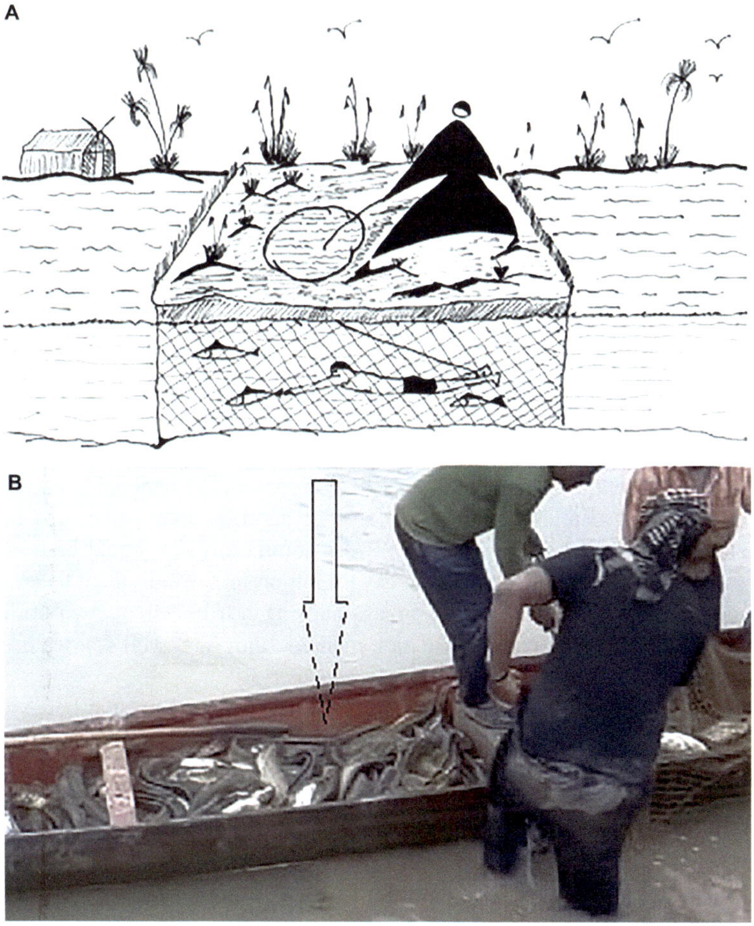

Fig. 23.10 (**a**) Al-Tawamees fishing method (fishing by diving) (Image courtesy of Jawad 2006); (**b**) separating catfish Silurus triostegus specimens from the rest of the catch (Image courtesy of Laith Jawad)

what purpose it could be used. The common name of the plant that they usually used as a poison to kill fish is "Neem" (*Azadirachta indica*, family: Meliaceae). They usually crush the plants and mix them with dough and spread it on the surface of an enclosed water area. Fish of different sizes and types will be affected by the poison and float to the surface of the water killed or temporarily anaesthetized. Most poisons affect the gills of the fish and the flesh is generally safe to eat, although when pesticides are used residues may accumulate in the fish flesh to toxic levels. The number and size of the fish depend on the concentration of the chemical present in the plant. The catch from this method of fishing usually fetch higher prices because the flesh is undamaged and very difficult for the non-expert people to tell whether the fish had been poisoned or not. Marshland fish caught with chemicals are easily recognized by their reddened eyes and swollen abdomens. Sometimes fishermen agitate the mud at the bottom of the marsh to liberate hydrogen sulphide gas, which suffocates fish and makes it easy to catch them by hand or by encircling them with net or a piece of cloth.

23.6 Aspects of the Fisheries in the Southern Marshes of Iraq

Motivation, behaviour and attitude of fishermen are always important when the assessment and the management of fishing are addressed, although social aspects of fishing management are often neglected. Fishermen usually rank the desirability of prey according to energetic or nutritional value, and foraging decisions will be based on a cost-benefit balance in which the currency of the benefits is energetic or nutritional reward (Emlen 1966). On the other hand, aquatic organisms became valuable as a means of earning money or as a commodity that could be traded for other goods. Thus, in the marshlands, it is very important for individual fishermen to know the boundaries of their fishing areas, since each tribe has its own traditional area. Although area boundaries are not marked, they are well known to local fishermen.

During the transition from subsistence to a market economy, the behaviour of fishers often changes because they realize greater overall benefits from selling fish than eating them. This case is quite clear in the villages distributed throughout the marshlands. Here all of the fish caught by villagers close to the fish markets of big cities are usually sold, but for villages located far away from major centres and not within easy range of a fish market, only about 10% of the catch is sold. The former villages use the income from fishing to pay for new motors for their boats, for fuel and for village improvement projects and to upgrade their lifestyle, while the latter villages have a much smaller income that pays for little more than some new boats and fuel (Al-Khait 1978).

The behaviour of fishermen within their own fishing grounds is influenced by the relative importance of the relevant market economy, which also explains differences in behaviour between fishermen on separate fishing grounds. Within each fishing area, there is a tendency to catch species of higher value by those fishermen who plan to sell their catch. In both the marshlands and the marine waters of lower Mesopotamia, areas with high-value species are usually owned by the stronger tribe (in the case of the marshlands) or exploited by companies that use better fishing vessels (Al-Khait 1978). Moreover, even though there are no major differences in the structure of the fish communities between grounds, fishermen in the proximity of the main city centres tend to target high-value species, while rural villagers are still targeting many species that they fish traditionally rather than dramatically changing their fishing behaviour to suit the whims of the fish markets (Jennings and Paulin 1996).

Throughout history, man worshipped the water. Ancient Mesopotamians considered water as the means of their existence and the eternal source of their food and even worshipped water gods such as Enki, god of the sweet water (Parrot 1961). Not surprisingly, fishing communities are often closely linked to religion, and this in turn affects their fishing behaviour.

Religion also influences fishing in the marshlands of southern Mesopotamia. Here all the fishermen are Shiite Muslim, a sector of Islamic religion that forbids eating fish without scales. Thus, all species of catfish, *Silurus triostegus*, *Mystus pelusius* and *Heteropneustes fossilis*, and the freshwater eel, *Mastacembelus halepensis*, found in the marshlands (Al-Hassan et al. 1989) are not included among the targeted species (Fig. 23.10b). Any of these species that do get caught are usually thrown back again into the water or left on the riverbank. Large catches of these nontarget species will thus lead to huge losses in income. By-catch may also include several species of freshwater crabs, which are also forbidden as food items by religion and thus represent an additional income loss.

Islamic beliefs play an important role in the timing of fishing activities in the lower Mesopotamian plain. For example, there must be no fishing on Friday or at the end of the Ramadan celebration (3 days) or during the Big Eid (4 days). In addition, during the month of Ramadan, hours of daily fishing are usually reduced due to the religious fasting tradition. Fishermen in the marshlands do not fish on certain days of the Arabic month of Muharram (first month in the Arabic lunar year) that commemorates the death of the Shiite Imam (religious leader) Hussain, grandson of Prophet Mohammad. The implications of such factors will be in favour of the health of the fisheries. Any damage might have occurred to the fish stock will be restored through the period of non-fishing.

23.7 Fisheries Communities' Conflicts and Solutions

Many fishing areas are viewed as common property and some are entirely open-access.

As a result, there is little to be gained by one fisherman trying to conserve fish because someone else will simply catch the fish. In other words, the race to fish occurs because it is better to catch a fish today than to leave it in the water. The consequences of the race to fish have been particularly dramatic in the marshlands where management systems and resources have collapsed. For many poor Marsh Arabs, and those living at the edges of the marshlands, the marsh area is the only potential source of food and income. Due to the activities of road building that lead to easy access to the marsh area, increasing numbers of people have entered the fishery of the area, which further contributes to the decline in fish catch. The newcomers have initiated a wholesale destruction of the resource base to maintain their livelihood. This may involve fishing with explosives and poisons that damage the ecosystem, killing most of the nontarget species and compromising any possibility of sustaining yields in the future.

Competition between fishers using different methods is clearly occurring in certain areas that are permanently closed to certain fishing practices. The race for using different fishing gear has led to a potential conflict that is not yet settled, and there is close competition between strong tribe-supported fishermen using preferred gear. In the marshlands, fishermen belonging to some strong local tribes usually control a fishing area and do not let fishermen from other tribes operate in their areas (Al-Khait 1978). Most of the time, such fishermen consistently use one type of gear that is suitable for catching target species. This selectivity in fishing gear will cause a rise in the stock of other fish species not caught by the same method. To solve this problem, a partition of resources between interested parties needs to be implemented. This can be achieved when the tribes supporting fishermen sit and discuss these problems together (such discussion usually took place in the *Mudhifs*, the large guest houses). The negotiation always ends when the weak tribe conforms to the ways of the strong tribe.

23.8 Fish Wars

Fish might be behind several undocumented clashes between strong tribes in the marshlands, who used whatever weapons were available in order to get control over rich fishing areas. Strong well-supported tribes usually win the conflicts and spread their fishing rules in the area. The distribution of fish played such a role throughout history and is still doing so (Kurlansky 1997).

23.9 Traditional Management Systems

Traditional systems management focuses on resolving gear use or allocation problems.

Access control is enforced by fishermen and by local moral and political authority. Elements of contemporary strategies applied to fishermen who violate them are the major issues of many traditional systems applied in the marsh area by tribes. Supernatural sanctions are probably the most effective punishment for poachers. As Shiite Muslims, the Marsh Arabs have among them a number of individuals termed "Syeds" (the religious masters), who claim descent from the Prophet Mohammad. They usually wear green scarves on their heads, the colour standing for peace for Muslims through the centuries. Therefore, any fishing area put under their names will be guarded by their supernatural power, and the poachers will face damage or disappearance of gear, or the threat of sickness or death for them or for their families. Clearly, illegal fishing was unlikely to be rewarding when it was against Islamic teaching.

In the lower Mesopotamian plain, however, the colonial intervention gave more power to tribes and more support to the traditional management systems. The colonial intervention achieved two aims: first, they can control the whole area by monitoring the heads of tribes; second, they do not need to involve any personnel from their side that might get killed in an unexpected conflict with the Marsh Arabs. Moreover, the colonial forces introduced what is known as customary freshwater tenure systems where certain families of a tribe control each fishing area. The tribe usually appoints some people to manage a fishing operation, and the individual fisherman then works within a group in a certain fishing area that belongs to a known family within the tribe.

References

Al-Hamed MI (1966) On the reproduction of three cyprinid fishes of Iraq. Minist Agric Tech Bull Iraq 136:1–26

Al-Handel A, Hu C (2015) MODIS observations of human-induced changes in the Mesopotamian Marshes in Iraq. Wetlands 35:31–40

Al-Hassan LAJ, Miller PJ (1987) Rhinogobius brunneus (Teleostei: Gobiidae) in the Arabian Gulf. Jpn J Ichthyol 33:405–408

Al-Hassan LAJ, Hussain NA, Soud KD (1989) A preliminary, annotated check list of the fishes of Shatt al-Arab River, Basrah, Iraq. Polskie Archiwum Hydrbiologii 36:283–288

Al-Khait KS (1978) An economic study on the fishing industry in Iraq. Publications of Centre for the Arab Studies No. 22, University of Basrah, Basrah, Iraq, 196 pp

Baeteman C, Dupin L, Heyvaert V (2005) Geo-environmental investigation. The Persian Gulf Shorelines and the Karkheh, Karun, and Jarrahi Rivers: a geo-archaeological approach. a joint Belgo-Iranian Project. First Progress Report. Akkadica *125*(2):2005

Banister KE (1980) The fishes of the Tigris and Euphrates Rivers, pp 95–108. In: Róska J (ed) Euphrates and Tigris, Mesopotamian ecology and destiny. Monographiae Biologicae, 122 pp

Botros GA (1968) Fishing strategy. Dar al-Maraf, Egypt, 373 pp

Coad BW (1991) Fishes of the Tigris-Euphrates Basin: a critical check-list. Syllogeus 68:49

Coad BW (2010) Freshwater Fishes of Iraq. Pensoft Publishers, Sofia/Moscow

Contenau G (1954) Everyday life in Babylon and Assyria. Edward Arnold (Publications) Ltd., London, 324 pp

Eger AA (2011) The Swamps of Home: Marsh Formation and Settlement in the Early Medieval Near East. J Near East Stud 70:55–79

Emlen JM (1966) The role of time and energy in food preference. Am Nat 100:611–617

Garstecki T, Amr Z (2011) Biodiversity and ecosystem management in the Iraqi Marshlands–screening study on potential World Heritage Nomination. IUCN (International Union for Conservation of Nature), Amman, p 191

Hadid II, Al-Mahdawi KJ (1977) Fishing gears and methods in Iraq. Ministry of Agriculture, Iraq, 67 pp

IUCN (2010) The IUCN Red List of threatened species. Retrieved 2 February, 2018, from www.iucnredlist.org

Jawad LA (2006) Fishing gear and methods of the lower Mesopotamian plain with reference to fisheries management. Marina Mesopotamica 1(1):1–39

Jawad LA (2012) History of the study of the fish fauna of Iraq. Water Res Manag 2:11–20

Jennings S, Paulin NVC (1996) Fishing strategies, fishery development and socioeconomics in traditionally managed Fijian fishing grounds. Fish Manag Ecol 3:335–347

Jennings S, Kaiser MJ, Reynolds JD (2001) Marine fisheries ecology. Blackwell Science, Oxford, 379 pp

Khalaf KT (1961) The marine and freshwater fishes of Iraq. Al-Rabitta Press, Baghdad, 164 pp

Kurlansky M (1997) Cod: a biography of the fish that changed the world. Walker, New York, 348 pp

Lawler A (2002) Archaeology: Report of oldest boat hints at early trade routes. Science 296:1791–1792

Mahdi N (1962) Fishes of Iraq. Ministry of Education, Baghdad, 82 pp

Nicholson E, Clark P (eds) (2002) The Iraqi marshlands: a human and environmental study. The Amar Appeal International Charitable Foundation.

Parrot A (1961) Nineveh and Babylon. Thames and Hudson Publishers, London, 380 pp

Richardson CJ, Hussain NA (2006) Restoring the Garden of Eden: An ecological assessment of the marshes of Iraq. Bioscience 56:477–489

Richardson CJ, Reiss P, Hussain N, Alwash A, Pool D (2005) The restoration potential of the Mesopotamian marshes of Iraq. Science 307:1307–1311

Saggs HWF (1987) Everyday life in Babylonia and Assyria. Dorset Press, New York, 207 pp

Sahrhage D (1999) Mesopotamia: fishing and fish cult in ancient Mesopotamia. Peter Lang GmbH, Frankfurt, 241 pp. (In German)

Sainsbury JC (1996) Commercial fishing methods: an introduction to vessels and gears. Fishing News Books, Oxford, 475 pp

Sanlaville P (2002) The deltaic complex of the lower Mesopotamian plain and its evolution through millennia. In: Nicholson E, Clark P (eds) The Iraqi Marshlands: a human and environmental study. Politico's Pub, London, pp 133–150

UNEP (United Nations Environment Programme) (2001) The Mesopotamian marshlands: demise of an ecosystem, early warning and assessment. Early warning and assessment. Division of Early Warning and Assessment, United Nations Environment Programme, Nairobi, Kenya

UNEP (ed) (2005) Assessment of environmental "Hot Spots" in Iraq. Nairobi, UNEP

UNEP-DTIE-IETC, C (2009) UNEP-UNESCO project and World Heritage Programme: Project overview. Kick-off meeting: natural cultural management of the iraqi marshlands as world heritage. Amman, Jordan.

Von Brandt A (1984) Fish catching methods of the world. Fishing News Books, Blackwell Science, Oxford, 325 pp

Welcomme RL (2001) Inland fisheries: ecology and management. Fishing News Books, Blackwell Science, Oxford, 358 pp

Ziyadi MSF, Jawad LA, Al-Mukhtar MA (2018) *Halicampus zavorensis* Dawson, 1984 (Syngnathidae): new record for Iraqi marine waters and for the Arabian Gulf area. Cahiers de Biologie Marine 59:121–126

Chapter 24
The Artisanal Fishers in the Southern Marshes of Iraq and the Traditional Ecological Knowledge

Laith A. Jawad

Abstract Forty-five fishing areas were recognized and georeferenced conferring to fishermen's traditional ecological knowledge (TEK) in the marshes located north of Basrah City, Iraq. These marsh areas are included within the greater Al-Hammar Marsh located on the Euphrates River. Nearly 75% of the descriptions of the fishing areas were acquainted by fishers only and are registered for the first time. Fishermen identify fishing areas according to depth, certain shapes of reeds, presence of some aquatic plants, gradation of water transparency, and customary use of definite areas. This study discloses the wealth of data attained by local fishers, explains their hard straightforward link with the natural habitats where they live, and displays the possible usages of TEK for fisheries administration in the future.

Keywords Fishing methods · South of Iraq · Wetlands · Lifestyle · Income · Professional

24.1 Introduction

Fishermen own a thorough information of constituents and changing aspects of natural assets, environment, and fishing processes that can be of unlimited value for fishery valuation and administration. In the inland waters of Iraq and in the southern marshes, notwithstanding the presence of artisanal fisheries for years, fishermen's traditional ecological knowledge has not been appreciated by the scientific community, and it has not been regularly evaluated. Such case is also observed in other countries around the world such as Brasil (Schafer and Reis 2008; D'Incao and Reis 2002). Traditional ecological knowledge (TEK) can be assumed as a unified system of information, knowledge, and confidence, conveyed over generations on the relationship between local community and environment; this data is snowballing and active, and it is grounded on attained skill (Berkes and Folke 1998).

L. A. Jawad (✉)
School of Environmental and Animal Sciences, Unitec Institute of Technology, Mt Albert, Auckland, New Zealand

L. A. Jawad (ed.), *Southern Iraq's Marshes*, Coastal Research Library 36,
https://doi.org/10.1007/978-3-030-66238-7_24

The conservative method to fishery administration is founded wholly on scientific means and data, overlooking the acquaintance and administration systems informally accepted by fishing societies (Schafer and Reis 2008). The application of old-style top-down managing systems is frequently complex, out of action, and exclusive since fishermen do not contribute in the conclusion construction procedure and, therefore, do not fulfill with the forced guidelines (Schafer and Reis 2008). Old-style (and local) information turn out to be significant in resource administration, considering that the ritual beliefs correct, having as significance a strong social identity (Hanna and Jentoft 1996). The inclusion of TEK into scientific data—or research-based knowledge (RBK)—(Wilson et al. 2006) signifies only one of the constituents of the debate process of the inclusion of the artisanal fisheries management in the region (Reis and D'Incao 2000). Fishermen's acquaintance has been thoroughly explored lately, particularly concentrating on how to sufficiently relate it to several features of fishery administration (Davis and Wagner 2006; Satria et al. 2006; Grant and Berkes 2007; Hall and Close 2007).

Fishing locations, as conventionally recognized by fishermen, are a valued device for fishery administration. This study presents the first information given about the traditional ecological knowledge available with the artisanal fishers in the southern marshes of Iraq. It is about knowing the fishing locations by fishers through their experience over the years and where are they located, how to reach them, and what sort of fishing gears should be used in each of them.

24.2 Methodology and Attaining Data

This study was performed in a village located in the marsh area north of Basrah City, where most of the inhabitants are involving in fisheries business.

In the methodology, the documentation and mapping of the fishing areas as conventionally known and visited by fishermen were based on participative mapping used on rural environment. Conferring to the technique, fishermen identify the main fishing areas using the following method: experienced fishermen of each fishing society were asked to take one or more of the surveying team to their usual fishing areas for identification and perimeter mapping. The region involved in this study covers all the marshes in north of Basrah City, which belong to Al-Hammar Marsh. Ten fishing societies were designated as origin points to the fishing areas to prevent surveying problems and to minimize sailing time.

The main problems in attaining information data collection differed from the unwarranted admission to the fishing societies to uphold data collection while fishers performing their catch, thus, were unfilled to contribute to this study; and to opposing climatic settings, that delayed navigation.

Motorized boats of different size and motor engines appropriate for getting both deep and shallow areas were used to reach the fishing areas. The perimeters of the fishing areas were mapped using a GPS. During the course plotting, while the fisherman sailed along the area according to his knowledge, the perimeter points were registered. Geographical references such as the main shallow and deep areas were recorded. At each field trip, around 100 GPS points were recorded that joint together that permitted the identification of each fishing area. It took about eight working hours to map the perimeter of one specific area of the estuary.

Data recorded by GPS were moved to a computer, using TrackMaker TM software. The data was then introduced to the Microsoft WordPad where the original data were accustomed to allow a clear and impartial export to Microsoft Excel that allowed a direct plotting of the points registered by GPS to Surfer TM. This interchange between different softwares stopped contribution mistakes.

24.3 Results Attained

Forty-five fishing areas were identified and georeferenced conferring to the fishermen's; ten sites useful to locate some of these areas were also recorded, the so-called reference sites; these were located either inland, represented by certain shapes of reed's plants, or a rising landscape.

Nearly 75% of the descriptions of the fishing areas were known by fishermen only and have been listed for the first time; these are found in the inner part of the marsh areas.

The size of the fishing areas varied from very small (0.14 km^2) to large (5 km^2). A high density of small fishing areas is found in the inner regions of the marsh area.

Fishermen recognize fishing areas by means of four individual aspects or a mixture of them. These are (a) areas with different depths that are gathered in three categories, those less than 1m deep, the places up to 3m deep, and places deeper than 3m; (b) the existence of a certain shape of reed's plantation or certain aquatic plants; (c) different gradation of water transparency; and (d) the frequent (traditional) use of certain areas. Fishing areas are often named conferring to the landmarks nearby, or to present or past facts/situations.

Multiple names may be allocated to a fishing area liable on the society that fishermen are from. Extra field journeys to the fishing societies were performed to check the double titles for the same areas. The use of the same name for diverse fishing areas was also perceived.

24.4 Remarks

The location and description of the fishing areas of the marshes located north of Basrah City, which are part of Al-Hammar Marsh, are solidly recognized among the fishers and have been delivered through the generations for centuries, signifying the true TEK with historical and multigenerational charismas (Berkes et al. 2001).

The great number of fishing areas acknowledged in this study from the information of the artisanal fishers (45) discloses the fruitfulness of data kept by fishers as well as their close affiliation with the natural environment where they live, and from where they take the attitudes for documentation of the fishing areas. For assessment, the nautical chart, conventionally used as reference for location of the fishing areas, presents 20 areas only, named distinctly from TEK.

Several generic names were used in locating fishing areas that have caused unclear circumstances, particularly in those instances where inexact location refers to large areas. For instance, several generic names were used by fishers to designate the size of the fishing areas, and these names differ between fisher's societies inhabiting different areas.

Fishers, in contrary, are talented to recognize 21 different fishing areas using this generic location, making a difference between areas conferring to a mixture of features.

The exact site of the fishing areas and their link with other information on fishing action such as total catch, type of fishing gear used, and type of boats, among others, can provide passable harvest models and situations that will envisage catch policies and help the upkeep of the fishing stocks (Schafer and Reis 2008). These information are also vital when studying fish population dynamics, particularly when evaluating reproduction and nursery areas, which may ultimately be the emphasis of conservation. To reduce the decline of the fish stocks ill-used by the artisanal fisheries, forthcoming studies are predictable to use the fishing areas according to the knowledge kept with the fishers.

Accepting restriction in fishing areas at different regions of the marshes, accompanied with minimum mesh size and the species minimum landing size, would increase the likelihoods of the native fish species inhabiting the marsh areas to reach a maintainable catch size. Furthermore, fishermen move freely in the marsh area looking for the target species. If fishermen's movements were controlled to certain regions to their societies, a more impartial portion of capture, and therefore turnover, for each fisherman may be attained. Yearly disparities on the richness of the target species conferring to the fishing areas would be stable in a medium term, helping to decrease struggles between fishers' societies.

Locating the precise site of fishing areas diminishes discrepancies when dealing with the data in relation to the geographical distribution of fishing effort, fishing, and fishermen migration, among others. This will also enable the selection of zones limited to some or all fishing gears for means upkeep and that can be directly agreed by fishermen. The acceptance of fishing areas according to the knowledge of the fishermen and fisheries managers is predictable to haste up the application of management events as well as to improve the administration itself.

To deliberate the TEK in the region as a vital foundation of data for fishery administration produced in some positive features, particularly because fishermen understand that their knowledge is appreciated, (1) fishermen have sureness in contributing in the management verdicts are supported; (2) struggles among investors are curtailed; and (3) fishermen are stimulated to fulfill with the guidelines that they assisted to set up.

A mutual difficulty with fisheries of developing countries is the absence of data regarding the construction of the fishery itself (number of boats and fishermen, type of gears, among others), as well as elementary statistics on landings and fishing effort. Frequently, total landings are not logged correctly, fishing work data is lacking, landings by diverse gears, areas, and periods are not distinguished, and all these information may have numerous levels of consistency.

For furthermost artisanal fisheries, no formal long-term data sets occur. This status is similar to fisheries areas in other countries (Johannes et al. 2000). Nevertheless, there might, in fact, be more supplies of data than are usually esteemed and, more, that there may be the suitable procedure for at least their initial examination (Pauly 1979). The absence of decent data is a customary condition and can be a hindrance in attaining a consistent stock valuation. Regularly, there are two choices only in this case. One is to tackle the data that is existing. The other is to delay the analysis for a well infrastructure and sampling situations, which permit a decent system of data gathering to be recognized.

In the latter case, the hazards are huge, as these settings may last for a long time to be attained (if ever) that the stocks will maybe be shattered by then (Reis 1992). Though fishermen's data frequently gets laid off for being personal (Ames 2003) or unreliable (Johannes et al. 2000), the use of TEK as matching to scientific knowledge (Mackinson and Nøttestad 1998) is a third possible choice. Outside that the most passable administration events in the area are deliberated by fishermen, scientific community, fishery managers, and other investors in the light of scientific knowledge backed by TEK.

Often fishermen's data is not considered owing to researchers' incapability to comprise it in administration systems, and ecosystem modeling or even to show it in a satisfactory approach. It is problematic to show it sufficiently mostly since data is very precise and unusual to the region where it came from and it may not seemingly be valuable to give chief instructions from it.

In this study, the prominence to link the TEK with scientific knowledge to improve the data class is utilized in fishery administration in the marsh area north of Basrah City, Iraq. As an alternative of showing instances where disregarding TEK results with negative costs for the natural assets, the present study strained to show how considering TEK may assist in improving information quality and stop failed fishery administration.

References

Ames T (2003) Putting fishermen knowledge to work: the promise and pitfalls. In: Haggan N, Brignall C, Wood L (eds) Putting fishers' knowledge to work: conference proceedings fisheries Centre research reports, vol 11, pp 184–188

Berkes F, Folke C (1998) Linking social end ecological systems. Management practices and social mechanisms for building resilience. Cambridge University Press, Cambridge

Berkes F, Mahon R, Mc Conney P, Pollnac R, Pomeroy R (2001) Managing small-scale fisheries. Alternative directions and methods. Ottawa, International Development Research Centre

D'Incao F, Reis EG (2002) Community-based management and technical advice in Patos lagoon estuary (Brazil). Ocean Coast Manag 45:531–539

Davis A, Wagner J (2006) A right to fish for living? The case for coastal fishing people's determination of access and participation. Ocean Coast Manag 49:476–497
Grant S, Berkes F (2007) Fisher knowledge as expert issue: a case from the longline fishery of Grenada, the eastern Caribbean. Fish Res 84:162–170
Hall GB, Close CH (2007) Local knowledge assessment for a small scale fishery using geographic information systems. Fish Res 83:11–22
Hanna S, Jentoft S (1996) Human use of the natural environment: an overview of social and economic dimensions. In: Hanna S, Folke C, Maler K-G (eds) Rights to nature: ecological, economic, cultural, and political principles of institutions for the environment. Island Press, Washington, DC
Johannes RE, Freeman MMR, Hamilton RJ (2000) Ignore fishers0 knowledge and miss the boat. Fish Fish 1:257–271
Mackinson S, Nøttestad L (1998) Combining local and scientific knowledge. Rev Fish Biol Fish 8(4):481–490
Pauly D (1979) Theory and management of tropical multispecies stocks: a review, with emphasis on the southeast Asian demersal fisheries. ICLARM Stud Rev 1:1–35
Reis EG (1992) An assessment of the exploitation of the white croaker Micropogonias furnieri (Pisces, Sciaenidae) by the artisanal and industrial fisheries in coastal waters of southern Brazil. PhD thesis, University of East Anglia
Reis EG, D'Incao F (2000) The present status of artisanal fisheries of extreme southern Brazil: an effort towards community-based management. Ocean Coast Manag 43:585–595
Satria A, Matsuda Y, Sano M (2006) Solution to the tragedy of property right in coastal fisheries. Mar Policy 30:226–236
Schafer AG, Reis EG (2008) Artisanal fishing areas and traditional ecological knowledge: the case study of the artisanal fisheries of the Patos lagoon estuary (Brazil). Mar Policy 32(3):283–292
Wilson DC, Raakjaer J, Degnbol P (2006) Local ecological knowledge and practical fisheries management in the tropics: a policy brief. Mar Policy 30:794–801

Section VII
The Marshes and Human Health

Chapter 25
Fish Consumption of Mothers and Their Children in Fishermen Society in the Marsh Areas

Laith A. Jawad

Abstract Inland fisheries whether large or small scale can subsidise to the food and nutrition safekeeping of the freshwater-dependent society. Fishers can allocate part of their daily yield for family usage or sell the catch and use the income to buy food items other than fish. Among the important food for human and especially for the growth of children is fish. On the other hand, women can play a vital role in supplying healthy food for the members of the family through controlling the household issues. With such accessibilities, food is still unsecured for several middle- and lower-income families in the developing countries. This chapter provides evidences obtained through examination of society inhabiting a remote village in the southern marshes of Iraq from the point of view of food and nutrition insecurity and concentrating on the eating habits and how fish is included in their daily food. The results showed that both mothers and children are facing less than the international level of the dietary diversity.

Keywords Fish eating · Digestive system · Kids · Housewives · Lifestyle · South of Iraq

25.1 Introduction

The current studies have proved that there is a positive link between small-scale fisheries whether it is marine or freshwater and reduction in micronutrient deficiency mainly between women and children (Gibson et al. 2020; Bene et al. 2015; Bogard et al. 2017; Thilsted et al. 2016). Such relationship may be introduced in three different ways as Kawarazuka and Bene (Kawarazuka and Bene 2010) suggested it: firstly, a revenue source, where fishermen first secure the need of their family from the fish catch before selling it or indirectly through the money earned from selling

L. A. Jawad (✉)
School of Environmental and Animal Sciences, Unitec Institute of Technology, Mt Albert, Auckland, New Zealand

L. A. Jawad (ed.), *Southern Iraq's Marshes*, Coastal Research Library 36,
https://doi.org/10.1007/978-3-030-66238-7_25

the catch afterward (Fabinyi et al. 2016), and secondly, a feeding way, where the habit of eating fish is higher in the fishers communities than other human gathering, where the rate of animal protein reaches up to 90% per capita (Charlton et al. 2016; Cisneros-Montemayor et al. 2016). With the high variable nutritional value of the fish, fishers' households get the nourishment better than those individuals who have less contact with the aquatic resources; thirdly, a supply way, the participation of women in the fishers' societies enables women to earn an extra income that they usually used to purchase food items other than fish such as vegetable, and through their caring for the children, the health of the households remains upright (Smith et al. 2003).

Women usually need greater nutritional food owing to the calls of pregnancy and lactation, with research stressing the significance of presumption health and nutrition (Stephenson et al. 2018). As to children, a nutritious food in their early life especially within the first 2 years is important in delivering satisfactory nutrition and balancing foods, subsidising in the crucial stages of foetal neurodevelopment and child growth (Development Initiatives 2018).

Nevertheless, food uncertainty, as shown by less nutrition, rests predominantly in low- and middle-income fish-producing countries regardless of the relative approachability of fish and the economic chances obtainable through sharing in small-scale fisheries (Thilsted 2012). Such notwithstanding determinations over the last few years to develop food security by growing the obtainability of fish through enhanced administration of fisheries and the use of landscape-scale methods (Bene et al. 2016; Foale et al. 2013; Steenbergen et al. 2017). Several factors such as gender, age and class among others also disturb access to food (Quisumbing et al. 1995) and, together with intra-household disparities in access to and supply of food, are frequently ignored (Burchi and De Muro 2016).

This chapter reports on an investigation related to the input of fish and small-scale fisheries' daily events to food and nutrition security. The study was conducted in some remote villages in the southern marshes of Iraq, where the daily life of the society is mainly based on fishing activities. This investigation adds to the literature by examining intra-household distribution of nutrient-dense foods, viz. fish, and how the gender became a main part player in access to and exploitation of food.

25.2 Methodology

The southern marshes of Iraq are a huge lowland covered with water all year round. In this area, reed's houses is distributed where the marsh Arabs live. These people are living mainly on the resources of the marshes, which comprise fish, birds and any vegetable that they grew on the small floating island that they have their hut on. The reed inhabitants are gathered in small villages, some of which are far away from the edges of the marshes where urban residents were located. The inhabitants of those villages rely in their daily food on fish that they catch. There is no previous data on the rate of protein consumption in the rural area in Iraq in general and the southern marshes in particular.

The inland fisheries from the southern marshes of Iraq support the main cities located in the south of Iraq. This means that the productivity of this area is high. Recently, the yield of inland fish catch from the marsh areas has dropped down due to several factors; among these are the significant decrease in the amount of water received by the marshes from both Euphrates and Tigris Rivers, human activities represented by pollution, overfishing, illegal fishing (bomb and poison fishing) and environment changes, i.e. increasing salinity.

The area selected for this case study was one of the villages located in the Chibayish marshes, south of Thigar Province, south of Iraq. The inhabitants of this village are very poor, and no assessment was made for their standard of living, and they seem to live at the lowest living standard. From the living style that the inhabitants have, it seems that there are some indications for high levels of food insecurity and undernutrition together with the admission and utilisation dimensions of food security: poor dietary intake, inappropriate child feeding methods and poor water, sanitation, and hygiene situations. The society contained households recognising as of Madan's or the Marsh Arabs engaged in several small-scale fisheries' livelihood events containing catching fishes, hunting birds and reed's related handmade.

The methods used to obtain data are variable ranging between indoor surveys, groups meetings, and a market survey. In the household survey, 80 households contributed for this. The sample includes qualified households, which comprise women of reproductive age and a children aged between 6 months and 5 years.

The data obtained contain the following features: (a) mother's age, education, eating of food during the 24-h recall method, living activities, sharing in gathering and household decision-making and health and nutrition knowledge; (b) child's age and sex and feeding practices (including breastfeeding, age at introduction of complementary foods, eating of food during the 24-h recall method); and (c) aspects of the household (house building and admittance to enhanced water, sanitation and electricity), ownership of livestock and production of food crops.

Interviews were conducted covering the following subjects:

1. Recognising the seasons and associated change or differences in daily life activities, food and water availability and health and nutrition matters
2. Historical and current food eating and meal patterns; matters with food admittance with taboos or intra-household distribution methods; child feeding ways; and insights and appreciative of healthy foods
3. Women's roles and tasks within the household and at the community level

Interviews were performed by the assistance of the head of the tribe or anybody he can allocate for this task and with volunteer women from the community. Interviews were conducted in the huts of the participants and lasted for less than an hour.

An additional questions were added to the questionnaires about the eating of fish (species and whether consumed fresh or dried), and other oils and fats, savoury and fried snack foods, sweet snacks and sugar-sweetened beverages. Fortified foods, infant formula and milk products and all snack foods were classified in accordance with the relevant guidance manuals.

Data on the height and weight for mothers and children were collected in accordance with standardised procedures (United Nations 1986). Maternal body mass index (BMI) was calculated and grouped as underweight (<18.49), normal (18.5≤24.9), overweight (>25≤29.9) and obese (≥30) (WHO Expert Consultation 2004).

Household wealth was assessed by a Material Styles of Life scale based on the building materials of the house, admittance to electricity (none, purchased or own source), admittance to enhanced sanitation and the presence or absence of household assets (fan, television, mobile phone, tablet, fridge, couch/sofa set, washing machine, scooter) (Pollnac and Crawford 2000; Filmer and Pritchett 2001). The scale was weighted by the number of household members, and households were assembled into three wealth levels: low (lowest 40%), middle and high (highest 20%).

25.3 Results Attained

The average age of the women was 32.8 ± 7.5 years (Table 25.1). All the women are illiterate and did not attend school in their lives. Seventy-five percent of the women sampled had a BMI within the normal range, while 6.25% of women were classed as overweight or obese (6.25%).

Over third of the children (31.25%) were 24 months and older. Small percentage of the children had been breastfed (12.5%), with only 2% of the children never breastfed. No children were observed to have breastfeeding beyond 2 years of age. Overall, the stunted growth showed a percentage of 5–20%, with less percentage has been observed in children over 24 months.

Households with higher wealth were only few (6.25%) as the poorness is the general outlook in the village. On the other hand, all households own their hut as they build it themselves. As to the land, there is a specific owner of the dry pieces of land in the marsh areas in Iraq.

The water, sanitation and hygiene environments are in a very bad situation, and proper toilets or any health measures are followed by the inhabitants. There is no source of freshwater; everyone in the village drinks from the clear area of the marsh.

Maternal dietary variety was low: nearly all the women of the village less than consumed foods of certain variety, i.e. fish, birds, vegetables and home-made bread. It is interesting to note here that Marsh Arabs are not used to eat snack food such as biscuits, cakes and sweets. The sweet items they have in their food are mainly dates and Khrait (a sweet made from the pollen grain of reed).

Table 25.1 Socio-demographic features and nutritional status of women and children in fisher's society in the southern marshes of Iraq

Features	Mean ± SD or *n* (%)	N
Maternal features		
Maternal age, years	32.8 ± 7.5	80
Highest level of schooling completed		80
Some primary	10 (100)	
Completed primary	0	
Some or completed secondary or further education	0	
Mother has income-generating livelihood activity	30 (3.75)	
Nutritional status		80
Underweight (BMI <18.49)	10 (12.5)	
Normal (BMI ≤ 18.5 <24.9)	60 (75)	
Overweight (BMI ≤25 <30)	5 (6.25)	
Obese (BMI ≥ 30)	5 (6.25)	
Child characteristics		
Sex		80
Male	38 (47.5)	
Age		80
6–11 months	10 (12.5)	
12–17 months	20 (25)	
18–23 months	25 (31.25)	
24+ months	25 (31.25)	
Breastfeeding, by age category		80
6–11 months	9 (100)	
12–17 months	15 (81)	
18–23 months	18 (83)	
24+ months	20 (45)	
Minimum acceptable feeds		
6–11 months	7 (18.75)	
12–17 months	10 (25)	
18–23 months	15 (18.75)	
24+ months	20 (3.75)	
Stunting		20
6–11 months	2 (10)	
12–17 months	4 (20)	
18–23 months	1 (5)	
24+ months	1 (5)	
Household		
Material styles of life scale (wealth)		
Lowest wealth	50 (62.5)	
Middle wealth	25 (31.25)	
Highest wealth	5 (6.25)	
Access to improved sanitation	0	
Dual burden households (mother overweight and child has stunted growth)	1 (1.25)	

As with the adults, the dietary diversity of children was also low, and they usually eat the same variety of food that the adult did. While the diets of mothers and children were considerably alike, mothers reported consuming animal-source foods and 'fruits and vegetables' more frequently than their children.

Fish was the most recurrently eaten animal-source food by mothers, with around 99% of mothers eating fish all months of the year. Among those, about 75% consume dried fish especially in the winter season when the catch drops. The habit of fish consumption is also true for children, with 87% of children consuming fish in all seasons. There was no difference in the rate of consumption of fish among seasons of the year. However, consumption of fish was low among infants: only 20.1% had consumed fish regularly. Children who did eat fish were more probable to have meals set from fresh fish than dried fish.

Intra-household variances in the consumption of nutrient-dense foods, especially fish, were sometimes affected by some social restrictions about suitable foods for infants and young children and pregnant and lactating women. The greatest disparity in mother-child diets occurred with respect to the consumption of fish: 85% of mothers do not consumed fish when they are pregnant. The analysis showed that both women and children have no variety of food to choose from to consume.

Women were responsible for the attainment of food, scheduling of meals and preparation of foods. Household duties were suggested by local culture that all household-related issues will be taken care of by the wife and the daughter(s) if there are any in the age that they can help in the house matters. Women also share men in the house living budget. However the diversity of foods eaten during meals – that is, the addition or inclusion of fish, other animal-source foods and vegetables – was reliant on the management ability of the women.

Household eaten of fish – and the obtainability of money to purchase foods – be contingent on the type and achievement of men's fishing activities. Daily decisions were made as to whether the catch of the day was sold fresh, dried for sale or retained for home consumption.

The income of the family is related to the availability of fish, which in turn is affected significantly with the changes in the weather. Fishing activities may cease due to stormy wet days or during very cold seasons that stop men going for fishing.

Other nutrient-dense foods, such as eggs, certain fruits such as dates, fruits, vegetables and milk were available at the site of each hut as each family owns at least one water buffalo, one cow and a number of chickens and plants some vegetables in what left of the dry land in the premises of the hut over the floating island. Nevertheless, without having refrigeration there is an expected loss of more than quarter of their produce to spoilage before it could be sold or consumed.

The load and strain of confirming that a family had adequate food are the responsibility of women. Women need to assist husband in daily income. In most cases, women share husband in earning some extra income by making and selling some milk products, making rugs from the reed and making dresses for other women in the village.

These events intended to secure the adequacy of food to the households.

25.4 Discussion

This chapter has shown the outcomes from a survey of how fish and small-scale fisheries' actions subsidise to food and nutrition security, converging on the eating habits in rural villages in the southern marshes of Iraq.

The information attained showed that the dietary diversity of both mothers and their children was under the minimum suggested for a diet with micronutrient adequacy, with the majority of women having inadequate dietary diversity during the year. Dietary variety is a vital issue of dietary quality, with an assorted diet addition of nutrient-dense animal-source foods and fruits and vegetables more probable to deliver the micronutrients important for the health, work capability of women and long-term health of their children (Torheim et al. 2010; Martin-Prevel et al. 2015; Gibson et al. 2020). Several investigations have shown similar studies, when women restricted their diet on certain food items, which cause a significant deficiency in micronutrient (Torheim et al. 2010). Women and all other adults in the village investigated have shown to have at least two daily meals based on rice and fish. Such restriction has been noted by Campbell et al. (2009) when investigating Indonesian women of reproductive age. They found that those women are at risk of vitamin A deficiency and a possible anaemia. This proposes that the diets of women in the investigation may be insufficient to meet their micronutrient needs.

Similarly, children's dietary variety was low, with children having less diet diversity in all seasons of the year. The dietary variety of infants (6–11 months) and young children (12–17 months) was normally lower than older children, indicating the need for introduction of balancing foods until children were given household food. The WHO suggested that breastfeeding of infants should last for the first 6 months of age and then continued breastfeeding be complemented by the introduction of low-cost locally available foods which are sufficiently energy- and nutrient-dense, and diverse, until a child reaches 2 years of age (WHO 2018). Such recommendation is not applied in the present example as the village is rural and complementary baby food is not known among the society. The shortage of this kind of food may be the reason behind the cases of stunting observed in the village (Victora et al. 2008).

Fish is progressively known as contributing great impending for addition in nutrition-sensitive food-based policies to upsurge the quality of children's diets (Bogard et al. 2017; Thilsted et al. 2016). Fish meal has large nutrient values, including iron and zinc, which are missing in breastmilk, and ground dried small fish could provide a nutritional improvement and are consumed in small quantities, given children's limited stomach capacity [Thilsted et al. 2016; Bogard et al. 2015; Gibson et al. 2020]. The results at hand showed that fish was the most usually eaten animal-source food by mothers and young children. This finding agrees with that of Bandoh and Kenu (2017), who found a higher level of fish eaten among children in fishing communities in Ghana, but differs with Thorne et al. (Thorne-Lyman et al. 2017) study and Gibson et al. (2020).

Sociocultural practices including some food taboos are found to affect women from having certain food items in certain days of the year. This can particularly disturb pregnant and lactating women and young children (Pachon et al. 2007).

Several investigation programmes in south and south-east Asia (e.g. Bangladesh [60]) and sub-Saharan Africa (e.g. Kenya (Konyole et al. 2012), Malawi (Hotz and Gibson 2005)) have shown that the use of home-made dried small fish powders will assist in the growth of children. Children in the southern marshes of Iraq eat fish twice a day. Therefore, they do not need an extra food item supplemented by fish. Other investigators were cautious about fish-supplemented food without giving another type of food enriched in iron (Dewey 2007).

The ability of the inhabitants of the investigated village to attain fridges to keep their food in is not present, and the families use their food in the same day and cannot save any for the next day due to the heat and the unhealthy environment. Therefore, the only way to preserve fish is through salting and drying.

This food environment includes the obtainability, budget availability, suitability and attraction of foods (Herforth and Ahmed 2015). In the present study, the obtainability of nutrient-dense foods was restricted by the isolation of the society of the village from markets and the absence of cold storage which caused high rates of decomposition.

References

Bandoh DA, Kenu E (2017) Dietary diversity and nutritional adequacy of under-fives in a fishing community in the central region of Ghana. BMC Nutr 3(2):1–6. Epub January 3, 2017

Bene C, Barange M, Subasinghe R, Pinstrup-Andersen P, Merino G, Hemre GI et al (2015) Feeding 9 billion by 2050—putting fish back on the menu. Food Secur 7(2):261–274

Bene C, Arthur R, Norbury H, Allison EH, Beveridge M, Bush SR et al (2016) Contribution of fisheries and aquaculture to food security and poverty reduction: assessing the current evidence. World Dev 79:177–196

Bogard JR, Hother AL, Saha M, Bose S, Kabir H, Marks GC et al (2015) Inclusion of small indigenous fish improved nutrition quality during the first 1000 days. Food Nutr Bull:1–14

Bogard JR, Farook S, Marks GC, Waid J, Belton B, Ali M et al (2017) Higher fish but lower micronutrient intakes: Temporal changes in fish consumption from capture fisheries and aquaculture in Bangladesh. PLoS One 12(4) Epub April 6, 2017

Burchi F, De Muro P (2016) From food availability to nutritional capabilities: advancing food security analysis. Food Policy 60:10–19

Campbell AA, Thorne-Lyman A, Sun K, de Pee S, Kraemer K, Moench-Pfanner R et al (2009) Indonesian women of childbearing age are at greater risk of clinical vitamin a deficiency in families that spend more on rice and less on fruits/vegetables and animal-based foods. Nutr Res 29:75–81

Charlton KE, Russell J, Gorman E, Hanich Q, Delisle A, Campbell B et al (2016) Fish, food security and health in Pacific Island countries and territories: a systematic literature review. BMC Public Health 16:26

Cisneros-Montemayor AM, Pauly D, Weatherdon LV, Ota Y (2016) A global estimate of seafood consumption by coastal indigenous peoples. PLoS One 11(12):16

Development Initiatives (2018) 2018 global nutrition report: shining a light to spur action on nutrition. Bristol, Development Initiatives

Dewey KG (2007) Increasing iron intake of children through complementary foods. Food Nutr Bull 28(4 Supp):S595–S609

Fabinyi M, Dressler WH, Pido MD (2016) Fish, trade and food security: moving beyond 'availability' discourse in marine conservation. Human Ecol 12. Epub December 29, 2016

Filmer D, Pritchett LH (2001) Estimating wealth effects without expenditure data—or tears: an application to educational enrolments in states of India. Demography 38(1):115–132

Foale S, Adhuri DS, Alino P, Allison EH, Andrew N, Cohen P et al (2013) Food security and the coral triangle initiative. Mar Policy 38:174–183

Gibson E, Stacey N, Sunderland TC, Adhuri DS (2020) Dietary diversity and fish consumption of mothers and their children in fisher households in Komodo District, eastern Indonesia. PloS One 15(4):e0230777

Herforth A, Ahmed S (2015) The food environment, its effects on dietary consumption and potential for measurement within agriculture-nutrition interventions. Food Secur 7:505–520. Epub May 7, 2015

Hotz C, Gibson RS (2005) Participatory nutrition education and adoption of new feeding practices are associated with improved adequacy of complementary diets among rural Malawian children: a pilot study. Eur J Clin Nutr 59:226–237

Kawarazuka N, Bene C (2010) Linking small-scale fisheries and aquaculture to household nutrition security: an overview. Food Secur 2:343–357

Konyole SO, Kinyuru JN, Owuor BO, Kenji GM, Onyango CA, Estambale BB et al (2012) Acceptability of amaranth grain-based nutritious complementary foods with Dagaa fish (*Rastrineobola argentea*) and edible termites (*Macrotermes subhylanus*) compared to corn soy blend plus among young children/ mother dyads in Western Kenya. J Food Res 1(3):111–120

Martin-Prevel Y, Allemand P, Wiesmann D, Arimond M, Ballard T, Deitchler M et al (2015) Moving forward on choosing a standard operational Indicator of Women's dietary diversity. Rome, Italy

Pachon H, Simondon KB, Fall ST, Menon P, Ruel MT, Hotz C et al (2007) Constraints on the delivery of animal-source foods to infants and young children: case studies from five countries. Food Nutr Bull 28(2):215–229

Pollnac RB, Crawford BR (2000) Assessing behavioural aspects of coastal resource use, Contract No.: Coastal Resources Centre Coastal Management Report #2226. University of Rhode Island, Rhode Island

Quisumbing AR, Brown LR, Feldstein HS, Haddad LJ, Pena C (1995) Women: the key to food security. Washington, DC

Smith LC, Ramakrishnan U, Ndiaye A, Haddad L, Martorell R (2003) The importance of women's status for child nutrition in developing countries, Contract No.: Research report 131. IFPRI

Steenbergen DJ, Clifton J, Visser LE, Stacey N, William MC (2017) Understanding influences in policy landscapes for sustainable coastal livelihoods. Mar Policy 82:181–188

Stephenson J, Heslehurst N, Hall J, Schoenaker DA, Hutchinson J, Cade JE et al (2018) Before the beginning: nutrition and lifestyle in the preconception period and its importance for future health. Lancet 391:1830–1841. Epub April 16, 2018

Thilsted SH (2012) Improved management, increased culture and consumption of small fish species can improve diets of the rural poor. In: Burlingame B, Dernini S (eds) Sustainable diets and biodiversity: directions and solutions for policy, research and action. FAO/Biodiversity International, Rome, pp 176–181

Thilsted SH, Thorne-Lyman A, Webb P, Bogard JR, Subasinghe R, Phillips M et al (2016) Sustaining healthy diets: the role of capture fisheries and aquaculture for improving nutrition in the post-2015 era. Food Policy 61:126–131

Thorne-Lyman A, Valpiani N, Akter R, Baten MA, Genschick S, Karim M et al (2017) Fish and meat are often withheld from the diets of infants 6 to 12 months in fish-farming households in rural Bangladesh. Food Nutr Bull 38(3):354–368

Torheim LE, Ferguson EL, Penrose K, Arimond M (2010) Women in resource-poor settings are at risk of inadequate intakes of multiple micronutrients. J Nutr 140:2051S–2058S

United Nations (1986) How to weigh and measure children: assessing the nutritional status of young children in household surveys, New York
Victora CG, Adair L, Fall C, Hallal PC, Martorell R, Richter L et al (2008) Maternal and child undernutrition: consequences for adult health and human capital. Lancet 371:340–357. Epub January 17, 2008
WHO (2018) Infant and young child feeding: WHO; 2018 [cited 2018 July 3, 2018]. Available from: http://www.who.int/news-room/fact-sheets/detail/infant-and-young-child-feeding
WHO Expert Consultation (2004) Appropriate body-mass index for Asian populations and its implications for policy and intervention strategies. Lancet 363:157–163

Chapter 26
Ingestion of Fish Bones: Clinical Cases from the Marsh Areas of Iraq

Laith A. Jawad

Abstract Fish bone is one of the most frequent foreign bodies that is inserted in the upper digestive tract due to swallowing events during eating fish meal, regularly positioned in the tonsil, epiglottis, pear-shaped fossa, and oesophagus, where it might be easily found using regular examination procedure and removed. In case of fish bone insertion, the classical advice is to eat either big bite of bread or big lump of dates. Such food could cause the movement of the fish bone from the pharynx, throat or oesophagus to the nearby tissues.

In this chapter, three clinical cases of ingestion fish bone were reported from people of different ages living in villages situated in the southern marshes of Iraq. These incidences represent insertion of fish bone in different parts of the digestive system and also brought about different impacts. Fish bone ingestion may cause a series of complications and endanger a patient's life. Incidences need great consciousness and attention on the part of the first physician to diagnose and treat the case, and suitable health teaching should be delivered to the patient.

Keywords Fish bones · Digestive system · Clinical case · Diagnosis · Clinical case

26.1 Introduction

In the regions where fish-eating habit is common especially cooking the whole fish without filleting, the small bones often get swallowed during the process of eating and trapped in the upper digestive tract (Akazawa et al. 2004; Dereci et al. 2015). The usual areas where the fish bone can be inserted are the palatine tonsils, tongue base and pyriform sinus (Nandi and Ong 1978). Doctors are usually able to locate and remove such bones without any complications, though fish bones inserted in the oesophagus can be very problematic to recognize and eliminate especially if they are

L. A. Jawad (✉)
School of Environmental and Animal Sciences, Unitec Institute of Technology, Mt Albert, Auckland, New Zealand

L. A. Jawad (ed.), *Southern Iraq's Marshes*, Coastal Research Library 36,
https://doi.org/10.1007/978-3-030-66238-7_26

small and transparent in colour. The causes that these bones can generate in the oesophagus region could range from mucosal ulceration and inflammation both locally and in neighbouring structures to oesophageal perforation, which in turn can lead to solemn medical difficulties such as deep neck abscess, mediastinitis, lung abscess and aortic fistulae (Ngan et al. 1990). Therefore, accurate diagnosis, quick removal and further treatment are recommended in such instances. In several instances it is not permanently conceivable to envisage a fine inserted fish bone on an X-ray film (Nandi and Ong 1978; Singh et al. 2018).

The difficulty, then, rests on how to deal with the patients whose X-ray films are negative, but having a case of swallowing a fish bone. Effective uncovering of inserted fish bones has lately been verified with computed tomography (CT) (Dereci et al. 2015).

The incidences of children ingested fish bones while they are eating are very common in countries where fish represents the main food item. Such cases are also affected by the method preparing the fish in. It might be difficult cases for physicians at the hospitals as it is quite problematic to obtain eating history for a child. To make the case worse, the event of ingesting fish bone might not be observed by an adult to assist in reporting the case to the hospital (Wong et al. 2019). In general, the incidence of swallowing fish in adult is different from that in children from the surgical and clinical point of views (Reilly et al. 1995; Singh et al. 2018; Wong et al. 2019). Given these thoughts, there is importance in assessing the indication concerning the characteristics, giving symptoms and administration choices for fish bone ingestion in children (Wong et al. 2019).

The present chapter offers some clinical cases of ingestion of fish bone by adults and children living in villages at the edge of the southern marshes of Iraq. The aim of the information given in this chapter is to raise awareness of the public about the dangerous of ingesting fish bone during consuming fish and to take extra care and look for any small bone before eating fish. Extra care should also be given to the children who usually eat by themselves without the supervision from the adults. Due to the complicated clinical terms, no full clinical descriptions were given for the cases mentioned in this chapter. Also, due to the restrictions about using clinical images of the patients, no photos of the cases were permitted by the medical centres where the cases were reported.

26.2 The Incidences

26.2.1 Case of Child Swallowed Fish Bone

A 10-year-old boy was introduced to a medical centre in small city at the edge of the southern marshes near Thigar Province. The boy was rushed to the medical centre soon after the incidence of swallowing fish bone happened while he was eating his lunch with his family. The symptom that the boy has is pain located in the neck and the whole oral cavity. In addition, the boy has a foreign body sensation. After a preliminary examination, the boy undertook an X-ray to show the position of the fish bone. The fish bone was located to impact the palatine tonsil. Using tongue depressor, headlight and forceps without requiring general anaesthetic, the fish bone was removed safely. No noteworthy problems were recognized in this cohort of patients. No children required a revision or second-look procedure, and the boy was discharged in the same day.

26.2.2 Case of Migration of an Ingested Fish Bone to the Submandibular Gland

A 46-year-old man, an inhabitant of one of the villages in southern marshes of Iraq close to the City of Amarah, was rushed to a medical centre located in a small city at the edge of the marsh area. This man swallowed a fish bone while he is eating fish for his lunch. The patient admitted eating one of the common fish found in the area and known locally as "Samtti" (species of carp family known also as common carp, *Cyprinus carpio*). The man tried several ways to release the bone from his throat, where it seemed to be inserted, such as eating a piece of bread or big piece of dates, but no instant release of the bone was reached. At the medical centre, an examination was done for the oral cavity and pharynx, and nothing was found. The patient was discharged from the centre after feeling relief in the oral cavity. He returned to the medical centre after 2 days suffering from an unusual pain in the right submandibular gland. An X-ray for the neck was made, and a foreign body was noted ingested in the right submaxillary gland. The man underwent a general anaesthetic for removal of the fish bone from the submandibular gland. After taking the fish bone out of the patient's neck, it revealed having 1.7 cm long, with an uncinated end. The patient was discharged from the medical centre the next day, and the incision healed after 1 week.

26.2.3 Case of Small-Bowel Perforation Caused by Fish Bone

A 42-year-old man, with no preceding abdominal illnesses, was introduced to a medical centre in a small city at the edge of the southern marshes of Iraq near the City of Amarah. The patient was suffering from severe abdominal pain for the last 3 days. The pain was all over his abdomen, but later on was localised in the lower right part of the abdomen. Later, the patient experienced high body temperature. The abdominal area showed some tenderness in the lower right part. Blood tests showed elevation in some parameters, while urine analysis was normal. The sonar for the whole abdomen showed fluid gathered over the lower right side. The preliminary diagnosis showed appendicitis and an appendectomy was arranged. During the operation, the appendix appeared to be normal, but the loop of the ileum in the right lower abdomen was swollen and showed an indication of an erythematous change. With the examination of the small bowel, a tiny sharp object was seen inserted in the wall in the inflamed area. The foreign body, which was revealed later as fish bone, was taken and removed. There were some complications due to the inflammation caused by the insertion of the foreign body in the small intestine. A closed-wound suction drainage tube was placed in the pelvis. The patient was discharged 5 days after the procedure.

26.3 Remarks

Fish bones are one of the most frequent foreign bodies swallowed necessitating for otolaryngology examination and surgical involvement among adult's patients especially in societies where fish diet is preferred (Wong et al. 2019).

The incidences of ingestion of fish bones and their insertion in the oral cavity and laryngopharynx were reported from different age groups (Kim et al. 2015). The physiological features of the oesophagus in the old people (>40 years) may make it easier to swallow a fish bone (Sheth and Diner 1988; Kim et al. 2015). It is essential to identify the presence of the inserted fish bone in the oesophagus or anywhere in the oral cavity quickly and precisely, because such foreign bodies infrequently can cause serious complications, such as perforation of the oesophagus, deep neck abscess, mediastinitis, lung abscess and aortic fistula (Akazawa et al. 2004).

The method of tracing the presence of fish bone in the oral cavity is tricky in most of the times. As fish has less calcium than humans, then parts of its skeleton would be too hard to detect by plain X-ray, and therefore, CT scan is recommended for this job as it is with high sensitivity (Lue et al. 2000; Evans et al. 1992; Sundgren et al. 1994).

To manage the incidence of fish bone insertion in the oesophagus, several treating methods can be used which depend on the site of the bone; age and clinical state of the patient; the size, shape and type of the bone; and technical skill of the medical physician performing the endoscopy (ASGE 1995). Previously, oesophageal alien bodies were detached by surgery, but elimination using an endoscope has been conceivable since the development of the inflexible endoscope in the 1930s. This type of surgery became easier with the discovery of the flexible endoscope (Kim et al. 2015).

In some cases, the fish bone reaches the gastrointestinal tract and causes perforation in this area, which is an uncommon incidence (Hsu et al. 2005). In addition to the fish bone, Pinero Madrona et al. (2000) suggested that chicken bones are the most frequent foreign bodies producing gastrointestinal tract perforation, but when fish bone is ingested and reaches the gastrointestinal tract, the ileum will be the most common site of perforation (Chu et al. 1998).

References

Akazawa Y, Watanabe S, Nobukiyo S, Iwatake H, Seki Y, Umehara T, Tsutsumi K, Koizuka I (2004) The management of possible fishbone ingestion. Auris Nasus Larynx 31(4):413–416

American Society for Gastrointestinal Endoscopy (ASGE) (1995) Guideline for the management of ingested foreign bodies. Gastrointest Endosc 42(6):622–625

Chu KM, Choi HK, Tuen HH, Law SY, Branicki FJ, Wong J (1998) A prospective randomized trial comparing the use of the flexible gastroscope versus the bronchoscope in the management of foreign body ingestion. Gastrointest Endosc 47:23–27

Dereci S, Koca T, Serdaroğlu F et al (2015) Foreign body ingestion in children. Turk Pediatri Arch 50:234–240

Evans RM, Ahuja A, Rhys Williams S, Van Hasselt CA (1992) The lateral neck radiograph in suspected impacted fish bones: does it have a role? Clin Radiol 46(2):121–123

Hsu SD, Chan DC, Liu YC (2005) Small-bowel perforation caused by fish bone. World J Gastroenterol: WJG 11(12):1884

Kim JP, Kwon OJ, Shim HS, Kim RB, Kim JH, Woo SH (2015) Analysis of clinical feature and management of fish bone ingestion of upper gastrointestinal tract. Clin Exp Otorhinolaryngol 8(3):261

Lue AJ, Fang WD, Manolidis S (2000) Use of plain radiography and computed tomography to identify fish bone foreign bodies. Otolaryngol Head Neck Surg 123(4):435–438

Nandi P, Ong GB (1978) Foreign body in the oesophagus: review of 2394 cases. Br J Surg 65:5–9

Ngan JH, Fok PJ, Lai EC, Branicki FJ, Wong JOHN (1990) A prospective study on fish bone ingestion. Experience of 358 patients. Ann Surg 211(4):459

Pinero Madrona A, Fernandez Hernandez JA, Carrasco Prats M, Riquelme Riquelme J, Parrila Paricio P (2000) Intestinal perforation by foreign bodies. Eur J Surg 166:307–309

Reilly JS, Walter MA, Beste D, Derkay C, Muntz H, Myer CM, Friedman E, Seibert R, Todd NW, Milbrath M, Andrews T (1995) Size/shape analysis of aero-digestive foreign bodies in children: a multi-institutional study. Am J Otolaryngol 16(3):190–193

Sheth N, Diner WC (1988) Swallowing problems in the elderly. Dysphagia 2(4):209–215

Singh N, Chong J, Ho J, Jayachandra S, Cope D, Azimi F, Eslick GD, Wong E (2018) Predictive factors associated with spontaneous passage of coins: a ten-year analysis of paediatric coin ingestion in Australia. Int J Pediatr Otorhinolaryngol 113:266–271

Sundgren PC, Burnett A, Maly PV (1994) Value of radiography in the management of possible fishbone ingestion. Ann Otol Rhinol Laryngol 103(8 Pt 1):628–631

Wong E, Cheng AT, Aggarwala S, Chong J, Duvnjak M, Azimi F, Cope D, Singh N (2019) Paediatric fish bone ingestion: a 10-year analysis at a tertiary paediatric hospital. Aust J Otolaryngol 2

Section VIII
Environmental Challenges

Chapter 27
The Status of Pollution in the Southern Marshes of Iraq: A Short Review

Nader A. Salman, Hamid T. Al-Saad, and Faris J. Al-Imarah

Abstract A short review of the water quality and status of pollution in the waters of the southern marshes of Iraq was discussed in this chapter. It starts with a hydrographical description of this important area for its biodiversity richness and historic cultural activities. Previous and recent scientific articles regarding water quality criteria, productivity and pollution with heavy elements and hydrocarbons were reviewed.Water quality criteria include dissolved oxygen, hardness, salinity, alkalinity and nutrients contents. The marsh water was found hard to very hard, rich in oxygen, phosphorus, and nitrogen. Such characters are favorable for fisheries and aquaculture activities. Special attention has been focused on the status of pollution with heavy metals, hydrocarbons and other toxic materials. The concentrations of the dissolved phase of some heavy elements in the marshes were below the permissible limits of drinking water. However, reported values indicate that the Al-Hammar marshland is rich in petroleum hydrocarbons as compared with other marshes. Factors affecting the pollution status and water quality deterioration were critically discussed.

Keywords Hydrocarbons · Toxic materials · Heavy metals · Wetlands · South of Iraq

27.1 Introduction

Marshes in the middle and lower basin of the Tigris and Euphrates Rivers in Iraq are the most extensive wetland ecosystems in the Middle East and Southeast Asia and covered more than 15,000 km^2. In their lower courses, these two great rivers have

N. A. Salman (✉)
Department of Medical Laboratory Technique, Al-Manara College for Medical Sciences, Misan, Iraq
e-mail: naderabed@uomanara.edu.iq

H. T. Al-Saad · F. J. Al-Imarah
College of Marine Sciences, University of Basrah, Basrah, Iraq

L. A. Jawad (ed.), *Southern Iraq's Marshes*, Coastal Research Library 36,
https://doi.org/10.1007/978-3-030-66238-7_27

created a vast network of wetlands, which is known as the Mesopotamian Marshes. These wetlands comprise a complex of interconnected shallow freshwater lakes, marshes, and seasonally swamped floodplains extending from the region of Basrah to within 150 km of Baghdad (Bedair et al. 2006).

Iraqi southern marshes form a large triangular region bounded by three major southern cities: Nasiriyah to the west, Amarah to the northeast, and Basrah to the south. Their vast area covers 20,000 square kilometers of open water and includes both permanent and seasonal marshes. Most of the largest wetlands within this complex are the Hor Al-Hammar and its associated marshes (350,000 ha) south of the Euphrates; the Central Marshes (300,000 ha), a vast complex of permanent lakes and marshes north of the Euphrates and west of the Tigris; and Hor Al-Huweza marsh and its associated marshes (220,000 ha) extending east from the Tigris into Iran.

Less than 10% of area remained as functioning marshland by year 2000 systematic plan to drain the marshes of southern Iraq (Partow 2001; Brasington 2002). The only remaining marsh of small size was the portion of Al-Huweza which straddles the border between Iran and Iraq. The other two marshes, Central (also locally known as the Qurna marsh with the largest lake) and Al-Hammar, were virtually destroyed by 2000. The remaining Al-Huweza was only 35% of its 1977 size of 3076 km^2 by 2000 (Richardson and Hussein 2006). Although the Mesopotamia marshes had been completely destroyed, it become clear on first inspection that they were restorable since they are true (river of grass) wetlands fed by river and dominated by the aquatic grass *Phragmites australis* (Alwan 2006). At the beginning of 2003, only 7% of the original marshlands remained; after June 2003 refolding of the marshes by people in the area, the opening of gates by local government officers, and release of water by Iran to the east were occurring. Field analysis concluded that water quantity and quality were sufficient to restore some area of the marshes and that rapid reestablishment of native plant species was occurring in some area (Richardson et al. 2005). Satellite photos indicate that by 2005, a percentage of 39% of the destroyed marshes had standing water and vegetation cover was expanding at 800 km^2/year (UNEP 2005).

The marshes were famous for their biodiversity and cultural richness. They were the permanent habitat for millions of birds and flyway for millions more migrating between Siberia and Africa (Maltby et al. 1994, Evans 2002). Sixty-six bird species may now be at risk. Other populations are thought to be in a serious decline. Coastal fisheries in the Arabian Gulf used the marshland for spawning, and they served as nursery grounds for shrimp and marine fish. Now fish catches have been decreased (UNEP 2003). Such specific wetlands of the southern part of Iraq play a vital role in the maintenance of biodiversity in the Middle East (UNEP 2001) primarily because of their large size, the richness of their aquatic vegetation (Alwan 2006), and their isolation from other comparable systems. The values of the marshes are numerous, including rich flora and fauna, livestock-grazing fields, fish, and other wildlife breeding places (Hussain and Ali 2006). Marshes are also known to be a farmland for rice, and cultivation areas for some other crops. The Mesopotamian marshlands are important for economic, social, and biodiversity values characterized by frequency of water flows, accumulation of nutrients and organic matters, and the production of commercially important vegetation and fish. Species lists, classifications, and specific characteristics, water chemistry, and some physical factors of the area can be found in a number of studies (Bedair et al. 2006).

Marshes are considered as biological filters of the Earth, and are often called as "Earth's kidney." They are considered as an intermediate region between aquatic and terrestrial systems which play a significant role in purifying the polluted water for centuries. The majority of marshes have been exploited for their natural cleansing capacity for assimilating different contaminants (Joyce 2012). In the natural system, marsh sediments accumulate types of chemicals that enter the water bodies through natural and anthropogenic activities (Li et al. 2006; Adekola and Eletta 2007; Lu et al. 2011). Marsh system can act as a sink of various compounds through either sedimentation or bioaccumulation. The Mesopotamian marshlands are known as a sink for different compounds (Mitsch and Gosselink 2007).

27.2 Water Quality

27.2.1 Dissolved Oxygen (DO)

Dissolved oxygen in the water of the southern marshes ranged between a minimum value of DO (5 mg/l) and maximum value of 12.25 mg/l (Douabul et al. 2013; Al-Mahmood and Hmood 2006). This referred to high levels of dissolved oxygen due to aerated, high water levels, well mixing, and density of phytoplankton (Al-Zubaidi 1985), while the lower values may be due to low water level and degradation of organic matter (AL-Saad et al. 2008; Al-Imarah et al. 2006). Optimal concentrations of dissolved oxygen for many organisms are over 5 mg/l (Lagler 1971). Conditions may be considered oxygen deficient at less than 0.5 mg/l, but some organisms can survive down to 0.1 mg/l. According to the Eden Master Plan Survey (2006), the waters of the Huweza marsh is well oxygenated. Only in some rather confined areas with high organic matter production was oxygen depletion observed. Except in these cases, nearly all of the measurements demonstrated aerobic conditions and typical oxygenation (around 10 mg/l).

27.2.2 Salinity

Salinity of the southern marsh waters is initially controlled by its source from the Tigris, and its branches which are the least-saline sources compared with Euphrates and Shatt al-Arab. In general, the lower overall salinities of some southern marshes such as Huweza marsh revealed that the marsh has achieved "hydraulic flow-through" conditions. Parts of the marsh that exhibit higher salinity and higher ranges of salinity are those where the water enters and stagnates (FAO 2010).

Seasonal variation was noticed in the southern marsh, being lower during the period from July to September (0.3 and 0.9 ppt) and higher during February to April (0.6 and 2.0 ppt) (Al-Mahmood and Hmood 2006; Douabul et al. 2013). The recorded levels may be attributed to the flooding and water exchange dynamics in addition to the uptake of salts by newly flourishing plants like *Phragmites* sp. and *Typha* sp. The salinity of water in the marshes is important for several reasons:

1. The possibility of using the water for drinking restrains the salinity level to less than 0.5 ppt.
2. It is important to evaluate whether salinity is appropriate to support wildlife, primarily birds and fish.
3. Salinity controls the type of vegetation present.

27.2.3 Total Hardness (TH)

High values of total hardness were recorded in the marshes, reflecting a common phenomenon in Iraqi inland water (Saeed 1997). According to Reid scale, waters of marshlands in Iraq were hard to very hard, and ranges of total hardness were between 570 and 1930 mg as CaCO3 mg/l. Spatial variations in total hardness showed large differences among different sites in the marshes. The higher values of total hardness were possibly related to high water level and discharge rate (Al-Falawi 2005) or related to evaporation which increase and concentrate the available cations (Toma 2002) or due to high precipitation and thus high soil leaching and high current velocities (Al-Mousawi et al. 1994), while the lower values are related to dilution factor due to precipitation or high water level (Toma 2002). In Al-Mahmood and Hmood (2006) study, fluctuation in the level of total hardness was related to change in water discharge rate.

27.2.4 Total Alkalinity

Total alkalinity of the southern marsh water is mainly due to bicarbonate ions which are abundant in Iraqi inland water (AL-Saad et al. 2008; and Al-Lami et al. 2002). The bicarbonate values recorded in the marshes were within the range of 244.04–854 mg/l and agreed with the other studies (Al-Saadi et al. 2001; Al-Lami et al. 2002). High levels of bicarbonate values recorded were probably related to phytoplankton activity as photosynthesis (Witton 1975) or due to the organic matter degradation and thus of calcium carbonate (Weiner 2000).

27.2.5 Nutrients

The waters of southern Iraqi marshlands in general are rich in nutrient especially nitrate and phosphate which enhance their suitability for growth and well-being of aquatic phytoplankton and aquatic plants which are necessary for primary productivity and food chains in marshland water.

27.2.6 Phosphate

The FAO (2010) survey in Huweza marsh referred to high concentration of total phosphorus recorded in April 2008, while the low values were recorded during July 2007. This is possibly due to additional input from agricultural land nearby or due to the activity of phytoplankton, human, and industrial effluent loads (Al-Imarah et al. 2006; Al-Shawi 2006; AL-Saad et al. 2008). It has been concluded that the waters of southern Iraqi marshlands are rich in phosphate which enhance their suitability for growth and well-being of aquatic plants and phytoplankton which are necessary for primary productivity in marshland water and food chain (Douabul et al. 2013).

27.2.7 Nitrite and Nitrate

The recorded nitrites in Huweza marsh were low in the range of 0.047–3.39 μg at $N\text{-}NO_2/l$ at Um-Al-Niaj, while nitrates characterized by high values were in the range of 0.240 at Lesan Ajerda to 20.01 μg at $N\text{-}NO_3/l$ at Al-Safiya (FAO 2010). Unlike phosphorus, nitrogen occurs in a number of very distinct inorganic forms in aquatic ecosystem; N_2, NH_4, NO_2, and NO_3 also occur in a variety of organic forms in aquatic plants (Wetzel and Likens 2000). The high concentration of total nitrogen was recorded during the winter, while the lower value was obtained during the autumn. This is possibly related to the change in water temperature and degradation processes (AL-Saad et al. 2008). Also the fluctuation in the concentration of total nitrogen depends on nitrogen uptake by phytoplankton and auto-chemotrophic bacteria.

27.2.8 Total Nitrogen/Total Phosphorus (TN:TP) Ratio

TN:TP ratios, the ratio of total nitrogen to total phosphorus by mass, are an indicator of nutrient conditions that define factors potentially limiting water productivity. Data presenting the ratios of TN:TP at wet (22.1–29.1) and dry (17.1–18.1) seasons at

Huweza marsh waters tend to be phosphorus limiting since the TN:TP ratio exceeded the theoretical level of 16:1 (Redfield ratio).

Typically organic matter of aquatic algae and macrophytes contains phosphorus, nitrogen, and carbon in approximately the ratio of 1P:7N:40C per 100 dry weight (Hillebrand and Sommer 1999), so decomposition of this organic matter will add more nutrients to marsh waters. In a phosphorus-limited system, an increase in phosphorus will result in an increase in productivity (Horne and Goldman 1994); this phenomenon occurred during dry season at southern marshes especially in Huweza waters, which showed an increase in chlorophyll (FAO 2010). The low TN:TP ratio during dry season stimulates the growth of different species belonging to Cyanobacteria.

27.2.9 The Heavy Elements

Ranges recorded previously by Al-Saad et al. (2009) and Salman (2011a, b) in the southern marsh water are as follows: Co (1.13–3.68), Mn (0.16–1.37), Ni (0.66–2.37), Fe (0.28–1.51), and Cu (0.10–0.28) mg/l. Certain areas from Al-Hammar marsh contained highest levels of heavy elements in their water. Lower values were recorded at Al-Huwaiza marsh (Salman 2011a). Trace elements enter the southern Iraqi marshes from both natural and anthropogenic sources (Abaychi and Al-Saad 1988). Natural sources include dust fall, erosion or crusted weathering, and dead decomposition of the biota in the water, whereas the anthropogenic sources include sewage wastes, pesticides, and fertilizers coming through irrigation and industrial effluent.

According to the Al-Atbee (2018) study, the concentrations of the dissolved phase of Cd, Cr, Ni, and Pb in Al-Chibayish marsh at all seasons were below the permissible limits of drinking water according to the WHO limit (2011). Most of the studies that have been carried out on the southern marshes revealed a decreasing trend of heavy element concentrations (Al-Khafaji et al. 2012; Al-Awady et al. 2015). This may be due to many factors such as the low flow rate of water in the marshes, quantities of plankton and suspended material that load in the water, complexion with organic matter and precipitation to the sediment, and accumulation of element in aquatic plant (Salman 2011a).

As for the particulate phase of Cd, Ni, and Pb concentrations, Al-Atbee (2018) found an increase in summer compared to spring. This may be related to high growth rate of phytoplankton in summer which has the ability to accumulate heavy elements in their bodies. In addition, high pH value leads to increased heavy elements in the particulate phase (Byrd et al. 1990; Fang and Lin 2002). This is consistent with many previous studies (Al-Haidarey 2009; Qzar 2009; Al-Awady et al. 2015).

27.2.10 The Heavy Elements Pollution Index (HPI)

Al-Atbee (2018) found that the HPI value in Chibayish marsh is less than the maximum threshold value of 100 as proposed by Reza and Singh (2010). This result indicates that the influence of the four studied elements on the marsh water quality is not alarming because it's below the permissible limits of drinking water according to the WHO (2011). Al-Hejuje (2014) has reported an alarming value of HPI at Shatt al-Arab river which was higher than the permissible limits of drinking water according to the WHO (2011). This means that the river was polluted with heavy elements, more than the marsh. Shatt al-Arab river received many pollutants with high levels of heavy elements through its branches at Basrah city (Al-Hejuje et al. 2017). Ewaid (2016) recorded the HPI value (98.6) in Al-Gharraf river water slightly below the critical value of 100 for drinking water.

27.2.11 Total Petroleum Hydrocarbons (TPHs)

TPH concentrations in the southern marshes varied between 0.6 and 46.6 μg/l (Al-Imarah et al. 2006). On the contrary, much lower values were recorded by Al-Saad et al. (2009) being 0.012–0.037 μg/l. Values at various stations in Al-Huweza marsh were recorded by Al-Khatib (2008) as 1.005–11.965 μg/l. TPHs in Al-Hammar marsh ranged between 0.41 and 0.82 μg/l, as shown by Talal (2008). Values in the range of 1.91–6.79 μg/l were seen in Al-Chibayish marsh by Al-Atbee (2018).

Previous reports have indicated that the Al-Hammar marshland is rich in petroleum hydrocarbons as compared with other marshes (Al-Imarah et al. 2006; AL-Saad et al. 2009). Spatial variations in hydrocarbon concentrations at Al-Hammar marsh water were noticed (Salman 2011a). Two stations (Al-Baghdadia and Al-Nagarah) recorded the highest levels (36.59 and 22.35 μg/l, respectively). The other two stations (Al-Bargah and Um Al-Ward) recorded significantly lower values (16.68 and 17.43 μg/l, respectively). The percentage of total organic carbon (TOC) reflected nearly the same trend. The values of all studies are within the normal limits for freshwater environment (less than 50 μg/l) and are within the safe limits of human usage, as recommended by the World Health Organization (WHO 2011).

There are two factors affecting the variation of petroleum hydrocarbons in water of the southern Iraqi marshlands, namely, temperature and water flow (AL-Saad 1995). It seems that the phenomenon of water flow is the more effective factor governing the random variation of petroleum hydrocarbons in the southern Iraqi marshlands more than seasonal temperature variations (DouAbul and Al-Saad 1985). High levels of TPHs were shown in Al-Chibayish marsh during winter (Al-Atbee 2018), which may be attributed to the increase fall of airborne compounds during rain and use of wood and hydrocarbons for heating in winter (AL-Saad 1995; Al-Khatib 2008; Jazza 2015).

27.2.12 Polycyclic Aromatic Hydrocarbons (PAHs)

PAH values in the Huwaiza marsh ranged between 1 and 50.8 μg/l (Al-Khatib 2008), while another range was recorded by Al-Atbee (2018) in Al-Chibayish marsh being 2.4–37.8 μg/l. To distinguish between pyrogenic and petrogenic sources of PAH in the marsh water, the result of Al-Atbee (2018) showed the ratio was less than one at all stations of Chibayish marsh and in all seasons. This has confirmed the pyrogenic origin which derived from incomplete combustion of fuel, except one station where PAH comes from petroleum origin. This result was in agreement with Al-Khatib (2008).

27.3 Pollution and Toxicity

The southern marshy habitats suffered from various types of pollution especially that related to pesticides, hydrocarbons, or heavy metal toxicants (Salman 2011a, b). Water pollution from agricultural and domestic sources seriously affects water quality in the southern marshes. They suffer from salinity through saltwater intrusion from the Shatt al-Arab and lack of throughflow. Irrigation return flows are discharged directly to the marshes. Industrial waste and raw sewage significantly impair water quality (Stevens and Salman 2015). A higher chemical pollution in the marshes and rivers was previously reported (DouAbul et al. 1988; Saeed et al. 1999).

The levels of pollution in marshes have increased substantially in recent years (Al-Imarah et al. 2006). Both biotic and abiotic factors can adversely be affected by the presence of toxicants even in trace concentrations. The introduction of motorboats to the deeper areas of the marshes has led to noticeable and frequent oil pollution along the heavily used waterways between the main villages. The distribution of polycyclic aromatic hydrocarbons (PAHs) in the Al-Hammar marsh sediments was studied by Al-Saad and Al-Timari (1989). In addition to their fatal effect on aquatic organisms, toxicants may directly or indirectly affect human health.

The uncontrolled flooding of marshes now presents potential problems and challenges regarding the quality of water because of the release of toxins from reflooded soils that are contaminated with chemicals from mines and military ordnance (Richardson and Hussein 2006). According to Richardson et al. (2005), the analysis of surface water and soils for organochlorine pesticides, polychlorinated biphenyls, and polycyclic aromatic hydrocarbons (PAHs) showed no detectable concentrations of any of these xenobiotics. However, recent surveys in the Abu Zarag marsh have shown low molecular weights of PAHs in the Abu Zarag soils, probably as a result of the severe burnings in the region (DouAbul et al. 2005).

According to Salman (2011a, b), a moderate level of pollution with petroleum hydrocarbons and traces of heavy metals has been detected in some sites of the Al-Hammar marshes. Lower levels were also found in the Al-Huweza marsh.

27.4 Impact of Water Quality on Fisheries and Aquaculture Activities in the Marshes

Three of the most important parameters which have a direct impact on fisheries and aquaculture status in the southern marshes are considered. Waters of the marshes, with total hardness of 300–380 mg/l during dry season and 600–1300 during wet season are considered hard to very hard according to Reid scale which ranges between 570 and 1930 mg as CaCO3 mg/l. The marsh fishes are well adapted to such hardness especially in the wet season. Some are good osmoregulators especially when hardness was associated with relatively high salinity (more than 1.5 ppt) as seen in previous laboratory studies. Fisheries of some sensitive species such as *Barbus* sp. may be affected adversely by waters of high hardness values, but many other species can tolerate such hardness, as seen in their relative abundance (FAO 2010). Phosphorus PO_4 concentration in Huweza marsh varied between 0.75 and 3.40 μg at P/l during dry season and from 0.64 to 3.40 μg at P/l during wet season. The marsh is considered rich in phosphorus. As for nitrogen values ranged between 0.24 and 7.35 in dry season and 6.94 and 20.01 during wet season that make the marsh water rich in this nutrient. It has also been noticed that relatively high phosphorus and nitrogen concentrations coincide with the spawning period of most of fish species in the marshes (May to February). Values of phosphates and nitrates are considerably high, reflecting nutrient richness of the marsh waters, which can act positively with fisheries and aquaculture potentials, being highly productive water.

References

Abaychi J, Al-Saad H (1988) Trace element in fish from the Arabian gulf and Shatt Al-Arab River. Iraq Bull Environ Contam Toxicol 40:226–232

Adekola FA, Eletta OAA (2007) A study of heavy metal pollution of Asa River, Ilorin, Nigeria; trace metal monitoring and geochemistry. Environ Monit Assess 125:157–163

Al-Atbee RS (2018) Assessment of some heavy elements and hydrocarbons in the water, sediments and dominant aquatic plants at Al-Chibayish marshes.MSc. thesis, Basrah University

Al-Awady A, Maktoof A, Al-Khafaji BY, Abid NM (2015) Concentration of some heavy metals in water, sediment and two species of aquatic plants collected from the Euphrates river, near the center of Al-Nassiriyia city, Iraq. Marsh Bull 10(2):161–172

Al-Falawi HJJ (2005) Ecological study Euphrates River between Al-Hindiya dam and Al-Kefel city-Iraq. MS.c thesis, University of Babylon

Al-Haidarey MJS (2009) Assessment and sources of some heavy metals in Mesopotamian Marshes. Ph.D. thesis, College of Science for Women, University of Baghdad, Iraq, 158p

Al-Hejuje MM (2014) Application of water quality and pollution indices to evaluate the water and sediments status in the middle part of Shatt AlArab River. Ph.D. thesis, Biology Department, College of Science, University of Basrah, 239p

Al-Hejuje MM, Hussain Najah A, Al-Saad HT (2017) Applied heavy metals pollution index (HPI) as a water pollution Indicator of Shatt Al-Arab River, Basrah-Iraq. Int J Mar Sci 7(35):353–360

Al-Imarah FJM, Al-Shawi IJM, Al-Mahmood HK, Hmood AY (2006) Study of some physical and chemical characterizations of water from southern Iraqi marshlands after rehabilitation. Marsh Bull 1(1):82–91

Al-Khafaji BY, Dawood YT, Maktoof AA (2012) Trace metals distribution in fish tissues (Cyprinus carpio and Barbus luteus) and sediment from Al-Massab Alamm river near the center of Al-Nassiriyia city. J Thi-Qar Sci 3(2):22–30

Al-Khatib FM (2008) Determination the concentrations, origin and distribution of hydrocarbon compounds in water, sediments and some biota of Hor Al-Howaiza, south of Iraq and their sources. Ph.D. thesis., University of Basrah, College of Science, Biology Department, 228p (in Arabic)

Al-Lami AA, Rathi AG, Al-Dylymi AA, Rasheed RS, Hassan AA (2002) The study of some ecological factors of four lotic aquatic system with different salinity, Middle of Iraq-Tikrit. J Pure Sci 5:14–25

Al-Mahmood H, Hmood A (2006) Study of some physical and chemical characterizations of water from southern Iraqi marshlands after rehabilitation. Marsh Bull 1(1):82–91

Al-Mousawi AHA, Al-Saadi HA, Hassan FM (1994) Spatial and seasonal variations of phyto-plankton population and related environments in Al-Hammar marsh. Iraqi Basrah G.Sci B 12(1):9–20

AL-Saad HT (1995) Distribution and sources of hydrocarbons in Shatt Al-Arab estuary and NW of the Arabian Gulf. Ph.D. thesis, College of Science, Basrah University, Iraq, 186p

AL-Saad H, Al-Timari A (1989) Distribution of polycyclic aromatic hydrocarbons (PAHs) in the marsh sediments. Iraq Bull Environ Contam Toxicol 43:864–869

AL-Saad HT, Al-Hello MA, Kareem S, DouAbul AAZ (2008) Water quality of Iraqi Southern marshes. Marina Mesopotamica 23(1)

AL-Saad HT, Al-Taein SM, Al-Hello MAR, DouAbul AAZ (2009) Hydrocarbons and trace elements in water and sediments of the marsh land of southern Iraq. Mesopotamian J Mar Sci 24(2):126–139

Al-Saadi HA, Sulaiman NA, Ismail AM (2001) On some limnological characters of three lotic water system, middle of Iraq. Ibin-Haitham. J Appl Pure Sci 2:5–12

Al-Shawi IJ (2006) Comparative study on some physic-chemical characteristic for northern Al-Hammar marsh water before destroyed and after rehabilitation. Marsh Bull 204:127–133

Alwan AR (2006) Past and presence status of the aquatic plants of the marshlands of Iraq. Marsh Bull 2:140–153

Al-Zubaidi AJM (1985) Ecological study on the algae (phytoplankton) in some marshes near Qurna (southern Iraq)M.Sc. thesis, University of Basrah, Iraq, 236p

Bedair HM, Al-Saad HT, Salman NA (2006) Iraqs southern marshes something to be conserved. A case study. Marsh Bull 1

Brasington J (2002) Monitoring marshes degradation using multispectral remote sensed imagery. In: Nicholson E, Clark P (eds) The Iraqi Marshlands: a human and environment study. Politico's, London, pp 151–168

Byrd JT, Lee KW, Lee DS, Smith RG, Windom HL (1990) The behavior of trace metals in the Geum estuary, Korea. Estuar Coasts 13(1):8–13

DouAbul A, Al-Saad H (1985) Seasonal variations of oil residues in water of Shatt Al-Arab river, Iraq. Water Air Soil Poll 24:237–246

DouAbul A, Al-Saad H, Al-Timari A, Al-Rekabi H (1988) Tigris-Euphrates delta: a major source of pesticides to the Shatt Al-Arab River (Iraq). Arch Environ Contam Toxicol 17:405–418

DouAbul AZ, Mahamed SS, Warner B, Abaychi JK, Alwash A (2005) Restoration processes in Abu Zarag– central marsh: a case study. 9th annual meeting of the ecological Society of America, special session 9: restoration of Mesopotamian marshes of Iraq, Montreal, Canada
Douabul AA, Al-Saad HT, Abdullah DS, Salman NA (2013) Designated protected marsh within mesopotamia: water quality. Am J Water Resour 1(3):39–44. http://pubs.sciepub.com/ajwr/1/3/4 © Science and Education Publishing. https://doi.org/10.12691/ajwr-1-3-4
Eden Master Plan (2006) New Eden master plan for integrated water resources management in the marshland area. Volume I, Overview of Present Conditions and Current Use of the Water in the Marshland Area, Book 4, Marshland. The Italian Ministry for the Environment and Territory and Free Iraq Foundation, Italy – Iraq. 256 pp
Evans ML (2002) The ecosystem. In: Nicholson E, Clark P (eds) The Iraqi marshlands: A human and environment study. Politico's, London, pp 201–219
Ewaid SH (2016) Water quality evaluation of Al-Gharraf river by two water quality indices. Appl Water Sci 26:1–7
Fang TH, Lin CL (2002) Dissolved and particulate trace metals and their partitioning in a hypoxic estuary: the Tanshui estuary in northern Taiwan. Estuar Coasts 25(4):598–607
FAO (2010) Base line survey of Huwaiza marsh, Final project report. FAO-Iraq & Basrah University, Basrah, 137pp
Hillebrand H, Sommer U (1999) The nutrient stoichiometry of benthic microalgal growth: Redfield proportions are optimal. Limnol Oceanogr 44(2):440–446
Horne AJ, Goldman CR (1994) Limnology. McGraw-Hill, New York
Hussain NA, Ali TS (2006) Tropical nature and feeding relationship among Al-Hammar marsh fishes, southern Iraq. Marsh Bull 1(1):9–18
Jazza SH (2015) The state of hydrocarbon compounds pollution of water, sediments and some aquatic biota in Al-Kahlaa river-Missan Province/Iraq. Ph.D. thesis, University of Basrah, College of Science, Biology Department, 137p
Joyce C (2012) Preface: wetland services & management. Hydrobiologia 692:1–3
Lagler KF (1971) Fresh water fishery biology. WM. Publishing Co, Iowa
Li Y, Yu Z, Song X, Mu Q (2006) Trace metal concentrations in suspended particles, sediments and clams from Jiaozhou Bay of China. Environ Monit Assess 121:491–501
Lu Q, He ZL, Graetz DA, Stoffella PJ, Yang X (2011) Uptake and distribution of metals by water lettuce (Pistia stratiotes L.). Environ Sci Pollut Res Int 18(6):978–986
Maltby E et al (1994) An environmental and ecological study of the marshlands of Mesopotamia: draft consultative bulletin, wetland ecosystems research group, University of Exeter. AMAR Appeal Trust, London
Mitsch WJ, Gosselink GJ (2007) Wetlands, 4th edn. Wiley, Hoboken, New Jersey
Partow H (2001) The Mesopotamian marshlands: demise of an ecosystem. Nairobi (Kenya): division of early warnings and assessment. United nations environment Programme. UNEP publication UNEP/DEWA/TR.01-3
Qzar EAA (2009) Concentration of trace metals in environment and some Gastrophoa (Mollusca) in east Hammar marsh. M.Sc. thesis, University of Basrah
Reza R, Singh G (2010) Heavy metals contamination and its indexing approach for river water. Int J Environ Sci Technol 7(4):785–792
Richardson CJ, Hussein NA (2006) Restoring the Garden of Eden: An ecological assessment of the marshes of Iraq. Bioscience 56(6):477–489
Richardson CJ, Reiss P, Hussain NA, Alwash AJ, Pool DJ (2005) The restoration potential of the Mesopotamian marshes of Iraq. Science 307:1307–1311
Saeed SH (1997) The study on water quality of Al-Kasa chai river in Kirkuk city. MS.c thesis. Tikrit University
Saeed T, Al-Ghandban N, Al-Shemmari H, Al-Mutair M, Al-Hashash H (1999) Preliminary assessment of the impact of draining of Iraqi marshes on Kuwait's northern marine environment, part II: sediment associated pollutants. Water Sci Technol 40:89–98

Salman NA (2011a) Assessment of environmental toxicity in Iraqi Southern marshes using fish as bio-indicators. EKOLOGIJA. T 57(1):21–29. © lietuvos mokslų akademija
Salman NA (2011b) Suggested methods for enhancement of aquaculture in the southern marshes of Iraq. Iraqi J Aquacult 8(2):157–177
Stevens M, Salman NA (2015) Application of international water low in Eden; Environmental protection of Mesopotamian marshes in Southern Iraq. Wetland Sci Pract
Talal AA (2008) A study for the seasonal and regional variations of hydrocarbon levels and origin of n-alkanes in water, sediments and some species of biota in Hor Al-Hammar Marshes. Ph.D. thesis, College of Science, University of Basrah, 166p
Toma JJ (2002) Limnological study of Dokan lake, Kurdistan region, Iraq. MS.c thesis, University of Salahaddin-Arbil
UNEP (2001) The Eden again project. Restoration of the Mesopotamian Marshlands
UNEP (2003) Environment in Iraq: UNEP progress report. UNEP, Geneva
UNEP (United Nations Environment Programme). (2005) UNEP project to help manage and restore the Iraqi marshlands. (6April, 2006 http://marashlands.unep.or.jp)
Weiner ER (2000) Application of environmental chemistry. Lewis Publisher, London/New York
Wetzel RG, Likens GE (2000) Limnological analysis, 3rd edn. Springer
WHO (2011) Guidelines for drinking-water quality, 4th edn., Geneva. http://www.who.int.
Witton BA (1975) River ecology. Black Well Scientific Publication, Oxford/London

Chapter 28
The Effects of Man-Made Noise on the Fishes in the Marshes of Iraq

Laith A. Jawad

Abstract Recently there was a cumulative interest among fisheries managers in regard to the man-made sound and its effects on the freshwater fish; this is because of the upsurge in the surrounding noise. In the literature, there is a dearth of information about the impacts of the sound on the freshwater fish species in comparison with the other aquatic animals such as marine mammals and fishes. The impact of human-generated sound on freshwater fishes can be seen in two main attributes, the physiological and behavioural. In the present chapter, a short description for both effects was given, and in a separate section, a discussion for the man-made sound present in the southern marshes of Iraq was given. This chapter aims to draw awareness of the scientific community to increase research on the impacts of the man-made sound on the freshwater fishes in the southern marshes of Iraq in order to reduce the injury that these fishes experiencing in their environment.

Keywords Boats · Guns · Illegal fishing · Physiological changes · Behavioural changes · South of Iraq

28.1 Introduction

Among the sensory stimuli to fishes is sound because it may be able to be used energetically for several biological activities that the fishes can go through during their life (Fay and Popper 2000; van der Sluijs et al. 2011; Mickle and Higgs 2018). Sound spreads very powerfully in deep water but not as such in shallow habitats owing to interface with surfaces and sediments (Kuperman and Ingenito 1998; Akyildiz et al. 2005); nevertheless it is a serious sensory incentive in most habitats (Popper and Fay 1973).

Numerous fish species are mainly dependent on sound as a method of interaction (van der Sluijs et al. 2011; Mickle and Higgs 2018), particularly as visual signals can

L. A. Jawad (✉)
School of Environmental and Animal Sciences, Unitec Institute of Technology, Mt Albert, Auckland, New Zealand

L. A. Jawad (ed.), *Southern Iraq's Marshes*, Coastal Research Library 36,
https://doi.org/10.1007/978-3-030-66238-7_28

be clogged in dark or turbid environments (Heuschele et al. 2012; Fischer and Frommen 2013). A few sounds in underwater habitats are further damaging than they are helpful, chiefly anthropogenic noise, which is a common human-made commotion for aquatic life (Popper and Hastings 2009; Radford et al. 2014; Solan et al. 2016; Mickle and Higgs 2018).

Man-made sound is chiefly instigated by urban expansions, the increase in shipping transportation systems, underwater resource removal, and seismic survey appliances and has been cumulative recently (Hildebrand 2009; Frisk 2012; Solan et al. 2016; Vazzana et al. 2017; Mickle and Higgs 2018). Such causes of sound are conjectured to upset acoustic interactions and have a significant impact on aquatic species (Wysocki et al. 2006; Popper and Hastings 2009). Most aquatic investigations have focused on an acute noise causes such as sonar, airguns, and pile driving owing to the unequivocal harm they can inflict on animals (Popper and Hastings 2009); nevertheless, shipping or movement of mechanised boat is the most overriding cause of man-made sound that spreads at low underwater frequencies and overlays with the hearing range of several aquatic species (Ross 1976; Dyndo et al. 2015; Solan et al. 2016; Mickle and Higgs 2018).

Most of the literature related to the impact of man-made sound on aquatic animals have dedicated their interest on perceiving noticeable behavioural variations in these animals, containing deviations to their foraging competence (Purser and Radford 2011; Sabet et al. 2015; McLaughlin and Kunc 2015; Mickle and Higgs 2018), or ensuing in physiological variations, such as cumulative impact levels, or instigating a hearing deficiency (Smith et al. 2004; Wysocki et al. 2006; Nichols et al. 2015; Mickle and Higgs 2018).

The impact of the man-made sound on marine animals was well investigated through the studies on marine mammals (Weilgart 2007; Heide-Jorgenson et al. 2013; Dyndo et al. 2015; Mickle and Higgs 2018); however, those investigations on freshwater species are very few (Popper 2003; Slabbekoorn et al. 2010). The space that the sound travels in within the marine environment is much bigger than that of the freshwater (Mickle and Higgs 2018). Therefore, freshwater systems may be less competent at sound transmission than marine environments and only comprise 1% of the water on the globe; nevertheless, they lodged an excessively high amount of earth's biodiversity (Combes 2003). In addition, diversity of aquatic biota is particularly susceptible to human-tempted environmental alteration owing to the high human populations near freshwater habitats together with their high species variability (Abell 2002; Dudgeon et al. 2005).

Fishes are an imperative inhabitants of freshwater environments and signify over half of all of the vertebrate species on the globe (Thomson and Shaffer 2010) and main worldwide aquaculture yield (Radford et al. 2014), underlining their prominence to humans and the need for additional studies (Mickle and Higgs 2018). Sound contamination investigations in marine habitats have been documented broadly indicating in general the influences of sound which can cause effects ranging from a behavioural alteration in an animal to death (Weilgart 2007; Popper and Hawkins 2012; Mickle and Higgs 2018). It is possible to apply such investigations as an indicator and recommendation for future freshwater sound pollution study.

The present chapter showed the possible noise effects on the freshwater fish in general, and besides it discusses the current noise pollution that exists in the southern marshes of Iraq. The present chapter offers information about the effects of human-made noise on freshwater fish. In its sections, the chapter does not intend to give all possible effects of noise on fish that might be found in the literature rather than arising an awareness about such effects on fish. The information offered in the present chapter also initiate further research and put conceivable guidelines on satisfactory levels of man-made sound in freshwater habitats.

28.2 The Effects of Sound Pollution on Fish

There are several causes that resulted from sound pollution in the freshwater environment. The main two of these causes are the physiological and the behavioural changes in the fish body. In addition, some anatomical changes might also become an outcome of such man-made pollution.

28.2.1 Physiological Changes

Stress is the main outcome of the effect of noise pollution on fish (Mickle and Higgs 2018). There are several examples obtained from experiments done on different fish species and showed significant cases of stress due to noise pollution. For instance, goldfish (*Carassius auratus*) display a change in hearing verge and hiding of sounds when exposed to four different kinds of filters in aquaria; nevertheless, there was no modification in threshold when goldfish were kept in ponds (Gutscher et al. 2011). On the other hand, Graham and Cooke (2008) exposed largemouth bass (*Micropterus salmoides*) to three dissimilar boat sound turbulences and found that fish experienced canoe paddle noise has augmented in their heart rate 29%, 44% when faced an electric trolling motor, and 67% when exposed to a combustion motor.

Finding how an animal responds to a stress response is not at all times an easy task, as it is imperative to decide the "setting, harshness, and period" of the impact and in this case sound when demonstrating if the animal is certainly affected (Bronson 1995). For instance, if the stress reaction of the animal continues for only 1 h, it is not clear at this stage whether the growth rate or appropriateness is actually influenced. Upcoming studies should contain the collection of GC levels (glucocorticoids are used as a gauge of stress in a wide selection of animals) at different time interludes to decide a stress against time that would also indicate if acclimatisation has happened. Moreover, it is significant to decide if the stress reaction is a consequence of natural diurnal or seasonal changes in GC levels, as contrasting to the stressor. In order to decide sound influences on fish, it is also conceivable to examine cardiac output as a measure of stress, as it has comparable programme to humans (Graham and Cooke 2008).

28.2.2 Behavioural Changes

It has been suggested that during the assessment of the impact of man-made noise on fish, it is preferable to use an integrative method that would better evaluate the accurate influences of noise as a possible stressor (Pickering 1992; Ellis et al. 2004; Huntingford et al. 2006; Mickle and Higgs 2018).

The majority of published impacts of sound on fish concentrated on precise behavioural or physiological features of a species, for instance defining the influence of sound on Mauthner-mediated startle responses (Zottoli 1977) or the effects of sound on hair cell impairment in goldfish (Smith et al. 2006). This is vital as it upsurges our understanding about the topic of sound pollution; nevertheless, the widely held of these investigations missing integration within their design (Mickle and Higgs 2018). Forthcoming investigations should include collaborative aspects originated from types of studies related to noise pollution and its effect on freshwater fish species in order to decide the degree to which sound disturbs them (Mickle and Higgs 2018). For instance, when examining the effects of stress on a local freshwater species, it may be helpful to examine behavioural features such as foraging efficiency and evading reaction but also to look at physiological replies such as GC levels.

Additional investigations are also required to decide the hearing ability of freshwater fish species and background sound intensities in the freshwater habitats in which they exist in to better apprehend conceivable man-made impacts (Mickle and Higgs 2018). Amoser et al. (2004) studied the hearing ability of some fish species both with and without known hearing knowledges in a freshwater lake (Lake Traunsee) and conclude sound levels during boating events to envisage effects of this sound on these species. Sound from boats overlays within the most sensitive hearing range of cyprinids in Lake Traunsee, therefore feasibly hiding noises present in their natural environment and damaging signal discovery (Amoser et al. 2004). In addition to the determination of influences of anthropogenic-generated sound, it is imperative to designate the background sound level at the first hand (Codarin et al. 2009).

In general, a stress reaction can be envisaged through behavioural features implanted in a change in searching for food efficiency, evasion reaction, startle–shelter reaction, or changes in metabolic rate (Mickle and Higgs 2018). It is important to mention here that all of the methods used to decide man-made effects on aquatic animals contain negative and positive sides in respect to their effects; consequently, to generate a more robust examination and evade confusing variables, it must be corporate procedure to contain the integration of numerous methods within each examination.

28.3 The Anthropogenic Sound Sources in the Southern Marshes of Iraq

At the first sight for any visitor to the southern marshes of Iraq, a complete noiselessness is dominating this lovely environment. But such situation will not last long when a noise of variable sorts breaks the natural silence.

It is possible to list the man-made noise in the southern marshes of Iraq as follows.

28.3.1 Sound Generated by Motors of Boats

In the last few decades, Marsh Arabs have upgraded their lifestyle by introducing oil-operated motors for their boats. Such introduction has changed the life pace in the southern marshes and introduces with it a new noise pollution that the marshes have never known before. Previously, oil engine-operated boats are usually used by Marsh Arabs for shopping from the nearest city or village, where the boats reach a certain area inside the marsh, where the depth of water allows these boats to go in. Later, the oil-operated engines became affordable for Marsh Arabs, and most of the boats that locals used in their movement within the marsh area are now fitted with this kind of engines. Some locals went even further and attained high capacity engines for faster communication. The general scene of the marshes before was a boat driven by Marsh Arab using a wooden pole, but this scene now has been changed and replaced by large boats with high capacity engine cutting the marsh water surface.

As mentioned in the section of the impacts of man-made noise on fish, motors of boats create a significant level of noise pollution in the area. Certainly, with the introduction of new high capacity motors, the level of impact of noise on fish has been upsurged dramatically.

28.3.2 Sound of Bombs Used in Illegal Fishing

Illegal fishing methods are always used by locals and visitors in the southern marshes of Iraq. Among these methods is the use of home-made and army bombs to catch fish. The sound these bombs produce can damage completely the hearing system of the fish that are found in the area where the bomb is used, while those found further away from the area of fishing will have a severe stress due to the transmission of sound wave through water. In addition to the mass killing of all aquatic animals in the vicinity of the bomb fishing event, severe stress resulted for the survival of this inhumane action.

28.3.3 Sound Generated from Gun Used to Hunt Birds

Every Marsh Arab adult individual owns one or more rifle for bird hunting. These arms vary in their made and they are efficient, but all of them produce huge noise that affects both the land and aquatic biota. In addition to the local's Marsh Arabs, regular bird's hunting enthusiasts are visiting the marsh area to practise their hobby especially in the spring and summer seasons. With this additional number of hunters, the level of noise produced from the guns upsurges significantly. In some shallow areas of the marshes and when bird's hunter is found, you can see through the water how the fish are frightened and disperse in different directions once the guns are used.

28.3.4 Sound Generated from Arm Shooting by Locals During Celebration Events

In addition to the guns that are used for hunting birds, each individual or at least each family owns a proper army machine gun. The Marsh Arabs usually prefer the Russian-made machine gun, which is available for them through those who run away from the military service or through agents who bring the arms to the marsh area regularly.

Among the habits of the Marsh Arabs in their events whether the sad or the happy occasion is to shoot live ammunition in the air. The shooting usually lasts for sometimes, and several machine guns are shared in this event. Shooting habit represents how important the person that they shoot is in his/her event, i.e. marriage, death, coming back from visiting the Shia holy shrine in Najaf and Karbala, or birth events. The shooting usually happened in the resident's area, where the living huts are located, and with such large number of shooting, a huge noise is usually created that certainly affects not only the aquatic animals but the land creatures too. It is not uncommon to have fatal incidence in the community of the Marsh Arabs as a result of shooting in the air; in addition several incidences of deafness happened mainly in children who usually attend the events.

28.3.5 Sound Generated by Machines Operating in the Marshes

In the oil exploration industry and in the inland oil fields that are out of reach of water supply for cooling of the machine, water quantities are usually brought to the oil field sites by pipelines (Van Hinte 2005). Water quantities can be drawn either from large rivers near to the oil fields or from a marine coastal area. In both cases, pipelines will cross lands of different ecological natures to reach their destination to the oil field. Among the habitats that the pipelines will cross are aquatic environments such as lakes, marshes, and rivers of different sizes (Van Hinte 2005).

The oil fields in the southern Iraq are recognised for being inland, away from any main water resources and neighbouring to the marsh area. Therefore, any pipelines need to be installed in this region that would cross the critically productive fish and other aquatic organism habitats of the lower reaches of Mesopotamia. If the proposed pipeline was built, it would extend over several kilometres stretched end to end. It would cross a number of rivers and streams in some of Iraq's most fish and other aquatic organism productive habitats. In building such project, the aquatic fauna and their ecosystems may be negatively impacted by its construction, operation, and potential failures.

Temporary noise impacts would occur during the construction phase. Noise levels associated with the construction of oil pipeline project would be variable in frequency and duration. Heavy machinery, the major source of noise in construction, is constantly accelerating and decelerating and/or moving in unpredictable patterns. This variation at the noise source leads to fluctuating noise levels. Noise levels would be elevated above "normal" in areas where installation work is being conducted. Wildlife would move away from the area of construction to avoid noise and/or noise sources, while the aquatic will have no option, but to stay in their environment.

Potential noise impacts from water construction cause serious hearing injuries and acoustic disturbance to adult and juvenile fish. Such disturbances will lead some fish species to leave the construction area, and hence there will be a loss in the fish stock and the local biodiversity in general.

The effect of underwater noise sources on the behaviour of aquatic animals is complex and depends on factors such as (1) hearing capability of individual species; (2) level and nature of noise exposure experienced; (3) habituation to man-made noise; and (4) background ambient noise.

References

Abell R (2002) Conservation biology for the biodiversity crisis: a freshwater follow-up. Conserv Biol 16(5):1435–1437

Akyildiz I, Pompili D, Melodia T (2005) Underwater acoustic sensor networks: research challenges. Ad Hoc Networks (Elsevier) 3(3):257–279

Amoser S, Wysocki LE, Ladich F (2004) Noise emission during the first powerboat race in an Alpine lake and potential impact on fish communities. J Acoust Soc Am 116(6):3789–3797

Bronson F (1995) Seasonal variation in human reproduction: environmental factors. Q Rev Biol 70(2):141–164

Codarin A, Wysocki L, Ladich F, Picciulin M (2009) Effects of ambient and boat noise on hearing and communication in three fish species living in a marine protected area (Miramare, Italy). Mar Pollut Bull 58(12):1880–1887

Combes S (2003) Protecting freshwater ecosystems in the face of global climate change. In: Hansen LJ, Biringer JL, Hoffman JR (eds) Global climate change and freshwater ecosystems. University of Washington, Seattle, pp 177–216

Dudgeon D, Arthington A, Gessner M, Kawabata Z, Knowler D, Leveque C, Naiman R, Prieur-Richard A, Soto D, Stiassny M, Sullivan C (2005) Freshwater biodiversity: importance, threats, status and conservation challenges. Biol Rev 81(2):163–182
Dyndo M, Wisniewska DM, Rojano-Donate L, Madsen PT (2015) Harbour porpoises react to low levels of high frequency vessel noise. Sci Rep 5:11083
Ellis T, James JD, Stewart C, Scott AP (2004) A non-invasive stress assay based upon measurement of free cortisol released into the water by rainbow trout. J Fish Biol 65(5):1233–1252
Fay RR, Popper AN (2000) Evolution of hearing in vertebrates: the inner ears and processing. Hearing Res 149(1–2):1–10
Fischer S, Frommen JG (2013) Eutrophication alters social preferences in three-spined sticklebacks (*Gasterosteus aculeatus*). Behav Ecol Sociobiol 67(2):293–299
Frisk GV (2012) Noiseonomics: the relationship between ambient noise levels in the sea and global economic trends. Sci Rep 2:437
Graham AL, Cooke SJ (2008) The effects of noise disturbance from various recreational boating activities common to inland waters on the cardiac physiology of a freshwater fish, the largemouth bass (*Micropterus salmoides*). Aquat Conserv Mar Freshw Ecosyst 18(7):1315–1324
Gutscher M, Wysocki LE, Ladich F (2011) Effects of aquarium and pond noise on hearing sensitivity in an otophysine fish. Bioacoustics 20(2):117–136
Heide-Jorgenson MP, Hansen RG, Westdal K, Reeves RR, Mosbech A (2013) Narwhals and seismic exploration: is seismic noise increasing the risk of ice entrapments? Biol Conserv 158:50–54
Heuschele J, Salminen T, Candolin U (2012) Habitat change influences mate search behaviour in three-spined sticklebacks. Anim Behav 83(6):1505–1510
Hildebrand JA (2009) Anthropogenic and natural sources of ambient noise in the ocean. Mar Ecol Prog Ser 395:5–20
Huntingford FA, Adams C, Braithwaite VA, Kadri S, Pottinger TG, Sandøe P, Turnbull JF (2006) Current issues in fish welfare. J Fish Biol 70(4):332–372
Kuperman WA, Ingenito F (1998) Attenuation of the coherent component of sound propagating in shallow water with rough boundaries. J Acoust Soc Am 61(5):1178–1187
McLaughlin KE, Kunc HP (2015) Changes in the acoustic environment alter the foraging and sheltering behaviour of the cichlid *Amatitlania nigrofasciata*. Behav Process 116:75–79
Mickle MF, Higgs DM (2018) Integrating techniques: a review of the effects of anthropogenic noise on freshwater fish. Can J Fish Aquat Sci 75(9):1534–1541
Nichols TA, Anderson TW, Sirovic A (2015) Intermittent noise induces physiological stress in a coastal marine fish. PLoS One 10(9):e0139157
Pickering AD (1992) Rainbow trout husbandry — management of the stress response. Aquaculture 100(1–3):125–139
Popper AN (2003) Effects of anthropogenic sounds on fishes. Fisheries 28(10):24–31
Popper AN, Fay RR (1973) Sound detection and processing by teleost fishes—a critical review. J Acoust Soc Am 53(6):1515–1529
Popper AN, Hastings MC (2009) The effects of anthropogenic sources of sound on fishes. J Fish Biol 75(3):455–489
Popper AN, Hawkins A (2012) The effects of noise on aquatic life. Springer, New York
Purser J, Radford AN (2011) Acoustic noise induces attention shifts and reduces foraging performance in three-spined sticklebacks (*Gasterosteus aculeatus*). PLoS One 6(2):e0017478
Radford AN, Kerridge E, Simpson SD (2014) Acoustic communication in a noisy world: can fish compete with anthropogenic noise? Behav Ecol 25(5):1022–1030
Ross D (1976) Mechanics of underwater noise. Pergamon Press, New York
Sabet SS, Neo YY, Slabbekoorn H (2015) The effect of temporal variation in sound exposure on swimming and foraging behaviour of captive zebrafish. Anim Behav 107:49–60

Slabbekoorn H, Bouton N, van Opzeeland I, Coers A, ten Cate C, Popper AN (2010) A noisy spring: the impact of globally rising underwater sound levels on fish. Trends Ecol Evol 25(7):419–427

Smith ME, Kane AS, Popper AN (2004) Noise-induced stress response and hearing loss in goldfish (*Carassius auratus*). J Exp Biol 207(3):427–435

Smith ME, Coffin AB, Miller DL, Popper AN (2006) Anatomical and functional recovery of the goldfish (*Carassius auratus*) ear following noise exposure. J Exp Biol 209(21):4193–4202

Solan M, Hauton C, Godbold JA, Wood CL, Leighton TG, White P (2016) Anthropogenic sources of underwater sound can modify how sediment welling invertebrates mediate ecosystem properties. Sci Rep 6:20540

Thomson RC, Shaffer HB (2010) Rapid progress on the vertebrate tree of life. BMC Biol 8:19

van der Sluijs I, Gray SM, Amorim MCP, Barber I, Candolin U, Hendry AP, Krahe R, Maan ME, Utne-Palm AC, Wagner HJ, Wong BBM (2011) Communication in troubled waters: responses of fish communication systems to changing environments. Evol Ecol 25(3):623–640

Van Hinte T (2005) Managing impacts of major projects: an analysis of the Enbridge gateway pipeline proposal (Doctoral dissertation, School of Resource and Environmental Management-Simon Fraser University)

Vazzana M, Celi M, Arizza V, Calandra G, Buscaino G, Ferrantelli V, Bracciali C, Sarà G (2017) Noise elicits haematological stress parameters in Mediterranean damselfish (*Chromis chromis*, Perciformes): a mesocosm study. Fish Shellfish Immunol 62:147–152

Weilgart L (2007) A brief review of known effects of noise on marine mammals. J Comp Physiol 20:159–168

Wysocki LE, Dittami JP, Ladich F (2006) Ship noise and cortisol secretion in European freshwater fishes. Biol Conserv 128(4):501–508

Zottoli SJ (1977) Correlation of the startle reflex and Mauthner cell auditory responses in unrestrained goldfish. J Exp Biol 66:243–254

Chapter 29
Heavy Metals in Wetlands in Turkey

Onur Can Türker and Jan Vymazal

Abstract Wetland areas are one of the most precious ecosystem because they contain rare ecological assets. The total area of wetlands of Turkey is 9.861 km^2 and wetlands cover 1.2% of total area of Turkey. Unfortunately, most wetland areas of Turkey face great challenges due to various pollution factors including heavy metals pollution caused by rapid industrialization and urbanization. In this chapter, current heavy metal pollution status of wetland areas, especially Ramsar Sites in Turkey, was assessed by more than 60 articles dealing with concentrations of heavy metals in sediment, water, and aquatic organisms. The results from the literature demonstrated that anthropogenic activities were closely related to heavy metal pollution causing a significant threat to wetlands including some Ramsar sites. In this respect, Ramsar areas of Lake Uluabat and Gediz Delta were severely contaminated by heavy metals. Moreover, heavy metal pollution in other Ramsar sites such as Kızılırmak Delta, Sultan Marshes, Lake Manyas, Lake Burdur, and Seyfe Lake had moderate levels. Low potential heavy metal pollution risk was reported for a Ramsar site of Akyatan Lagoon. On the other hand, high levels of heavy metals were found in various wetland areas of Turkey, and heavy metal concentrations in water of Lake Hazar, Yeniçağa Lake, Terkos Lake, and Beyşehir Lake were higher than the permissible levels for drinking water suggested by WHO (World Health Organization). Plankton, zoobenthos, and fishes which lives in polluted wetland habitats had also higher concentrations of heavy metals in their tissues, and thus heavy metal pollution can pose a potential risk for food chain both in wetland and terrestrial ecosystems of Turkey.

Keywords Heavy metals · Pollution · Wetlands · Turkey · Health risk assessment · Ramsar sites

---

O. C. Türker
Faculty of Science and Letters, Department of Biology, Aksaray University, Aksaray, Turkey

J. Vymazal (✉)
Faculty of Environmental Sciences, Czech University of Life Sciences Prague, Praha, Czech Republic
e-mail: vymazal@fzp.czu.cz

L. A. Jawad (ed.), *Southern Iraq's Marshes*, Coastal Research Library 36,
https://doi.org/10.1007/978-3-030-66238-7_29

29.1 Introduction

29.1.1 Wetlands in Turkey

Wetlands are defined as the most precious ecosystems and natural wetlands around the world provide fifteen percent of value of the world's ecosystems services (Costanza 2006; Costanza et al. 1997; Mitsch and Gosselink 2000). Correspondingly, ecosystem services from wetlands present 33 trillion USD of natural capital to human beings per year (Costanza et al. 2014). Wetland ecosystems provide important ecosystem services including carbon sequestration, water quality rehabilitation, and flood control, provision of staple food, biomass reservoir, mitigation of climate changes, recreation, and education, but also it presents shelters for various vertebrates and biodiversities (Mitsch et al. 2009; Tinner et al. 2008; Vymazal 2011). Among the wetland functions, the water quality improvement plays a crucial role, for example, wetlands can remove/retain more than 130 kg/ha of phosphorus, and 1000 kg/ha of nitrogen can be retained in wetland matrix every year (Meng et al. 2017).

Turkey is located in the center of old-world continents, and it consists of two peninsula surrounded by sea on three sides. It has 814.578 km^2 terrestrial area, and this area presents huge differences for the topographical viewpoint between the continents of Europe, Asia, and Africa (Karadeniz et al. 2009). In this case, it has important migration routes which can be used by both human society and various migratory birds throughout the historical periods (Çalışkan 2008; Horns and Şekercioğlu 2018). At the same time, these topographical changes have been approved to one of the crucial parameters for rich plant diversity in Turkey, which has diverse climatic conditions and assorted habitats for organisms. For example, around 11,000 flowering plants (34% of species defined as endemic) have been identified in Turkey. Owing to country's biogeographical characteristics, one of the main components associated with the rich biodiversity in Turkey is wetland habitats. In this case, Turkey's wetlands are defined as one of the most precious ecosystems which contain rare ecologic assets.

Wetlands in Turkey include seven types according to the classification of the European Community, covering estuaries and deltas, freshwater swamps, lakes, rivers and floodplains, peatlands, and artificial wetlands (Hughes 1995). More importantly, Turkey is the richest European country except Russian in terms of the existing wetlands (Karadeniz et al. 2009; Kuleli 2010). According to the Ministry of Agriculture and Forestry, the total wetland area in Turkey is approximately 3×10^6 ha, whereas the total area of registered wetlands of Turkey is 958,883 ha and accounts for 1.2% of Turkey's territorial area. Moreover, Turkey has 76 designated wetland areas, out of which 14 are wetlands of international importance, i.e., Ramsar sites with the area of 184,487 ha, 53 areas are wetlands of national importance with the total area of 764,107 ha, and 9 areas are wetlands of national importance with the area of 10,289 ha. The total peat wetland reservoirs in Turkey are 190×10^6 tons, and the peat wetlands stored 1300 tons of carbon per hectare in their matrix.

Wetland distribution in Turkey has obvious regional characteristics, and factors of structure and geomorphological, climatic, hydrological, and biological factors were influential in the formation of wetlands (Berberoglu et al. 2004; Somay and Filiz 2003). Wetlands are mainly distributed in the near-coastal area of Turkey and Anatolian plateau. Wetlands near the coastal area have higher water availability, plant and nutrient richness, and climate conditions to wetland animals such as birds. However, wetlands in Anatolian plateau have more harsh climatic conditions in winter time due to freezing of lakes, and the local bird populations migrate to wetlands in the coastal area (Ayas 2007). In this case, wetlands in coastal areas have many important breeding areas for many threatened bird species such as *Pelecanus crispus* and *Oxyura leucocephala* (Karadeniz et al. 2009). More importantly, Turkey is quite rich compared to its neighbors in terms of lake structures. Even if temporary ponds and swamps are not taken into account, the number of continuous lakes reaches 300, and the total area reaches the width of the Marmara Sea. This total area is 9.861 km^2 and covers an area of 1.2% in Turkey.

The Ministry of Agriculture and Forestry (MoAF) is directly responsible for the protection of Turkey's wetlands. In this respect, management wetlands in Turkey are being protected according to the national and international criteria regulated by MoAF. Manyas (Kuş) Lake is known as the first conservation wetland area with National Parks Law enacted in 1983 (Karadeniz et al. 2009). After this Law, many wetland habitats have been taken under protection in the scope of these protection statuses in Turkey. In addition, Turkey is part to various international conventions which are aimed at creation of conservation strategy for Turkey's wetlands. Among several important legal conventions, "conservation of natural areas for wetland protections" which is called the Bern Convention presents a crucial regulation to protection of wetland's species.

29.2 Current Status and Distribution of Ramsar Sites in Turkey

In order a wetland to be declared an internationally important wetland, it must meet certain criteria (McInnes et al. 2017). Correspondingly, if a wetland is rare, contains a considerable number of endangered animal and plant species, and presents permanent habitats to wetland birds up to 20,000 species, this wetland can be declared as an internationally important wetland or Ramsar site (Davidson et al. 2019; Geijzendorffer et al. 2019; McInnes et al. 2017). The Ramsar sites have been taken under protection with an intergovernmental treaty called as "Ramsar Convention" which was signed on February 2, 1971, in the city of Ramsar in Iran. Briefly, this convention imposes some obligations to member countries including production of ecological characters of international important wetland habitats and wise or

Fig. 29.1 Ramsar sites in Turkey (Ramsar 2019)

sustainable management strategy of wetlands in their territories (Geijzendorffer et al. 2019; Matthews 1993). Turkey has become a member to the Ramsar Convention in 1994, and the National Wetland Commission proposed that there are more than 20 wetlands to Ramsar Convention (Karadeniz et al. 2009). However, only 12 wetland sites have been declared as Ramsar sites among 20 international wetland sites in 2004. In 2009 and 2013, Kuyucuk Lake (Kars) and Nemrut Caldera (Bitlis) have been declared as new Ramsar wetland sites, and thus the number of the Ramsar sites in Turkey increased to 14 (Fig. 29.1). In this respect, total surface area of Turkey's Ramsar sites is 184,487 ha (Ramsar 2019). The largest Ramsar site in Turkey is Lake Burdur (southwest Turkey) with 24,800 ha (accounted for 13.4% of the total Ramsar sites in Turkey). Kızılırmak Delta (North Turkey) and Kuş Lake (North West Turkey) belong to larger Ramsar sites with areas of 21,700 ha (11.7%) and 20,400 ha (11%), respectively. These areas are also the most development regions in Turkey, meaning that they are exposed to high anthropogenic activities causing a significant threat to Ramsar sites (Kuleli et al. 2011). The intensive human disturbance and pollution have created a growing pressure on natural resources obtained from wetland habitats, and thus both wetlands and Ramsar sites in Turkey are in moderate shortage (Çalışkan 2008; Degirmenci et al. 2019; Karadeniz et al. 2009). Correspondingly, 1.3 million hectares of wetland areas have lost their crucial ecological and economic properties and functions due to land drying, filling, and intervention in water regimes according to the World Wide Fund for Nature (WWF). This lost area corresponds to almost half of the wetland areas in Turkey. Most Ramsar sites in Turkey have also lost area for controlling malaria, obtaining farmland, acquiring new construction land, as well as climate change (Gürlük and Rehber 2008; Karadeniz et al. 2009). For example, Sultan Marshes, one of the most important Ramsar areas in Turkey, is facing with significant threats (Dadaser-Celik et al. 2009). Specifically, this area provides habitat for some endemic medicinal aromatic plants, bulbous ornamental species, edible mushrooms, and various wetland birds. In this respect, biovalue rate of Sultan Marshes including national park area and buffer zone has been calculated as 24 133 272 USD by the Ministry of Agriculture and Forestry (MoAF). However, significant contraction has occurred in hydrological regime and grassland components of Sultan Marshes in recent years (Aksoy et al. 2005; Dadaser-Celik et al. 2009). The areal contraction of both Sultan Marshes and their surrounding regions was accelerated with intensive water use from wetland matrix, as well as storing surface water in agricultural dams (Karadeniz et al. 2009). However, it was very well known that some of the water which is used for irrigation purpose should be given to wetland matrix again in order to protect and improve wetland habitats (Meng et al. 2017). In this case, water regime of Sultan Marshes and water sources are decreased due to mostly anthropogenic activities, and the destruction of 1 ha of Sultan Marshes will result in a decrease of biovalue rate of 617.898 USD.

Although wetlands have gained an international status, the status sometimes cannot make sure that the wetland habitat is protected, even if the protection status came from the Ramsar Convention. For example, almost the total area of Seyfe Lake, which was declared as Ramsar area in 1994, is currently dry (Celik et al. 2008; Karadeniz et al. 2009). However, the lake is an important hatchery and breeding ground for waterfowl or other water birds. The reason of degradation of Seyfe Lake was a result of natural and artificial factors, and the excessive groundwater use is the main factor of lake drying (Karadeniz et al. 2009). Accordingly, the lake water is not used for agricultural activities due to its salinity and presence of soda; however, the groundwater which is added to the lake body is used as irrigation water in agricultural lands. Therefore, some activities such as illegally drilled wells result in excessive use of underground waters, and this practice causes to rapidly decreasing the resources which are used for feeding of the Seyfe Lake (Karadeniz et al. 2009). More importantly, a large part of the lake is drying in the summer times, and salt spilled from the salt layer constitutes an important threat for both ecological components and human health. Along with the drought period, the yield in agriculture zones decreased, reeds' stands were seriously damaged, and poplar and willow species dried up and disappeared. Also, Meke Maar (Karapınar, Konya) that was declared in 2005 as Ramsar area had nearly dried due to climate change and high excavation to obtain sand (Karadeniz et al. 2009).

Like most Turkey's Ramsar sites, Kızılırmak Delta, which is located at Northern Turkey, is facing the anthropogenic and climatic threats. This Ramsar site is one of the largest and most significant wetland habitats in Turkey, and it has been able to provide habitats and protect indirectly biological resources in Black Sea Coast to date. Recently, Kızılırmak Delta has been exposed to human activities such as land-use decisions for highway construction or dam construction, and this activity leads to wetland fragmentation and habitat degradation (Ersayın and Tağıl 2017; Karadeniz et al. 2009). It was very known that wetland fragmentation is defined as the first step of wetland degradation; it damages the normal energy flow and nutrient cycling. In this respect, approximately 11% of Kızılırmak Delta is defined as a high sensitivity and risk zone because of anthropogenic activities (Ersayın and Tağıl 2017). The fragmentation in Kızılırmak Delta matrix, habitat degradation, and decreasing genetic diversity are found for plants and animals, and thus biodiversity is significantly deteriorating (Degirmenci et al. 2019).

One of the important Ramsar sites that faces the danger of drying problems is Lake Burdur, which is the largest Ramsar site in Turkey just mentioned above. Correspondingly, the total area of Lake Burdur has been reduced from 230 km^2 to 131 km^2 with lake volume decreased from 6.475 km^3 to 3.927 km^3 between 1975 and 2016 (Davraz et al. 2019). At the same time, the water level has been decreasing approximately from 855 to 841 between 1975 and 2016 due to evaporation, groundwater use, and agricultural irrigation (Davraz et al. 2019).

The proportional decrease of the lake area was determined as 16% from 2000 to 2016. The main reasons of water level decreasing are construction of dams, groundwater withdraws, and climatic conditions (Sarp and Ozcelik 2017). Illegally drilled wells from farmers, residents which live around the lake, and the private sector in the urban area also contribute to the decrease in the volume of water due to excessive water use (Adaman et al. 2009). In addition, wastewater comes from Burdur city center and organized industry zone, as well as excessive pesticides and fertilizers used in agriculture zone are responsible for the pollution in Lake Burdur (Davraz et al. 2019).

Pollution is defined as one of the main factors in Ramsar wetland degradation in Turkey, and thus wastewater discharges from agriculture, industry, and aquaculture are directly responsible for threats to Ramsar ecosystems. Because large amounts of domestic, municipality, and industrial wastewater are discharged into the wetland habitats, many Ramsar sites have been seriously polluted and degraded (Celik et al. 2008; Dadaser-Celik et al. 2009; Davraz et al. 2019; Degirmenci et al. 2019). Most wetland ecosystems in Turkey have been polluted with nutrients and heavy metals, and thus eutrophication and heavy metal levels in Ramsar sites are continuously increasing. A number of factors including urbanization, infrastructure construction, and agricultural activities have caused an accelerated loss of Ramsar wetlands. Specifically, commonly used pesticides in agricultural zones constitute a crucial factor to damage of wetland habitats in Turkey because they contain heavy metals (Muhammetoglu et al. 2019). The agrochemical use is very widespread in Turkey's agricultural zones, and 30,000 tons of agrochemicals are used every year by farmers who quite often do not care about the correct practices (Karadeniz et al. 2009). More importantly, heavy metals in agrochemical contents can easily diffuse into the wetland matrix due to their transport through the flowing water bodies (Yılmaz and Parlak 2011). In this respect, wetland ecosystems are more sensitive to heavy metals than terrestrial ones. Thus the current pollution status of Ramsar areas in terms of heavy metals should be investigated in order to understand the general status of Ramsar sites, and the evaluation needs to be undertaken in order to enhance the conservation of Ramsar sites.

In this chapter, the current heavy metal concentrations in Turkey's Ramsar wetlands were discussed based on researches on heavy metal pollutions, and we also reviewed the effect of heavy metals on wetland components in Ramsar sites.

29.2.1 *Heavy Metal Pollution in Selected Ramsar Sites of Turkey*

29.2.1.1 Lake Manyas (Kuş)

The Ramsar area of Kuş Lake, which is located in Western Turkey, is defined as the bird-of-paradise, and it provides important hosting areas to migrant birds which use an immigration way between the continents of Europe and Africa (Alemdaroğlu et al. 2003). Kuş Lake's water body is fed by four rivers, namely, as Dutlu stream, Kocaçay, Akıntı stream, and Sığırcı stream, and Kuş Lake water flows into Marmara Sea through Susurluk River that just flows near Gemlik Bay. It was reported that the Sığırcı and Kocaçay streams were seriously polluted by industrial effluents which can be easily carried to the lake. Specifically, Sığırcı stream is the most polluted stream feeding the lake body because it comes from Bandırma district with many factories operated in various industrial sectors, and thus industrial pollution is heavy in this region.

The average concentrations of Mn, Fe, Cu, Pb, Ni, and Zn in sediment of Kuş Lake were found to be 1042, 32.8, 50, 217.4, 43.8, and 253.8 μg g^{-1} DW, respectively, in a 1996 study (Alemdaroğlu et al. 2003). In the same study, the highest levels of Mn, Fe, Cu, Pb, Ni, and Zn were mostly determined in sediment from Sığırcı stream and its lake junction as follows: 993, 32.9, 59.8, 124, 62.1, and 246 μg g^{-1} DW. Additionally, sediments from Kocaçay had the highest concentrations of Zn. These results indicate that heavy metals are transported through streams, especially Sığırcı and Kocaçay stream, into the Kuş Lake, and they are mainly deposited at the middle of the lake. Therefore, the amounts of heavy metals in lake body were found higher than in the Lake Manyas tributaries.

In parallel to these results, Çiçek et al. (2009) reported that concentrations of Mn, Fe, Cu, Pb, Ni, and Zn in the bottom sediment were 557.4, 27,848, 36.6, 176.4, 28.6, and 237.3 μg g^{-1} DW, respectively, and they demonstrated that Mn, Pb, Ni, and Zn concentrations in fish tissues were found higher than the critical limit values. The results of this study concluded that Kuş Lake has significant heavy metal pollution.

Türkmen (2018) demonstrated that the main factors for heavy metal pollution in Kuş Lake are industrial wastewater, agricultural effluents, wastewater from olive oil and tomato paste factories, and some effluents from boron mining. Unfortunately, there is no scientific report on heavy metal concentrations in Kuş Lake until 2009; thus it is not possible to make any comparisons between current and historical values of heavy metals in the sediments.

29.2.1.2 Lake Uluabat

The Lake Uluabat is one of the significant "Ramsar site" and "living lake" members; therefore it is an important ecosystem not only for Turkey but also for Asia and Europe. The Lake Uluabat is a shallow and eutrophic lake at an altitude of 9 m above sea level with a surface area of 156 km^2 (Magnin and Yarar 1997). The lake is mainly fed by the Mustafakemalpaşa River from the southwest and has its only outlet in the northwest where it rains to Kocaçay River (Arslan et al. 2010a). Deposits of incoming silt from the Mustafakemalpaşa River have formed an inland delta covering an area of 3747.6 ha that is under agricultural use (Salihoğlu and Karaer 2004). Lake Uluabat is located near the city of Bursa with high population, dense industry, and developed economy. With the rapid economic development and increasing population within the city of Bursa in recent decades, a high amount of uncontrolled industrial and domestic effluents has been drained into the rivers which enter the Lake Uluabat, and then transported into the lake body. Therefore, the Lake Uluabat has been contaminated by domestic and industrial discharge for many years. Specifically, region of Lake Uluabat has had an agricultural-based economy, and excessive pesticide usage and the continuous application of inorganic and organic fertilizers which contain heavy metals. In this respect, significant heavy metal pollution has been determined by various studies.

Barlas et al. (2005) reported that the average concentrations of Fe, Mn, Cu, Zn, Cr, Pb, Ni, Cd, and Co in the sediment of the Lake Uluabat were determined as 18.21, 6.57, 0.31, 0.74, 0.83, 0.43, 1.84, 0.03, and 0.24 mg kg^{-1} DW in November 2001. The same researchers found an increasing heavy metal pollution trend within the lake body in 2002. Accordingly, the highest Fe, Mn, Cu, Zn, Cr, Pb, Ni, Cd, and Co concentrations in the lake sediment were found as 27.03, 23.92, 1.19, 8.36, 4.88, 2.39, 8.93, 0.12, and 1.81 mg kg^{-1} DW, respectively, in late spring period of 2002.

Kazancı et al. (2010) determined heavy metal concentrations in Lake Uluabat surface water and in a river just flowing the lake body. Correspondingly, the mean concentrations of Mn, Cu, Pb, Ni, Cr, Fe, Al, Cd, Hg, and As in surface water were measured as 2.17, 2.83, 1.21, 12.51, 4.31, 41.4, 9.27, 0.18, 0.44, and 25.9 μg L^{-1} respectively, whereas the same heavy metal concentrations were found to be 4.56, 18.75, 3.09, 71.05, 3.64, 42.05, 84.3, 1.09, 0.59, and 22.81 μg L^{-1}, respectively, in river water. The highest amount of Al, Cu, Fe, Mn, Ni, Pb, and Hg in sediment of lake were also determined as 72.01, 440, 74, 38, 813, 337, 249 ppm, and 460 mg kg^{-1}, respectively.

Katip et al. (2016) found that concentrations of Fe, Mn, Cu, Zn, Cr, Pb, and Ni in surface sediment of Lake Uluabat were 27,664, 781, 30.4, 132.2, 131.6, 33.3, and 220 mg kg^{-1}, respectively. Accordingly, the researchers demonstrated that industrial effluents and the use of the agrochemicals around the lake basin were the main reason for heavy metal pollution in lake sediments.

On the other hand, the various studies indicated that the heavy metal pollution is mostly transported to the lake organisms through food chain in the Lake Uluabat. For example, Elmaci et al. (2007) investigated the heavy metal concentrations in water, sediment, and plankton of the Lake Uluabat, and they found elevated heavy metal content in water and plankton samples. However, the researchers demonstrated that some heavy metals such as Cu and Cr can be shown mobilizable fractions in lake sediment. According to the results of this study, an increasing trend for some heavy metal pollution also continued in 2003 and 2004, and consequently the average concentrations of Cu, Pb, Cr, Cd, Ni, and Zn in the sediment amounted to approximately 12, 13, 9, 2, 8, and 1 mg kg^{-1} DW, respectively. More importantly, higher and significant concentrations of Zn, Ni, Cr, Cd, and Cu were recorded for plankton samples. It can be concluded that heavy metals were released into the water body, and then finally entered the food chain through the planktonic organisms.

Arslan et al. (2010a) determined concentrations of Cd, Cr, Pb, Cu, Ni, and Zn in water, sediment, and Oligochaeta-Chironomidae of Lake Uluabat during the period of August 2004 to July 2005. They found that concentrations of Zn and Pb were highest among the heavy metals in lake body and the concentrations of these two heavy metals exceeded the CMC (criterion maximum concentration) limits declared by the EPA. Moreover, sediment analysis results of this study indicated that heavy metal pollution in sediment was still increased in 2004 and 2005, indicating that domestic and industrial effluents were discharged to the lake body. Also, the researchers reported that the levels of Cd, Cr, Pb, Cu, Ni, and Zn in the sediment reached the values of 0.699, 57.9, 110.7, 119.2, 209.4, and 171 mg kg^{-1} DW, respectively. Excessive Pb and Cu concentrations were recorded for Oligochaeta and Chironomidae, indicating that heavy metals entered the food chain through adsorption processes. Additionally, Katip et al. (2014) reported that there are temporal changes of some heavy metal concentrations (Cr, Pb, and Ni) in surface water of Lake Uluabat between 2003 and 2004 and 2008 and 2009. The authors found concentrations of 0.028 and 0.085 mg/L for Cr, 0.026 and 0.036 mg/L for Pb, and 0.022 and 0.03 mg/L for Ni during the time periods of 2003–2004 and 2008–2009, respectively. This result indicated that concentrations of Cr, Pb, and Ni had increased between 2003 and 2009, and elevated heavy metal level in surface water can exhibit a threat to the food chain in the lake body (Table 29.1).

The heavy metal pollution level in the Lake Uluabat was determined using some quality criteria and limit values (geoaccumulation index and severe effect level). For example, Katip and Karaer (2011) reported that water quality of Lake Uluabat was low in terms of the heavy metal concentrations in water, and they found that Cd and Cr concentrations exceeded water quality criteria in surface water issued by the Turkish Government during the summer times.

Katip et al. (2013) also found that the portions of As, Cd, Cr, Ni, and Pb in exchangeable and reducible fractions of sediment from Lake Uluabat were below 10% of the total concentrations; however, total concentrations of Cr and Ni in the sediment were close to the severe effect level (SEL) limit values (2.144% and 1.301% for Cr and Ni, respectively). The same researchers reported that the highest concentrations of Fe, Mn, Zn, Ni, Cr, Cd, Cu, As, and Pb amounted to 15.7, 0.17, 1.341, 0.358, 1.95, 0.38, 0.116, 0.09, and 0.5 mg L^{-1}, respectively, between May 2008 and May 2009 (Katip et al. 2013). Moreover, the heavy metal amount in sediment samples was found higher during the summer; the mean concentrations of Fe, Mn, Cu, Zn, Cr, Pb, Cd, Ni, and As in lake sediment were recorded as 2039, 50, 5.3, 26.87, 13.4, 7.6, 1.16, 18.6, and 0.9 mg kg^{-1}, respectively.

Hacısalihoğlu and Karaer (2016) demonstrated that lake body of Uluabat was moderately polluted by Cr and Ni, and they found that dissolved concentrations of Cr and Ni in the water sample were 0.022 and 0.012 mg L^{-1}, respectively. Moreover, total concentrations of Cr and Ni in the sediment were determined as 119.7 and

Table 29.1 Wetlands of national importance in Turkey according to the Ministry of Agriculture and Forestry

No.	Wetland name	Area of wetland (ha)	Region	Registered date	No.	Wetland name	Area of wetland (ha)	Region	Registered date
1	Acıgöl	55,095	Afyonkarahisar-Denizli	08.04.2015	**28**	Ladik lake	1836	Samsun	08.04.2015
2	Ahlat Marshes	243	Bitlis	08.04.2015	**29**	Nazik Lake	11,164	Bitlis	08.04.2015
3	Akgöl	1203	Van	08.04.2015	**30**	Putka Lake	4181	Ardahan	08.04.2015
4	Aktaş Lake	5847	Ardahan	08.04.2015	**31**	Sarısu Ovası wetlands	10,092	Ağrı	08.04.2015
5	Aras Karasu floodplain	9090	Iğdır	10.06.2016	**32**	Tödürge Lake	4340	Sivas	10.06.2016
6	Arin Lake	4322	Bitlis	10.06.2016	**33**	Turna (Keşiş) Lake	3045	Van	08.04.2015
7	Avlan Lake	10,062	Antalya	10.06.2016	**34**	Ulaş Lake	7994	Sivas	10.06.2016
8	Aygır Lake	1034	Kars	08.04.2015	**35**	Yarışlı Lake	13,219	Burdur	10.06.2016
9	Bendimahi Delta	27,177	Van	10.06.2016	**36**	Yazır Lake	2705	Burdur	10.06.2016
10	Bulanık wetlands	3496	Muş	10.06.2016	**37**	Yeniçağa Lake	8224	Bolu	09.04.2015
11	Çalı Lake	391	Kars	10.06.2016	**38**	Yüksekov reeds	21,533	Hakkâri	09.04.2015
12	Çelebibağ reeds	1337	Van	10.06.2016	**39**	Tol lake	1414	Ankara	19.04.2017
13	Çıldır Lake	27,058	Ardahan	08.04.2015	**40**	Tortum Lake	2709	Erzurum	19.04.2017
14	Çorak Lake	7892	Burdur	10.06.2016	**41**	Akşehir ve Eber Lakes	117,779	Afyonkarahisar; Konya	19.04.2017

15	Doğubeyazıt reeds	22,179	Ağrı	10.06.2016	**42**	Gölbaşı Lake	792	Hatay	19.04.2017
16	Dönemeç Delta	5945	Van	10.06.2016	**43**	Gölmarmara lake	24,893	Manisa	12.06.2017
17	Erçek Lake	22,269	Van	10.06.2016	**44**	Ekşisu reeds	8736	Erzincan	12.06.2017
18	Gölhisar lake	5877	Burdur	10.06.2016	**45**	Dipsiz lagoon	1035	Mersin	12.06.2017
19	Gönen Delta	9770	Balıkesir	10.06.2016	**46**	Efteni Lake	8314	Düzce	30.05.2018
20	Güney Keban dam	41,424	Elazığ	08.04.2015	**47**	Kocaçay Delta	17,025	Bursa	13.08.2018
21	Hazar Lake	28,846	Elazığ	08.04.2015	**48**	İznik lake	61,606	Bursa	13.08.2018
22	Heybeli (Norşin) Lake	53	Bitlis	08.04.2015	**49**	Acarlar Longoz forest	17,528	Sakarya	07.02.2019
23	Hürmetçi marshes	15,713	Kayseri	08.04.2015	**50**	Balıkdamı Lake	14,147	Eskişehir	07.02.2019
24	Işıklı Gökgöl wetlands	33,693	Denizli	10.06.2016	**51**	Karakuyu reeds	12,625	Afyonkarahisar	07.02.2019
25	İron marshes	13,746	Bitlis; Muş	08.04.2015	**52**	Azap Lake	2183	Aydın	07.02.2019
26	Karasu Delta	339	Van	08.04.2015	**53**	Gökçeada lagoon	3491	Çanakkale	07.02.2019
27	Karkamış floodplain	27,396	Gaziantep; Şanlıurfa	08.04.2015	**Total**		**764,107**		

196.2 mg kg^{-1}, respectively. The researchers also used the geoaccumulation index (Igeo) in order to assess the degree of contamination from heavy metals in the sediment. The Igeo value for Ni and Cr indicated that sediment of the Lake Uluabat was moderately polluted by Ni, whereas the value from Cr Igeo index demonstrated relatively uncontaminated sediment by Ni pollution.

Consequently, the reports from various studies demonstrated that the Lake Uluabat body was mostly polluted by Cu, Cr, and Ni, and these heavy metal concentrations were attributed to discharges from domestic and industrial wastewater between 2001 and 2014. At the same time, agriculture runoff and the use of pesticides were also dominant pollution sources. Analysis of chemical fraction of some heavy metals in the sediment from Lake Uluabat demonstrated that Pb, Cd, and Ni existed in readily available forms including reducible forms, carbonate, and exchangeable. The remaining fraction of Cu, Cd, and Ni existed in oxidable and residual forms which are rather stable in the lake environment.

29.2.1.3 Gediz Delta

The Ramsar area of Gediz Delta wetland ecosystem basically consists of freshwater and saltwater marches located from north to south, namely, Kırdeniz Lagoon, Homa Lagoon, Çamaltı Salt water, Çilazmak Lagoon, Ragıppaşa Dalyan, and northern Gediz Delta. The Gediz Delta contains various important bird areas for breeding populations of many bird species. Unfortunately, the delta is heavily polluted by domestic and industrial wastewaters, and agricultural drainage water, because three big cities (İzmir, Manisa, and Usak) are just located near the delta. Therefore, significant heavy metal pollution has been reported by various researchers for this region (Table 29.2).

Akcay et al. (2003) investigated heavy metal pollution in the Gediz River which was the part of the Gediz Delta between 1996 and 1998, and they found a significant Pb, Cr, Mn, and Zn pollution in the river body. The highest concentrations of Co, Cr, Cu, Fe, Mn, Ni, Pb, and Zn in river water were determined as 0.011, 0.028, 0.011, 1.1, 0.05, 0.011, 0.040, and 0.076 mg L^{-1}, respectively. In addition, the total concentrations of Co, Cr, Cu, Fe, Mn, Ni, Pb, and Zn in the sediment of Gediz River were 38, 200, 140, 25,500, 510, 106, 128, and 160 mg/kg, respectively. Specifically, they also suggested that high concentrations of Cr, Mn, and Zn in the Gediz River body indicated anthropogenic activities including industrial effluents from industrial zones and pesticides used in the agricultural area.

Parlak et al. (2006) reported that the highest concentrations of Cr, Co, Cu, As, Ni, Pb, and Cd in Gediz River sediment in 1998 were determined as 1192, 9.09, 19.9, 11.15, 42.94, 40.89, and 0.18 mg kg^{-1} DW, respectively. In this respect, the authors suggested that there is no significant heavy metal pollution in Gediz River except Cr, and a substantial amount of Pb may be potentially harmful in the river body in the future.

Table 29.2 Heavy metal concentrations (mg kg^{-1}) in sediments of the Lake Uluabat between 2001 and 2014 (Arslan et al. 2010a; Barlas et al. 2005; Hacısalihoğlu and Karaer 2016; Katip et al. 2016; Kazancı et al. 2010)

Heavy metal (mg kg^{-1} DW)	The min and max concentrations (mg kg^{-1} DW) between 2001 and 2002	The min and max concentrations (mg kg^{-1} DW) between 2003 and 2005	The min and max concentrations (mg kg^{-1} DW) between 2008 and 2009	The min and max concentrations (mg kg^{-1} DW) between 2013 and 2014
Fe	18.21–37.83	–	0–27,664	–
Mn	6.57–813	–	0–781	–
Cu	0.31–74	0.3–289	0.1–1029	–
Zn	0.74–8.36	0.07–2	0–213	–
Cr	0.83–4.88	0.8–132.7	0–190	60–180
Pb	0.42–249	0.06–372	0–49	–
Ni	1.84–337	1–303	0–353	10–240
Cd	0.02–0.12	0.5–395	0.5–1.6	–
Co	0.24–1.81	–	–	–
Al	72.01	–	–	–
Hg	0.046	–	–	–
Ba	440	–	–	–
As	–	–	0–2	–

Dora et al. (2007) demonstrated the presence of heavy metals in the sediments in Homa Lagoon, which is part of Gediz Delta, between 2003 and 2004. They found that Cu and Cr concentrations in sediment samples increased between 1998 and 2004 and the highest concentrations of Cd, Cr, Cu, and Pb were 0.0043, 0.849, 0.041, and 0.0191 mg kg^{-1} DW, respectively. They indicated that local anthropogenic inputs, particularly domestic, industrial, and agricultural, were the main factors affecting the spatial distribution of Cu and Cr in this region. On the other hand, the heavy metal concentration in polychaete was higher than sediment concentrations, indicating that heavy metals entered the food chain through adsorption processes.

Kucuksezgin et al. (2008) investigated the distribution of heavy metals in water, particulate matter, and sediment samples of Gediz River. The maximum heavy metal concentrations in river water body were found in the summer period because of industrial and agricultural activities, and the highest Hg, Pb, Cu, Zn, Mn, Ni, and Fe concentrations were 0.81, 1.5, 1.6, 2.9, 170, 9, and 687 μg L^{-1}, respectively. The highest heavy metal concentrations of Hg, Cr, Pb, Cu, Zn, Mn, Ni, and Fe in the sediment of the Gediz were 0.49, 814, 198, 148, 196, 1371, 175, and 72,387 mg kg^{-1} DW, respectively. Moreover, contamination factor (C_r) derived by Hakanson (1980) suggested that the river body was highly polluted by Pb and Cr, and moderately polluted by Hg, Cu, Zn, Ni, and Fe, and the lowest C_r value was found for Mn. The reasonable explanation of Pb pollution in the Gediz River is the effect of the industrial and agricultural activities in Muradiye region. Sediment analysis results of study indicated that heavy metal pollution in the sediment was still increasing.

Taş et al. (2009) determined some heavy metal concentrations including Cd, Cu, Zn, Pb, Cr, and Fe in *Hediste diversicolor* (polychaete worm) and sediment of the Homa Lagoon in Gediz Delta between 2002 and 2004. In this respect, the highest Cd, Cr, Cu, Pb, Zn, and Fe concentrations were measured at 0.062, 1.135, 3.857, 9.69, 14.503, and 202 μg g^{-1} DW, respectively. The results from this study are in agreement with the reports by Dora et al. (2007), who demonstrated significant Cu and Cr pollution in the lagoon body. In this respect, Taş et al. (2009) found a significant correlation between accumulated Cu and Cd by *Hediste diversicolor* and their concentrations in Homa Lagoon sediment, and significant Cr pollution value (90 μg g^{-1}) was obtained according to SQG (sediment quality guide) in the sediment. The high concentrations of Cu, Cd, and Cr in the sediment indicated that significant heavy metal pollution was still increasing because of used pesticides in agricultural activities and textile wastewater which comes from factory in the Gediz Delta.

Uluturhan et al. (2013) investigated heavy metal concentrations in sediment of Homa Lagoon between 2005 and 2006, and they assessed heavy metal pollution level with SOQ in lagoon body. The highest concentrations of Hg, Cd, Pb, Cr, Cu, Mn, Ni, Zn, Fe, Al, and Li were 0.48, 0.19, 17.2, 129, 25.8, 729, 108, 91.9, 30,234, 42,637, 37.5, and 4.2 mg kg^{-1} DW, respectively. The authors also found that sediment of lagoon was mostly enriched by Cr and Ni, indicating that the lagoon was polluted by agricultural effluents and industrial and domestic wastewater. This fact was also supported by a comparison with SOG values. On the other hand, it was found that the concentrations of Cr, Hg, and Cu were calculated above TEL (threshold effect level) values, whereas the contamination of Pb and Zn in sediments was classified as non-polluted according to TEL values.

Tas and Sunlu (2013) reported heavy metal concentrations in surface sediments from Homa Lagoon in 2012. The highest Cd, Cu, Pb, Zn, Cr, and Fe concentrations were determined as 0.8, 24, 30, 88, 220, and 28,300 mg kg^{-1} DW, respectively. The researchers discussed the contamination levels according to the previous years. The results indicated that there are no significant temporal trends for heavy metal concentrations in the sediment of the lagoon. For Cd pollution, the decreasing trend was shown until 1998, and the concentrations reached minimum values of 0.19 mg kg^{-1} DW in 2006, and then the Cd concentration increased again until it reached 0.8 mg kg^{-1} DW in 2012. For Pb concentrations in the lagoon sediment, decreasing trend was shown along the years, and then it reached 30 mg kg^{-1} DW in 2012. The Zn concentration in the sediment increased between 2002 and 2004; the Zn pollution reached a maximum value of 95.5 mg kg^{-1} DW in 2004. Moreover, Zn concentrations in sediment steadily decreased and reached 88 mg kg^{-1} DW in 2012. Conversely, Cr concentration in sediment was continuously increasing until 1998, and it reached a maximum level as 220 mg kg^{-1} DW in 2012. The researchers suggested that the heavy metal pollution was mostly transported by Gediz River which is continuously polluted by industrial and domestic wastewater, as well as by

Table 29.3 Heavy metal concentrations (mg kg^{-1}) in sediment of Gediz Delta between 1996 and 2012 (Akcay et al. 2003; Dora et al. 2007; Parlak et al. 2006; Tas and Sunlu 2013; Taş et al. 2009; Uluturhan et al. 2013)

Heavy metal (mg kg^{-1} DW)	The min and max concentrations (mg kg^{-1} DW) between 1996 and 1998	The min and max concentrations (mg kg^{-1} DW) between 2002 and 2004	The min and max concentrations (mg kg^{-1} DW) between 2005 and 2006	The min and max concentrations (mg kg^{-1} DW) in 2012
Fe	9–72,387	182–253	17,054–30,234	16,500–28,300
Mn	15–1371	–	410–729	–
Cu	0.39–148	3–41	10.3–25.8	9–24
Zn	1–192	13–15	46.2–91.9	38–88
Cr	10–494	0.67–1.396	83.9–129	92–220
Pb	1–198	8–10	2.43–17.2	6–30
Ni	6–175	–	58.1–108	–
Cd	0.01–0.39	0.05–0.43	0.06–0.19	0.5–0.8
Co	0.5–24	–	–	–
Al	–	–	12,663–42,637	–
Hg	0.25–0.49	–	0.22–0.48	–
Ba	–	–	–	–
As	0.02–30.98	–	–	–
Ch	5.87–488	–	–	–

effluents from agricultural zones. The concentrations of heavy metals in sediments of Gediz Delta between 1996 and 2012 are summarized in Table 29.3.

29.2.1.4 Lake Burdur

Lake Burdur is one of the biggest and deepest lakes in Turkey. The Lake presents important habitats for birds such as *Oxyura leucocephala*, so the lake represents a significant Ramsar area (Arslan et al. 2016). The maximum depth of Lake Burdur is 110 m, and a few fish species can live in the lake body due to its salinity (Davraz et al. 2019). However, the main factors contributing to heavy metal pollution are urbanization and industrialization in lake area. The highest Zn, Pb, Cu, As, and Fe concentrations in water samples of Lake Burdur between 2004 and 2006 were 12.35, 1.45, 0.26, 0.13, and 0.25 mg L^{-1}, respectively. Moreover, Ni, Cr, and Zn concentrations in water samples increased between 1986 and 2006 from 0.06–0.26 mg L^{-1} to 3.2–5.3 mg L^{-1}, from 0.10–0.15 mg L^{-1} to 0.65–1.84 mg L^{-1}, and from 0.3–0.6 mg L^{-1} to 1.3–4.5 mg L^{-1}, respectively (Davraz et al. 2019). These results indicated that the heavy metal pollution increased along with the years. However, some researchers also reported that there were no significant differences between concentrations of Cu, Fe, and Cd in water in 1986 and 2006 (Beyhan et al. 2007; Davraz et al. 2019).

29.2.1.5 Kızılırmak Delta

Kızılırmak Delta has been protected by Ramsar Convention since 1993, and it is defined as the most significant and largest wetland area in Turkey. At the same time, the delta also contains the part of Kızılırmak River which is the longest river (1355 km) within the borders of Turkey. The origin of the heavy metal pollution was determined as both natural such as geological weathering and anthropogenic. In this respect, the highest concentrations of Fe in the river sediment were recorded at 686 mg kg^{-1} DW in July 2012, whereas the lowest sediment heavy metal concentrations were reported for Co (10.24 mg kg^{-1} DW) and Pb (19.06 mg kg^{-1} DW), respectively (Engin et al. 2017). The heavy metal concentrations in river sediment were found in the order of Cr>Zn>Pb>Cu>Co>Hg>Br in August 2008 (Pulatsü and Topçu 2015). At the same time, they reported that Pb, Ni, Cu, Zn, and Cd concentrations in the river sediment were higher than TEL value, whereas Cr and Hg pollution in the sediment exceeded both TEL and PEC (probable effect concentration) values. Results from this study indicated that the reason of the significant heavy metal concentrations in aquatic animals such as fish and carps was high heavy metal concentrations in sediment and aquatic plants. This result emphasized that heavy metals easily enter the food chain. Therefore, it can be suggested that if the river food is consumed by people, attention should be paid to potential toxicity.

29.2.1.6 Sultan Marshes

Sultan Marshes are located in central Anatolia between Develi, Yahyalı, and Yeşilhisar in Kayseri province. It is Turkey's most visited Ramsar area which includes very rich marsh plants such as *Phragmites australis*, *Typha* sp., and *Carex* sp. In Sultan Marsh, a total of 462 bird species exist (Yalcin et al. 2007). According to the Ministry of Agriculture and Forestry (MoAF), the calculated biovalue of the Sultan Marshes is 24,133,273 USD. The main factors contributing to the heavy metal pollution in this region are domestic wastewaters, industrial effluents, and pesticide use in the agricultural zones. In a study by Aksoy et al. (2005), it was found that the Marshes' body is mainly contaminated by Cd, Cr, Pb, and Ni and the plants had also higher concentrations of Cd, Pb, and Cr than sediment samples. Demirezen and Aksoy (2004) measured concentrations of heavy metals (Cd, Pb, Cr, Ni, Zn, and Cu) in the sediments, in water, as well as in plants *Typha angustifolia* and *Potamogeton pectinatus*. The results indicated that the wetland system was polluted with Pb and Cd and partly with Cu and Zn. In general, the concentration of heavy metals in all studied samples was more or less elevated as compared to control, unpolluted sites. A strong positive correlation was found between concentration of lead in water and plant biomass. The authors pointed out that higher heavy metal concentrations were found in *T. angustifolia* as compared to *P. pectinatus* (Demirezen and Aksoy 2004). They also found a significant correlation between concentrations of Cd, Pb, and Cr in plants and heavy metals in sediment

(Demirezen and Aksoy 2006). Result from the correlation analysis obtained from this study emphasized that heavy metal concentrations in plants are the most crucial factors governing the heavy metal content in the aquatic animals such as fish. The Cd, Pb, and Cr pollution in Sultan Marshes was also confirmed in 2007 (Yalcin et al. 2007), and the highest Cd, Pb, and Cr concentrations in sediment were found as 8.1, 50.1, and 7130 mg kg^{-1}, respectively. This result suggested that domestic wastewater, effluents from leather factory, and industrial waste drainage into the surface water increased the heavy metal concentrations in Sultan Marches. Yalcin et al. (2007) also found relatively high concentrations of the elements Fe, Pb, Zn, Sb, W, Mo, Co, Cu, Hg, Ni, Cr, Mn, and Cd in Sultan Marsh. They pointed out that the major sources of pollution for the Sultan Marsh are surrounding rocks (geogenic sources), mines of Fe and Pb/Zn, industrial facilities, residential and agricultural areas, and major traffic routes. The highest concentrations in the Sultan Marsh soils were found for Al (75,300 mg kg^{-1}), Fe (73,950 mg kg^{-1}), Cr (7130 mg kg^{-1}), Mn (2567 mg kg^{-1}), and Zn (552 mg kg^{-1}).

29.2.1.7 Seyfe Lake

Seyfe Lake is located in the middle Kızılırmak region, and it hosts 167 aquatic bird species and more than a half million birds (Bölükbaşı and Akın 2016). The heavy metal concentrations in sediment samples were in the order of Zn > Pb > As > Cr > Ni > Cu > Cd > Co in a study conducted by Bölükbaşı and Akın (2016). The researchers reported that the sediment was classified as high in terms of SQG values and high in terms of geoaccumulation index for As and Zn, which come from an intensive agricultural activity. The heavy metal pollution in the sediment was classified as anthropogenically "highly polluted," whereas it was classified "moderately to highly polluted" in terms of geochemical status.

29.2.1.8 Summary on Heavy Metals in Ramsar Sites in Turkey

The heavy metal concentrations for Ramsar areas in Turkey emphasized that the heavy metal concentrations in sediment of Lake Uluabat and Gediz Delta were found to be extremely high and the water sources from these lakes create a risk associated with food contamination to local people. However, the heavy metal pollution in Kızılırmak Delta, Sultan Marshes, Lake Manyas, Lake Burdur, and Seyfe Lake sediments was defined as moderately polluted by both natural and anthropogenic factors. On the other hand, heavy metal pollution in other Ramsar areas in Turkey was also reported. For example, Ni, Cd, and Pb pollution which comes from pesticide use in agricultural areas was reported for Göksu Delta (Ayas et al. 1997; Ayas and Kolankaya 1996; Ergene et al. 2007). Low potential heavy metal pollution risk was detected for Mn, Ni, Pb, and Zn in Akyatan Lagoon. Moreover, no heavy metal toxicity risk was recorded for aquatic organisms in Akyatan Lagoon (Türkmen et al. 2012). Unfortunately, there are no reports associated with heavy metal pollution in Ramsar area of Meke Maar, Kızören Obrouk, Lake Kuyucuk, Nemrut Caldera, and Yumurtalık Lagoons in Turkey.

29.2.2 Heavy Metal Pollution in Selected Wetland Area of Turkey

Lake Hazar is defined as a national importance wetland habitat in Eastern of Anatolia (Elazığ, Turkey), and various native fish species such as *Aphanius asquamatus* live in the lake body. Heavy metal concentrations of water, sediment, and fish samples were reported by various studies so far (Nergiz and Şamat 2019; Özmen et al. 2004). The highest heavy metal concentrations in water and sediment samples were recorded for Fe and Mn, whereas the lowest values were measured for Pb both in water and sediment. Moreover, Cr, Cu, and Ni concentrations in sediment exceeded the TEL and PEC values. In a study conducted by Nergiz and Şamat (2019), the order of heavy metal concentrations in Hazar Lake water was found to be Fe > As>Zn > Ni > Cd > Cu > Cr > Pb > Mn > Se. The Cd, Cu, Zn, and Cr concentrations were higher than the permissible levels for drinking water suggested by the WHO (World Health Organization), thus posing a serious threat for drinking water use. At the same time, the authors measured heavy metal concentrations in feather of Armenian gull and concluded that feathers are a useful monitoring tool for assessing heavy metal contamination.

Yeniçağa Lake (Bolu, Turkey) is one of the national important wetland areas in Turkey because of its localization on the migration route for migratory birds and fishing activities in lake body. The heavy metal pollution was detected in the lake body, with Mn, Cd, Pb, and Cr being observed in water, sediment, and organisms of the lake. The mean concentrations of Mn, Cd, Pb, and Cr in sediment were 1143, 0.8, 16, and 92.8 mg kg^{-1}, respectively. At the same time, the concentrations of Cd and Ch in sediments exceeded the TEL values, and Al, As, Mn, Pb, and Fe concentrations in water were higher than the permissible levels for drinking water suggested by the WHO (World Health Organization) (Saygı and Yiğit 2012). Tunca et al. (2013) also reported that Yeniçağa Lake is polluted by As and Ni in the lake body according to the criteria set by the United States Environmental Protection Agency.

Lake Aktaş is the second soda lake in Eastern Anatolian highlands of Turkey, and defined as very shallow lake body, varying in depth between 1.5 and 2 m. A study performed by Kükrer (2017) on Lake Aktaş indicated that the heavy metal concentrations were ranked in the order of Al>Fe>Mn>Zn>Ni>Cr>Cu>Pb>As>Cd>Hg. More importantly, EF (enrichment factor) values and geoaccumulation index verified minimal Hg pollution, whereas no potential contamination risk was found for other metals according to PER (potential ecological risk) index. At the same time, a moderate ecological contamination risk was detected for Hg.

Terkos Lake is the most important drinking water reservoir for İstanbul, the biggest city in Turkey (Kurun et al. 2010). In a study conducted by Kurun et al.

(2010), Al, Cu, Mn, Pb, Cd, Fe, Zn, Cr, and Ni were detected in the lake sediment and an aquatic crayfish (*Astacus leptodactylus*) tissues in May 2008. The high enrichment factors (EF) were also found for Zn, Cr, Cd, and Pb in sediment, indicating that anthropogenic sources including industrial and domestic effluent flow into the lake body. At the same time, current concentrations of Mn and As in Terkos Lake can be considered as possible risk originating from fertilizer use in the agricultural zones (Kükrer et al. 2019).

Beyşehir Lake is located in the southwestern Turkey, and the water from lake is used for drinking and agricultural purpose. However, the lake body was contaminated by various heavy metals including Cd, Pb, Hg, and Cr, and the heavy metal concentrations in plankton tissues were found higher than in the sediment samples (Altındağ and Yiğit 2005). This result indicated that heavy metals have been released into water body, and then finally entered the food chain through a trophic level with plankton metabolisms. The water of Beyşehir Lake was also containing extremely high with Cd and Pb according to the permissible level of drinking water. Specifically, the highest heavy metal concentration in sediment was measured for Pb, whereas the lowest concentration was found for Hg. Moreover, it was found that Cd concentration in sediment was higher than TEL and PEC values and the mean concentration of Hg was also higher than TEL value.

Kavak Delta is a saltmarsh area located on the eastern coast of the Gulf of Saros in the northwest part of the Aegean Sea. The delta covers the area of about 20 km^2 and extends along the Çanakkale-Istanbul highway, which constitutes the eastern border of the delta (Özcan et al. 2009). Sungur and Özcan (2015) monitored the concentrations of heavy metals in kaval Delta and concluded that Pb and Se were mainly derived from anthropogenic sources while Ba, Cd, Cr, Cu, LI, Ni, and Zn were derived by a combination of anthropogenic and lithospheric sources while Sr came from mainly lithospheric sources. Geoaccumulation index, enrichment factor, and contamination factor indicated that Cd, Li, Ni, Pb, and Se accumulate in the sediments and may create an environmental risk and impact existing species, namely, halophytic plants and bird species found in the delta.

Lake Sapanca is one of the few lakes which provide drinking water in Turkey, but it is exposed to heavy urbanization because of its natural beauty and its proximity to the metropolitan Istanbul (Duman et al. 2006). The water surface is 45 km^2, and it is surrounded with major highways and railway connecting Europe and Asia. Duman et al. (2006) focused on the heavy metal concentrations in *Potamogeton lucens* growing in the littoral zones of the lake.

The results revealed that the concentrations in the water and sediments decreased in the order of Zn>Cr>Ni>Pb>Mn>Cu>Cd and Mn>Zn>Ni>Cu>Mcr>Pb>Cd respectively. The concentrations of heavy metals in plant biomass followed closely the order found for sediments.

Güleryüz et al. (2008) monitored wetland vegetation along the Nilüfer stream in heavily industrialized Bursa City. The authors measured concentrations of Cr, Cu, Mn, Ni, and Zn in sediment as well as in aboveground and belowground biomass of *Rumex obtusifolius*, *Polygonum lapathifolium*, *Urtica dioica*, and *Xanthium strumarium*. The results suggested that all monitored species may be considered bioaccumulation species for Cr, Cu, Ni, and Zn, and, therefore, these plants may play a significant role in remediation of polluted sites.

Sasmaz et al. (2008) evaluated concentrations of Zn, Ni, Cu, Pb, Co, Mn, Cd, and Cr in sediments of the polluted Kehli stream (Elazığ, central Turkey) as well as in biomass of *Typha latifolia*. The stream is mainly polluted by sewage from small villages. For all heavy metals, the highest concentrations were found in roots followed by concentration in either leaves or sediments. The transfer factor (leaf/root) was the highest for Mn (>1). For other metals the transfer factor was lower than 1 indicating low movement of heavy metals from roots to leaves.

References

Adaman F, Hakyemez S, Özkaynak B (2009) The political ecology of a Ramsar site conservation failure: the case of Burdur Lake, Turkey. Environ Plan C: Gov Pol 27(5):783–800
Akcay H, Oguz A, Karapire C (2003) Study of heavy metal pollution and speciation in Buyak Menderes and Gediz river sediments. Wat Res 37(4):813–822
Aksoy A, Demirezen D, Duman F (2005) Bioaccumulation, detection and analyses of heavy metal pollution in sultan marsh and its environment. Wat Air Soil Pollut 164(1–4):241–255
Alemdaroğlu T, Onur E, Erkakan F (2003) Trace metal levels in surficial sediments of lake Manyas, Turkey and tributary Rivers. Int J Environ Stud 60(3):287–298
Altındağ A, Yiğit S (2005) Assessment of heavy metal concentrations in the food web of Lake Beyşehir, Turkey. Chemosphere 60(4):552–556
Arslan N, Koç B, Çiçek A (2010a) Metal contents in water, sediment, and Oligochaeta-Chironomidae of Lake Uluabat, a Ramsar site of Turkey. Sci World J 10:1269–1281
Arslan N, Ayık Ö, Şahin Y (2010b) Diversity and structure of Chironomidae (Diptera) limnofauna of Lake Uluabat, a Ramsar site of Turkey, and their relation to environmental variables. Turk J Fish Aquat Sci 10:315–322
Arslan Y, Diğdem T, Bulgök U, Bodur S, Kendüzler E (2016) Burdur Göl suyu ve çamurunda metal tayini. Mehmet Akif Ersoy Üniversitesi Fen Bilimleri Enstitüsü Dergisi 7(2):144–156
Ayas Z (2007) Review on DDT and its residues in Turkey's wetlands. J Environ Biol 28(4):707
Ayas Z, Kolankaya D (1996) Accumulation of some heavy metals in various environments and organisms at Göksu Delta, Türkiye, 1991-1993. Bull Environ Contam Toxicol 56(1):65–72
Ayas Z, Barlas NE, Kolankaya D (1997) Determination of organochlorine pesticide residues in various environments and organisms in Göksu Delta, Turkey. Aquatic Toxicol 39(2):171–181
Barlas N, Akbulut N, Aydoğan M (2005) Assessment of heavy metal residues in the sediment and water samples of Uluabat Lake, Turkey. Bull Environ Contam Toxicol 74(2):286–293
Berberoglu S, Yilma KT, Özkan C (2004) Mapping and monitoring of coastal wetlands of Cukurova Delta in the eastern Mediterranean region. Biodivers Conserv 13(3):615–633
Beyhan M, Şahin Ş, Keskin M, Harman B (2007) Burdur gölü uzun periyotlu seviye değişiminin su kalitesi ve ağır metaller üzerindeki etkisi. SDÜ Fen Bilimleri Enstitüsü Dergisi 11(2):173–179
Bölükbaşı V, Akın BS (2016) Agriculturally induced heavy metal accumulation in Seyfe lake. Turkey Bull Environ Contam Toxicol 96(3):401–407
Çalışkan V (2008) Human-induced wetland degradation: a case study of Lake Amik (Southern Turkey). Balwois (Ohrid, Republic of Macedonia). 27:1–10
Celik M, Ünsal N, Tüfenkçi OO, Bolat S (2008) Assessment of water quality and pollution of the Lake Seyfe basin, Kırşehir, Turkey. Environ Geol 55(3):559–569
Çiçek A, Emiroglu Ö, Arslan N (2009) World Lake conference. Wuhan, China, Wuhan
Costanza R (2006) Nature: ecosystems without commodifying them. Nature 443(7113):749

Costanza R, d'Arge R, De Groot R, Farber S, Grasso M, Hannon B, Limburg K, Naeem S, O'Neill RV, Paruelo J, Raskin RG, Sutton P, Van der Belt M (1997) The value of the world's ecosystem services and natural capital. Nature 387(6630):253–260

Costanza R, De Groot R, Sutton P, Van der Ploeg S, Anderson SJ, Kubiszewski I, Farber S, Turner RK (2014) Changes in the global value of ecosystem services. Glob Environ Chang 26:152–158

Dadaser-Celik F, Coggins JS, Brezonik PL, Stefan HG (2009) The projected costs and benefits of water diversion from and to the Sultan Marshes (Turkey). Ecol Econ 68(5):1496–1506

Davidson N, Dinesen L, Fennessy S, Finlayson C, Grillas P, Grobicki A, McInnes R, Stroud D (2019) A review of the adequacy of reporting to the Ramsar Convention on change in the ecological character of wetlands. Mar Freshw Res 71(1):117–126

Davraz A, Sener E, Sener S (2019) Evaluation of climate and human effects on the hydrology and water quality of Burdur Lake. Turkey J Afr Earth Sci 158:103569

Degirmenci FO, Acar P, Kaya Z (2019) Consequences of habitat fragmentation on genetic diversity and structure of *Salix alba* L. populations in two major river systems of Turkey. Tree Genet Genom 15(4):59

Demirezen D, Aksoy A (2004) Accumulation of heavy metals in Typha angustifolia (L.) and Potamogeton pectinatus (L.) living in sultan marsh (Kayseri, Turkey). Chemosphere 56(7):685–696

Demirezen D, Aksoy A (2006) Common hydrophytes as bioindicators of iron and manganese pollutions. Ecol Indic 6(2):388–393

Duman F, Obali O, Demirezen D (2006) Seasonal changes of metal accumulation and distribution in shining pondweed (Potamogeton lucens). Chemosphere 65:2145–2151

Dora EÇ, Sunlu U, Ergen Z (2007) Heavy metal concentrations in *Hediste diversicolor* (Polychaeta) and sediments from Homa Lagoon (Izmir Bay-Turkey). Rapp Comm Int Mer Médit 38:253

Elmaci A, Teksoy A, Topac FO, Ozengin N, Kurtoglu S, Baskaya HS (2007) Assessment of heavy metals in Lake Uluabat, Turkey. Afr J Biotechnol 6(19):2236–2244

Engin MS, Uyanik A, Cay S (2017) Investigation of trace metals distribution in water, sediments and wetland plants of Kızılırmak Delta, Turkey. Int. J Sediment Res 32(1):90–97

Ergene S, Çavaş T, Celik A, Köleli N, Kaya F, Karahan A (2007) Monitoring of nuclear abnormalities in peripheral erythrocytes of three fish species from the Goksu Delta (Turkey): genotoxic damage in relation to water pollution. Ecotoxicology 16(4):385–391

Ersayın K, Tağıl Ş (2017) Ecological sensitivity and risk assessment in the Kizilirmak Delta. Fresenius Environ Bull 26(11):6508–6516

Geijzendorffer I, Beltrame C, Chazee L, Gaget E, Galewski T, Guelmami A, Perennou C, Popoff N, Guerra C, Leberger R, Jalbert J, Grillas P (2019) A more effective Ramsar Convention for the conservation of Mediterranean wetlands. Front Ecol Evol 7:21

Güleryüz G, Arslan H, Çelik C, Güćer Ş, Kendall M (2008) Heavy-metal content of plant species along Nilűfer stream in industrialized Bursa City, Turkey. Water Air Soil Pollut 195:275–284

Gürlük S, Rehber E (2008) A travel cost study to estimate recreational value for a bird refuge at Lake Manyas, Turkey. J Environ Manag 88(4):1350–1360

Hacısalihoğlu S, Karaer F (2016) Relationships of heavy metals in water and surface sediment with different chemical fractions in Lake Uluabat, Turkey. Polish J Environ Stud 25(5):1937–1946

Hakanson L (1980) An ecological risk index for aquatic pollution control—a sedimentological approach. Water Research 14:975–1001

Horns JJ, Şekercioğlu ÇH (2018) Conservation of migratory species. Curr Biol 28(17):R980–R983

Hughes J (1995) The current status of European wetland inventories and classifications. In: Finlayson M, van der Valk A (eds) Classification and inventory of the world's wetlands. Springer, Dordrecht, pp 17–28

Karadeniz N, Tınl A, Baylan E (2009) Wetland management in Turkey: problems, achievements and perspectives. Afr J Agric Res 4(11):1106–1119

Katip A, Karaer F (2011) Uluabat gölü su kalitesinin türk mevzuatina ve uluslararasi kriterlege göre degerlendirilmesi. Uludağ Üniversitesi Mühendislik-Mimarlık Fakültesi Dergisi, Cilt 16. Sayı:2

Katip A, Karaer F, İleri S, Sarmaşik S (2013) Uluabat Gölü'nde iz metallerin askıda katı madde ile su arasındaki dağılımının araştırılması. Gazi Üniversitesi Mühendislik-Mimarlık Fakültesi Dergisi 28(4):865–874
Katip A, Özengin N, İleri S, Elmacı A, Karaer F (2014) Uluabat gölü'nde sudaki iz element kirliliginin zamasal degisiminin (2003–2004 VE 2008-2009) izlenmesi. 2014. Uludağ Üniversitesi Mühendislik Fakültesi Dergisi, Cilt 19, Sayı 2
Katip A, Karaer F, Hacısalihoğlu S (2016) The role of different kinds of particles on heavy metals in Lake Uluabat sediments. Environ Forensic 17(3):263–273
Kazancı N, Leroy S, Öncel S, İleri Ö, Toprak Ö, Costa P, Sayılı S, Turgut C, Kibar M (2010) Wind control on the accumulation of heavy metals in sediment of Lake Ulubat, Anatolia, Turkey. J Paleolimnol 43(1):89
Kucuksezgin F, Uluturhan E, Batki H (2008) Distribution of heavy metals in water, particulate matter and sediments of Gediz River (eastern Aegean). Environ Monit Assess 141:213–225
Kükrer S (2017) Pollution, source, and ecological risk assessment of trace elements in surface sediments of Lake Aktaş, NE Turkey. Human and Ecological Risk Assessment 23(7):1629–1644
Kükrer S, Çakır Ç, Kaya H, Erginal AE (2019) Historical record of metals in Lake Küçükçekmece and Lake Terkos (Istanbul, Turkey) based on anthropogenic impacts and ecological risk assessment. Environ Forensic 20(4):385–401
Kuleli T (2010) City-based risk assessment of sea level rise using topographic and census data for the Turkish coastal zone. Estuar Coasts 33(3):640–651
Kuleli T, Guneroglu A, Karsli F, Dihkan M (2011) Automatic detection of shoreline change on coastal Ramsar wetlands of Turkey. Ocean Eng 38(10):1141–1149
Kurun A, Balkıs N, Erkan M, Balkıs H, Aksu A, Erşan MS (2010) Total metal levels in crayfish Astacus leptodactylus (Eschscholtz, 1823), and surface sediments in Lake Terkos. Turkey Environ Monit Assess 169(1–4):385–395
Magnin G, Yarar M (1997) Important bird breeding areas in Turkey, İstanbul
Matthews GVT (1993) The Ramsar convention on wetlands: its history and development. Ramsar Convention Bureau Gland
McInnes RJ, Simpson M, Lopez B, Hawkins R, Shore R (2017) Wetland ecosystem services and the Ramsar Convention: an assessment of needs. Wetlands 37(1):123–134
Meng W, He M, Hu B, Mo X, Li H, Liu B, Wang Z (2017) Status of wetlands in China: a review of extent, degradation, issues and recommendations for improvement. Ocean Coast Manage 146:50–59
Mitsch WJ, Gosselink JG (2000) The value of wetlands: importance of scale and landscape setting. Ecol Econ 35(1):25–33
Mitsch WJ, Gosselink JG, Zhang L, Anderson CJ (2009) Wetland ecosystems. Wiley, Hoboken
Muhammetoglu A, Keyikoglu R, Cil A, Muhammetoglu H (2019) Integrated management of pesticides in an intensive agricultural area: a case study in Altinova, Turkey. Environ Monitor Assess 191(9):599
Nergiz H, Şamat A (2019) Assessment of heavy metal concentration in feathers of Armenian gull (Larus armenicus Buturlin, 1934) and water samples of Hazar Lake, Turkey. Appl Ecol Environ Res 17(4):10221–10227
Özcan H, Akbulak C, Kelkit A, Tosunoğlu M, Uysal İ (2009) Ecotourism potential and management of Kavak Delta (Northwest Turkey). J Coast Res 25(3):781–787
Özmen H, Külahçı F, Çukurovalı A, Doğru M (2004) Concentrations of heavy metal and radioactivity in surface water and sediment of Hazar Lake (Elazığ, Turkey). Chemosphere 55(3):401–408
Parlak H, Çakır A, Boyacıoğlu M, Arslan ÖÇ (2006) Heavy metal deposition in sediments from the Delta of the Gediz River (Western Turkey): a preliminary study. Su Ürünleri Dergisi 23(3):445–448
Pulatsü S, Topçu A (2015) Review of 15 years of research on sediment heavy metal contents and sediment nutrient release in inland aquatic ecosystems, Turkey. J Water Resour Protect 7(2):85
Ramsar (2019) www.ramsar.org

Salihoğlu G, Karaer G (2004) Ecological risk assessment and problem formulation for Lake Uluabat, a Ramsar state in Turkey. Environ Manag 33(6):899–910
Sarp G, Ozcelik M (2017) Water body extraction and change detection using time series: a case study of Lake Burdur. Turkey J Taibah Univ Sci 11(3):381–391
Sasmaz A, Obek E, Hasar H (2008) The accumulation of heavy metals in Typha latifolia L. Grown in a stream carrying secondary effluent Ecol. Eng. 33:278–284
Saygı Y, Yiğit SA (2012) Heavy metals in Yeniçağa Lake and its potential sources: soil, water, sediment, and plankton. Environ Monitor Assess 184(3):1379–1389
Somay MA, Filiz S (2003) Hydrology, hydrogeology and hydrochemistry of wetlands: a case study in Izmir bird paradise, Turkey. Environ Geol 43(7):825–835
Sungur A, Özcan H (2015) Chemometric and geochemical study of the heavy metal accumulation in the soils of a salt marsh area (Kavak Delta, NW Turkey). J Soils Sediments, 15:323–331
Tas EÇ, Sunlu U (2013) Distribution of some heavy metals in surface sediments fro the Homa Lagoon (Izmir Bay, Turkey). 40:275
Taş EÇ, Ergen Z, Sunlu U (2009) 2002-2004 Yılları Arasında Homa Lagünü'nden (İzmir Körfezi) Toplanan Hediste diversicolor'da ve Yaşadığı Sedimentte Ağır Metal Düzeylerinin (Cd, Cu, Zn, Pb, Cr, Fe) Araştırılması. Su Ürünleri Dergisi 26(3):179–185. (in Turkish)
Tinner W, Bigler C, Gedye S, Gregory-Eaves I, Jones RT, Kaltenrieder P, Krähenbühl U, Hu FS (2008) A 700-year paleoecological record of boreal ecosystem responses to climatic variation from Alaska. Ecology 89(3):729–743
Tunca E, Üçüncü E, Kurtuluş B, et al. (2013) Accumulation trends of metals and a metalloid in the freshwater crayfish Astacus leptodactylus from Lake Yeniçağa (Turkey). Chem Ecol 29(8):754–69
Türkmen A, Tepe Y, Türkmen M, Ateş A (2012) Investigation of metals in tissues of fish species from Akyatan lagoon. Fresenius Environ Bull 21(11c):3562–3567
Türkmen EI (2018) Manyas Kuş Gölü'nün zooplankton kommunite yapısı. Balıkesir University, Balıkesir/Turkey
Uluturhan E, Kontas A, Can E (2013) Sediment concentrations of heavy metals in the Homa Lagoon (Eastern Aegean Sea): assessment of contamination and ecological risks. Mar Pollut Bull 62(9):1989–1997
Vymazal J (ed) (2011) Water and nutrient management in natural and constructed wetlands. Springer, Dordrecht
Yalcin MG, Battaloglu R, Ilhan S (2007) Heavy metal sources in Sultan Marsh and its neighborhood, Kayseri. Turkey Environ Geol 53(2):399–415
Yılmaz DD, Parlak KU (2011) Changes in proline accumulation and antioxidative enzyme activities in Groenlandia densa under cadmium stress. Ecol Indic 11(2):417–423

Chapter 30
Locals' Awareness of Ecotourism in the Southern Marshes of Iraq

Laith A. Jawad

Abstract The investigation delivered in this chapter intends to discover out the apparent economic, environmental, and sociocultural influences of ecotourism by locals who live in village in the heart of the southern marshes of Iraq near Chibayish area, south of Dhi Qar Province, south of Iraq. The investigation is qualitative and experiential which emphasizes on the locals' approaches to the economic, environmental, and sociocultural variations owing to ecotourism. A total of 350 participants were questioned from the locals using a conveyed questionnaire containing 9 demographic features and 30 influence articles categorized into economic, environmental, and sociocultural classes. The study displays the influence articles concerning economic and socio-cultural features that attained higher ranks. Contrariwise, articles linked with the environmental feature attained lower ranks. Such results signify that locals prefer both the economic and sociocultural influences of tourism as they certainly supposed both influences, whereas the environmental influence was least preferred by the locals. Also, it was observed that local's approaches to ecotourism differ with the disparities in revenue, job, and gender. This investigation resolved that locals recognize economic and sociocultural assistances carried by ecotourism without forgoing environmental and aesthetic features. Additional investigations are suggested to reveal the tendencies of ecotourism and its influences on locals and the marshes.

Keywords Lifestyle · Marsh Arabs · South of Iraq · Tourism · Socio-economic

30.1 Introduction

The ecotourism business is growing globally as the request for tourism amenities, knowledges a sturdy rise (Drumm and Moore 2005; Mree et al. 2020). Travel has turned out to be a significant economic event, particularly in the developing

L. A. Jawad (✉)
School of Environmental and Animal Sciences, Unitec Institute of Technology, Mt Albert, Auckland, New Zealand

L. A. Jawad (ed.), *Southern Iraq's Marshes*, Coastal Research Library 36,
https://doi.org/10.1007/978-3-030-66238-7_30

countries. It is considered as one of the wildest growing businesses worldwide, making about two trillion USD annually (about 12% of the world's economy) (Fennell 2004). Ecotourism is the new aspect of travelling well thought-out as low-impact nature tourism backing the care of species and ecosystems either directly through preservation or indirectly by providing the local society with adequate income, which prods them to appreciate and keep the wildlife inheritance as a supply of revenue (Weaver 2008; Mree et al. 2020). Within a specific society, ecotourism can deliver a context for motivating economic growth by income making along with the growth and progress of local businesses and industries. Therefore, ecotourism is linked mainly with the small-scale society-organized action and long-term social security and health (Reichel et al. 2008; Mree et al. 2020).

As ecotourism is chiefly connected with environmental protection and society growth, including the local society or inhabitants is one of the essential features of ecotourism. The contribution of the locals should be trained widely in scheduling, rules-making, and implementation by incorporating their insights, standards, and benefits (Manu and Kuuder 2012; Vincent and Thompson 2002). As ecotourism impacts locals substantially, it can be supposed as encouraging or adverse by the locals relying on their participation with ecotourism (Haddle 2005; Manu and Kuuder 2012; Mree et al. 2020).Tourism insights by locals have attained specific consideration lately, and the significance of their views for scheduling topics, in regard to the maintainable administration, has been accredited (Dyer et al. 2007). It has been documented that the assessment of locals' acuities could be an appreciated factor in recognizing and gauging the influences of ecotourism (Cordero 2008; Getz 1994). A study (Mathieson and Wall 1982) required to examine the supposed effects of tourism within a theoretical framework converging on the merger and links of different kinds of aspects. Many levels or agendas have been established to evaluate locals' arrogances on several facets of tourism influences.

There are several areas around the world comprising numerous of the world's highest ecotourism lures (Drumm and Moore 2005). In regard to land use, prearranged ecotourism can be supportably accomplished for preserving natural outfit (Kourandeh and Fataei 2013; Mree et al. 2020). The southern marshes of Iraq is an exclusive area, where the beauty of the nature combines with the simple life of the Marsh Arabs to produce a blend that no one can stand without a huge appreciation. In this area, birds from the northern part of the north hemisphere pay a regular visit to spend the cold winter of North Europe in a warm and moderate weather. In addition, there are the aquatic fauna and flora, which represent a unique selection of animals and plants. Furthermore, the exclusive life style of the Marsh Arabs, the locals of the marsh areas, their food and way of life and much more are unlimited attractions for an excellent ecotourism. Recently, little tourist activities were introduced into the marsh areas, where some tourist companies were established in southern cities of Iraq. They advertised for 1-day stay in different southern marshes in a hope for such activities to boom and attract tourist from outside Iraq.

No studies have been performed on the apparent influences of ecotourism by the locals of the southern marshes of Iraq. Such studies should be established as they have been done in similar area in the world (Debashish et al. 2013; Mree et al. 2020), with objectives of measuring the socioeconomic influences of ecotourism on locals in the area overlooking the environmental effects. Other studies when they took into consideration areas similar to the southern marshes of Iraq have considered the environment issues (Jahan and Akhter 2018) and shown the influence of ecotourism on the environment, community, and ethos of the region. However, no investigations endeavoured to assume a tourism influence scale to show out the locals' insights concerning the effect of ecotourism. Also, these examinations didn't deliberate the link between demographic components and apparent influences of ecotourism by locals. Therefore, in addition to the lack of studies that concentrate on the local's perceptions for the ecotourism in the southern marshes of Iraq, this chapter also tries to find out the apparent economic and sociocultural influences of ecotourism bearing in mind environmental impressions using the tourism influence level anticipated by Ap and Crompton (1998) as well as to correlate apparent influences of ecotourism with demographic components.

30.2 Methodology and Data Collection

In this chapter of methodology, the design of the study was given in the form of a set of questionnaires, and methods of data collection and analysis will be shown.

A set of questionnaires was established conferring to the influence objects scale advanced by Ap and Crompton (1998). According to these authors, 35 influence items were chosen from an initial pool of 147 items and were grouped into economic, environmental, and sociocultural categories. The objects comprise both encouraging and adverse influences alleged by locals. The specific scale was preferred owing to the scale comprising each influence article that exhibited the most statistically dependable results on ecotourism influence insight. The scale includes an assessment scheme of explicit ecotourism influences on two features of each criterion: confidence variable and assessment variable.

The objects were hierarchically arranged regarding an index of apparent ecotourism influence. The confidence variable was measured by asking each participant to assess the level of variation linked with each item. A five-point score system was used, ranging from 1 to 5. An additional notation "DK" was used for a "don't know" reply. The symbolization of the values was as follows: 1 = large decrease; 2 = moderate decrease; 3 = no change; 4 = moderate increase; and 5 = large increase. The assessment variable of the index also shadowed a similar rating system, measuring the local's level of like or dislike for each item by five points ranging from 1 to 5. The ranks with notation were as follows: 1 = dislike; 2 = somewhat dislike; 3 = neither like nor dislike; 4 = somewhat like; and 5 = like.

Main data were gathered via face-to-face interviews steered through first-hand visits in the villages situated in Chibayish area, south of Dhi Qar Province, south of Iraq. The sampling method was random but without fixed bias. Interviews of 350 participants were steered from the society. A survey, grounded upon the scale mentioned in Ap and Crompton (1998), was used for attaining data. Beside with influence insight data, nine demographic components were also reported. As the area was never studied, no data were available so far for comparison.

Nine population-related components were considered in this survey, containing gender, age, marital status, education, occupation, length of residency, job type, income, and visiting urban areas. Percentages were computed conferring to the reactions to the variables made by participants. For data analysis, IBM SPSS version 21.0 was used. The local's reactions to the influence articles embodied in the questionnaire were verified. Standard deviations were also recorded for estimating the extent of variances among the participants.

30.3 Results Attained

Reactions were attained from a village near Al-Gibayish area, south of Dhi Qar Province, south of Iraq. A total of 350 participants were questioned following the conveyed questionnaire according to Ap and Crompton (1998). There were 320 males and 30 females questioned. The population structure of the participants is shown in Table 30.1. The participant's age groups were also diverse, with about 28.6% being between 31 and 40 years and another 22.9% between 21 and 30 years. People aging between 41 and 50 years of constituted about 20% of the participants, and people above the age of 50 years made a percentage of about 21.4%. About 75.7% of the participants were married while the remaining 24.3% were unmarried.

Table 30.1 Demographic profile of participants in a study in the southern marshes of Iraq

Age (year)	11–20	21–30	31–40	41–50	>50
	25	80	100	70	75
Marital status	Married			Unmarried	
	265			85	
Residency	<10	10–15	16–20	>20	
	0	0	0	350	
Education	Illiterate	Primary	Secondary	Others	
	340	7	3	0	
Job Type	Tourism based			Non-tourism based	
	348			2	
Occupation	Agriculture	Civil servant	Student	Worker	
	250	0	2	98	
Income	<50,000 ID	50,000–100,000	100,000–150,000	>150,000	
	250	45	5	0	
Travel outside the village	Yes			NO	
	200			150	

All participants were living in the village vicinity for the whole of their life. About 97.1% of the respondents were illiterate, and another 2% finished primary education. About 0.9% educated up to secondary school level, while no one finished higher secondary education.

Out of 350 respondents, about 71.4% had monthly incomes (i.e. Iraqi Dinar = ID) below 50,000/ month, while another 12.9% earned between 50,000 and 100,000 ID per month and 1.4% have 100,000–150,000 ID per month. Most of the participants are farmers (71.4%), worker (25.4%), and student (2%).

Locals' approach to supposed influences of ecotourism was analysed. In this analysis, 30 influence articles were separated into three classes (economic, environmental, and sociocultural impacts or aspects) and each item was given a rank. Among the three groups, the items linked with economic impacts attained higher ranks. Economic factor objects, namely "local economy improvement", "contribution to income and standard of living", and "increases employment opportunity" ranked first, second, and third correspondingly as these three items were mostly favoured by the locals. The least attracted entry in this category was "improves public utility infrastructure" which ranked 33rd. Besides sociocultural factor items were more selected by locals, with articles "improves the quality of life" and "improves quality of police protection". The items regarding the general setting of the community also graded high in the ranking system, namely, "positive attitude of locals towards tourists", "pride of locals", and "community spirit among locals". Perversely, articles related to environmental influences were slightly favoured by the participants, with items "preservation of the natural environment", "overcrowding", "improvement of the area's appearance", and "increased noise pollution and waste".

These observations show that the participants were positive about the economic benefits transferred by ecotourism events, but they were also worried about the environmental and aesthetic sides. This agrees with the results of Tatoglu et al. (2002).

30.4 Remarks

The locals living in the village near Al-Gibayish area, south of Dhi Qar Province, south of Iraq in the heart of the southern marshes of Iraq were found as enduring inhabitants as they are living here for several generations (more than 20 years). This shows that the society in this area is more or less homogenous within the time frame. Some locals of the village were settlers, coming mainly from other villages in the marsh areas after the drying of the marshes by Saddam Hussein, and now are considered as permanent locals. Nevertheless the deeper marsh area was unacquainted to the public for a long period due to the political unrest, and it has become a popular tourist site, with a hope of an ever-cumulative number of tourists visiting each year due to government and private sector reporting.

Recently, private sector initiatives were introduced, where several tourism companies were established to instate the business of tourism in the southern marshes of

Iraq to include restricted activities. Such events will encourage other investors to establish more tourist businesses with wider tourist activities, with probably building hotels and restaurants in the area for tourists to rest for few days. Other investors may be interested to establish different kinds of water sports or amusement cities for recreations in the marshes.

The study discloses about 35% of the total participants were involved in certain type of ecotourism activities such as driving boats hired for visitors from the big cities for recreations during the holidays. Their views inclined to denote the economic and sociocultural profits that came with ecotourism events. Locals, who were not intricate in ecotourism, inclined to emphasis on the failing influences of ecotourism on the environment.

The locals commonly alleged the absence of public utility delivery and infrastructure development in an undesirable sense, suggesting they were not pleased with the level of public amenities delivered by the government. The study shows objects of the economic influence group such as "improvement of the quality of life" were favoured by the locals. This finding is also supported by results of Debashish et al. (2013) and Mree et al. (2020), which exposed a positive approach of locals concerning tourism in the marsh area. According to Crouch and Ritchie (1999), tourism action represented a significant issue in defining the class of life of the local society of the marshes. Williams and Lawson (2001) suggested that the type of life of the locals in any society could bring changes to the apparent influence and approach of the locals, which was also backed by (Zhang et al. 2006).

Study at hand provokes a motivated approach of locals to the positive influences of ecotourism in the case of economic and sociocultural factors, and locals were also found worried about the negative influences of ecotourism on the surrounding ecosystem. It reflects locals were attentive of environmental deprivation instigated by ecotourism events in the area in case this business became widely dispersed, which backs the results of other investigation (Jahan and Akhter 2018). The results showed that variations in revenue, occupation, and gender influenced the apparent bearings of ecotourism by the locals. However, the investigation reveals that participants who were intricate in ecotourism reinforced the growth of ecotourism further.

References

Ap J, Crompton JL (1998) Developing and testing a tourism impact scale. J Travel Res 37(2):120–130

Cordero JC (2008) Residents' perception of tourism: a critical theoretical and methodological review. Ciencia Ergo-sum, Revista Científica Multi Disciplinaria de Prospectiva 15(1):35–44

Crouch GI, Ritchie JB (1999) Tourism, competitiveness, and societal prosperity. J Bus Res 44(3):137–152

Debashish R, Gope MC, Alam MD (2013) Socioeconomic impact of tourism to the local residents: a case study on Ratargul swamp forest. J Socioecon Res Dev 10(5):1557–1562

Drumm A, Moore A (2005) Ecotourism development: a manual for conservation planners and managers, vol l, 2nd edn. The Nature Conservancy, Arlington, p 85

Dyer P, Gursoy D, Sharma B, Carter J (2007) Structural modeling of resident perceptions of tourism and associated development on the sunshine coast, Australia. Tour Manag 28(2):409–422

Fennell DA (2004) Ecotourism: an introduction. Routledge, London

Getz D (1994) Residents' attitudes towards tourism: a longitudinal study in Spey Valley, Scotland. Tour Manag 15(4):247–258

Haddle JB (2005) Community residents' perceptions of ecotourism impacts and conservation issues in rural creole Belize: a case study of crooked tree wildlife sanctuary. Doctoral dissertation, University of Florida

Jahan KM, Akhter H (2018) Impact of ecotourism on the environment, society and culture of ratargul swamp forest in Sylhet, Bangladesh. Asian J Environ Ecol 8(1):1–8

Kourandeh HH, Fataei E (2013) Estimation of tourism carrying capacity of fandoqloo forest in Ardebil province, Iran. Bull Env Pharmacol Life Sci 2(12):64–70

Manu I, Kuuder WC-J (2012) Community-based ecotourism and livelihood enhancement in Sirigu. Ghana Int J humanit Soc Sci 2(18):97–108

Mathieson A, Wall G (1982) Tourism, economic, physical and social impacts. Longman, London

Mree CL, Das S, Ray TK, Chowdhury P, Saha N (2020) Residents' perception of ecotourism in Ratargul Freshwater Swamp Forest of Bangladesh. Asian J Res Agric For:1–11

Reichel A, Uriely N, Shani A (2008) Ecotourism and simulated attractions: Tourists' attitudes towards integrated sites in a desert area. J Sustain Tour 16(1):23–41

Tatoglu E, Erdal F, Ozgur H, Azakli S (2002) Resident attitudes toward tourism impacts: the case of Kusadasi in Turkey. Int J Hosp Tour Admin 3(3):79–100

Vincent VC, Thompson W (2002) Assessing community support and sustainability for ecotourism development. J Travel Res 41(2):153–160

Weaver D (2008) Ecotourism, 2nd edn. Wiley, London

Williams J, Lawson R (2001) Community issues and resident opinions of Tourism. Ann Tour Res 28(2):269–290

Zhang J, Inbakaran RJ, Jackson MS (2006) Understanding community attitudes towards tourism and host-guest interaction in the urban-rural border region. Tour Geogr 8(2):182–204

Chapter 31
The Effects of Thermal Pollution on the Aquatic Life in the Southern Marshes of Iraq

Laith A. Jawad

Abstract A huge amount of heat are released into freshwater habitats from power stations using once-through cooling systems, frequently causing an increase in temperature that disrupts aquatic ecosystems and biota living in. Internationally, 56% of the heat released to the freshwater system included in the hot water discharge from the thermal power plants appeared to be the source of thermal pollution in the area where it has been rejected in. In the literature today, numerous studies were giving evidences that the heated water from the thermal power stations is responsible for the huge environmental changes that lead to an unfavourable condition and affect directly the freshwater biota. In Iraq, all the power plants are thermal stations using the once-through cooling systems as a cooling system. Such procedure has approved its deleterious effects on the freshwater environment where the hot water is rejected to. In the present chapter, a short review is given about the basic idea of the thermal power stations and their effects on the freshwater environment and its biota. Also, in this chapter examples on the effect of the heated water rejected from the thermal power plants in Iraq were given. At the end of the chapter, a discussion to the future of the development of the electricity industry in Iraq is given, which reports on establishing further power plants in the western desert area of Iraq to serve as electricity supply source to be transferred to European countries. If this project sees the light regarding the harmful effects of these power plants on the environment, then thousands or millions of freshwater biota will never see the light again.

Keywords Water temperature · Aquatic life · Fishes · Power plants · Electricity generators

---

L. A. Jawad (✉)
School of Environmental and Animal Sciences, Unitec Institute of Technology, Mt Albert, Auckland, New Zealand

L. A. Jawad (ed.), *Southern Iraq's Marshes*, Coastal Research Library 36,
https://doi.org/10.1007/978-3-030-66238-7_31

31.1 Introduction

Internationally, the freshwater habitats experiencing manifold stresses owing to man-made activities like dams, channelization, deforestation, irrigation and many more (Brooker 1985; Bunn and Arthington 2002; Sweeney et al. 2004; Raptis et al. 2016). These events have significant joined physical and chemical influences on aquatic ecosystems (Malmqvist and Rundle 2002). Among the serious physical impact on freshwater system is thermal pollution, an influence that has two sides of effects on the environment, one on the biota and other on the chemical pollution (Heugens et al. 2002; Holmstrup et al. 2010; Raptis et al. 2016). Therefore, it is important to recognize the regions where the thermal pollution is significantly impacted (Raptis et al. 2016).

Among the main sources of thermal pollution in the freshwater system is the thermoelectric power area (Hester and Doyle 2011). Electrical power stations are usually located near freshwater sources such as rivers and lakes so they can use a method known as once-through and recirculation (tower) cooling. With this method, the heat attained by the cooling water during the steam cycle is straight excluded back into the river. Some of the heat absorbed is usually eradicated through the evaporation and dissolution into the atmosphere (Stewart et al. 2013). The three basic kinds of cooling systems—once-through, closed-cycle and dry cooling—vary intensely in their water usage, with once-through cooling being the most water-rigorous and environmentally damaging method (Fleischli and Hayat 2014).

In rivers, particularly those with large thermal release centres, the water temperature upsurges as a consequence of cooling water releases, which can be extensive (Madden et al. 2013). In many parts of the world, there are certain legislations that control the release of hot water directly to the freshwater environment. Among such countries is the United States of America, where thermal power stations are commonly found across the country. The laws that govern such practices state that numerous power stations are required to limit their electricity output, often at times of peak request, and are expected to do so even more under climate change situations (Förster and Lilliestam 2009; van Vliet et al. 2012). Precisely, for the reason of keeping the aquatic habitats, many US states impose an upper temperature limit of 32 °C for surface water (Madden et al. 2013), while in the European Union, temperature of the disposed water should not go beyond 1.5 °C–3 °C above natural temperatures (or 21.5 °C–28 °C) (European Parliament and Council of the European Union 2006).

Water removals for thermoelectric power production were assessed in 2005 to be 201 billion gallons per day—the uppermost use of any manufacturing (Fleischli and Hayat 2014). Furthermost of that water goes for the cooling process. Power factories boil water to yield steam, which is used to turn the turbines that create electricity.

Preceding investigations on the influences of cooling water thermal discharges have concentrated on the effects either from a single power plant (e.g. Wu et al. 2001, Contador 2005, Verones et al. 2010) or from multiple factories over a large freshwater habitat (e.g. Stewart et al. 2013, Pfister and Suh 2015). In a current investigation, the cooling systems for the most common of thermal power plants worldwide were identified (covering 92% of the global thermoelectric power installed capacity), and the thermal releases for those plants with once-through cooling systems were modelled (Raptis and Pfister 2016).

The use of once-through cooling systems can lead to austere environmental influences such as killing billions of fish, damaging aquatic ecosystems and upsurging the temperature of rivers, lakes and even ocean waters. Power stations using once-through cooling also need to introduce more intervals of shutdowns or limitations during times of drought and extreme heat (Fleischli and Hayat 2014). Any legislation that put forward in regard to protecting the freshwater environment from the extra heat generated by the power plants should take into consideration that the electricity sector is acting in the direction of cleaner and more water-smart future by substituting old-fashioned and habitat-damaging once-through cooling systems with contemporary, less water-rigorous machineries (Fleischli and Hayat 2014).

The aim of the this chapter is to give a short review about the effects of the thermal power stations on the freshwater environment and then discuss how such effects have already been observed in Iraq in general and the southern marshes. The chapter also discusses the future impacts of the thermal power stations in Iraq, with the forthcoming plans to build extra thermal power stations using both gas and solar energy to heat water and freshwater water in the once-through cooling systems.

31.2 The Background of the Electrical Power in Iraq

Iraq of a long history going back to more than 5000 years BC is a Middle Eastern country with a population of about 27 million and stretching over an area of 437,000 km^2 (Reda et al. 2006). For the Middle East region, Iraq was at the leading edge during the 1970s, with developed technology, education and medical resources and quickly emerging economy (Reda et al. 2006). Nevertheless, for the previous 25 years, it has undergone from political unrest, economic restriction and uprising.

Electric power was introduced into Iraq in 1917 during the First World War. This service started as small mobile DC sets, which assisted precise local loads. Later, a big power station was first introduced in 1932 encompassing two units of 2.5 MW. Iraq went further and developed the electric power sector in the late 1950s, when the first three thermal power plants, working on heavy fuel oil, were fitted in south of Baghdad City (Reda et al. 2006). During the periods 1960s and 1970s, economic growth in Iraq went faster, and the power system was extended with the introduction of 132 kV and 400 kV networks, and sufficient power production to satisfy the 6000 MW request (Reda et al. 2006).

During the periods 1980s and 1990s, the wars that have covered the heaven of Iraq have led to a significant decrease in the growth and even deterioration of the status of what have been built in the electric power development. With the economic sanction that is applied on Saddam Hussein regime, several technical companies desisted from supplying spare parts or any services to the existing power plant stations in Iraq (Reda et al. 2006; Library of Congress – Federal Research Division, 2006).

Iraq reached its utmost development in regard to electric power generation industry by 1990, where the total connected producing bulk in Iraq was about 10,000 MW with a peak demand of about 5100 MW. The production of electricity mix contained 50% thermal, 28% gas and 22% hydro, and nearly 90% of the population had access to electricity (Reda et al. 2006; Library of Congress – Federal Research Division, 2006). Due to a significant decline in the electric power generation in Iraq in the following years and as a consequence of wars and unsettle politic status, numerous generation power stations were rigorously spoiled causing a huge reduction in supply of electricity. This situation of poor power supply in Iraq continued until the present time, with slight enhancement for the electric supply service was resumed after 2003 (Ministry of Planning 2010).

The main feature of electric power generation in Iraq is the presence of the thermal power plants that depend in their operation on both oil and natural gas as fuel (Austin et al. 2005). This might shed light on the significant impacts that these power stations can inflict on the freshwater environment of Iraq.

31.3 Impacts of Thermal Power Plants on the Freshwater Habitats

Radical fluctuations in release and temperature linked with thermal pollution may modify the biotic society as well as connected ecosystem courses such as organic matter dynamics (Lakly and McArthur 2000). The quantity and quality of organic matter contributions and invertebrate communities all these may propose alterations in ecosystem task (Cummins et al. 1983; Benke et al. 1984; Hauer 1989). Alterations in channel morphology, valley riparian vegetation and organic matter additions as results of thermal release may cause a swing in the stream energy base (Lakly and McArthur 2000). Characteristically, terrestrially originated organic matter (i.e. allochthonous) structures macroinvertebrate societies in forested basins that interact widely with their floodplain. Contrariwise, open, channelized systems such as grassland and desert streams may obtain most of their energy from instream algae or aquatic plants (i.e. autochthonous) (Minshall 1996). Original indication for a shift in energy base can be found in a systems change in primary instream structural constituents and production, conveyance of organic matter, retaining and sedimentation within a stream reach and possibly the organic energy flow between trophic levels in the system (Lakly and McArthur 2000).

Several studies have shown that the impact on the freshwater biota starts when hot water is discharged to the habitat from the power plants. Raptis et al. (2016) found that changes in the freshwater habitats can happened when water with temperature of 9.5–10 °C above ambient stream temperatures during the summer from 1996 to 2005, while Hester and Doyle (2011) indicated that considerable aquatic ecosystem influences can happen with temperature upsurges of 5 °C or higher, which proposes that the highest temperature releases during the summer could disturb aquatic ecosystems and biota living in.

The following types of impacts have been suggested by Bera (2016) to affect the physical features of the water of the rivers and lakes due to the input of hot water from the power plants.

31.3.1 Impacts on Water Quantity

Thermal-based power plants chiefly rely on water for cooling requirements. This allows them to continue high production competences, but also shows that they use an incredible amount of water every day (McDermott and Nilsen 2014). Circulating system using cooling tower loses a significant amount of continuous loss of water. Therefore, there is about 20,000 m^3/day for the case of Plant on Hugli River, Hugli, West Bengal (Bera 2016). Cooling towers evaporate a large amount of water into the atmosphere, which condensate and return to water and fall in another area other than the stream it came from. Diminished stream base flow can unfavourably impact stream morphology, habitat and aquatic plants (Bera 2016).

31.3.1.1 Temperature

Thermal release may cause a disruption in physicochemical components of water body, impacting species constituent (Kulkarni et al. 2011). Bera (2016) showed that temperature of hot water releases in Hugli River, Hugli, West Bengal, is 27 °C and such temperature is above the upstream and downstream temperature with an average of 24 °C and 25 °C, respectively. So, this case shows that passage release water temperature 30 °C higher than upstream temperature of river Hugli. This hot water release could damage the native aquatic biota, chiefly temperature-subtle plants, fish and microbial activities and also influence the chemical and physical reactions in the water where the disposal outlet channel encounters the river Hugli (Bera 2016). Owing to the fact that water can engross thermal energy with only small deviations in temperature, therefore, the majority of aquatic organisms have established enzyme systems that function in only narrow arrays of temperature. These stenothermic organisms can be exterminated by abrupt temperature deviations that are outside the tolerance bounds of their metabolic systems (Pokale 2012).

Furthermore, thermal pollution may upsurge the metabolic rates of aquatic animals leading to an increase in foraging rate in an environment that did not change but has an increase in its water temperature (Fleischli and Hayat 2014). Accordingly, such an activity will cause food shortages, which result in the movement of the biota from their own area looking for food. On the contrary, event of immigration could happen due to changes in water temperature, where fish species or other biota that prefer warm water temperature could move in and bring with them several drawbacks to the new area such as competition on food and niches and parasites. All of these ecological influences linked with thermal pollution can reach to a noteworthy variation in aquatic biodiversity (Fleischli and Hayat 2014).

31.3.1.2 Dissolved Oxygen (DO)

Dissolved oxygen is vital for biota in aquatic environment and is one of the most significant quality factors. Usually, dissolved oxygen found in surface water in a range of 8–10 mg/L, and such a level is deliberated decent quality of water. Bera (2016) in his study on the effect of thermal pollution in Hugli River, West Bengal, showed that outlet released water from the thermal power station with an average DO value of 1.87 mg/L, which is very low and lower than the value from the upstream and downstream that have shown an average of 9.2 mg/L and 9.1 mg/L, respectively. In addition, it was lower than the standard limit (BIS: 6 mg/L) of DO in surface water. The effect of warm water can be seen in decreasing the solubility of oxygen in water and disturbs the habitat of healthy aquatic ecosystem (Bera 2016). The aquatic biota found it very stressful when the level of dissolved oxygen falls below 2.5 mg/L, and it is very difficult to endure and could cause asphyxia. On the other hand, the level of pH of water may also be considered as a factor when is it too high or too low, which in turn can cause strain for aquatic organisms, and it can also cut down on species diversity in freshwater habitats (Bera 2016). Slight variations of pH levels in water can upsurge the solubility of elements and compounds. Particularly more sensitive aquatic species are influenced by slight changes of pH levels in water.

31.3.1.3 Total Dissolved Solids (TDS) and Electrical Conductivity (EC)

The measurement of inorganic and organic substances in the freshwater environment is known as total dissolved solid (TDS). Electrical conductivity (EC) is a measure of capability of water to pass an electrical current flow. Both TDS and EC linked together in their function. Electrical conduction in water is usually upsurged in the presence of more dissolved solid ions. Bera (2016) found that water of Hugli River, West Bengal, has an average values for TDS and EC of the released water from the thermal power plant of 240 mg/L and 375 μs/cm, which are considerably augmented if related to the upstream and downstream samples with an average of 180 mg/L and 281 μs/cm and 190 mg/L and 297 μs/cm, respectively. On the other hand, EC is also related to water temperature as it increases by 2–3% for an increase of 1 °C of water temperature (Iyasele and Idiata 2015).

31.3.2 Impact on Fishes

Through the process of suctioning water into a cooling system, individuals of different sizes and of different freshwater taxa are ruined and stuck counter to screens at the opening of an intake structure (Fleischli and Hayat 2014). This is referred to as 'impingement'. Moreover, early-life-stage fish, eggs and larvae are frequently slurped into the cooling system, where they are impaired by heat, pressure, mechanical stress and/or chemicals used to clean the cooling system before being discarded back into a water body (Fleischli and Hayat 2014). This is known to as 'entrainment'. In the USA, the damage of fish due to the effects of thermal power plants is considered higher than that of the fishing industry (Fleischli and Hayat 2014). The following are examples on the number of fish individuals being killed per year at the site of the thermal stations in the USA. At the Salem Nuclear Plant in New Jersey, the toll estimated was 1.12 million weakfish and 842 million bay anchovies, which represents fourfold more than caught by commercial fishermen (Summers et al. 1989); the 16 California power plants by means of once-through cooling systems combined have the capability to draw about 14.5 billion gallons of seawater every day. Each year they eradicate about 2.4 million fish and 17.5 billion larvae (California State Water Resources Control Board 2010); the combined effects of the five power plants on New York's Hudson River have exterminated as many as 79% of all the fish born in a single species in a single year (Boreman and Goodyear 1988); the Bayshore coal power plant in Ohio killed more than 60 million adult fish and more than 2.5 billion fish eggs and larvae in 2008 (Ager et al. 2008).

31.3.3 Impact on Plankton

31.3.3.1 Phytoplankton

Vinitha et al. (2010) studied the effects of chlorine on some species of phytoplankton entrained into the cooling water circuit in a thermal power plant in India. They reported on the stress that the phytoplankton experienced during the entrainment process. They showed that there is a substantial reduction in chlorophyll *a* as water passed from the intake point to the outfall. Also, their results revealed that growth rate, chlorophyll *a* and primary productivity of chlorine-handled diatoms reduced, liable to the quantity.

The phytoplankton travelling through the power plant cooling system suffer the joined consequence of chemical, thermal and mechanical strains. Some of the previous investigations have reduced the likelihood of mechanical pressure being a likely cause for noteworthy death of phytoplankton (Bienfang and Johnson 1980). Other investigations concentrated on temperature as the major impact leading to phytoplankton lessening (Morgan and Carpenter 1978; Briand 1975). Reports put forward by Choi et al. (2002) and Poornima et al. (2006) highlight that chlorine might be the chief pressure agent to entrained organisms.

31.3.3.2 Zooplankton

Zargar and Ghosh (2007) in their study showed that the percent growth rate of the zooplankton was greater at 26 °C compared with 31 °C, 33 °C and 36 °C. They also showed that the growth rate of *C. reticulata* was unfavourably influenced at 36 °C after day 21 of the experimentation; it augmented at 33 °C. In moderate habitat, Martin and Novotny (1975) suggested a negative influence of increasing temperature on the survival of *C. cuspidatus*. Generally, studies intended to determine the influences of different temperature levels on species of zooplankton have shown that the growth rates of these organisms were delayed at higher temperatures.

31.3.4 Impact on Invertebrates

The early investigations regarding the effect of increased water temperature on freshwater biota have started in the USA (Gaufin and Hern 1971). Later, this subject became familiar, and several studies were published on the effect of heated water on different invertebrate taxa (Stewart et al. 2013). Among aquatic insects, mayflies (Ephemeroptera) and stoneflies (Plecoptera) were revealed to be mainly sensitive (e.g. Ward and Stanford 1982; Quinn et al. 1994), which were considered as a biomarker for water quality ecologically. On the other hand, some other taxa have shown greater acceptance levels and increased water temperature as species of the insect orders Coleoptera and Odonata. The Odonata have shown a wide range of thermal tolerances than several other constituents of the freshwater inhabitants (Martin and Gentry 1974). For the planarians, Claussen and Walters (1982) recommended that the high thermal tolerances of these animals are linked with their prevalent and eurythermic dispersals.

In many parts of the world, rising water temperature owing to climate changes or any other factors such as water discharge from thermal power plants has caused the movement of species to cooler zones either at higher latitudes or altitudes (Jacobsen et al. 1997; Davies 2010). Stewart et al. (2013) suggested that sublethal impacts of heated water might also be imperative and can cause fluctuations in community structure. Such sublethal influences could contain altered behavioural reactions. For instance, for caddisflies with upper thermal limits around 31 °C, Gallepp (1977) discovered a reduction in filtering rate at temperatures above 24 °C and recommended that these larvae were questionable to survive and reproduce at higher water temperatures.

31.4 Effects of Heated Freshwater Owing to the Thermal Power Plants: The Iraqi Case

Studies on the effects of the discharge of heated water by the thermal power plants in Iraq are very scarce. The only available investigations were that of Nashaat et al. (2017) on one of the thermal power stations operating on Tigris River near Baghdad City and Jebure and Meshjel (2019) on a similar power plant located on the Tigris River in Wasit Province, south of Baghdad City.

Nashaat et al. (2017) investigated the invertebrate's community in the vicinity of the thermal power plant on the Tigris River at Baghdad City, and their results showed that there were variations in species composition between the area near the hot water discharge and areas further away from this point. The authors suggested also that the heated water discharge area is dominated by certain invertebrate species and suggested that such dominance is due to the wide range of tolerance of these species. In addition, there were seasonal variations in the abundance of the invertebrate species in the different months of the year. Nashaat et al. (2017) suggested that the decrease in the abundance may be due to the high discharge of heated water from the power plant.

Jebure and Meshjel (2019) in their study have shown that water temperature at the site of the thermal power station investigated is falling in a range of 15–38.8 °C. They showed that the heated water of the thermal power plant was the causative agent for the increase of electrical conductivity (EC) and total dissolved solid contents as 656–1790 μs/cm and 588–1145 mg/L, respectively. Due to the increase of dissolved solids, Jebure and Meshjel (2019) claimed that such factor has contributed to the restriction in light penetration and increase in the level of turbidity, which inevitably affect the value of the dissolved oxygen. According to these changes in the environment variables, the invertebrate groups that Jebure and Meshjel (2019) examined have shown a significant decrease in both the abundance and diversity in localities neared the heated water discharge from the thermal power plant. Among the invertebrate species that Jebure and Meshjel (2019) found with high level of tolerance to the changing environment is the different species of worms and in particular the annelids, *Limnodrilus hoffmeisteri* and *Branchiura sowerbyi*, which are usually found living in water thermal-polluted areas (Nashaat 2010).

The work of Al-Aboodi et al. (2018) on the impacts of the two thermal power plants located in southern of Iraq on the banks of Shatt al-Arab River could be the best example for this study as it deals with heated water discharged in an area near to the southern marshes of Iraq. The results of this study showed that there is an increase in the temperature of water up to 30% during summer season and up to 45% during winter season in the site of the power plant compared with other sites far away from the heated water discharge point. High values of electrical conductivity (EC) were observed, but they suggested that such an increase is due to the high salinity. This reason may not be correct as high water temperature increase dissolved solids, which in turn increase ions that upsurge the electric conductivity. Al-Aboodi et al. (2018) concluded that the thermal power stations located on two sites on Shatt al-Arab Rivers have contributed significantly to the deterioration of the water quality of this river.

Although there are only a limited studies on the impacts of the thermal power plants in Iraq, the results obtained from the available literature are enough evidences for the deterioration of both the environment and the freshwater biota in Iraq. Therefore, a perquisite attention is required for this issue in order to save the environment.

31.5 The Future of the Effects of Thermal Power Plants in Iraq

The future of using thermal power plants in Iraq looks to last forever as there are plenty of project plans for building more plants using once-through, closed-cycle cooling, which needs water for cooling the turbines.

Al-Kayiem and Mohammad (2019) in their study were talking about a major project of developing electrical power in Iraq using solar energy to heat water to produce steam. They concentrated on the following benefits that will be gained from implementing such megaproject:

1. Reduction of greenhouse gas discharges.
2. Reductions in fuel use.
3. The development of the local industry and new job openings.
4. The attainment of the aim set by the increase in electrical power produced by renewable energy supplies which is approximately 2000 MW in 2030.

They propose the following as means to replace the fuel usage in heating water to produce steam:

1. By integration with thermal power plants.
2. As stand-alone solar power plants.

If this proposal is approved, the desert area of Iraq that is neighbouring the two main rivers, Euphrates and Tigris, will be a forest of power-generated plants that operate on solar energy, but using water from the rivers for cooling the turbines. If this project is approved, then the freshwater life will be vanished, and the aquatic biodiversity of the two great rivers will disappear forever.

In his study, Sharif (2012) is talking about another mega electricity project to be implemented in Iraq to serve European countries. This project also uses solar energy to produce electricity and be transmitted to Europe via Turkey. The background of this project is to use solar power during daytime plus natural gas during night time. In both cases, thermal power stations will be established that need water to cool down the turbines. Although the project is very ambitious in making available more than 6000 MW, it is on the other hand very damaging to the environment and the biota living in the freshwater system that this project will take water from to cool down the turbines of the power plants.

The reader of the proposal of this project can directly pin the point of the damaging side of the project on the environment on the biodiversity. If such project is approved and sees the light, then the freshwater life in Iraq will diminish and can see no light forever. Policy makers in Iraq should think twice and probably hundreds of time before accepting such project. European countries will have electricity power given to them on a plate of gold not caring about how this power was generated and how many millions of aquatic creatures disappear in order for people in Europe to enjoy their lives.

References

Ager D, Marttila D, Patrick P (2008) Bay Shore power plant cooling water intake structure information and I&E sampling data. Kinectrics Report 112026-005

Al-Aboodi AH, Abbas SA, Ibrahim HT (2018) Effect of Hartha and Najibia power plants on water quality indices of Shatt Al-Arab River, south of Iraq. Appl Water Sci 8(2):64

Al-Kayiem HH, Mohammad ST (2019) Potential of renewable energy resources with an emphasis on solar power in Iraq: An outlook. Resources 8(1):42

Austin C, Borja R, Phillips J (2005) OPERATION SOLAR EAGLE: a study examining photovoltaic (PV) solar power as an alternative for the rebuilding of the Iraqi electrical power generation infrastructure. Naval Postgraduate School, Monterey

Benke AC, Van Arsdall TC, Gillespie DM (1984) Invertebrate productivity in a subtropical black water river: the importance of habitat and life history. Ecol Monogr 54:25–63

Bera A (2016) Impact of Bandel Thermal Power Plant (BTPS) on Hugli River, Hugli. West Bengal Geogr 5(4)

Bienfang P, Johnson W (1980) Response of subtropical phytoplankton to power plant entrainment. Environ Pollut Ecol Biol 22(3):165–178

Boreman J, Goodyear P (1988) Estimates of entrainment mortality for striped bass and other fish species inhabiting the Hudson River Estuary. Am Fish Soc Monogr. Exh. 8

Briand FJP (1975) Effects of power plant cooling systems on marine phytoplankton. Mar Biol 33 (2):135–146

Brooker MP (1985) The ecological effects of channelization. Geogr J 151:63–69

Bunn SE, Arthington AH (2002) Basic principles and ecological consequences of altered flow regimes for aquatic biodiversity. Environ Manag 30:492–507

California State Water Resources Control Board (2010) Water Quality Control Policy on the Use of Coastal and Estuarine Waters for Power Plant Cooling Final Substitute Environmental Document, May 4, 2010, pp 33–34

Choi DH, Park JS, Hwang CY, Huh SH, Cho BL (2002) Effects of thermal effluents from a power station on bacteria and heteromorphic nano-flagellates in coastal waters. Mar Ecol Prog Ser 229:1–10

Claussen DL, Walters LM (1982) Thermal acclimation in the freshwater planarians *Dugesia tigrina* and *D. dorotocephala*. Hydrobiologia 94:231–236

Contador JFL (2005) Adaptive management, monitoring, and the ecological sustainability of a thermal-polluted water ecosystem: a case in Sw Spain Environ. Monit Assess 104:19–35

Cummins KW, Sedell JR, Swanson FJ, Minshall GW, Fisher SG, Cushing CE, Petersen RC, Vannote RL (1983) Organic matter budgets for stream ecosystems problems in their evaluation. In: Barnes JR, Minshall GW (eds) Stream Ecology. Plenum Press, New York, pp 299–354

Davies PM (2010) Climate change implications for river restoration in global biodiversity hotspots. Restor Ecol 18:261–268

European Parliament and Council of the European Union (2006) Directive 2006/44/EC of the European Parliament

Fleischli S, Hayat B (2014) Power plant cooling and associated impacts: the need to modernize US power plants and protect our water resources and aquatic ecosystems. Natural Resources Defense Council. https://www.nrdc.org/sites/default/files/power-plant-cooling-IB.pdf

Förster H, Lilliestam J (2009) Modeling thermoelectric power generation in view of climate change. Reg Environ Chang 10:327–338

Gallepp GW (1977) Responses of caddisfly larvae (Brachycentrus spp.) to temperature, food availability and current velocity. Am Midl Nat 98:59–84

Gaufin AR, Hern S (1971) Laboratory studies on tolerance of aquatic insects to heated waters. J Kansas Entomol Soc 44:240–245

Hauer FR (1989) Organic matter transport and retention in a black water stream recovering from flow augmentation and thermal discharge. Regul Rivers Res Manag 4:371–380

Hester ET, Doyle MW (2011) Human impacts to river temperature and their effects on biological processes: a quantitative synthesis. J Am Water Resour Assoc 47:571–587

Heugens EHW, Hendriks AJ, Dekker T, Van SNM, Admiraal W (2002) A review of the effects of multiple stressors on aquatic organisms and analysis of uncertainty factors for use in risk assessment. Crit Rev Toxicol 31:247–284

Holmstrup M et al (2010) Interactions between effects of environmental chemicals and natural stressors: a review. Sci Total Environ 408:3746–3762

Iyasele JU, Idiata DJ (2015) Investigation of the relationship between electrical conductivity and total dissolved solids for mono-valent, Di-Valent and Tri-Valent metal compounds. Int J Eng Res Rev 3(1):40–48

Jacobsen D, Schultz R, Encalada A (1997) Structure and diversity of stream invertebrate assemblages: the influence of temperature with altitude and latitude. Freshw Biol 38:247–261

Jebure HA, Meshjel MH (2019) Impact of Wasit power plant effluents on biodiversity of benthic fauna in Tigris River, Province Wasit, Iraq. Iraqi J Agric Sci 50(3):960–971

Kulkarni VA, Naidu VS, Jagtap TG (2011) Marine ecological habitat: A case study on projected thermal power plant around Dharamtar creek. India J Environ Biol 32(2):213–219

Lakly MB, McArthur JV (2000) Macroinvertebrate recovery of a post-thermal stream: habitat structure and biotic function. Ecol Eng 15:S87–S100

Library of Congress – Federal Research Division (2006, August) Country Studies. Retrieved September 2011, from Library of Congress: http://lcweb2.loc.gov/frd/cs/profiles/Iraq.pdf

Madden N, Lewis A, Davis M (2013) Thermal effluent from the power sector: an analysis of once-through cooling system impacts on surface water temperature. Environ Res Lett 8:035006

Malmqvist B, Rundle S (2002) Threats to the running water ecosystems of the world. Environ Conserv 29:134–153

Martin WJ, Gentry JB (1974) Effect of thermal stress on dragonfly nymphs. In: Gibbons JW, Sharitz RR (eds) Thermal ecology. Atomic Energy Commission, Oak Ridge

Martin, D.B. and Novotny, J.F., 1975. Studies to determine methods for culturing three freshwater zooplankton species. National Environmental Research Centre

McDermott GR, Nilsen ØA (2014) Electricity prices, river temperatures, and cooling water scarcity. Land Econ 90(1):131–148

Ministry of Planning of Iraq (2010) National development plan for the Years 2010–2014. Ministry of Planning of Iraq, Baghdad

Minshall GW (1996) Organic matter budgets. In: Hauer RF, Lamberti GA (eds) Methods in stream ecology, vol 99. Academic Press, San Diego, pp 591–607

Morgan RP III, Carpenter EJ (1978) Biocides. In: Schuber JR, Marcy BC Jr (eds) Power plant entrainment: a biological assessment. Academic Press, New York, pp 96–113

Nashaat MR (2010) Impact of AL-Durah power plant effluents on physical, chemical and invertebrates biodiversity in Tigris River Southern Baghdad. PhD. Theses, College of Science, Baghdad University

Nashaat MR, Radhi AG, Mohammad AA, Reassn KH (2017) Impact of Al-Rasheed Power Plant Effluents on Biodiversity of Benthic Fauna in Tigris River, Southern Baghdad. Ibn AL-Haitham J Pure Appl Sci 30(1):315–324

Pfister S, Suh S (2015) Environmental impacts of thermal emissions to freshwater: spatially explicit fate and effect modeling for life cycle assessment and water foot printing. Int J Life Cycle Assess 20:927e36

Pokale WK (2012) Effects of thermal power plant on environment. Sci Rev Chem Commun 2(3):212–215

Poornima EH, Rajadurai M, Rao VNR, Narasimhan SV, Venugopalan VP (2006) Use of coastal waters as condenser coolant in electric power plant: impact on phytoplankton and productivity. J Therm Biol 31:556–564

Quinn JM, Steele GL, Hickey CW, Vickers ML (1994) Upper thermal tolerances of twelve New Zealand stream invertebrate species. N Z J Mar Freshw Res 28:391–397

Raptis CE, Pfister S (2016) Global freshwater thermal emissions from steam-electric power plants with once-through cooling systems. Energy 97:46–57

Raptis CE, van Vliet MT, Pfister S (2016) Global thermal pollution of rivers from thermoelectric power plants. Environ Res Lett 11(10):104011

Reda M, Saied N, Khaldi A, Abdul-Hussain MA, Ekwue A (2006) Iraq electrical power infrastructure-Past, present and future developments and expectations. In: 2006 IEEE PES power systems conference and exposition. IEEE, pp 1792–1795

Sharif SF (2012) The project; Iraq Desertec is the future second electricity supplier to Europe through Turkey. Int J Sci Eng Res 3(11):1–8

Stewart BA, Close PG, Cook PA, Davies PM (2013) Upper thermal tolerances of key taxonomic groups of stream invertebrates. Hydrobiologia 718(1):131–140

Summers, J. K., Holland, A. F., Weisberg, S. B., Wendling, L.C., Stroup, C.F., Dwyer, R. L., Turner, M.A. and Burton, W., 1989. Technical review and evaluation of thermal effect studies and cooling water intake structure demonstration of impact for the Oyster Creek Nuclear Generating Station: Revised final report

Sweeney BW, Bott TL, Jackson JK, Kaplan LA, Newbold JD, Standley LJ, Hession WC, Horwitz RJ (2004) Riparian deforestation, stream narrowing, and loss of stream ecosystem services. Proc Natl Acad Sci USA 101:14132–14137

Van Vliet MTH, Yearsley JR, Franssen WHP, Ludwig F, Haddeland I, Lettenmaier DP et al (2012) Coupled daily streamflow and water temperature modelling in large river basins. Hydrol Earth Syst Sci 16:4303e21

Verones F, Hanafiah MM, Pfister S, Huijbregts MAJ, Pelletier GJ, Koehler A (2010) Characterization factors for thermal pollution in freshwater aquatic environments. Environ Sci Technol 44:9364–9369

Vinitha E, Veeramani P, Venugopalan VP (2010) Chlorination for power plant biofouling control: potential impact on entrained phytoplankton. Int J Environ Stud 67(4):515–530

Ward JV, Stanford JA (1982) Thermal response in the evolutionary ecology of aquatic insects. Annu Rev Entomol 27:97–117

Wu J, Buchak EM, Edinger JE, Kolluru VS (2001) Simulation of cooling-water discharges from power plants. J Environ Manag 61:77–92

Zargar S, Ghosh TK (2007) Thermal and biocidal (chlorine) effects on select freshwater plankton. Arch Environ Contam Toxicol 53(2):191–197

Section IX
Conservation

Chapter 32
What Ecological Principles Required for a Proposed Establishment and Management of National Parks in the Southern Marshes of Iraq

Laith A. Jawad

Abstract Ecological principles are important criteria in the process of establishing protected areas. Regularly, these features of the area are neglected, and conservation measures related to the biodiversity area usually considered. It appears the ecological facets of any landscape are vital to determine the validity of the protected area and its future endurance. The lack of experience of the managers that have no knowledge of ecological science can cause severe damage to the protected areas that are selected without considering the ecological issues. Therefore, several prevailing protected regions around the world facing problems in attaining their upkeep goals endure a vital choice for upholding and improving the conservation of biota diversity and habitat functions.

This chapter discusses briefly the important ecological principles that should be considered in the process of establishing new protected areas in Iraq in general and in the southern marshes region. Failing to consider these factors will lead definitely to the collapse of the protected area/s and malfunctioning in the future. With such bad consequences, biodiversity will certainly be affected. Therefore, the policymakers in Iraq are urged to consider the ecological side when establishing any protected area in Iraq and especially in the southern marshes region.

Keywords Ecology · Restoration · Protected areas · Biodiversity · Management · Aquatic life

32.1 Introduction

There are several restrictions and rules when any area needs to be allocated as a protected region for conservation of the biodiversity in this zone. Such limitations are embodied in some environmental characters known as ecological features

L. A. Jawad (✉)
School of Environmental and Animal Sciences, Unitec Institute of Technology, Mt Albert, Auckland, New Zealand

L. A. Jawad (ed.), *Southern Iraq's Marshes*, Coastal Research Library 36,
https://doi.org/10.1007/978-3-030-66238-7_32

(Polunin and Eidsvik 1979). In the early days when the first national park was established in the USA in 1872, the scientists at that time did not pay much attention to the new branch of knowledge known as "ecology", and it may be a coincidence that both science of ecology and the first national park were found at the same period of time (Polunin and Eidsvik 1979).

In the present, the main part that ecology can play when establishing any area for protection is generally familiar that it is nearly absurd that any natural protected area should be built without taking into consideration the ecological rules and the local condition (Polunin and Eidsvik 1979). In spite of these familiarities, establishment of new national parks and reserves still lacks the congruence among several aspects such as what should be done and where and when, to establish and maintain a suitable network of such protected areas in diverse regions of the world (Polunin and Eidsvik 1979). These problems are increased and accumulated several folds by the recurrent requirement to work with a fully interdisciplinary group comprising sociologists and economists as well as ecologists (Polunin and Eidsvik 1979).

Allocated protected areas and national parks are deliberated important to endeavours to preserve the biota diversity globally (Gaston et al. 2006). Hitherto, numerous present protected areas are at this time facing problems in attaining their conservation goals, and their maintainability on the long run has been in jeopardy (Bruner et al. 2001).

Criticizers have appealed that the protected areas cannot be predictable to service the load of keeping biodiversity (Eagles 1993), that they do not essentially keep biotic uprightness within their limits (Terborgh 2004; Bruner et al. 2001; Salafsky and Wollenberg 2000), or that they have been weakly positioned from the view of biota diversity upkeep (Scott et al. 2001; Rodrigues et al. 2004). The worldwide system of protected areas presently is even so the chief choice for preserving and improving biodiversity upkeep; but there are several methods that need to be located and assigned to reinforce those that are weakening (Terborgh and van Schaik 2002) and to comprehend and duplicate those that are consequent.

Any habitat has a normal and natural life when its main ecological features (composition, structure, function) and courses happen within their natural arrays of disparity and can endure and improve from disquiets caused by natural environmental or human factors (Parrish et al. 2003). In order to assess the level of integrity in any protected area, certain aspects that related to the ecological features of the area need to be checked such as intactness, species viability, ecological processes and functioning, and the threats and stressors facing a protected area (Ervin 2003).

In the present study, the chief ecological features of any protected areas are discussed and made obvious for any future study that considers to allocate a protected area/s in the southern marshes of Iraq to take into consideration. Such ecological factors are the key points in the flourishing of any protected areas to be assigned. Therefore, the aim of this chapter is direct the policymakers to set their goals for the better ecological aspects of any area before finalizing their decision about allocating the protected region/s.

32.2 Ecological Codes Relating to Selection of Protected Area and Preliminary Founding

The principal measure to be deliberated in selecting a location for a protected area is the distinct (rather exclusive) natural feature that is limited within its borders. This more or less elite natural class may be conveyed through a great variety of species or by vast numbers of a specific species or, on the opposite side, by little examples of a threatened species (Polunin and Eidsvik 1979). In addition, the area may be categorized by an exclusive geological attribute, landform, or marine environment. Exclusivity is the class that differentiates a certain area as meriting the protection which can be manage to pay for by a distinct legal term.

If the choice is offered, one would normally give favourite to an area with high biota diversity rather than smaller diversity. At the time of the natural values of the area recognized, then one needs to meticulously assess the composite social, economic, and political features that will have an effect on the final decision on whether it can be efficiently secured for future generations and certainly in eternity (Polunin and Eidsvik 1979). In taking the final decision about allocating any area for protection, ecological values will be put first in the process of such consideration. Therefore, it is imperative to keep and widen public awareness and teaching to upsurge gratitude of "natural history" and the meaning of conservation, for which the desirability of diversity and change (including decay) can be of excessive assistance (Polunin and Eidsvik 1979). In the meantime, it is important to specify an area of enough size to be ecologically feasible conferring to the latest standards. Such an area will act as a separating zone between the protected area and the neighbouring region, which is known as a buffer zone. In addition to considering the ecological terms in selecting the areas for protection, the management team should also be selected so they show ecological knowledge and prudence (Polunin and Eidsvik 1979).

The relationship between the ecological principles of any protected area is strongly linked to conservation issue, and any decision that needs to be taken about allocating any area for protection should also consider the issue of conservation as much as the ecological principles. Regrettably, many cases approved today in regard of choosing areas for protection overlooked the conservation issues such as several aspects of the community of the fauna and flora and other environmental aspects (Polunin and Eidsvik 1979).

The ecological principles have a direct link with the area to be protected. Each piece of land on this earth is open for natural changes that might be small or big, fatal or slight. In such cases, the need for the intervention of human to reset the area to its previous natural form is required. In doing so, humans need to know the ecological features of the land that they are going to deal with.

One of the major ecological issues that the managers should know when looking after any protected area is that the surface of earth is covered with thousands or millions of communities that are sharing neighbouring niches. Within each of these ecosystems of variable sizes, there are individuals of organisms, which have an allocated job to perform to support the community it lives in. Therefore, extra care should be paid to all organisms living in the selected area for protection and for a targeted group of animals or plants.

To control the ecological integrity in any protected area, Woodley (1993) has suggested an agenda to fulfil this aspect. His proposal contains three common constituents – biodiversity, ecosystem processes, and stressors (Timko and Innes 2009). Several protected areas around the world use these criteria such as the parks in Canada.

From the information given in this chapter, it is clear that the process of allocating an area for protection should follow a rigorous examination and analysis of several issues among which are the ecological principles that appeared to be very important for any protected areas. This chapter is only a preliminary study for the subject of selecting a protected area, but it is at the same time a baseline study for policymakers in Iraq to follow in order to allocate one or more areas in the southern marshes as protected areas for the preservation of the biodiversity.

References

Bruner AG, Gullison RE, Rice RE, da Fonseca GA (2001) Effectiveness of parks in protecting tropical biodiversity. Science 291:125–128

Eagles PFJ (1993) Environmental management in parks. In: Dearden P, Rollins R (eds) Parks and protected areas in Canada: planning and management. Oxford University Press, Toronto, pp 154–183

Ervin J (2003) Protected area assessments in perspective. Bioscience 53:819–822

Gaston KJ, Charman K, Jackson SF, Armsworth PR, Bonn A, Briers RA, Callaghan CS, Catchpole R, Hopkins J, Kunin WE, Latham J (2006) The ecological effectiveness of protected areas: the United Kingdom. Biol Conserv 132(1):76–87

Parrish JD, Braun DP, Unnasch RS (2003) Are we conserving what we say we are? Measuring ecological integrity within protected areas. Bioscience 53:851–860

Polunin N, Eidsvik HK (1979) Ecological principles for the establishment and management of national parks and equivalent reserves. Environ Conserv 6(1):21–26

Rodrigues ASL, Andelman SJ, Bakarr MI, Boitani L, Brooks TM et al (2004) Effectiveness of the global protected area network in representing species diversity. Nature 428:40–643

Salafsky N, Wollenberg E (2000) Linking livelihoods and conservation: a conceptual framework and scale for assessing the integration of human needs and biodiversity. World Dev 28:1421–1438

Scott JM, Davis FW, McGhie RG, Wright RG, Groves C et al (2001) Nature reserves: do they capture the full range of America's biological diversity? Ecol Appl 11:999–1007

Terborgh J (2004) Reflections of a scientist on the World Parks Congress. Conserv Biol 18:619–620

Terborgh J, van Schaik C (2002) Why the world needs parks. In: Terborgh J, van Schaik C, Davenport L, Rao M (eds) Making parks work: strategies for preserving tropical nature. Island Press, Washington, DC, pp 1–14

Timko JA, Innes JL (2009) Evaluating ecological integrity in national parks: case studies from Canada and South Africa. Biol Conserv 142(3):676–688

Woodley S (1993) Monitoring and measuring ecological integrity in Canadian National Parks. In: Woodley SJ, Francis G, Kay J (eds) Ecosystem integrity and the management of ecosystems. St. Lucie Press, Boca Raton, pp 155–176

Chapter 33
The Possible Difficulties and Outcomes of the Biodiversity Conservation of the Southern Marshes of Iraq

Laith A. Jawad

Abstract The health of any biological system can be estimated by studying its biodiversity, which can be defined as the disparity in the biota in a given habitat over certain time limit. Biodiversity is the factor that supports the living earth that human cannot live if it vanished. Human-induced impacts have threatened rigorously the diversity of both the terrestrial and aquatic lives. As a result of such threat, several niches have shown fragmentation and species became on the edge of extermination. If humans continue harming the environment that he is living in, the biodiversity will be the first to be affected, and the future of the life of human on the surface of this earth will be threatened too. In an attempt to evade such unfavourable conditions or at least decrease it to bearable level by the biota, several actions were put forward to perform so to reinstate the normal status of the biodiversity on earth.

This chapter aims to discuss the main causes that lead to deterioration of the biodiversity on earth including the freshwater system, and in another section of this chapter, special concern was given to the biodiversity of the southern marshes of Iraq and its status. At the end of this chapter, a list of solutions were given that can restore biodiversity in freshwater system together with the possible implementation of these factors in Iraq.

The idea behind this chapter is to draw the attention of the policy makers in Iraq about a major threat to the biodiversity of the freshwater system in general and the southern marshes in particular that might arrive in short period of time. Prompt actions are prerequisite by the policy makers in Iraq to save the unique habitat of the southern marshes from vanishing forever.

Keywords Protected areas · Aquatic life · Freshwater environment · Biodiversity · South of Iraq

L. A. Jawad (✉)
School of Environmental and Animal Sciences, Unitec Institute of Technology, Mt Albert, Auckland, New Zealand

L. A. Jawad (ed.), *Southern Iraq's Marshes*, Coastal Research Library 36,
https://doi.org/10.1007/978-3-030-66238-7_33

33.1 Introduction

A quickly augmented international inhabitant has affected the biodiversity severely. For example, the South Asian region, one of the world largest population gathering, has been facing population impact on its environment that backing high biodiversity (Bhandari 2018). Such an increase in human population worldwide was in the form of expansion of agricultural lands towards the forests, additional roads constructed, and cities expanded. Investigations have revealed that habitats comprising several plant species in certain area are more fruitful than those holding only one of those species (Fargione et al. 2008).

The opinions are that keeping high biodiversity may assist encountering the requirements for the increasing population while reinstating habitat services. Nevertheless, humans' straight and unintended impacts have been irrevocably altering the biodiversity in this region (MEA 2005). Direct impacts directly influencing biodiversity comprise, but not restricted to, the building of dams and roads and eradicating of forests for farming and city growth. Unintended impacts are more or less hard to realize. For instance, it may acquire a longer time of investigation to comprehend how a rising population or cultural acceptance might adversely impact the rate at which ecosystem procedures undo (Bhandari 2018).

On the other hand, environmental strain, environmental instabilities, great environmental settings, austere restriction of assets, introduction of exotic species, and geographic separation of numerous species have been the chief issues that cause the losses of biodiversity and climate variations. These features have enhanced the damages of the environment via the overutilization of appreciated flora and fauna (Bhandari 2018). Consequently, investigations have resolved that human events are encouraging powerful natural courses, such as flood, and a change in plant production system owing to climate variations that influence biodiversity (Bhandari 2018).

The examples given above can be perfectly applied to the environment of the southern marshes of Iraq. In addition, in those areas from around the world, continuous management for the affected environments is in progress, while no such conservation measures were on hand in the marsh areas in Iraq. If this status continues, several natural habitats in the freshwater system of Iraq in general and the southern marshes in particular will be fragmented from extreme human impacts. Probably many species are in the verge of extinction, but no one knows who and how many are they. As additional pressures to biodiversity are endured, it is probable that the southern marshes of Iraq could face extraordinary ecosystem penalties of different kinds. In reaction to these conceivable exceptional fluctuations, efforts have been made to support and recover the ecosystem abilities over the conservation of biodiversity, though all these events have been inadequate to restock the lost environmental wealth that has been damaged. In this chapter, a crucial requirement of official arrangement to distribute an extremely consistent and expressive data scheme at diverse levels is lost in Iraq at the moment unfortunately.

33.2 The Biodiversity of the Southern Marshes of Iraq

The southern marshes of Iraq are a unique environment in the sense of comprising wide range of fauna and flora that belong to large number of taxa. The animal kingdom taxa, for example, range from varieties of protozoa to vertebrate animals. As to the taxa of plants, there are thousands of both aquatic and land species that some of them are native to the marsh area. In several chapters of this book, listing of the major taxa of the aquatic biota that inhabit the marshes of Iraq is given. Therefore, this section is too small to accommodate such information and it will be a repetition of information.

Transboundary biodiversity is present in the southern marshes of Iraq, where both animal and plant species are found across the boundaries separating Iraq and Iran. Such connection might sometimes be considered as a support for the survival of the aquatic biodiversity as the Iranian depth provides additional niches for the Iraqi species if they face any threat. At the same time, this cross boundary connection might affect negatively the aquatic biodiversity in Iraq through the dispersal of exotic species or transmission of parasites and other diseases.

33.3 Biodiversity Hotspots

Hotspots denote to rich biodiversity regions which are under the continuous pressures with the jeopardy of dropping significant gene pools, if not sheltered straightway. This notion was first suggested by Myers (1988, 1990) and later was reviewed by Mittermeier et al. (2000). There are numerous biodiversity hotspots around the world in both fresh and marine environments in addition to the terrestrial locations. These hotspots contain important populations of a significant species of different kinds both animal and plants. The Himalaya is considered one of the largest biodiversity hotspot in the world, and it is suitable as an example of hotspots in this section. The Himalaya area covers over 3000 km of northern Pakistan, Nepal, and Bhutan and the northwestern and northeastern states of India. It embraces all of the world's mountain peaks higher than 8000 m. Due to its extreme height, mountains present varied environments (IUCN 2003; UNEP 2001; Bhandari 2012). Such hotspot area is home for rare animal and plant species.

In spite of the importance of the hotspot areas, their biodiversity is facing deterioration and the populations of the different species experience fragmentation (FAO 1999; WRI 2010). International organizations such as the Global Biodiversity Strategy (WRI/IUCN/UNEP 1992) and the United Nations Economic and Social Commission for Asia and the Pacific (ESCAP)-ADB (2000) have recognized six important causes of biodiversity losses in the hotspots in South Asia. These contain (1) unmaintainable high levels of natural resource depletion by the rising human population; (2) progressive reduction range of transacted goods from agriculture and forestry; (3) introduction of non-native species linked with agriculture, forestry, and fisheries; (4) financial schemes and strategies that were unsuccessful to assist the environment and its wealth; (5) unfairness in possession and admission to natural properties, as well as the assistances from use and saving of biodiversity, and insufficient knowledge and incompetent use of data; and (6) feeble legitimate and official schemes that were unsuccessful to keep against unbearable misuse of wealth (ESCAP-ADB 2000).

We do not know whether there are biodiversity hotspots in the southern marshes of Iraq owing to the dearth in environmental research and studies that should be taken into consideration this issue. In case, there are any of such biodiversity hotspots, to which animal or plant groups they belong? How big these hotspots are? What are their geographical limitations? No answers are available for such questions, and consequently, we as human through our inhuman activities are destructing these hotspots without knowing what we are doing and what would be the outcomes of our impacts on the environment.

Biodiversity hotspots in all over the world have been identified and their geographical locations were known. Accordingly, special legislations to manage and protect them were put forward by the governments that these hotspots belong too. In Iraq, a similar action is needed in order to save whatever left of the biodiversity hotspots in the southern marshes of Iraq.

33.4 Biodiversity Conservation Problems

It has been agreed that biodiversity is indispensable to deliver ecosystem facilities for the human populations, and it is very problematic to set an explicit theoretical framework of preserving the biodiversity that is appropriate to all the countries (Escobar 1998).

Among the suggestions put forward to conserve biodiversity on global level is that suggested by Poiani et al. (2000). They claimed for a widespread agenda centred on the focal environments, species composition, and supporting natural processes within their natural geographical ranges as the first step in the direction of conserving biodiversity.

A danger to the biodiversity owing to shattering several niches followed by effect of capitalist budget and globalization, industrial growth, and urban expansion is totalling extra griefs to keep biodiversity. In many examples, the addition of industrial discharge onto freshwater has contaminated waterbodies impacting the environments of aquatic flora and fauna.

The expansion of human populations and building new urban areas have led to the destruction of both the terrestrial and aquatic biodiversity on a global standard. Such growths have disturbed the forest coverage that is responsible for the water cycle in the nature and considered as the lung of mother nature (Cleugh 1998).

Natural tragedies also have endangered biodiversity, for example, flood, fire, and drought. Examples from around the world are many about these issues (FAO 1988; UNISDR and World Bank 2009).

Pollution of different kinds, i.e. chemical, thermal, sound, plastic debris, and air, is all affecting terrestrial and aquatic biodiversity in different levels. All kinds of pollutions are human-originated, and their consequences on the environment are not reversible in many cases unfortunately.

The other important problem that is facing biodiversity is the budget allocation by governments. In many instances, governments are slack in allocating budget to manage the biodiversity in their countries and consider such expenditure as not among the prompt needs for the country. Examples were given by several international organizations on cases reported by many countries (WWF 1997).

Some political and economic assemblies and policies approve supply change as new economists contend; nevertheless, there is no worldwide method to assist biodiversity and maintainable administration schemes (WWF 1997).

33.5 Conservation of the Freshwater Biodiversity and Its Implementation in Iraq

Notwithstanding the inclusive unhappy scenario for freshwater biodiversity, there are chances for preservation act and operative administration (Reid et al. 2019). Developing facilities and skills will be indispensable in extenuating some developing pressures (Jackson et al. 2016). Some current methods could also aid upkeep biodiversity protection while sufficing human requirements; these, nevertheless, have been achieved with variable degrees of accomplishment and, hitherto, give vision into the efficiency of diverse freshwater protection policies (Reid et al. 2019). In this section, selective strategies and practices that have significance to freshwater protection will be discussed. At the end of each strategy, information of the applicability of implementing this strategy in Iraq will be stated.

33.5.1 Environmental DNA

Deoxyribose nucleic acid (DNA) from organisms living in the freshwater system is found in the water as secretions, cells, tissues, faeces, or gametes and is conveyed through natural networks. Pieces of this environmental DNA (eDNA) can be insulated from organic matter in water samples and process through the molecular techniques to allocate the identity of the species using metabarcoding (Elbrecht and Leese 2017). The possible protection uses of eDNA methods are considerable in perceiving infrequent and threatened freshwater species whose occurrence cannot be established simply by more conservative methods (Jerde et al. 2013; Laramie et al. 2015; Bellemain et al. 2016) and for observing the colonization of new niches by possibly invasive species or pathogens (Rees et al. 2014). This directed or "active" scrutiny directed to discovery of eDNA for a single species of attention can be compared with "passive" observation, using high amount molecular technique, while sampled eDNA is used to evaluate population configuration and unscrupulously discloses the occurrence of a species of concern (Simmons et al. 2016). However, when joined with next-generation molecular methods, collection of eDNA conveyed in river networks provides a geographically unified way to evaluate the species abundance (both aquatic and terrestrial) of whole drainage basins and could well convert biodiversity data gaining in the future (Deiner et al. 2016).

33.5.1.1 A Possible Implementation in Iraq

There is a possibility of this tool to be applied in the freshwater of Iraq including the southern marshes, but not immediately like tomorrow. The scientific team should get special training on the different methods of molecular analysis before attempting to do such investigation. There might be some difficulties facing the application of this tool in Iraq owing to the severe shortage of equipment and technical facilities required for such application.

33.5.2 Environmental Flows

One tactic to extenuating the influence of current regulation on freshwaters is the exercise of water distributions (environmental flows or e-flows) to keep or reinstate habitats (Reid et al. 2019). Scientifically, the e-flows should deliver water levels or release that similar to natural hydrologic patchiness and include a variety of flows indispensable to upkeep operative ecosystems (Arthington et al. 2016). Depending on the unevenness of hydrographs, e-flows allow connection longitudinally along rivers and laterally within floodplains; this is vibrant in allowing familiarized reactions by the riverine biota to the threats of living in a warmer world, allowing transportation among possible immigrants as environmental settings modified. The e-flows have attracted ample research to answer the question "how much water does a river (or stream or wetland) require"? The answer for this question can fit the cases in different parts of the world (for instance, 37% of mean annual flow) (Pastor et al. 2014), but other scientists were not sure about this answer as the general nature of the answer is doubtful to accommodate all ecologically imperative features of flow inconsistency. As an alternative, the achievement of river safety and renovation will rely on the precisely modelling relations between hydrological designs and ecological reactions, followed by application of water distributions within a choice set by the flexibility of these habitats (Poff and Zimmerman 2010). Result of the analysis must be vital in any administration involvement: a meta-analysis by Palmer et al. (2008) shown inadequacies in the broadly used "if you build it, they will come" tactic of reinstating physical habitat and flows in rivers if other impacts endure to affect ecological reclamation. Such disappointments are common given that many freshwater niches may face multiple cooperating impacts (Craig et al. 2017).

33.5.2.1 A Possible Implementation in Iraq

There is possibility of implication of this tool in Iraq as the governmental offices that are responsible to perform such research are not up to the standard of applying such project owing to other projects in progress that they think are more important than this job.

33.5.3 Aliens and Aquaculture

Non-native species have imposed intensely destructive influences on receiver freshwater habitats (Gallardo et al. 2016). Nevertheless, it must not be overlooked that some exotic species can now have a vital part in the ecological functions in human-changed habitats, such as backup lake food webs (Twardochleb and Olden 2016) and riverine ecosystem functions (Moore and Olden 2017). Species have been frequently and intentionally included in areas outside their natural ranges with the aim to upkeep food safekeeping, leisure chances, and ecosystem restoration (Reid et al. 2019). Where the protection of the freshwater habitats is no more a genuine choice, the view of improving ecosystem functions through bringing in an exotic species might become a choice. Some researchers have claimed that exotic species might, under definite situations (Gozlan 2008), subsidize to conservation aims by delivering habitat or accomplishing needed ecosystem roles (Schlaepfer et al. 2011). However, others powerfully differ to such suggestion, claiming that the hazards of exotic introductions compensate any useful parts they could act in improving ecosystem functions (Reid et al. 2019).

Reducing natural yield of freshwater fishes (comparative to aquaculture) is an important issue as it offers the equal of all nutritional animal protein for 158 million people, with poor and undernourished populations chiefly dependent on these fisheries comparable with marine or aquaculture supplies (McIntyre et al. 2016). Aquaculture can cause an increase in parasite incidences, diseases, and species introductions, in addition to polluting recipient waters with trashes and medical-related materials (FAO 2016). Taking these drawbacks into consideration, cultivated fishes may not be an ample supernumerary for capture fisheries (Reid et al. 2019).

33.5.3.1 A Possible Implementation in Iraq

The reform of the present practices of introducing non-native species and widespread aquaculture ventures in Iraq is out of question. Although there are some rules and legislations to control such practices, they are completely ignored by the public. Exotic species are introduced to the country through different ways such as the aquarium business, where large numbers of freshwater species were introduced intentionally or non-intentionally. For aquaculture, every member of the public has an access to a riverine or near the marsh area establishes aquaculture facility to cultivate fish to increase the income of the family regardless of the negative outcome of such practice on the environment. Furthermore, the establishment of such projects is processed without getting any licence for the governmental office responsible for building aquaculture premises.

33.5.4 Fish Abstractions and Dam Elimination

Investigations regarding building ways that allow fish to cross dams (in both directions) are required immediately, as several dams are missing such conveniences or they have connected structures that fail sufficiently to pass native species (Pelicice et al. 2015). Valuations of the efficiency of different method to assist fish to cross dams and their designs and types are directly required, especially in the tropics (Silva et al. 2018). The general scientific agreement suggests for the complete removal of all dams. Such decision was implemented in the USA, and about 1200 dams have been removed in the last 40 years, and the process of removal is in progress since then (Bellmore et al. 2016). Such actions can show a success in conservation aspects, where migration courses are reopened. One good thing about our mother nature is that rivers can restore their nature rapidly after dam removal, eroding and redistributing sediment and returning to pre-impoundment conditions within a few years, rather than decades (O'Connor et al. 2015).

33.5.4.1 A Possible Implementation in Iraq

The implementation of such tool is out of question in Iraq in spite of the major shortage of freshwater. One thing can be made that might solve the problem and assist in conservation the biodiversity, and that is to build passages to allow fish individuals to pass both ways across the dam.

33.5.5 Climate Modification and Accomplished Moving of Species

The promptness of climate alteration is forecast to surpass the capability of many freshwater species to acclimatize or to scatter to more climatically suitable environments (Brook et al. 2008; Loarie et al. 2009). Protection of these species might need achieved moving of individuals to sites where the likelihood of their future perseverance is probable to be high, but where the species is not known to have found formerly (Olden et al. 2011). However, there is moral reason to question whether such action of moving individual of species is a practical conservation policy. For instance, such movement encourages the dispersal increase of species and therefore might have unwanted impacts on other species or ecological courses. Choices concerning such movement of species are obviously difficult simply because the influences of bringing a freshwater species to a new environment are indeterminate (and possibly calamitous); consequently, the necessity for moving individuals must be evaluated in contrast to the likelihood of species damage related with doing nothing (Olden et al. 2011).

33.5.5.1 A Possible Implementation in Iraq

Since there are some difficulties in implementing this tool, its application is certainly not feasible as people responsible for the conservation of the freshwater system in Iraq have no previous experience in such meticulous job.

References

Arthington AH, Dulvy NK, Gladstone W, Winfield IJ (2016) Fish conservation in freshwater and marine realms: status, threats and management. Aquat Conserv Mar Freshwat Ecosyst 26:838–857

Bellemain E, Patricio H, Gray T, Guegan F, Valentini A, Miaud C, Dejean T (2016) Trails of river monsters: detecting critically endangered Mekong giant catfish *Pangasianodon gigas* using environmental DNA. Glob Ecol Conserv 7:148–156

Bellmore JR, Duda JJ, Craig LS, Greene SL, Torgersen CE, Collins MJ, Vittum K (2016) Status and trends of dam removal research in the United States. Wiley Interdiscip Rev Water 4:e1164

Bhandari MP (2012) Environmental performance and vulnerability to climate change: a case study of India, Nepal, Bangladesh, and Pakistan. In: Climate change and disaster risk management. Series: Climate change management. Springer, New York/Heidelberg, pp 149–167. ISBN 978-3-642-31109-3

Bhandari MP (2018) The problems and consequences of the biodiversity conservation: a case study from Bangladesh, India, Nepal, and Pakistan

Brook BW, Sodhi NS, Bradshaw CJA (2008) Synergies among extinction drivers under global change. Trends Ecol Evol 23:453–460

Cleugh HA (1998) Effects of windbreaks on airflow, microclimates and crop yields. Agrofor Syst 41:55–84

Craig LS, Olden JD, Arthington AH, Entrekin S, Hawkins CP, Kelly JJ, Kennedy TA, Maitland BM, Rosi EJ, Roy AH, Strayer DL, Tank JL, West AO, Wooten MS (2017) Meeting the challenge of interacting threats in freshwater ecosystems: a call to scientists and managers. Elementa Science of the Anthropocene 5:1–15

Deiner K, Fronhofer EA, Mächler E, Walser J-C, Altermatt F (2016) Environmental DNA reveals that rivers are conveyer belts of biodiversity information. Nat Commun 7:12544

Elbrecht V, Leese F (2017) Validation and development of COI metabarcoding primers for freshwater macroinvertebrate bio assessment. Front Environ Sci 5:11

Escobar A (1998) Who's knowledge, whose nature? Biodiversity, conservation, and the political ecology of social movements. J Polit Ecol 5

FAO (1988) Agro-ecological Regions of Bangladesh. FAO, Rome

FAO (1999) Forest Resources of Nepal Country Report (FRA 2000). Forest Resources Assessment Program Working Paper 16. Food and Agriculture Organization of the United Nations, Rome

FAO (2016) The state of world fisheries and aquaculture. FAO, Rome
Fargione J, Hill J, Tilman D, Polasky S, Hawthorne P (2008) Land clearing and the biofuel carbon debt. Science 319:1235–1238
Gallardo B, Clavero M, Sánchez MI, Vilà M (2016) Global ecological impacts of invasive species in aquatic ecosystems. Glob Chang Biol 22:151–163
Gozlan RE (2008) Introduction of non-native freshwater fish: is it all bad? Fish Fish 9:106–115
IUCN (2003). IUCN in Asia: Growing Together – Ecosystems and Livelihoods. IUCN, Bangkok/Gland
Jackson M, Loewen CJG, Vinebrooke RD, Chimimba CT (2016) Net effects of multiple stressors in freshwater ecosystems: a meta-analysis. Glob Chang Biol 22:180–189
Jerde CL, Chadderton WL, Mahon AR, Renshaw MA, Corush J, Budny ML, Mysorekar S, Lodge DM (2013) Detection of Asian carp DNA as part of a Great Lakes basin-wide surveillance program. Can J Fish Aquat Sci 70:522–526
Laramie MB, Pilliod DS, Goldberg CS (2015) Characterizing the distribution of an endangered salmonid using environmental DNA analysis. Biol Conserv 183:29–37
Loarie SR, Duffy PB, Hamilton H, Asner GP, Field CB, Ackerly DD (2009) The velocity of climate change. Nature 462:1052–1055
McIntyre PB, Reidy Liermann CA, Revenga C (2016) Linking freshwater fishery management to global food security and biodiversity conservation. Proc Natl Acad Sci U S A 113:12880–12885
Millennium Ecosystem Assessment (2005) Ecosystems and human well-being: biodiversity synthesis. World Resources Institute, Washington, DC. Retrieved from http://www.Millenniumassessment.org/documents/document.354.aspx.pdf
Mittermeier RA, Myers N, Mittermeier CG (2000) Hotspots: earth's biologically richest and most endangered terrestrial eco-regions. Conservation International. ISBN 978-9686397581
Moore JW, Olden JD (2017) Response diversity, nonnative species, and disassembly rules buffer freshwater ecosystem processes from anthropogenic change. Glob Chang Biol 23:1871–1880
Myers N (1988) Threatened biotas: hotspots in tropical forests. Environmentalist 8:178–208
Myers N (1990) The biodiversity challenge: Expanded hot spots analysis. Environmentalist 10:243–256
O'Connor JE, Duda JJ, Grant GE (2015) 1000 dams down and counting. Science 348:496–497
Olden JD, Kennard MJ, Lawler JJ, Poff NL (2011) Challenges and opportunities in implementing managed relocation for conservation of freshwater species. Conserv Biol 25:40–47
Palmer MA, Reidy Liermann CA, Nilsson C, Fl¨orke M, Alcamo J, Lake PS, Bond N (2008) Climate change and the world's river basins: anticipating management options. Front Ecol Environ 6:81–89
Pastor AV, Ludwig F, Biemans H, Hoff H, Kabat P (2014) Accounting for environmental flow requirements in global water assessments. Hydrol Earth Syst Sci 18:5041–5059
Pelicice FM, Pompeu PS, Agostinho AA (2015) Large reservoirs as ecological barriers to downstream movements of Neotropical migratory fish. Fish Fish 16:697–715
Poff NL, Zimmerman JKH (2010) Ecological responses to altered flow regimes: a literature review to inform the science and management of environmental flows. Freshw Biol 55:194–205
Poiani K, Aren A, Brian D, Richt ER, Anderson MG, Richter Holly E (2000) Biodiversity Conservation at Multiple Scales: Functional Sites, Landscapes, and Networks. BioScience 50(2):133–146. Retrieved from http://science.Natureconservancy.ca/salishsea/documents/Background/general/ Poiani%20et%20al%202000.pdf
Rees HC, Maddison BC, Middleditch DJ, Patmore JRM, Gough KC (2014) The detection of aquatic animal species using environmental DNA – a review of eDNA as a survey tool in ecology. J Appl Ecol 51:1450–1459
Reid AJ, Carlson AK, Creed IF, Eliason EJ, Gell PA, Johnson PT, Kidd KA, McCormack TJ, Olden JD, Ormerod SJ, Smol JP (2019) Emerging threats and persistent conservation challenges for freshwater biodiversity. Biol Rev 94(3):849–873
Ricciardi A, Rasmussen JB (1999) Extinction rates of North American freshwater fauna. Conserv Biol 13:1220–1222

Schlaepfer MA, Sax DF, Olden JD (2011) The potential conservation value of non-native species. Conserv Biol 25:428–437

Silva AT, Lucas MC, Castro-Santos T, Katopodis C, Baumgartner LJ, Thiem JD, Aarestrup K, Pompeu PS, O'Brien GC, Braun DC, Burnett NJ, Zhu DZ, Fjeldstad H-P, Forseth T, Rajaratnam N et al (2018) The future of fish passage science, engineering, and practice. Fish Fish 19:340–362

Simmons M, Tucker A, Chadderton WL, Jerde CL, Mahon AR (2016) Active and passive environmental DNA surveillance of aquatic invasive species. Can J Fish Aquat Sci 73:76–83

The United Nations Economic and Social Commission for Asia and the Pacific (ESCAP) and ADB (2000) State of the environment in Asia and the Pacific 2000. Economic and Social Commission For Asia, Asian Development Bank, Bangkok

Twardochleb LA, Olden JD (2016) Non-native Chinese mystery snail (*Bellamya chinensis*) supports consumers in urban lake food webs. Ecosphere 7:e01293

UNEP (2001) South Asia: State of the Environment 2001. United Nations Environment Program, United Nations Environment Program, Regional Resource Centre for Asia and the Pacific (UNEP RRC. AP). Outreach Building, Asian Institute of Technology, Klong Luang, Pathumthani, Thailand. Retrieved from http://www.rrcap.unep.org/pub/soe/sa_cover.pdf

UNISDR and World Bank (2009) Global facility for disaster reduction and recovery. Disaster Risk Management Programs for Priority Countries, Global Facility for Disaster Reduction and Recovery GFDRR Secretariat, Washington, DC. Retrieved from http://unisdr.org/files/14757-6thCGDRMProgramsforPriorityCountrie.pdf

World Resource Institute (2010) Forests, grasslands, and dry lands — Pakistan; Earth trend country profile. WRI, USA. Retrieved from http://earthtrends.wri.org/pdf-library/country-profiles/for_cou_586.pdf. Accessed on 21 May 2010

WRI, IUCN and UNEP (1992) Global biodiversity strategy. World Resources Institute, Washington, DC

WWF (1997) Socioeconomic root causes of biodiversity loss: an analytical approach paper for case studies (by Pamela Stedman-Edwards) For the Macroeconomics for Sustainable Development Pro-gram Office World Wide Fund for Nature December 1997, WWF-MPO Washington, DC. Retrieved from http://assets.panda.org/downloads/analytic.pdf

Chapter 34
Co-management Scheme to Protect the Southern Marshes of Iraq: A Proposal

Laith A. Jawad

Abstract The development of co-management schemes for several protected areas in many parts of the world has gone through several steps of legislations and law establishment until this practice has settled down in its present form that we can see in many South Asian countries. In the introduction of this chapter, basic ideas about the context of the co-management were given, and in the following sections, a short history of the development of the co-management practice in South Asian countries was allocated. The South Asian countries were chosen as an example of implementing the co-management because these countries have struggled in establishing such well-designed program that suits both locals of any protected areas and the government. In the later section, the protected area was discussed and the concentration was given for the southern marshes of Iraq as a possible candidate for implementing the co-management practice. In the last section, the experiences of the South Asian countries that implemented the co-management practice were reviewed shortly, and the possibility of applying such practice in Iraq and on the southern marshes was discussed. The aim of this chapter was to make available for the policymakers in Iraq a program that both the locals of the southern marshes of Iraq and the government can share in order to manage and protect the marsh areas.

Keywords Freshwater management · Biodiversity · Protected areas · Marsh Arabs · Legislation

34.1 Introduction

A protected area is recognized by IUCN (the World Conservation Union) as "a zone of land and/or sea particularly devoted to the security and upkeep of variety of biota, and of natural and linked cultural means, and maintained over permissible or other

L. A. Jawad (✉)
School of Environmental and Animal Sciences, Unitec Institute of Technology, Mt Albert, Auckland, New Zealand

L. A. Jawad (ed.), *Southern Iraq's Marshes*, Coastal Research Library 36,
https://doi.org/10.1007/978-3-030-66238-7_34

actual means" (Sharma 2011). Cooperative management – or co-management – is denoted as a condition in which two or more social performers discuss, outline, and pledge between themselves a fair distribution of the supervision purposes, rights, and duties for a defined terrain, area, or set of natural assets (Borrini-Feyerabend et al. 2004). A reasonable distribution of aids and costs of protected areas guard and running among the investors is, then, a significant feature of co-managing any protected area. Such localities may be co-managed for a range of causes comprising scientific research, wilderness safety, conservation of species and genetic variety, upkeep of habitat functions, keeping of exact natural and cultural topographies, tourism and recreation, education, maintainable use of means from natural environments, upkeep of cultural and traditional aspects, etc. (Sharma 2011). The administration of any areas in the tropical and subtropical zones is considered among the jobs that experience several difficulties. In such areas, the biodiversity is very rich and at the same time they are facing human-made impacts (Baird and Dearden 2003).

In South Asian part of the world, diverse sorts of methods are followed in protecting any area showing in need for protection from the environment point of view. For example, three kinds of protection such as national parks, wildlife sanctuaries, and game reserves have been recognized in Bangladesh, while in Sri Lanka there are six types of ways of protection implemented, i.e. strict natural reserves, national parks, nature reserves, jungle corridors, intermediate zones, and national wilderness areas (Sharma 2011).

The present chapter analyses pertinent co-management involvements at policy level in some countries in South Asia in order to attain vital lessons for a possible implementation of such management system in the southern marshes of Iraq. The information obtained in this chapter is based on the local peoples' perspective of the countries where the studies were performed.

34.2 Short History of Co-management Practised in South Asian Countries

The main areas in South Asian countries that need to be protected are the forest lands, and since ancient time, the procedure of protecting such areas was run through involving the locals in the process of the management of the land (Sharma 1994, 1995). The locals in return drive their interest in the environment and protect it from the different kinds of religions that people have faith in. With such practice, the normal use of biodiversity was ruled by the outmoded schemes of reserve use and protection based on religion, culture, and folklore (Sharma 2011).

Thousands of years may mark the beginning of legislation, or similar act may be created in the Indian subcontinent by Hindu Rulers such as King Ashoka (early AD) (Sharma 2011). In these rules, which were applied to the forest areas, suggested that the approach of forest management realised that local people be subject to for their livelihood cannot be managed in separation of the cultural and religious does of neighbouring communities (Sharma 2011). Later and during the Moghul rules, the situation remains as is the new rules respected the local customs and do not impede the local performs.

Later the dissipating of forests to create lands for agriculture was cheered by the governments in the South Asian countries. For instance, the first official effort was made in India in 1793, when stable society of estates including forest land was performed with local leaders (designated as *'Zamindars* and *Talukdars),* who became inherited amassers of the land revenue in lieu of a fixed annual royalty (Sharma 2011). This system of land tenancy established conspicuously in the eastern Indian subcontinent as related to *Ryotwari* system (ownership rights with farmers) of land tenancy that developed in the western Indian subcontinent (Sharma 2011).

These disparities in land tenancy schemes were later revealed in the higher approval of social forestry initiatives (during 80s and 90s) by farmers in *Ryotwari* regions of western India as compared to *'Zamindari* regions of eastern India (Sharma 2011).

The laws of keeping the forests were completely changed during the British colonial rules of the Indian subcontinent, and several forest areas were allocated as hunting areas for royalties. In the amended laws, certain types of trees were listed under conservation such as teak, which is reserved entirely for the use in royal navy (Bawa et al. 2011).

Progressively, the traditional/customary rights of local people on the nearby forests were unrestricted, thus estranging local communities from previously open access forest resources. Instead, Forest Department implemented control over forests by awarding harvesting rights through acquisition contract system, which a buyer was permitted to chop any tree over and above a certain girth. Local community's requirements for forest yields were thus overlooked (Sharma 2011).

Things were changed again after independence, and new rules were introduced in which forests can be manage by *Zamindars* and native rulers and international agencies such as the World Bank have presented aids to protect forest areas in the Indian subcontinent since 1992. The environmental services of forests and the protected areas were being gradually known during this period, both nationally and internationally (Sharma 1994).

Later different countries in the South Asian regions have declared different acts that deal with protection of the wildlife. Such movement started during the 1970s (Sharma 2011). These laws paved the way for the introduction of a new law to deal with the preservation of any protected areas after which several protected areas were designated in different South Asian countries.

In the projects of preserving protected areas, eco-development scheme was adopted, which can be denoted as a policy to overwhelmed unmaintainable and unsuited forest use by reliance on populations for their living in and around the protected areas via controlled use and substitutes. The policy goal was aimed to protect the diversity of biota through dealing with the impact of local people on the protected areas and vice versa (Sharma 2011).

At the moment, the system in use in management of the forest lands in the Indian subcontinent or at least in India is the traditional and community-based forest resource use systems, which replace the centralized government systems during colonial times. Later, several countries have gone further in their legislation and introduced co-management plans for preserving any designated protected land. Otherwise certain areas will be destroyed completely if such movement will not be approved by governments (Wells et al. 2004). Therefore, kinds of associations have been introduced and created between managers that are responsible for protection of lands and the locals as the co-management tactic is progressively legalized. As a result, conservation assistances to the locals have happened in those countries where co-management performs have been in practice.

Although the example given above is not drawn from the aquatic environment, it shows the changes through time for the development of laws that govern the management of any area that needs to be protected for environment purposes. The best results were attained when a conclusion came across the co-management scheme that the present chapter is proposing to be used in the management of the southern marshes of Iraq.

34.3 Protected Areas in Iraq

In Iraq, there are no protected areas as known internationally. The southern marshes of Iraq could be designated perfectly for such position. They have a significant part in the freshwater system of Iraq and the whole region as they hold vital roles represented in ecological, historical, and cultural significance, being the foundation of Sumerian, Akkadian, Assyrian, and Babylonian civilizations (Hashim 2017). The characteristic features of the southern marshes of Iraq are different from any other wetlands in the world in being different water movement, the amount of dissolved oxygen, the wealth of plants, the diversities of its biota, its endurance during history, and vigorous efficiency are fundamentally (Omar 2012).

The importance of the southern marshes of Iraq can be seen in preserving the ecosystem of both Euphrates and Tigris's river basins through several aspects such as keeping the currents of the two rivers and the tidal estuaries, removing harmful pollutants like metals and organic matters from water, and analysing organic wastes and decreasing their destruction (Salman 1994). On the other hand, the environment of these marshes is deteriorated over the years owing to several factors that are caused by climate changes and anthropogenic variations of the habitats (Ministry of Water Resources 2010). Due to the significance of the southern marshes of Iraq, they were included in the world heritage by the UNESCO (Barrara 1985).

Therefore, the inclusion of the southern marshes of Iraq in the world heritage by UNESCO has triggered another issue which is the possible protection of these marshes and their inhabitants both human and the biota (terrestrial and aquatic) against climatic changes, natural disasters, and man-made impacts. Such views were also suggested by others (e.g. Du Saussay 1985).

34.4 The Experience of Co-managing of Protected Areas of Some Countries and Applicability to the Case of the Southern Marshes of Iraq

In this section, a short review is given about the practices and experiences of some countries about implementation of co-management scheme in protecting any area. The aim of this review is to observe the possible application of this scheme in the future in the southern marshes of Iraq.

Originally, the co-management schemes have historically been applied in the forest environment in India that was open to local societies as common pool natural means for their livings. The laws were changed later as the ability of locals to manage such huge area failed, and for this, state land rights were established as a result. The co-management schemes were established in other countries of South Asia, where locals have started cooperative moves to meet the degrading forests and environment. Such ingenuities have been extra reinforced under donor subsidized sharing forestry and biodiversity projects in these countries.

Several co-management efforts have been successful in some South Asian countries such as Nepal, India, and Sri Lanka. During the implementation of these practices, many pertinent teachings have been attained for co-managing of any protected area. The preservation of the maintainable diversity of biota and use schemes is to be progressively permitted to the dwellers any areas that need to be protected, and in case of Iraq, the locals of the southern marshes are needed to be involved in such scheme. There are some important issues that the locals need to be involved in, and their responses should be considered before any management to go ahead such as landscapes. According to the co-management agreements, specific roles and responsibilities should be set for the gainful partnerships of the locals. Natural renewal and eco-renovation are to be stimulated by linking local societies. Improvement implanting of native species of shrubs and trees may be taken in those areas where recovering rootstock and/or mother trees do not occur. In the case of the southern marshes, for example, several native bird species can be reintroduced, and native fish species that disappear from the marsh area since a long time may attempted to be reinstated in the marsh again and many other native animal and plant species that the locals used to acclimatized their life with in the marsh area. Such efforts can be designated as explicit co-management purposes.

The future accomplishment of co-management of the southern marshes of Iraq would depend on effectively applying the lessons in developing profitable corporations with key investors, who are authorized by passing and applying permitting strategies and laws. Socio-environmental meanings and services from any protected areas need to be sufficiently esteemed by policymakers. To implement a better co-managing scheme, this program should have the ability to put in place pertinent official and financial tools, and sustainability efforts are equally imperative for maintainable biodiversity conservation in the southern marshes of Iraq.

The experience of the countries that have implemented the co-management practice and succeeded in this aspect has established several laws that paved the way for the general acceptance of the locals. For the case of the southern marshes of Iraq, the policymakers in Iraq should follow the same path of the other countries that have gone and prepare for the responses of the locals and solve the obstacles that might rise accordingly. After approval of such legislations, another set of laws should be discussed and put in place that allow participation of the locals in the management of the marsh areas through the co-management scheme.

After the implementations of the co-management, a certain percentage from the revenue of any commercial projects that the government establish should go as a share of the locals to assist in the development of their environment and also to assist in establishing some private projects that help to increase the family income of the locals, in case of the southern marshes of Iraq, projects to bloom the tourist industry in the marshes like building small restaurants, motels, coffee shops, and even amusement centres that contain water sports. Small factories for traditional hand-made crafts might be a good idea to build in the marsh area to make new revenue to the locals, and at the same time it can involve the young workers to teach them the traditional craft of their culture, and at the same time they earn extra income that helps them for their daily life.

In this respect, the scientific community at the universities in the cities near the marsh area can contribute in building science centre that contains both exhibition and deposition locations for natural history specimens representing the fauna and flora of the marsh area. Such science centre will be the attractive tourist's monument to visit and know more about the biota of the marsh area.

To protect the biodiversity, locals need to be enrolled in some sort of sessions that teach them the basic information about the diversity of the biota around them and their importance to life and how to protect them. Such teaching will give its fruitful results once the first few groups attained the lessons and assist effectively in the process of protecting their environment.

References

Baird IG, Dearden P (2003) Biodiversity conservation and resource tenure regimes: a case study from Northeast Cambodia. Environ Manag 32(5):541–550

Barrara J (1985) Lausche. Guidelines for protected areas legislations. (UNEP) and (IUCN), no 1, IUCN Environmental policy and Law Paper, no 16. p 34

Borrini-Feyerabend G, Michel P, Farvar MT, Kothari A, Renard Y (2004) Sharing Power. Learning by doing in co-management of natural resources throughout the world. IIED and IUCN/CEESP/CMWG, Cenesta, Tehran

Bawa KS, Rai ND, Sodhi NS (2011) Rights, governance, and conservation of biological diversity. Conserv Biol 25(3):639–641

Du Saussay C (1985) Legislation on wildlife and protected areas, United Nations Legal studies no. 20. FAO, Rome, p 22

Hashim SJ (2017) International legal regulation of Iraq Wetlands. Revista Publicando 4(13 (3)):802–809

Ministry of Water Resources (2010) Animation in the centre of the marshes. Introduction to the study of Maysan conference. Lmcharaa restoration of Wetlands, p 20

Omar AQ (2012) Criminal protection of terrestrial and aquatic flora and fauna in Iraqi law, Baghdad, p 43

Salman NA (1994) Legal opportunity to exploit the marshes of southern Iraq (problems and solutions), environmental studies, the University of Basra, Marine Scientific Centre, p 23

Sharma RA (1994) Learning from the Experiences of Joint Forests Management. For Trees People 24:36–41

Sharma RA (1995) Participatory forest management in India. Ambio 24(2):131–133

Sharma RA (2011) Co-management of protected areas in South Asia with special reference to Bangladesh. Asia-Pac J Rural Dev 21(1):1–28

Wells MP, McShane TO, Dublin HT, O'Connor S, Redford KH (2004) The future of integrated conservation and development projects: building on what works. Getting biodiversity projects to work: Towards more effective conservation and development. Columbia University Press, New York, 1:397–422

Chapter 35
Eradication as a Method to Manage the Non-native Fish in the Marsh Environment: A Proposal

Laith A. Jawad

Abstract The non-native fishes that enter a new environment show a significant hazard of different grades on the new environment and on the economy of the country. Attempts to remove those invasive fishes have not well attracted the attention of researchers globally. In the present chapter, a short review about the methods of removal of invasive fishes was given in aim to make this information available for the policy makers in Iraq to start thinking about finding a solution to the dilemma of the invasive species in its freshwater system. At the moment, there are several invasive species swarming the inland waters of Iraq mainly carp species. The procedures discussed in this chapter are those used and implanted by other countries from around the world experiencing the same problem of invasive fish species, which they approved their practicality in reducing significantly the number of individuals of the non-native fish species. Once these practices are introduced in Iraq, further review and modification for them should be applied in order to improve their efficacy that may be more discriminating and less harmful to nontarget species.

Keywords Invasive species · Management · Aquatic life · Biodiversity · Economy · Environment

35.1 Introduction

Biological invasion is one of the opposing ecological factors that can bring about drastic changes to the aquatic environment and especially the freshwater habitats (Gozlan et al. 2009; Ricciardi and MacIsaac 2011; Vitousek et al. 1997; Donaldson and Cooke 2016) and enforce substantial economic charges (Pimentel et al. 2005), allocating augmented stresses on policy makers, supply executives, and scientists (Simberloff et al. 2013). The non-native species have several route to invade a new

L. A. Jawad (✉)
School of Environmental and Animal Sciences, Unitec Institute of Technology, Mt Albert, Auckland, New Zealand

L. A. Jawad (ed.), *Southern Iraq's Marshes*, Coastal Research Library 36,
https://doi.org/10.1007/978-3-030-66238-7_35

environment, among these are shipping routes (Ruiz et al. 1997; Levine and D'Antonio 2003), seepages from aquaculture, aquaria and ornamental trade (Padilla and Williams 2004), fish individual supplying (including authorized and unauthorized efforts), entertaining vessels, and live food occupation, which may be intentional or unplanned. Furthermore, the inferior dispersal of invasive species creates contests for source executives (Fredenberg 2002; Lintermans 2004; Vander Zanden and Olden 2008).

Choices for handling invasive fish species can comprise leaving them as is, regulate and/or suppression, population destruction, and/or species abolition (Varley and Schullery 1995). Control, such as effecting barriers, is typically the most required method to stop the spread of the invasive species into new habitats (Fausch et al. 2006; Finnoff et al. 2007; Peterson et al. 2008; Britton et al. 2011). Nevertheless, where control is not conceivable or has not been choice for handling biological invasions (Rinne and Turner 1991; Genovesi 2005), eradication is the removal of whole fish populations or fish species from distinct areas or bodies of water (Gresswell 1991). The means of this method incline to be directed, for example, by using susceptible periods in the life cycle (Buhle et al. 2005; Syslo et al. 2013) or by concentrating on regions of high great quantity (Lodge et al. 2006).

The kinds of fish administration practices obtainable to source executives to use eradication plans can vary extensively (Donaldson and Cooke 2016). They may comprise chemicals, harvest rules, physical elimination, or biological regulations (Meronek et al. 1996). The efficiency of chemical eradication (e.g., rotenone, Fintrol) rests on the environmental settings (e.g., water temperature, depth, pH, release, target fish species, hydrology, substrate configuration, regions of groundwater renew Johnson and Bobrovskaya 2015); some worries of security impairment may accompany the chemical eradication when nontarget species are influenced by chemical conducts (Vinson et al. 2010). Yield rules can contain different fishing methods of target species (Paul et al. 2003; Syslo et al. 2011; Gaeta et al. 2015) or alteration of fishing guidelines. Physical exclusion practices can consist of traps, electrofishing, and/or netting plans, while biological managing can comprise the introduction of predators, intraspecific handling, or targeted pathological responses (Davis and Britton 2015). When applying fish management agendas, risk study is necessary to assist in decision of when administration policies should be applied, what plan should be selected, and what the probabilities of achievement of diverse approaches are (Britton et al. 2011). The risk investigation comprises identification and valuations of threats, containing forecasting the possibility and sternness of opposing impacts (Koel et al. 2010).

The achievement of invasive fish control tactics can differ largely liable to the controlling aims for the scheme: whether regulation, eradication, elimination, or repression (among others) was the final objective of the project (Donaldson and Cooke 2016). As can be anticipated given the difficulties of the natural environment, accomplishment can be problematic to enumerate, and some methods can be failed even though best determinations were involved (Meronek et al. 1996). Disappointment of invasive fish eradication systems can happen owing to a number of issues, containing (but not restricted to) unsuccessful catch practices (e.g., size-specific efficiencies), environment intricacy (e.g., areas of sanctuary, plant thickness), species-specific features (e.g., size, habitat preferences), and physical water setting (e.g., water chemistry, temperature, water depth; Koel et al. 2010). Defining the results of administration interferences, particularly when refurbishment of freshwater ecosystems is an objective (i.e., to eliminate invasive aim fish species from a precise water body), needs long-term assessment and appraisal in relation to attaining the purposes (Britton and Brazier 2006). At the end, program assessment and appraisal are essential not only to control the efficiency of practices but also to search the cost-efficacy and cost/profit of each policy.

The present chapter discusses the possibility of implementing the policy of eradication as a method to remove the invasive fish species from the southern marshes of Iraq. In the first part of the chapter, a background information was given about the eradication method, and in the second part different fish eradication methods were suggested. All these tactics were applicable to the inland waters of Iraq and specifically in the southern marsh areas. In giving such information, the policy makers should be aware about which method they need to choose in order to remove the invasive fish species and save the native ones.

35.2 Eradication Approaches

Generally, procedures for eradication of non-native fishes can be grouped into three classes: chemical, physical, or biological (Nico and Walsh 2011). A unified approach is often chosen, using multiple systems in grouping (Lee 2001; Diggle et al. 2004; Kolar et al. 2010). Several non-native fishes have great reproductive potential, and the endurance of even one adult pair can possibly lead to thousands of progenies (Nico and Walsh 2011). Consequently, spawning grounds are often the main target of both eradication and control endeavors (Diggle et al. 2004).

35.2.1 Chemical Methods

Fish toxicants (i.e., ichthyocides, piscicides, or fish poisons) are the main technique for eradicating non-native fishes, which are available globally in more than 40 different chemical kinds (Kolar et al. 2010). The majority of these chemicals have not been entirely industrialized or verified, several are not permitted for fish administration, and only a few are broadly and reliably used (Dawson 2003; Clearwater et al. 2008; Cailteux et al. 2001; Kolar et al. 2010). The utmost normally ichthyocides in use are rotenone and antimycin-A (Fintrol®). Strategies for the active and safe planning and implementation of projects using rotenone are broadly obtainable (Finlayson et al. 2000; Moore et al. 2008).

Rotenone is found in nature as a component in the cells of the plants belonging to the family Leguminosae and is the active constituent in some plants used by early Pacific islanders as a venom in the harvest of food fish (Morrison et al. 1994). Fish biologists in North America have used rotenone to kill unwanted fish species since the 1930s (Rinne and Turner 1991; McClay 2005). At the moment, considerable studies were available about its usage as an eradicator (Wydoski and Wiley 1999; Cailteux et al. 2001; McClay 2005).

The other piscicide is the antimycin, which is a fungal antibiotic known for its possible use in fish control since the early 1960s (Melo et al. 2015; Moore et al. 2008). Rotenone and antimycin are both common piscicides, but liable on the environment and fish species to be managed, they have occasionally been used discriminately (Willis and Ling 2000; Moore et al. 2008). For instance, scaled fish and some rotenone-resilient species are frequently vulnerable to antimycin (Melo et al. 2015). Owing to efficacy rest on water and habitat features (e.g., pH, water flow, and amount of leaf litter), antimycin is occasionally used in small streams whereas rotenone is used in large, deep lakes (Melo et al. 2015). Usage of each chemical characteristically includes discharge of diluted liquid solutions straight into the water, though rotenone powder is frequently used (Nico and Walsh 2011). These two chemicals can come as eatable pellets (Mallison et al. 1995; Kroon et al. 2005).

The chief benefits for antimycin are its efficiency at lower concentrations and non-noticeability by fish, whereas rotenone has the leads of broad range of toxicity to all species of fish and efficiency under a large array of pH conditions (Melo et al. 2015). Both chemicals destroy comparatively rapidly into harmless composites and are neutralized by potassium permanganate (Moore et al. 2008). Depending on water temperature and sunlight contact, distraction may be within days or weeks for rotenone or within hours or days for antimycin (Dinger and Marks 2007). Contingent on concentration, both chemicals can be harmful to aquatic invertebrates, especially those with gills (Nico and Walsh 2011). Nevertheless, considerably fewer information is known regarding the nontarget impacts of antimycin (Melo et al. 2015; Dinger and Marks 2007).

A few investigations, possibly beneficial toxicants comprise a varied group of plant-derived saponins or triterpene glycosides, containing definite produces given in the literature as teaseed cake and Mahua oilcake (Clearwater et al. 2008). Further auspicious ichthyocides comprise squoxin, chosen counter to northern pike minnow (*Ptychocheilus oregonensis*), and several others owing to their deceptive discrimination, low toxicity to nontarget organisms, ease of usage, safety to humans, tenacity in the habitat, low inclination to bioaccumulate, and low cost (Dawson 2003).

Utmost chemicals that used as piscicides have the drawback of non-explicity, instigating death or injury not only to directed invasive fish but also nontargeted native fishes and aquatic invertebrates (Nico and Walsh 2011). Many invasive species are less subtle to piscicides than the nontarget species (Schofield and Nico 2007; Schreier et al. 2008). The piscicides that kill native species usually need replacing to counterbalance their influences, though in some tropical insular Pacific environments this is frequently needless since native fishes and macroinvertebrates invade the area again naturally from coastal regions or nearby inland waterfalls. However, attention is essential particularly since concurrent chemical conduct of more than a few streams might eradicate nonmigratory stream invertebrates from the entire region such as islands.

35.2.2 Physical Methods

Several usual fishing gears including the harmful ones are included under the physical eradication methods to remove the non-native fish species. However, the majority of these have restricted possibility for eradication (Roberts and Tilzey 1996; Wydoski and Wiley 1999; Mueller 2005; CDFG 2007; Kolar et al. 2010).

Nets and trap when used in the eradication of the non-native fishes showed a restricted efficiency in isolated water bodies or portions of drainages. For example, exhaustive usage of a seine net during 1976–1978 supposedly removed all invasive sheepshead minnow (*Cyprinodon variegatus*) and its hybrids from a trivial stream system in Texas, USA (Minckley and Deacon 1991). On the other hand, using gill net showed a possibility to remove invasive trout from high mountain lakes in California, USA (Knapp and Matthews 1998; Vredenburg 2004), and Banff National Park in Canada (Parker et al. 2001). Small traps were found useful in removing invasive fishes from an isolated pool in Mexico (Lozano-Vilano et al. 2006). In contrary, gill nets in New Zealand ponds (Neilson et al. 2004) were unsuccessful to remove or mange rudd (*Scardinius erythrophthalmus*).

A mixed results were obtained when a backpack electrofishing gear was used in eradication of the invasive fish in a small upland streams in North America (Moore et al. 1986; Thompson and Rahel 1996; Kulp and Moore 2000). Electrofishing may be beneficial for management but not removal in larger or more multifaceted water areas. For instance, electrofishing on boat has been arrayed frequently since 2001 to remove Asian swamp eels (Synbranchidae) from canals in South Florida, USA (Nico and Walsh 2011). Almost 1400 swamp eels were eliminated the first year, but results seemed to have little preliminary impact on whole population size or size-length structure (L. G. Nico, unpublished data, cited in Nico and Walsh 2011).

The harmful fishing method, underwater explosives such as detonation cord, is able to kill or harm fishes (Teleki and Chamberlain 1978; Keevin 1998) but is costly and principally unsuccessful for removal of non-native species (CDFG 2007). Substantial difference occurs in blast impacts liable on charge sort and habitat characteristics (Keevin 1998). Susceptibility to explosives also differs between fish species. Fish with gas bladders suffer great harm, whereas those that lack gas bladders often endure underwater blasts (Goertner et al. 1994).

Fishes may show variation in their patience toward temperature, but handling of water temperature to eliminate or control invasive fish is rarely possible (Nico and Walsh 2011). In rare instances, Stauffer et al. (1988) suggested that the lower lethal temperature of non-native blue tilapia (*Oreochromis aureus*) in the Susquehanna River of Pennsylvania (USA) was about 5 °C. The local tilapia population overwintered in the thermal effluent of an electric power plant, so the plant temporarily dropped the water temperature during winter. This deceptively removed the local population, but the tilapia continued owing to the presence of other thermal discharges in the area.

Complete removal of water from any pond or water body to eliminate invasive fish populations has been projected for some large reservoirs (CDFG 2007) but has mostly been restricted to small water bodies, frequently aquaculture ponds (Alvarez et al. 2003; Mueller 2005). The water level of lakes or reservoirs is occasionally dropped in combination with the use of fish toxicants (CDFG 2007), thus dropping the amount of toxicant required and comprising directed fish within smaller and more unprotected locations.

Augmented yield impact as a technique of managing non-native species or undesirable fishes can include alteration of guidelines to encourage fishing and commercial collecting (Lee 2001). Nevertheless, since fishes differ in their vulnerability to catch, the means used by fishers and commercial fishermen are characteristically size and species designed. Accordingly, the probability of eliminating the whole population by an augmented catching is normally little (Thresher 1996, Yonekura et al. 2007).

35.2.3 Biological Methods

The set free of predators to feed on unwanted or non-native species as a form of biological controller has long been applied though it is not normally used contrary to non-native fishes (Nico and Walsh 2011). Similar to what happens in the environments on land, this tactic to invasive fishes could have unintentional concerns (Fuller et al. 1999). Transmissible diseases such as koi herpes virus or KHV have potential to be used to kill invasive fishes but are debatable owing to the possible injury that linked to the targeted species (Gilligan and Rayner 2007) and probable problems with amending accidental results. Furthermore, enduring fish could have resistance to the disease, making the technique impractical once one application has been done. Yet, brining in the environment of the invasive fish species of a highly specific transmissible disease may assist if joint with other means.

Genetic managements which have been suggested include (1) chromosome set handlings encompassing construction and discharge of triploid sterile invasive fish with the commitment of decreasing the population size of directed naturalized individuals and (2) recombinant DNA approaches comprising transgenic practices intended to yield sterile fish or disperse harmful transgenes (i.e., "Trojan horse" genes) to a target invasive species (Gilligan and Rayner 2007; Thresher 2008). In Australia, there have been studies on the use of "daughterless genetic technology" to fight non-native fish, particularly common carp. This includes generating a transmissible gene that subdues the production of female offspring, causing a reduction in the irritant population over succeeding cohorts (Gilligan and Rayner 2007). Some genetic handlings have been verified in the field such as that related to the discharge of sterile males to help control sea lamprey (*Petromyzon marinus*) in the Laurentian Great Lakes (Bergstedt and Twohey 2007).

An auspicious and possibly friendly biological management technique under progress is to use pheromones, which are natural chemicals secreted by many fish and are significant in impacting their behavior. Up till now, advancement of this technique has been focused at the management of sea lamprey in North America

(Sorensen and Hoye 2007). Practical tests showed that pheromone indications drive sea lampreys into traps. However, the project to control sea lamprey in the Laurentian Great Lakes – although offering supported base for procedures of possible advantage for removing other species – has been exhaustive, taking a long time, and costly (Kolar et al. 2010).

35.3 Remarks

As non-native species are damaging and can create destruction for both the environment and the economy, eradication rests a vital controlling choice. Nevertheless, comparable to other non-native animals and plants, non-native fishes can be problematic and costly to eliminate especially on the mainland. Besides, the situation of information on fish removal is active, and, as each elimination project has its own exclusive set of difficulties, answers may be site or species specific (Nico and Walsh 2011).

Eradication plans aiming non-native fishes are frequently provocative, somewhat due to the probability of the damage to native species (Britton et al. 2008) and particularly when non-specific fish chemicals, such as rotenone, stay one of the few operative implements.

In the inland waters of Iraq and especially in the southern marshes, there are several non-native species, i.e., the different carp species. These invasive species have a significant impact on the environment of the freshwater ecosystem and on the native fish fauna living in. It is clear now that most of the native fish species have disappeared from the marsh area and any native species has left is represented by low abundance. This is due to the impacts that the invasive species exert on the native fish species through several aspects among them is the competition on the niches and food. Therefore, it is the time for the policy makers to sit and think about a policy to eradicate the non-native species from the freshwater ecosystem in Iraq. For this, the present chapter has offered the practices that are usually followed in countries from around the world experiencing problems of invasive fish species.

References

Alvarez JA, Dunn C, Zuur A (2003) Response of California red-legged frogs to removal of non-native fish. Trans West Sect Wildl Soc 38/39:9–12

Bergstedt RA, Twohey MB (2007) Research to support sterile-male-release and genetic alteration techniques for sea lamprey control. J Great Lakes Res 33(Special Issue 2):48–69

Britton JR, Brazier M (2006) Eradicating the invasive topmouth gudgeon, *Pseudorasbora parva*, from a recreational fishery in northern England. Fish Manag Ecol 13:329–335

Britton JR, Brazier M, Davies GD, Chare SI (2008) Case studies on eradicating the Asiatic cyprinid *Pseudorasbora parva* from fishing lakes in England to prevent their riverine dispersal. Aquat Conserv Mar Freshwat Ecosyst 18:867–876

Britton JR, Copp GH, Brazier M, Davies GD (2011) A modular assessment tool for managing introduced fishes according to risks of species and their populations, and impacts of management actions. Biol Invasions 13:2847–2860

Buhle ER, Margolis M, Ruesink JL (2005) Bang for buck: cost effective control of invasive species with different life histories. Ecol Econ 52:355–366

Cailteux RL, DeMong L, Finlayson BJ, Horton W, McClay W, Schnick RA, Thompson C, C. (eds) (2001) Rotenone in fisheries: are the rewards worth the risks? American Fisheries Society, Bethesda

CDFG (California Department of Fish and Game) (2007) Lake Davis pike eradication project: Final environmental impact report/environmental impact statement (EIR/EIS). California Department of Fish and Game and U.S. Forest Service, Plumas National Forest. Available from http://www.dfg.ca.gov/lakedavis/EIR-EIS/. Accessed 11 Mar 2010

Clearwater SJ, Hickey CW, Martin MI (2008) Overview of potential piscicides and molluscicides for controlling aquatic pest species in New Zealand. Science for Conservation Series 283. New Zealand Department of Conservation, Wellington, 74 p

Davis GS, Britton JR (2015) Assessing the efficacy and ecology of biocontrol and biomanipulation for managing invasive pest fish. J Appl Ecol 52:1264–1273

Dawson VK (2003) Successes and failures of using piscicides. In: Dawson VK, Kolar CS (eds) Integrated management techniques to control nonnative fishes. Upper Midwest Environmental Sciences Center, U.S. Geological Survey, La Crosse, pp 33–38

Diggle J, Day J, Bax N (2004) Eradicating European carp from Tasmania and implications for national European carp eradication. Inland Fisheries Service, Moonah, 67 p

Dinger EC, Marks JC (2007) Effects of high levels of Antimycin A on aquatic invertebrates in a warmwater Arizona stream. N Am J Fish Manag 27:1243–1256

Donaldson LA, Cooke SJ (2016) The effectiveness of non-native fish eradication techniques in freshwater ecosystems: a systematic review protocol. Environ Evid 5(1):12

Fausch KD, Rieman BE, Young MK, Dunham JB (2006) Strategies for conserving native salmonid populations at risk from nonnative fish invasions: trade-offs in using barriers to upstream movement. General Technical Report RMRS-GTR-174. USDA Forest Service, Rocky Mountain Research Station, Fort Collins

Finlayson BJ, Schnick RA, Cailteux RL, De Mong L, Horton WD, McClay W, Thompson CW, Tichacek GJ (2000) Rotenone use in fisheries management: administrative and technical guidelines manual. American Fisheries Society, Bethesda, 212 p

Finnoff D, Shogren JF, Leung B, Lodge D (2007) Take a risk: preferring prevention over control of biological invaders. Ecol Econ 62:216–222

Fredenberg W (2002) Further evidence that lake trout displace bull trout in mountain lakes. Int J Sci 8:143–152

Fuller PL, Nico LG, Williams JD (1999) Non-indigenous fishes introduced to inland waters of the United States. American Fisheries Society, Bethesda, 613 p

Gaeta JW, Hrabik TR, Sass GC, Roth BM, Gilbert ST, Vander Zanden MJ (2015) A whole-lake experiment to control invasive rainbow smelt (*Actinoperygii, Osmeridae*) via overharvest and food web manipulation. Hydrobiologia 745:433–444

Genovesi P (2005) Eradication of invasive alien species in Europe: a review. Biol Invasions 7:127–133

Gilligan D, Rayner T (2007) The distribution, spread, ecological impacts and potential control of carp in upper Murray River. New South Wales Department of Primary Industries Fisheries Research Report Series 14, Cronulla, 25 p

Goertner JF, Wiley ML, Young GA, McDonald WW (1994) Effects of underwater explosions on fish without swim bladders, Technical Report NSWC TR 88-114. Naval Surface Warfare Centre, Silver Spring

Gozlan RE, Whipps C, Andreou D, Arkush K (2009) Identification of a rosette-like agent as *Sphaerothecum destruens*, a multi-host fish pathogen. Int J Parasitol 39:1055–1058

Gresswell RE (1991) Use of antimycin for removal of brook trout from a tributary of Yellowstone Lake. Trans Am Fish Soc 11:83–90
Johnson ME, Bobrovskaya L (2015) An update on the rotenone models of Parkinson's disease: their ability to reproduce the features of clinical disease and model gene–environment interactions. Neurotoxicology 46:101–116
Keevin TM (1998) A review of natural resource agency recommendations for mitigating the impacts of underwater blasting. Rev Fish Sci 6:281–313
Knapp RA, Matthews KR (1998) Eradication of nonnative fish by gill netting from a small mountain lake in California. Restora Ecol 6:207–213
Koel TM, Arnold JL, Bigelow PE, Ruhl ME (2010) Native fish conservation plan/ environmental assessment. National Park Service, Yellowstone National Park, Wyoming
Kolar CS, Courtenay WR Jr, Nico LG (2010) Managing undesired and invading fishes. In: Hubert WA, Quist MC (eds) Inland fisheries management in North America, 3rd edn. American Fisheries Society, Bethesda, pp 213–259
Kroon FJ, Gehrke PC, Kurwie T (2005) Palatability of rotenone and antimycin baits for carp control. Ecol Manag Restor 6:228–229
Kulp RA, Moore SE (2000) Multiple electrofishing removals for eliminating rainbow trout in a small southern Appalachian stream. N Am J Fish Manag 20:259–266
Lee DP (2001) Northern pike control at Lake Davis, California. In: Cailteux RL, DeMong L, Finlayson BJ, Horton W, McClay W, Schnick RA, Thompson C (eds) Rotenone in fisheries: are the rewards worth the risks? American Fisheries Society, Bethesda, pp 55–61
Levine JM, D'Antonio CM (2003) Forecasting biological invasions with increasing international trade. Conserv Biol 17:322–326
Lintermans M (2004) Human-assisted dispersal of alien freshwater fish in Australia. N Z J Mar Freshw 38:481–501
Lodge DM, Williams A, MacIsaac HJ et al (2006) Biological invasions: recommendations for US policy and management. Biol Invasions 16:1035–2054
Lozano-Vilano ML, Contreras-Balderas AJ, Garcia-Ramirez ME (2006) Eradication of spotted jewelfish, *Hemichromis guttatus*, from Poza San Jose del Anteojo, Cuatro Cienegas Bolson, Coahuila, Mexico. Southwest Nat 51:553–555
Mallison CT, Hestand RS, Thompson BZ (1995) Removal of triploid grass carp with an oral rotenone bait in two central Florida lakes. Lake Reservoir Manage 11:337–342
McClay W (2005) Rotenone use in North America (1988–2002). Fisheries 30(4):29–31
Melo KM, Oliveira R, Grisolia CK, Domingues I, Pieczarka JC, de Souza Filho J, Nagamachi CY (2015) Short-term exposure to low doses of rotenone induces developmental, biochemical, behavioral, and histological changes in fish. Environ Sci Pollut Res 22(18):13926–13938
Meronek TG, Bouchard PM, Buckner ER, Burri TM, Demmerly KK, Hatleli DC, Klumb RA, Schmidt SH, Coble DW (1996) A review of fish control projects. N Am J Fish Manag 16:63–74
Minckley WL, Deacon JE (eds) (1991) Battle against extinction: native fish management in the American West. University of Arizona Press, Tucson, 519 p
Moore SE, Larson GL, Ridley B (1986) Population control of exotic rainbow trout in streams of a natural area park. Environ Manag 10:215–219
Moore SE, Kulp M, Rosenlund B, Brooks J, Propst D (2008) A field manual for the use of Antimycin A for restoration of native fish populations. National Resource Report NPS/NRPC/NRR-2008/033, U.S. National Park Service, Fort Collins, Colorado, USA, 149 p. Available from http://www.Nature.Nps.gov/water/ FisheriesReports/NPS_Antimycin_SOP_2008_reformat.pdf. Accessed 11 Mar 2010
Morrison J, Geraghty P, Crowl L (eds) (1994) Science of Pacific Island peoples. Volume 3: fauna, flora, food and medicine. Institute of Pacific Studies, Suva, 237 p
Mueller GA (2005) Predatory fish removal and native fish recovery in the Colorado River mainstem: what have we learned? Fisheries 30(9):10–19

Neilson K, Kelleher R, Barnes G, Speirs DA, Kelly J (2004) Use of fine-mesh monofilament gill nets for the removal of rudd (*Scardinius erythrophthalmus*) from a small lake complex in Waikato, New Zealand. N Z J Mar Fresh Res 38:525–539

Nico LG, Walsh SJ (2011) Non-indigenous freshwater fishes on tropical Pacific islands: a review of eradication efforts. In: Island Invasives: eradication and management. Proceedings of the International Conference on Island Invasives, vol 97107. International Union for Conservation of Nature, Gland

Padilla DK, Williams SL (2004) Beyond ballast water: aquarium and ornamental trades as sources of invasive species in aquatic ecosystems. Front Ecol Environ 2:131–138

Parker BR, Schindler DW, Donald DB, Anderson RS (2001) The effects of stocking and removal of a non-native salmonid on the plankton of an Alpine lake. Ecosystems 4:334–345

Paul AJ, Post JR, Stelfox JD (2003) Can anglers influence the abundance of native and non-native salmonids in a stream from the Canadian Rocky Mountains? N Am J Fish Manag 23:109–119

Peterson DP, Rieman BE, Dunham JB, Fausch KD, Young MK (2008) Analysis of trade-offs between threats of invasion by non-native trout brook trout (*Salvelinus fontinalis*) and intentional isolation for native west slope cutthroat trout (*Oncorhynchus clarkii lewisi*). Can J Fish Aquat Sci 65:557–573

Pimentel D, Zuniga R, Morrison D (2005) Update on the environmental and economic costs associated with alien-invasive species in the United States. Ecol Econ 52:273–288

Ricciardi A, MacIsaac HJ (2011) Impacts of biological invasions on freshwater ecosystems. In: Fifty years of invasion ecology: the legacy of Charles Elton., pp 211–224

Rinne JN, Turner PR (1991) Reclamation and alteration as management techniques, and a review of methodology in stream renovation. In: Minckley WL, Deacon JE (eds) Battle against extinction: native fish management in the American West. University of Arizona Press, Tucson, pp 219–244

Roberts J, Tilzey R (eds) (1996) Controlling carp: exploring the options for Australia. CSIRO Land and Water, Canberra, 133 pp

Ruiz GM, Carlton JT, Grosholz ED, Hines AH (1997) Global invasions of marine and estuarine habitats by non-indigenous species: mechanisms, extent, and consequences. Am Zool 37:621–632

Schofield PJ, Nico LG (2007) Toxicity of 5% rotenone to non-indigenous Asian swamp eels. N Am J Fish Manag 27:453–459

Schreier TM, Dawson VK, Larson W (2008) Effectiveness of piscicides for controlling round gobies (*Neogobius melanostomus*). J Great Lakes Res 34:253–264

Simberloff D, Martin J, Genovesi P, Maris V, Wardle DA, Aronson J, Courchamp F, Galil B, García-Berthou E, Pascal M (2013) Impacts of biological invasions: what's what and the way forward. Trends Ecol Evol 28:58–66

Sorensen PW, Hoye TR (2007) A critical review of the discovery and application of a migratory pheromone in an invasive fish, the sea lamprey *Petromyzon marinus* L. J Fish Biol 71:100–114

Stauffer JR, Boltz SE, Boltz JM (1988) Cold shock susceptibility of blue tilapia from the Susquehanna River, Pennsylvania. N Am J Fish Manag 8:329–332

Syslo JM, Guy CS, Bigelow P, Doepke PD, Ertel BD, Koel TM (2011) Response of non-native lake trout (*Salvelinus namaycush*) to 15 years of harvest in Yellowstone Lake, Yellowstone National Park. Can J Fish Aquat Sci 68:2132–2145

Syslo JM, Guy CS, Cox BS (2013) Comparison of harvest scenarios for the costeffective suppression of lake trout in Swan Lake. Montana. N Am J Fish Manag 33:1079–1090

Teleki GC, Chamberlain AJ (1978) Acute effects of underwater construction blasting on fishes in Long Point Bay, Lake Erie. J Fish Res Board Can 35:1191–1198

Thompson PD, Rahel FJ (1996) Evaluation of depletion-removal electrofishing of brook trout in small Rocky Mountain streams. N Am J Fish Manag 16:332–339

Thresher RE (1996) Physical removal as an option for the control of feral carp populations. In: Roberts J, Tilzey R (eds) Controlling carp: exploring the options for Australia, Proceedings of a workshop, 22-24 October 1996, Albury. CSIRO Land and Water, Canberra, pp 58–72

Thresher RE (2008) Autocidal technology for the control of invasive fish. Fisheries 33(3):114–121
Vander Zanden MJ, Olden JD (2008) A management framework for preventing the secondary spread of aquatic invasive species. Can J Fish Aquat Sci 65:1512–1522
Varley JD, Schullery P (1995) The Yellowstone Lake crisis: confronting a lake trout invasion: a report to the Director of the National Park Service. Yellowstone Centre for Resources, National Park Service
Vinson MR, Dinger EC, Vinson DK (2010) Piscicides and invertebrates: after 70 years, does anyone really know? Fisheries 35:61–71
Vitousek PM, Mooney HA, Lubchenco J, Melillo JM (1997) Human domination of Earth's ecosystems. Science 277:494–499
Vredenburg AT (2004) Reversing introduced species effects: experimental removal of introduced fish leads to rapid recovery of a declining frog. Proc Nat Acad Sci 101:7646–7650
Willis K, Ling N (2000) Sensitivities of mosquitofish and black mudfish to a piscicide: Could rotenone be used to control mosquitofish in New Zealand wetlands? N Z J Zool 27:85–91
Wydoski RS, Wiley RW (1999) Management of undesirable fish species. In: Kohler CC, Hubert WA (eds) Inland fisheries management in North America, 2nd edn. American Fisheries Society, Bethesda, pp 403–430
Yonekura R, Kariya T, Fujii R, Kumazaki H, Saito K, Kumazaki T, Kuwada T, Hara T, Tokuhara T, Kageyama T (2007) Control of a bluegill population by angling. Nippon Suisan Gakkaishi 73:839–843

Chapter 36
Freshwater Commercial Bycatch in the Southern Marshes of Iraq: The Unexploited Aquatic Wealth and Understated Conservation Problem

Laith A. Jawad

Abstract While bycatch is well investigated in marine fisheries, relatively little is recognized about bycatch in freshwater fisheries. Yet basic data on bycatch configuration and mortality in freshwater is unobtainable, given that a few inland authorities need recording of bycatch.

Similar scarcity is also noted in the amount of researches being done on the bycatch resulted from the freshwater fisheries worldwide. As there are a large number of aquatic species inhabiting freshwater environment, the lack in researches on the bycatch is alarming. The results of studying the bycatch resulted from the marine environment fortunately can be used as a model for the freshwater habitats and can lead to quantifiable improvements in the conservation of freshwater ecosystems.

In the southern marshes of Iraq, a wide range of fisheries activities are performed. With these activities, a huge amount of bycatch is produced that is usually left to desiccate in the sun on the marsh banks. Several issues that govern the noninclusion of the different bycatch items in the food menu of Iraqi people attained from the marshes in Iraq were discussed. The problem of the freshwater bycatch in the southern marshes of Iraq is arising, and if no solution is found, an ecological disaster can happen.

Keywords Freshwater · Bycatch · Environment · Conservation · Invertebrates

36.1 Introduction

Appreciated resources can be obtained through the commercial fisheries both the marine and freshwater as these resources supply food for humans and support the economic around the globe (Raby et al. 2011). In some instances, fisheries are

L. A. Jawad (✉)
School of Environmental and Animal Sciences, Unitec Institute of Technology, Mt Albert, Auckland, New Zealand

L. A. Jawad (ed.), *Southern Iraq's Marshes*, Coastal Research Library 36,
https://doi.org/10.1007/978-3-030-66238-7_36

mishandled, ensuing in unmaintainable extreme harvesting of aquatic wealth, which cause an enduring loss of biodiversity, or modification of ecosystem construction (Hilborn et al. 2003). Consequently, the manipulation of the commercial fisheries is known as one of the main dangers to aquatic biodiversity (Agardy 2000). Recently, attention was directed toward the non-targeted aquatic animals and the impact of fisheries on these groups (Hall 1996; Cox et al. 2007). The non-targeted aquatic animals that are caught with the commercial catch are known as a bycatch (Crowder and Murawski 1998).

The bycatch in commercial fisheries is a rising worry and a vital conservation matter (Alverson et al. 1994; Hall and Mainprize 2005; Harrington et al. 2005; Kelleher 2005). The issue of the bycatch has made the fisheries not only worried with the availability of targeted organisms but also start to deliberate whether catches of non-targeted species are at maintainable quantities (Hall et al. 2000). Such interest was echoed on research-based interest, and the number of studies that take into consideration the bycatch has been increased consequently (Soykan et al. 2008; Larocque et al. 2012). There was a drawback in these investigations as they concentrated mainly on marine fisheries, dropping freshwater bycatch matters comparatively uninvestigated (Raby et al. 2011). Such disturbing event puts biodiversity of the freshwater environment in a declining position waiting for further investigations (Dudgeon et al. 2006). As with the marine environment, where bycatch shares the process of the degradation of marine habitats (Crowder and Murawski 1998; Hall et al. 2000; Lewison et al. 2004), bycatch is possible to have similar influences on freshwater ecosystems. Consequently, there is a necessity to control the extent and significances of bycatch in freshwater commercial fisheries. It is important to note here that when the animal (e.g., fish, birds, reptiles, mammals, invertebrates) was released alive from a bycatch fishing gear, it is usually experiencing late mortality or sublethal injuries (e.g., behavioral changes or injury), which can decrease its viability and health over the long term (Davis 2002).

In Iraq, the fisheries regulations are not enforced although there are laws approved by the government to preserve the natural life. In addition to the silence that the government gives to the continuous breaching of the laws of fisheries and the conserving of the nature in Iraq, the issue of the bycatch forms a separate problem that the government in Iraq needs to introduce specific laws about conserving this aquatic wealth before the time is too late and most of the aquatic biodiversity vanish.

In the southern marshes of Iraq, there are a wealth of aquatic biodiversity that include a wide range of animal and plant species. Some of the animal species are considered commercial as they are included in the diet of the Iraqi people, while other biota, which contain a large number of species belonging to several groups, are not favored as food.

The noncommercial aquatic taxa, at least on the Iraqi people's standard, are usually caught by the daily commercial fishing events in the rivers and in the marsh areas in different parts of Iraq. A discard for these animals usually happened for all the bycatch represented by these animal groups whatever the amount of catch was.

In the present chapter, an overview about the bycatch in the freshwater fisheries is given, and a highlight on the bycatch generated from the commercial fishing activities in the marsh area is delivered at the end of the chapter. The information given in this study is considered as a baseline study for further investigations in the future to evaluate the status of the bycatch resulted from the freshwater fisheries in the marsh area and other freshwater bodies in Iraq. The policy makers in Iraq should consider the information given in this study when they plan to develop the fisheries industry in the inland water in Iraq.

36.2 Significant Contribution of the Bycatch in the Freshwater Fisheries

As with marine fisheries, bycatch in freshwater commercial fisheries has shown a significant input to local economics (Raby et al. 2011). In addition, the inland fisheries bycatch has also conjoined with different levels of impacts on the (Crowder and Murawski 1998), but unfortunately the scientific investigations have ignored such fact (Raby et al. 2011). In all parts of the world where there are freshwater environments, commercial fisheries are present. Such trade is ancient and evolves with the social evolution of humans on this planet. Among the largest freshwater commercial fisheries are found in the Amazon River in South America (McDaniel 1997), in the Great Lakes of North America (Johnson et al. 2004), in Africa at lakes Victoria and Malawi in Kenya (Preikshot et al. 1998), in Asia at the Yangtze and Mekong rivers in China (Allan et al. 2005; Turvey et al. 2007), and in minor inland waters (e.g., Lowry et al. 2005; Siira et al. 2006; Scholten and Bettoli 2007; Fratto et al. 2008; Hyvarinen et al. 2008). In Iraq, inland fisheries are performed in the two large rivers, Euphrates and Tigris, and their tributaries, in Shatt al-Arab River and its two tributaries, and in the southern marshes. The latter area represents one of the most economic productivities regarding the fisheries trade.

Allan et al. (2005) proclaimed that worldwide, inland fisheries are being desecrated. If this statement is effective, then it is reasonable to propose that bycatch of fish, bird, mammal, and reptile species may be ordinary, given the indication from marine fisheries (Soykan et al. 2008). Though the global bycatch of marine commercial fisheries has been assessed (Alverson et al. 1994; Kelleher 2005), no such approximations have been given for freshwater fisheries. In the absenteeism of assessments for freshwater, we can use goal catch as a gauge of the possible scale of bycatch, since this gauge catch is positively linked with bycatch in most fisheries (Kelleher 2005). The enormous amount of the world's known freshwater commercial catch happens in the developing world; only 5% of the global catch takes place by commercial fishers in industrialized economies, where the focus has moved to recreational fishing (Arlinghaus et al. 2002; Allan et al. 2005, FAO 2009). Nevertheless, more than 10 million metric tons of freshwater commercial catch was stated internationally for 2006, a 12.8% upsurge from each of the previous 2 years (FAO 2009).

It has been documented that Asian countries experience the highest percentage for inland fisheries (66.9% of the total global capture) and then African states (23.5%), followed by the Americas (5.9%), Europe (3.5%), and Oceania (0.2%) (FAO 2009). These figures represent only those given by countries and published by the FAO, but still there are some regions in the world that achieve a high level of inland fisheries, but their records were not reported in any of the international fisheries organizations (Allan et al. 2005). As to the bycatch, in some developing countries such as Iraq, the bycatch mortality reaches 100% as the catch is left outside water instead of being rejected alive. This means that many of these countries may not produce bycatch in a legal sense (i.e., there is no regulation against saving the bycatch for a personal usage on any or all species caught). Nevertheless, fish species vary widely in economic value, and some species can always be regarded as bycatch. Moreover, small-sized individuals of the aimed species, as well as mammals, birds, and reptiles, may have no market or artisanal value and can be considered bycatch.

36.3 Research Chances for Freshwater Commercial Bycatch

Conservation schemes need to be implemented quickly and in a cheap cost resolution in the freshwater environment (Raby et al. 2011). Luckily, numerous bycatch difficulties can be resolved using knowledge of what bycatch is happening where and when and an appreciation of the behavior, spatial ecology, and life histories of the species at risk.

In order to put a conservation plan for the inland fisheries bycatch, information regarding the distribution of bycatch needs to be gathered by observer programs, which have been sometimes critical in recognizing bycatch species and rates in the marine area (Hall 1996). Applying the observer program in the developing countries might not be the correct choice as inland fisheries performed on a small scale (Raby et al. 2011). Therefore, the option is to perform a self-reporting of bycatch rates, but such system might experience some difficulties as the inland fisheries in the developing countries are unregulated, which can render the process of recording the bycatch obscure. However, defining bycatch rates is a clear initial point for the identification of possible problems that need a study and administration action. In artisanal fisheries in developing states, it is probable that most of the bycatch is eaten somewhat than rejected alive, but there is little quantifiable data to support that possibility (Raby et al. 2011). Therefore, observer-reporting programs are an important first step to enumerating bycatch and should be applied in inland commercial fisheries where species of that raising alarm may be impacted by the fishery. There are two main goals that such programs can attain: (1) recognize bycatch problems of worry and (2) gather both quantitative and qualitative information on those matters.

As soon as this basic step starts to gather data, simple resolutions that decrease bycatch without influencing target catch should be required (Raby et al. 2011).

The other type of research that needs to be included in the investigation process to reduce bycatch in the inland waters is to develop the selectivity of the gear being used (e.g., Barlow and Cameron 2003). This research area characterizes a second source of effort for bycatch administration. While facing and then escaping fishing gear can be harmful to animals, it should be deliberated better to the requirement that fishers pull their gear and release the animals, a procedure that has the possibility to intensify impact and injury among discards (Raby et al. 2011). A few of the freshwater commercial fisheries bycatch investigations until now deliver examples of the research chances in this area. With a grasp of the morphology and behavior of non-targeted species, which raise alarm, alterations to current net designs can be attained (Lowry et al. 2005; Gessner and Arndt 2006). Several types of fishing gear used in commercial fisheries in marine systems are matching those used in freshwater fisheries; consequently, most of the solutions obtained for marine fisheries should be applicable to freshwater.

The last issue in the freshwater bycatch strategy, but not the least, should be to grow a consideration of the destiny of discards and animals that have runaway fishing gear. Among the adaptive administration procedures for bycatch programs, examining discards offers the chance to govern whether specific gear alterations or handling methods recover the attainability of populations. By monitoring environmental factors and gear types and handling practices, researchers can recognize the issues affecting discard destiny and implement that information to fishing performs (e.g., Olla et al. 1998). Knowing discard fortune for marine species so far has been largely restricted to investigating direct mortality in the lab (Davis 2002). Due to their smaller size, freshwater environments offer an original chance to attain a thorough consideration of the destiny of discarded animals through the usage of biotelemetry (Cooke et al. 2004). Makinen et al. (2000) and Fox et al. (2002) were among the pioneer researchers to use biotelemetry in the freshwater habitats.

Several studies have targeted the bycatch resulted from the inland fisheries in the developing countries, and most of them have reached the following two conclusions: (1) there is sufficient chance for research on freshwater bycatch in developed countries, and (2) commercial bycatch is particularly poorly investigated in the inland waters of these countries. Therefore, there is an urgent need to upsurge the determinations to study the problem of the bycatch resulted from the inland fisheries in the developing countries, which require collaborations between local governments and NGOs to implement such reforming program.

36.4 What Aquatic Animals Fall in the Category of Bycatch and Usually Caught in the Southern Marshes of Iraq

In the inland waters of Iraq and especially in the southern marshes, fisheries activities are very active. The results of such activities are in the form of large freshwater fish species landings that support human populations in the southern part of Iraq.

Through the fishing activities in the southern marshes, several fishing gears and methods are used. The artisanal fishing gears are the predominant followed by other illegal methods and gears such fishing using bombs and poison. Even the artisanal fishing gears have included some illegalities that can be seen in the use of an illegal mesh size.

In all these methods, bycatch is present, and its rate increases with the different types of fishing gear. The rate is very high in both illegal methods, the fishing with bomb and poison. In those two methods, a large number of non-targeted species and small-sized specimens of targeted species fall as a bycatch that left to die on the marsh bank instead of returning it alive to the water.

In addition to the fishing gears used, there are two social issues that affect very much the rate of the bycatch. These are:

1. The religious issue: Nearly all the southern marsh inhabitants belong to the Shia Muslim sector. The people of this sector do not eat and/or even touch the species belonging to the catfish groups, i.e., members of the order Siluriformes. The silurid fish species are very abundant in the marsh area and represent 30% of fishing yield performed by any fishing gear. Individuals of silurid fish species have seen desiccating in the sun on the banks of the marsh as the fishermen through them after removing them from their nets.
2. The specificity of the food habits of the Iraqi people: The range of the freshwater fisheries products is very wide in Iraq, and there are several items that can be offered as food to the consumers, but due to the narrow selectivity of the Iraqi people in choosing freshwater food items, a large number of items are rendered useless and discarded once they are caught. Among the groups that Iraqi people never think to include them among their food items are the different species of mollusks, crustaceans other than shrimps, certain species of insects, small-sized fish species, and other invertebrate species. In other parts of the world, mollusks and crustaceans of any species are considered among the delicacy and can be sold for high prices.

   An education programs are needed for the people of the marshes of Iraq to learn about two main things: first, they need to be aware about the environment and the biodiversity issues and to discard alive whatever animals they caught and that do not fall in the list of their preferable food list, and, second, if they continue to catch reasonable quantities of those unpreferred food items, they can be delivered to a collection station(s) that can take their catch and process as freshwater food item products for export.

Until such understanding of the environment awareness for the Marsh Arabs, the rate of the freshwater bycatch can upsurge and probably reaches to a point where certain animal groups might be vanished.

References

Agardy T (2000) Effects of fisheries on marine ecosystems: a conservationist's perspective. ICES J Marine Sci 57:761–765

Allan JD, Abell R, Hogan Z, Revenga C, Taylor BW, Welcomme RL, Winemiller K (2005) Overfishing of inland waters. Bioscience 55:1041–1051

Alverson DL, Freeberg MH, Murawski SA, Pope JG (1994) A global assessment of fisheries bycatch and discards. United Nations Food and Agriculture Organization. Fisheries Technical Paper no. 339

Arlinghaus R, Mehner T, Cowx IG (2002) Reconciling traditional inland fisheries management and sustainability in industrialized countries, with emphasis on Europe. Fish Fish 3:261–316

Barlow J, Cameron GA (2003) Field experiments show that acoustic pingers reduce marine mammal bycatch in the California drift gill net fishery. Mar Mamm Sci 19:265–283

Cooke SJ, Hinch SG, Wikelski M, Andrews RD, Kuchel LJ, Wolcott TG, Butler PJ (2004) Biotelemetry: a mechanistic approach to ecology. Trends Ecol Evol 19:334–343

Cox TM, Lewison RL, Zydelis R, Crowder LB, Safina C, Read AJ (2007) Comparing effectiveness of experimental and implemented bycatch reduction measures: the ideal and the real. Conserv Biol 21:1155–1164

Crowder LB, Murawski SA (1998) Fisheries bycatch: implications for management. Fisheries 23:8–17

Davis MW (2002) Key principles for understanding fish bycatch discard mortality. Can J Fish Aquat Sci 59:1834–1843

Dudgeon D et al (2006) Freshwater biodiversity: importance, threats, status and conservation challenges. Biol Rev 81:163–182

Food and Agriculture Organization (2009) The State of World Fisheries and Aquaculture 2008. In: FAO Fisheries and Aquaculture Department

Fox DA, Hightower JE, Parauka FM (2002) Estuarine and nearshore marine habitat use by Gulf sturgeon from the Choctawhatchee River system, Florida. Am Fish Soc Symp 28:111–126

Fratto ZW, Barko VA, Pitts PR, Sheriff SL, Briggler JT, Sullivan KP, McKeage BL, Johnson TR (2008) Evaluation of turtle exclusion and escapement devices for hoop-nets. J Wildlife Manag 72:1628–1633

Gessner J, Arndt GM (2006) Modification of gill nets to minimize by-catch of sturgeons. J Appl Ichthyol 22:166–171

Hall MA (1996) On bycatches. Rev Fish Biol & Fisher 6:319–352

Hall SJ, Mainprize BM (2005) Managing bycatch and discards: how much progress are we making and how can we do better? Fish Fish 6:134–155

Hall MA, Alverson DL, Metuzals KI (2000) By-catch: problems and solutions. Mar Pollut Bull 41:204–219

Harrington JM, Myers RA, Rosenberg AA (2005) Wasted fishery resources: discarded by-catch in the USA. Fish Fish 6:350–361

Hilborn R, Branch TA, Ernst B, Magnusson A, Minte-Vera CV, Scheuerell MD, Valero JL (2003) State of the world's fisheries. Annu Rev Env Resour 28:359–399

Hyvarinen P, Leppaniemi V, Johansson K, Korhonen P, Suuronen P (2008) Stress and survival of small pike-perch *Sander lucioperca* (L.) after trawling and chilling. J Fish Biol 72:2677–2688

Johnson JE, Ebener MP, Gebhardt K, Bergstedt R (2004) Comparison of catch and lake trout bycatch in commercial trap nets and gill nets targeting lake whitefish in Northern Lake Huron. Michigan Department of Natural Resources. Fisheries Research Report 2071, Lansing

Kelleher K (2005) Discards in the World's Marine Fisheries: an update. Food and Agricultural Organization. Fisheries Technical Paper no. 470

Larocque SM, Colotelo AH, Cooke SJ, Blouin-Demers G, Haxton T, Smokorowski KE (2012) Seasonal patterns in bycatch composition and mortality associated with a freshwater hoop net fishery. Animal Conserv 15(1):53–60

Lewison RL, Crowder LB, Read AJ, Freeman SA (2004) Understanding impacts of fisheries bycatch on marine megafauna. Trends Ecol Evol 19:598–604

Lowry MB, Pease BC, Graham K, Walfourd TR (2005) Reducing the mortality of freshwater turtles in commercial fish traps. Aquatic Conserv: Marine & Freshwater Ecosystem 15:7–21

Makinen TS, Niemela E, Moen K, Lindstrom R (2000) Behaviour of gill-net and rod-captured Atlantic salmon (*Salmo salar* L.) during upstream migration and following radio tagging. Fish Res 45:117–127

McDaniel J (1997) Communal fisheries: management in the Peruvian Amazon. Hum Organ 56:147–152

Olla BL, Davis MW, Schreck CB (1998) Temperature magnified post capture mortality in adult sablefish after simulated trawling. J Fish Biol 53:743–751

Preikshot D, Nsiku E, Pitcher T, Pauly D (1998) An interdisciplinary evaluation of the status and health of African lake fisheries using a rapid appraisal technique. J Fish Biol 53(suppl. A):381–393

Raby GD, Colotelo AH, Blouin-Demers G, Cooke SJ (2011) Freshwater commercial bycatch: an understated conservation problem. Bioscience 61(4):271–280

Scholten GD, Bettoli PW (2007) Lack of size selectivity for paddlefish captured in hobbled gillnets. Fish Res 83:355–359

Siira A, Suuronen P, Ikonen E, Erkinaro J (2006) Survival of Atlantic salmon captured in and released from a commercial trap-net: Potential for selective harvesting of stocked salmon. Fish Res 80:280–294

Soykan CU, Moore JE, Zydelis R, Crowder LB, Safina C, Lewison RL (2008) Why study bycatch? An introduction to the theme section on fisheries bycatch. Endanger Species Res 5:91–102

Turvey ST, Pitman RL, Taylor BL, Barlow J, Akamatsu T, Barrett LA, Zhao X, Reeves RR, Stewart BS, Wang K, Wei Z (2007) First human-caused extinction of a cetacean species? Biol Lett 3(5):537–540

Section X
Socio-economic Aspects

Chapter 37
Ornamental Fish Farming: A Proposal for a Successful Small-Scale Aqua Business Project in the Southern Marshes of Iraq

Laith A. Jawad

Abstract The southern marshes of Iraq comprise a huge potential for ornamental fish industry on both small scales run by locals and on larger scales run by investors or government. Although there are a few ornamental fish species inhabiting the marsh areas of Iraq, such presence reflects the suitability of both the water quality and the climate for breeding further ornamental fish species in the frame of monitored ornamental fish industry. In the present chapter, a review of the idea of establishing ornamental fish industry centre on a large scale financed by investors or government or several small breeding centres run by families of the locals is discussed. In case any of the options was chosen, certain steps of keeping the sustainability of the project and the health of the environment should be considered.

Keywords Fishes · Environment · Socio-economic · Lifestyle · Aquatic management

37.1 Introduction

Ornamental fish is one of the imperative articles among the numerous types of commercially significant fishes sold nationally and internationally (Mandal et al. 2007). Ornamental fishes, commonly known as "aquarium fish", have a great commercial market in some countries like India, where these fishes are shipped to 27 countries, which reached to 2568 Mt. (0.86% of the total marine export) in terms of quantity and US$14 million (0.50% of total marine export) in terms of value (MPEDA 2007).

Ornamental fish raising turned out to be as prevalent as an easy and stress-releasing hobby. For example, about 7.2 million houses in the USA and 3.2 million in the European Union have an aquarium, and the number is cumulative continuously globally, and accordingly, ornamental fish farming is also mounting to attend

L. A. Jawad (✉)
School of Environmental and Animal Sciences, Unitec Institute of Technology, Mt Albert, Auckland, New Zealand

L. A. Jawad (ed.), *Southern Iraq's Marshes*, Coastal Research Library 36,
https://doi.org/10.1007/978-3-030-66238-7_37

this request. Asia is the largest spot for export of ornamental fish species in the world, but the USA, Europe and Japan are the main importers of these fishes.

Iraq is among the countries that own both suitable environment and climate to start this type of industry and can be considered as one of the main national revenues as it did with other countries in Asia. The nominated area for developing such industry in Iraq is the southern marshes, where the amount and quality of water are available and cheap labour and easy distribution make the marsh area suitable for ornamental fish culture.

The present chapter reviews the idea of establishing centres across the southern marshes of Iraq for ornamental fish culture. Such an idea is cheap to implement, but it is profitable otherwise. This proposal will be available for policy makers in Iraq to adopt and probably for the private sectors as it is easy and cheap to implement.

37.2 The Ornamental Fish

The ornamental fish species are diverse fishes with a variety of shapes and colours, which attracted the people for a long time. These fishes have spectacular colours and other features that make them suitable to keep as pets in houses (Mahapatra 2018).

Ornamental fish is a general term that includes species designated as ornamental species as they are sold in aquarium's shops as pets and other species, which are not included among the ornamental fish species, but they are usually cultured in aquarium due to their small size and the success in their culture in small confinements (Mahapatra 2018).

The reasons behind calling the fish species ornamental fishes owe to their colourful and shape diversification and being able to breed in the aquarium (e.g. *Tetraodon cutcutia*, *Colisa lalia*), stripes and banding pattern (e.g. *Botia dario*, *Botia striata*), attractive appearance (e.g. *Notopterus chitala*), transparent body (e.g. *Pseudambassis baculis*, *P. ranga*) and beautiful jumping behaviour (e.g. *Esomus danricus*).

37.3 Diversified Ornamental Fish Group

In the world of ornamental fish species, there arise several species that are well known for breeders and hobbyists. In the following, a short review is given for the most well-known species of ornamental fishes that on deal between breeders and hobbyists.

37.3.1 Catfish Group

Species of this group can be freshwater or marine and distinguished in having whisker-like barbels around the mouth, and they are mainly bottom feeders. They belong to the order Siluriformes, which comprises a large number of species inhabiting freshwater and marine habitats globally (Mahapatra 2018).

Examples on species of this group are *Clarias batrachus*, *Heteropneustes fossilis*, *Pangasius pangasius*, *Ompok bimaculatus*, *O. pabda*, *O. malabaricus*, *Wallago attu*, *Mystus seenghala*, *M. aor*, *Gagata gagata*, *G. cenia*, *Hara serratus*, *H. jerdoni*, *H. hara* and *Olyra longicaudata*.

37.3.2 The Barbs

A barb is one of numerous ray-finned fish species belonging to the genus *Barbus* and *Puntius* of the family Cyprinidae. Those species belonging to the subfamily Barbinae have barbels around their mouth (Mahapatra 2018).

Famous examples of this group are *Puntius sophore*, *P. stoliczkanus*, *P. guganio*, *P. chola*, *P. sarana sarana*, *P. shalynius*, *P. terio*, *P. gelius*, *P. conchonius*, *P. ticto*, *P. bizonatus*, *Rasbora daniconius* and *R. rasbora*.

37.3.3 Danionin

The genus *Danio* comprises many of the species that are acquainted by aquarists. The common name "danio" is used for members of the genera *Danio* and *Devario*. The name "danio" originates from the Bangla name "dhani", meaning "of the gouramis or gouramies", which are a group of freshwater perciform fishes that encompass the family Osphronemidae (Mahapatra 2018). The native geographical distribution of these species is in Asia, from Pakistan and India to the Malay Archipelago and north-easterly towards Korea. The name gourami, of Japanese origin, is also used to call species of the family Anabantidae (Mahapatra 2018).

Among the known species of this group are *Brachydanio rerio*, *B. acuticephala*, *Devario devario*, *D. aequipinnatus*, *Danio regina*, *D. dangila*, *D. naganensis*, *D. yuensis*, *Colisa sota*, *C. labiosus*, *Trichogaster fasciata*, *T. lalius*, *T. chuna* and *Ctenops nobilis*.

37.3.4 Loach

The members of this group are energetic and living near the bottom as they are scavengers in their feeding habit. Therefore, they are suitable for community aquarium (Mahapatra 2018). This group includes some well-known species that are usually found in the aquaria and pet shops. They are recognised in having beautiful colours, and they are semi-aggressive in their behaviour especially when they are placed in a larger group of the same species.

Examples on species of this group are *Balitora brucei*, *Acanthocobitis botia*, *A. pavonaceous*, *Schistura manipurensis*, *S. scaturigina*, *S. barapaniensis*, *S. sikmaiensis*, *Botia berdmorei*, *B. dario*, *B. rostrata*, *B. histrionica*, *Lepidocephalichthys guntea* and *L. annandalei.*

37.3.5 Cichlid

Members of this group are belonging to the family Cichlidae of the order Perciformes. These species are territorials and have aggressive habits (Mahapatra 2018). Therefore, caution needs to be taken when establishing aquarium and when selecting species to include in.

There are many species of this group that fall under ornamental fish species designation, and *Etroplus suratensis* is one of them.

37.3.6 Eels

Species of this group is belonging to the order Anguilliformes which contains 4 suborders, 20 families, 111 genera and about 800 species (Mahapatra 2018). The majority of eels are predators. The term "eel" can be referred to different species of fish that has a snake-like body shape such as electric eels and spiny eels, but these are not members of the Anguilliformes order.

Examples on species of this group are *Anguilla bengalensis*, *A. bicolor*, *Monopterus cuchia*, *Macrognathus pancalus*, *M. morehensis*, *M. aral*, *Mastacembelus armatus*, *M. alboguttatus*, *Chaudhuria khajurial* and *C. indica*.

37.3.7 Snakehead

The species belonging to this group are of the family Channidae, a residence of parts of Africa and Asia. They take free air through their gills while breathing, which permits them to travel short distances over land (Mahapatra 2018). The 2 existing genera are *Channa* in Asia and *Parachanna* in Africa, containing about 35 species.

Among the well-known species of this group are Channa punctatus, C. striatus, C. stewartii, C. orientalis, C. marulius and *C. barca.*

37.4 The Ornamental Fish Species of the Southern Marshes of Iraq

Recently, several species of ornamental fish groups were established in the water of the southern marshes of Iraq owing to the ornamental fish industrial activity that has been active in Iraq after 2003. In this section, a short identification of these species is given to make the readers familiar with the species that might be considered to be included in the future ornamental fish industry that might be established in the marsh areas.

1. *Carassius auratus* (Linnaeus, 1758)

Goldfish
Order: Cypriniformes
Family: Cyprinidae

Habitat: Freshwater; brackish; benthopelagic (Riede 2004).

Distribution: This species is found in Asia from central Asia and China (Kottelat et al. 1993) and Japan (Kailola et al. 1993), and it has been introduced throughout the world.

Maximum length, 480 mm TL (Murdy et al. 1997); common length, 100 cm TL (Frimodt 1995).

Ecology and Biology: Individuals of this species live in rivers, lakes, ponds and ditches (Man and Hodgkiss 1981; Etnier and Starnes 1993), with still or slow-flowing water (Billard 1997). They are found in eutrophic waters, in rich vegetated ponds and canals (Kottelat and Freyhof 2007). Plankton and benthic invertebrates are the main food items preferred by this species (Kottelat and Freyhof 2007). The eggs of this species are placed on submerged vegetation, while females spawn several times during the year (Teletchea et al. 2009). Individuals of this species follow the oviparous method of laying eggs, with pelagic larvae. They are considered as ornamental fish for ponds (for big-sized individuals) and aquaria (for small-sized individuals). No more than five individuals should be placed in one aquarium of 1000 mm (aquarium keeping: in groups of five or more individuals; minimum aquarium size 100 cm) (BMELF 1999).

2. *Poecilia latipinna* (Lesueur, 1821)

Sailfin molly
Order: Cyprinodontiformes
Family: Poeciliidae

Habitat: Marine; freshwater; brackish; benthopelagic (Allen et al. 2002).

Distribution: North America from Cape Fear drainage in North Carolina, USA, to Veracruz, Mexico. This species has been transferred to several countries.

Ecology and Biology: Individuals of this species reach maximum length of 150 mm TL (Page and Burr 1991), with common length of 34 mm TL (Hugg 1996). This species prefers areas with still or gently flowing warm water of small creeks and drains (Allen et al. 2002). It is rich in vegetation (Page and Burr 1991) and also in coastal waters (Robins and Ray 1986). Individuals of this species have algae as a main food item, and also contain in its food are rotifers, small crustaceans (such as copepods and ostracods) and aquatic insects (Hassan-Williams et al. 2007).

3. *Gambusia holbrooki* (Girard, 1859)

Eastern mosquitofish
Order: Cyprinodontiformes
Family: Poeciliidae

Habitat: Freshwater; brackish; benthopelagic (Riede 2004).

Ecology and Biology: Individuals of this species reach a maximum length of 47 mm TL (Tarkan et al. 2006). Adults of this species prefer areas of standing to slow-flowing water, mainly found in vegetated ponds and lakes, backwaters and quiet pools of steams (Page and Burr 1991). This species feeds on small terrestrial insects usually in the drift and among aquatic plants, vigorously picking very small prey (Arthington 1989), and is approved as mosquito larvae eradicator (Gabrielyan 2001). Individuals of this species have been distributed worldwide in an aim to combat malaria by eradicating the larvae of the mosquito that transmit malaria parasite (Kottelat and Freyhof 2007; Allen et al. 2002). Individuals of this species reach maturity at 4–6 weeks of their lives, with 3 generations can be produced in 1 year. Gestation lasts 3–4 weeks and the number of young can reach up to 354, but commonly it is about 40–60 (Riehl and Baensch 1991).

Although the number of the ornamental fish species is very few in comparison with the other Asian countries, it is clear that the environment of the southern marshes of Iraq is suitable for breeding more species of ornamental fishes in a commercial level in the future.

37.5 Livelihood Opportunities Through Ornamental Fisheries

The job and hobby of breeding ornamental fish are attracting acceptance in different Asian countries including Iraq especially after 2003 when the import was open for the private sector to import ornamental fish species from around the world. In Iraq and as in other countries, the percentages of houses that own an aquarium have increased significantly, but there is no survey showing such an increase. The trend of increase in the number of houses that own an aquarium can be noticed through the sale of the aquarium shops in the main cities in Iraq. Central Baghdad City is the main area for ornamental fish trade as in that area, there are dealers that import fishes from around the world and traders from other cities usually visit this area in Baghdad to buy their stock to sell it in their shops that are distributed in cities around the country. Therefore, the chance to attain a job in the ornamental fish industry such as production, marketing and exports is moderately high especially in Baghdad City, the capital. Breeding for commercial purposes of ornamental fish species on a small scale is present in all large cities in Iraq, where families perform this job using small glass or plastic aquaria or even small earth ponds. It looks that those families are happy of their additional business as they increased their income nearly by ⅓ of what they usually earn from other businesses.

37.6 Main Daily Jobs That the Ornamental Fish Industry Requires in Case of the Establishment in the Southern Marshes of Iraq

In case of establishing ornamental fish breeding in the southern marshes of Iraq, there will be several different jobs that people can work in. These jobs definitely will absorb most of the unemployed personnel available in the area. At the same time, the ornamental fish industry even on small scales will bring an additional income to the families of the Marsh Arabs. The different jobs are:

1. *Wild caught*: This type of job can be divided into two parts: first, the workers will be involved in collecting fish specimen alive from the wild and bring them to the breeding centre; second, they need to get in contact with aquarium shops in the neighbouring cities to bring adults of species that are not found in the marsh environment for breeding.
2. *Rearing*: To master the profession of rearing the ornamental fish species, a special training programme should be set for those who would like to establish such a business. Usually, this type of job is suitable for young people men or women. The rearing of ornamental fish can raise a quite good income as related to collection of fish from wild catch owing to their well quality and lower risk of death during transportation.
3. *Live feed collection, culture and trade*: This job is about providing live food for the cultured ornamental fishes. Either this food is collected from the environment such as different kinds of crustaceans, insect larvae and worms or these organisms are cultured in small aquaria and made available for the cultured fish species. This job is important as the live food is vital for larval stages of the ornamental fishes. People who will do this job should have the experience and good training in how and where to collect the live food.
4. *Artificial feed manufacture and trade*: Prepared food has a similar value as the live food. Therefore, separate business can be set up in the vicinity of the marsh area to manufacture the artificial food from ingredients obtained locally. The

manufacture of such food does not need heavy or advanced machinery, but on the other hand, household equipment will be suitable for such a project. The artificial food can include some items that did not require any machinery such as sun-dried small crustaceans like shrimps or sun-dried worms. Producing such food can be exported to the larger cities to be used by shop aquaria and hobbyists keeping aquaria at their house.

5. *Aquarium plant breeding and trade*: The aquatic plants are an important constituent in setting up an aquarium, and the marsh area is the best place for this type of business. In the marshes, there are a large number of aquatic plant species that can be easily obtained and grow in a small aquarium or even in earthed ponds. The centres that will take this business will support not only the ornamental fish breeding centres at the marsh areas but the whole country wherever fish aquaria are set up. The aquatic plants are important for the fish aquaria as they provide a natural habitat and environment inside the aquarium.
6. *Aquarium and accessories maker:* The business of making aquarium accessories is another side job that ornamental fish trade relies on and needs it. There are several things that housewives and elderly can produce while they are at home. Among these are small fishnets, gravel vacuum, air-stones, decorative wooden pieces, colourful stones, etc.
7. *Ornamental fish export*: This job will come last, and in case of covering the local markets with ornamental fish species and there is a surplus of stock, then it is feasible to get in contact with traders in the neighbouring countries to export such a stock.

37.7 Aspects of Maintainability and Living Security

In order to secure the business of the ornamental fish species, the following aspects are required to be available:

A. The breeding areas of the wild ornamental fish species should be identified and be secured of any vandalism.
B. The official sustenance shall be comprehensive to diagnose breeding and fishing grounds; to estimate production, fishing potential, etc.; and also to control the whole procedure of collection and marketing.
C. Only trained fishermen who have endured a particular skill progress shall be allowed to collect fish. No pesticides should be used in collecting ornamental fish from the wild, and it is recommended to use only customary gears.
D. The required number of fish and species should be cached, and any bycatch should be released live into the water.
E. Fish collected from the wild should pass through a period of quarantine to reveal any diseases before using them as a stock for breeding.
F. Public awareness programme should be established in order to make public that there is no fish harvesting at the stations where wild ornamental fish species are located.

G. Encouraging the breeders of the ornamental fish to include more exotic fish species as they are fetching more income.
H. Stoppage and management of aquatic pollution in the breeding grounds.
I. Biological information about the feeding and breeding of ornamental fish species is required for all staffs that are involved in the different steps of the ornamental fish business.

References

Allen GR, Midgley SH, Allen M (2002) Field guide to the freshwater fishes of Australia. Western Australian Museum, Perth, 394 p

Arthington AH (1989) Diet of *Gambusia affinis holbrooki*, *Xiphophorus helleri*, *X. maculatus*, and *Poecilia reticulata* (Pisces: Poeciliidae) in streams of southeastern Queensland, Australia. Asian Fish Sci 2:193–212

Billard R (1997) Les poissons d'eau douce des rivières de France. Identification, inventaire et répartition des 83 espèces. Delachaux & Niestlé, Lausanne, 192p

Bundesministerium für Ernährung, Landwirtschaft und Forsten (BMELF) (1999) Gutachten über Mindestanforderungen an die Haltung von Zierfischen (Süßwasser). Bundesministerium für Ernährung, Landwirtschaft und Forsten (BMELF), Bonn, Germany, 16 p

Etnier DA, Starnes WC (1993) The fishes of Tennessee. The University of Tennessee Press, Knoxville

Frimodt C (1995) Multilingual illustrated guide to the world's commercial warmwater fish. Fishing News Books, Osney Mead, Oxford, 215 p

Gabrielyan BK (2001) An annotated checklist of freshwater fishes of Armenia. Naga ICLARM Q 24(3&4):23–29

Hassan-Williams C, Bonner TH, Thomas C (photogr.) (2007) Texas freshwater fishes. Texas State University-San Marcos: Biology Department/Aquatic Station

Hugg DO (1996) MAPFISH georeferenced mapping database. Freshwater and estuarine fishes of North America. Life Science Software. Dennis O. and Steven Hugg, 1278 Turkey Point Road, Edgewater, Maryland, USA

Kailola PJ, Williams MJ, Stewart PC, Reichelt RE, McNee A, Grieve C (1993) Australian fisheries resources. Bureau of Resource Sciences, Canberra, 422p

Kottelat M, Freyhof J (2007) Handbook of European freshwater fishes. Publications Kottelat, Cornol and Freyhof, Berlin, 646 pp

Kottelat M, Whitten AJ, Kartikasari SN, Wirjoatmodjo S (1993) Freshwater fishes of Western Indonesia and Sulawesi. Periplus Editions, Hong Kong, 221 p

Mahapatra BK (2018) Ornamental fishery resources in India: diversified option for livelihood improvement. In: Mahapatra BK, Patra SB (eds) National seminar on recent trends in fishery and ecological science. Organised on 19th May 2018, pp 33–44p

Man SH, Hodgkiss IJ (1981) Hong Kong freshwater fishes. Urban council. Wishing Printing Company, Hong Kong, p 75

Mandal S, Mahapatra BK, Tripathi AK, Verma MR, Datta KK, Ngachan SV (2007) Agribusiness opportunities of ornamental fisheries in North-Eastern region of India. Agric Econ Res Rev 20(347-2016-16835):471–488

MPEDA (2007) Export performance of marine products during 2005–06. http://www.mpeda.com

Murdy EO, Birdsong RS, Musick JA (1997) Fishes of Chesapeake Bay. Smithsonian Institution Press, Washington and London, 324 p

Page LM, Burr BM (1991) A field guide to freshwater fishes of North America north of Mexico. Houghton Mifflin Company, Boston, 432 p

Riede K (2004) Global register of migratory species – from global to regional scales. Final report of the R&D-Projekt 808 05 081. Federal Agency for Nature Conservation, Bonn, 329 p

Riehl R, Baensch HA (1991) Aquarien Atlas. Band. 1. Melle: Mergus, Verlag für Natur-und Heimtierkunde, Germany, 992 p

Robins CR, Ray GC (1986) A field guide to Atlantic coast fishes of North America. Houghton Mifflin Company, Boston, 354 p

Tarkan AS, Gaygusuz O, Acipinar H, Gürsoy C, Ozulug M (2006) Length-weight relationship of fishes from the Marmara region (NW-Turkey). J Appl Ichthyol 22:271–273

Teletchea F, Fostier A, Kamler E, Gardeur J-N, Le Bail P-Y, Jalabert B, Fontaine P (2009) Comparative analysis of reproductive traits in 65 freshwater fish species: application to the domestication of new fish species. Rev Fish Biol Fish 19:403–430

Chapter 38
Migration of Youth to the Big Cities from the Southern Marshes of Iraq: Solutions to Keep Locals with an Upgrading for Their Skills and Livelihood

Laith A. Jawad

Abstract There are indications available showing that intra-rural disparity is a chief reason of rural-urban movement. Such migration has significant effects on the socio-economic aspects of the rural areas.

In the area of the southern marshes of Iraq, such movement of young individuals happened and has increased recently with the development of the technology and with the reduction in the economic aspects of the families in the marsh area. In order to stop such migration out of the marshes, some solutions need to be implemented so the young people can be attracted to such options and stay in their lands. In this chapter, several options for businesses and jobs are suggested that ensure the young individuals can remain in their homeland and in addition they can improve their income and enhance the style of their living. To achieve such goal, government has to intervene in such mega project so the young people can start their projects, which in the end the benefits of these projects will be within the national economy of the country in general.

Keywords Youth · Professions · Jobs · Technology · South of Iraq · Wetlands · National economy

38.1 Introduction

Henderson (2002) defined migration as the movement of people from one place to another across the limitations of the area that living in to settle and live in a new are. Such movement could happen between two different countries or within different places of the same country, i.e. rural to urban areas.

Economic development leads to structural transformation, and as a result, the share of agriculture sector declines while that of industrial sector increases in the

L. A. Jawad (✉)
School of Environmental and Animal Sciences, Unitec Institute of Technology, Mt Albert, Auckland, New Zealand

L. A. Jawad (ed.), *Southern Iraq's Marshes*, Coastal Research Library 36,
https://doi.org/10.1007/978-3-030-66238-7_38

country's gross domestic product (GDP). As the industrial output increases, employment opportunities at the urban centres also increase, and people begin moving from the rural subsistence agriculture sector to the urban areas in search of better employment opportunities and better living conditions (Sadaf et al. 2010). This phenomenon increase in the urban population as a consequence of the mass movement of people to urban centres is commonly referred to as urbanization. However, movement of people from the rural subsistence agricultural sector is not the only cause of urbanization; the higher population growth rate especially in the developing world is also a major reason of the growing trend of urbanization. The process of urbanization, on the one hand, provides opportunities of better standard of living, while it is also having certain negative effects such as congestion, environmental pollution, and an increase in crimes and so on (Imran et al. 2013). Urbanization is of many developing countries and Pakistan is no exception. During the last 63 years, the total population of Pakistan in general and of Pakistani cities in particular increased manifolds (Imran et al. 2013). The overall population increased by more than 52.5% during 1951 and 2010. In 1951, 82.26% of the total population lived in rural areas, whereas this figure dropped down to only 66% in 2008. The current estimated population growth rate of Pakistan is 1.513%, while the population growth rate of cities is 3% which clearly points out that rural-urban migration is nearly 2% annually (Anonymous 2011). On a smaller scale, migration from rural areas denotes the leaving of individuals or households, for more than a week or so, from the small, primarily agricultural community in which they live (Lipton 1980). Such local movement of people within the country may be designed for several aspects such as harvest, leaving for good, marriage, or study.

The resent study discusses the problem of migration of the youth from the southern marsh areas of Iraq to the cities for different purposes mainly for better work opportunities and higher income. Such movement out of the marsh areas will have several drawback effects on the socio-economic aspect of the marsh area. To stop or at least to reduce such movement out of the marsh areas, there should be some solutions that keep the young people attracted to their homeland. Some of the proposed solutions contain business ideas that can upgrade the skills of the young people and at the same time enhance their income and finally to keep them in their homeland.

38.2 Effects of Migration and Their Incentives

The expected effects of the migration of the youth out of the southern marshes areas in Iraq can be seen in the following: (a) effect on the number of worker, (b) effect on man-hour level, (c) effect in regard to efficiency-weighted unit of rural labour, and (d) effect on unit of price-weighted total factor (Imran et al. 2013).

On the other hand, there are several factors that cause motivations for the young individuals to leave their homeland (Imran et al. 2013). Among these are the following:

1. Sociocultural factors, which include social discrimination to some extent, family expansion, crime, and social Injustice
2. Political factors such as political instability, ethnic conflict, and propaganda
3. Economic factors like poverty, unemployment, low wages, land tenure issues, landlessness, depleting resources, and lack of infrastructure
4. Environmental factors such as environmental degradation, natural disasters, food security, disease, climate change, and water scarcity

38.3 Preparatory Measures Before Setting the Solutions for the Problem of the Migration of the Young People out of the Marsh Areas in Iraq

Before starting with any upgrading program for the socio-economic status of the young people in the marsh areas of Iraq, it is necessary to put plans to upgrade the infrastructure of the marsh area in order to receive the socio-economic upgrading plans later on. Among such programs to upgrade the infrastructure of the marsh areas are:

A. **Improving Roads That Connect Between Marsh Areas and the Neighbouring Cities**

There is a network of roads that the government of Iraq needs to be paved between main cities and villages and the marsh areas making accessibility to these areas from the cities easy and possible. Also, small roads that can connect those rural areas of the marshes with each other and with main area of the marshes should be built.

In addition, water taxies in the form of small boats designed to take about ten passengers need to be introduced in the marsh area to facilitate the movement of Marsh Arabs between the different parts of the marshes. To make the movement of boats possible in the marshes, small floating petrol stations should be installed in several points in the marshes, to overcome the problem of shortage of fuel.

B. **Water Purification**

Probably, the water purification step is one of the preliminary solutions for the settlement of the Marsh Arabs. Government should install stations for water purification operated by solar energy in different parts of the marsh area so the inhabitants of these areas can live healthy life. Nowadays, there are several models of water purification that come in different shapes and sizes for single or multiple family usage. Such equipment need to be subsidized by the government and offered for sale for locals of the marshes in low prices so they can afford buying them.

C. **Treatment and Removal of Human and Household Wastes**

The inhabitants of the marsh areas usually have no proper toilets to use. Instead, they use the marsh as their toilet. In this way, the human wastes will go directly to the environment of the marshes. In order to teach the marsh inhabitants the habits of using toilets, the government need to introduce the portable toilets that are equipped with solar-operated heater to provide hot water for washing. These toilets may be connected to what are called clever tanks that treat the waste and return them to the environment or may be used for agriculture businesses.

D. **Establishing Medical Centres**

The health of the inhabitants is the main goal in any socio-economic rehabilitation of the marsh areas in Iraq. To achieve such aim, small medical centres should be established in several locations inside the marshes. To make the process of building such centres easy, prefabricated or plastic constructions may be used, which can be provided with electricity generated by solar energy and water purification units.

38.4 Solution Measures to Stop Young People from Leaving the Marsh Areas in Iraq

In Sect. 38.2 of this chapter, the main motivations for migration from the marsh areas that the young people may take have been given briefly. The main jobs in the marsh areas depend on the amount of water available in the marshes, where the availability of water will sustain the obtainability of fish and other water biota, which are the basic food for the inhabitants of the marshes. Also, water areas will attract more birds to visit the marshes for wintering, which again can be a source of food for the people of the marshes. Moreover, the water buffaloes that the Marsh Arabs usually bred are using water during their life. Water buffaloes are the source of milk that the Marsh Arabs can use for several dairy products. Therefore, if water level decreases in the marsh area as one of the factors behind the migration of the young people, all the economic life in the marsh area will be reduced.

In this section, several options will be discussed regarding upgrading the skills of the young people and involving them in projects that they can earn better income and enhance their livelihood. The solutions cover both the environment and the Marsh Arab inhabitants.

A. **Enhancing the Environment of the Marsh Areas**

There are several ways of enhancing the environment of the marsh area so they will continue to be suitable for living of the Marsh Arabs in on its resources. Among these ideas that might be implemented in the area are the following:

1. Planting date palms and other trees in all the dry lands of the marsh area, so they will be a source of food and possible agriculture business for the inhabitants
2. Increasing the green coverage of the edges of the marsh areas with different kinds of indigenous aquatic plants so they will form suitable habitats for different small aquatic animals, which in turn will become food source for the bigger animals such as fish, amphibian, and reptiles

3. Increasing the plantation of both reeds and papyrus in those areas that lack these plants. These two species of plants are considered among the sources that the Marsh Arabs based most of their livelihood on in making different handcrafts and as a food source of the buffaloes and cows

B. **Establishing Some Projects as a Solution to Stop the Migration of the Young People from the Marsh Areas in Iraq**

As it has been mentioned earlier in this chapter about the effects of the migration of the young people from their homeland, the marsh areas to the cities and other urban regions, in this section, suggestions for some projects will be proposed that the young people can involve in to upgrade learning and upgrade their skill and at the same time to increase their income. Such projects will have a direct benefits to the marsh area inhabitants first and then the whole region in the second place. These project ideas are:

1. **Cage Fish Farming**

Cage fish farming is one of the fast-mounting sectors in aquaculture and usually maintained unceasingly by the government. Around the world, this type of aquaculture industry usually makes a major contribution to provide the increasing request for animal protein due to the human population increase (Iliyasu et al. 2015).

This method of culturing fish is easy to handle and look after especially if the species chosen are among those species that have wide adaptability to the marsh environment. This project contains two parts: the first part is training, where the young people can attend session arranged by the government to teach them the basics of the fish culture. The second part starts at the end of the teaching period; government can supply those who attend the courses with all the required equipment to set their projects in the areas near their living areas so they do not need to travel long way to look after the cages. Also, the government can supply the attendees of the course with the fingerlings of the species of fish that are suitable to be cultured in the marsh environment. The cost of the equipment supplied by the government can be returned after one of establishment of the project and once the project starts to give a steady income.

2. **Prawn Farming**

Freshwater prawn farming is currently one of the most significant subdivisions of the national economy of many countries, and during the last two decades, its development has attracted considerable attention because of its export potential (Ahmed et al. 2008). In Bangladesh, for example, the prawn and shrimp sector as a whole is the second largest export industry making US$380 million annually and 5.6% of the total value of exports (Department of Fisheries (DOF) 2006). There are 1.2 million people employed in prawn and shrimp production, and a further 4.8 million household members are linked with the sector (USAID 2006).

This project is similar to that of cage fish farming, but the technique is different. The basic ideas are the same and the government approaches are the same in providing the training and assisting in setting up the project. In both the fish cage culture and the prawn culture projects, government can assist in marketing the yield outside the marsh area. In this project, several species can be cultured such as those suitable for human consumption and the small-sized species that can be used as fodder for the chicken and other animals.

3. **Non-commercial Fish Species Processing**

In the marsh areas, there are many fish and other aquatic animals that are not considered as food items for the Marsh Arabs or the Iraqi people in general. Among these are the catfish species, where there are more than three species that are found in high abundances. Among the invertebrates are the different species of molluscs and still to add the several species of amphibians that are found in abundance in the marsh area.

The idea is to establish centres that can culture these aquatic animals and process their meat and make available for international markets. Probably this idea looks strange at first, but investigating the aquatic resource markets especially in Southeast Asia and India and Japan, such food items are considered among the delicacy that can fetch high income.

The role of the government here is to give the training for the young people, to build small factories for processing the aquatic resources, and to assist the projects owner in marketing their products.

4. **Establishment of Methane Gas Production**

Alternative energy supplies are being investigated through the world to win the battle of energy calamity (Yasin and Wasim 2011). Among numerous substitutes, methane production from animal waste (biogas) is one of the significant sources of energy (Yasin and Wasim 2011). In rural regions of developing countries, energy supplies are partly met by burning animal dung as cooking fuel due to shortage of wood and other fuel. In order to reduce the level of pollution created by burning animal wastes as fuel, the crop, animal, and poultry wastes can be used as feeding materials for the biogas plants (Ahmad 1995; Yasin and Wasim 2011). The material can aid double drive by producing energy and offer valued crop manure with increased nitrogen due to anaerobic fermentation.

The biogas plants of various types and sizes were installed in several Asian countries such as Pakistan, and there were some mistakes that made this project inefficient (Chaudhry and Farooq 1984), but later such problems were resolved and enhanced the enactment of the plant (Ghaffar et al. 1986).

With the presence of large number of buffaloes and cattle in the marsh area, the amount of waste produced by these cattle is high. At the moment, the Marsh Arabs dry the buffalo's dung used as fuel. Simple building to produce methane gas from the animal waste can be established in certain centres distributed in the marsh area. These centres can receive amounts of animal waste for processing to generate methane gas. This gas can be pressed in cylinders and sold locally or even in the big cities in the neighbourhood of the marsh area.

5. **Dairy Products Factories**

Dairy, fisheries, and forestry are other constituents of agriculture with countless unfamiliar prospective (Ghosh and Maharjan 2001). So there is a prerequisite to consider them among the different agriculture activities. Dairy farming can be the feasible substitute to improve the economic settings of the farmers. For example, the Bangladesh Bureau of Statistics (1994) has revealed that a very high percentage of cattle (50.9%) are possessed by small farmers as compared to the medium farmers (37.3%) and large farmers (10.2%). In this setting, dairy farming is capable to bring the security to the enormous mainstream of the poor societies (Ghosh and Maharjan 2001). Dairy farming is a labour-intensive creative work, which can produce employment chances for the rural poor, and this is one of the main purposes of rural development. Dairy offers a feasible secondary job for the jobless rural poor so as to increase their income earning aptitudes (Ghosh and Maharjan 2001). Therefore, any methodical step to dairy progress can be considered as main event in enhancing rural economy of the developing countries (Ghosh and Maharjan 2001).

This is the most promising project that can be built in the vicinity of the marsh area. The large number of buffaloes and cattle in the marsh area makes this area very rich in producing milk, which in many places is covering more than the actual local needs.

The idea of the project is to establish a number of small dairy factories at the edges of the marsh area designed to process milk of buffaloes, sheep, and goat. Inside the marsh areas, small centres should be established responsible to collect milk from the locals and deliver it to the dairy factories, which in turn can process it and produce different kinds of dairy products such as cheese, yoghurt, and cream. The young people can involve in both the factory business and the milk collecting centres.

6. **Poultry Farming**

Rural poultry farming using native breeds has been in usage in many developing and underdeveloped countries throughout the world (Guèye 1998; Khan 2008; Bett et al. 2011). The significance of native birds for rural economy is huge in different countries (Barua and Yoshimura 1997). Backyard farming has over the years been underwritten to a great degree to the agricultural economy of different countries (Padhi 2016). In the same way, rural backyard poultry assembly can be considered as one of the pushing factors in the growing economy. It offers living security to the family besides keeping the sustainability of food. Unemployed youth and women can also earn an income through poultry farming (Padhi 2016). Indigenous breeds are well known for their tropical adaptability and disease resistance, while their plumage colour helps in protecting themselves against predators. The first importance of today's rural poultry farmer is not only having birds which lay just more eggs but also having birds which lay eggs with an optimal size as well as birds which grow to the best body weight with plumage colour similar to indigenous birds (Padhi 2016). Producers therefore have an option out of a number of native breeds.

With the increasing dry lands at the middle of the marsh areas, poultry farming becomes a permissible project to establish. This project can be performed on the family bases or even bigger if investors are interested in such project. In this project, eggs, chickens, and hatchlings can be provided for marketing in the big cities, where there is a great demand on these products. Integration between this project and the dry aquatic resources food project can be made, so the fodder produced in the later can be used for the poultry farming project.

7. **Beekeeping Project**

Beekeeping or apiculture is the skill, science, and or trade of rearing and managing honeybees for the drive of manufacturing honey, beeswax, and other bee yields for personal eating and industrial usage and producing honeybee hives for selling to other beekeepers (Khan and Anjum 2016). Honeybee keeping is a profitable business in most Asian countries such as Pakistan, where about 400,000 honeybee colonies are in business producing 10,000 metric tons of honey annually (Khan and Anjum 2016). Beekeeping supports the livings of rural societies and helps them to become less susceptible to different jolts and decreases the hazard that they will fall into emergency (Nicola, 2009).

Running this project is an easy task, but it needs an initial training, which can be done by the government. After attaining the training, the attendees of these courses may be able to initiate their own project to produce honey. All equipment should be supplied by the government, and their cost will be paid back after 1 year when the project starts to give benefits.

8. **Establishing Handcraft Centre**

The handicraft sector is considered among the chief factor in increasing the revenue and employment creation and has also been known globally as a way to reduce poverty (Allal and Chuta 1982; Jones et al. 2012; Yang et al. 2018). It is an approach of maintaining and encouraging cultural and artistic conducts, such as numerous methods and talents of traditional crafts being conveyed from generation to generation (Yang et al. 2018). For many countries, the important exclusive cultural tradition is remembered in their handicrafts.

Consumers of handicraft goods diminish as industrial yields become more stylish, which forces the artists to leave their occupation and move to the cities for earning income. Consequently, several old-style crafting talents and methods have vanished as fewer younger generation are concerned in learning the tradition. Thus, the handicrafts stonewashed, and no place in the society was left for them owing to the superiority of the mechanized handcraft products (Leesuwan 2010).

In the marsh areas, there are many handcrafts that are made by housewives, young girls, and elderly women. These products have good commercial market in both the marsh areas and the big cities. The idea of this project is to allocate a large reed's hut in each centre of the Marsh Arabs inhabitants designated for production of the handcrafts. The products of these centres can be marketed by middle men that can take the products to big cities for distribution. In these centres, several handcraft ideas can be done, and not only the Iraqi traditional ones, but it is possible to introduce ideas of other countries and produce them in the Iraqi way.

References

Ahmad Y (1995) Production of methane by anaerobic digestion of animal and poultry wastes. MSc thesis, Univ., Agric., Faisalabad

Ahmed N, Demaine H, Muir JF (2008) Freshwater prawn farming in Bangladesh: history, present status and future prospects. Aquac Res 39(8):806–819

Allal M, Chuta E (1982) Cottage Industries and Handicrafts; some Guidelines for Employment Promotion. International Labour Organization, Genève. ISBN 9221030296

Anonymous (2011) Punjab Development Statistics. Bureau of Statics Government of Punjab, Lahore

Bangladesh Bureau of Statistics (1994) Survey on livestock and poultry in Bangladesh, 1988–89. Bangladesh Government, Dhaka

Barua A, Yoshimura Y (1997) Rural poultry keeping in Bangladesh. Worlds Poult Sci J 53(4):392–394

Bett HK, Peters KJ, Bokelmann W (2011) Hedonic price analysis to guide in breeding and production of Indigenous chicken in Kenya. Livest Res Rural Dev 23(6):2011. http://www.lrrd.org/lrrd23/6/bett23142.htm

Chaudhry MA, Farooq MA (1984) Present status of biogas plants in Punjab Province, Report no. 65. AMRI, Multan

DOF (2006) Fishery statistical yearbook of Bangladesh 2004–2005. Fisheries Resources Service System, Department of Fisheries (DOF), Ministry of Fisheries and Livestock, Dhaka, 42pp

Ghaffar M, Yasin M, Farooq MA (1986) An improved and efficient biogas plant. Paper presented at PSAE Annual Convention

Ghosh AK, Maharjan KL (2001) Impacts of dairy cooperative on rural income generation in Bangladesh. J Int Dev Coop 8(1):91–105

Guèye EHF (1998) Village egg and fowlmeat production inAfrica. Worlds Poult Sci J 54(1):82–86

Henderson V (2002) Urbanization in Developing Countries. World Bank Res Obs 17(1):89–112. Oxford University Press

Iliyasu A, Mohamed ZA, Hashim M (2015) Productivity growth, technical change and efficiency change of the Malaysian cage fish farming: an application of Malmquist Productivity Index approach. Aquac Int 23(4):1013–1024

Imran F, Nawaz Y, Asim M, Hashmi AH (2013) Socio-economic determinants of rural migrants in urban setting: a study conducted at City Sargodha, Pakistan. Acad J Interdiscip Stud 2(1):71

Jones E, Smith S, Wills G (2012) Women producers and the benefits of collective forms of enterprise. Gend Dev 20:13–32

Khan AG (2008) Indigenous breeds, crossbreds and synthetic hybrids with modified genetic and economic profiles for rural family and small scale poultry farming in India. Worlds Poult Sci J 64(3):405–415

Khan W, Anjum F (2016) Bee keeping's impact on sustainable livelihoods development in Bajaur agency Pakistan. Int J Appl:213–217

Leesuwan V (2010) Folk handicrafts: local identity. Silpakorn Univ J 30:163–182

Lipton M (1980) Migration from rural areas of poor countries: the impact on rural productivity and income distribution. World Dev 8(1):1–24

Nicola B (2009) Bees and their role in forest livelihoods a guide to the services provided by bees and the sstainable harvesting processing and marketing of their products. Food and Agriculture Organization of The United Nations, Rome

Padhi MK (2016) Importance of indigenous breeds of chicken for rural economy and their improvements for higher production performance. Scientifica 2016:156–159

Sadaf M, Khan IA, Khan AA, Shahbaz B, Akhtar S (2010) Role of international migration in agricultural development and farmers' livelihoods: a case study of an agrarian community. Pak J Agric Sci 47(3):297–301

USAID (2006) A pro-poor analysis of the shrimp sector in Bangladesh. The United States Agency for International Development (USAID), Development and Training Services, Arlington, 93pp

Yang Y, Shafi M, Song X, Yang R (2018) Preservation of cultural heritage embodied in traditional crafts in the developing countries. A case study of pakistani handicraft industry. Sustainability 10(5):1336
Yasin M, Wasim M (2011) Anaerobic digestion of buffalo dung, sheep waste and poultry litter for biogas production. J Agric Res 49:73

Chapter 39
The Daily Life of the Marsh Arab of the Southern Marshes of Iraq Through a Camera Lens

Laith A. Jawad

Abstract Iraq is considered ecologically unique among the other countries of the Middle East. Such exclusivity can be seen in the presence of a vast are of wet lands located in the southern part of the country and known as marsh areas. The marsh areas are lowlands that filled with freshwater originated mainly from the Euphrates River. Inhabiting in these areas, a distinctive human population known as Marsh Arabs. They may be considered as a relics of the Sumerian that lived in south of Iraq more than 5000 years B.C. The life of the Marsh Arabs is very simple and the people living their used the natural products such as fish and birds besides some plantation that they usually grow. In addition, they breed water buffaloes, which it considered the icon of the southern marshes of Iraq. The Marsh Arabs have a series of daily jobs, where each members of any family has his/her own job to do. In this chapter, the lens of a camera has follow the Marsh Arabs in their daily life and recorded several aspects of this life in photos that the author used to comment on them.

Keywords Lifestyle · Social life · Food · Wetlands · Fishing · Hunting · Building · Landscape

39.1 Introduction

Unlike other countries in the Middle East, Iraq is characterised in having wet land areas located in the southern part of the country. This area is known as the marshes or is called in Iraq as "Al-Ahwar". The landscape of this area is fairly different form the middle and north parts of Iraq, and certain group of people inhabit this marsh areas and they have been living for thousands of years. They believed that they are the relicts of the famous Sumerian civilisation.

L. A. Jawad (✉)
School of Environmental and Animal Sciences, Unitec Institute of Technology, Mt Albert, Auckland, New Zealand

L. A. Jawad (ed.), *Southern Iraq's Marshes*, Coastal Research Library 36,
https://doi.org/10.1007/978-3-030-66238-7_39

39.1.1 Who Are the Marsh Arabs?

It is thought that the civilisation of the wetlands in Iraq can be mark out to the time of Gilgamesh who cited this area in his Epic while he was looking for an effective herb for perpetuity (Al-Tameemi 2016). The major group of people in the marshes is the Arabs. Some tribes of this group arrived in the area preceding the Islamic takeover, and some through and after it. Salim (1962) suggested based on anthropological evidence that there are two chief groups that have a clear power in both racial and cultural ways in the marsh areas. These two groups known as "the Eastern Group" and "the Western Group". The former contains the Ma'dan, Albu Muhammad and other communities in the Tigris marshes; they had associations with those human societies cross the border in Iran over and done with visiting and marriage. The latter group includes the non-Ma'dan Euphrates Marsh Arabs, and they had relations with the Bedouin societies of the Arabian Peninsula through the same ways as the eastern group has done with the Iranian societies (Al-Jubouri 2011).

Generally, the people living in the southern marsh areas of Iraq can be known by three names: *Ma'dan*, *Marsh Arabs* and *marsh dwellers*. The word *Ma'dan* has incompatible meanings. The Marsh Arabs consider that the word "*Ma'dan*" is initiated from "Ma'aidi" which represents the enemy, and another view proposes that it stems from "Mou'adat" which is the Arabic word for unfriendliness and dislike. There is a belief that the word "Ma'aidi" came in usage after the British forces have completed invasion of Iraq and reached the marsh areas and were confronted by a fierce opposition from the local inhabitants of the marshes (Salim 1962). Mustafa (2008) suggested another version of the word Ma'dan, which designates dwellers of the wetlands who rely on raising water buffalo and selling its products to earn income from. Here, at this point it is possible to differentiate between the two societies, the *Ma'dan* and other Arab tribes. In the *Ma'dan* society, women can work in trading animal products in the markets, and they have almost freedom of wondering around outside their living areas, on contrary to those of the other Arab tribes that forbid their women from trading outside their home area alone.

On the other hand, *Marsh Arabs* word expresses the people who dwell the wetlands. They are under the effects of the nature of their habitats that they live in. Therefore, their culture, social activities and economics are risen from such environment. Furthermore, the word "Marsh Arabs" is designated to define groups of people that do not raise buffalo and their women cannot wonder around lonely as *Ma'dan* do (Mustafa 2008).

The human societies that live in the marsh areas are known as *marsh dwellers*. Those societies have their own communal morals, conducts and ethnicities. This word is used mutually with the term *Marsh Arabs*, but it has more topographical and ecological implications to show the usual, communal and ethnic influence on the dwellers of wetlands (Mustafa 2008).

The inhabitants of the marsh areas in Iraq are followers of various religions starting from the ancient time. The majority of people in these areas are Muslims followed by Sabians and Jews, but no Christians. Evidences of this ancient belief can be seen in the wholly memorials of each religion that is present in the area. The interesting thing about these diverse religious groups is that they were living in peace and integration since the old time. In the last few decades and due to the political status in Iraq, the number of Sabians and Jews was reduced dramatically (Al-Khayoun 2003; Kubba 2011).

39.1.2 The Landscape of the Marsh Arabs

The wetlands of Iraq are found in the southern part of the country, near the convergence of Tigris and Euphrates Rivers. These areas have exclusive features that seldom can be encountered globally, which turns out to be one of the utmost imperative lowlands in the world (Partow 2001).

The marshes spread southwards from Amara on the Tigris and eastwards from Suq al Shuyukh on the Euphrates to Qurna, where the two rivers converge to form the Shatt al Arab River (Levy 1924). The Mesopotamian wetlands contain three chief localities 1): the northern Al-Hawizah, the southern Al-Hammar and the so-called Central Marshes all rich in both natural resources and biodiversity (Maltby 1994; Evans 2002). Nevertheless, throughout the second half of the twentieth century, an organised strategy of water alteration and pumping out severely diminished the size of the Iraqi lowlands. This ecological disaster forced marsh dwellers of the dried areas to evacuate their homes (Al-Zahery et al. 2011). Nevertheless, owing to the affection to their regime, marsh inhabitants have been reverted to their land as soon as the renovation of marshes started (2003) (Richardson et al. 2005).

39.1.3 The Dwellings of the Marsh Arabs

Al-Tameemi (2016) studied the type of housing built by Marsh Arabs and group these inhabitants into four categories depending in her classification on several factors such as the topography of the islands, weather settings, types of plantation and water quality:

1. *Dwellings on the boundaries of the marshes*. These inhabitants are usually constructed in a location that lifted slightly above the water surface and not far inside the marsh area. The Marsh Arabs called it "*Salaf*" or communities. In the vicinity of Euphrates River, such dwelling is known as "*Nazl*". Each community contains around 100 and 300 small huts, reliant on the position and the number of people living in this village. Each village usually has one or more guest houses "*Mudhif*", where the place of the head of the tribe is located and where all the social affairs of the villagers are discussed and solved. The inhabitants at the edge of the marshes incline to be larger than those dwellers far inside the marshes (Salim 1962; Al-Tameemi 2016).

2. *Dwelling on environmental isles*. In the marsh areas, islands can be either naturally formed or man-made islands from reeds. The inhabitants of the nature-created isles, with their hard base, are able to breed several types of animals mainly buffalo, sheep and goats. This type of naturally created isle is known as "*Ishan*" and could comprise not more than 40 huts (Jadran 2010; Al-Tameemi 2016).
3. *Dwelling on everlasting non-natural isles*. Marsh Arabs usually gather plantation and mainly reeds and deposit in shallow areas of the marshes in an aim to build islands. Once such isles are established, people start to build their inhabitants. This is common in several marsh areas and Marsh Arabs are forced to create such isles due to the lack of drylands in the marsh areas. They called such isles as "*Chibasha*" characterised as having a base made of sheets of mud and plantation and can hold up to 25 small dwellings (Jadran, 2010; Al-Tameemi 2016).
4. *Moveable false isle dwelling*. This is known as "*Dibin* or *Dubun*" and involves basement made of reeds and soil and in a shape of small boat. Such dwellers are suitable for one hut only and a few buffalo to keep (Jadran 2010; Al-Tameemi 2016).

39.1.4 Socioeconomic Life of the Marsh Arabs

It is pretty much that the topography and the terrain setting have direct effect on the socioeconomic activities of the marsh dwellers. There are several agricultural-related events that the Marsh Arabs performed in their daily lives. These include crop assembly, animal rearing, bird shooting and local trades like weaving of reed mats (Ibrahim 2007; Al-Tameemi 2016). The marsh dwellers usually grew rice, corn and wheat in addition to leafs' vegetable on a restricted scale (Kubba 2011).

Animal rearing forms the chief supply of earnings for the people of the wetlands of Iraq. Types of animals that they reared differed reliant on the kind of action performed: water buffalo propagating is the most familiar doings in the wetlands, and it was one of the basic events that go back to thousands of years. Milk can be obtained from the buffaloes and form milk, and several by-products can be made in addition to the meat. Other than buffalo, bird shooting and fishing are two main jobs that are performed by the people of the lowlands in Iraq (Alwan 2005).

Indigenous businesses in the marshes are grounded on hand-made work that used the naturally present materials. Such trades are reed cutting, mat weaving and boat-building from reeds identified as "*mash-huf*", which is the utmost significant way of movement in the lowlands. It is little kayak that can simply be controlled through the thick plantation in the marshes and is usually constructed from wood and coated with tar on the outside (Al-Tameemi 2016).

39.1.5 Scope of the Chapter

The present chapters deal with daily life of the Marsh Arabs in different ways that other similar publication have dealt with. Here, the events of life of the people inhabiting marsh areas of Iraq have been narrated through the photos that depict such events. This chapter is divided into sections related to the different aspects of the life of the Marsh Arabs.

39.2 The Housing at the Marshes

In general, the dwellings of the people living in the marshes are simply constructed from reeds and mud. Their sizes vary depending on the number of the people in the family and on the type of island that the dwelling is building on. The architectural art in the ordinary dwelling is simple and in many cases represents a cover for the people against cold and rain in winter and hot weather in summer, but nowadays and since electricity became available in most areas either through main overhead lines or through small generators, the basic design of the marsh dwelling has changed.

Al-Mudhif will remain the landmark of the southern marshes of Iraq. It is designated by the unique architecture that native people of the marshes has inherited from their predecessors "the Sumerians" and continued to apply such talent in the building of their guesthouse, Al-Mudhif. Therefore, it is imperative to describe the steps involved in building of Al-Mudhif supported by photos to aid the readers to comprehend the basis of the plans of building such an incredible construction designed and built by simple people with using no modern mechanisation. Also, the narration on the photos will reveal some facts behind some features of Al-Mudhif that remain encrypted among the marsh people only.

The values behind the building of mudhifs are linked to those of the clans of the marshes, with a generosity, hospitality and simple life. They signify the village mansion and the local tavern in our present time, with one major difference that everything will be paid by the tribe to anybody entering the mudhif (Al-Jumeily et al. 2018). The rules of Al-Mudhif state that whoever the outsider or visitor can reside at the mudhif and will receive full board hospitality to spend his/her night with food and shelter during the night that he/she is staying in.

The interior of mudhif as it is quite big gives you the sentiment that you are in a holy place. The only furnitures found in the mudhif are coffee pots of different shapes and sizes, the water containers, the area where the fire is located to make the coffee and situated at the middle of mudhif the colourful carpets placed on thin cushions located around the place (Al-Jumeily et al. 2018). The construction of mudhif is similar to a historical building in the ancient city of Ur (south of Iraq); archaeologists located clay artefacts and cylindrical seals with engravings similar to the mudhif, verifying that these reed buildings are identical to those constructed by

Fig. 39.1 Clay artefacts showing image of the Sumerian *mudhif*. The structure of this ancient building is similar to that of the present time *al-mudhif* built by the Marsh Arabs. (Image courtesy of Laith Jawad)

the Sumerians thousands of years ago (*The Illustrated London News*, 1956 in Al-Jumeily et al. 2018) (Fig. 39.1).

39.2.1 Constructing the Mudhif

The basic design of the mudhif is constructed on a structure made of recycled and natural materials. The well-developed reed plants are usually collected by the Marsh Arabs to start the process of building their mudhif. The reeds' plants should be clean of leaves as these may cause weakening and instability of the bundle later on. The bundles are locally called "Shabeh" (Al-Jumeily et al. 2018).

Some facts about the construction of mudhif given by Al-Jumeily et al. (2018) are shown below:

1. The chief element in the body of mudhif is the huge bundle of reeds that stand on its sides in a parabolic shape, with length exceeding 6 m sometimes.
2. These bundles have a narrowing shape, with 0.6 to 1.25 m in diameter at the base, while the height varies from 3 to 10 m.
3. The wide end of each bundle is usually buried in the mud, and doing so, a pair of bundles normally stick opposite to each other. The tapering end of each bundle overlaps with that of the opposite bundle, and they usually fasten together to give a parabolic shape for the formed arch. Once the outside frame of mudhif is completed, it is covered with mats woven from reed on both its top and the sides (Al-Jumeily et al. 2018).
4. The bundles are positioned into the ready-made ditches, with an angle of about 300 (vertical)_ (to the outside). This location aids the constructers to generate the exclusive elliptical arch of the mudhif.
5. Bundles are located in two corresponding rows; the distance between these rows governs the width of the reed house that is frequently falling between 3 and 6.5 m.

   The distance between each two consecutive bundles, within the same row, varies between 1.5 and 2 m (centre to centre). The final total number of arches is regularly odd number. This fact has a religious origin as there is only one God "Allah" and has a social meaning as there is only one head of tribe "Sheikh" at a time.
6. It is important to indicate that the bundles are very light in weight in contrast with concrete and brick ones. The usual weight of these bundles with base diameter of 1 m and height of 6 m is around 628 kg, whereas the weight of analogous concrete and brick pillars is 15,072 kg and 12,560 kg, respectively.
7. The following phase is the spreading of long lateral reed bundles, 10–20 cm in diameter, on the elliptical roof at distance about 30 cm (centre to centre).
8. Next, the vertical sides of the mudhif are enclosed with holed reed walls.
9. The last stage in the building of the mudhif is making the roof, which is done by means of laying mats made of reed that is lightweight and known as "Baryah". These mats are manually woven using fresh and well-developed reed plants.
10. At the end of the construction of the mudhif, Marsh Arabs burn the remaining of the construction materials and other dry plants so the smoke will seal all small vents present in the roof and the side of the mudhif and in turn making it a rainproof structure.

39.2.2 Rituals of the Mudhif

The door of *Al-mudhif* is always open, as a sign of generosity and welcome. There is a divine and emotive connection between Marsh Arabs and *Al-mudhif* (Al-Tameemi 2016). Marsh Arabs usually respond to the sound of grinding coffee beans in the mortar or what is known as "Hawan" telling them that their attendance is needed at the mudhif to discuss some important issues.

Fig. 39.2 Two Marsh Arab men chatting at *al-mudhif* wearing their formal cloths, which consisted of a head cover held by a head band and an abaya placed on their shoulder to cover their bodies. (Image courtesy of Laith Jawad)

Within *al-mudhif*, everybody has to perform respectfully, use appropriate words and be truthful, because dishonesties will not be allowed (Suhair 2013). To attend a meeting at *al-mudhif*, normally, men dress in their best formal clothes (Fig. 39.2) which contain the "*abaya*", a body cover made of thin wool material having various colours, but mainly brown and black as the dominant colours. Their heads are covered with "*kaffiyah*", which is made of cotton material and with design similar to the fish net. This design is similar to that found on the head cover of Gudea of Lagash, the Sumerian ruler in the third dynasty of Ur (2112 BC–2004 BC) (Fig. 39.3). On the top of the cloth head cover, a black and wool-made headband known as "*Akal*" is usually worn to keep the cloth head cover in its place. This headband is a characteristic of the Arabs in general, but that worn by Marsh Arabs is hand-made woven by women of the marshes.

Seating in *al-mudhif* is among the rituals that attendees should follow and never break them, which is based on the communal place of the person in the marsh society. In the primary places sit the respected people, such as the sheikh. These places are renowned in having extravagant rugs and cushions. This important place changes with the seasons; in winter it moves to the middle of *al-mudhif* near the fireplace (Fig. 39.4), while in summer it is located to the end of *al-mudhif* opposite the main entrance and near the small openings in the sides of *al-mudhif* to receive fresh air (Fig. 39.5) (Rapoport 1982; Suhair 2013).

There are some rituals that need to be followed as a person at *al-mudhif*. These are related to the drinking of coffee and tea. Coffee is usually served in *al-mudhif* in

Fig. 39.3 Statue of Gudea the ruler of Lagash in southern Mesopotamia who ruled c. 2144–2124 BC. Note the pattern of the head cover is similar to that used at the present by Marsh Arabs. (Image courtesy of Pinterest.nz)

small cups and in small quantity (Fig. 39.6, a, b). After the first serve of coffee, the person serving the coffee will wait for the guest to give him a sign to stop serving more and more cups of coffee. The sign is to shake the cup with your hand denoting you do not want any more coffee. On the other hand, tea is served black with sugar and no milk in small glass containers known as "*Istikan*" and sits in a small plate with small spoon to stir the tea to mix the sugar (Fig. 39.7). The person serving the tea should be very careful in not putting the little spoon in the "*Istikan*". Putting the little spoon in the "*Istikan*" means that the person served has an issue with his honour. Therefore, it should be taken when serving tea.

Among the other rituals of *al-mudhif* is that you need to take your slippers at the doorstep of *al-mudhif* (Fig. 39.8).

Fig. 39.4 Winter location of the Sheikh of the tribe (head) in *al-mudhif* at the centre and away from the entrance. (Image courtesy of Laith Jawad)

Fig. 39.5 Summer location of the sheikh of the tribe (head) in *al-mudhif* near the entrance, with adequate ventilation and fresh air. (Image courtesy of Laith Jawad)

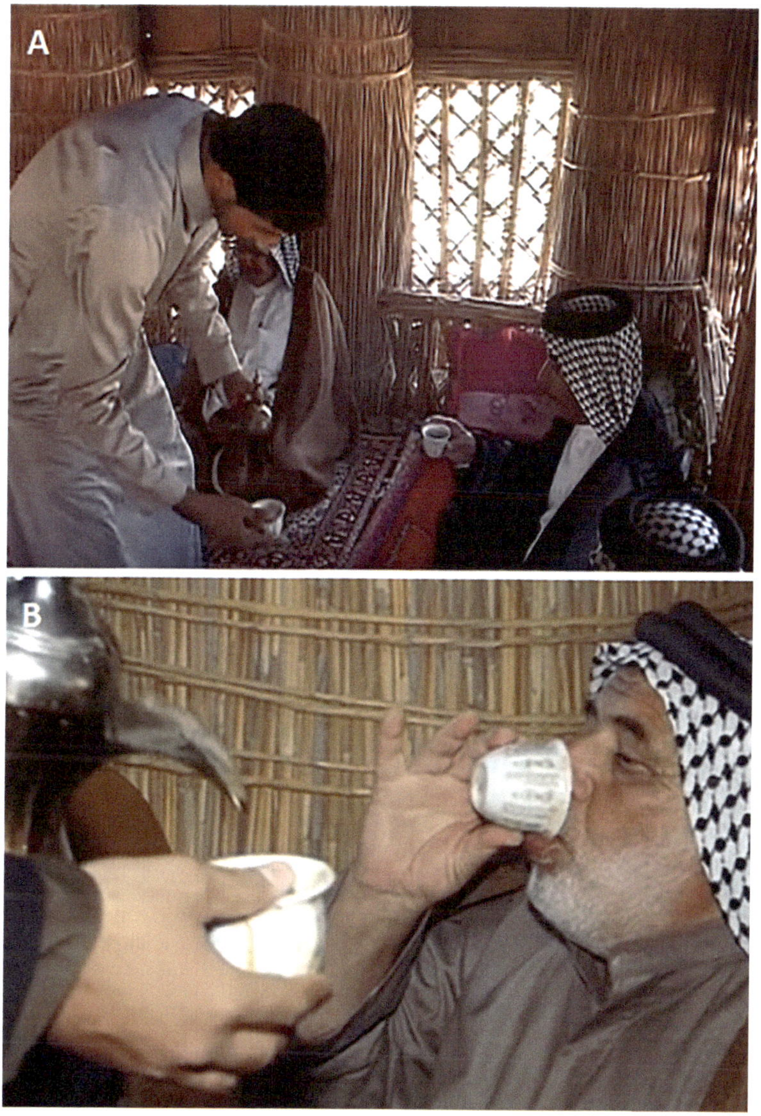

Fig. 39.6 Serving coffee at *al-mudhif* in small distinguished cups. (**a**) the guest drinking coffee and the server waiting to fill another; (**b**) coffee served to a visitor. (Images courtesy of Laith Jawad)

39.2.3 Steps of Building al-mudhif *as Shown in Photos*

In this section, the steps involved in constructing *al-mudhif* are shown in photos, with narration on each of them.

Fig. 39.7 Serving tea in small glass container "*Istikan*". Tea is distributed in tray to visitors. (Image courtesy of Laith Jawad)

Fig. 39.8 Among the rituals of al-mudhif is to remove and leave your slipper at the entrance. (Image courtesy of Laith Jawad)

The process of building *al-mudhif* starts with digging ditches in the place chosen for *al-mudhif* to be built. To inaugurate this event, the first spade in the ground should be done by a respected person from all the tribe and in this case is the *sheikh* of the tribe (head of the tribe). In Fig. 39.9, a *sheikh* of one of the tribes inaugurates the process of building *al-mudhif* in the outskirt of Al-Nasiriya City, south of Iraq (Fig. 39.9). Later on, volunteers from the tribe start to deepen the planned ditches prior to installing the bundles of reed (Fig. 39.10).

Making the bundles of *al-mudhif* requires selecting healthy and matured reed plants (Fig. 39.11a). After this process they move the assembled bundle to its final place in *al-mudhif* planned site (Fig. 39.11b). After fetching the assembled bundles to the site of the future *al-Mudhif*, they lifted off the ground prior to set the parabolic arch and keep them straight in their final position (Fig. 39.12a, b). Workers will tie

Fig. 39.9 A *sheikh* of one of the tribes inaugurating the process of building *al-mudhif* in the outskirt of Al-Nasiriya City, south of Iraq. (Image courtesy of Laith Jawad)

each bundle from one side with that from the opposite side to form the parabolic arch, which will be covered with reeds' mats. In addition longitudinal bundles are usually put by an expert in this job (Fig. 39.13a, b).

On a nearby site, piles of reeds' mats are ready to go as a cover to the arches of *al-mudhif* (Fig. 39.14). In this process, ropes made of reed are used and special knots are applied which might be inherited from the ancestors of the Marsh Arabs (Figs. 39.15a, b, c and 39.16a, b). Different knots were also used to fasten several kinds of bundles in the constructed *al-mudhif* (Fig. 39.17a, b).

The inside view of the completed *al-mudhif* is fascinating, with its windows covered with reeds' sheets knitted perfectly leaving diagonal openings for light and for ventilation (Fig. 39.18a). The completed *al-mudhif* is a piece of art entirely built from recycling environment-friendly materials. As a ritual, the bundles are built in an odd number and in the *al-mudhif* shown in Fig. 39.18b having 19 arch bundles including the entrance and back side arches.

Fig. 39.10 Voluntary men are digging the required ditches as foundation for the reeds' bundles. (Image courtesy of Laith Jawad)

In the present time and after the electricity supply reaches to the marshes in Iraq, the state of *al-mudhif* has also changed accordingly. In the modern *al-mudhif*, you can see light, fans and even television and air-conditioning, with the floor paved with tiles. The seating area is furnished with fancy furniture to receive visitors from both the local marsh area and nearby villages and cities (Fig. 39.19a, b). The mudhif is also the place to settle down and solve the conflicts that occur within the tribes or between neighbouring tribes. The sheikh will be the judge in such case. Special rituals will be followed in settling down any dispute/s between tribes. As a ritual and in the reception at the *al-mudhif*, men shake hands and exchange kisses on the cheek and touches on the shoulders (Fig. 39.20).

Among the chief landmarks inside *al-mudhif* is the place where the coffee is made and where the man serving it is sitting. This is usually located at the centre of *mudhif*. The person serving coffee should be expert in his job and no untrained person can do this type of work. Coffee serving job might run in the family and pass from one generation to another to keep the standard of preparing and the way of serving it. Different mudhifs have different tastes of coffee depending on the person preparing the coffee. The first thing that attracts the attention of the visitor is the large coffee pots that are in odd number as in the photo shown in Fig. 39.21, where there are seven pots in front of the coffee maker. Those pots on the sides usually contain hot water. The coffee is normally served to the guest with the smallest pot. Once this

Fig. 39.11 (**a**) Selecting matured reed plant for making bundles to build *al-mudhif*; (**b**) moving the assembled bundle to its final place in *al-mudhif* planned site. (Image courtesy of Laith Jawad)

pot runs out of coffee, the coffee man will fill it up from the pot next larger in size. He continues refilling the pots from the largest to the smallest (Fig. 39.21).

The majority of the Marsh Arabs belong to the Shia Muslims who respect Ali to a high degree, the cousin of Prophet Mohamed, his sons, Al-Hassan and Al-Hussain, and their descendants. In Fig. 39.22, an imaginary photo of Ali, cousin of Prophet Mohamed, was put at the entrance of *Mudhif* to protect the place from the evil eyes and to announce their strong belief in Shia subdivision of Islam.

Fig. 39.12 (**a**) Bundles of reeds being lifted off the ground prior to set the parabolic arch; (**b**) bundles of reed standing straight in their final position. (Image courtesy of Laith Jawad)

39.2.4 Hut Building

The building materials used in building huts are all available at the environment of the marshes such as mud and reeds. Such a method is similar to that used thousands of years ago by the descendants of the Marsh Arabs (Ochsenschlager 1998).

Fig. 39.13 (**a**) the finished parabolic arches prior to their covering with reeds' mats; (**b**) man laying and fastening the longitudinal bundles as a support for the arched bundles. (Image courtesy of Laith Jawad)

Ordinary people do not used mud bricks in the building of their dwellings due to the high cost of materials and the need for professional builders. On the other hand, an ordinary person can build a mud hat within a day with help of members of his family.

Fig. 39.14 Piles of reeds' mats ready to go as a cover to the arches of *al-mudhif.* (Image courtesy of Pinterest.nz)

The hut is composed of four walls, with no roof. Two sides of the hut have a triangular-shaped top. Once the mud structure dries in the sun for a day or two, bundles of reed are put longitudinally, joining the two triangular ends of two sides of the hut. Later on, smaller bundles from reed are laid across the large middle bundle and fasten with ropes made of reed. The final step of building the hut is to lay mats made of reed on the top of reed structure and fasten together with ropes. Sometime, mud is laid on the top of the reed mats to make it waterproof and cold during summertime.

Preparing the mud for building a hut requires gathering soil and mixing it with water to form a soft mud that is spreadable and can be shaped in layers on the top of each other. The mud is usually mixed with hay or small fragments of reeds to strengthen the wall after being dried. Figure 39.23 shows a man building a hut with help of one member of his family. The man is shown sitting on the new wall receiving lumps of mud mixed with hays and reeds from a young fellow who is probably his son. Note a rectangular opening was left in the side wall of the hut as a window.

39.3 The Daily Works

All members of the family of the Marsh Arabs started their day very early and probably with first light as they do their morning prayer, eating their breakfast and then each one going to his/her daily routine work.

Fig. 39.15 Methods of fastening the bundles of *al-mudhif*. (**a**) from the front; (**b**) from the side top; (**c**) from the base. (Image courtesy of Pinterest.nz)

Fig. 39.16 (**a**) man laying reed mat on the top of the arched structure of *al-mudhif*, using ropes made of soft reeds; (**b**) special and very ancient type of knot used to fasten the bundles of reeds. (Image courtesy of Pinterest.nz)

The breakfast usually is composed of hot bread that the women usually bake it. Here, it is important to mention that there are several dairy products used by Marsh Arabs and all of them were made from buffalo milk. With the bread people eat butter, fresh cream or yoghurt.

Fig. 39.17 Types of tying the different shapes of reed bundles. (**a**), binding bundles at the entrance of *al-mudhif*; (**b**), enlarged shot showing the type of ties at the entrance of *al-mudhif*. (Image courtesy of Pinterest.nz)

The daily works that all members of the family at the marshes perform are very hard and extend through all the day until darkness. Each member of the family has his/her designated job/s. Men, for example, are responsible mainly for bringing food for the family. This food can be fish collected from the marshes or birds hunted from the nearby field. Other jobs that men can do are gathering reeds for several purposes and herding buffaloes and taking them to their daily swimming activities. On the other hand, women have more to do than men. In addition to looking after all

Fig. 39.18 Inside view of the final setting of sides of *al-mudhif*; (**a**), the reed sheets are very well knitted to form diagonal opening for light penetration and for ventilation; (**b**), completed *mudhif*. Note the odd number of parabolic arches. (Image courtesy of Pinterest.nz)

members of the family, they prepare food for lunch and dinner, bake the bread, milk the buffalo and cows if there are any, wash dishes, wash clothes, help in collecting reed and, if the dwelling living in is near to a village of small city, take the fish catch and sell it there in the market. Helping women are young girls, who usually take containers to fill them with water and fetch them to the living area for washing and drinking. On the other hand, teenager boys usually accompany men in their daily fishing work, bird hunting or reed gathering to learn the skills of these jobs. The followings are detailed daily jobs performed by different members of the family accompanied with photos.

Fig. 39.19 Modern *mudhif*. (**a**), *Mudhif* supplied with electricity as shown with light distributed in its roof; (**b**) *Mudhif* supplied with roof fans. (Image courtesy of Pinterest.nz)

39.3.1 Food Making

39.3.1.1 Bread Making

Making bread is a daily job for each family. Women are usually responsible for this type of job. There are different types of bread and there are two types of ovens that are made of mud for the women to use to make the bread. The common oven is known as "Tanoor" and the other type known as "*Sadj*". Before describing the method of preparing the dough of the bread and baking procedure, it is important to know how the Marsh Arabs build their mud-made ovens.

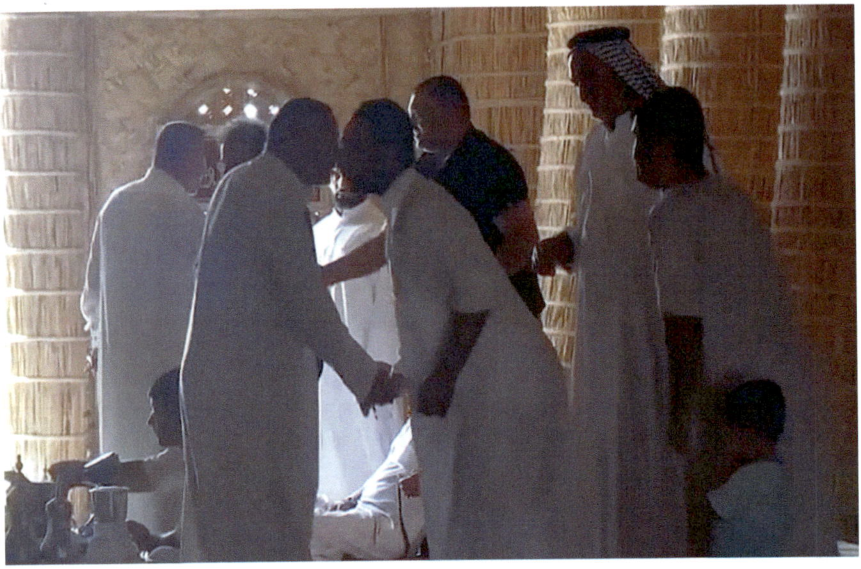

Fig. 39.20 Receiving members of the tribes at mudhif to discuss several social issues of the tribe or to solve conflicts that might occur within the tribes or between neighbouring tribes. (Image courtesy of Laith Jawad)

Fig. 39.21 The coffee serving area, which is located at the centre of *Mudhif*, with the person serving coffee sitting behind the coffee pots. (Image courtesy of Pinterest.nz)

Fig. 39.22 An imaginary photo of Ali, cousin of Prophet Mohamed, was put at the entrance of *Mudhif*. (Image courtesy of Laith Jawad)

39.3.1.2 Tanoor Building

Ancient Mud "Tanoor"

The mud oven or what is known as "Tanoor" is an ancient household fitting that is vital for every dwelling in the marsh area and villages at the edge of the marshes. In this "Tanoor" bread is baked daily by women for the rest of the members of the family. The Marsh Arab and all Iraqis reckoned that the bread baked in a mud-made "Tanoor" is more delicious than any bread baked in any type of ovens. In their daily life and as a tradition, Iraqis are sticked to the habit of eating bread made in "Tanoor", but with the modernisation another type of "Tanoor" was introduced that does not need to use reeds as a fuel, but gas. Such "Tanoor" is made of aluminium and mainly used in the cities.

To make the traditional "Tanoor", women in the marshes are expert in choosing the appropriate type of mud for making "Tanoor". Prior to starting the process of building the "Tanoor", women go searching for the best type of soil, which has high percentage of clay. Then the soil is mixed with a certain amount of water and definite quantity of hay or reed to harden the walls of the "Tanoor" afterwards (Fig. 39.24). The soil and water are mixed very well so as to get a homogenised lump of mud (Fig. 39.24c). Later and with both hands, the mud is placed in a thin circular row and left to dry completely in the sun for 2–3 days (Fig. 39.24d). After getting really dry, another row of mud is put on the top of the first row (Fig. 39.25a, b). Women continue adding rows of mud on top of each other until the required size of "Tanoor" is reached (Fig. 39.26a). To quicken the process of drying the mud, fire is set in small quantity of reeds and put near the wet mud so it gets dry quickly and another layer can be laid on. Since there is a demand on the "Tanoor" in villages and small cities to

Fig. 39.23 Man and his son building mud hut. (Image courtesy of Pinterest.nz)

use it in small bakery shops, an order to make large "Tanoor" is usually received by women to make such large size "Tanoor" (Fig. 39.26b).

Metal "Tanoor"

This type of "Tanoor" is used mainly in the urban areas, where the availability of getting dry plantation as a fuel is nearly nil. In addition the metal oven is smoke-free and suitable for the inhabitants of houses in the metropolitan areas. Moreover, the Iraqi people like to eat the bread baked in the traditional way of baking and the metal oven can provide them with such bread similar to the mud-made "Tanoor" to some extent.

Fig. 39.24 (**a**) Mixing hays with soil prior to making mud; (**b**) making mud required to make Tanoor; (**c**) process of building Tanoor starts with making one by one mud rows; (**d**) completed mud Tanoor. (Images courtesy of Laith Jawad)

The metal "Tanoor" is composed of two aluminium cylinders, with different diameters (Fig. 39.27a, b) inside each other and kept separated by small distance. The base of the inner cylinder is sealed with an aluminium plate supplied with a ring of holes at its centre. Beneath this ring of holes and in the distance between the two cylinders, pipes of gas in a ring shape are present, which in turn is connected to a horizontal pipe leading to a high-pressure cylinder that supplies the gas to the two ring pipes under the central plate and in between the two hollow cylinders. Once the gas is released from the high-pressure cylinder, the person using the metal "Tanoor" will ignite the gas and fire starts to run in the ring pipes. In this way the inner cylinder will receive heat from two sources, from the large ring pipe outside its wall and from the small ring pipe at its centre. The metal "Tanoor" has a metal cover that should be put on to keep the "Tanoor" hot enough to stick the dough of the bread on its inner walls. Figure 39.28a shows a factory manufacturing metal "Tanoor" and Fig. 39.28b shows a housewife using this type of "Tanoor" to make Iraqi traditional bread. It is clear form this photo the connection of the metal "Tanoor" with high-pressure gas cylinder as a source of fuel. Also it appears in this photo that inhabitants of urban areas are using this "Tanoor" and they used it in the front of the house for safety reasons.

Fig. 39.25 (**a**, **b**) Processes of making mud Tanoor. (Images courtesy of Laith Jawad)

39.3.1.3 Making Iraqi Bread

The Iraqi traditional bread differs from any other bread in its shape and the ingredients. The usual shape is large round and slightly thin at the centre and thick at the edges, but it can be done in an elliptical shape too. It characterised in high different sizes of bubbles formed while the dough is baking. It has a distinctive

Fig. 39.26 (**a**, **b**) Process of making commercial-size mud Tanoor. (Images courtesy of Laith Jawad)

Fig. 39.27 (**a**, **b**) Metal Tanoor. (Images courtesy of Laith Jawad)

attractive smell especially when hot. The ingredients are as follows: 2 Kg of wholemeal flour. White plain flour is not suitable for this kind of bread as it will melt once it is put in the "Tanoor". Sometimes, white flour is used, but only if mixed with wholemeal flour as support, two large spoons of yeast, 1½ large spoon salt and one large cup of warm water (Fig. 39.29a–d).

In the process of preparing the dough of the bread, part of the warm water is poured over the flour and other ingredients (Fig. 39.30a), and then gently all ingredients are mixed with water to make the dough (Fig. 39.30b). At this stage, water should be added in steps while mixing is taking place until the dough is formed (Fig. 39.30c, d). After the dough has completely become homogenised, it needs to be left over for half an hour to rest and for fermentation. After that, the dough is separated into small balls and left over a small quantity of flour so it will not stick

Fig. 39.28 (**a**) factory manufacturing metal Tanoor; (**b**) women using metal Tanoor to bake Iraqi bread. Note that the Tanoor is outside the house for safety reason. Also, note the connection of butane cylinder gas to the Tanoor. (Images courtesy of Laith Jawad)

to each other (Fig. 39.31). During this time, the "Tanoor" needs to be started. In this process, dry reeds or dry date palm leaves are usually gathered from the nearby area and ignited inside the "Tanoor". The dry plantation will burn vigorously and a lot of smoke is formed accordingly, but later, fire slows down (Figs. 39.32a–d). The sign that "Tanoor" is ready for baking bread is when the sides of "Tanoor" facing fire become white and clear of smoke. The next step is to spread the balls of the dough in turn on rounded cushion (Fig. 39.33a) in a way that does not stretch the dough too much; otherwise it will break and holes form in it and cover the whole surface of the cushion (Fig. 39.33b,c). Then women take the cushion and with strong hit sticks the dough on the inside wall of "Tanoor" (Fig. 39.33d). The number of dough sticked in "Tanoor" depends on its size, but on an average four doughs can be sticked at the same time giving four breads afterwards. Women who bake bread daily became professional in this job. Therefore, they can take out the ready-baked bread from the hot "Tanoor" with their bare hand. The non-professional women use special metal

Fig. 39.29 Ingredients of the traditional Iraqi bread. (**a**) white and brown flour; (**b**) water; (**c**) salt; (**d**) yeast. (Images courtesy of Jawad)

Fig. 39.30 Process of making the Iraqi traditional bread. (**a**) mixing water with flour; (**b**) making the dough; (**c**) homogenising the dough; (**d**) leaving the dough to rest or to complete fermentation. (Images courtesy of Laith Jawad)

tongue with a long handle to pick up the hot bread (Fig. 39.34a). The hot baked bread will have irresistible smell and with bubbles on the surface (Fig. 39.34b).

Marsh Arab men when they go on their daily job for fishing, bird hunting and gathering reeds, they usually take with them some flour and basic utensils so they will be able to make bread to eat for their lunch while they are away from home. Together with the bread, they grill some of the fish they collect. The men follow a simple way of baking bread that does not need "Tanoor تنور". In this method, a small

Fig. 39.31 Dough made in small balls after the complete fermentation and prior to baking. (Images courtesy of Laith Jawad)

Fig. 39.32 Process of preparing Tanoor for baking the dough to make the traditional Iraqi bread. (**a**) burning the reed and the dry leaves of date palms; (**b**) baking should wait until the fire inside Tanoor drops down and no more flames come out. Also, the inside panel of Tanoor needs to be red hot; otherwise the dough will not stick; (**c**) reed and dry date palm leaves burning inside Tanoor; (**d**) ideal settled fire for bread making to start. (Images courtesy of Laith Jawad)

amount of flour is mixed with water to make the dough and spread in a circular shape and left on the fresh reed (Fig. 39.35a, b, c). In the same area that is intended to be used for baking the bread, fish is spread open by sticks and positioned in the ground

Fig. 39.33 Process of baking the Iraqi bread. (**a**) women taking one dough ball at a time and with her hand she makes it circular and thin as much as possible. Later, she will stick it on the circular convex pillow prior to sticking it on the hot inside plane of Tanoor. The professional women stick the dough with their bare hand inside the Tanoor. (Image courtesy of Timeline photos); (**b**) method of laying flat the dough on the circular and convex pillow; (**c**) stretching the dough to make circular and with thick edges; (**d**) bread sticked inside Tanoor in the process of baking. (Images courtesy of Laith Jawad)

facing the fire. A small bundle of reed is selected to be used as a torch to bake both the fish and the bread (Fig. 39.35d). Men continue making circular soft doughs as they require and put them on the fresh reed (Fig. 39.36a). The small bundle of the reed will be ignited and bring as close as possible near the bread and the fish controlled by the feet of one of the men until the bread is baked and the fish is grilled (Fig. 39.36b, c, d).

The profession of making Iraqi bread "*Tanoor Bread* خبز تنور " is an ongoing job in the villages and cities due to the great demand on the traditional Iraqi bread. Therefore a special bakery is usually found mainly in the old parts of the cities and in the villages that bake the Iraqi traditional bread (Fig. 39.37a–d). In these bakeries, large-sized "Tanoor تنور" are usually used to take more than ten breads at a time. Since the bakery is designed for commercial usage, the owner of the bakery usually fits 2–3 "Tanoor تنور", and there are three or more people making the bread, one person specialised in preparing the dough balls "شنگه", another person for sticking the dough on the inner side of "Tanoor تنور" and a third person to take out the bread after being baked. When there is a big demand on bread, two people will be assigned to stick bread in the "Tanoor تنور".

Fig. 39.34 (**a**) Ready-baked bread usually taken out of Tanoor by bare hand by professional women, but metal tongue is used by others; (**b**) ready-baked Iraqi traditional bread, with its distinctive baked bubbles on the surface. (Images courtesy of Laith Jawad)

39.3.1.4 Making "*Tabak and Siah* طابك و سياح" Bread

The origin of these types of bread is the Sumerian, who used to make such bread. In contrary to the Iraqi bread described above, the ingredient in these two types of bread is rice, water and salt. For each of these two types of bread, a special technique is required in the process of making them.

Tabak Bread خبز طابك

For this type of bread, a specific oven is needed that can be made at the time of making the bread. This unique oven is a hot plate made of mud known as "*Tabak* طابك". The word comes from the Arabic (Tabak طبك), which means large rounded

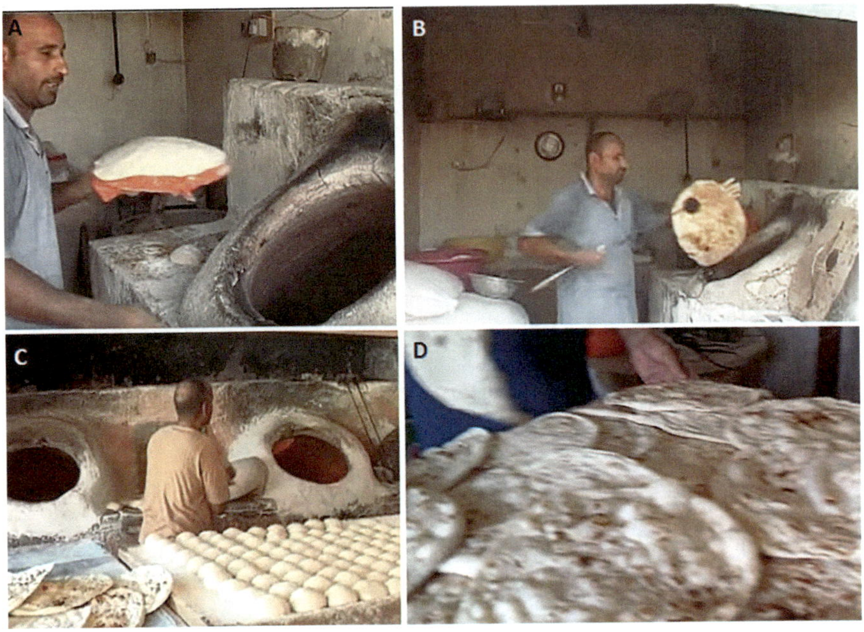

Fig. 39.35 Baking Iraqi traditional bread commercially. (**a**) man sticking the bread inside the hot mud Tanoor using the circular convex pillow; (**b**) man taking out baked piece of bread using metal tongue; (**c**) man sticking dough inside the hot mud Tanoor. Note the dough balls sitting on the table waiting to be baked; (**d**) ready-baked bread. (Images courtesy of Laith Jawad)

Fig. 39.36 Process of making Tabak bread at the field. (**a**) dough made of mixture of rice flour and water is made and spread on back of metal tray; (**b**) completion of spreading the rice flour dough to become thin and circular; (**c**) the rice four dough is ready for baking; (**d**) process of igniting the red to bake the rice flour dough "Tabak bread". (Images courtesy of Mohammed Al-Darwesh, Iraq)

Fig. 39.37 In the process of making Tabak bread. (**a**) aiming the burned reed against the surface of the bread; (**b**) holding the burning red by feet; (**c**) Tabak bread starts to bake; (**d**) directing the fire from the burning reed against the bread. (Images courtesy of Mohammed Al-Darwesh, Iraq)

plate or tray usually made of reeds or date palm leaves. In the process of making the mud-made hot plate, seeds of reed are collected as a first step (Fig. 39.38a) and then soft mud gathered from the area where the bread is going to be baked (Fig. 39.38b,c). The seeds of the reed mixed with soft mud hold parts of mud hot plate together. The mud containing reed seeds is mixed very well until it becomes homogenised (Fig. 39.39a). The mud dough is then put in a circular shape, with thickness of a few centimetres and its surface smooth as much as possible by moving wet hand on the top of the circular mud piece (Fig. 39.39b, c). The final step in making the mud-made hot plate is baking it over strong fire using dry reeds (Fig. 39.39d). The mud-made hot plate is left baking until its surface becomes completely dry.

After the required mud hot plate is prepared, the next step is to prepare the dough of the bread. Flour of grinded rice is usually used for this type of bread (Fig. 39.40a) mixed with water and some salt until soft dough is obtained (Fig. 39.40b, c). Then the mud hot plate taken is off the fire and put on some stones to make it slightly higher than the ground (Fig. 39.40d). On the red hot top of the hot plate, the watery mixture of rice flour and water is poured and levelled by hand. Now the bottom of the laid mixture of rice flour and water is touching the hot plate and in a process of baking. To bake the top of the rice flour-water mixture, a dry buffalo dung is brought and put on the top of the rice flour-water mixture to bake the top of the future bread. Usually the whole surface of the spread rice flour-water mixer is covered with the dry buffalo dung in order to bake the whole surface of the bread (Fig. 39.41a-d). Once the surface of the bread is baked, the burned buffalo dung is by holding the side of

Fig. 39.38 Process of making Tabak bread at the field away from home. (**a**) selecting seeds of reed's plants; (**b**) digging to get suitable soil to make mud; (**c**) taking wet soil to make mud; (**d**) mixing reed's seeds with mud to make it hard. (Images courtesy of Mohammed Al-Darwesh, Iraq)

the mud-made hot plate (Fig. 39.42a) and lifted so all the dung will fall on the floor. The next step is cleaning the surface of the bread from any remaining parts of the burned buffalo dung (Fig. 39.42b) and then removing the sticked bread from the mud-made hot plate (Fig. 39.42c). Finally, the bread is cut into quarters as shown in Figs. 39.42d and 39.43a, b). This method of cutting the *Tabak Bread* خبز طابك is characteristic of the people of the marshes, and both women and men follow the same technique.

Grilled fish of different species is usually eaten with Tabak Bread خبز طابك. In the following images, a method of preparing grilled fish while people are in the field and away from home is shown. The fish species they used is what is known as "*Zoori* زوري". People in south of Iraq in general and the Marsh Arabs call any small-sized fish species as "*Zoori* زوري". The fish species that is shown in the images discussed in this section is muglid fish, *Planiliza abu*, family Mugilidae. This species is also called "*Abu Khraiza* أبو خريزه ". The word "*Khraiza*" in Arabic is small bead, and the fish was given this name relative to the shape of its stomach that has a bead shape. After catching fish, it is washed and fresh hard reed is sticked in the whole fish together with its viscera (Fig. 39.44a, b). Several fish stick together and the sticks carrying the fish are grilled using buffalo dung as a fuel (Fig. 39.44c, d). Once the

Fig. 39.39 Finishing making the mud hot plate. (**a**) rolling the mud to homogenise it; (**b**) making circular mud hot plate; (**c**) making the surface of the mud hot plate smooth; (**d**) baking the mud hot plate to dry and to make red hot prior to putting the rice flour-water mixture on top of it. (Images courtesy of Mohammed Al-Darwesh, Iraq)

fish is grilled, the presumed table on the floor is set. The table in such occasions is composed of green leaves of date palms or any plantation (Fig. 39.45).

The description of making *Tabak Bread* خبز طابك is given for people who are out in the field and away from home and where there are no utensils to use. In the following images, the process of making *Tabak Bread* خبز طابك takes place at home and the person doing it is a housewife. Here, at the home, the "*Tabak* طابك" is usually found and no need to make it (Fig. 39.46a) and to make it red-hot, dry tree stems are used instead of reed (Fig. 39.46b). The mixture of rice flour and water is prepared in proper metal utensil (Fig. 39.46c), and then poured over the surface of the hot "*Tabak* طابك" (Fig. 39.46d) and levelled with the hand to give it a circular shape (Fig. 39.47a). Once the bread gets dry, the whole mud hot plate "*Tabak* طابك" is tilted to its side so the bread will face the flame so to get baked (Fig. 39.47b) and remains in this position until the surface of the bread gets roasted (Fig. 39.47c). Later on, the hot "*Tabak* طابك" is brought down and the bread is taken off and served hot (Fig. 39.47d).

Siah Bread خبز سياح.

The constituent of this type of bread is also rice flour mixed with water and small amount of salt; it differs from *Tabak Bread* خبز طابك in being thin in its configuration and it is usually baked on hot metal plate and in this case people use the back of a metal tray or metal bowl turned upside down (Fig. 39.48a), and then the watery mixture of rice flour and water is poured on the hot metal plate and levelled by hand

Fig. 39.40 Process of preparing rice flour dough in the field. (**a**) pouring the rice flour in a bowl; (**b**) mixing the rice flour with water; (**c**) making the dough very thin; (**d**) taking the mud hot plate of the burning reed. (Images courtesy of Mohammed Al-Darwesh, Iraq)

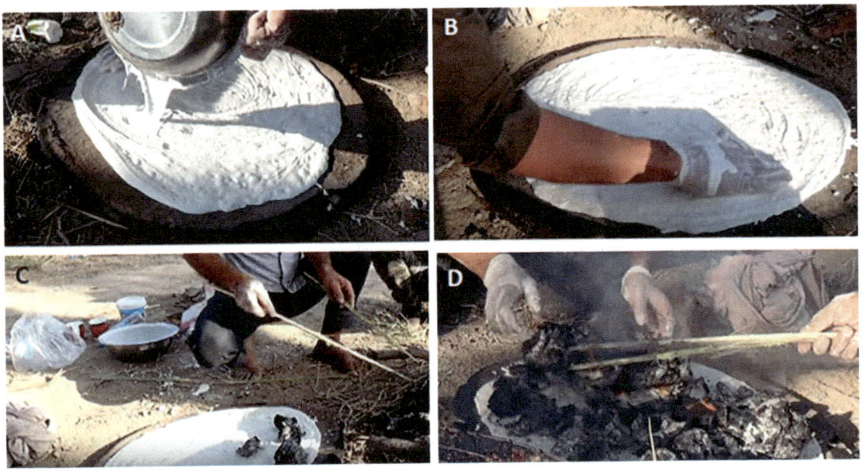

Fig. 39.41 Process of baking Tabak bread. (**a**) pouring the thin rice flour-water mixture over the red hot mud hot plate; (**b**) levelling the mixture in a form of thin circular shape; (**c**) putting burned buffalo dung on the top of the rice flour-water mixture; (**d**) keeping the surface of Tabak bread covered fully with burned buffalo dung. (Images courtesy of Mohammed Al-Darwesh, Iraq)

Fig. 39.42 Final steps of baking Tabak bread. (**a**) removing burned buffalo dung from the surface of the bread; (**b**) removing the remains of any burned buffalo dung on the surface. Note the bread is completed baked and ready to serve; (**c**) removing the bread from the red hot mud hot plate; (**d**) traditional way in cutting Tabak bread. (Images courtesy of Mohammed Al-Darwesh, Iraq)

to form a rounded thin layer of bread (Fig. 39.48b, c) and covered (Fig. 39.48d). The final product is a thin layer of bread (Fig. 39.49).

39.3.1.5 Making Kharait خريط.

Kharait خريط is a kind of sweet that the Sumerians have made and enjoyed eating it. This kind of sweet is made of the pollen of the reed when they appeared during spring time. The technique of making Kharait خريط has been transferred from one generation to another over the thousands of years from the Sumerian in ancient Mesopotamia to the Marsh Arab of the present time. The people of the marshes has retained the method making this sweet and they enjoy eating it these days. Moreover, there is a great demand on Kharait خريط not in the marsh areas, but in the urban areas of the southern cities of Iraq during spring time when it is available. Kharait خريط is distinguished in having bright yellow colour, with no smell and with powder configuration. It is usually made in thick disc-shaped or in a shape of small balls.

Fig. 39.43 (**a**, **b**) Completing the process of breaking the ready-baked bread. (Images courtesy of Mohammed Al-Darwesh, Iraq)

Fig. 39.44 Process of baking Zoori fishes to be eaten with Tabak bread. (**a**) selecting fishes; (**b**) threading the fish in fresh reed sticks; (**c**) sticks with fish thread through them waiting to be baked; (**d**) sticks with fish thread through them on fire baking. (Images courtesy of Mohammed Al-Darwesh, Iraq)

Process of Making Kharait خريط.

In spring time (March–May), the reed plants in the marshes start to flower and produce their pollen, which is usually large in size. The first step in making Kharait خريط is to collect the pollen of the reed plants (Fig. 39.50a, b). Later, dry leaves of date palms are collected to make fire, where Kharait خريط will be cooked on (Fig. 39.50c). The pollen of reed is usually collected with some parts of the reed plant itself. Therefore, the pollen is required to be sieved in order to get clean pollen only (Fig. 39.50d). Boiling water is needed in the process of making Kharait خريط. In a small pan, a small amount of water is put and it usually comes from the marsh directly and left over the fire to boil (Fig. 39.51a). The collected pollen of the reed is put in another small pan and inserted inside the pan with the boiling water and the top of the small pan is wrapped with part of the cloth holding the pollen (Fig. 39.51b). The top of the large pan with the boiling water is put on, and its edges are sealed with mud to keep the heat inside to cook the pollen using the steam coming from the boiling water (Fig. 39.51 c, d). The pollen is left for about 30 min to be cooked on the steam inside the small pan. Later, the small pan is taken out from

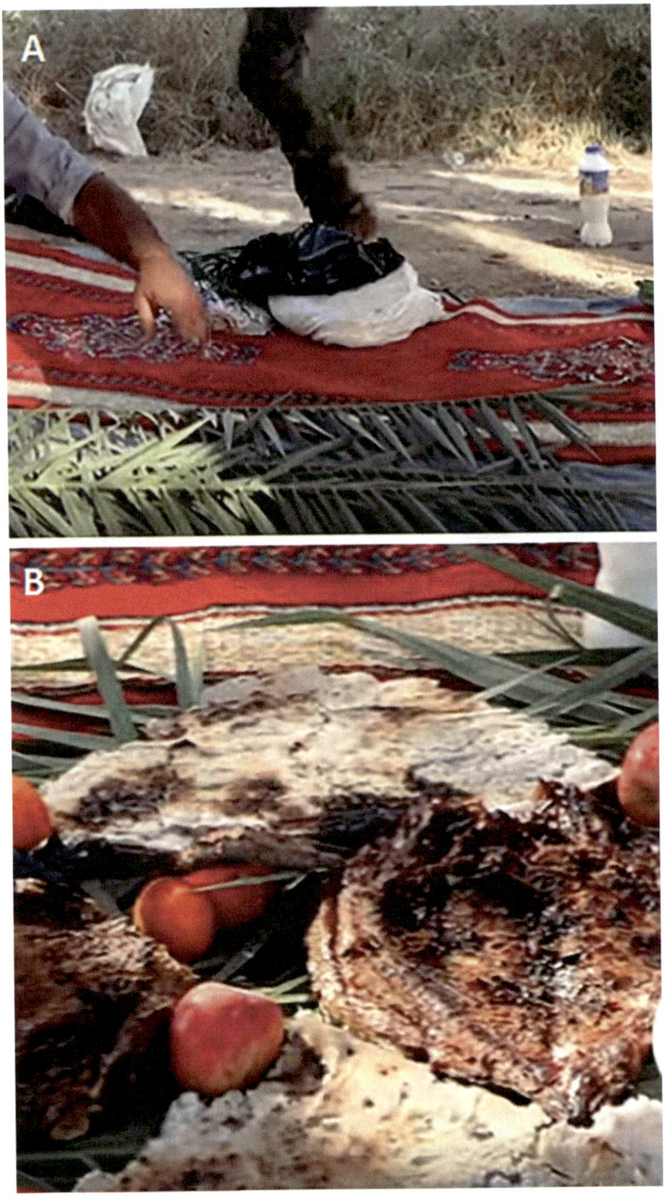

Fig. 39.45 (**a**, **b**) Once the Tabak bread and Zoori fishes were baked and ready to eat, the table is set on the floor for lunch. Note the use of fresh date palm leaves as a mat to put food on. (Images courtesy of Mohammed Al-Darwesh, Iraq)

the larger pan (Fig. 39.52a) and put aside (Fig. 39.52b), and the cloth wrapping is opened to get the Kharait خريط (Fig. 39.52c), which is ready to serve and eat (Fig. 39.52d). During spring time, Kharait خريط is offered for sale in markets of the urban area in the main cities in southern Iraq (Fig. 39.53a, b).

Fig. 39.46 Making Tabak bread at home. (**a**, **b**) putting already made mud hot plate on fire to get hot; (**c**) preparing the dough of Tabak bread from rice flour and water; (**d**) pouring the mixture of rice flour and water on top of the red hot mud hot plate. (Images courtesy of Laith Jawad)

Nasrallah (2013) related Kharait to the food used in the festivals of Jewish community where they used to live in Iraq. The story related to the prophet Moses where he found among the reeds and since Kharait is made of the pollen of this plant, then it may have a religious significance for the Jewish community. As they settled down at the city of Babylon in ancient time, they kept looking for the reed plant and then they figure it in the southern marshes of Iraq and noticed the process of manufacturing the sweet from the pollen of this plant.

39.3.1.6 Making Kaimer عمل القيمر.

Kaimer قيمر is the Iraqi term for a very thick cream usually obtained from the milk of a buffalo characterised with a high percentage of fat in it (Fig. 39.54). Iraqi across the country have Kaimer قيمر for their breakfast and it is very popular in Iraq, together with the traditional Iraqi bread and date molasses or home-made jam. In the marsh areas, Kaimer قيمر is usually made as there are a surplus of buffalo milk and also the

Fig. 39.47 Final steps of making Tabak bread at home. (**a**) levelling the rice flour-water mixture on top of the red hot mud hot plate; (**b**) tilting the mud hot plate to face the burned reeds; (**c**) nearly baked bread; (**d**) removing the ready-baked bread from the mud hot plate (Images courtesy of Laith Jawad)

Marsh Arabs believe that the buffalo milk is a natural medication for several illnesses especially for young people and kids.

The first step in the process of making Kaimer قيمر and after the buffalo milk is collected is warming the milk, and it is usually put in a large container and left to get warm for a period ranging between 5 and 10 h (Fig. 39.55a). Once the milk is heated adequately, the fat contents of the milk are mobilised by mixing it for few minutes so the fat will not aggregate at the bottom (Fig. 39.55b, c). After getting mixed while the milk is warm, it is usually kept warm by covering the container with thick cover such as blankets (Figs. 39.55d). After nearly 10–15 h, all the fat present in the milk will float on the surface making a thick layer of cream. At this stage, the thick layer of the cream is usually broken using a special fine needle (Fig. 39.56a). The purpose of breaking the thick layer of the cream is to make a way for the cold milk to go through, which will be added in the next step. A small amount of milk is usually mixed with small pieces of ice and then poured over the warm milk so any remaining dissolved fat will coagulate and float to the surface (Fig. 39.56b, c), which is then taken in pieces and rolled and put in plates ready to serve (Fig. 39.56d).

Fig. 39.48 Process of making Siah bread. (**a**) putting the metal circular and convex plate on burning reeds; (**b**) levelling the poured thin mixture of rice flour-water on top of the metal hot plate; (**c**) waiting for the bread to get baked; (**d**) cover of the baking bread is removed to check whether the bread is ready or not. (Images courtesy of Laith Jawad)

39.3.1.7 CheeseMaking عمل الجبن.

Cheese is another milk product that the Marsh Arab make in their daily life. The process of making cheese is simpler from that of making Kaimer قيمر. In the process, buffalo, cow and sheep milks are used and the cheese coming out of each of those milks has its distinctive taste. For making the cheese, the milk needs to be warmed and then a starter that contains a special yeast only used for cheese is added to the milk once it becomes slightly cold. The milk is left for a while with the starter so the protein present in the milk will coagulate and form chunks. The women making the cheese will continue mixing the milk to ensure all the protein in the milk get in contact with the starter. The next step is putting the chunks of the coagulated protein in a form of broad thread and bread it such as breading of the hair (Fig. 39.57a–d).

Fig. 39.49 Ready-made Siah bread. (Images courtesy of Laith Jawad)

Fig. 39.50 Process of making Kharait خريط. (**a**, **b**) collecting pollen of reed; (**c**) gathering dry date palm leaves for making fire; (**d**) sieving the collected pollen to remove any plant remains. (Images courtesy of Mohammed Al-Darwesh, Iraq)

Fig. 39.51 The second stage of making Kharait خريط at the field. (**a**) boiling water; (**b**) putting the collected pollen of the reed in a small pan and inside a bigger pan containing boiling water so it will cook on the steam; (**c**) pollen wrapped in a piece of cloth covered; (**d**) the cover of the pan sealed with mud to ensure steam will remain in and cook the pollen. (Images courtesy of Mohammed Al-Darwesh, Iraq)

Fig. 39.52 The final steps of making Kharait خريط at the field. (**a, b**) taking the small pan containing the cooked pollen; (**c**) opening the wrapping to examine the cooked pollen; (**d**) cooked pollen or Kharait خريط is ready to serve. (Images courtesy of Mohammed Al-Darwesh, Iraq)

Fig. 39.53 (**a**, **b**) Kharait خريط is offered for sale in marketplaces in the urban areas of the major cities in south of Iraq

Fig. 39.54 The Iraqi thick cream known locally as "Kaimer قيمر". Note how thick the cream is and how its pieces have been rolled. (Images courtesy of Laith Jawad)

Fig. 39.55 Process of making Kaimer قيمر. (**a**) warming the buffalo milk; (**b**, **c**) after warming, the milk is mixed to mobilise the fat and stop it from gathering at the bottom of the pan; (**d**) covering the milk to keep it warm (Images courtesy of Laith Jawad)

Fig. 39.56 Final steps of making Kaimer قيمر. (**a**) cutting the thick fat layer to give way for the poured cold milk to go through; (**b**) ice crashed into small pieces to chill the added milk; (**c**) process of pouring the chilled milk over the warm milk; (**d**) rolling the thick layers of fat prior to serving Kaimer قيمر. (Images courtesy of Laith Jawad)

39.3.1.8 Butter Making عمل الزبد.

Butter is usually made by Marsh Arabs. The process of making this milk product is very simple and every housewives can do it. The essential constituent in making butter is the thin yoghurt that forms when a small quantity of yoghurt is mixed with milk and kept warm by wrapping with thick cover such as blanket. The thin yoghurt is made by mixing the thick yoghurt with certain amount of water and stirred until the mixture becomes homogenised. For making butter, a special container is designed for this purpose known as Makatha مخضه (Fig. 39.58a–c). It is a metal horizontal cylinder with a small cylinder connected 90° at the top centre. This vertical cylinder is used to pour the yoghurt through. The whole butter maker cylinder needs to be hanged so it is easy to move it forwards and backwards. The process of making butter depends on the idea that if the fatty yoghurt is mixed very well by moving it forwards and backwards, all the fat contents will gather together in lumps. To achieve such performance, the butter cylinder is usually fastened at their ends with two ropes and those ropes are connected to a tripod made of dry stem of any tree. The women making the butter will sit with the tripod holding the butter cylinder in front of her. She keeps moving it forwards and backwards. Every now and then, the women check for any clustered fat or butter is formed. There is no certain time to follow, and women in the marsh areas know by experience the time needed for the butter to form. Once the women are certainly sure that no more butter will be formed, they stop the process and take the amount of butter formed with small amount of thin yoghurt and then it is ready to serve (Fig. 39.58b). In the old

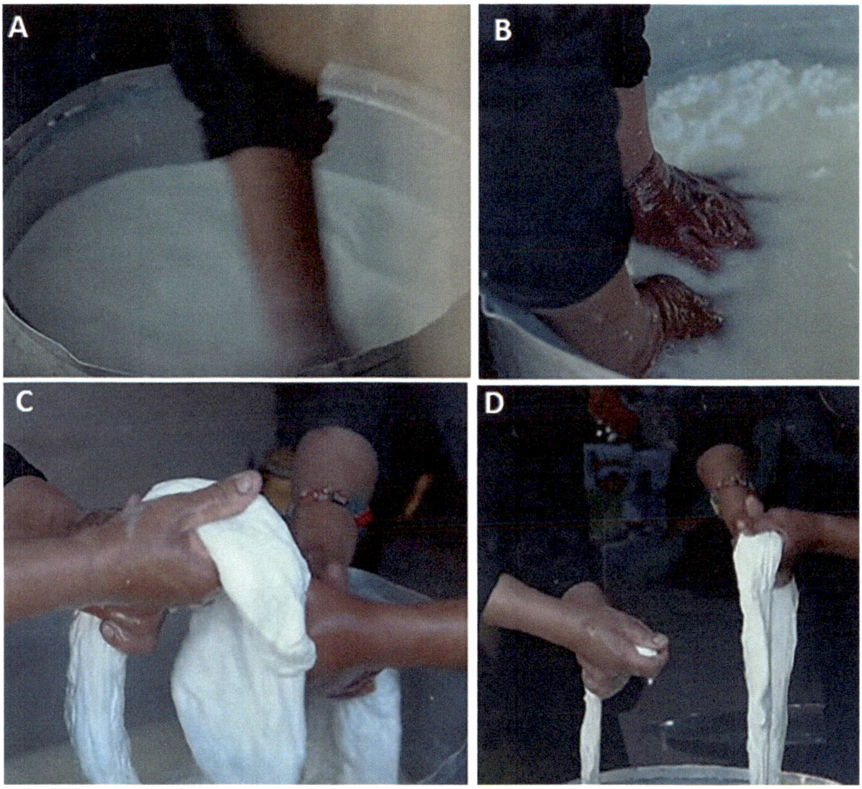

Fig. 39.57 Process of making cheese الجبن. (**a**) adding the starter to the warm milk; (**b**) mixing the milk with the starter to ensure all protein gets in contact with the starter. Note the lumps of protein coagulated; (**c**, **d**) forming the lumps of cheese in threads and then breading them. (Images courtesy of Laith Jawad)

days and before the civilisation reaches the marsh area, Marsh Arabs used to make butter in cylinder made of skin of sheep or a goat (Fig. 39.58b).

39.3.1.9 Making Flour عمل الطحين.

Making flour is one of the essential and important jobs that women and mainly the old ones do. In saying flour, it means wheat and rice. The flour that originated from wheat will be used in making the Iraqi traditional bread, while the flour that comes from rice will be used for Tabak and Siah breads. For grinding wheat and rice, a special mortar is used, and this is made of two discs made of stones sitting one on the top of each other. At the edge of the top disc, there is a hole where a long piece of wood is inserted to turn the upper disc (Fig. 39.59). At the centre of the upper disc, there is a small hole that is used to feed in the rice or wheat grains to be grinded. The ground grains come out from a space located between the two discs and fall on a

Fig. 39.58 Making butter. (**a**) women making butter using metal cylinder. The cylinder is hanged by ropes and tied to a wooden tripod; (**b**) hand-made butter mixed with thin yoghurt. It also shows the old-days cylinder used to make butter which is made of animal skin; (**c**) butter clumps floating on the top of thin yoghurt; (**d**) ready to eat butter mixed with thin yoghurt. (Image courtesy of Pinterest.nz)

piece of cloth usually put under the mortar. The stones that the grinding mortar is made of are not available at the environment of the marshes. People of the marshes usually obtain this mortar from villages at the edge of the marsh areas and have connected with urban areas.

39.3.1.10 Food-Related Activities

Habit of Drinking Tea

Tea and coffee are two heavily and daily type of drinks that Marsh Arabs take. Unlike coffee which is usually drunk at *al-mudhif* المضيف, tea is normally drunk anywhere in the marsh areas, at *al-mudhif* المضيف, homes, and at the field. The black tea is the one preferred by people at the marshes and they like it strong with lots of sugar. The tea is usually served in small glass container known as *Istikan* استكان and sits on small porcelain saucer with small spoon to mix the sugar. The habit

Fig. 39.59 Women using stone mortar to grind grains of rice and wheat to make flour. Note that mortar is composed of two heavy discs one on the top of the other, with the top disc turned to grind grains falling between the two discs. (Image courtesy of Pinterest.nz)

of drinking tea by people in the marsh areas is the same as that used by villagers and even some people in the urban areas of big city in Iraq. This habit, the person drinking tea, starts by stirring first the tea to mix the sugar, then leaving the spoon inside the *Istikan* استكان, and pouring some of the tea in the saucer while still holding the spoon inside the *Istikan* استكان (Fig. 39.60a, b). Sometimes, a piece of bread is usually dipped in the tea especially when the people are in the field working. The habit of drinking tea described above is distinctive to the people of Iraq and especially those living in south of Iraq, rural, and the marsh areas.

Fig. 39.60 Habit of drinking tea. (**a**) men drinking tea while they are standing; (**b**) old man drinking tea at the field during a break. (Image courtesy of Pinterest.nz)

Fig. 39.61 Fetching water from the clear nearby marsh for domestic usage. (**a**) women and young holding containers heading to the edge of the marsh to fetch water; (**b**) women dipping small container to get water from the marsh; (**c**) women, with covered face, filling large container with water; (**d**) women helping each other with containers filled with water. (Images courtesy of Laith Jawad)

Providing Drinking Water

Drinking water is not a problem as the Marsh Arabs are used to drinking water directly from the marsh. They usually drink from the surface area of a clear water. Although this water is not suitable for human drinking, the people of the marshes have no option. To get over the problem of heat during summertime, people at the marshes as other Iraqi do in the old days before the introduction of fridges keep water in a large mud container, with conical shape. The pointed end is directed downwards. This container is known as Hib الحب.

The mud has some degree of porosity in its structure. This will enable a seepage of small amount of water to the outside and then it evaporates. With the evaporation, the temperature of the mud container will drop and water gets slightly cooler (Fig. 39.61). A smaller version of the Hib الحب is also used in Iraq and is known as Hobana حبانه .

Among the daily job that women and young girls perform is fetching water from the clear water area in the nearby marsh. In doing so, they use different containers and pans to fill water and take back to their huts (Fig. 39.61 a–d).

39.3.2 Jobs Related with Reeds

The Marsh Arabs perform several jobs where reed is involved in. Before describing these jobs, it is worth to shed a light on the plants known as reed that inhabit the marsh area. There are two species of plants that fall under the common name reed, and these are the southern cattail *Typha domingensis* known in Iraq as *Bardi* بردي and the less common which is *Typha lugdunensis*. Both these two species of reed are well established in the southern marshes of Iraq and since the ancient time, and they manage to re-establish again after the drying of the marshes during the late Saddam Hussein regime.

Typha domingensis is found in several regions of the world such as the temperate and tropical zones (KWCSPF 2020). This species of aquatic plant is a very energetic, herbaceous perennial plant. Using the rhizomatous rootstock, it can grow and disperse in large cluster of plants having stems with branched leaves (UTP 2020).

Among the agricultural characteristics of this species are massive roots that are backing the bed of the marshes and the banks of the rivers, and the thick roots of this plant can purify the water from the pollutants (Huxley 1992). *Typha domingensis* has several medicinal usage, and this fact is probably already known by the ancestors of the Marsh Arabs who lived in the marsh area thousands of years ago. Each part of the plant has a specific medicinal use. For example, the leaves have an analgesic, antioxidant and diuretic effect (Duke and Ayensu 1985). In addition, the leaves have shown a possibility of treatment for Alzheimer's patients in containing nootropic criteria (Kumar et al. 2014). On the other hand, pollens have amazing features as they are astringent, desiccant, diuretic, haemostatic and vulnerary (Duke and Ayensu 1985). Also, it showed to stop nosebleeds, haematemesis, haematuria, uterine bleeding, dysmenorrhoea, postpartum abdominal pain and gastralgia (Yeung 1995).

Cutting, gathering and transporting the reeds are daily jobs for both men and women. The reeds are food for buffaloes and some of it are left to dry as a food source for the animal in the winter. It is not unusual to see women of all ages doing all the jobs related with the reeds (Fig. 39.62a–d).

Transporting reed is another hard daily job that Marsh Arabs perform especially women. The reed is cut either by the man or woman in the boat and is laid directly on the surface of the boat, or they have to come off the boat on to a small dry island to cut the reed and then store it in one or two stations near their boat. After that they start to upload the big bunches of the reed on the boat. After loading the reed, the journey back home starts. In many cases the driver/s of the boat put a lot of force in pushing the wooden pole in the bottom of the marsh so the boat can move forwards. Women in many instances do this job by themselves (Figs. 39.63a–d and 39.64a–c).

Fig. 39.62 Cutting the reed. (**a**) two men cutting reed in the field; (**b**) two women cutting reed in the field; (**c**) women cutting reed while she is on the boat; (**d**) old women carrying large bunch of reed. (Images courtesy of Laith Jawad)

39.3.3 Dung Patty Making

Although dung is cow's waste which is known in the marsh area as Mattal مطال, it is worth nothing in the urban areas; however it is an asset or fortune in the marsh areas. It is exactly like petrol for people living in the cities. Dry dung is the only fuel that people at the marshes rely on. It is better than the dry reed as the latter burns quickly and does not give too much heat.

In the process of recycling cow's dung, women mainly perform this job daily. Every day, the dung is collected from the area where the buffaloes are staying, which is next door to the living area of the people. Dung is usually gathered in a plate and then carried on the top of the head of the women to the collection station (Fig. 39.65a–d). The next step in processing the dung is mixing it and making patties of it, which look like a thin disc (Fig. 39.66a) after which it is piled in different shapes in the sun to dry (Fig. 39.66b–d).

Fig. 39.63 Carrying the reed to stations near the boat. (**a**) young women carrying bunch of reed on her head to deliver it to the collecting station; (**b**) two women carrying reed to the collecting station; (**c**) young lady driving a boat in which she is transporting reed to her house; (**d**) women covered her face driving boat loaded with several bunches of fresh reed. (Images courtesy of Laith Jawad)

The dry dung is usually used for household cooking and not for Iraqi traditional bread, i.e. not to be used in mud Tanoor. Among the daily cooking events are using dry dung in making Siah and Tabak breads and making tea (Fig. 39.67a, b).

Fig. 39.64 Transporting reed. (a) boat waiting to be loaded with reed before heading to the living area; (b) man and women struggling in pushing their boat forwards. Note the boat is fully loaded with reed; (c) man downloaded the catch of reed. (Images courtesy of Laith Jawad)

The buffalo waste is different in shape and colour from that of the cow. The former is black in colour, circular in shape and slightly harder. It is known as Al-Jilla الجله (Fig. 39.68a, b).

39.3.4 Milking Animals

Milking animals such as cows, buffalo and sheep is a job performed by both men and women, but for buffalo, men mainly do the milking. For cows and sheep, the milking

Fig. 39.65 Process of gathering buffalo dung. (**a**) young lady gathering dung from the area where cows stay; (**b**, **c**) young lady carrying the dung on top of her head to dump in the collection area; (**d**) dung dumping area at the living space of Marsh Arabs. (Images courtesy of Laith Jawad)

process is simpler than that of milking buffalo as you need to be professional in milking and you need to become a friend to the buffalo (Figs. 39.69a–d, 39.70a–c). There are special facts about the buffalo the milking person should know before attempting to milk a buffalo. Among these are as follows: if the buffalo has a calf, then the mother buffalo will never let you milk her until her calf comes besides her on one side and the milking man on the other. You need to know how to deal with the buffalo, i.e. you need to learn the simple meaningless words or sounds that you should direct to the buffalo while you are milking the animal. The Marsh Arabs think that the sound they produce while milking buffalo is recognisable to the buffalo in general, i.e. even from other areas or even from other countries. Finally you need to show your friendship to the buffalo before attempting to start milking (Fig. 39.71a–d).

The amount of milk that buffalo gives per female daily is far more than that given by cow (Fig. 39.72a, b). Therefore, there is always a surplus of buffalo milk available at any place in the marsh area. This surplus amount of milk is usually sold for an

Fig. 39.66 Piling dung patties for drying in the sun. (**a**) making dung patties as thin discs; (**b–d**) ways of piling dung patties for drying in the sun. (Images courtesy of Laith Jawad)

agent living at the edge of the marshes where a car can be driven. People from all small villages bring the surplus of the milk they have to the milk collection point, where the agent is present with car and the empty containers he usually bring with him to fill them with milk (Figs. 39.73a–d, 39.74a, b).

39.3.5 Feeding the Animals

Among the daily jobs that men and women do is feeding the cows and buffalo. This job seems easy, but it needs a lot of preparation as the food given for the cows and buffaloes in summer and winter is different. In summer, the fresh plants are available, while in winter, they become scarce. Therefore, a lot of effort is usually being put during summertime to collect plant that is used as food for cows and buffaloes and dry it for winter usage. The job of feeding cows is easier than that of feeding buffaloes. For the former, all you need is to put the fresh reeds in front of the animal and it will proceed in eating it (Fig. 39.75a, b). For the latter, two types of food need to be given, fresh and dry. In giving the fresh plant food, it is usually put on the floor

Fig. 39.67 Usage of dry dung patties as fuel in household cooking. (**a**) women igniting dry dung patty prior to start cooking; (**b**) women preparing to bake Siah bread using heat generated by dry dung patties; (**c**) kettle containing tea usually people drink after their meal; (**d**) women pouring the rice flour-water mixture on a hot tray heated by dry dung patties. (Images courtesy of Laith Jawad)

in front of the animal and then it will proceed in eating. In case of feeding dry food, it is usually given in a bag made of cloth with long ribbons so the bag can be tied along the neck of the buffalo. The process of fastening the food bag around the neck of the buffalo needs a professional person, who has already established a friendship with the animal (Figs. 39.76a–d).

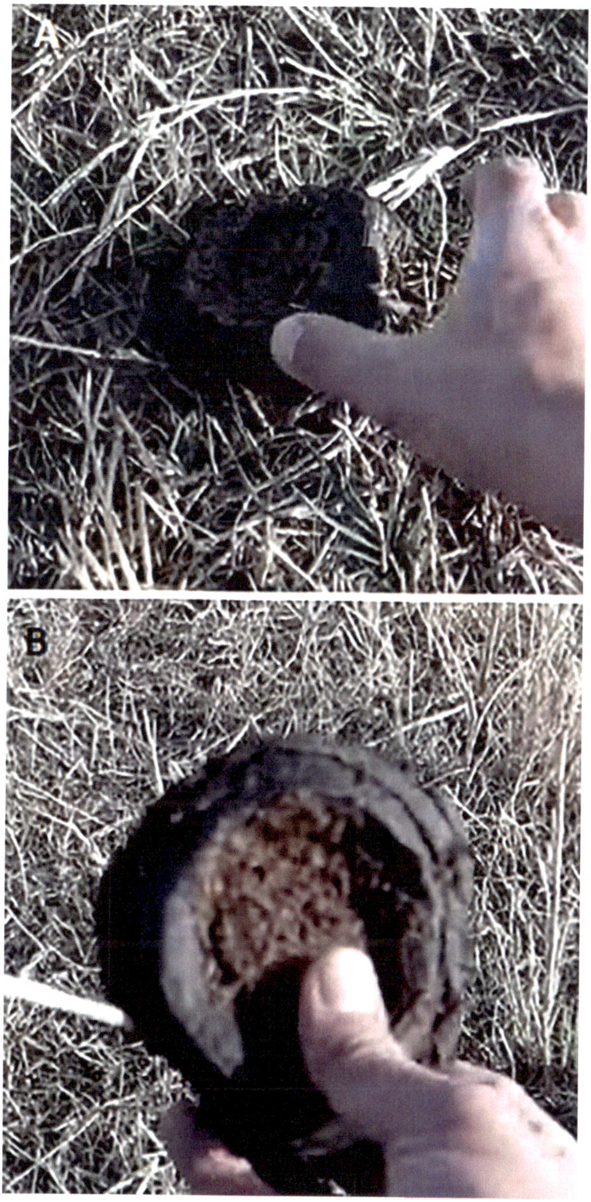

Fig. 39.68 Buffalo dry dung. (**a**, **b**) dry buffalo dung note its shape and colour, which is different from that of the cows. (Images courtesy of Mohammed Al-Darwesh, Iraq)

39.3.6 Dishwashing

Dishwashing is not ideal as they get washed in water obtained from a clear area of the marsh. In the old days when soap is not available in the marsh area, people of the marshes used to wash their dishes with the dust that is left from separating wheat

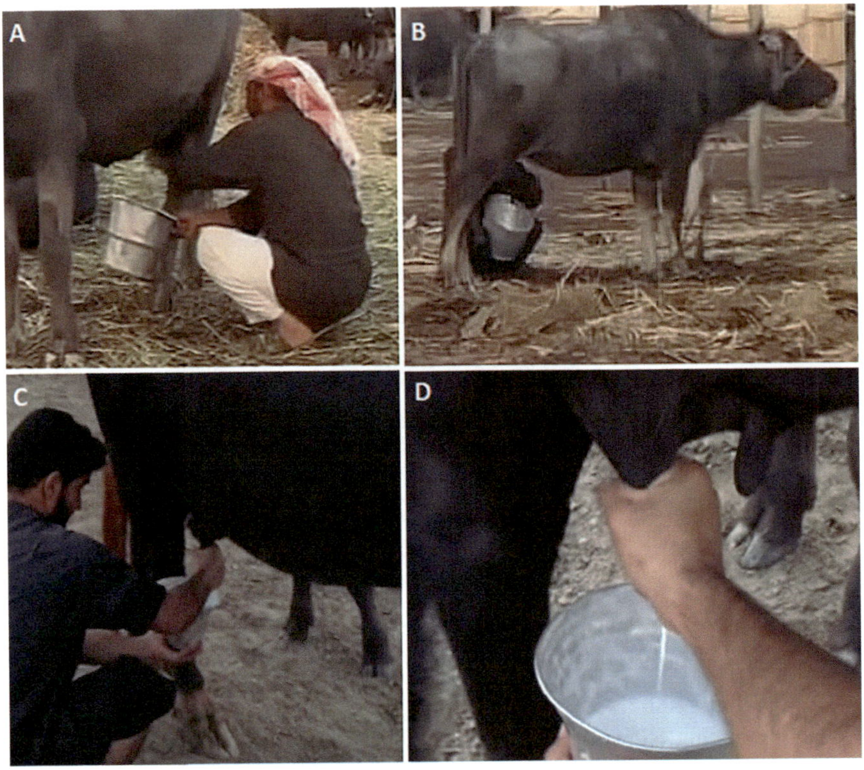

Fig. 39.69 Men collecting milk. (**a–d**) different men with the same technique of milking buffalo. (Images courtesy of Laith Jawad)

grain from their stems, but most of the people in the marshes are able to get some kind of soap to wash their dishes with from those villages at the edge of the marsh. Traditionally, water is usually fetched by containers or food pan from a clear water area of the marsh normally by women or by young girls. No men or young males are involved in this type of work as it is recognised as feminine job (Fig. 39.77a). Later, the housewives start washing the dishes outside the hut where they living in (Fig. 39.77b).

39.3.7 Taking Care of the Kids

Taking care of kids and supplying their needs is solely the responsibility of women in the community of the Marsh Arabs. One may think how the people in the marshes get their bodies washed. The answer to this question is critical and no one dares to ask anybody from the marsh area as it is considered shame. For men, they can get washed in a clear area of the marsh using soap, but in the old days, they used clay

Fig. 39.70 Women milking buffalo. (**a**) old women preparing to milk buffalo; (**b**, **c**) large amount of milk obtained from only one buffalo. (Images courtesy of Laith Jawad)

from the marsh bank. They usually cover their body with the clay and then dip in the marsh. For kids, the story is different. A wide cooking pan is used and put in the open air. In the pan, the kid usually sits in and the mother washes his/her hair and body using water from the marsh (Fig. 39.78a).

The other job that women do in regard to the care of the kids is to keep them tidy as far as possible. The level of cleanness and tidiness is not relevant to that known for people in the urban areas, but it is very basic. In the recent years, schools became in the reach of the Marsh Arabs, and they started to send their kids (boys and girls) to these schools. Kids going to the schools need to have a uniform and not all the families of the Marsh Arabs can afford to buy uniforms for their kids. Those who can afford this extra family cost usually like their kids to look tidy and smart as much as possible when mothers prepare them every day before leaving for school (Fig. 39.78b).

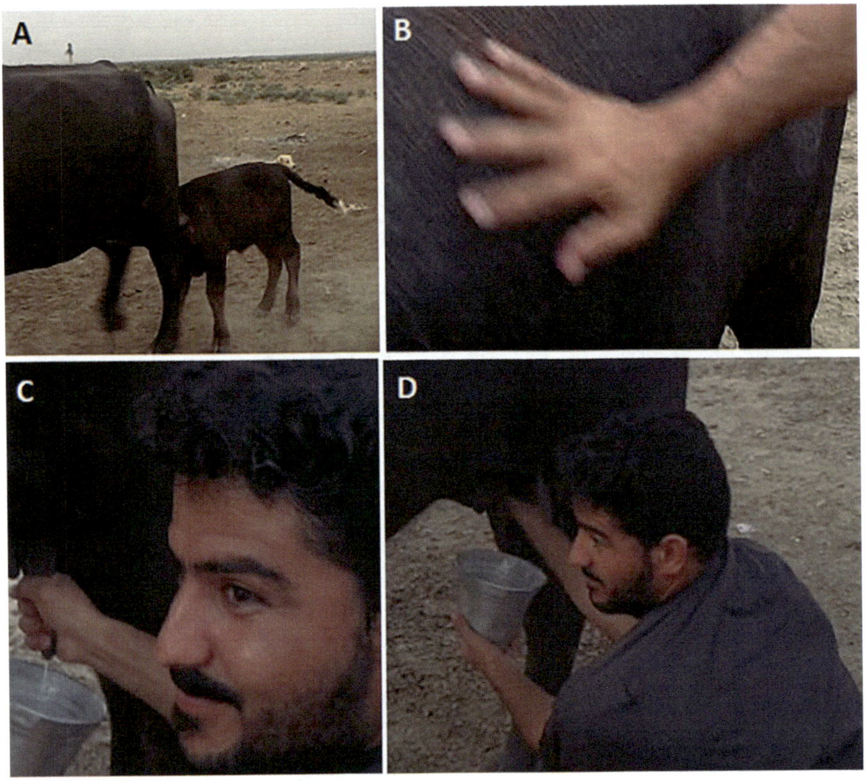

Fig. 39.71 Technique should be followed and learned before attempting to milk a buffalo. (**a**) buffalo with calf. It is dangerous to milk buffalo if she has a calf. The calf should be beside her before touching buffalo; (**b**) tapping on the body of buffalo to make it relaxed and show you are friend; (**c**, **d**) man saying meaningless words and sounds in a way to communicate with the buffalo and calm her while milking her. (Images courtesy of Laith Jawad)

39.3.8 Fishing

This a daily job for mainly men, but some women perform it too. A detailed information about fishing as daily activity is given in a separate chapter in this book. In this section, the marketing of the fish catch will be covered, which is considered a vital part from the economic perspective.

The people at the marshes stay the whole night beside their nets catching fishes. This time of the day is suitable for catching fishes especially for the technique in which they use the spears (Fig. 39.79). The technique of catching fishes during the night requires having light to attract fishes towards the surface so they will be easily caught by the spears or by the net. For this purpose, Marsh Arabs use oil lantern as light source to attract fishes to the surface (Fig. 39.80).

The sunset marks the end of fishing session for those of Marsh Arabs fishing during the day. Soon after gathering their fishing gears, they start to sort out their

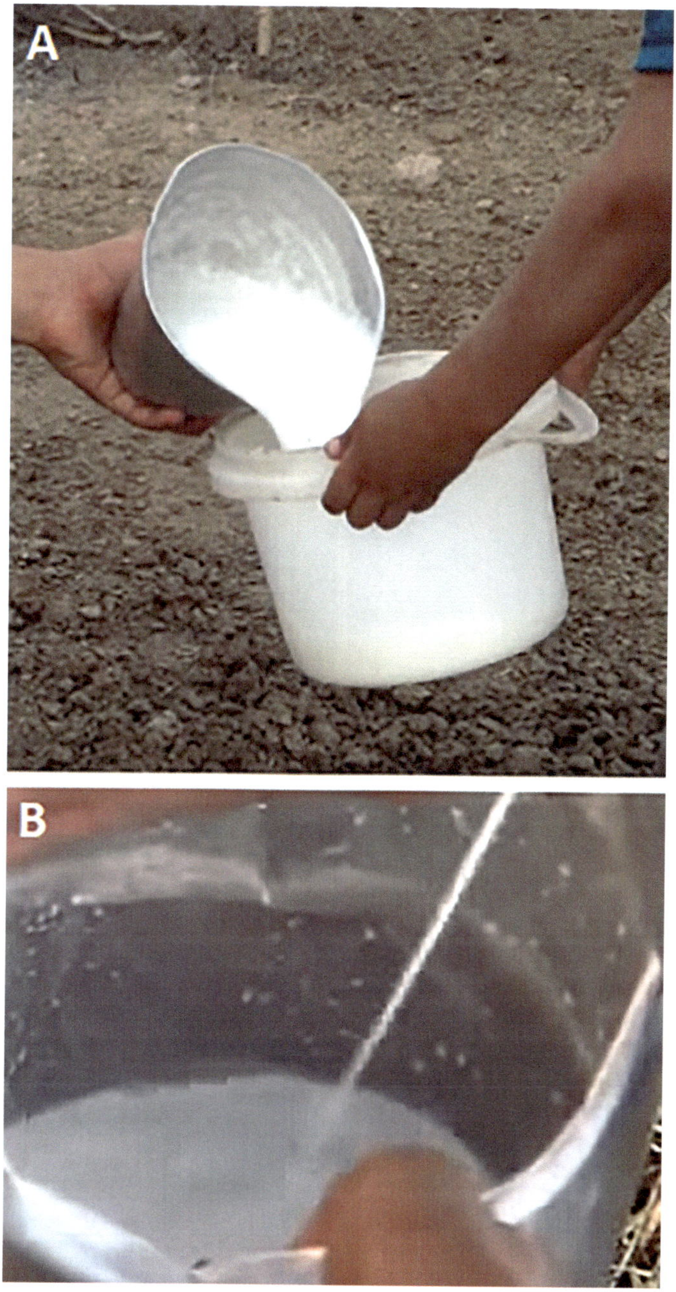

Fig. 39.72 (**a**, **b**) Large amount of milk obtained from only one buffalo compared with the amount obtained from one cow. (Images courtesy of Laith Jawad)

Fig. 39.73 Women taking surplus milk to the agent to sell it. (**a**) one old woman and young lady carrying milk containers over their head; (**b**) at the milk collection point, the agent measuring the milk brought by villagers; (**c**) unit of measuring milk; (**d**) milk bought by the agent loading on a car. (Images courtesy of Laith Jawad)

catch prior to taking it for the market in the villages at the edge of the marsh areas. Once they reach there, they will be entering in the process of a fish auction arranged by professional people in this aspect (Fig. 39.81a–d). Others prefer to sell their catch personally by taking their fishes to villages at the edge of the marshes and where cars can reach. There, they make for them a stall and sell their fish directly to people passing by (Fig. 39.82).

39.3.9 Fowl Hunting

This job is entirely men dominated and no women have been seen in the field hunting for bird. However, most women know how to use guns. Bird hunting is not an everyday job, but it is usually performed two to three times a week and whenever the stock for fresh meat is needed. Birds and fish are the two meat

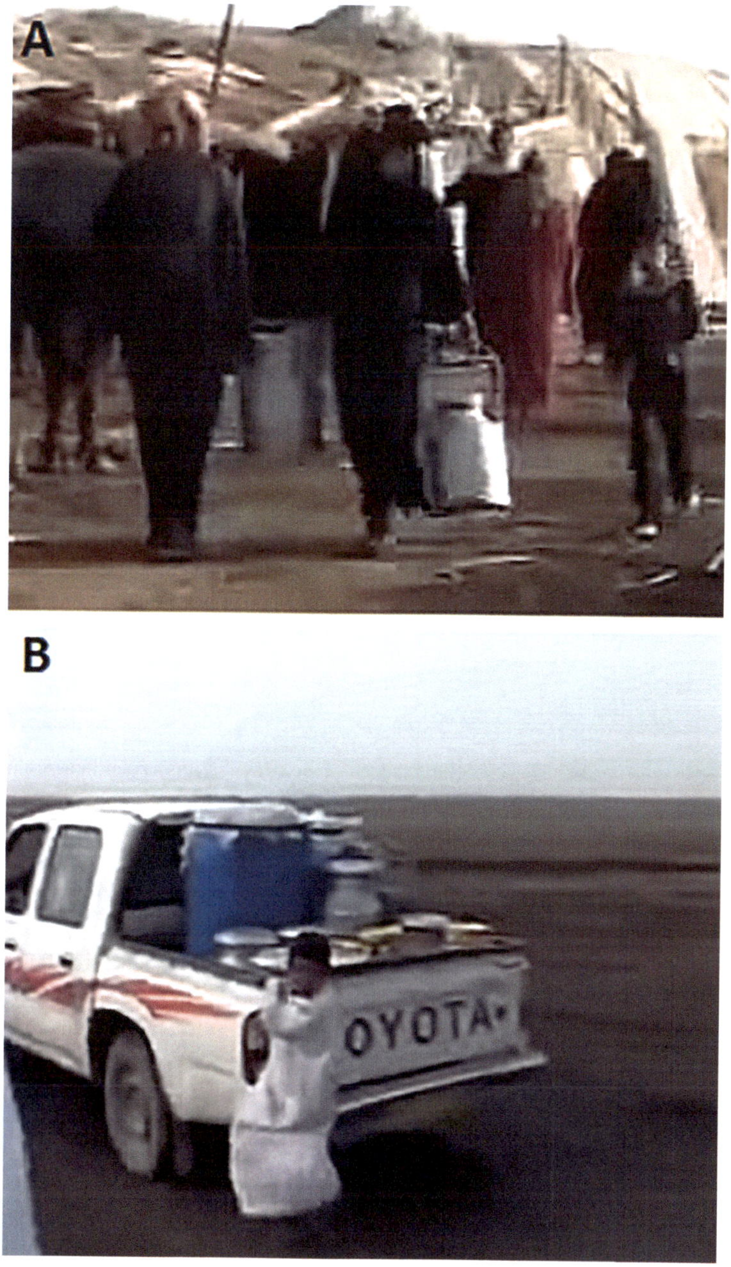

Fig. 39.74 Bringing more milk for sale. (**a**) man and woman helping each other in carrying milk containers; (**b**) car of the milk agent leaving the milk collection point heading for another point in another village. (Images courtesy of Laith Jawad)

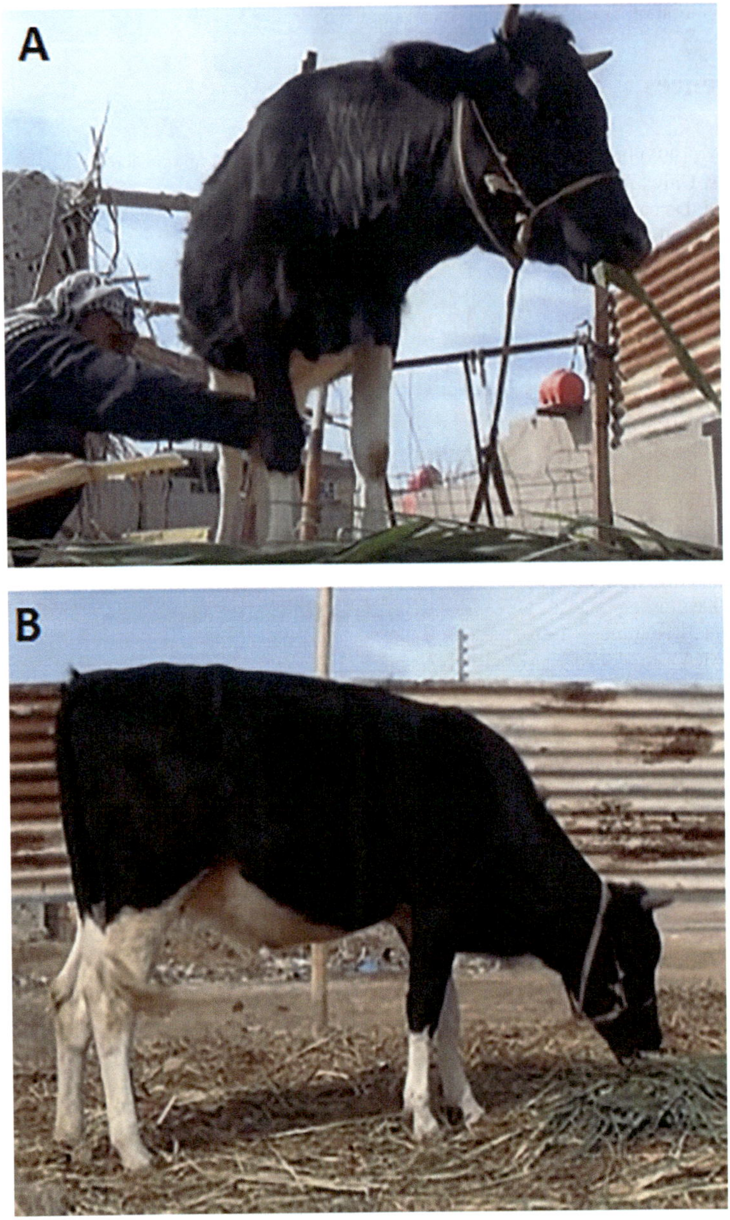

Fig. 39.75 Cow feeding. (**a**, **b**) fresh reed is offered for the animal to eat and put on the floor in front of them. (Images courtesy of Laith Jawad)

Fig. 39.76 Buffalo feeding. (**a**) bag containing dry food prior to fastening it around the neck of the animal; (**b**) fastening the bag around the neck of the buffalo; (**c**) buffaloes start eating from the bag containing dry food; (**d**) buffaloes eating fresh reed's plant. (Images courtesy of Laith Jawad)

resources for the Marsh Arabs. Cows, buffaloes and sheep if there are any are not slaughtered as they are kept as a source for milk and its products.

Usually men go for bird hunting using boats (Fig. 39.83a, b), but sometimes hunting is operated on large islands with dense vegetation that is rich in birds (Fig. 39.84). In the trip of bird hunting, several men usually go together for a whole day, or sometimes they stay overnight to hunt for birds in the early hours of the day. Most men are expert in shooting and seldom have they missed their target. Once the shotted bird falls in the water, the person driving the boat comes closer to the bird and another person collects the birds (Fig. 39.85–d). There are certain bird species that the Marsh Arabs look for and like to eat. It is one of the customs at the marsh areas that grilled hunted birds and fish are among the food offered for the guest.

Fig. 39.77 Dishwashing. (**a**) women getting water from a clear area in the marsh for dishwashing. Note the woman is struggling to put the heavy pan on the top of her head; (**b**) women cleaning dishes in an area outside the living hut. (Image courtesy of Pinterest.nz)

Fig. 39.78 Taking care of kids. (**a**) mother washing kid with water from the marsh. Note the kid is sitting in a cooking pan and the process takes place in the open air; (**b**) members of the family including the father looking tidy and clean up to the standard of the Marsh Arabs. Note the little girl is dressed in a school uniform ready to go to the school. (Image courtesy of Pinterest.nz)

Fig. 39.79 Man holding spear ready for fishing. (Image courtesy of Pinterest.nz)

In this section, a concise information about the biodiversity of waterfowl inhabiting the southern marshes of Iraq is appropriate to be given. Such information will give an idea to the readers about the importance of the marsh areas from the ecological point of view.

39.3.10 Boat Building and Repair

This type of job is performed only by men and not from any village in the marsh areas, but from those villages at the edge of the marshes and that have connection with large urban centres or cities. The Marsh Arabs usually visit the boat making centres in the villages that their inhabitants are professional in making boat and put

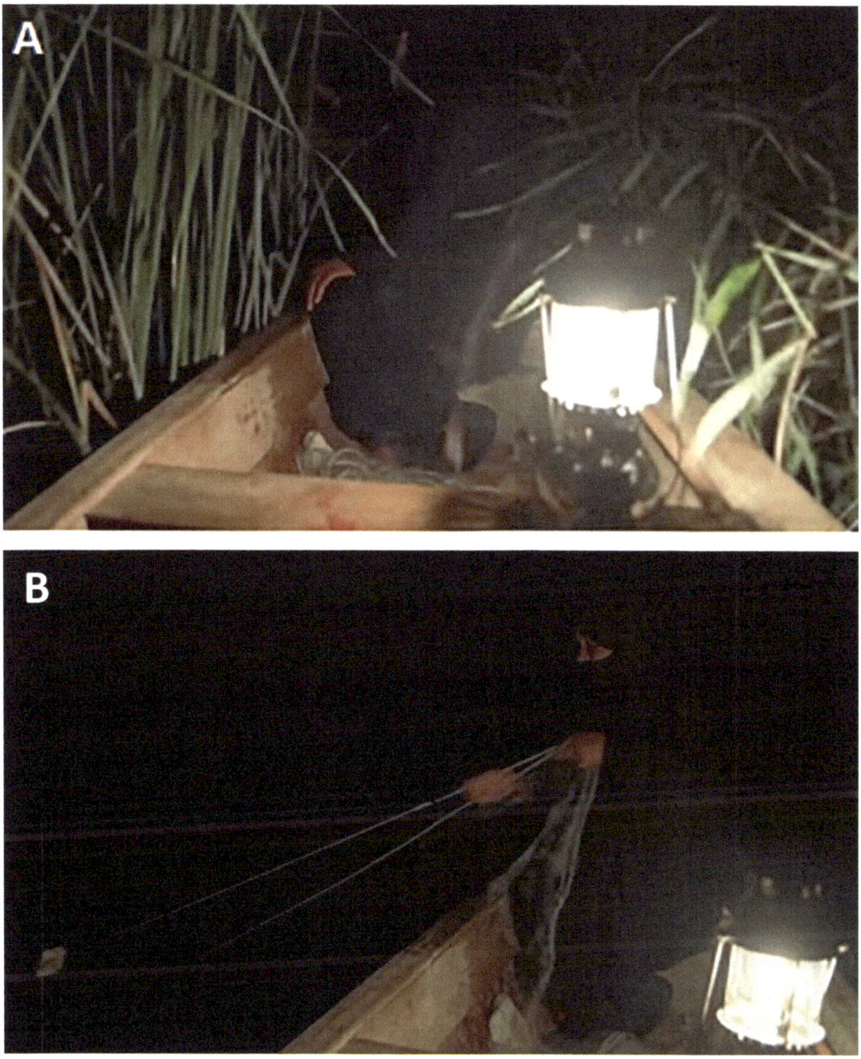

Fig. 39.80 (**a**, **b**) Oil lantern used in the night fishing session by Marsh Arabs. (Images courtesy of Laith Jawad)

their orders for making boat according to the criteria that is suitable for the usage of the boat, for example, if the boat is needed for bird hunting, collecting reeds, fishing or for transportation.

There are several types of boats "*Mashhoof* مشحوف" the Marsh Arabs used in their daily life and cannot do anything without them. The boat in its different kinds is the centre of life in the marshes. The boat or "*Mashhoof* مشحوف" is a way of transport in the marsh area that has been in practice since the time of the Sumerian 5000 BC (Fig. 39.86a, b).

Fig. 39.81 Activities after catching fish at night. (**a**) sorting the catch; (**b**) packing the catch; (**c**) heading to the auction ground; (**d**) auction ground in a village at the edge of the marsh area. Note the cars of the sellers who buy the fish catch from the marsh areas and sell it in urban fish market at nearby large cities. (Images courtesy of Laith Jawad)

The main three types of boats that are commonly used by Marsh Arabs are:

1. Boats "*Mashhoof* 7 "مشحوف feet long. This type of boat is large and usually used for reed's gathering (Fig. 39.87).
2. Al-Mukhait المخيط. This boat is slim and light, without long neck at its rear end and used in bird hunting (Fig. 39.88).
3. Al-Mudakum المدكم . It is a large boat and reaches more than 7 feet in length and is usually used for fishing and transportation (Fig. 39.89).
4. Tarada طراده . This boat is very small and is designed to take one person only. It is usually used in hunting birds as it is very low and can hide easily among the reeds (Fig. 39.90).

In making the boats "*Mashhoof* مشحوف" simple hand tools are used, which represent those being used in ancient time (Fig. 39.91). Once the professional boat builder receives the order to make "*Mashhoof* مشحوف ", he makes on the ground using white rope the design and full length of the boat. Later, the wood will be cut and curved according to the size and design requested. Once the boat is completed, its external base will be covered with a thick layer of tar, which is usually brought to

Fig. 39.82 Others prefer to sell their fish privately. Man setting his stall on the side of the main road at village at the edge of the marshes. (Images courtesy of Laith Jawad)

the manufacturer from northwestern part of Iraq. Tar is usually brought in a semihard state and it will be made a thick syrup by boiling it. A thin layer of tar is spread on the outside bottom of the new boat and on its sides. Tar will protect the boat from leaking (Figs. 39.92, 39.93, and 39.94).

Repairing boats is another job that men only do and it is usually done at the spot where the boat is found. The common damage is a leak at the base of the boat or changing a piece of wood (Fig. 39.95).

39.4 Creative and Artistic Works

The Marsh Arabs are creatives in some artefacts that they do either for their usage or for selling to improve their income. In this section the talk will go around some creative skills performed by different members of the family.

Fig. 39.83 Fowl hunting. (**a**) man sitting in a boat ready to shoot a bird; (**b**) man standing in a boat aiming for a bird in a distance. (Images courtesy of Laith Jawad)

39.4.1 Date Palm Leaves as Artistic Products

The date palm is characterised among plants as the tree that is completely useful from the roots to the tip of the leaves. For example, the roots if they are exposed can be used as fuel for making Iraqi traditional bread in the mud Tanoor, the stem is used for bridging and roofing of the huts, the fresh leaves are used for artistic artefacts, while the dry leaves are used as fuel, roofing for houses, mats حصران and other

Fig. 39.84 Fowl hunting on an island. Man hiding behind high vegetation to shoot birds. (Images courtesy of Laith Jawad)

Fig. 39.85 During a bird's hunting trip. (**a**) men got ready for bird's hunting equipped with food and guns; (**b**, **c**, **d**) man collecting shotted birds. (Images courtesy of Laith Jawad)

Fig. 39.86 The boat *Mashhoof* مشحوف. (**a**) Sumerian time 5000 BC. (Image courtesy of Pinterest.nz); (**b**) present time used by Marsh Arabs. (Images courtesy of Laith Jawad)

artefacts. The following are some of the artefacts that can be made of the date palm leaves.

39.4.1.1 Making Baskets سلال, Tabak طبگ

Different types of baskets are used for various household purposes. Therefore, they are important and can replace the household utensil that is made of glass or plastic. They are good, unbreakable and washable. Tabak on the other hand is a flat rounded plate used to hold Iraqi traditional bread in or vegetables.

Fig. 39.87 Boat 7 feet long. (Images courtesy of Laith Jawad)

Fig. 39.88 Boat Al-Mukhait المخيط. (Images courtesy of Laith Jawad)

The first step in making baskets is collecting date palm leaves and staining them with different colours using natural colour materials found in the plants grown in the marsh environment. Later on, the weaving of the basket starts with a few leaves and then increased gradually to form the body of the basket (Fig. 39.96a–d). The weaving of the date palm leaves is usually made in squarish crosses of leaves of different colours (Figs. 39.97 and 39.98).

Fig. 39.89 Boat Al-Mudakum المدكم. (Images courtesy of Laith Jawad)

Fig. 39.90 Boat Tarada طراده. (Images courtesy of Laith Jawad)

The process of making Tabak طبگ starts with making small circle from a bunch of fine threads of date palm leaves. The women doing Tabak طبگ continue adding more circles using sometimes coloured bundles of shredded date palm leaves. This process continues until the required size of Tabak طبگ is reached. The tricky part of making Tabak طبگ is when reaching the required size, how to end the process of adding circles and how to make a handle (Figs. 39.99 and 39.100). The final products are of different colourful-shaped baskets to be used for everyday jobs (Fig. 39.101a, b).

39.4.2 Reed Plant's Artistic Products

The main product that Marsh Arabs usually do is the matt that is made of reed plant. In the production of this artistic article, the reed is usually collected from the marshes and left to dry in the sun (Fig. 39.102a–c). Since the reed plant is in a form of rounded sticks and mats need a flattened material, then the dry sticks of the reed will be smashed to make them flat either manually and or by hitting them with piece of

Fig. 39.91 Primitive hand tools used in making boat *Mashhoof* مشحوف . (Images courtesy of Laith Jawad)

wood (Fig. 39.103a–c). Once the reed's sticks become flat and before we start weaving them, they are usually dipped in water for a few hours to get them soft and be able to bend them during the weaving process. The weaving job can be done by both women and men and sometimes young girls are involved so they learn the job. One person can do 1–3 reed mats a day (Fig. 39.104a, d). Making reed mats is a job the can be done on small or wide scales, and there are some agents that take the reed's mat product and sell them in the villages at the edge of the marshes or even in the cities as it serves many purposes. Therefore, a small number of ready-made mat can come from a single hut, but they usually store their products before selling it (Fig. 39.105a, d). At the collecting point where the agent, who will buy the mats and take to the market, he usually bring large truck and upload the mats in a peculiar way that looks completely unsafe to drive the truck with (Fig. 39.106a, c). In addition the reed mat is normally used as cover on the floor of the hut where the Marsh Arabs usually live, and it is also included in all building constructions such as huts and Mudhifs (Fig. 39.107a, b).

The other artefacts that can be made from the reed plant are baskets that come in different shapes and sizes and for different purposes (Fig. 39.108).

39.4.3 Weaving and Making Rugs

The skill of weaving, which includes wool spinning and threading, that the Marsh Arabs have gained is related to those of their ancestors and transferred through the

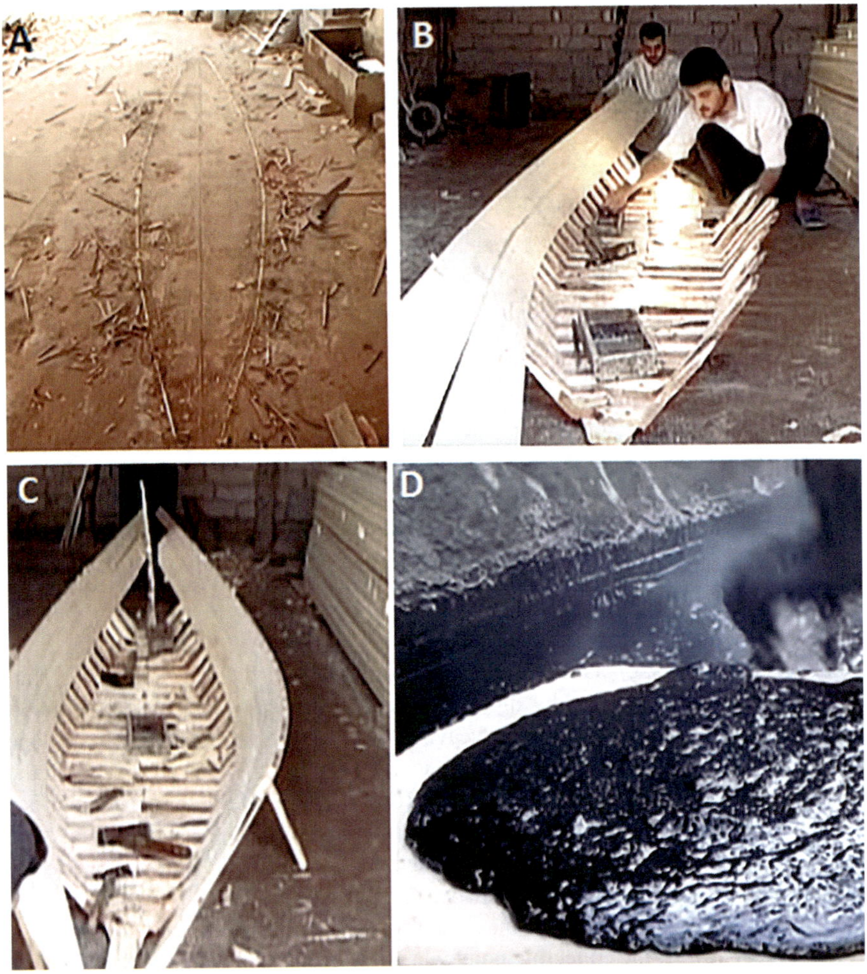

Fig. 39.92 Process of making boat *Mashhoof* مشحوف. (**a**) make a full length design on the floor using white rope; (**b**, **c**) first step in building the boat; (**d**) asphalt material used to cover the outside bottom and sides of the boat to prevent leaking. (Images courtesy of Laith Jawad)

long history to them. In order to reveal the similarity in the process of weaving and making rugs between the Marsh Arabs and their ancestors, it is important to shed a light on the history of wool spinning and weaving that started in the ancient Mesopotamia for the first time in the history of humankind thousands of years ago.

Old Babylonian literature from the first half of the second millennium BC related to trade frequently proposed that Mesopotamia is the land of wool. The history of the Ancient Near East comprises a huge chronological setting, from the first symbolic texts of the late fourth millennium to the invasion of Alexander the Great in 333 BC (Breniquet and Michel 2014). During these millennia, different groups advance in a

Fig. 39.93 The final steps in making the boat. (**a**) boiling the asphalt to make spreadable on the body of the boat; (**b**, **c**) spreading the asphalt on the bottom and sides of the boat; (**d**) after spreading the asphalt, the outside bottom of the boat is covered with clay to cool down the hot asphalt. (Images courtesy of Laith Jawad)

Fig. 39.94 The finished boat ready to use. (Images courtesy of Laith Jawad)

changing landscape where sheep (and their wool) has always shown an imperative economic part (McCorriston 1997).

Wool was initially utilised for weaving and perhaps managed by the first urban societies, but, as an original material, it also became a form of payment, given to workers as such and swapped within the outline of a basic market economy

Fig. 39.95 (**a**, **b**) Men repairing their boats locally at the bank of an island inside the marsh

(Breniquet and Michel 2014). Mesopotamia in the third millennium BC has been called the birthplace of wool; this is due to the local socioeconomic growths that directed to its augmented production, as well as enhancements in methods for the production of woollen threads and textiles. In this area, factories first began to yield textiles and clothing in an extraordinary amount (Breniquet and Michel 2014).

The utilisation of wool started as an importance of the taming and selective breeding of livestock. In southern Mesopotamia, with the beginning of socialisation in the late fifth and fourth millennia BC, archaeozoological remains propose that taking care of farm animals such as pigs and cattle started to change towards nomadic goat and sheep-herding (Breniquet 2006; Sudo 2010).

According to cuneiform literature, large-scale fabric manufacture started throughout the second half of the third millennium. Local making went on, but only women wove at home; spindles and spindle whorls became a female gender marker amongst artefacts that were retrieved from graves (Lambert 1961; Maekawa 1980).

The use of sheep for meat and wool is characterised as having a rich background in the Near East. Zooarchaeological and genetic indication show that sheep were tamed about 9000 BC (11,000 and 10,500 cal. BP) through a region that extended from Iran to Turkey (Zeder 2008, 2009). When and where special strains of sheep for wool were established is less certain. Although spindle whorls were learned about 7000 BC (9000 cal. BP), it is uncertain whether they were for spinning linen fibre or wool. By Ur III, textile manufacturing had become an industrial action (Pomponio 2010; Waetzoldt 1972, 2010).

From the above short history of humankind and the discovery of wool and its usages, it is clear that the people of ancient Mesopotamia were the first who learned the manufacturing of wool and making fabrics out of it, while the people in other parts of the world were still using fabrics to cover their bodies made from plant

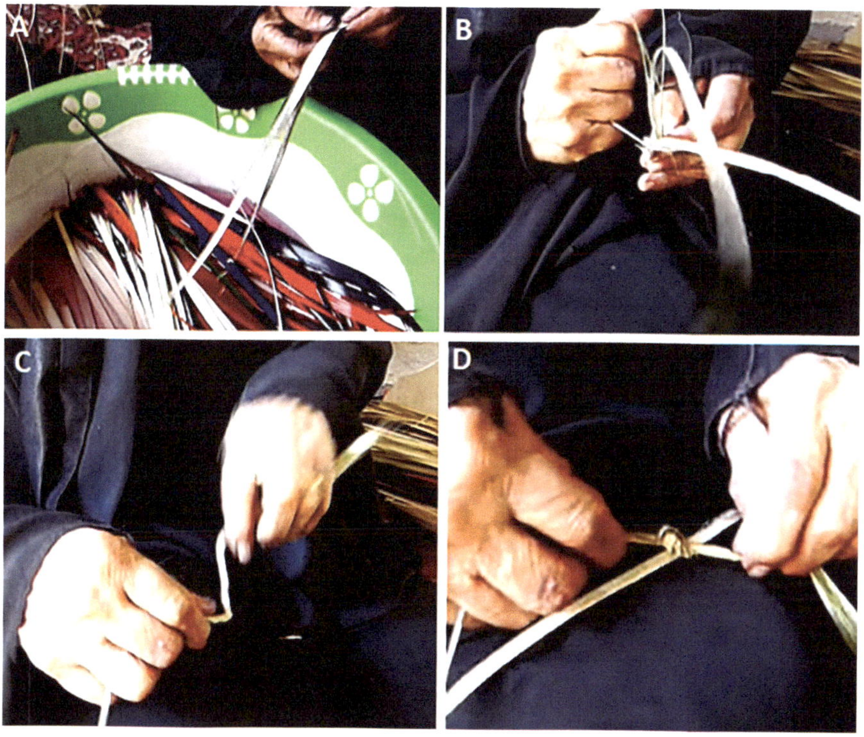

Fig. 39.96 Processes of making baskets from the date palm leaves. (**a**) dying the date palm leaves with natural colours; (**b**) the starting step in making the basket, which is small knot; (**c**) making ropes made of date palm leaves to fasten the coloured leaves; (**d**) the starting knot has been made. (Images courtesy of Laith Jawad)

sources. In the southern part of ancient Mesopotamia, the first loom was invented (Fig. 39.109) and the ancestors of the Marsh Arabs from thousands of years have used the simple wool spinners to change the wool into threads prior to start of the weaving process. In addition to fabrics, ancient Mesopotamian have created the rug making and used threads made of wool that they got from the domesticated sheep in their area.

Carpet making is one of the oldest professions that human has learned. The process of making rugs includes several steps starting with making threads from the wool, cleaning, dying and weaving. The oldest surviving carpet (Fig. 39.110) was found enclosed in ice in ancient Scythian (or perhaps Achaemenidian) tombs. The tomb is situated under a Pazyryk burial area, in the Pazyryk Valley of the high Altai Mountains in Siberia (a crossroad of Chinese, Russian and Kazakh people). It measures 6′6“ x 6’0” and framed by a border of enormous griffins. It dates back to approximately 2500–2000 BCE (Silk Road Treasure 2020).

This design is now reproduced in modern rug manufacturing in Iraq (Fig. 39.111).

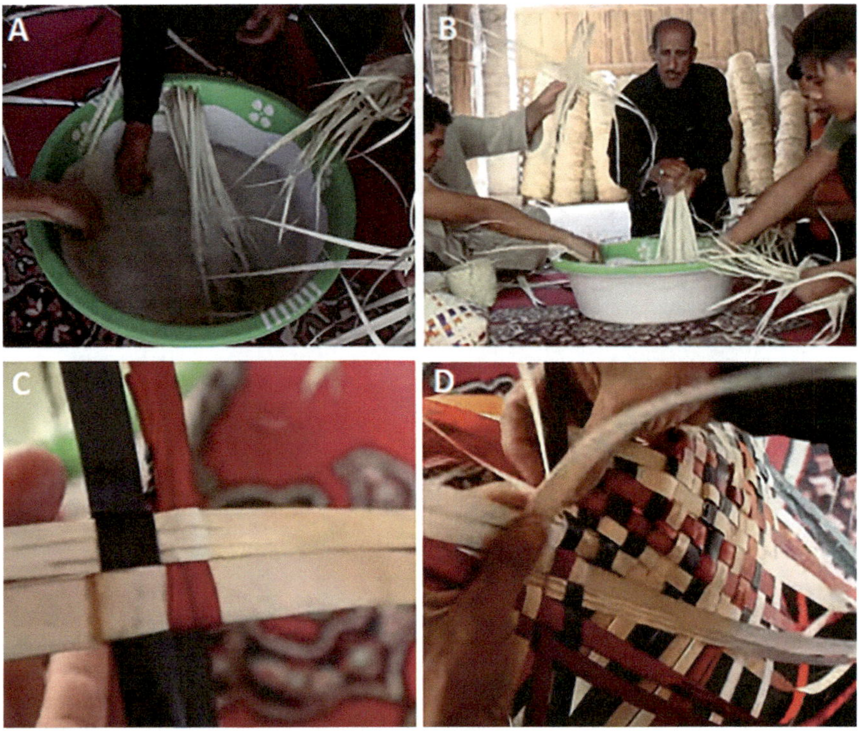

Fig. 39.97 Detailed processes of making baskets. (**a**) immersing date palm leaves in water to make them soft and easy to be woven; (**b**) group of men making baskets in a hut similar to Mudhif; (**c**) making the first cross from coloured leaves; (**d**) nearly finished basket (Images courtesy of Laith Jawad)

Rug making starts with the step of spinning the wool by using an old-style spindle and then changing it to threads. This job is usually done by women (Fig. 39.112a), but men sometimes do it too (Fig. 39.112b). The next step is cleaning the wool and dying it. After that the step of weaving starts when the loom is used. There are two types of looms that the Marsh Arabs usually use, the horizontal type, which is similar to that used by Sumerian (Fig. 39.113a) and usually used by women. The other type is vertical (Fig. 39.113b) and used normally by men. The patterns of the mats made by Marsh Arabs are very attractive as they are colourful (Fig. 39.114). In addition to the mats, the Marsh Arabs do also cushion covers, which is used often in Mudhif (Fig. 39.115).

Fig. 39.98 Making baskets commercially. (**a**) men making baskets of different shapes; (**b**) old women making large basket for sell. (Images courtesy of Laith Jawad)

Fig. 39.99 Process of making Tabak. (**a**) the starting point of making the first circle made of bundle of shredded palm date leaves; (**b**) adding circles of bundles; (**c**) joining the circles one to each other with broad shred of date palm leaves; (**d**) nearly finished Tabak. (Images courtesy of Laith Jawad)

39.5 Other Skills

The boat as we have seen from the previous sections plays a vital role in the life of the Marsh Arabs. Therefore, everyone should learn to drive the boat even the kids (Fig. 39.116). The boat is usually driven by using a long piece of wood, where the driver sticks it in the bottom of the marsh and pushes the boat forwards. The other device used in driving the boat is the oar, which can be in different shapes and sizes depending on the size of the boat. Generally, the oar is used in driving small boats that take one person only. Recently, the mechanisation has entered the marsh area and the people of the marshes started to follow the development in technology. In doing so, they fitted engines in their large-sized boat to make them faster and reach their destination in a shorter time. Petrol is used to operate these engines. The Marsh Arabs learned how to use these engines and became expert in driving their boats using such engines. In some cases, the driver stands while driving instead of sitting and holding the steering bar of the engine. With his foot, he learned skilfully how to

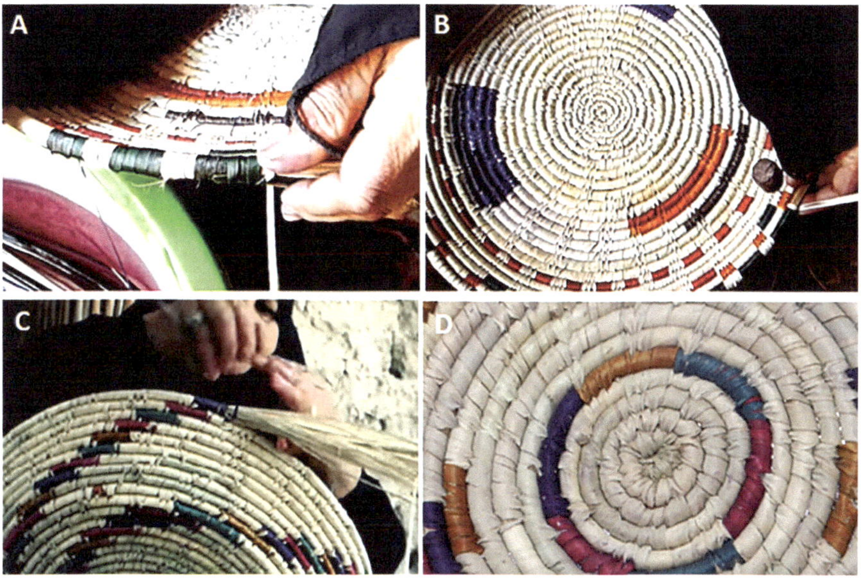

Fig. 39.100 Finish making Tabak. (**a**, **b**) making the outer circles; (**c**) making the handle; (**d**) ready-made Tabak. (Images courtesy of Laith Jawad)

control the boat by moving the steering bar with his foot. Moreover, he usually drives very fast and can manoeuvre his boat to the left and right (Fig. 39.117a, b).

39.6 Buffalo

About 150 thousands of buffaloes are raised in Iraq (Ministry of Agriculture, 2006) out of 170 million heads reported worldwide (FAO 2004). They are mainly dairy stocks that yield thick butter fat (Kaimar) and are not included in the farm services, and they, therefore, give high rates in milk and meat conversion from poor-quality feeds such as reed (*Phragticum communis*) and bardi (*Typha latifolia*) with other marsh vegetation (Juma 1997). Notwithstanding the advantage amongst cattle, buffaloes in Iraq have been disused for a long time and affected by many influences that cause severe drop in population and production (Al-Saedy 2007).

The Asian buffalo comprises two subspecies known as the river and swamp types, the outer characteristics are different as are the genetics. The river buffalo has 50 chromosomes of which five pairs are submetacentric, while 20 are acrocentric: the swamp buffalo has 48 chromosomes, of which 19 pairs are metacentric (Al-Fartosi n.d.).

In Iraq, according to results of studies attained by Borghese and Mazzi (2004), there was a total of 98,000 river Khuzestani or Iraqi buffaloes, with 40,000 adult females with 1320 kg as medium lactation milk yield, in a 270-day lactation period.

Fig. 39.101 Large Tabak and basket making. (**a**) old women finish making large Tabak by adding a handle; (**b**) final products representing baskets of different shapes and colours for everyday jobs. (Images courtesy of Laith Jawad)

Fig. 39.102 Stocking reeds for making mats. (**a**) women and young girl sitting sorting out reed sticks; (**b**) young man stacking reed stick bundles; (**c**) heap of dry reeds. (Images courtesy of Laith Jawad)

Buffaloes are raised in the marshes and swim far and wide to feed on papyrus, reeds, common ash and other plants (Al-Fartosi n.d.). Rice hulls are also given as food for buffaloes when available. The buffaloes in villages at the outskirt of big cities rarely feed on natural pasture; they are fed frequently on concentrates, green feed, straw and agricultural by-products. These animals are large in size and weight, reaching up to 900 kg. The body length is about 115.2–128 cm and the chest girth is 207.2–223.8 cm. Additionally, water buffaloes are black or grey in colour, with white sporadically shaped spots that are sometimes present on the chest, legs or tail (Abid and Fazaa 2007).

Cockrill (1981) stated that it is well established that the buffalo was tamed very early in history, but when and where is unknown. The water buffalo appeared in East Asia and mainland Southeast Asia. It extends in north and west of China and to the Indian subcontinent (Lau et al. 1998). Two contradictory theories occur concerning the source of buffalo in Iraq. The first theory assumes that wild buffalo lived in Mesopotamia and dispersed from there to other regions. Others consider that the buffalo was brought to the marshes of southern Iraq from India 13 centuries ago.

The archaeological evidences showed that the wild buffaloes have domesticated as work animal in Mesopotamia about third millennium BC (Zeuner 1963; Sousa 1983). These evidences are depicted in the cylindrical seals and ivory materials that were discovered in Ur at a royal cemetery particularly the black marbled seal

Fig. 39.103 Breaking reed stick prior to be used in making mats. (**a**) young man hitting reed sticks with heavy piece of wood; (**b**) young women flattening reed sticks; (**c**) man flattening reed sticks manually. (Images courtesy of Laith Jawad)

comprising artefacts showing Gilgamesh with water buffaloes in the rivers of Tigris and Euphrates with typical crescent horn of swamp buffalo (Aldessouky 1987) (Fig. 39.118).

The genetic investigations showed that Iraqi buffalo originated in Iraq, not from India as Indian theory appealed. And there are three main clusters: the first one included Basra, Baghdad and Al-Qadisyia, the second included Kirkuk and Missan

Fig. 39.104 (**a–d**) Weaving reed mat. (Images courtesy of Laith Jawad)

and the third included Babylon and Mosul, with highest polymorphism (Jaayid and Dragh 2013).

39.6.1 Varieties of Buffaloes

The Food and Agriculture Organization (FAO) suggested that there are two general kinds of Asian water buffalo, wild and domestic. Wild buffalo approximations are usually concealed due to the trouble of differentiating between domestic, feral and truly wild buffalo (Abid and Fazaa 2007). Global wild buffalo approximations total nearly 4000 animals and shelter is provided in special areas of Bhutan, India, Nepal and Thailand for parts of this wild population. Unluckily, the number of wild buffalo has dropped intensely, owing to interbreeding with domestic and/or feral buffalo, hunting and habitat loss.

Fig. 39.105 Carrying ready-made reed mats to the collection point for sale. (**a**) women carrying reed mat on her head; (**b**) old man happy finishing one reed mat; (**c**) ready-made reed mats rolled ready to go for sale; (**d**) ready-made reed mats on boat in their way for sale. (Images courtesy of Laith Jawad)

There are two forms of domestic Asian buffalo: swamp buffalo (*Bubalus carabanesis*), present in the eastern half of Asia, and river buffalo (*Bubalus bubalis*) which is inhabiting the western half of Asia, Egypt and Europe. *Bubalus bubalis* typically is black or dark grey with curved horns, weighing about 300–1000 kg.

Abid and Fazaa (2007) suggested that there are several kinds of buffaloes in Iraq, and these are black buffalo, white-spotted buffalo and white spotted-tail and leg buffalo. This system of recognition of types of buffalo does not comply with that introduced by Marsh Arabs, where the locals can recognise six forms of buffaloes according to the shape of their horns. These forms are as follows:

1. *Al-Hauka* الحوگه. Buffaloes of this form are characterised as having the area of the white of their eyes to be large (Fig. 39.119a).
2. *Al-Nasba* النصبه. The horns of this group are uniformly curved and their tips directed forwards (Fig. 39.119b).
3. *Al-Kora* الگوره. The horns of this form are curved and directed upwards (Fig. 39.119c).
4. *Al-Sebbla* السبله. The horns of the buffaloes of this type are slightly curved and directed to the back (Fig. 39.119d).

Fig. 39.106 Reed mats at the collection point. (**a**) ready-made reed mats rolled waiting to be loaded on the truck; (**b**) men loading truck with reed mats. (Images courtesy of Laith Jawad); (**c**) truck fully loaded with reed mats. Note the unique way in loading the truck. (Image courtesy of Pinterest.nz)

5. *Al-Gadsha* الگطشه. The horns of these buffaloes are curved and directed backwards (Fig. 39.120a).
6. *Al-Shattha* الشطحه. The horns in these buffaloes are straight and directed backwards (Fig. 39.120b).

   In some Asian countries, the recognised forms of buffaloes are according to the shape of their horns.

Sohel and Amin (2015) were able to identify three types of buffaloes in Mymensingh district, Bangladesh, according to their external features including the shape of their horn. Therefore, the method that the Marsh Arabs are following is not a haphazard one, and their system goes back to their ancestors, the Sumerians, who lived in the area thousands of years ago.

39.6.2 The Horns of the Buffaloes

The males of all species of buffaloes have enormous curving or straight large horns, with deep circular grooves on their surface ridges that sit on both upright corners of their heads (Surakamhang and Sangsubun 2015). The horns of the buffaloes are not of bony origin and do not have blood or nerve supply. Therefore, when they break

Fig. 39.107 (**a**, **b**) Constructed buildings at the marsh areas, with reed mats included in their structures. (Images courtesy of Laith Jawad)

due to physical action, they will not grow again. There are no nerves and they do not bleed when fractured (Tombolato et al. 2010). The horn of a water buffalo consists mainly of two parts, a bony core and a keratinous horn sheath. Inside the horn, there is a bony core that has blood and nerve supply and consists only of a small part of the horn itself. The remaining large part of the horn consists of substance similar to the hair of mammals known as "keratin" (Currey 2010). The majority hard part of the horn is linked with the defence and communication behaviour of the buffalo and also has some physiological functions such astemperature and water regulation (Li et al.

Fig. 39.108 Men making large baskets from reed plant for several daily jobs. (Images courtesy of Laith Jawad)

2010). As to the air temperature, buffaloes know when the weather is cold or hot. During the cold days of winter time, buffaloes do not go in the water as the water temperature is too low for them to swim. In Iraq, and during January, there is usually a week that the air temperature usually drops down and the weather becomes very cold. These days are known in Iraq as "the blue cold", referring to the human skin becoming nearly blue in colour due to the dropping of air temperature. The buffaloes can tell the air temperature through their horns and they do not go in the water. The Marsh Arabs can tell whether the 7 days of the blue cold has finished when they see the buffaloes swim in the water marking such event.

The hardness of the horns of the buffalo resulted from the mineralisation of the keratin with calcium and other salt contributes (Fraser and Macrae 1980). The horns are involved in several decoration and jewellery industries in addition to using them as handle for knives and other kitchen utensils owing to their hard structure (Surakamhang and Sangsubun 2015). In Iraq, the usage of buffalo's horns is limited in the industry of making handles for knives used in butchery and in basic agricultural tools such as the sickle knife used in cropping the agricultural yields. Among the common scenes in the marsh areas is the buffalo's fight, where two males fight over gaining females for mating (Fig. 39.121). This fight could be dangerous for the spectators as one of the buffaloes might go astray and hit the people or the shepherd that might stand watching the fight.

Owing to the hard nature of the horn of buffaloes, several engineering ideas were proposed to check their mechanical properties and degree of hardness for future

Fig. 39.109 Sumerian women weaving. (Image courtesy of De Agostini/Getty Images, Swisshippo/Getty Images)

design of toughening synthetic materials such as engineering ceramics, metals and polymers (Surakamhang and Sangsubun 2015). Since in Iraq there is a reasonable number of buffaloes, then it might be feasible to think about introducing an economic idea to manufacture hard biological materials from the horns of the buffaloes. Such materials will be environment friendly and will generate a new national revenue.

39.6.3 Herding the Buffaloes

During the herding of the flock of the buffaloes to and from the marsh area where they wallow themselves in, the shepherd usually takes one or more dogs with him to

Fig. 39.110 The oldest surviving carpet. (Image courtesy of Silk Road Treasure. https://silkroadtreasures.tumblr.com/

guard the herd from any danger and usually holds a long stick with him. It is important to mention here that buffaloes can recognise the person that they usually deal with and they don't easily let the person touch them, milk them or even ride on its back. It is not an easy job for a newcomer to the area to touch or ride on the back of the buffalo. There is a special technique that the person should follow in order to get the friendship of the buffalo before it allows you to ride on its back. The person should approach the buffalo from the back and start to put his hand on its back and rub gently. If the buffalo continues to feed without objection of having the hand of the person on its back, then this is a sign that the friendship has been approved. The person can then ride on the back of the buffalo, but gently (Figs. 39.122 and 39.123). Usually buffaloes are friendly with kids and allow them to ride on their back and swim with them in the marsh (Fig. 39.124). Also, some boys can take buffalo herd across the marsh to the feeding area (Fig. 39.125).

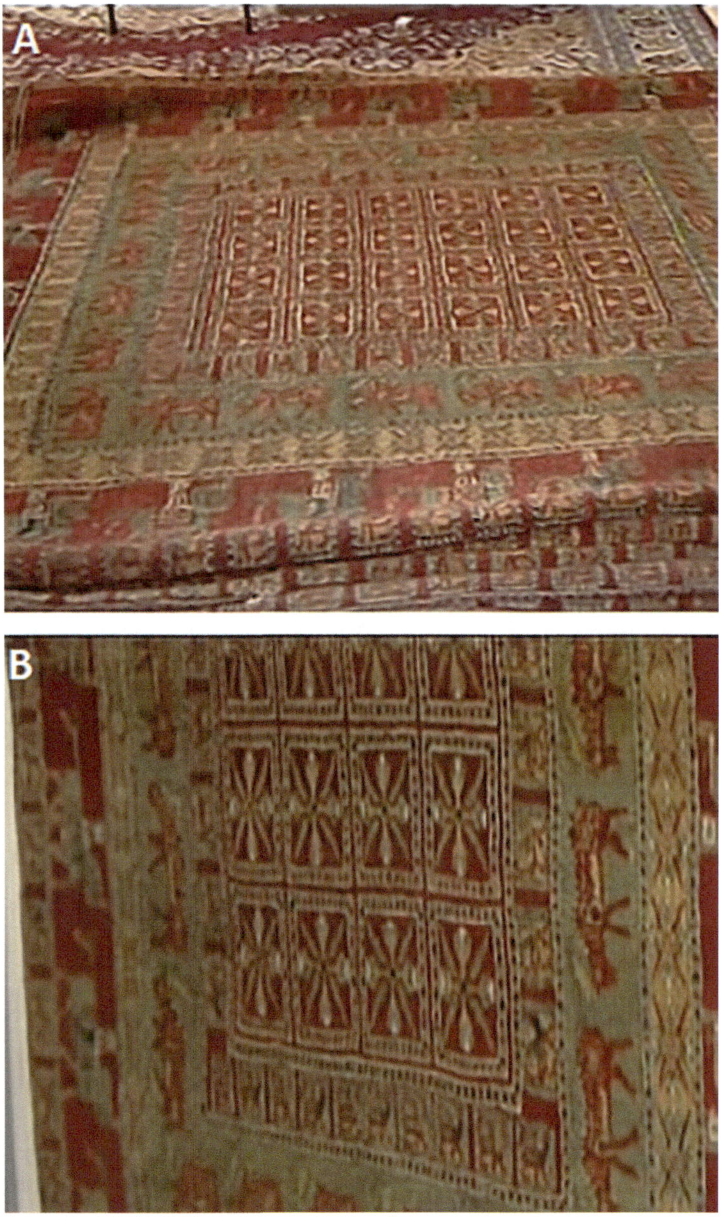

Fig. 39.111 (**a**, **b**) Duplicate design of the oldest carpet in the world made by rug's manufacture in Iraq. (Images courtesy of Laith Jawad)

Fig. 39.112 Using the old-style spindle to wool spinning. (**a**) young women using the spindle; (**b**) man using the hand spindle. (Image courtesy of Pinterest.nz)

39.7 Teaching in the Marsh Area

Due to the nature of the marshes, the teaching process is not similar to that of the cities or even in the villages bordering the marshes. Here in the marsh areas, there is no proper building for schools, where the students could sit and receive learning from the teachers. Instead, the teaching rooms are nothing but a hut larger than those the students live in. This study room as we can call it as such seldom has desks for the students to sit on; hence they sit and write on the floor. The teaching hut or the classroom is lucky if it has a blackboard that the teacher can write on. During Saddam's time, the photo of the president "Saddam" should be put on the side of the teaching hut and above the head of the teacher before any other furniture can be placed in (Fig. 39.126a). When the weather becomes hot, the students and their teachers take an open air teaching session (Fig. 39.126b). In the deep areas of the marshes, the students attend their classes with their clothes that they usually have during the day, i.e. there is no uniform to wear. The curriculum the schools in the marshes follow is the same as what other schools in different parts of Iraq follow. Therefore, there are several essential requirements the students at the schools in the marshes are missing such as the illustration facilities that are used in several subjects such as science.

Although the teaching in the marsh areas is not up to the standard, the students still show enthusiasm to learn and educate themselves (Fig. 39.127a, b). The students

Fig. 39.113 The loom. (**a**) women using the horizontal type of loom in front of kids to teach them the profession; (**b**) man using the vertical type of loom. (Image courtesy of Pinterest.nz)

find a great opportunity to change the atmosphere of their daily lives with something different. Probably, the students like very much the physical exercise classes as they spend 45 min every day stretching their bodies and having fun during this short period of time (Fig. 39.128). The other entertainment in the school that the students like is the event of raising the Iraqi flag at the morning of each Thursday, where they recite the national anthem of Iraq and the headmaster of the school gives a short talk and probably mentions the names of those students that have done very well in their

Fig. 39.114 (**a–d**) Different colourful patterns of rugs usually made by Marsh Arabs. (Images courtesy of Laith Jawad)

Fig. 39.115 Colourful cushion with attractive pattern made by Marsh Arabs. (Images courtesy of Laith Jawad)

Fig. 39.116 Young girl driving large boat using the long piece of wood that is usually used by adults only. (Image courtesy of Pinterest.nz)

study (Fig. 39.129). With the reach of electricity to the marshes and the introduction of television to the huts where the people are living, the students become more eager to learn, write and read so they can follow up the advancement in technology and understand what is happening in the world outside the marshes that they know only.

The students cannot find any assistance from the parents and grown-up at home to do homework as their parents are illiterate, so they struggle in coping up with their education.

At the edge of the marshes and where the students are able to attend schools at somehow large villages, the status of the schools is much better than those in the deep areas of the marshes. Here, separate schools for girls and boys are found, where female teachers are serving. In addition, in these schools, students usually wear uniform. Also, the schools in these villages are equipped with teaching facilities that students of the deep marshes are not able to use.

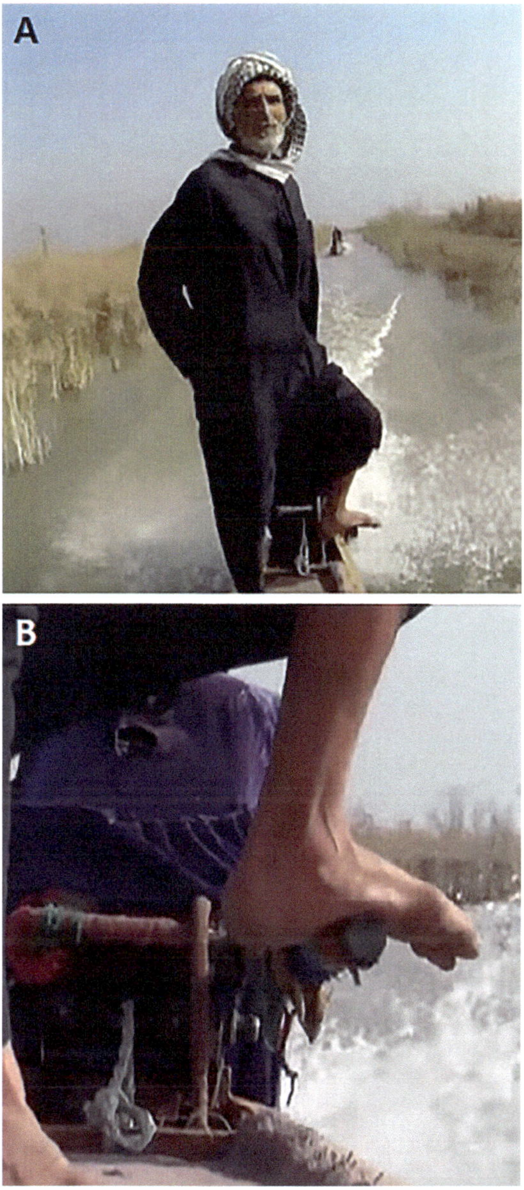

Fig. 39.117 High skill in driving a boat. (**a**) man standing while driving a boat instead of sitting; (**b**) man controlling the boat while driving with his foot. (Images courtesy of Laith Jawad)

39.8 Myths and Legends of the Marsh Arabs

As in any other places around the world, there are a lot of legends and myths of different kinds in the marsh areas that go around between people and move from one generation to another. These legends vary between the reality and illusion and

Fig. 39.118 Cylinder seal of Shar-Kali-Sharri, King of Akkad, Mesopotamia, C. 2340–2100. (Image courtesy of collection, Louvre, Paris, France)

Fig. 39.119 Buffalo type. (**a**) Al-Hauka الحوگه; (**b**) Al-Nasba النصبه; (**c**) Al-Kora الكوره; (**d**) Al-Sebbla السبله. (Images courtesy of Laith Jawad)

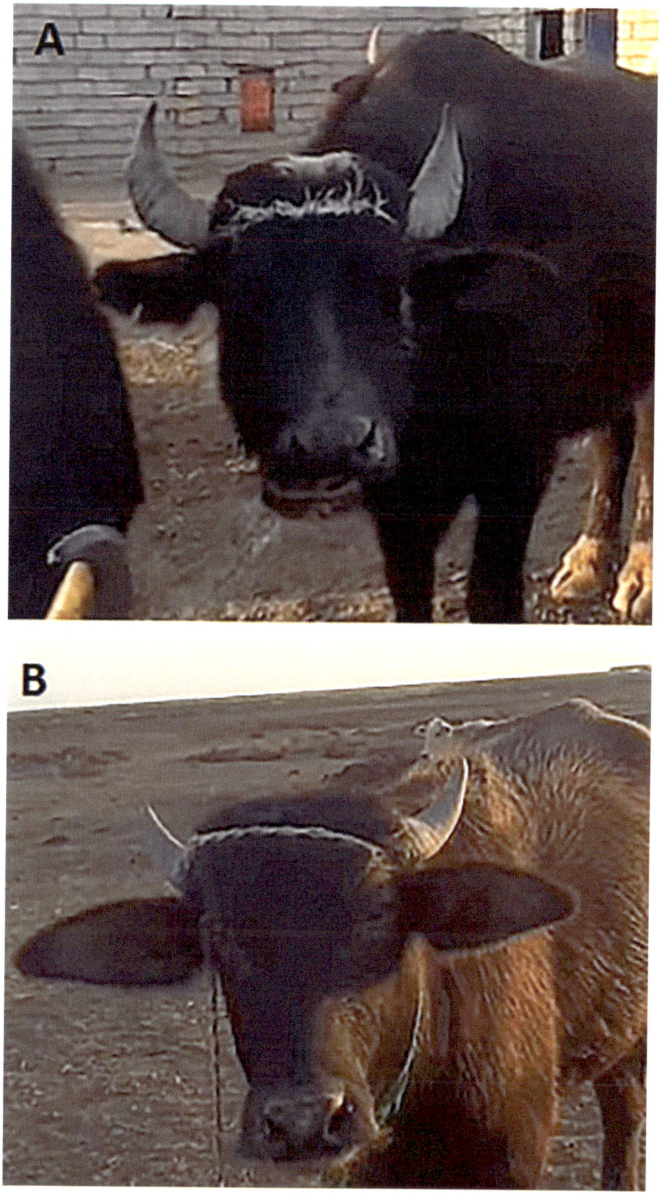

Fig. 39.120 Buffalo types. (**a**) Al-Gadsha الگطشه; (**b**) Al-Shattha الشطحه. (Images courtesy of Laith Jawad)

fiction. Adding to these is the religious inheritance of the society. With the case of the myths and legends in the marsh areas, another factor can also be involved and that is the ancient civilisations.

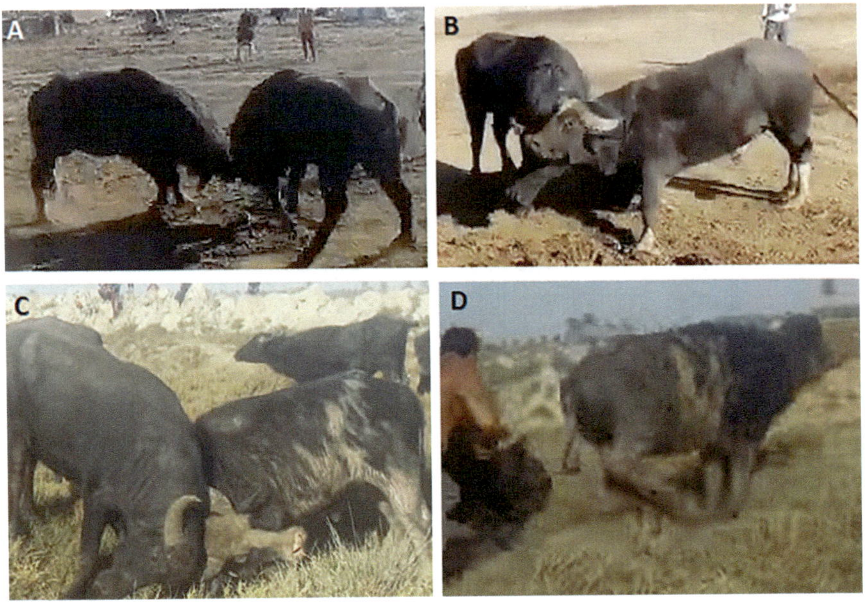

Fig. 39.121 (**a–d**) Buffaloes fighting. Note in figure "D" the defeated buffalo runs away followed by the strong buffalo. (Images courtesy of Laith Jawad)

The southern marsh areas of Iraq are located in an area that is used to be the land of great civilisation such as Sumer, Akkad and Babylon. Epics, legends and myths dominated the Sumerian literature (Kramer 1972), which is later transferred to the Assyrians and Babylonians. Therefore, it is not unusual to find a great deal of legends and myths in the marsh areas as the people of this area have inherited such stories from their ancestors.

Among the famous legends that are common among the people of the marshes are Ihfaid إحفيظ, Um Al-Hanna أم الحنه and Al-Tantal الطنطل. In the next sections, an account about each of these legends will be given to throw a light on the bases of these legends and myths and their possible origin.

39.8.1 The Legend of Ihfaid إحفيظ

Ihfaid is a "Yashen" يشن (from the Sumerian the high land) usually found in in the middle of the marsh area. Archaeologists suggested that these are remains of ancient cities that are ruined by time and then have been taken by the Marsh Arabs as inhabitants. The folklore or the stories of the marsh people say something else. They believe that these highlands were living places for people who were non-believer in God "Allah" الله and therefore, and under their feet has turn over and buried them underneath.

Fig. 39.122 An attempt of Mohammed Al-Darwesh from Missan province, Iraq, to ride on the back of the buffalo. (a) Mohammed getting nearer the buffalo herd; (b) Mohammed stopped by barking guard dog; (c) finally the dog let him to come nearer; (d) young boy, the shepherd, showing Mohammed the technique to become a friend with the buffalo. (Images courtesy of Mohammed Al-Darwesh, Iraq)

The Ihfaid Yashen now is nothing but heaps of bricks, broken pottery and dilapidated walls that used to be the fence that encloses the ancient cities. The legends say that whoever visit this Yashen will not come back and get killed definitely as this area is haunted with demons الجن. Among the interesting stories that talks about Ihfaid Yashen is that there is a bright light that comes out of this land every Friday night of the week and people who saw this light narrated that strange creatures look similar to human, but very tall, and anybody trying to reach there will be found dead in the morning (Figs. 39.130 and 39.131).

The other stories say that Ihfaid Yashen is a land full of trees, fruits, gold and precious stones, but all this is guarded by demons that do not allowed anybody to reach there and take anything away. This land is a remnant of some kingdoms known as "Al-Akar" العكر, "Abu Shathir" ابوشذر, "Ihfaid" إحفيظ and "Al-Wajif" الواجف. People in these kingdoms used to live happily and were very rich having all the resources of food and water, but later on they became non-believers and so God destroyed their kingdoms. This story is somewhat similar to the account of the Epic of Gilgamesh.

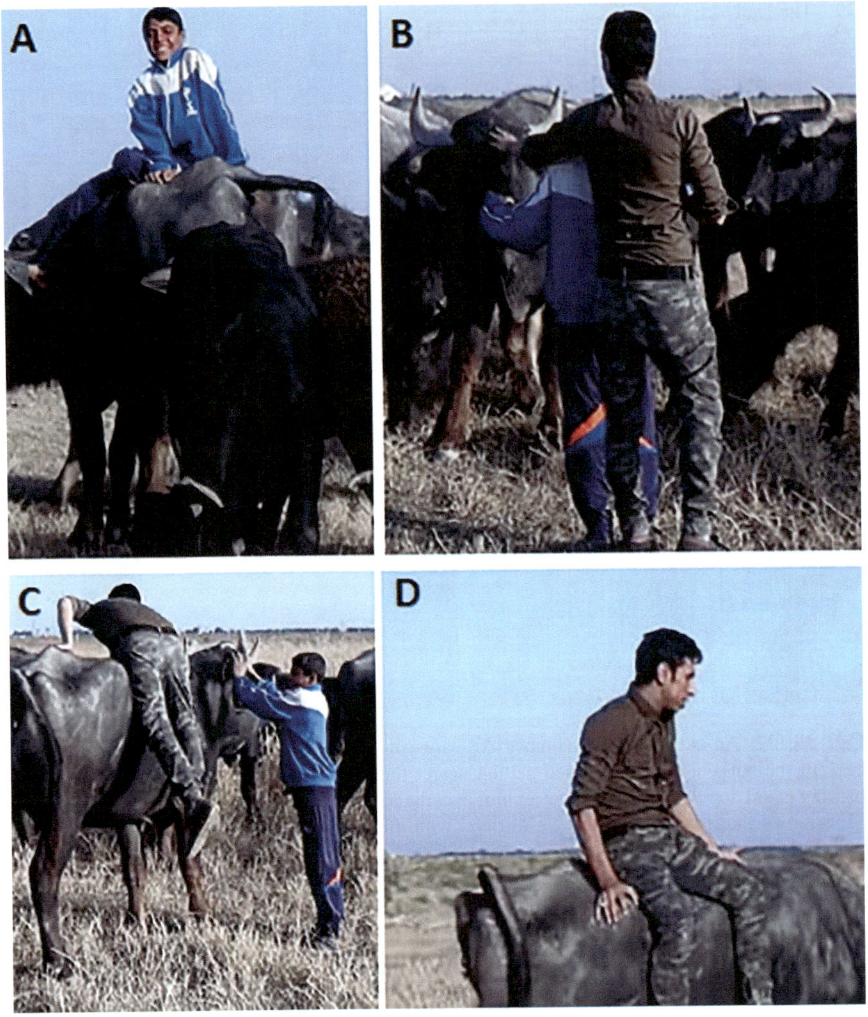

Fig. 39.123 The attempt continued. (**a**) the young shepherd managed to get on the back of the buffalo and showed Mohammed the way; (**b**) Mohammed trying cautiously with the help of the young shepherd; (**c**) buffalo accept the friendship of Mohammed and he is trying to ride the buffalo; (**d**) finally, Mohammed on the back of the buffalo. (Images courtesy of Mohammed Al-Darwesh, Iraq)

39.8.2 Um Al-Hanna أم الحنه

This legend or as it is also known as Yashen Um Al-Hanna is not common as Ihfaid Yashen. The stories about this legend say that a light comes out of this highland in the marsh area during the night and people can hear strange noises. The people of the marshes tell interesting stories about this legend. They said that during the night, they can see strange looking people doing their work and everything disappears once

Fig. 39.124 Buffaloes like kids. (**a**) kid playing with the calf of buffalo; (**b**) swimming with buffaloes. Kids swimming with the buffaloes in the marsh; (**c**) young boy taking his little sister for a ride on the back of buffalo; (**d**) jumping on the back of buffaloes. Note skilful are the kids as they continue jumping from the back of one buffalo to another without falling down. (Image courtesy of Pinterest.nz)

the first light of the sun appears. The Yashen Um Al-Hanna is located in an area which is used to be a cemetery where people who pass by the area can hear the noises of the dead, and when they dig in this highland, several artefacts used as torturing tools were found. No substantial evidences were obtained so far from this Yashen.

As with Ihfaid Yashen no one dares to come close to this area of the marsh as it is dreadful. Nothing has been given in the literature so far about the image of a very thin old woman that usually shows whenever Yashen Um Al-Hanna is mentioned (Fig. 39.132). In addition no stories were said about this woman.

39.8.3 Al-Tantal الطنطل

This legend is about an imaginary creature that looks like human and is very tall. This creature can take the shape of dog, donkey or a cat and is able to produce

Fig. 39.125 (**a**, **b**) Young boy herding buffaloes and taking them to the feeding ground. (Images courtesy of Laith Jawad)

different noises to frighten people. Iraqi people were the only nation among the Arabs that have this legend in their folklore history.

The name Al-Tantal الطنطل could be originated from the Greek "Tantalous", which is one of the underworld inmate in Greek mythology. The origin of the

Fig. 39.126 Hut as a teaching class. (**a**) teacher at one of the schools in the deep marsh area. He is teaching math subject. Note the photo of Saddam Hussein hanging on the top of the hut. This is an obligatory issue to put the president photo in every class of any school in Iraq. (**b**) an English Language teacher giving his lesson in an open air session during the hot season. (Image courtesy of Pinterest.nz)

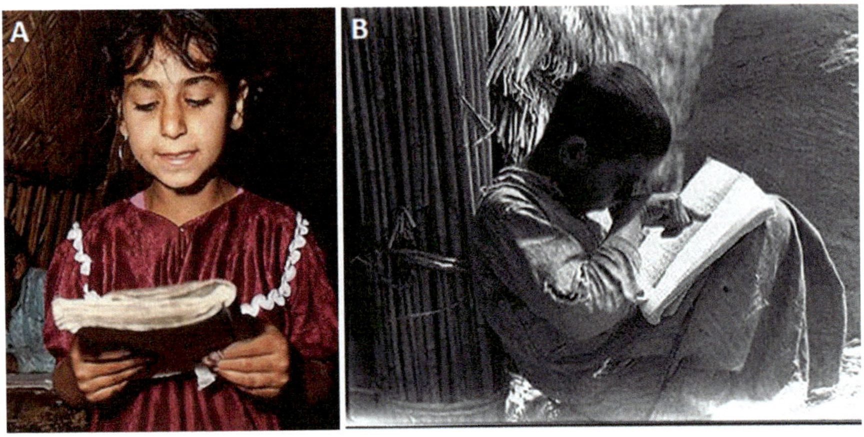

Fig. 39.127 Keenness to learn. (a) young girl doing her homework by reading; (b) young boy struggling to read. (Image courtesy of Pinterest.nz)

Fig. 39.128 Physical exercise at the school. PE teacher doing some exercises with his students during the PE class. (Image courtesy of Pinterest.nz)

name is not properly fitting for the name of the legend found in the Iraqi mythology, which might come from the ancient Mesopotamian mythology. There are two stories that shed light on the origin of the name of this legend. The first story originates from the time of the invasion of the Mongols to Baghdad in 1265 under the command of Hulagu Khan. The granddaughter of this Mongol leader decided to dread people of Baghdad and put them under control. In doing so, she created a story about a giant

Fig. 39.129 Raising the Iraqi flag on Thursday morning every week. (Image courtesy of Pinterest.nz)

creature that can take the shape of different animals and come out during the night and kill people. There is no special name given for the creature that the granddaughter of the great leader has invented. The other story, which seems more acceptable logically, happened during the invasion of Iraq by the British troops. When the British troops completely controlled the cities of Iraq, they moved to the marsh area

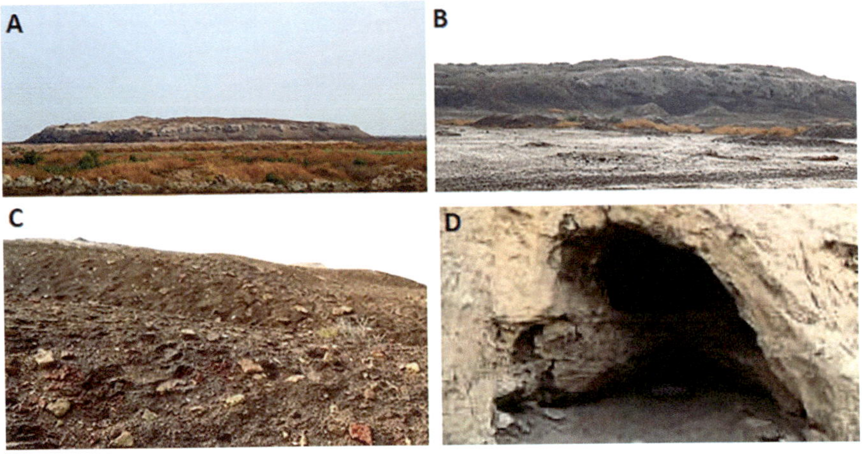

Fig. 39.130 (**a–d**) General outlook of Ihfaid Yashen. (Images courtesy of Between Two Rivers TV Channel)

to put it under their control. As with any invaders, the British built military camps on the drylands near the marsh areas as a military point to serve as places where the store of arms is kept. Once the Marsh Arabs came to know about these stores of arms, they started to steal the arms from the British Army stores. To stop the locals of the marshes was a difficult issue. Therefore, the army commanding officer decided to invent a story that scared the marsh people and stop them from stealing the arms. He asked some soldiers to circulate that there is a tall creature that comes out at night and kills anybody found in his way. This creature has ten tails. The word "Ten-Tails" went around the locals of the marshes and was finally changed to "Tantal", which is more likely the origin where it comes from. Others think that the name "Tantal" has a connection to the ancient Mesopotamian myths and legends and in particular the Sumerians, but there is no a profound evidence to support this suggestion.

In Iraq, both the people of the cities and marsh areas know the legend of "Tantal", but the creature in each case is different. Although the creature in this legend is dangerous, in the marsh area it is ferocious. To keep the Tantal away from you while you are going out at night or keep it away from the house, you need to take a long needle with thread and a small amount of salt. Housewives usually spread the salt in the corner of the house to keep Tantal away. No explanations are given to the use of the needle, the thread and the salt in keeping away the Tantal. Since Tantal is imagined to be very tall, interesting stories related to this character that the people call the very tall person Tantal (Fig. 39.133).

Fig. 39.131 (**a**, **b**) Ruins of an old city as seen in Ihfaid Yashen. (Images courtesy of Between Two Rivers TV Channel)

39.8.4 Yashen Abu Al-Thahab يشن أبو الذهب

This legend is narrated in the marsh areas of Al-Huwaiza, Al-Salam and Al-Mejar. People talking about this legend say that during the night, a boat full of gold appears on the surface of the marsh and stays there until the sun rises. Several people tried to get to the boat, but they failed and died accordingly.

Fig. 39.132 The old and very thin women supposed to b Um Al-Hanna. There are no stories supporting the place where this object is found. (Images courtesy of Al-Istora TV Channel)

Fig. 39.133 The Tantal, an artistic imagination. (Images courtesy of Laith Jawad)

39.9 Social Life of the Marsh Arabs

The social life in the marsh area does exist, but in a different way to that of the cities and even the large villages depending on the availability of equipments and the need of the people. For example, for many years, the people in the marsh areas lived without electricity, which means that all electric equipments were not part of their lives. For instance, the televisions and radio that are considered the main two equipments that control the lifestyle of the people in the cities are not available in the marsh areas even for some years ago. Instead, the people got used to living without them and continue to perform social life especially at night through meeting, drinking tea and having chat and sometimes they hear songs from people with nice voices singing different kinds of songs mainly about lovers.

Women are usually getting together and discussing feminine subjects away from the men. Joining such meeting are young girls and kids. In such meeting, women who have young male adults in the age of marriage usually discuss this subject with the other women that have young girls in the age of marriage and arrange for any future marriage. They usually drink tea and some old women smoke cigarettes (Fig. 39.134a). When women go to meetings, they usually put their best dresses and head cover. Women in the marsh areas have a specific way of covering their heads. They put the black cloth material that they cover their heads with in a way similar to a turban. The way they do this cover is comparable to that done by women in ancient Mesopotamia (Fig. 39.134 b, c).

The women's meeting may receive visitors from the nearby hut gathering. Those visitors took a boat to travel and paddle it by themselves (Fig. 39.135).

Men, on the other hand, also have their own meetings which usually take place in either Al-Mudhif if there is a visitor coming from another village or they meet to discuss one another's tribe issues with the head of the tribe "Sheikh" الشيخ. To attend such meeting, men usually dress in their formal dress (Fig. 39.136).

39.9.1 Carrying String Beads "Subhah" سبحه *by Men*

The "Subhah" سبحه is used by men and women across the Islamic countries for religion purposes. The Arabic name "Subhah" سبحه is originated from the Arabic word "Tasbeah" تسبيح, which means mentioning the name of God "Allah" الله daily and as many as possible. Marsh Arabs are no different about this habit and their men and women are using "Subhah" سبحه, and each person carries one in his pocket and starts using it once he or she sits.

The "Subhah" سبحه or string of beads has shown to be used in the ancient Mesopotamia in the period 883–856 BC. The sculpture in the palace of King Ashur-nasir-pal II of Assyria (reigned 883–859 BC) shows one of the goddesses who appears to hold a string of beads with her finger shown to be moving the beads. This goddess might be Ishtar (Inana) (Green 1992) (Fig. 39.137).

Fig. 39.134 (**a**) women doing their meeting in the sun outside their huts probably during Friday. Women are drinking tea. (Image courtesy of Pinterest.nz); (**b**) statue of head of women wearing a turban headdress as an ancient Mesopotamian style. This statue is located in the Museum of Fine Arts, Boston, MA, USA. (Image courtesy of Babylon Chronicle); (**c**) present-time Marsh Arab women wearing headdress made in as similar way to that of the ancient Mesopotamian women. (Image courtesy of Pinterest.nz)

The “Subhah” سبحه of Muslims is composed of either 33 or 99 and usually 99 as this number resembles the number of names of Allah الله. If the “Subhah” سبحه contains 99 beads, then a small bead of different shapes is usually inserted after the bead no. 33 and another one inserted after the bead number 66. At the end of the

Fig. 39.135 (**a**, **b**) women dressed to attend a women's meeting somewhere in one of the hut gatherings in the marsh area. (Images courtesy of Laith Jawad)

"Subhah" سبحه, a longer bead is usually inserted representing minaret. This piece is sometimes made of silver or even gold depending on how wealthy the person who owns the "Subhah" سبحه. The circular shape that the "Subhah" سبحه makes represents unity and integrity of the universe in Islamic philosophy.

Fig. 39.136 Men wearing their best dresses to attend a formal meeting. (Image courtesy of Pinterest.nz)

The "Subhah" سبحه can be made from any material such as wood, plastic or precious stones. This will depend on the ability of the person to attain any of those string of beads.

In the marsh areas and mainly in the Al-Mudhif, men usually use "Subhah" سبحه while they are sitting, chatting and drinking tea and coffee. The colour and the material that "Subhah" سبحه are made of are different from one person to another (Fig. 39.138).

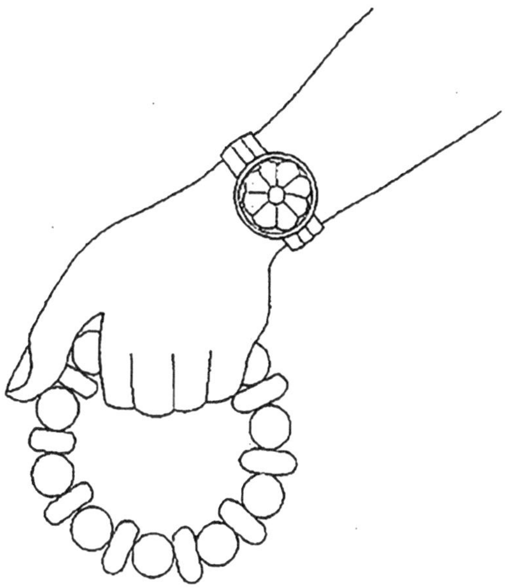

Fig. 39.137 A string of beads, held in the hand of a goddess. This image appeared in the wall relief at the royal palace of the Assyrian king Ashurnasirpal (reigned 883-859 BC). (Image courtesy of Green 1992)

39.9.2 Wearing Rings

Since the ancient civilisations, wearing rings is mainly done by women. This habit has extended to cover men too. In the Islamic world, wearing gold by men is religiously not favourable, and men if they would like to wear rings, they should wear those rings made of anything but gold. The Shia sector of Islam believes that God "Allah" الله has informed his Prophet Mohamed via the messenger Gabriel to wear a ring with sapphire عقيق stone from Yemen and asked him to tell his cousin Imam Ali to do so too. Therefore, nearly all men of Shia Muslims are wearing one or two rings with large stone in their hands.

The gemstones that were used in the rings worn by men in Iraq in general including the marsh area are amber; diamond; Yemeni sapphire, which can be red, yellow and blue, aquamarine, eye of the cat; amethyst; emerald; corundum and malachite stone (Fig. 39.139).

The aim of men wearing rings is versatile as it includes getting better health, being successful in the daily life, attracting the favourable reception of others and for spiritual purposes such as to avoid the evil eye. The Yemeni red sapphire is believed to be good for some circulatory system disorders such as stopping haemorrhage in the body and reducing heart beat and blood pressure. The yellow sapphire on the other hand is believed to stop palpitation, while the malachite green stone is for extracting the kidney stone.

Recently in Iraq, the commodity of men's rings found a wide market as the majority of Shia Muslims were inclined to wear one or more rings in their hands;

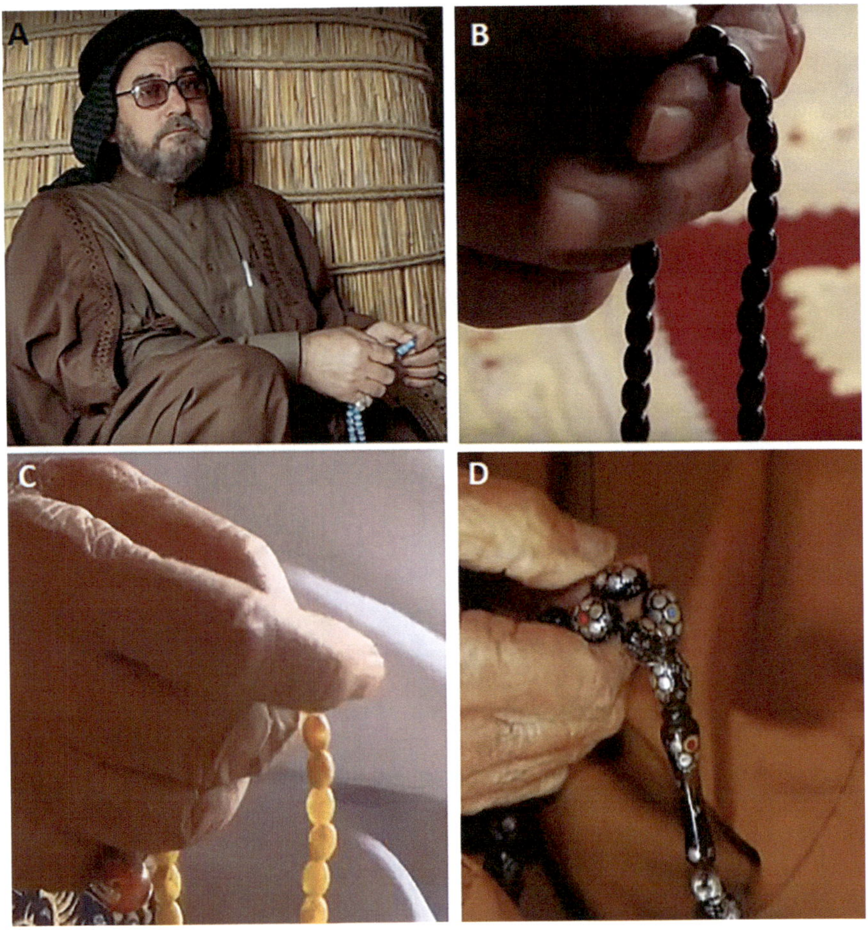

Fig. 39.138 (**a–d**) the "Subhah" سبحه in the hands of different people sitting at Al-Mudhif. Note the differences in the stones of the beads. (Images courtesy of Laith Jawad)

hence special shops or even stalls on the side of the road were created in all cities in Iraq offering a wide range of men's rings for sale (Fig. 39.140).

Men of the Marsh Arabs seen usually wear their rings of different colours, shapes and stones every day while they are working, visiting, sitting or chatting in Al-Mudhif (Fig. 39.141).

39.9.3 Carrying Arms

Carrying arms or different guns in the marsh areas is a sign of strength and manhood. The sheikh of any tribe usually carries a full set of arms, and sitting in his Al-Mudhif

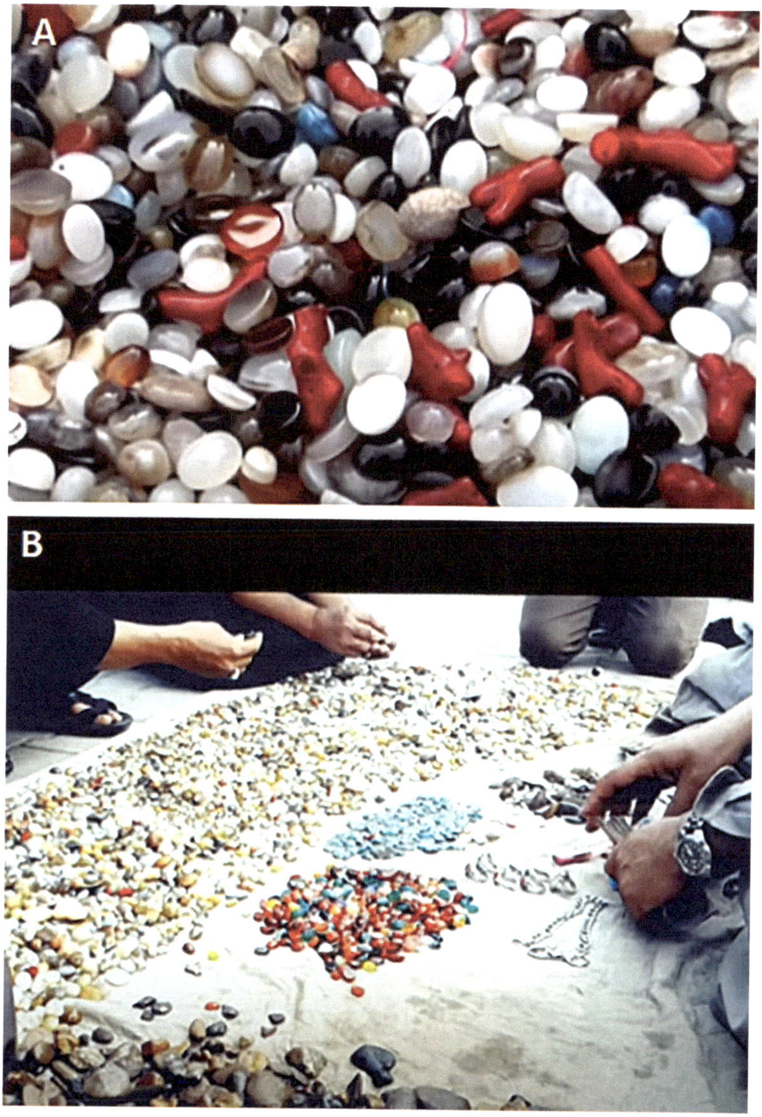

Fig. 39.139 (**a**,**b**) gemstones of different colours and shapes ready to be fit on men's ring. (Images courtesy of Laith Jawad)

is usually seen in the marshes (Fig. 39.142). On the other hand normal Marsh Arabs normally own one or more pieces of arms at home and take it with him whenever they go to work away from their living area. Shooting using different arms is usually seen in several social events such as marriage, death or at the reception of an important person (Fig. 39.143).

Fig. 39.140 Men's rings. (**a**, **b**) men's rings on display for sale in big shops in the large cities in Iraq; (**c**, **d**) gemstones and men's rings offered for sale on stalls at the side of the main road in big cities in Iraq. (Images courtesy of Laith Jawad)

The arms' market has flourished in Iraq after 2003 when the Saddam regime has collapsed. Although there is a licence to carry weapon that the person needs to obtain from the Iraqi government, majority of the people own arms without licence especially in the marsh areas. Most people carrying weapons have no knowledge about how to use it, look after it and how to keep himself and the members of his family safe away from these arms (Fig. 39.144).

39.9.4 Sleeping

The hut with whatever it has been built from does not give a room for the members of the family to use proper beds to sleep on for two reasons, because it is too small to fit several beds in and second beds will be problematic when Marsh Arabs decided to move to another place. To solve the problem where and how to sleep, the marsh locals usually use mattresses filled with cotton, pillows and blankets to use as beds. During the day, these convertible beds are folded and stacked in one corner of the hut (Fig. 39.145).

Fig. 39.141 Men's rings. (**a–c**) men wearing different shapes of rings with different colour stones sitting at Al-Mudhif; (**d**) man wearing ring with red Yemeni sapphire stone. This man wearing the ring even in his daily work. (Images courtesy of Laith Jawad)

39.9.5 The Daily Formal Meal

Marsh Arabs gather together for a formal meal during the night only, when man and woman finish their daily work. In the evening, the housewives usually start preparing the food, which consists of a kind of bread, grilled fish or birds, date, yoghurt or milk. As the fruit will be eaten after the meal, date is the one and only fruit usually eaten and no other fruit. If area where the hut is located somehow is large, then vegetable might be cultivated and eaten with the meal. For this meal, all members of the family will gather together around a big round metal tray or they use reed's mat, where they put their food on. This depends on the size of the family, and if there is a visitor, then the mat or what is known as "Suffra" سفره will be bigger and several kinds of food will be included (Fig. 39.146).

39.9.6 Social Night

Parties or social events are not common in the society of the Marsh Arabs. Instead, they might get together in certain occasions such as engagement, circumcision, birth,

Fig. 39.142 Photo of one of the tribe's standing in the doorway of Al-Mudhif. Note the way he is carrying ammunitions in. (Image courtesy of Pinterest.nz)

nights of Ramadan and the days of Eid (the days that follow the month of Ramadan, the month where people fast from sunrise to sunset).

When they get together, men and women will sit at the same hut, but usually women and kids sit on one side of the hut and men sit on the other. Among the habits of visiting is the visitor women should kiss on the check the of host woman and the man shakes hands with host man. The gathering usually takes place after the night meal, which is usually after the sunset. Daily life events will be discussed, and if

Fig. 39.143 People carrying arms in the street in one of the large cities in Iraq. (Images courtesy of Laith Jawad)

someone has a nice voice, then he/she can start singing and the rest of the people clap (Figs. 39.147 and 39.148).

39.9.7 Celebrating Eid

Eid days are very important events in Islamic countries. During the year, there are 7 days of Eid that people celebrate. Three days of the 7 days are those for small Eid that comes after the month of Ramadan and known as "Eid al-Fitr" عيد الفطر and 4 days of "Eid Al-Adha" عيد الأضحى. In the cities and villages and in those days, people usually wear new clothes and shoes and eat the best food and kids go to a playground. In the marsh areas, not much of what people at the cities is present and the only things that Marsh Arabs can do in these days are probably not doing their daily jobs and eating special food. Talking about food, the special meal of the Eid is called "Masmuta" مصموطه. This special meal consists of dry and salted fish. Prior to the Eid, people get large-sized fish, gutted, salted and hang on the line in the sun to dry. To prepare this meal, the fish is usually boiled in a water and some salt and some spices will add to it. Once the fish is ready, they pour it with the boiled water that contains all the nutrients on pieces of bread and then it will be ready to eat

Fig. 39.144 (**a**) At the shop of selling arms. Man buying a weapon to protect himself and his family; (**b**) man cleaning at home a set of personal weapon. (Images courtesy of Laith Jawad)

Fig. 39.145 Mattresses, pillows and blankets stacked at the corner of the hut during the day. (Images courtesy of Laith Jawad)

(Fig. 39.149). The meal has a strong fishy smell that people cannot stand other than those who are used to eating it.

39.9.8 Generosity

Among the characters of the Marsh Arabs is generosity. No visitor can leave their place without having food and drink as much as he/she wants. If the visitor needs to stay overnight, then he will sleep in Al-Mudhif. People of the marshes offer to the visitors the best food they have and offer them all the help and assistance in their jobs that they came about. In such occasion, when several visitors arrived, then people prepare big meal for lunch that includes several grilled fishes and several grilled birds and sometimes a full grilled sheep. In addition, there will be all sorts of bread (Fig. 39.150).

There are some rituals that the visitors should follow while they are attending a meal held at Al-Mudhif or at the hut of any of the marsh people. The visitors should eat whatever food and drink offered to him/her. If the visitors refuse to eat, then this means an insult to the host or the tribe that he is sitting in their Mudhif.

Fig. 39.146 Evening meal. (**a**) women preparing evening meal for the family; (**b**) meal prepared in a rounded metal tray; (**c**) members of the family taking their evening meal. (Images courtesy of Laith Jawad)

39.10 Childhood in the Marshes

Unlike the children in the cities or even in the villages, the children of the marsh areas are not living their childhood time. They are missing several essential things that other kids have. There is no proper teaching, they wear shabby clothes, their personal cleanness is not what it should be, most of them do not know how to wear shoes and finally the toys that other kids play with are something from the outer

Fig. 39.147 Social evening in a hut. (**a**) Marsh Arabs gathering in a hut around kerosene lamp; (**b**) man talking to the people in the hut; (**c**, **d**) some women and kids from those who attend the social event. (Images courtesy of Laith Jawad)

space to them. The young girls in the cities or in the villages normally wear gold earrings, but in the marsh areas, young girls put in the ear piece of thread to keep the hole in their ears open (Fig. 39.151). Therefore, once the kids reach the age of 10, they boys will assist men and learn their daily jobs, while girls follow their mothers and learn all women's jobs that they do during the day. Among the things that kids can do is taking a ride in a small boat in the marsh (Fig. 39.152a), running in the small land where their hut is located or playing with the calf of the buffalo or the cow (Fig. 39.152b–d). In the villages at the edge of the marsh area which has a sort of connection with the big cities, young girls usually sit at the main road selling fresh reed, where some passing people in their cars buy from them (Fig. 39.153a). Recently and after the technology reaches those marshes near the villages, young boys became to know about mobile phones and how to use it. The most comfortable place to sit and watch the mobile is on the back of a buffalo (Fig. 39.153b).

Fig. 39.148 In the social event. (**a**, **b**) young man singing; (**c**) young women enjoying the song and clapping expressing her enjoyment; (**d**) women kissing each other on cheeks. This is the way of greeting between women in the marshes and other parts of Iraq. (Images courtesy of Laith Jawad)

39.11 Women's Beauty in the Marsh Areas

Beauty among women of the Marsh Arabs is natural. Unlike the women of the cities and the villages, women in the marsh areas do not use makeup. Many women and even men have whiter skin colour than normal Arab people (Fig. 39.154). Moreover, they have yellow hair and coloured eyes. These characters suggest that they inherited some features that are not usually found among the Arab people through their long history.

Probably the most beautiful woman from the marsh area that her photo was dispersed widely in Iraq in general is "Bint Al-Muaidi" (daughter of the marsh dweller) بنت المعيدي. The story of this very beautiful young lady happened during the British invasion to Iraq and when the British troops centred in the marsh area. One day, one of the British military officers operating from a camp in the marsh area has

Fig. 39.149 The special meal for Eid. (**a**) women preparing the special meal for the lunch of the first day of Eid. Dry fishes were put in the pan for boiling; (**b**) water being added to fish prior to boiling; (**c**) heating the pan to boil the fish; (**d**) finally, the special meal is ready and members of the family stated to eat. (Images courtesy of Laith Jawad)

seen a very beautiful young lady. He decided to marry her, but due to the customs of the marsh people that they do not let their daughters marry foreigners from outside the area, the request of the British officer was refused. He insisted to marry her and 1 day he kidnapped her and took her with him to England. In her new home, she was very sad and missed her previous life at the marshes and her family. Her husband and, in a humane step, asked one of the best painters in England to draw his beautiful wife and send the photo to her family to let them know that their daughter is living as a princess in her new home in England. This photo spread among the people of the marshes very quickly and people started to put it on the wall in their houses even in the cities (Fig. 39.155). Recently, the marsh areas became one of the tourist attraction sites. Young women started to visit the marshes and wear the traditional clothes of women of Marsh Arabs as fashion (Fig. 39.156).

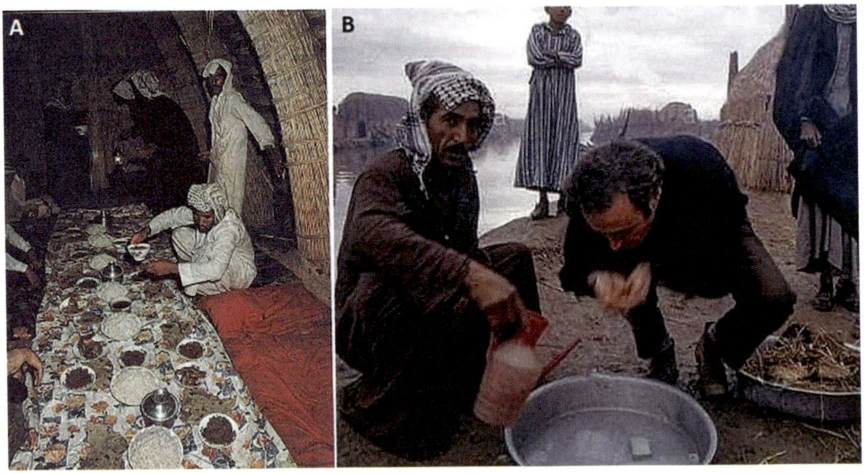

Fig. 39.150 Generosity of the Marsh Arabs. (**a**) big meal prepared for a visitors in Al-Mudhif. (**a**) a long mat holding varieties of food as a meal in honour of some visitors; (**b**) Marsh Arab holding container of water for the guest to wash his hand after taking his meal. (Images courtesy of Laith Jawad)

39.12 Boy or Girl, a Crucial Matter in the Marsh Society

The issue of preferring boys over girls has been on debate for a long time. In most parts of the world including the Middle East, people usually favour boys over girls. In such cases, when a person has several daughters and then has a boy, he gives all his attention to the boy and when he gets older he has all the right to do anything, while the girls are not allowed to do so. In the marsh area, men when they get married always wish to have boys rather than girls. In the Middle East, there is a habit of calling a person after the name of his son. For example, if a man has a boy named Ali, then people will call him "Abu Ali" (father of Ali). If the man has boys and girls and the girls are older than the boy, the people still call him "Abu Ali" even if Ali, his son, is younger than the girls, which the man called accordingly.

Among the things that Marsh Arabs do to keep away the evil eyes off their kids and mainly boys is to give them strange names such as "Zubala" (means rubbish) or "Kashia" (mean tile) and so on. The idea behind giving such names is to keep the evil eye away from them. Moreover, they usually attach to their hair some amulets to keep the evils away. The amulets can be a ring with seven holes which is known as "Umm Sabbaa Eyoun" (mother of seven eyes) أم سبع عيون; the hand, which is known as "Hamsha or Khamsaa" خمسه; or the blue eye (Fig. 39.157). Recently, these amulets have experienced good market around the Middle East, and special shops are opened to sell them in different shapes and sizes (Fig. 39.158).

An amulet or talisman in Arabic "Talism" تلسم is an item either natural or synthetic which is thought to enjoy magically defensive influences, to fetch good affluence or to prevent evil (or both) (Green 1992). Amulets are kept by the person or positioned

Fig. 39.151 (**a–d**) Kids from the marsh areas. Note the condition of their clothes and their hair. (Image courtesy of Pinterest.nz)

at the site of the anticipated magical upshot. Regularly they are thought to originate their power from a sympathetic magic ensuing from their contacts with nature, from religious links or from the favourable time of, or the rituals intricate in, their creation (Green 1992).

Amulets were known by both the Sumerians and Akkadians, and a number of items that were used as amulets, for instance, seals, in their times (Green 1992) were found in the graves. Among the famous amulet used by the ancient Mesopotamians and still being used in the present time particularly in Iraq and the other Arab countries is the "Umm Sabbaa Eyoun" (mother of seven eyes) أم سبع عيون. It is a hexagonal shape disc with seven holes, six holes are surrounding the seventh hole, which is located in the middle. People usually stain this artefact with blue colour as they believe that the blue colour is able to receive and destroy the power of the evil eye.

Fig. 39.152 (**a–d**). Kids playing and enjoying themselves. (Image courtesy of Pinterest.nz)

The Umm Sabbaa Eyoun probably has a long history dating back to the prehistoric times, and the symbol of the seven dots (or globes) is first known in clear form, which became common in the Neo-Assyrian and Neo-Babylonian periods. From early on (at least from Middle Assyrian times), the seven dots seem as a symbol in close link with other clearly astral symbols such as the solar disc and the crescent (green 1992). In Assyrian and the Babylonian beliefs, the seven dots have a connection with the seven gods that have the power against the demons. These seven holes represent the seven gods collaborating together to defeat the evil (green 1992). The habit of placing the hexagonal disc with seven holes in the entrance of the house is an old ritual experienced by ancient Mesopotamians and still being performed by the present inhabitants of Iraq and the neighbouring countries, which have been included under ancient Mesopotamia.

Going back to the issue of close attendance of men to their male kids, it is important to speak about a ritual that Marsh Arabs perform. This habit is teaching young boys how to wear the head "Scarf and Aqal" جفيةو العكال. Once the boys reach the age of 6 or 7, fathers usually take them with him to sit in Al-Mudhif. In sitting there, the young boy learns gradually all the rituals related to Al-Mudhif starting from wearing the proper head's cover. In the Mudhif, the father teaches the young boy how to wear the head's cover properly (Fig. 39.159).

Fig. 39.153 How the kids in the marsh areas spend their time. (**a**) young girl selling fresh reeds on the side of the road in a village at the edge of the marsh (Image courtesy of Pinterest.nz); (**b**) young boy busy playing with his mobile on the back of a buffalo. (Image courtesy of Hayder Ismael, photographer, Iraq)

Fig. 39.154 (**a–d**) Beautiful women form the marsh area. (Image courtesy of Pinterest.nz)

39.13 Death and Burial Ceremony

Death is one of the major social events in the life of people in the Islamic world and especially in Iraq, where there are many practices and rituals that need to be done and followed. Although the marsh areas are cut off from the villages and the cities, the marsh people are still doing and following the same practices and rituals that people in the cities do.

Before describing the death and burial rituals followed by people of the marsh areas, which seem comparable to those of the ancient Mesopotamians, the sorrow for the deceased person was showed by beating on the chest and stridently weeping. This action is in support to the Epic in Gilgamesh, when Gilgamesh was very sad for the death of his friend Enkidu. He left his hair uncombed, took off his jewellery and wore a mourning dress which could be black (Crawford 2013). The inscribed tables of the Sumerians indicated that the mourning will last for a few days probably more than two (Crawford 2013).

Fig. 39.155 Photos of "Bint Al-Muaidi" (daughter of the marsh dweller) بنتالمعيدي. (**a**) the photo hanged on the wall of the hut; (**b**, **c**) on the wall in houses in many cities of Iraq. (Images courtesy of Laith Jawad)

Once an incidence of death happened in anywhere in the marsh area, the family of the deceased person will be alerted and start promptly to prepare for the burial of the corpse, which should be done before the sunset if the incidence occurred during the day, but if it happened during the night, then the preparation should wait until the morning as it is impossible to do anything during the night. Before taking away the

Fig. 39.156 Ladies from the cities wearing the costumes of the women of the marshes as a recent fashion. (Image courtesy of Pinterest.nz)

corpse, a decision should be made by the family of the deceased person whether the burial should take place in the City of Najaf, where all Shia Muslims bury their relatives, or in a nearby village or city. The corpse is usually put in a coffin made of wood and then loaded in a boat to be taken to the nearest village that can be loaded on a car to take the coffin either to the City of Najaf or to a nearby area cemetery depending on the financial ability of the deceased family. In both cases, one or two members of the deceased family will accompany the deceased person to his final journey for burial.

The formal mourning will not take place only after the corpse is buried. The rituals of mourning continue for 3 days according to the Islamic practices during which the whole Quran will be recited by the person reading the Quran, but

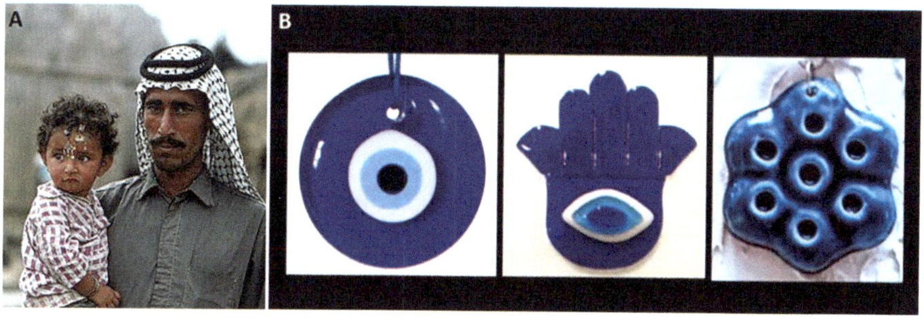

Fig. 39.157 Boys or girls. (**a**) young man carries his son, which seems very proud of having a boy. Note the amulet fastened to the front hair of the young boy (Images courtesy of Laith Jawad); (**b**) common amulets, which are in use by people in Iraq and other Arab countries. (Image courtesy of Pinterest.nz)

Fig. 39.158 Amulets on display for sale in special shops in the Middle East. (Image courtesy of Focal point Wikipedia retrieved April 2020)

nowadays and since there is electricity, a tape playing the Holy Quran will run for 3 days from 7: 00 AM until the sunset. This event usually takes place in Al-Mudhif, where the members of the deceased family sit near the entrance of the Mudhif on the left. The mourners normally enter and say Al-Salam Alaikum (means peace on you) and sit and then one of them if they are more than one asks the whole people sitting in the Mudhif to read the first verse of Quran "Al-Fatiha". Once everyone is finished

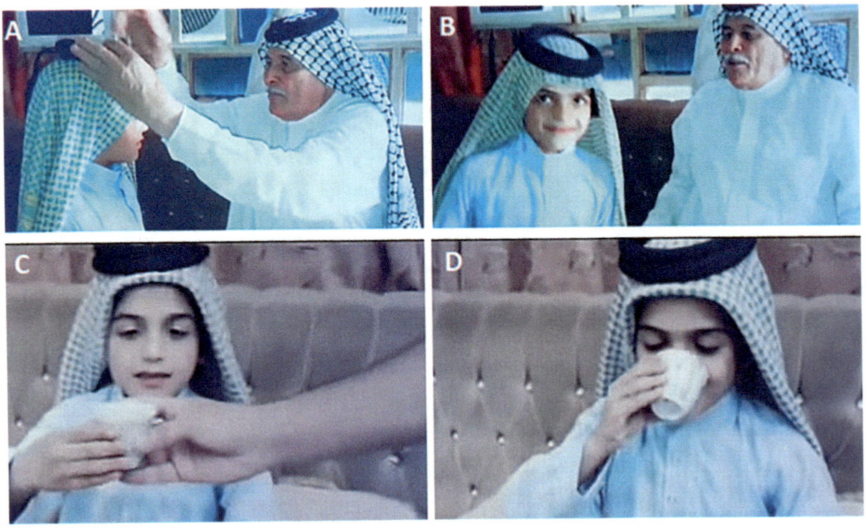

Fig. 39.159 Teaching the boys. (**a**, **b**) man teaching his son on how to wear properly the head cover; (**c**, **d**) boy learning how to drink black coffee in Mudhif. (Images courtesy of Laith Jawad)

reading the verse of Quran, each of the sittings will salute the mourners, which they stay silent or read some verses from the Holy Quran available in Mudhif. The mourners will be served black coffee and they will stay for about 10–15 min and then leave.

On the other hand, the women also have their own mourning gathering, where female relatives of the deceased are usually crying and weeping loudly. This action is normally accompanied by hitting hard their faces, chest and arms (Fig. 39.160). Outside the hut and Mudhif, women and men are gathering to prepare food for all the mourners that came and stayed until the lunch time or in the evening session until the sunset. Food preparation process normally lasts for 3 days, with a big banquet done on the evening of the third and last day of mourning. The cost of all the food will be paid by the deceased family, but there is a habit that Muslims usually follow in the death occasion. Each mourner will bring with him something to assist in the cost of the food. Some people give money to the deceased family, while others give dry food like rice, cooking oil, vegetable, tea, sugar, cigarettes and even sheep so they can be slaughtered and used in the food preparation that takes 3 consecutive days.

Outside Al-Mudhif, flag of the tribe that the deceased person belongs is raised. Also, some members of the deceased family and other will make "Hawssa" هوسه (people singing some slogans and dancing in circle) (Fig. 39.161). The dancer usually carries guns and they shoot while they are dancing. This will alarm inhabitants of other gathering living away from the place, and it is also an invitation for them to attend the funeral. The time the dancers spent in singing their slogans is relative to the social rank of the deceased. If he/she was a very important person, then people from several tribes join the dance and last for a long time. Visiting the tomb of

Fig. 39.160 Women in a session mourning. (Images courtesy of Laith Jawad)

the deceased person is normally done after 40 days of the date of his/her death and not before.

All female relatives of the deceased person will wear black clothes for 1 year. If the deceased is a man and has a wife, then his wife will not go out of her house and meet males for 3 months. This is an Islamic practice.

From the description of the death and mourning events given above, it is possible to say that there are some similarities with the events occurring during the time of the ancient Mesopotamians. The similarities can be seen in the way the relatives of the deceased person perform the mourning, with crying and screaming in loud voices, hitting their faces and bodies and wearing special mourning dresses, which is black in this case. In addition to what the present Iraqis add to all these, the grief is declared by the family of the deceased for 1 year.

39.14 Day of Ashura

The month of Muharram is the first month in the Islamic calendar. This month is also known as month of Ashura عَاشُورَاء relative to the event that happened years ago in the tenth day of this month, Yawm Ashura (Arabic: يَوْم عَاشُورَاء). This month is very important for all Muslims and in particular for the Shia sector of Muslims. Ashura designates the pinnacle of the Commemoration of Muharram (Facts and details 2020), the annual tribute of the death of Hussain and his family and followers at

Fig. 39.161 Men doing "Hawssa" هوسه (people singing some slogans and dancing in circle) shooting, dancing and carrying flags of the tribes in mourning session. (Image courtesy of Pinterest.nz)

the Combat of Karbala on 10 Muharram in the year 61 AH (in AHt: 10 October 680 CE) (Madelung 2014). Grief for the occasion commenced nearly directly after the fight. Widespread requiems were inscribed by poets to remember the Battle of Karbala during the Umayyad and Abbasid era, and the original public mourning ceremonies happened in 963 CE during the Buyid dynasty. In the Islamic countries where there are many Shia Muslim population, Ashura has become a national holiday (Wikipedia 2020a, b).

For the whole month of Muharram, Shia Muslims wear black clothes especially in the first 10 days. This ritual is common in all parts of Iraq including the marsh areas (Fig. 39.162). During the tenth day of Ashura, most of the Shia Muslim sector cook special food including wheat and meat. They spend the whole night cooking

Fig. 39.162 Month of Muharram. (**a**) Modern Mudhif equipped with electricity, light and fan. Note the black sheets on the wall signifying the mourning status. People sitting listening to speakers talking about the battle of Karbala, where Imam Al-Husain was killed by the Umayyad army in the year 61 AH (in AHt: 10 October 680 CE); (**b**) people wearing either black or dark clothes during month of Muharram signifying their grief and mourning. (Images courtesy of Laith Jawad)

Fig. 39.163 Month of Muharram. Cooking during the tenth day of month Muharram. Note the amount of food prepared and offered for free for anybody and designated for the poor people. (Images courtesy of Laith Jawad)

this food as they cook large amount of food so they can distribute it to the poor people (Fig. 39.163).

39.15 Tattoo in the Marsh Area

A tattoo is a form of body alteration where a drawing is made by introducing ink, dyes and pigments, either permanent or transitory, into the dermis layer of the skin to change the pigment. The purpose of making tattoos can be grouped into three broad classes: purely decorative (with no explicit meaning), symbolic (with a precise meaning related to the wearer) and pictorial (a representation of an exact person or item) (Wikipedia 2020a, b).

Tattooing has been used over time across many nations, carrying a broad selection of significances and purposes (Deter-Wolf 2013). With the invention of writing in the late fourth millennium onwards and with dispersal of cuneiform writing system, it became easy to signify possession over certain items including human bodies. At the time of ancient Mesopotamian, the main function of tattooing was a way of marking slaves and temple workers for purposes of recognizing a human or divine owner to stop the loss of human possessions. Secondly, tattooing was used as disciplinary action against fugitives or otherwise disobedient slaves (Ditchey 2017).

Through the ancient Mesopotamian literature, there are three elementary techniques that human can make an enduring pattern, and these are tattooing, cauterisation (branding) and cicatrisation (scarification) (Giessler and Pientka-Hinz 2012). Tattooing is described as making patterns on the body with the aid of placing pigment into the dermis. A brand is made by the sealing of the skin using a very hot iron, usually in the shape of a symbol, so that the subsequent wound heals in a way that carries that symbol. Scarification is the making of patterns on the skin by means of cicatrisation, or cutting.

Indications on body marking continuously appeared through the evidences obtained from the third millennium and first millennium texts from Babylonia (Ditchey 2017). These signs show an association between temple devotion and marked persons and animals, while disciplinary tattooing for runaways and the tattooing of prisoners persist through the second and first millennia (Ditchey 2017).

The archaeological evidence indicates that the ancient Mesopotamians were the first who promoted the usage of tattoo in the form of branding the slaves. This happened in the early third millennium Mesopotamia (Green and Nissen 1987). Nevertheless, there are some oldest samples of tattooing and brand marks found in other cultures (Deter-Wolf 2013). A famous example is "Ötzi the ice man", a 5300-year-old natural mummy found on the border of Austria and Italy in the Ötztal Alps. Conserved by the freezing nature of the area, the mummy showed to have tattoos in the form of parallel lines along the wrist, back, ankles and calves, with a cruciform tattoo on the right knee (Fleckinger and Steiner 2002). The ancient Mesopotamians might practise tattooing even earlier than the date of the ice man of Austria lived in, but due to the climate of the area where they used to live in and the burial procedures they used to follow, it was not suitable to conserve the skin of the human long enough as in the case of the freezing weather. Nevertheless, porcelain items from sixth millennium sites in Mesopotamia show stand-up images decorated with marks that may propose tattooing (Oates 1978).

Using tattoos as an adornment by women could start in the same period when it was used as branding for the slaves. In the third millennium, women start tattooing her body starting from the back of the hand, the wrist, around the mouth and on the brows. This is usually done by using a pointed device to embed the stain or what is usually used is the black soot found at the bottom of the cooking pans. Once it introduces under the skin, it turns green and it will be permanent.

In Iraq, tattooing is a widespread exercise. It is identified locally as Daqq or Dagg دگ from a root connotation to strike or knock, and as the name suggests, it is making a pattern by puncture (Smeaton 1937). Tattooing is a tradition which already shows signs of disappearing during the last few decades, but it returns in another form and considered fashion in the cities. In old days, women from the lower social class used to do tattooing, but in these days, tattooing became a fashion for women in the cities (Figs. 39.164 and 39.165).

People in Iraq usually do tattooing in two different ways; it is ornamental or decorative and tattooing applied for magic or therapeutic reasons (Smeaton 1937). Generally, all tattooing imply some magic-religious meaning along the history of this practice. The most common kind of curative tattooing is for sprains. Another is tattooing against headache and eye disease. The tattooing is smeared on the fore or the sides of the head or near the eye. These areas are designated as a pain locations (Smeaton 1937).

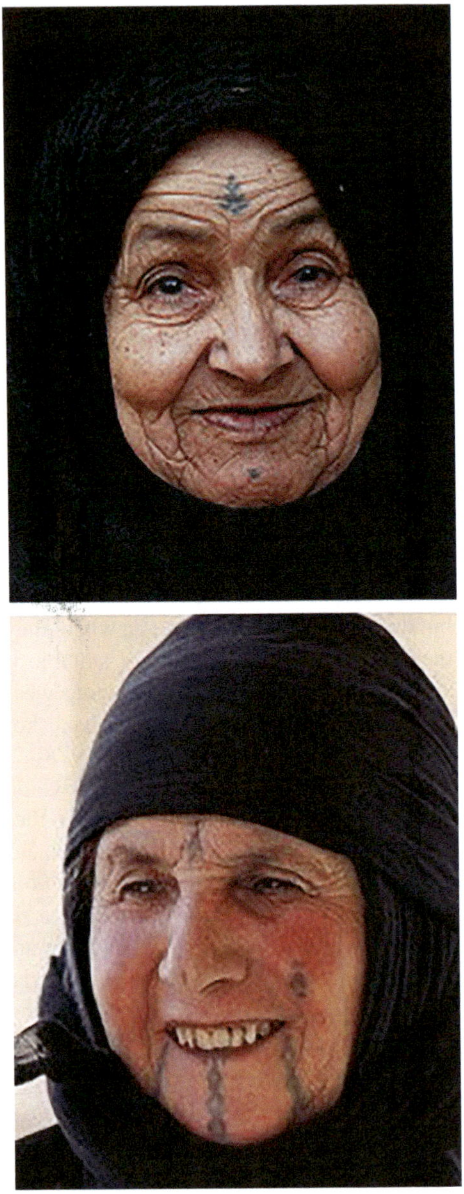

Fig. 39.164 Face tattooing. (Images courtesy of Laith Jawad)

Tattooing is also done to fulfil magical purposes and it is a type that women usually look for. Among the most common types of tattooing are tattoo designed to induce pregnancy, a matter of pronounced anxiety to Arab women; the drive of guarding children, especially boys, against death; and to obtain charms for love or against other magic (Smeaton 1937).

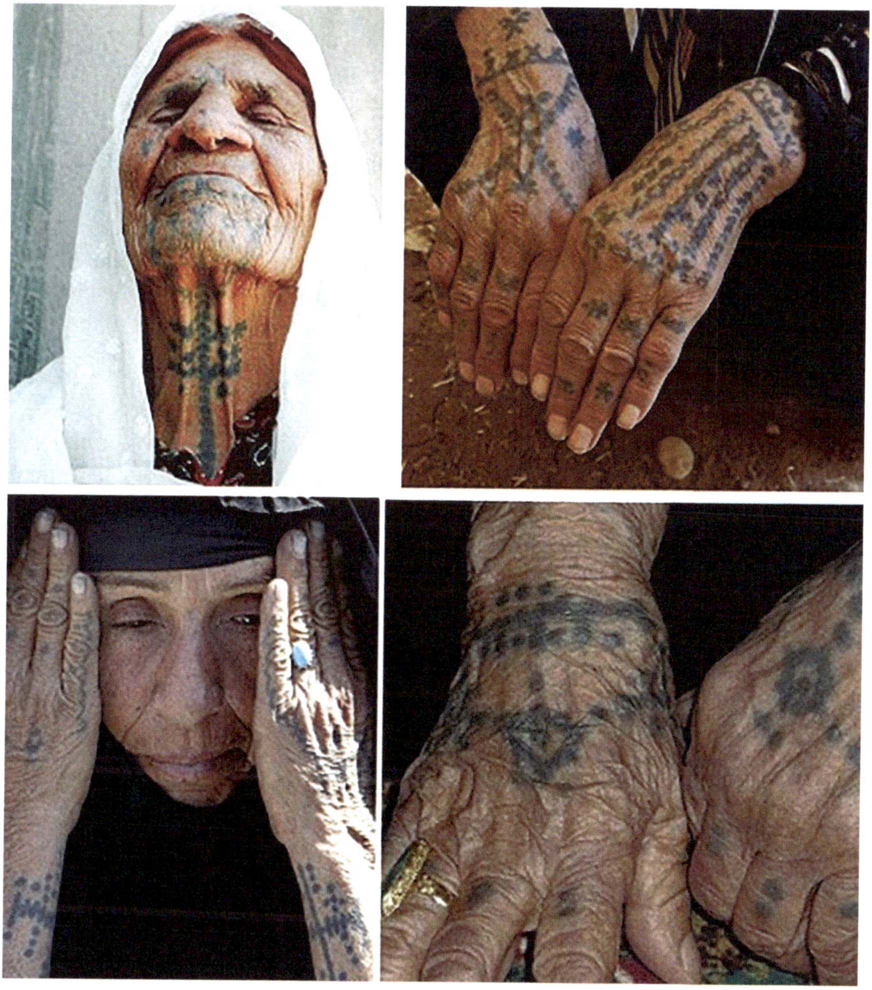

Fig. 39.165 Hand tattooing. (Images courtesy of Laith Jawad)

Tattooing in different tribes of different cities in Iraq may show different patterns and may be done in different body regions. This is clear in tribes originating from Nasiriya and Amarah cities. In the former, men usually have one dot on the tip of the nose and another on the back side of both their hand just under the wrist. In the latter, men have small cross on the sides of their head and small dots above and below the edges of their lips. The latter design is related to the eagerness of the women to have her boy child healthy and continue to live.

With such a wish, they make the boy child look like the girls.

As an adornment, tattooing is used by all women especially those in the age of marriage. Men in the rural areas such the marshes prefer to marry woman with tattoo rather than woman with no tattoo. Therefore, all young women cover their body with tattoo of different shapes. This information has been documented in words of a traditional song in Iraq. The words of the song say something is relative to the type and beauty of tattooing on certain places of the woman's body. The words say "... if see the Daqq (tattooing) under my vail, you will get mad as it is charming....".

There are several designs where tattoos can be drawn in. Some of these designs were inherited from the ancestors. The patterns are uniform or artificial. Normally they contain a mixtures of dots and lines, particularly zigzag and cross-hatched lines, circles, crescents, chevrons, triangles, stars and crosses. A motivating and primitive pattern, which was done by the majority of the social sectors, is the cross. It is constantly done on arms of equal length, and often there is a dot in each angle. The cross has attained peculiar magic meaning connection to it. The sign of the cross, or what is known as the four-sided, is the strongest design and gives power to the place where it has been drawn on (Smeaton 1937).

39.16 Tourism in the Marsh Area

The southern marshes of Iraq have experienced hard time during Saddam Hussein's era as it faced an intentional devastation and endured years of war, have converted to haven for its inhabitants and probably for the tourists. The marshes, which used to extend over 9000 km^2 back in the 1970s, had shrunk to just 760 km^2 by 2002 before regaining some 40% of the original area by 2005.

Relaxation through tourism in Iraq's marshes aims to familiarise guests to a genuine Iraq. The ferocity of war during the past years made tourism in Iraq's marshes impossible. The main purpose of planning tours in the marshlands is to attract more people to Iraqi wetlands and put them in direct interaction with the life of Marsh Arabs in addition to discovering the cultural heritage of the marsh inhabitants. One of the setbacks that can delay the flourishing of the ecotourism in the marsh area is the lack of resting places and recreation facilities including hotels and restaurants, and the government of Iraq should pay more attention to this area and to encourage tourism in order to assist the economy of Iraq and to develop the abilities of the marsh people and support them.

Recently, several tourist companies have established in the main cities road to the marshes such Amara, Nasiriya and Basrah. These companies arrange regularly trips to the marsh areas, where the tourists are usually taken by tourist buses equipped with excellent facilities so the tourists will have a comfortable trip. Reaching the villages at the edge of the marsh area, the tourists leave the buses and take the small boats arranged by the tourist company to take them deep in the marshes, where they can meet the Marsh Arabs and speak with them face to face and discover their daily lives and taste their daily food (Figs. 39.166 and 39.167).

Fig. 39.166 Tourists in boats in a daytrip in the marshes. (Images courtesy of Laith Jawad)

39.17 The Museum of the Marshes

Recently, a special museum was built in Chibayish, Nasiriya, south of Iraq. This museum is devoted to contain the heritage of the people of the marshes. The museum is known as "the museum of the marshes of Iraq", and it was designed and founded by southern Iraq local Mr. Raad Habib Al-Assady (Fig. 39.168), the head of Al-Chibayish organisation for the ecotourism in the marsh area.

Fig. 39.167 Tourists taking their lunch in a hut in a daytrip in the marshes. (Images courtesy of Laith Jawad)

The area of the marsh museum is about 300 m^2 and it is designed as a Mudhif building (Figs. 39.169, 39.170, 39.171, and 39.172). It contains four galleries; the first is assigned to the different types of boats and fishing gears used in the marsh area, the second for the daily life, the third for the nature of the marshes and the

Fig. 39.168 Mr. Raad Habib Al-Assady the founder of the museum of the marshes in Chibayish, south of Iraq. (Images courtesy of Raad Al-Assady, Iraq)

Fig. 39.169 Steps of building the museum of the marshes. (**a**) levelling the ground; (**b**) preparing the location of the museum; (**c**) making the parabolic arches prior to covering them with reed; (**d**) men preparing reed to make the large main piles. (Images courtesy of Asia Cell)

Fig. 39.170 Steps of building the museum of the marshes. (**a**) man arranging reeds in a shape of large bundle; (**b**) men fastening the large bundle of reed; (**c**) locating the large bundles of reed on the parabolic arch; (**d**) parabolic arches completely covered with bundles of reed. (Images courtesy of Asia Cell)

fourth is considered as an educational hall (Figs. 39.173, 39.174, 39.175, 39.176, 39.177, and 39.178).

The museum already proved an excellent step towards educating tourist about the daily life of the Marsh Arabs and their heritage. As the director of the museum, Mr. Raad Al-Assady said it is not possible to convey the idea about the life of the Marsh Arab in words. Therefore, a decision was made to make the idea of this museum a reality.

The area where the museum was built on was owned by the family of Mr. Raad Al-Assady and donated to be used as a museum. Mr. Raad Al-Assady has put a great effort in attaining more than 400 different artefacts on display in the museum. He has done a huge search and contacts with locals to gather the different items that related to different purposes.

In this section, selected photos showing the artefacts on display in the museum will be given together with short description for each of them.

Fig. 39.171 Steps of building the museum of the marshes. (**a**) covering the top and sides of the museum, which is built in shape of Mudhif; (**b**) skilful man doing reed's work to cover the sides of the museum's building; (**c**) laying the longitudinal reed's bundles over the parabolic arches; (**d**) Raad Al-Assady talking about the museum after it has been built. (Images courtesy of Asia Cell)

Fig. 39.172 Initiation of the museum of the marshes. Mr. Raad Al-Assady, the founder and director of the museum on the right of the official VIP, who inaugurated the museum. (Images courtesy of Asia Cell)

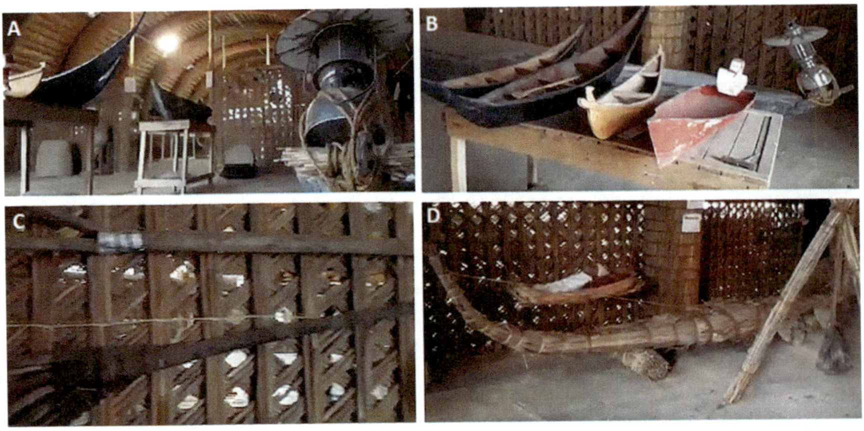

Fig. 39.173 Artefacts displayed in the museum of the marshes. (**a**, **b**) boats; (**c**) fishing gear. Note the wooden spear used for fishing; (**d**) boat made of reed. (Images courtesy of Asia Cell)

Fig. 39.174 Artefacts displayed in the museum of the marshes. (**a**) pots for making coffee and tea; (**b**) different shapes of coffee pots; (**c**) coffee pot and small containers for drinking; (**d**) metal mortar for grinding seeds. The large one is designed for big occasions such as marriage. (Images courtesy of Asia Cell)

Fig. 39.175 Artefacts displayed in the museum of the marshes. (**a**) kerosene lamp and burners of different shapes; (**b**) kerosene lamps of different shapes; (**c**) stone mortars for grinding seeds; (**d**) wood-made mortars. (Images courtesy of Asia Cell)

Fig. 39.176 Artefacts displayed in the museum of the marshes. (**a**) wide pan and pitcher known as "Ibriq" إبريق. This pitcher and the pan are used for washing hands after eating. Marsh Arabs usually eat with no spoon and with their bare hands; (**b**) different types of stone mortars for grinding seeds and for making a dough; (**c**) large pan used for fetching water from the marsh for domestic usage and for boiling water; (**d**) different types of trays made of copper. (Images courtesy of Asia Cell)

Fig. 39.177 Artefacts displayed in the museum of the marshes. (a) hand-made artefacts. Hand fan and Tabak طبگ used for putting different types of bread; (b) animal skin used for making butter. The milk is usually put in this animal skin bag and shaken for a period of time, and then butter will gather at the surface of the milk, which will be collected. Note step up is hanged from a tripod made of reed; (c) baby bed made of reed and hanged from both ends so it can rock; (d) old riffles. Also, a piece of wood covered at one of its end with tar. It is known as "Meqwar" مگوار. (Images courtesy of Asia Cell)

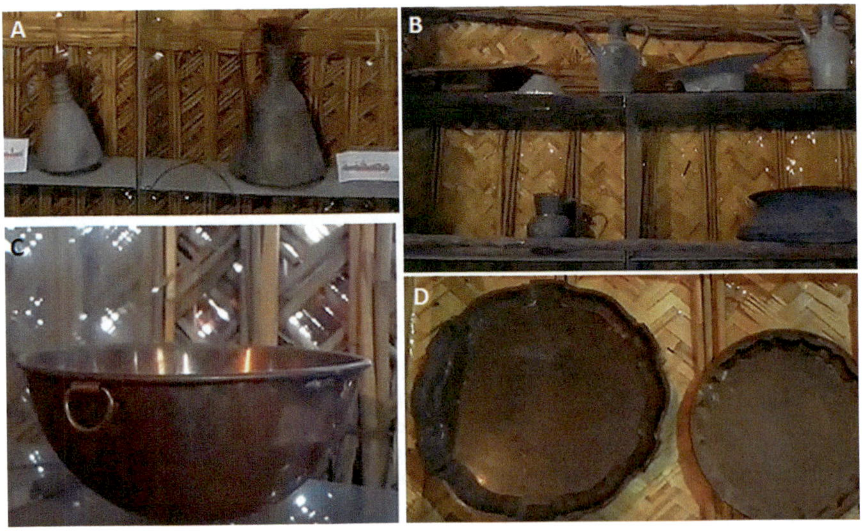

Fig. 39.178 Artefacts displayed in the museum of the marshes. (a) special containers for fetching and boiling water. They are called "Masekhnaa" مصخنه; (b, c) different household utensils; (d) copper trays. (Images courtesy of Asia Cell)

References

Abid HS, Fazaa NA (2007) Water Buffalo in the Iraqi marshes. Nature Iraq Status Report, 29p

Aldessouky FI (1987) Water buffalo in Iraq. Report, UNDP/FAO 1987

Al-Fartosi K. (n.d.) Preliminary study on water buffaloes (*Bubalus bubalis*) in marshes of Iraq. Report to the Department of Biology, College of Education, University of Thi-Qar, Iraq, 10p

Al-Jubouri MS (2011) Southern Iraqi marshlands between rehabilitation and drying. Al-Rafidain Centre for Strategic Studies and Research (Arabic)

Al-Jumeily D, Hashim K., Alkaddar R, Al-Tufaily M, Lunn J (2018) Sustainable and environmental friendly ancient reed houses (inspired by the past to motivate the future). In: 2018 11th International Conference on Developments in eSystems Engineering (DeSE), September, IEEE, pp 214–219

Al-Khayoun R (2003) Religions and sects in Iraq. Al-Kamel Verlage, Koln

Al-Saedy JKH (2007) Iraqi Buffalo now. Ital J Anim Sci 6(Sup 2):1234–1236

Al-Tameemi R (2016) Reading the cultural specificities of the Iraqi marsh Arabs from their landscape. Landscape research record no. 5. Council of Educators in Landscape Architecture (CELA), Salt Lake City, Utah, pp 39–51

Alwan WA (2005) Iraqi's marshlands – Eden again. Environmental Tourism Approach. Sage, Beverly Hills

Al-Zahery N, Pala M, Battaglia V, Grugni V, Hamod MA, Kashani BH, Olivieri A, Torroni A, Santachiara-Benerecetti AS, Semino O (2011) In search of the genetic footprints of Sumerians: a survey of Y-chromosome and mtDNA variation in the Marsh Arabs of Iraq. BMC Evol Biol 11 (1):288

Borghese A, Mazzi M (2004) Buffalo population and strategies in the world. In: Borghese A (ed) Buffalo production and research (FAO, 2005), pp 1–39

Breniquet C (2006) "Dans le mouton, tout est bon". Remarques sur les usages socio-économiques des animaux en Mésopotamie de la préhistoire récente au IIIe millénaire avant J.-C. In Michel C, Lion B (eds) De la domestication au tabou: le cas des suidés dans le Proche-Orient ancien, Paris, pp 247–255

Breniquet C, Michel C (2014) Wool economy in the ancient near East and the Aegean. In: Wool economy in the ancient Near East and the Aegean: from the beginnings of sheep husbandry to institutional textile industry, 17. Ancient Textiles Series, 9781782976318. halshs-01226052. Oxbow Books, pp 1–11

Cockrill WR (1981) The water buffalo: a review. Br Vet J 137(1):8–16

Crawford H (ed) (2013) The Sumerian world. Routledge/Tylor and Francis Group, London, 659pp

Currey JD (2010) Mechanical properties and adaptations of some less familiar bony tissues. J Mech Behav Biomed Mater 3(2010):357–372

Deter-Wolf A (2013) The material culture and middle stone age origins of ancient tattooing. In: Casa PD, Constanze W (eds) Tattoos and body modifications in antiquity, Zürich studies in archaeology, vol 9. Chronos Verlag, Zürich, pp 15–25

Ditchey M (2017) Body language: tattooing and branding in ancient Mesopotamia. J Anc Near East Hist 3(1):1–24

Duke JA, Ayensu ES (1985) Medicinal plants of China. Reference Publications

Evans MI (2002) The ecosystem. In: Nicholson E, Clark P (eds) The Iraqi Marshlands: a human and environmental study. Politico's, London, pp 201–219

Facts and Details (2020) Shiite history beliefs and differences between Sunnis and Shiites: Muslim sects and Sunnis. http://factsanddetails com/world/ cat55/ sub359/ item 1439 html. Retrieved 22 April 2020

Fleckinger A, Steiner H (2002) Otzi, Der Mann Aus Dem Eis: Alles Wissenwerte Zum Nachslagen Und Staunen. Folio-Verlag, Wien

Food and Agriculture Organization --FAO (2004) Production yearbook 2004. Rome, FAO

Fraser RDB, Macrae TP (1980) Molecular structure and mechanical properties of keratin. In: Vincent JFV, Currey JD, (eds) The mechanical properties of biological materials, symposium of the society of experimental biology. Cambridge University Press, Cambridge, pp 211–246

Giessler J, Pientka-Hinz R (2012) Tätoweirung. Reallexikon Der Assyriologie Und Vorderasiatischen Archäeologie Band 13: 399–400. De Gruyter, Berlin

Green A (1992) Gods, demons and symbols of ancient Mesopotamia: an illustrated dictionary. British Museum Press for the Trustees of the British Museum

Green MW, Nissen HJ (1987) Zeichenliste Der Archaischen Texte Aus Uruk. Mann Verlag, Berlin

Huxley A (1992) The new RHS dictionary of gardening. Macmillan Press, London

Ibrahim MH (2007) The role of culture and education in the development of the Marshland People's life. Sada al-Ahwar Magazine, year I, issue 2, Dhi Qar University, April. (Arabic). Issue 18, July–August

Jaayid TA, Dragh MA (2013) Genetic diversity among Iraqi buffalo using micro satellites markers. J Agric Sci Technol A 3:297–301

Jadran AA-ZT (2010) The pollution of Marshland's water – a Survey of the water of hwar Abu-Zark (Unpublished Graduation Project of Diploma in Arabic). The Higher Institute of Urban and Regional Planning: University of Baghdad, Baghdad, Iraq

Juma KH (1997) Present status in buffalo production in Iraq. Buffalo J 2:103–113

Kramer SN (1972) Sumerian mythology, vol 47. University of Pennsylvania Press, Philadelphia, 65pp

Kubba S (2011) The Iraqi marshlands and the Marsh Arabs: The Ma'dan, their culture and the environment. Trans Pacific Press

Kumar KA, Kumar MS, Sravanthi K, Teja VA (2014) Evaluation of memory enhancement activity of leaf extract of Typha angustata. Int J Phytopharmacol 15:218–220

KWCSPF (2020) Kew world checklist of selected plant families, *Typha domingensis*. https://wcsp.science.kew.org/namedetail.do?name_id=270970. Retrieved on 28 Mar 2020

Lambert M (1961) Recherches sur la vie ouvrière: les ateliers de tissage de Lagash. Archiv Orientalni 29:422–443

Lau CH, Drinkwater RD, Yusoff K, Tan SG, Hetzel DJS, Barker JSF (1998) Genetic diversity of Asian water buffalo (*Bubalus bubalis*): mitochondrial DNA D-loop and cytochrome b sequence variation. Anim Genet 29(4):253–264

Levy R (1924) A note on the Marsh Arabs of lower Iraq. J Am Orient Soc 44:130–133

Li BW, Zhao HP, Feng XQ, Guo WW, Shan SC (2010) Experimental study on the mechanical properties of the horn sheaths from cattle. J Exp Biol 213(2010):479–486

Madelung W (2014). Ṫisayn B. ‘Ali i. Life and significance in Shi'ism. Encyclopædia Iranica Online. Retrieved 22 April 2020

Maekawa K (1980) Female weavers and their children in Lagash: pre-Sargonic and Ur III. Acta Sumerologica 2:82–125

Maltby E (1994) An environmental and ecological study of the marshlands of Mesopotamia. Draft consultative bulletin. Wetland Ecosystem Research Group, University of Exeter, AMAR Appeal Trust, London

McCorriston J (1997) The fiber revolution. textile extensification, alienation and social stratification in ancient Mesopotamia. Curr Anthropol 38–4(5):17–549

Ministry of Agriculture (2006) Survey of animal resources. Central static's Bureau, Baghdad

Mustafa SMA (2008) The economic, architectural and social characteristics of the marshland. Sada al-Ahwar Magazine, year II, issue 5, Dhi Qar University, July. (Arabic)

Nasrallah N (2013) Delights from the garden of Eden: a cookbook and history of the Iraqi cuisine. Equinox Publishing, London, 458p

Oates J (1978) Religion and ritual in sixth-millennium B.C. Mesopotamia. World Archaeol 10:117–124

Ochsenschlager E (1998) Life on the edge of the marshes. Expedition 40(2):29–39

Partow H (2001) The Mesopotamian marshlands: demise of an ecosystem. Division of Early Warning and Assessment, United Nations Environment Programme. UNEP publication UNEP/DEWA/ TR. 01–3, Nairobi, Kenya
Pomponio F (2010) New texts regarding the Neo-Sumerian textiles. In: Michel C, Nosch M-L (eds) Textile terminologies in the ancient near East and Mediterranean from the third to the first millennia BC. Oxbow Books, Oxford, pp 186–200
Rapoport A (1982) The meaning of the built environment – a nonverbal communication
Richardson CJ, Reiss P, Hussain NA, Alwash AJ, Pool DJ (2005) The restoration potential of the Mesopotamian marshes of Iraq. Science 307:1307–1311
Salim SM (1962) Marsh dwellers of the Euphrates Delta. University of London/The Athlone Press
Silk Road Treasure (2020). https://silkroadtreasures.tumblr.com/. Retrieved 10 April 2020
Smeaton W (1937) Tattooing among the Arabs of Iraq. Am Anthropol 39(1):53–61
Sohel MMH, Amin MR (2015) Identification of types of buffaloes available in Kanihari buffalo pocket of Mymensingh district. Res Agric Livest Fish 2(1):109–115
Sousa A (1983) The history of Mesopotamian civilization in light of agriculture irrigation projects, recent archaeological discovers and historical sources, vol. 1. Al-Hurria Printing House, 572 p. (in Arabic)
Sudo H (2010) The development of wool exploitation in Ubaid-period settlements of North Mesopotamia. In: Carter R, Phlip G (eds) Beyond the Ubaid. Transformation and integration in the late prehistoric societies of the middle east. Chigaco, pp 169–179 (SAOC 63)
Suhair SA (2013) Iraqi guesthouse is the aesthetics of Sumerian art and the gate of generosity. Alwilaia Cultural Foundation July 1
Surakamhang P, Sangsubun C (2015) Structure and mechanical properties of water buffalo horns. Appl Mech Mater 804:247–250
Tombolato L, Novitskaya EE, Chen PY, Fred A, Sheppard J, McKittrick A (2010) Microstructure and elastic and fracture properties of horn keratin. Acta Biomater 6:319–330
UTP (2020) Useful tropical plants. http://tropical.theferns.info/viewtropical.php?id=Typha+domingensis. Retrieved 28 Mar 2020
Waetzoldt H (1972) Untersuchungen zur Neusumerischen Textilindustrie. Centro per le Antichita e la Storia Dell'Arte del Vicino Oriente, Studi Economici e Technologici -1, Rome
Waetzoldt H (2010) The colours and variety of fabrics from Mesopotamia during the Ur III period (2050 BC). In: Michel C, Nosch M-L (eds) Textile terminologies in the ancient near east and Mediterranean from the third to the first millennia BC. Oxbow Books, Oxford, pp 201–209
Wikipedia (2020a) Ashura. https://en.wikipedia.org/wiki/Ashura#cite_ref-Factsanddetails_4-1. Retrieved 22 April 2020
Wikipedia (2020b) Tattoo. https://en.wikipedia.org/wiki/Tattoo. Retrieved 22 April 2020
Yeung HC (1995) Handbook of Chinese herbs and formulas. Institute of Chinese Medicine
Zeder M (2008) Domestication and early agriculture in the Mediterranean basin: origins, diffusion, and impact. Proc Natl Acad Sci 105:11597–11604
Zeder M (2009) The Neolithic macro-(re) evolution: macroevolutionary theory and the study of culture change. J Archaeol Res 17:1–63
Zeuner FE (1963) A history of domesticated animals. Hutchinson, London

Chapter 40
Socio-economic Status Comparison of Fishermen Community in Two Marsh Areas in Southern Iraq

Laith A. Jawad

Abstract In the southern marshes of Iraq, the majority of males are involved in the occupation of fishing. Among the glitches experienced by fishermen in the marsh areas in Iraq as in similar areas around the world comprise poverty and minimal fishing activity due to the environmental and climate concerns, in which these both matters can cause a failure in the fishery yield and revenue. This study is performed to designate the social and economic signs of the fishermen and their family in two areas at the southern marshes of Iraq, one in the marshes near Thiqar Province on Euphrates River and the other near Mysan Province on the Tigris River.. A number of 120 and 130 of the total participants were interviewed in marshes near Thiqar and Mysan Provinces, respectively. The participants were interviewed arbitrarily designated by using Slovin's formula. In this study, the results showed that 14.9% and 19.6% of the wives in the marshes near Thiqar Province and Mysan Province, respectively, were verified working and backing to household income. The survey also revealed that in the area of the marshes near Thiqar Province, a majority of 36.2% fishermen are above 50 years old, with 77 years old being the oldest and 19 years old the youngest. On the other hand, 25.3% majority of the fishermen in the marshes near Mysan Province are in the group age of 31–40 years old with the 71 years old being the oldest and 19 years old being the youngest. Consequently, job-related chances for fishermen household should be accessible so that fishermen's wife and children can underwrite in growing their family revenue. On the other hand, introduction of modern fishing service to decrease fishermen's energy during fishing actions to improve catches and yield of the fishes have imapcted the income of fishermen by reducing the number of fishers.

Keywords Social life · Fishers · Daily income · Women · Jobs · Wetlands · Fishing activities

L. A. Jawad (✉)
School of Environmental and Animal Sciences, Unitec Institute of Technology, Mt Albert, Auckland, New Zealand

L. A. Jawad (ed.), *Southern Iraq's Marshes*, Coastal Research Library 36,
https://doi.org/10.1007/978-3-030-66238-7_40

40.1 Introduction

There are difficulties that frequently disturb and annoy the local fishermen including those in the southern marshes of Iraq. Such issues include environment-climate as well as income deficiency matters (Zainudin et al. 2019). Usually, climate issues can be seen in rainy or stormy days, and such aspects have been experienced by fishermen in many Asian countries such as Malaysia (Islam et al. 2016). In the southern marshes of Iraq, there are certain periods of time during the year where fishermen stayed away from fishing activities. In these days, strong wind and rain are dominating the climate, which prevent any fishing boat to practice fishing in the marsh area. Besides, the issue of poverty is considered the main factor that affects the socio-economic status of the fishermen in the marsh areas in Iraq. Similar circumstances were observed in other Asian countries such as Malaysia (Islam et al. 2016).

Studies on socio-economic of the fishermen can enable to enhance the fishermen's matters. Results of the present chapter will also permit the policy makers in Iraq to recognize the concerns and difficulties that are linked to fishermen's socio-economic status and also their solutions. Abdullah-Bin-Farid et al. (2013) wrote that fisherman socio-economic investigations are usually steered so that actions can be taken to enhance fisheries administration through the development of fishermen intricate. In support of this suggestion, Zuraini et al. (2017) recommended that the description of society's growing scheme particularly for those with lower income can be performed by investigating the socio-economic profile of the fishermen societies.

The study on the fishermen's socio-economic construction and their living status aims to investigate the socio-economic setting of groups such as education, income, poverty, and unemployment. In general, all these components are related to the quality of local communities' life features such as housing, health, telecommunications, social facilities, and others (Zuraini et al. 2017).

40.2 Data Attainment and Processing

A socio-economic study was conducted on the fishermen societies in two villages, one in the marshes near Thiqar Province located on the Euphrates River and the other near Mysan Province on the Tigris River to reveal the differences of the fishermen's socio-economic growth between those two populations.

Socio-economic measureable information and the main data were obtained using a structured questionnaire, interviews, and group discussion with the fishermen's societies. The sample size of fishermen as respondents was decided using Slovin's formula (Limi et al. 2017) at 90% confidence level. A number of 120 and 130 of the total participants were interviewed in marshes near Thiqar and Mysan Provinces, respectively. Random and purposive sampling method (Shettima et al. 2014) was also performed during the procedure of attaining the main information. The questionnaire on socio-economic status and features contained questions on fishermen's general data and essential information (Perumal et al. 2016). An analysis of the fishermen's socio-economic status and their livelihood examined the aspects of socio-economic position such as education, income, poverty, and unemployment. All these components appear to have links with the type of the local communities' life in the traits of housing, health, telecommunications, social facilities, and others (Zuraini et al. 2017).

40.3 Results and Remarks

The overall demographic structure of the two villages investigated is dominated by male fishermen representing 100% of the participants. This suggests that males are more appropriate to the occupation of fishing than women owing to the requirement of energy and facing the harsh climate and the environment. In addition women usually have other jobs, which are looking after Also, things like the technical aspects of running petrol boats and fishing gears require an extra skilled women might not have those (Zainudin et al. 2019). Conferring to the World Fish Centre (2018), the fishery area has long been recognized as a male-playing area, suggesting a sagacity of venture and jeopardy appreciated by men. Several studies on fishermen societies from around the world have confirmed the dominance of males over women in fishing occupation. Shettima et al. (2014) and Sudarmo et al. (2015) gave evidence on such finding from Lake Alau Borno of Nigeria and Tegal City of Indonesia, respectively. For women to be involved in fisheries is somehow restricted (Biswal 2015), but could be found in certain areas. In the southern marshes of Iraq, there are cases where a group of women established their own fishing venture, where they use some nets to set and catch fish and then later on they market the yield in the villages at the verge of the marsh areas.

In the present study on the two marsh areas, the role of women appeared confined to the household aspects and contributes to the income of the family. The survey showed that 14.9% and 19.6% of the wives in the marshes near Thiqar Province and Mysan Province, respectively, were verified working and backing to household income. The mainstream of the fishermen's wives in the marshes near Thiqar Province and Mysan Province (64.9% and 70.1%, respectively) are operated with others and become an employee, while only 31.1% and 27.2% of the wives have their own businesses, but on a small scale. Besides, 83.5% and 80.2% of wives of the fishermen in the marshes near Thiqar Province and Mysan Province were housewives.

Age is an imperative measure for socio-economic construction since it disturbs the fishermen's efficiency and output in regard to fisheries and family living as well as novelty getting (Zuraini et al. 2017). The youth among the fishermen have more energy to work and are more hospitable to agree on modernism involved in fishing unit or fishing administration in comparison with the aged fishermen who are contented with their classical method involved in their job. The present survey showed that in the area of the marshes near Thiqar Province, a majority of 36.2%

fishermen are above 50 years old, with 77 years old being the oldest and 19 years old the youngest. On the other hand, 25.3% majority of the fishermen in the marshes near Mysan Province are in the group age of 31–40 years old with the 71 years old being the oldest and 19 years old being the youngest. By relating the data attained, in the marshes near Mysan Province recorded the majority of productive and economic age participant who are competent to take vigorous duty linked to fishing and upsurge their revenue along with fish catch. Such a finding was supported by Olaoye et al. (2012) in Nigeria. Consequently, fishermen's fishing knowledge describes one's fishing talents since the more skill gained by the fishermen, the more capabilities have been educated and qualified (Shettima et al. 2014). With the talents, fishing knowledge also did echo the fishermen's age where older fishermen had more fishing involvement. In this study, fishermen's familiarity shows a negative statistically significant on fishermen's household revenue in both studied areas. With the coefficients of 0.007 and 0.006, a unit increase in fishermen's experience will reduce the fishermen's home revenue by 0.6% and 0.5% in the marshes near Thiqar Province and Mysan Province, respectively. Jeyrajah and Santhirasegaram (2015) showed comparable results related to the number of fishing experience increases; income per fishermen will decrease by USD 1.7 owing to the upsurge in age that causes the reduction in fishing action.

A high percentage of married fishermen were recorded which were 65.2% and 49.6% for the marshes near Thiqar Province and Mysan Province, respectively. It suggests that job-related movement could be decreased in both studied areas since each has economic tasks. Similar conditions were reported in Tegal City, Indonesia, (Sudarmo et al. 2015) and Bagan Lalang, Sepang (Perumal et al. 2016) in which 100% of 76.5% of married fishermen were reported.

Fishermen in both areas were retarded in concern of educational standing which are in need of higher education with specific teaching. In the marshes near Thiqar Province, a majority of 35% fishermen own education until their lower secondary level, while in Mysan Province, a majority of 34.3% went until upper secondary level. On the other hand, 13.3% and 33.1% of fishermen left school in both studied areas in the marshes near Thiqar Province and in Mysan Province, respectively. This condition infers that fishermen in both areas, especially in Mysan Province, might come from families that did not worry and care about education, at their time. This is because, in the past, children are dedicated more on having job talents for the sake of helping their families to find provisions. Most horrible circumstances were reported in southwest region of Bangladesh (Das et al. 2015) where 75% of fishermen were uneducated while more than 60% of this group of fishermen were reported in Ogun State, Nigeria (Olaoye et al. 2012). As clarified by Ahmed et al. (2013), some fishermen society did not have faith in welfares of having a good education standing and choose for their children to trail their parent's jobs as a fisherman. This is confirmed by the high percentage of fishermen's children in Mysan Province, which were school-leavers with 29.1%, while 19.2% had been recorded in Thiqar Province and in Mysan Province. Conferring to Mamun (2011), the main reasons of fishermen's children to leave school were the lack of education, seeking for attaining revenue, and economic and social problems in addition to family member's oblivion.

In addition, in relation to the deficiency of awareness about learning, 34.5% and 30.4% of fishermen were uneducated. Also, high percentage of information showed several problemes particularly smoking were reported with 91.0% and 81.1% in addition to the status about having secondary or any extra job were taken blithely as 91.0% and 82.3% of fishermen in in Thiqar Province and in Mysan Province, respectively, have no any extra job.

The mainstream of fishermen in both study areas have 5–8 number of children. In Mysan Province, 6.7% of fishermen's household is composed of more than 13 members, while none had been recorded in Thiqar Province. Conferring to Dzuhaimi et al. (2013) in the study of Malay-Muslim fishermen community in Selangor, a mainstream of them owe to the big household size and share the same inhabitant. This study displays that the fishermen's household number has a positive statistically significant link with the fishermen's household revenue. With the coefficients of 0.056 and 0.049, a unit upsurge in fishermen's household member raises the fishermen's household income by 4.6% and 4% in Thiqar Province and in Mysan Province, respectively. As specified by Mercado and Mercado (2016), some fishermen living in a rural area incline to relate the use of family jobs and restricted savings, which is as many as 12% and 30.4% of respondents in Thiqar Province and in Mysan Province, respectively, that uses family job's type for fishing events. This improves the cost decrease and upsurges the turnover during fishing action since there occurs full belief in numerous matters such as supervision fishing facility, fishing catch and turnover allocation that can be performed smoothly which enables the working conditions.

The kinds of fishing boats have an important effect in the fishing-covered distance in which progressive boat with engine allows fishermen to go further away from their home areas associated with the outdated hand-rowed vessel. Conferring to Zainudin et al. (2019), a fishing area with more fish population can be reached only by using motorboat as they are away from the living areas. As much as 55.2% of fishermen in Thiqar Province works from 15 to 20 hours per trip; 40.2% of fishermen in Mysan Province works from 9 to 15 hours per trip. Consequently, fishing effort displays a positive important link with fishermen's domestic revenue in Thiqar Province. As clarified by Islam et al. (2016), fishing effort which entails fishing hours in a month is important at 1% which infers that it extremely disturbs the fishermen's catch along with their revenue. The present survey showed that more than half of the fishermen were vessel possessors, which consists of 55.1% and 48.5% of ownership recorded in marshes near Thiqar Province and in marshes near Mysan Province, respectively. This suggests that vessel possession has a positive noteworthy link with fishermen's home revenue in marshes near Mysan Province.

For fishermen's kind of fishing methods used for fishing action, 80.34% of the fishermen in marshes near Thiqar Province uses a single gear. Differently, in marshes near Mysan Province, 70.18% of the fishermen use a single gear. The gear sorts used by fishermen display a positive noteworthy link with the fishermen's home revenue. In contrast with Kenya's South Coast, fishermen using old-style fishing methods favor using many fishing gears to catch numerous fish species (Tuda et al. 2016), and several gears are also reported to be used in San Miguel Island for multi-fishery drives (Nieves et al. 2009). This finding is also backed by Y and Antonia (2017), as the number of gear types increased the fishermen's revenue that will be upsurged by 44.7%.

Concerning the fishermen's accommodation circumstance profile, all the participants in both localities have declared that they own the reed's huts that they live in. The housing status in the southern marshes of Iraq is different from the fishermen's societies around the world, where inhabitants can build a hut and live in on any dry land they can locate in the marshes as these areas are nearly no-man's-land. In many Asian countries, house ownership in the rural areas is one of the problems that the fishermen experience and affects their income (Zaimah et al. 2015).

In regard to toilet facilities, 100% of the huts own no bathroom or toilet facilities in both study areas. Similar findings were also recorded in Andhra Pradesh, India, by Ismail (2014), where the fishermen use lands, fields, canals, bushy areas, hidden places, and latrines to answer the call of nature (Reza et al. 2015). In the contrary and in Malaysia, toilets are built in every house as it is a basic component in a house construction nowadays (Perumal et al. 2016).

References

Abdullah-Bin-Farid BS, Mondal S, Satu KA, Adhikary RK, Saha D (2013) Management and socio-economic conditions of fishermen of the Baluhar Baor, Jhenaidah, Bangladesh. J Fish 1(1):30–36

Ahmed N, Rahman S, Bunting WS, Brugere C (2013) Socio-economic and ecological challenges of small-scale fishing and strategies for its sustainable management: a case study of the Old Brahmaputra River, Bangladesh. Singap J Trop Geogr 34:86–10

Biswal LR (2015) Fishing is more than just a livelihood: well-being and small-scale bag net fisheries governance in Gujarat India. Master thesis, University of Manitoba

Das RM, Ray S, Kumar U, Begum S, Tarafdar RS (2015) Livelihood assessment of the fishermen community in the South West Region of Bangladesh. J Exp Biol Agric Sci 3(4):354–361

Dzuhaimi D, Turiman S, Mohamad RH, Jegak U, Jamilah O, Nobaya A, Nor AWA, Farid SABM, Mondal S, Satu AK, Adhikary KR, Saha D (2013) Management and socio-economic condition of fishermen of the Baluhar Baor, Jhenaidah, Bangladesh. J Fish 1(1):30–36

Islam NMG, Ali J, Zamhuri S, Viswanathan KK, Abdullah H (2016) Impact of subsidies on the economic and environmental conditions of small scale fisheries in Malaysia. Eur Proc Soc Behav Sci

Ismail M (2014) An analysis of socio-economic situation of fisherman in Visakhapatnam district of Andhra Pradesh India. Int Res J Mark Econ 1(3):24–35

Jeyrajah S, Santhirasegaram S (2015) Socio-economic factors influencing marine small scale fishers' income in the Batticaloa District of Sri Lanka. Int J Humanit Soc Stud 3(1):75–79

Limi AM, Sara L, Ola LT, Yunus L (2017) Environmental changes and fisherman. Welfare in coastal area of Kendari Bay. J Agric Forest Fish 6(1):20–25

Mamun AA (2011) Socio-economic conditions of fishermen in Kaptai Lake area. Master thesis, Bangladesh Agriculture University Mymensingh

Mercado OJ, Mercado ER (2016) Analysis of socioeconomic profile of rural fishers in Northern Part of Surigao Del Sur, Philippines. World J Fish Mar Sci 8(1):64–67

Nieves MP, Pelea RN, Bradecina GR, Pereyra AM, Morooka Y, Shinbo T, Rivero PCM (2009) Agriculture and the adaptive capacities of households and communities in San Miguel Island, Albay, Philippines in the Kuroshio Sphere of influence. Kursoshio Science 3(1):23–32
Olaoye OJ, Idowu AA, Omoyinmi GAK, Akintayo IA, Odebiyi OC, Fasina AO (2012) Socio-economic analysis of artisanal fisher folks in Ogun Water-side local government areas of Ogun State, Nigeria. Glob J Sci Front Res Agric Biol 12(4):9–22
Perumal C, Zaini S, Siti KZ (2016) Impak ekopelancongan terhadap komuniti tempatan di Malaysia: Kajian kes komuniti nelayan Bagan Lalang, Sepang, Selangor. Malays J Soc Space 12(4):94–108
Reza S, Hossain SM, Hossain U, Zafar AM (2015) Socio-economic and livelihood status of fishermen around the Atrai and Kankra Rivers of Chirirbandar Upazila under Dinajpur District. Int J Fish Aquat Stud 2(6):402–408
Shettima BG, Mohammed ST, Ghide AA, Zindam PL (2014) Analysis of socio-economic factors affecting artisanal fishermen around Lake Alau, Jere local government area of Borno State, Nigeria. Niger J Fish Aquac 2(1):48–53
Sudarmo PA, Baskoro SM, Wiryawan B, Wiyono SE, Monintija RD (2015) Social economics characteristics of coastal small-scale fisheries in Tegal City, Indonesia. Int J Sci Technol Res 4(1)
Tuda MP, Wolff M, Breckwoldt A (2016) Size structure and gear selectivity of target species in the multispecies multigear fishery of the Kenyan South Coast. Ocean Coast Manag 130:95–106
World Fish Centre (2018) Gender and fisheries: do women support, complement or subsidize men's small scale fishing activity. The World Fish Centre
Y AA, Antonia M (2017) Determinants influencing fishing income to the coastal households of Indian Ocean. Oceanogr Fish Open Access J 5(3):1–7
Zaimah R, Suhana S, Azima AM, Hussain YM, Samsudin M, Omar M, Yusoff I (2015) Kesejahteraan subjektif: Kajian kes nelayan di Sedili, Kota Tinggi, Johor. Subjective well-being: a case study of fishermen in Sedili, Kota Tinggi, Johor. Malays J Soc Space 11(2):67–94
Zainudin LM, Zein A, Idris MH, Luqman W (2019) Socio-economic profile comparison of fishermen community in Kuala Marang and Seberang Takir, Terengganu, Malaysia. Journal of Sustainability Science and Management 14(6):145–157
Zuraini A, Roseliza MA, Noorhaslinda KAR (2017) Profil sosioekonomi nelayan di Setiu. In: Transformasi inovasi sosial komuniti nelayan. Penerbit UMT, Malaysia, pp 61–80

GPSR Compliance

The European Union's (EU) General Product Safety Regulation (GPSR) is a set of rules that requires consumer products to be safe and our obligations to ensure this.

If you have any concerns about our products, you can contact us on ProductSafety@springernature.com

In case Publisher is established outside the EU, the EU authorized representative is:

Springer Nature Customer Service Center GmbH
Europaplatz 3
69115 Heidelberg, Germany

Batch number: 10732057

Printed by Printforce, the Netherlands